The series "Studies in Fuzziness and Soft Computing" contains publications on various topics in the area of soft computing, which include fuzzy sets, rough sets, neural networks, evolutionary computation, probabilistic and evidential reasoning, multi-valued logic, and related fields. The publications within "Studies in Fuzziness and Soft Computing" are primarily monographs and edited volumes. They cover significant recent developments in the field, both of a foundational and applicable character. An important feature of the series is its short publication time and world-wide distribution. This permits a rapid and broad dissemination of research results.

Indexed by SCOPUS, DBLP, WTI Frankfurt eG, zbMATH, SCImago.

All books published in the series are submitted for consideration in Web of Science.

More information about this series at http://www.springer.com/series/2941

Studies in Fuzziness and Soft Computing

Volume 406

Hanns-Werner Heister

Music and Fuzzy Logic

The Dialectics of Idea and Realizations in the Artwork Process

Springer

Hanns-Werner Heister
Rosengarten, Germany

ISSN 1434-9922 ISSN 1860-0808 (electronic)
Studies in Fuzziness and Soft Computing
ISBN 978-3-662-62905-5 ISBN 978-3-662-62907-9 (eBook)
https://doi.org/10.1007/978-3-662-62907-9

This Springer imprint is published by the registered company Springer-Verlag GmbH, DE part of Springer Nature.
The registered company address is: Heidelberger Platz 3, 14197 Berlin, Germany

To Inge Rubbert

Preface: Subject, Themes and Aims

With this book I pursue a twofold objective. On the one hand, it is about appropriating fuzzy logic even more intensively for the science of music and to make it as broadly as differentiatedly usable there: for the analysis of musical works and facts, for the expansion and specification of thinking about music and thinking in music. On the other hand, it is a concern to prepare the broad field of music, at least to a certain extent, in such a way that music becomes accessible for research from the side of fuzzy logic up to its practical application in an easier, more intensive, more detailed and more differentiated way. The book goes back to studies about fuzzy logic and music, which I started in 2009 mainly due to suggestions by Rudolf Seising (See Heister 2012).

Double Approach, Hybrid Perspective

In this book I deal with numerous further aspects, which go beyond of what I have already researched and published in this context. This includes as well such aspects, that have so far not received enough attention, by research, which takes Fuzzy Logic as a starting point. For here we extend the scope of research beyond the elementary, like acoustics (in the context of space and on perception), or systems of pitch or rhythm, of elementary problems of musical tuning systems, but also beyond, what in the realm of Fuzzy Logic so far has likewise been dealt with, like the complex matters of timbre simulation and styles of interpretation, of procedures for storage and automatic finding, and many more.

Complexity and Priority of Qualitative Questioning

Due to my disciplinary competence I generate primarily *qualitative* reflections. Starting from music and different arts, they are elaborated in this perspective. From now on, it has to be taken into account to analyze such qualitative facts, as far as possible, or as useful, with Fuzzy Logic' *quantitative*-mathematics, as it has already and in varied ways been undertaken. Thus we may achieve further and more far-reaching specifications, for instance, about problems of motivic-thematic affinities as special cases of similarities. By so to say hand-made music analysis, such similarities may be hardly, rarely and in some points even in principle never be reached.

In such cases reflected formalizing, mathematical logic, and modeling, on the basis of computerizing, just in the realm of "Soft Computing", offer plenty, new, and surely at the moment not fully foreseeable potentials of research (and practically applicable too). In the meantime, the series *Studies in Fuzziness and Soft Computing* has already published an array of relevant themes. Their main topics, such as storage and recall at databases, and similar, connected or related task areas. Automated recognition of patterns is probably one of its core areas.

Mutual Benefit for Science of Music and Fuzzy Logic: Tunnel Digging from Two Different Sides

On the other hand, experts of Fuzzy Logic may be provided by this book, depending on how far as they embark on it, with a deeper comprehension about what music means. While I have rather too little to tell, or hardly anything new and of actual relevance, about *direct practical* application of Fuzzy Logic and about practicing Soft Computing, I can tell surely a lot about music and its connections with Fuzzy Logic. I am concerned with the ontological, gnoseological, psychological, and (music-) aesthetical status and the relative importance of different phenomena of relationship between music and Fuzzy Logic.

This seems to be a journey full of risks: For specialists of music, this book contains to much of what is well-known in their department, while on the other side, specialists of Fuzzy Logic are commonly familiar with its content. Nevertheless, from a double access, a cross-wise, for not to say "hybrid" perspective, the combination of music analysis and fuzzy logical modes of thinking, might reveal plenty of new aspects in both directions. I took the risk of loosing the ground and falling into the abyss, but followed the path with my eyes wide-opened. But to travel on trodden routes only is unproductive. Without the risk of going astray, there will be no scientific progress. By the way, to make it clear, once and for all. What is presented in this book, remains always under the proviso "Under the actual state of affairs ..."—This refers to research in general, and I concede, that naturally my knowledge about it remains rudimentarily, but also refers to my own researches, even if they only mark attempts, but nevertheless, they are new. Therefore, the theory, based on empirical knowledge, as it is hereafter unfolded, contains hypothetical components. It can be enlarged, reduced, and partially rejected.

As it is unavoidable, I am dealing with matters, that concern various scientific disciplines. As far as possible, I challenge questions of a truly interdisciplinary dialogue, which surpasses declarations and lyrical applications for research grants. It should not be a conversation among deaf persons. And instead of being competitive, it has to be handled in a productive and cooperative way. As I had to acquire, as much as possible, concepts, presentations, questions, and results of Fuzzy Logic, conversely one should expect, that those, who are scientifically as well as practically engaged in various areas of Fuzzy Logic and its application on music, get involved with questions of music - at least, as far as they are dealing with music from the perspective of Fuzzy Logic. Facing the vast expertise of Fuzzy Logic's researchers in information science and mathematics, as well as in related scientific disciplines, I hold full respect, with all due modesty. But if someone takes 'science' only as something in combination with the utmost in mathematical formulas and logical formalizations, then such a person will be disappointed. Anyway, he or she must not continue to read. On the other hand, I cannot totally do without musical examples. Even if someone is unable to read the lines, they bear as an approach of understanding an aesthetical and sensual quality of charming, totally in the sense of Fuzzy Logic. Taken as a picture without any understanding of what is written, it somehow exposes the intended sound and purpose.

A work partly from the science of music, like this one, and partly from Fuzzy Logic, respectively, from soft computing, may be taken in such a way: We dig a tunnel from various directions. Whether we meet exactly in the middle, is however not quite sure. In case, that this may not be completely achieved, and if there were blurs, nevertheless yet undetected states of affairs will be laid bare, which were until now hidden in the darkness of ignorance. Thereby, for both sides the interests and tasks might get clearer, and it could further the studies with or about music and Fuzzy Logic.

About the Structure of This Book

The introduction gives an explanation for a number of important categories and terminologies in relation to Fuzzy Logic, focusing on their relevance to matters of music. Thereupon, correlatively and complementarily, basic categories of the musical process are outlined in the light of Fuzzy Logic. This encompasses its absorption in the individual's memory, or in external storage and forms of reflection like texts or pictures from the concept and first ideas to the different forms of storing. During this kind of process, a musical work of art is the centering focus for at least thirteen different forms of existence, with which I analyze in detail the manifold occurrences of fuzziness in the music process and thus reveal further questions for the field of fuzzy logic.

The Notion of '(Art)work': A structural, Genre-Comprehensive Concept, Not a Concept of Assessment

'Work' is not considered as a concept of assessment, but as a structural one. The characteristics of a 'work' are not confined to 'classical' music, contrary to common prejudices held by enthusiasts of either 'Pop' or 'Classics'. Here, it is meant in the sense of being free of any value judgement, just a specific aesthetic representation, taken in its concreteness and objecthood: a 'piece' of music with more or less complex qualities of character, being limited in space in its temporal and in concrete realization. In principle, such an objecthood is likewise valuable for popular music, or even for improvised, relatively completed productions, being 'pieces', or works.

On the other hand, there is no reason to dismiss the concept of work altogether, as it became a trend in modernity and post-modernity. If firmly outlined characters are missing during processes of art or actions, like in the case of a 'Happening', then the action, as a whole, stands for the 'Work'. No matter how much it is emphasized by the ideology of the 'Open Form' or similar concepts: Actually, even such actions may be reproduced, for example, by filming. Since the 1980s, this is done by 'Performance', which indeed is nothing but performance, only that most of the time there is nothing to be performed. Although in a regularly screened video-reproduction it becomes a work. As will be further explained below (see 2.), the work as a whole appears in and through its different forms of existence as a 'processing identity'; as a whole, that, in the end, may never be absolutely completed.

The main part of this book treats the four principles and five manifestations or phenotypes of Fuzzy Logic in musical process. The pivot rests basically in the dialectic, which is multi-dimensional, multi-layered, involved (encapsulated) and folded in its fuzziness and sharpening along with blurring, for intentionally compositional-artistic utilization of fuzziness in its different facets of art.

Towards a 'Philosophy of Fuzzy Logic'

A brief and final part recapitulates and outlines some further perspectives of future research, above all, interdisciplinary, especially for the realm of theories of art. Apart from that, I am generally tempted to draw some consequences from Fuzzy Logic's way of thinking for the world view in general, and specifically in a musically-artistic context. Seen from this angle, some reflections about analogies and isomorphies are even for those productive, who are hardly interested in music. May be, this book has in a direct way only little to contribute to the methods of Fuzzy Logic, but probably for its meta-theory. Anyway, the book offers a small contribution in this direction. Bluntly, it might even be labeled as being a piece of 'Philosophy of Fuzzy Logic'.

Rosengarten, Germany
September 2020

Hanns-Werner Heister

Acknowledgements

I am grateful to Rudolf Seising for his essential encouragement, as well as for his arrangement of contact to the series *Studies in Fuzziness and Soft Computing*, to Janusz Kacprzyk, the editor of the series *Studies in Fuzziness and Soft Computing* for the book's inclusion in the series and his generous, open minded acceptance of the ideas being developed here; to Hanjo Polk, Hans Krauss, Maren Köster for various suggestions; to Frank Böhme for the supply with books; to Gavin Bruce, Stefanie Everke Buchanan, Hilmar Heister, Elke Keilhofer-Schmidt, and Klaus Sonnendecker for their English translations. In the end of this book as process, Bernhard Rusam made a substantial contribution to giving the manuscript a homogeneous, coherent shape in both content and form. I am also grateful to my wife, Karin Heister-Grech and my children Georg in Kiel and Hanne in Braunschweig for having tolerated and accepted, especially from spring to autumn 2020, that I had to partially withdraw from family life, i.e. that husband, papa and grandpapa were not fully available. May the book make this renunciation worth it. Moreover, I am grateful to A. Gowrishankar and Jayarani Premkumar, project-coordinators and production managers at the Springer-Verlag, for the persistent reminders about delivering the book's manuscript. I am also grateful to Shukkanthy Siva for the final text processing and finally, to Leontina di Cecco, my editor, for her never ending, infinite patience, as she constantly had to remind me, with warm and friendly words, that she expected the completion of the manuscript—after all not in vain.

Last but not least I am grateful to the sociologist and visual artist Inge Rubbert. Mathematically gifted and from their own theoretical and practical projects interested in Fuzzy Logic, she gave me decisive impulses in the final phase of the production of the manuscript, February and March 2020: She commented the existing parts of the book and gave me valuable advices, for the structuring of the whole, and for redrafting, rewording and shortening parts, too. All this accelerated the completion. In the final phase, July to September 2020, she helped with revising, correcting, editing and polishing to ensure that the manuscript was really finished and ready for typesetting. Therefore the book is dedicated to her.

Contents

Bio-Bibliography of the Author

Hanns-Werner Heister

Prof. Dr. phil. and Scientist with focus on music, born in 1946 in Plochingen/Neckar (Germany, Baden-Württemberg)
1965–1977 Musicology, German studies and Indo-European studies/linguistics in Tübingen, Frankfurt a.M. und Berlin (University of Technology)
1977 Ph.D. in musicology at Technische Universität Berlin, Germany, supervised by Carl Dahlhaus. Dissertation published in 1983 with the Title: "The Concert. Theory of a Cultural Form" (2 volumes).
1993 Habilitation at the Carl von Ossietzky-University Oldenburg on the subject *Musical analysis (including methodological studies)*
1992–1998 Professor for history of music and music communication at the Musikhochschule „Carl Maria von Weber" in Dresden, Germany
1994/95 und 1995 Visiting Professor at the Musikhochschule "Franz Liszt" Weimar, Germany
1998–2011 Professor for musicology at the Hochschule für Musik und Theater Hamburg, Germany

Themes of research and publications:

Methodology of musicology; aesthetics, sociology, history, and anthropology (in particular music and human perception, origins of language and art) of music; political, popular music, new music, jazz; music and musical culture in Nazism, resistance movement and exile; aesthetics and history of music theatre; media/technology and institutions of music culture; music analysis; music and: other arts, psychoanalysis, play, maths, cybernetics, fuzzy logic; gardening.

Books

The Concert. Theory of a Cultural Form, 2 vols. (1983) *Jazz* (1983)
Of the Universal New. Analyses of Engaged Music: Dessau, Eisler, Ginastera, Hartmann (2006)
In/Finity. Encountering Görgy Ligeti (2008)

Background Sound Art (2009)
Heinz Gellrich—Times, Paths, Signs (2014)

Editor and Co-Editor of Series and Books (Selection)

Over/Tones. Music and Other Arts, series since 1995
Music/Society/History, series since 2005
History of Music in the 20th Century, Vol. III: 1945–1975 (2005)
Charles Ives, 1874-1954. American Pioneer of New Music (Atlantische Texte, vol. 23) (2004)
Layers, History, System. Geological Metaphors and Forms of Thought in Art Studies (*Over/Tones, New series* vol. 6), 2016
(with W.-W. Sparrer) *Contemporary Composers* (loose leaf lexicon, since 1992, till April 2020 66 deliveries, ca. 12.700 pages)
For more details see www.Hanns-Werner-Heister.de.

Chapter 1
Introduction: Fuzzy Logic's Four Art-Related Principles and Five Phenotypes in the Musical Process

Abstract The research program for the book is designed from both the FL side and the music side and aims at a specifically music-related appropriation of the FL. With reference to L. Zadeh, categories and ways of thinking of FL and its 'intra-interdisciplinary' approach, also under the sign of the paradox of "Precise fuzziness", the many and manifold relationships between musical process, artwork and FL are reflected. This applies initially to the relationship between modal, polyvalent and musical logic. thereby respecting the asymmetric relation between the infinite possibilities and finite reality. Four art-specific principles will be developed from FL's concept of fuzziness and precision: similarity, sharpening - as filtering and as crystallization -, blurring and variation. These principles are the guiding categories and analytical basis for the book and will be applied systematically and in detail during the course of the investigation, and in turn will be continuously concretized.

Keywords Aesthetics · Art and science · Art as anticipation · (Art)work' as a structural, genre-comprehensive concept · Craftsmanship & engineering · Fuzzy Logic's four art-related principles: similarity, sharpening as filtering and crystallization, blurring, variation · Infinite possibilities and finite reality · 'Intra-interdisciplinarity' of FL and musicology · Life sciences · Mathematical theory versus real matters · Modal logic · Music affine applications · Musical logic · Polyvalent logic · "Precise fuzziness" · Real and ideal · Realistically relativized absolute precision · L. Zadeh

Fuzzy Logic is treated here from the perspective of music and the musical process, particularly regarding the different forms of the appropriation of music. Fuzziness is necessary in an at least double sense: it is inevitable, and it is necessary for changes, developments, variations of the given.

Production of music, interpretation of music and appropriation of music can only be extendedly and deeply understood in the context of Fuzzy Logic and in the complementarity of fuzziness and sharpening. Appropriation of music is the comprehensive idea and it encompasses productive as well as receptive realization. It includes not only cognitive actions, but also practical and sensorial ones, like singing, playing, and listening. There is always a reciprocity between fuzziness and sharpening, in their respective forms.

H. Heister, *Music and Fuzzy Logic*, Studies in Fuzziness and Soft Computing 406,
https://doi.org/10.1007/978-3-662-62907-9_1

1.1 Realization, 'Appropriation'. Craftsmanship and Engineering as a Bridge Between Fuzzy Logic and Music

The idea of 'appropriation' is ambiguous. It signifies here: (1) Appropriation of the world, in its three realms of reality, thus, of nature, mind, and society, through or in music. (2) Appropriation of existing music, as a composition and as a work, in their different forms of existence, in a productive as well as a receptive realization. Realization means (1) 'Realization' of what is meant, thought, and written as a sensorial-material sound. (2) Its 'realization', again in a technically reproduced form. (3) Its 'realization' in the reception as a perception, remembrance etc., and (4) Any other 'realization' of these forms of existence in the arts of language and images.

'Musical process', as an expression, states the fact, that music is not a thing (See Chapter 2). It is embedded in structures, relations, and processes. Within them, it will respectively concretize in a representational way, as some piece of music, or rather a work of music. And again, this proceeds as a tonal realization in a temporal process. The material dimension, in the wider sense, of the musical process draws such an attention on Fuzzy Logic and its relation to music. For within this context the prevalence of the workmanship and the engineering over the mathematical and formally logical, of the sensorial-practical over the purely cerebral, is undeniable. And through the category of production, the art of engineering joins crosswise with the engineering of art.

Especially the poiesis-dimension of art, the making, producing, doing, connects music and engineering.[1] Even more so, since music as sound, and therefore a sound to be produced, has a material-bodily dimension, along with its cognitive dimension. Altogether, this unveils all kinds of parallels and references, which qualify Fuzzy Logic as a methodology of thought, ready for application.

An example: A musical formal arrangement in 1, 1, 2, 3, 5, 8 bars, the well-known beginning of the "Fibonacci-row", which can numerically be noted in an exact way.[2] Therefore 1 means = 1 etc. But the duration of bars in tonal realization (on condition that the same amount of beats is measured for each bar) differ more or less slightly. Perhaps, it remains imperceptible, as while listening, the proportions of duration are in conformity to sensuality and intensions adapted to the ears, but it can be measured: Fuzzy Logic instead of logic of identities. Fascinating, promising and essential from the point of view of a scholar of music and other arts, the realistic reference of Fuzzy Logic to necessary blurs in the reality of art production and generally in „real

[1]For instance, Peter Janich, in his contribution, *Handwerker und Mundwerker* (craftsmen and mouth-workers), points to the historical and logical primacy of real geometric structures over mathematical and geometrical idealizations, like 'Platonic bodies', see Wehr/ Weinmann 2005, pp. 271–292.

[2]During the 1910s, Charles Ives and Igor Stravinsky applied successive, fast changing series of measures. During the 1950s, Boris Blacher's "Variable Metres" systemized this. There sequence was calculated in accordance with mathematical proportions and regularities. Karl Amadeus Hartmann adopted this method for some of his works.

life problems" (L. Zadeh, 2013) is crucial. In the light of Fuzzy Logic, relations of music to reality, relations between various layers, between sensorial modalities, and between phases or forms of existence of music, receive new and precise contours during the musical process. The same applies also in regard to shared features and variations to relations between music and other arts. Thus, it serves to realizing a "mutual clarification of arts" (Oskar Walzel, 1917), a program, that was stated a century ago, but is still valid.

1.2 Logic of Difference and Identity. The Internal and External Interdisciplinary Subjects of Fuzzy Logic

The notion of Fuzzy Logic has to be understood extensively, in order to apply it more broadly and deeper to matters of music. Fuzzy Logic in regard to music has to be defined as an objective *and* at the same time methodological-analytic contradictory unity of the complementary interplay of blurs (along with subjectively produced, conscious 'slurring'), and their precision and clarification during the musical process. Components of Fuzzy Logic are both, features of a thing, of music, or rather the musical process, and also features of their analytical procedure. Like already mentioned in the preface, Fuzzy Logic is a real logic of things and a methodological logic of their research. For this double-faced definition we might even call it *Musical* Fuzzy Logic. The expression "*their* precision" (or sharpening) gives a hint, that this is no simple bipolar symmetry. Because being *blurry or fuzzy* takes primacy; it is the real, objectively essential. And even more than this: a certain amount of blurring is *necessary* for developments: in the development of respective works as "processing identities", and even in the development of European polyphony in and for the history of music as a whole—the latter will be outlined in case studies (See below 6.10 and 9.1). Starting from this asymmetrical fundamental inconsistency, and taking it as our main theme, the relevant and thematic phenomena, as they are presented here, can be intellectually developed and portrayed as a progressing explication for something, that is initially rather exposed in an abstract way.

Fuzziness is "necessary" in a double meaning: It may not only exist, but it must exist and it *is*, and it is *necessary* for a development. It is therefore not only inevitable, just like some inevitable trouble. Because no development will be achieved by mere repetition of identity. In the sense of a "Philosophy of Fuzzy Logic" we may therefore even include evolution or even cosmology. Because without any minor divergences, variations or mutations from an identical reduplication, the genome would be unable to select any material for a further development. By analogy, this is valid for the "Big Bang". (Possibly, or even probably, this was no sudden event, but rather something, that happened repeatedly in cycles throughout an unimaginably long period of time of about a hundred billions of years.[3] Whatever might have been the details, and how

[3] See Weinberg 1977.

things took place foregone times: without any of those tiny asymmetries, neither space nor our world would exist.

Against such a background friends of Fuzzy Logic are over and over again amazed and frustrated, that generally only a few know the word, let alone have any knowledge of Fuzzy Logic.[4] One of the implicit aims of this book is to supply some help, at least as far as music is concerned. At times, Fuzzy Logic appears unexpectedly, for instance, in a German collection of glossaries, which on the whole is quite instructive.

> *Fuzziness.* Also blurring; an expression used in semantics of prototypes. It hints to the fact, that features of categories can quite often be drawn in a clear-cut way. Therefore quite often the classification of elements to certain categories may only be responded conditionally with yes or no […].[5]

So far it is correct. But it continues:

> *Fuzzy Logic.* Term for the analysis of logical structures (arguments), which are under strictly logical consideration invalid or incorrect. Like a poorly defined statement with poorly defined arguments, such as: If apples are ripe, they taste sweet./ This apple is quite sweet./ Therefore. This apple is quite sweet.[6]

Here, "strictly logical" means indeed, formal logic in the conformist sense. Such a limitation retaliates. Instead of 'sweet', it should be: ripe. As in the present form it is merely a tautology, and not even a simple poorly defined or vague syllogism. So, this is truly a point, that such a vaguely quasi-critical description is wrong, and even according to formal logic.

1.2.1 Sources and Elements of Fuzzy Logic as Interdisciplinary Synthesis

Lotfi A. Zadeh, already around 1965, developed, in the context of electro-technics, "Fuzzy-Set-Theory", as a theory of fuzzy sets, with set theory amongst its linkings. The expression was marked by the linguist George Lakoff, following Joseph Goguen, one of Zadeh's doctoral candidates. With Fuzzy Logic Zadeh consciously faced a contradiction, that was often suppressed or denied:

> There is gap of insecurity between mathematical theory on the one hand, and the real matters or phenomena on the other hand. Precise conformity of experimental results with values, that were theoretically predicted within the range of observation, hardly ever occurs. In most cases they can only be given by approximation. There exists a range of 'insecurity', or 'inaccuracy', or 'blurring'.[7]

---

[4]Referring to everyday usage of something, like a thermostat or automatic anti-blurring in cameras, can forge a small link for some understanding.

[5]http://www.teachsam.de/deutsch/glossar_deu_p.htm#prototypensemantik, retrieved February 1, 2014.

[6]Ibidem.

[7]Seising 2005, Vorwort p. XII.

Zadeh alludes to Albert Einstein's well-known apodictic phrase:

> As far as the laws of mathematics refer to reality, the are not certain, and as far as the are certain, they do not refer to reality.[8]

Zadeh differentiates between blurred resp. *fuzzy* and *vague*:

> the term vague is not coextensive with fuzzy, although it is frequently used in this sense in the literature of philosophy and linguistics. Thus, in my perception, vague and fuzzy are distinct concepts, with vagueness pertaining to insufficient specificity whereas fuzziness relates to unsharpness of boundaries. For example, "I will see you sometime," is vague and fuzzy while "I will meet you at approximately 5 pm" is fuzzy but not vague.[9]

Zadeh's terminological differentiation including his example can be illustrated with a Yiddish joke:

> Two Jews from Chelmo meet. "So we'll meet tomorrow?" "Yes." "Where?" "Where you want." "And when?" "I do not care." "Good. But I ask you: Be punctual."

Vagueness in the sense outlined here is thus objective and subjective indeterminacy and uncertainty. By contrast, "fuzzy" is already more precise, defined in the literal sense, i.e. there are limits outlined. It moves within the bounds of uncertainty and fuzziness, even if the exact-sharp is not and cannot be precisely defined.

> Taking fuzziness into account complies with the calculations of technical engineering that include imperfections of reality and matter, knowingly allowing for and including "tolerances".[10]

On the other hand, it conflicts with mathematical-computational logic, which unfailingly demands that $1 \equiv 1$[11]—at least in principle.[12] However, in the context of Fuzzy

---

[8]Einstein 1921, p. 3f.

[9]Zadeh in: Kruse/ Gebhardt/ Klawonn 1994, p. VI.

[10]„For a long time, theory of probabilities or statistics served as unique tools for attempts to control this area. As Lotfi Zadeh developed his Theory of Fuzzy Sets […] another scientific tool was available to handle the 'incompatibility' between scientific and technical theory and practice. In 1973, he defined the often quoted 'principle of incompatibility': 'The closer one looks at a 'real world' problem, the fuzzier becomes its solution.'"(Seising 2005, preface, p. XII) Fuzzy Logic as such, is „an attempt, to close the gap between pure mathematics and technical application."(Seising 2005, p. 3).

[11]Modern mathematics, which already started with Newton's and Leibniz's differential and integral calculus, has moved certainly far away from any uncritical complacency about absolute exactitude. The latter presents the image or self-image of something, we could designate as "vulgar mathematics".

[12]That might lead to an, only seeming, false precision: "when exact numbers are used for notions that cannot be expressed in exact terms. For example, 'I am 90% sure he is wrong'." "In science and engineering, convention dictates that unless a margin of error is explicitly stated, the number of significant figures used in the presentation of data should be limited to what is warranted by the precision of those data. For example, if an instrument can be read to tenths of a unit of measurement, results of calculations using data obtained from that instrument can only be confidently stated to the tenths place". Converted in an joke: "A tour guide at a museum says a dinosaur skeleton is 100,000,005 years old, because an expert told him that it was 100 million years old when he started working there 5 years ago." (https://en.wikipedia.org/wiki/False_precision, 21.2.17, retrieved Febr. 25, 2017.

Logic $1 \neq 1$, or more precisely $1 \approx 1$, or rather equals 1 more or less. This corresponds with many circumstances in the field of music, where components of nature, mind, and society are interwoven. For a long time tone a^1 was according to international standard cycles per second (c.p.s.) 870 Hz; meanwhile it is forced up to 880, and with upward tendency. Nevertheless it remains in context of different levels, like pitch system, or depending on what key is used, or depending on the respective orchestra, or work etc., within a certain tolerance of a range of frequency identifiable as just that tone and no other one. And within a respective context of a work, which may also be considered as a 'system', while taking these kinds of fuzziness into account, and irrespective of concrete frequencies, even then $A = A$ prevails, because of the relations between elements *and* musical logic.

Even though exactness as measure and scale has necessarily and historically, as well as culturally been realized in very diverse manners, the numerical-mathematical exactness is only seemingly exact in the world of sounds. Even more fuzziness exists in the systematic interferences between music, culture, and society: These are even more complex and difficult to assess and understand than any yet-so-complex musical composition. On the other hand, taking into account and technically utilizing fuzziness in the context of Fuzzy Logic even by today's technical standards achieves confounding simulations of complex operations with the parameters of tonal sequences, computerized style analysis, up to remarkably deceptive imitations of personal music styles, sound characteristics and so on.

In particular, living systems and social processes do not evolve according to the logic of yes and no, and do not follow strictly determinist rules.[13] Lotfi Zadeh had "assumed that his new theory would lead to new developments in the humanities, i.e. in linguistics, philosophy, and economics", further stating "that fuzzy systems are predestined for the representation of living systems in the life sciences."[14] "Especially when he first put forward his theory, no one could have imagined, that in technical industries there would be a demand for so many applications of Fuzzy Sets, Fuzzy Logic, and Fuzzy Algorithms."[15]

The scope of procedures in Fuzzy Logic, as well as the scope of objects and subjects are continuously extended. Interestingly, L. Zadeh himself designed Fuzzy Logic (along with Fuzzy Set Theory and Soft Computing) as transdisciplinary right from the start, as R. Seising above all had worked out from history of science and in theory.[16] Just as it integrates various procedures of mathematical, computer, and

---

[13]A parallel from the physiology of senses: In the body, the receptors of smell (not alone in the nose, but in other organs as well) react sometimes weaker, and at other times stronger, in conformity with the idea of Fuzzy Logic. However initially the reaction of the neurons in the brain is binary, adopting all or nothing as their values. By repetition a qualitative differentiation is generated, which is more or less graded, also in accordance with the idea of Fuzzy Logic.

[14]Seising 2005, p. 306; cf. p. 220. And even in the so-called "exact" sciences, and also in technical sciences, as in technics itself—the decisive starting point for Lotfi Zadeh.

[15]Ibidem.

[16]"In the end of the 1960s Zadeh took his theory of Fuzzy Logic as part of the General Systems Theory, which could be used to master the so-called input-output analysis of systems." Seising 2005, p. 5.

engineering sciences, so should any of the concerned disciplines search for its applications beyond the limits of their immediate departments. Meanwhile, as it is generally known inside the Fuzzy Logic Community, there are numerous approaches and contributions from totally different scientific disciplines. All this points to a multi-lateral and productive importance.

In later years (1994) Zadeh imagined the following fields of applications:

> Application to applied fields such as systems theory, neural network theory, stability theory, pattern recognition, mathematical programming, etc., fuzzification leads to fuzzy systems theory, fuzzy neural network theory, fuzzy stability theory, fuzzy pattern recognition, fuzzy mathematical programming, etc. What is gained through fuzzification is greater generality, higher expressive power, *an enhanced ability to model real world problems and a methodology for exploiting the tolerance for imprecision and uncertainty* to achieve tractability, robustness and low solution cost. For these reasons, it is inevitable that, eventually, *fuzzy set theory will pervade most* scientific *theories* and will have a particularly strong impact in the realms of systems analysis, artificial intelligence, intelligent control, signal processing, numerical analysis, optimization techniques, diagnostics, *linguistics*, information processing, decision analysis, *cognitive science* and related fields.[17]

Most interestingly, Zadeh picks up upon a specifically musical expression, which was suggested by von Wolzogen in relation to the musical theater of Wagner:

> In a very basic sense, this [i.e. the wide field of applications] is the *leitmotiv* of the, Foundations of Fuzzy Systems.[18]

Zadeh's basic assumption including humanities was right. Fuzzy Logic helps musicology to expand its range by including processes and modes of thought regarding a profounder and improved understanding of music, further increasing our understanding of musical processes.2 (see Heister). This applies to all musicological issues, but also to all kinds of anthropological, sociological, aesthetic, as well as historical and ethnological research pertaining to music. In addition to the initially primarily technical applications of Lotfi A. Zadeh's Fuzzy Logic, since 1965 these fields have been extended to other contexts, such as artificial intelligence, automatic voice recognition, and more generally to all kinds of optical and acoustical pattern recognitions, as for example in the field of 'neuronal networks', which attempt to replicate cognitive processes with the help of computer technology.

From here on can easily make the connection to music, too. For nearly everywhere in music all kinds of fuzziness can be found: most basically on the level of pitches. Distribution of timbre and pitch are characteristic for processes of transient oscillation. The various tunings reveal the limits of exactness. In the context of music further

---

[17]Zadeh in: Kruse/Gebhardt/Klawonn 1994, p. VI. R. Seising gives a detailed account of the initial conception of Fuzzy Logic by Zadeh, see Seising 2005 and 2007. Cf. an early, but already nearly endless list, in Dubois/Prade 1980, content p. VII–XI and p. 362: Aid to creativity/Analysis of scientific literature/Applied Operations Research/Artificial Intelligence and Robotics Image/Biological and Medical Sciences/Control/Damage assessment of structures/Economics and Geography/Linguistics/Processing and Speech Recognition/Psychology/Semiotics/Sociology. (The list, a little bit unsystematically ordered and uneven, and long, but nevertheless perhaps gappy, is given here in alphabetical order.)

[18]Zadeh in: Kruse/Gebhardt/Klawonn 1994, p. VI.

areas for Fuzzy Logic's application are numerous and by far exceed those discussed in this book. Firstly, among others, this can be said of all questions regarding classification and typology, be it musical genres, forms, even styles, idioms, and so on. The classifications and designations of elements and quantities show direct parallels to other sciences: Such as the botanical or zoological taxonomy,[19] or even in the area of applied sciences, as for example anamnesis and diagnosis of illness (for which corresponding software applications are already in place). This also includes the vast area of the analysis of thematic correspondences, such as degrees and kinds of *similarity*, structural and/or genetic correspondences, either within a work or across works of individual composers, schools, groups, but also diastematic, rhythmic, melodic types in folk and popular music—too wide a subject to discuss it here in exhaust.

Meanwhile there are numerous areas of music-affinitive applications, like automation of music distribution, reproduction and production, storage and search, whose stress ratio between Fuzzy Logic's real spheres of activities and its non-material reflections were worked out by many authors in an excellent way.[20] Therefore my endeavor is to fill some gaps, even as I am not fully aware about all the gaps, that were closed, or which are the ones, that need still to be detected and mended. But I will not venture as far as Atanassov. "I realized that it would be virtually impossible to make a survey of all the others' results and contented myself with the presentation of my own works only." [21] Though I may reach the limits of my competency, time and possibly size of the book, I shall nevertheless try as far as possible—as already mentioned in the preface—to include and elaborate results of research hitherto achieved, in order to work on them further on. Not each of the numerous case studies, which I shall subsequently out-line and analyze, can already be presented in such a way, that they might instantly be further analyzed with methods of Fuzzy Logic. But it should suffice to demonstrate the vast field of efficiency of the principles of Fuzzy Logic.

In Fuzzy Logic whole bundles of related sciences and disciplinary sections are assembled in search for trustworthy and realistic values. Naturally, the bundle of disciples has to be selected, in accordance with the chosen subject or object. It is deemed unnecessary to present here any list in detail. Fuzzy Logic, in reverse, gets integrated in Soft Computing:

> The principal constituents of Soft Computing (SC) are Fuzzy Logic (FL), Neural Computing (NC), Evolutionary Computation (EC) Machine Learning (ML) and Probabilistic Reasoning (PR), with the latter subsuming belief networks, chaos theory and parts of learning theory. […] Soft computing is not a melange. Rather, it is a partnership in which each of the partners contributes a distinct methodology for addressing problems in its domain. In this perspective, the principal constituent methodologies in SC are complementary rather than competitive.[22]

---

[19] Cf., for instance Löther 2015: *The Control of Diversity. Philosophical Fundamentals of Taxonomy.*

[20] Only some names: Arcos, Carrión, Díaz-Báñez, León, Ligges, Mauch, Morchen, Müllensiefen etc.

[21] Atanassov 2016, *Preface*, p. VII.

[22] http://www.soft-computing.de/def.html, retrieved Febr. 17, 2017.

In the context of Fuzzy Logic and Music, relations to or inclusion of adjacent sciences and disciplinary sections—and even with the ones, that are not adjacent, are multiplied. If applied on music the interdisciplinary bundle is further enlarged, for instance by news theory, communications theory, game theory or the General Systems Theory. Thereby and moreover not at last, for research of generative deep structures neuroscience and cognition science, and also psychoanalysis receive enhanced significance.[23] "The *role model* for soft computing is *the human mind*."[24]

1.2.2 Heuristics. Exploring the Uncertain

The discussion inside Fuzzy Logic, as far as I have perceived it, has not so exhaustively as one would expect taken note of heuristics, although they operate with deliberately calculated fuzziness. Heuristics is defined as a "procedure for obtaining a solution of general problems, for which no clear strategy of solution is known."[25] A functional description of heuristics points at the search, hypothetical—and at best—of finding:

> A doctrine of finding any "true" statement has to be divided from a logic in search of giving reasons for true statements. The field of research is mental creativity, that is creativity, which by way of operating with mental elements reaches a target of thinking, respectively a solution for a problem, that seemed to be concealed before.[26]

For the sake of completeness, even if any further discussion shall be omitted, it has to be mentioned, that nowadays numerous researchers deem the aim of fixing truth, as complete as possible and adequate to reality, as non-existent, misguided or as unattainable. Any strive to find the truth will always only lead to approximations:

> Heuristic procedures serve for an approximate solution of complex issues of decision and optimization […]. Introductory procedures design a (first) permissible solution, while successive transformations of solutions during procedures of improvements lead to improved (locally optimal) solutions. Meta strategies direct procedures of improvement with the prospect of testing any of the most promising ranges of solutions, or to overcome (solely) local optimums.[27]

---

[23]This is an extension of Fuzzy Logic—from its point of view—or inclusion of Fuzzy Logic from the point of view of Soft Computing. Neural computing and genetic computing were included since the later 1990s. (http://www.soft-computing.de/def.html, retrieved Febr. 17, 2017).

[24]Furthermore, soft computing may be viewed as a foundation component for the emerging field of conceptual intelligence. (http://www.soft-computing.de/def.html, retrieved Febr. 17, 2017).

[25]Heuristik, in: Springer Gabler Verlag (ed.), Gabler Wirtschaftslexikon, see: http://wirtschaftslexikon.gabler.de/Archiv/49697heuristik-v8.html, retrieved Aug. 7, 2017.

[26]Klüsener, Sebastian 1998/1999, p. 4.

[27]Domschke/Scholl 2006, p. 2.

1.2.3 Realistic Limits of Absolute Precision

In the context of the engineering sciences and mathematical theories, as the set theory and others, the discovery and development of Fuzzy Logic picks up the thread of the traditions of differential and infinitesimal calculus. A long time ago they falsified the naïve idea of the absolute exactness as main feature of mathematics—an idea, which even after Gödel survives in the common sense mentality and seems ineradicable. In the late enlightenment, the scientist Georg Christoph Lichtenberg wrote:

> The big stratagem to reckon minor divergences from the truth as truth itself, whereupon is built the whole differential calculus, is withal the reason for our whitty cogitations wherein the whole would result, if we would take the deviations in a philosophical rigour.[28]

While determining differential and infinitesimal calculus this consideration of inexactness means, that in the end often only an approximation is possible in mathematics, as such, as there is no absolute identity between calculation and what is to be calculated. This may be interpreted epistemologically. In the scope of an object-related perception, there will be an objective truth. But there is no absolute truth. From a scientific point of view, that is chimerical. Instead, there is a relative truth, which may, however, advance in an asymptotical approach towards attaining (the telos of) absolute truth.[29] In fact, this has to be the case.

The problem of utmost precision and elimination of vagueness and blurring divides on the one hand, science as thinking and as theory, and on the other hand, arts and technics as practice. The aim of science, and likewise in Fuzzy Logic as theory, is to minimize asymptotes. Again, this happens within objective limits, basically those of perception, and particularly those, for instance, in an area of applications of probabilities, which are put up by reality and practice upon themselves. Engineering may often be content with approximations. As a rule it makes sense to minimize "allowances", instead as eliminating them. In so far as there is any chance, then usually no reasonable proportions of efforts and returns can be achieved. However, in most cases minimization is pointless, for instance, when making adjustments. Here, science aims at maximization, engineering at optimation, whereas one might say, arts or music are placed somewhere in the middle. While on the one hand, for instance, it is practically impossible, to intonate every pitch absolutely clear. On the other hand, an A should be intonated so precisely, that it may clearly be distinguished from an adjacent B. Always, within the limits and the system of historically changing

---

[28]Lichtenberg, 1765.

[29]Here, the objective truth is meant, not an absolute one. In nature there is an ontological parallel to this kind of gnoseological limit. For instance, thermodynamics demonstrates, "that a reversible reaction signifies an ideal case, which at times might be nearly achieved, but in reality could never happen."(2nd principle; see https://www.uni-ulm.de/fileadmin/website_uni_ulm/nawi.inst.251/Didactics/thermodynamik/INHALT/HS2.HTM, retrieved March 17, 2017). From the same section: „The 3rd principle of thermodynamics states, that no procedure may ever reach an absolute point zero of temperature by limited numbers of steps. An absolute point zero may be approached anyhow, but will be never attained."(https://www.uni-lm.de/fileadmin/website_uni_ulm/nawi.inst.251/Didactics/thermodynamik/INHALT/HS3.HTM).

Fig. 1.1 Franz Schubert: *Die Forelle,* 4. Version, Ith stanza, 2nd accolade without bass Eusebius Mandyczewski (Hrsg.): Schubert's Werke, Serie XX: Sämtliche einstimmige Lieder und Gesänge. Band 5. 1817-1818, No.327 (S 141-143) Breitkopf & Härtel, Leipzig 1895. Plate F.S. 670-673. IMSLP10602-SchubertD550d_Die_Forelle_4th_Version

tunings. Especially the 'idea' of the prevailing pitch sets the point of reference or aim, if it bears motivic importance.

1.2.4 "Precise Fuzziness"

The polarity blurring/sharpening, the hostile and friendly twins imprecision/"precisiation" (Zadeh) for its part is subject of a dialectic, whose pendulum swings back and forth, depending on its field of application or target. 'Similarity' (see Chapter 3) is one of the compromises.[30] In his supporting motif Schubert draws tiny whirls and splashes for his song *Die Forelle* (the trout; lyrics: Chr. Fr. D. Schubart) with an artistically adequate tonal and picturesque obviousness, as the fish streaks "like an arrow" through the waves of the water (Fig. 1.1).

This is not an "exact"'illustration'. Leonardo da Vinci e.g. grasped whirls more accurately, but visually too, however not quantifying, as it is done rather painstakingly in fluid mechanics, where the search for exact algorithms is obviously still going on. But again a different picture emerges, if one turns to the analysis of blurriness in sciences or epistemology *and* the realization of experiences in the arts. Here, blurriness has to be accepted as a given, and even a necessity of the matter, but while performing music or when analyzing facts that are blurred, than it is the task, to apprehend these as far as possible—or as far as it makes sense.

For such problems Wolfgang Eiden coined the brilliant dialectic formula of "precise fuzziness[31]" for Fuzzy Logic. As Fuzzy Logic is not content in stating blurs, it tries to apprehend methodically and as far as possible its real and blurred facts in their qualitative as well as quantitative exactness. Generally this demands an enormous effort of mathematics. But for applications to (cautiously: many of the) questions in relation with music, it is seemingly, at least for the time being, not necessarily required.

---

[30] Aristotle already had recognized that in representing an object one should not be more precise, than nature permits this object to be.

[31] See Eiden 2002.

1.2.5 *Ideology, Illusions, Realities. Vulgar Mathematics and Fuzzy Logic*

The mathematics in the background are just as fascinating—but here I remain somehow skeptical regarding the absolute claim to sole representation of authority and unquestionable exactness; in the realm of the mind, to which mathematics principally belong, it might apply, at least in principle.

That mathematics can be applied nearly universally to matters of nature, technics, society, mind is founded in its quality (the same with logic) of reflecting in a high degree of abstraction real relations, nexuses, proportions etc. Mathematics is in contrary to the common prejudice not a (natural) science but the discipline of the humanities ("Geisteswissenschaft") par excellence, along with philosophy, logic in particular. There is a place for shrewdness, for the principle of identity, and for a propensity towards absolute exactness. Common knowledge (not the self-image of someone, who is engaged in it) assigns to it, an undoubted, or even absolute exactness. In the pure realm of the mind, where initially mathematics essentially belongs to, this might generally be true, at least in principle. Qualified by "in principle", because, as far as I know, mathematics has at its borders like in the basics and in its logic of arguments some uncertainties and indeterminacies.

But taking into account reality, be it nature or society, material or artistic production, the totalizing use of mathematics as quantification lastly does not sum up to more than a seemingly (absolute) exactness, even if this is neither wanted nor meant.[32] Mathematics is not simply to formalize facts or to count correctly. Also a computer counts correctly. The ingenious mathematician Gauß claimed, that exactly excesses of accuracy while counting numbers show an incomprehension of mathematics. Mathematical precision means: being able to estimate, while consciously considering any of the mistakes, that may occur. As has been already stated, Fuzzy Logic's modes of thinking and procedures are better adapted to reality. Here, I could even quote one of the many ingenious mathematicians.

Genz, a physicist, remarked about the relation between mathematics and physics, which for him served as pars pro toto expression of material reality, "that fundamental progress in physics coincides with a replacement of assumptions, that were solely mathematically stated, by non-mathematical principles".[33] As a concrete evidence: The proportions of numbers used by Pythagoreans with reference to music had partly an empirical origin, for instance, by measuring resonating strings, partly in theoretical mathematics, partly in cosmological speculations. But their physical and acoustic substrates were discovered after more than a millennium later.

> "The mathematical description of pitches by Pythagoras and his school [… were] replaced in physics by the vibrating bodies and air column".[34]

[32]The "veneration of the number", as Ludwik Fleck, the sociologist, in 1935, attributed to the natural sciences, "can presently even be detected in everyday life" as "mental attitude" and belief in "quantification's promise of objectivity". (Zillien 2017, p. 17)

[33]Genz 2004, p. 233.

[34]Genz 2004, p. 235.

Projected onto the principles of art and its procedures the mathematics resemble the chiaroscuro, the clear and distinct demarcation of light and darkness in painting, whereas Fuzzy Logic appears more like a 'sfumato', the blending of contours and emphasis on transitions, or in a more general sense, they relate to each other like clock and cloud (see below apropos Ligeti in Chapter 7.4). In music it is interesting to observe the quantification of the divergence, but above all qualitative aspects, i.e. why and on the basis of which factors these divergences from the notation come to be the prescriptions, the blueprint for the performance.

1.2.6 Abstract Scientific and Concrete Artistic Anticipation

Perhaps what is most fascinating in this highly abstract intellectuality is the fact, that now, on the one hand, structures and processes of the material, of "nature", can be understood and became calculable, and on the other hand, one may generate from the inherent logic of science even calculations, algorithms, and formulas, for which facts and applications have still to be detected.

In 1930, Pauli considered the problem of beta decay. On 4th of December he sent a letter to Lise Meitner et al., with the opening, "Dear radioactive ladies and gentlemen", where he proposed the existence of a hitherto unobserved neutral particle with a small mass, no greater than 1% the mass of a proton, in order to explain the continuous spectrum of beta decay. In 1934, Enrico Fermi incorporated in his theory of beta decay the particle, calling it a neutrino. The neutrino was first experimentally confirmed in 1956 by Frederick Reines and Clyde Cowan, two and a half years before Pauli's death.[35]

This abstract anticipation shows analogies with the arts. About whether or not and to what extent procedures and results have proposed or form anything of relevance in advance for science would merit a special study. Anyway, it is immanent in art to create incessantly new procedures and detect new materials, which after a more formal and technical application, later on give evidence of being suitable for a better appropriation of reality.1 In 1978, Aribert Reimann arranged in his opera Lear (after Shakespeare) strains of terror and horror, fright and fear, which in 1984, in his antiwar opera Troades (after Euripides/Werfel), so to speak, get back on an even keel, or rather generalized: intensified and historically and socially set in concrete terms for the presentation of war atrocities, even after the war, here, surviving women and children get murdered and enslaved.

[35]https://en.wikipedia.org/wiki/Wolfgang_Pauli, 2017-02-24 retrieved Febr. 25, 2017. He commented on this: "Today, I committed something awful, something, that no theoretical physicist should ever venture to do. I proposed something, that may experimentally never be verified." http://www.spektrum.de/news/das-gewissen-der-physik/976922, retrieved Febr. 25, 2017.

1.3 Modal Logic, Polyvalent Logics and Musical Logic

Ultimately, in Fuzzy Logic there are infinitely many intermediate stages between 0 and 1, in accordance with the principle of something more or something less. In the concrete material realization of this 'infinity' in and as art, however, the infinitely many possibilities are always reduced to just one reality. Fuzzy Logic belongs also to the line of traditions of multivalent, non-Aristotelian logics, or rather modal logic, to which already Aristotle, himself, formulated some basic ideas. Also logics, like Georg Wilhelm Friedrich Hegel's dialectic logic, or such logics as the ones from Loyd (1906), Jan Łukasiewicz or Gotthard Günther and others belong to the further historical and systematical sphere of modal logic.

In multivalent logics the classical principle of the theory of extensionality is maintained without any restrictions, however, the principle of bivalence is given up. Supposing, that more than two values (possibly, even an infinite number) may occur as implications of predicates. With this in mind, in a multivalent logic the principle of the exclusion of a third (tertiary/tertium quid) constituent ceases to be valid. In a generalized sense, however, it retains its validity for n-valent logics as principle of the exclusion of the (n + 1) value.[36] The first mature systems of multivalent logics were presented, in 1920, by Jan Łukasiewicz (1878–1956), and in 1921, by Emil Post (1897–1954). In his three-valent logic, Łukasiewicz introduced between the value 1 (true) and 0 (false) the value ½ as *potential*, with that it received a modal interpretation.[37]

1.3.1 Possibilities and Reality, Coincidence and Necessity

From Fuzzy Logic's point of view, "potential" may also be (re-)interpreted as "indeterminate".

In music, this dialectic of possibilities and realities plays an important role for musical logic. Especially, on the part of inevitabilities at compositional conversion of principles and rules. Even under such a perspective there are manifold references to music. Multivalent logics and modal logic are especially relevant for compositional possibilities and ways—along with departures from conventionality. Such decisions are set within the limits of the dialectic of necessity and arbitrariness, or rather subjective decisions, of quasi-automatic guided motions and coincidence.

Particularly, the stringent regular "Palestrina-counterpoint", which was the "classical", and before the beginning of the 17th and after the beginning of the eighteenth century, normative control system for a polyphonic compositions, allowed at one's

[36]In the same sense as an abolishment of a higher, superior unity as also "classical physics includes partly quantum theory, as a limiting case", and, on the other hand to this, in analogy, one has to understand "the classical, Aristotelean Logic as a limiting case of quantum logic" (Heisenberg 1971, p. 176). (H.-W.H.)

[37]Stelzner 2010, p. 1442b; in-depth, see p. 1443b.

own discretion to start the melody with a leap, i.e. a quart, a quint, or even a larger interval. But then such a leap has to be homostatically compensated by some minor nests of intervals. There are some options of choices available, but they are limited. In the case of art, what is often vaguely called "coincidence", occurs mainly as a subjective composition of producers. Initially, it is or seems to be arbitrary. But unrestrained compositions are nevertheless frequently motivated, quite often involuntarily. Thus, the final set of Anton Weber's *Sechs Bagatellen für Streichquartett* op. 9 (1913) starts with note A. Within the limits of free tonality, that is not fixed to a key, he might have started with any note within the chromatic total of twelve-tone music. But a composer has to select one out of twelve possibilities, and it must be solely one. Which one, however, rests with him. An apparent arbitrariness of choice is nevertheless motivated: In this one, well as in other works from that time, Webern was mourning for his mother, as he explicitly wrote in his letters. Amalie, her name, obviously starts with A too.

Similarly, but a bit more complicated is the case of the final A in Gustav Mahler's *Das Lied von der Erde*, which was composed at about the same time (1907/1909). But it was totally set in another idiom, within the frame of traditional, major-minor keys, which were lead close to their limits. The tonic-triad C-E-G plus A, as sixte ajoutée, a sixth chord added to the triad, was manifoldly motivated. Though it may be regularly used as an extension of the subdominant chord, in major-minor keys it may not serve as permitted final tone. Thereby, Mahler suggests something like "die and be", as if after death of an individual, life continues: Earth is "eternally" (that is the last word of the lyrics) blossoming and steadily starting anew, with *E-D*. Secondly, the chord is exotically motivated, as an allusion to the Chinese pentatonic key system. Thirdly, just this semitone lacking, or so to say, anhemitonic pentatonic is something that is at the same time simple and elementary, as it is just alluding to 'earth'. Fourthly, in *Lied von der Erde*, A in particular, is like his unfinished 10th symphony biographically embedded. In those days, Mahler had to cope with the lost love of his wife.[38] Her name start and ends with the A of the final chord, which under the rules of major-minor tonality for "pure" music may not belong to this place: AlmA. Luigi Nono in his orchestral work *Per C.S. (Carlo Scarpa) architetto e suoi infiniti possibilità*, from 1984, gave proof of a dialectic of infinite possibilities and limited realities. Nono, in his homage to Scarpa, the architect, wrote the pitch with German letters, as they serve better for coding. In this bi-tonal piece he took advantage of "infinite possibilities", within, or between a—relatively—large interval of the minor third C and Es (Eb), as well as in the immediate vicinity of a particular pitch with numerous micro-tones (quarter tones etc.), for actually to create sounds of an infinite number of intermediate stages, totally in accord with the rules of Fuzzy Logic.

---

[38]Mahler wrote in his sketches to the 10th Symphony, among other things, an appellation to "Almschi" (that was his pet name for Alma).

1.3.2 *Infinite Possibilities and Finite Reality. Intermediate Values*

But even if there is an infinite number of intermediate pitches, their composed as well as their tonally realized numbers, which are not entirely identical, are limited. A composer, as well as the performers, have somehow to make a decision for something. A realized possibility, such as in this case, a commitment for A, and thus against non-A, eliminates for the time being any other possibility, that was available before, as now all of them became unreal. One way or another, any composition draws consequences from its starting point, which exceed subjective decisions. On the one hand, such consequences follow the logic of composition, while on the other hand, quite often even a certain inevitable automatism exists. For instance, in tonal music it would be natural to start a piece in A-major or A-minor, with a first chord on A, hardly without any choice to make. The same pitch has to appear in the final chord, as long as the composer stays in line with the traditional rules of major-minor tonality. Here, TINA "there is no alternative" is true, while otherwise it is questionable.

1.3.3 *Musical Logic*

Generally, there are in musical languages and in styles in each case more or less strict rules. Ambiguously, they define margins as rooms of possibilities: They create them, limit them, and restrict them. Rules determine, what could be, and what may not be. In case of art quite often: what *should* not be. Musicians may comply with rules or may deliberately violate against them, if this is motivated or justified by their need for expression or innovation, along with the yield. Unintentional breaches of the rules are merely errors. The limits of Modal Logic are in musical compulsory logic, as it is applied within historical and epochal systems of styles or genres, or even personal languages.

Already in 1788, Johann Nicolaus Forkel used the term „Musical Logic". It means both the "logicality of music" and the "participation of musical processes in the laws of reason".[39] Motives, themes, parts of movements, forms are supposedly related to each other: horizontally in time in a consequent and apparent manner, either as derived variation or as contrast, and vertically in (musical) space as an order of various voices simultaneously in the score. In both dimensions it should sound meaningful and beautiful, harmonically without dissolving contradictions (but with contrasts, even extreme ones), and following a guideline of consequence, development and intentional-teleological and dramaturgical finality.[40] That is valid even for improvisations. (Here, in the artwork, the traditional Aristotelian "causa finalis", chimerical in reality, outside of human/social contexts, as an inner principle of the artwork has

[39]In-Depth, see Adolf Nowak, *Musikalische Logik*, HmT – 38. Delivery, winter 2004/05.

[40]New tendencies among others: Nicolas 2002, pp. 89–112.

its place.) In a high level of abstraction this logic is universally valid. But in concrete music it depends on a complicated framework of implicit or explicit principles, rules, on the conditions of genres, forms, of period norms and styles etc.[41] And furthermore, it depends on the even more complicate interactions between artwork, realization, reception[42], and then, on the public's diversification in social strata like class, gender, generation etc., at last in individual differences too.

The logic of music however reaches even further. It includes a cultural and historical context. And it includes analogies, similarities, which at the utmost can be identifiable with mathematical methods and procedures, e.g. shaping forms with the proportions of durations, following the famous 'Fibonacci Row' and others types of rows for organizing musical measures (as Boris Blacher's "Variable Metren", (variable metres) with measure-sequences as e.g. 1/4, 1/4, 2/4, 3/4, 5/4, 8/4, 11/4, 7/4, 4/4 etc., even in retrograde mode), permutations on different levels of the material of music and the techniques of composition etc. On different levels of organization it includes analogies and similarities to methods and procedures of Fuzzy Logic.

Part of the musical logic and its realization in concrete artworks is the motivic-thematic work ("motivisch-thematische Arbeit"). Musical motifs, often combined to more expanded themes, form shapes ("Gestalten") with a fairly stable core and diverse and multiple varied elaborations in the course of the art work as process. Core can be interpreted as the inner kernel, partially subconscious, quasi also the 'generative' dimension. To the core as essence correlates a phenomenon, respectively phenomena in different variants, the 'performative' dimension. The 'contours' of these variants for their part are a generalization and a fuzzily identity-establishing commonality of the different various forms and appearances of the kernel. The interrelations between both these entities belong to the general dialectic of essence and appearance (the Hegelian 'Wesen' vs. 'Erscheinung')—nota bene a real dialectic, because both sides or dimensions participate in the constitution of the object, and both belong to the objects of Fuzzy Logic.

1.4 The Four Principles with Its Five Phenotypes: Similarity, Sharpening (Filtering and Crystallization), Blurring, Variation

In the following, I start from four principles, which in the light of Fuzzy Logic are essential for musical matters of fact. They appear in five phenotypes, which may also be designated as 'expressions', 'types', 'forms of appearance', or 'manifestations'. Here, like always, such a systematization is made on the proviso that it is tentative. There can be some other principles that matter, which until now had escaped the author's range of vision. Though in the context of music and Fuzzy Logic, very

[41]Cf. e.g. Chemillier 2002.

[42]Cf. e.g. Leman 2002, pp. 185–198.

probably, this concerns only a very limited number of principles, that have an effect. In any case, essential principles are surely treated here.

Within these four or five principles, the principle of similarity appears to be preconditional. Similarity is the resultant of equality and difference: What resembles is *such as* and *different from* this, in relation to something it refers to. On the one hand, this reference alleges a relation with music, or else to a work of art and the world. While on the other hand, those manifold, countless relations among its elements, for instance, in case of motivic relations. And thirdly, again on the other hand, its relation on those layered and nested mediations between them and matters of fact outside of art. For instance, the question how far or even whether or not impulsive interpretations are similar to what is to be interpreted, and if so, then how in particular.

Some considerations in this book, especially in chapter 4, are building bricks for a prospective 'Theory of Similarity'. Similarity is a result of the interaction of blurring and sharpening, fuzziness and "precisiation". The similarity principle presents what is relatively static, what is presupposed, and at the same time, however, what figures as the resultant, at the end of the processes. As such, it constitutes some kind of reference mark or benchmark as well. It virtually creates the basis for the development of the other principles. Thus, in a way, similarity represents a—relative—identity in the face of other diverging differences. On the other hand, the principle of variation contributes the dynamic as counterpart into the continuous process, where always something new or different has to be produced artistically. For their part the differences are due to the interplay of repetitions of something that is already given, and modification or opposition of something not similarly new. After all, all this is just constantly happening in the dialectic of fuzziness and sharpening, together with blurring as a countermove.

Hereupon the following principles are based on, or rather add to. Essentially, in similarity fuzziness has to be included. One may conceive relations of similarity also as relations of reflections. By reflections is meant, that something can be expressed within something else. This can be a prehistoric human footprint, solidified in volcanic mud of Laetoli—thus something organic being part of the inorganic. Since the beginning of the twentieth century, it may be even more complicated, as the differentiation of material elements underneath the level of atoms is steadily increasing. Thus, their artificial reproduction or reshaping, done mostly instinctively, could lead to a steadily increasing differentiation of elements of art, underneath the level of words, of figurative and representational shape, of motifs, and the twelve-tone music etc. What is in such a way reflected, will be partially identical—the shape of a foot, the view of the world or the mode of thinking, which progressively penetrates into constantly deeper layers—, but principally different. The core of Fuzzy Logic's central concern is represented by the principle of fuzziness. Already mentioned were amongst other examples, the systems of mood in matters of music. Contrast is the principle of sharpening. It is noticeable in two different basic or main forms. One may be called 'filtering'. Thereby, elementary sequences emerge out of a chaos, of non-compliance, or from coherent systems with more or less disparate elements. Examples may be found in the development of pitch systems and of rhythm systems.

"Information-narrowing" in cybernetics belongs to the category of "filtering" as well.[43] As for further processing some kind of abstraction is necessary, which is explicitly described as 'filtering' (see p. 74). This corresponds in outline to a brain's (subsequent) processing of sensory data. Likewise, within control systems "information-expansion" (see p. 89) is the compensative answer.

The second main form of sharpening is 'crystallization'.[44] Here, shapes of an already higher level of organization are formed from poorly or amorphous shapes, either consciously designed and produced, or as an emergence, just like a musical motif from a sequence of initially scattered and amorphous isolated tones and interval configurations. Thus crystallization is the formation of new, up till now non-existent shapes.

Compositional work with motifs is always both, may be not in its development, but surely in the advanced development, and completely in the variation: crystallization as a special sharpening and blurring, because motifs are necessarily altered—identity and difference—and they should be distinguishable from the starting point. And this takes place within a net of relations and a process of work, which finds only in its end the conclusion. And possibly, in may be even further continued beyond the particular work. Perhaps the principle of blurring is the rigid antithesis of sharpening in both forms. The basic idea of blurring is a fuzziness, that is deliberately and artificially produced. This is in some ways the return of objective, unintentional fuzziness on a higher stage. For generally, within this context, blurring bears a double-character: If it happens unintentionally, even against the will of participants and operators—it may be called 'objectively'. Attempts to sharpen fuzziness—or, in a second step, to use it either willingly or unwillingly, in an aesthetical and artistic way, are directed against this. This leads to the counterpart. For on the other side, it is willingly, deliberately done—'subjectively' intended.

Both are counter-rotating, even antagonistic. Both is specific to art, but the intended dimension is more of interest from arts production as a point of view. In opposition to the (unintended) *becoming* blurred or being blurred stands the (willful) deliberate *making* blurred of the blurring.

Thus again, in a reversal of the fuzziness/sharpening dialectic, blurring may take the effect of sharpening. For instance, in the case of graphic arts, the shape of outlines may be emphasized, together with hatchings, which simulate three-dimensionality. Thereby a higher stage of shape or shaping is achieved.

Blurring and sharpening, as polarity and complementarity, are seemingly identical with the coupling fuzzyfication/defuzzyfication, but at best only similar or analogue. For the time being they have no concern with such complex mathematical procedures. They are rather an

---

[43] Klaus 1973, p. 81.

[44] In this context both are new terms for illustrating matters of fact in a comprehensive way. Just like in mathematics, the Peano-*sieve* serves as a metaphor, here, I use it in both cases as a scientific image. In physical, or chemical, or biological research and in technical processes, filters of various grades of fineness serve for the differentiation and separation of particles. Crystallization, which is often quite nicely marked on minerals, is a process in inorganic as in organic chemistry between rocks und proteins, well-known, for instance, of ice and snow.

attempt to conceptualize matters of music, as well as arts in general, and furthermore even of reality, more distinctly. But certainly, they are indebted to Fuzzy Logic's modes of thought.

Finally, the principle of variation goes from filtering, as this step is already taken, and then to crystallization. It lifts the opposition between fuzziness and precision, again on a higher level of shaping. Therefore, it appears as a synthesis specific to art[45] of such contradictory interrelations and movements.

References

Atanassov, K.T.: Intuitionistic Fuzzy Logics. Studies in Fuzziness and Soft Computing, Springer, Berlin etc (2016)

Chemillier, M.: Ethnomusicology, ethnomathematics. The logic underlying orally transmitted artistic practices. In: Assayag, G., Feichtinger, H.G., Rodrigues, J.F. (eds.): Mathematics and music. a diderot mathematical forum, pp. 161–184. Springer, Berlin, Heidelberg, New York, Barcelona, Hong Kong, London, Milan, Paris, Tokyo (2002)

Domschke, W., Scholl, A.: Heuristische Verfahren (Jenaer Schriften zur Wirtschaftswissenschaft 08/2006)

Dubois, D., Prade, H.: Fuzzy sets and systems: Theory and applications (Mathematics in Science and Engeneering, ed. by Richard Bellman, vol. 144). Academic Press, New York etc. (1980)

Eiden, W.: Präzise Unschärfe. Informationsmodellierung durch Fuzzy-Mengen. Ibidem-Verlag, Stuttgart (2002)

Einstein, A.: Geometrie und Erfahrung. Erweiterte Fassung des Festvortrages gehalten an der Preußischen Akademie der Wissenschaften zu Berlin am 27. January 1921, p. 3 f.. Julius Springer, Berlin. archive.org (1921)

Genz, H.: Wie die Naturgesetze Wirklichkeit schaffen. Über Physik und Realität, Reinbek (2004) (1/2002)

Heisenberg, W.: Schritte über Grenzen. Gesammelte Reden und Aufsätze. R. Piper & Co., München (1971)

Heister, H.-W.: Invariance and variance of motives: A model of musical logic and/as fuzzy logic. In: Seising, R., Sanz, V. (eds.) Computing in humanities and social sciences (Studies in Fuzziness and Soft Computing, Vol. 273), pp. 423–450. Springer, Heidelberg (2012)

Janich, P.: Handwerker und Mundwerker. In: Wehr, M., Weinmann, M. (eds.) Die Hand. Werkzeug des Geistes, pp. 271–292. Spektrum Akademischer Verlag, Heidelberg (2005)

Klaus, G.: Kybernetik—eine neue Universalphilosophie der Gesellschaft? (Zur Kritik der bürgerlichen Ideologie, Vol. 27). Akademie-Verlag, Berlin (1973)

Klüsener, S.: Qualitative Heuristik. Strukturierendes Entdecken. Ruprecht-Karls-Universität Heidelberg, Geographisches Institut, Heidelberg. http://www.kluesener-net.de/Deutsch/Pdf/Qualitative_Heuristik.pdf, retrieved 7 August 2017 (1998/99)

Leman, M.: Expressing coherence of musical perception in formal logic. In: Assayag, G., Feichtinger, H.G., Rodrigues, J.F. (eds.) Mathematics and music. A diderot mathematical forum, pp. 185–198. Springer, Berlin, Heidelberg, New York, Barcelona, Hong Kong, London, Milan, Paris, Tokyo (2002)

[45] To this and not too far fetched, there are analogies in evolution. The origin, formation, and progressive differentiation of species as "varieties", can be interpreted in the interaction between mutation and selection as 'variation'. Thereby, selection in environment is intensified through the sexual selection, and a further interaction exists between environment and epigenesis, which creates additional variations. In reality this mechanism is somewhat more intricate.

Lichtenberg, G.C., Sudelbuch A. [A 1], Erstes bis fünftes Heft (1765). Georg Christoph Lichtenberg. Schriften und Briefe. Herausgegeben und kommentiert von Wolfgang Promies. Vol. 1. Hanser Verlag, München (1968)

Löther, R.: Die Beherrschung der Mannigfaltigkeit. Philosophische Grundlagen der Taxonomie, Digitalisat Max Stirner Archiv Leipzig – 17.5.2015, retrieved 7 February 2017 (2015)

Nicolas, F.: Questions of logic: Writing, dialectics and musical strategies. In: Assayag, G., Feichtinger, H.G., Rodrigues, J.F. (eds.) Mathematics and music. A diderot mathematical forum, pp. 89–112. Springer, Berlin, Heidelberg, New York, Barcelona, Hong Kong, London, Milan, Paris, Tokyo (2002)

Nowak, A.: Musikalische Logik, HmT – 38. Delivery (winter 2004/05)

Seising R.: Die Fuzzifizierung der Systeme. Die Entstehung der Fuzzy Set Theorie und ihrer ersten Anwendungen—Ihre Entwicklung bis in die 70er Jahre des 20. Jahrhunderts. Steiner, Stuttgart (2005) (English version Seising 2007)

Seising R.: Information—ein Begriff im Fluss. 60 Jahre 'Mathematische Theorie der Kommunikation'", Vortrag zum Symposion 1948–2008: 60 Jahre Kybernetik—eine nach wie vor junge Wissenschaft?!, Berlin, 21–22.11.2008 (not published)

Stelzner, W.: 3.2.2 Mehrwertige Logik, from: Logik. In: Enzyklopädie Philosophie. Three volumes, plus one CD-ROM. In cooperation with Dagmar Borchers, Arnim Regenbogen, Volker Schürmann and Pirmin Stekeler-Weithofer, ed. by Hans Jörg Sandkühler, p. 1442f. Felix Meiner Verlag, Hamburg (2010)

Weinberg, S.: Die ersten drei Minuten. Der Ursprung des Universums. dtv, München (1980) engl. ed.: The first three minutes. A modern view of the origin of the Universe. Basic Books, New York (1977)

Zadeh, L.A.: Foreword to: Kruse, R., Gebhardt J., Klawonn, F.: Foundations of fuzzy systems. J. Wiley & Sons, Chichester, p. VI (1994)

Zillien, N.: Ludwik Fleck und die »Verehrung der Zahl«. Beitrag zu einer Soziologie der Quantifizierung (6 Robuste Vagheit: Quantifizierung in der Wissenschaftsgesellschaft). In: Endreß, M., Lichtblau, K., Moebius, S. (eds.) Zyklos 3. Jahrbuch für Theorie und Geschichte der Soziologie, pp. 15–51. Springer, Wiesbaden (2017)

Chapter 2
The Structure of the Music Process and the Work's Thirteen 'Forms of Existence'. Identity in and as a Process, Against and Through Differences

Abstract The music process is an overarching whole from its motivation to its effect. Here, we analyze a section of it, the work process in its metamorphoses from the composer's initial idea through the performances to the quotation of the work. This work process is structured into its thirteen 'forms of existence' (H.-W. Heister): Conceptual, Sketch, Thought, Score, Anticipation, Sound, Perception, Phonographic, Recollection, Word, Picture, Imagination, and Quotation Form. The music process and the work process are a simultaneously and successively fanned out 'Segmented continuum', vertically organized in five anthropological layers, and horizontally in different, in tendency all sensorial areas with the 'Mimetic Ceremony' as background. The work develops its identity in and through differences: a sublation as 'processing identity'. Basic categories are thus provided for the following detailed analyses of all facets of the work relevant to Fuzzy Logic.

Keywords Arrow of time · Artwork · Dialectics: difference and identity · Five anthropological layers of the music process · 'Mimetic Ceremony' · Music as process and as a result of a process · 'Processing identity' · Segmentation, horizontal and vertical · Sensorial areas · Simultaneity · 'Sublation' · Thirteen 'Forms of Existence' · Three phases: production + composition, reproduction + interpretation, reception · Variance and "constancy of shape"

2.1 The Music Process as Overarching Entirety—The Artwork as Integral and Concentrate. Productive, Reproductive and Receptive Realization of the Idea

An important component of the musical process are the various modes of realization. That is, how an idea of a work is sensorially and tonally transformed, and thereby receives a material dimension of the non-material ideas. (see above, *Introduction*). Here, tensions and contradictions between idea and realizations are decided, and last but not least, those between sharpening and blurring; and generally, for each case solutions which include interactions between sensorial-tonal and sense have to be found.

H. Heister, *Music and Fuzzy Logic*, Studies in Fuzziness and Soft Computing 406,
https://doi.org/10.1007/978-3-662-62907-9_2

2.1.1 Music as Process—and as Result of the Process

Somehow music appears only in a certain abstraction from reality as a simple "thing". Even if it is communicated economically in form of a commodity, as it is nowadays almost universal. This means, the character of technically reproduced music bears at the same time the form of a thing (object). For sound storage mediums from Edison's phonograph to the compact disc this seems obvious, but is also valid, somehow less noticeable, for concretizations and storage in the internet (cloud) as well. In fact, music originates, exists and operate in an entire net of relations. As product and object it is a partial moment of the musical process. Or, to express this more radically: By nature music is a process passing in time. In the form of notation or sound storage mediums, then this process is only temporarily halted and at the same time concretized. Therefore, we could consider 'music' following the pattern of terms like 'composition'. Because such terms refer "in most of the western European languages" "to a process as well as to the result of this process".[1]

In an "art work" this musical process is concentrated in aesthetical and artistic, sensorial and mental concretizations. As already mentioned, "art work" or work as term means a structural fact, and not a value concept. The process is concretized in an object—also improvisations have a work-character, as they can more or less be sharply started and ended, along with a logical sequence in a wider sense. The work is the musical process's integral and concentrate as a structured and specific shape. There are reasons and purposes within the musical process that are resting on its outside. It comprises the whole range, which is timely and functionally quite sophisticated, ranging from idea and performance to impression and impact. The musical process itself is neither driving motif nor determinative reason. We play or listen to music, because each time this makes a certain sense or reason for us—and even if this exceeds to the limits, like killing time or cutting off aesthetic sensibility: This is also inner-musically neither reasonable nor justified. Although music, as well as art, is not a thing, but a process, that does not mean, it needs no concretization. Concept art is no artwork, only an instruction for art, which still has to be realized.[2]

Natural and historical-social, sensorial and mental, emotional and rational, individual and collective, practical and theoretical matters of fact enter in the musical process, in its genesis, content and character, along with function of music. Thus facts, from various areas and layers of reality, starting from physical-acoustical across the psychological till the social and ideological, are embraced in the musical process. The simpler gets superseded by what is more developed. But it will also be preserved and is further effective: In music, as in arts, it will be raised on a higher stage, without eliminating the "lower" ones. Yet, from the most primitive to the most sublime, generally even within every single piece of music, or musical process, the internal arrangement and the arrangement in layers of the musical process may be observed. This may be one explanation for the fact, that a piece of music can have many impacts

[1]Bellos 2011, p. 32. To this concept of music as process, cf. Heister 1999, Heister 2000, and Heister 2001.

[2]About the discussion see, among others, Siebert 1998, p. 199f.

or multifarious functions, depending on whatever side of the listener it touches upon, and whatever strings it lets there sound and resonate. The physiological and psychological side of human nature is incarnated objectively and representationally within the musical process, and on the other hand it resonates in object-subject-relation. Between the layers, and also inside of each layer, always principles and problems of Fuzzy Logic from sharpening to blurring are set in function.

2.1.2 *Vertical Segmentation: Simultaneity. The Five Anthropological Layers of the Music Process*

The layers are not strictly defined from one another, but are distinguishable. (For the terminology of such layers there is still a potential for development as for their specific study.) All of the layers are mediated with each other. As no one and no music consists only as a physical constitution, or only as an intellect, although each of the separate layers is distinct in its parts, weight and relevance. Therefore, objects as well as subjects, humans as well as music pieces, are varying according to context and situations. Hence, most of the layers are interacting in the concrete. Nevertheless, with regard to their quantitative share and qualitative weight, they are variable, depending on purpose, social location, how they are performed, their style or genre of music etc. The layers of the musical process, being the relevant category, include the physiological-biotic, the mental, the socio-psychological and the ideational-ideological, apart from the physical-acoustics, as a special layer, which carries the whole. Added to these is a timely perspective of depth with different dimensioning: longue durée for the physiological, somewhat shorter set up for the mental, and for the social and ideological more or less rapidly changing throughout history. Here too, we find the phenomenon, that the historical-chronological is quasi-spatially reflected. Once again, Adorno may also be cited in this case: Material as such is "sedimented mind", "a social matter, preformed by man's consciousness".[3]

2.1.3 *The Layers Between Physics and Philosophy*

We have the following layers:

1. physical-acoustic
2. physiological/somatical—unconscious
3. affective/emotive—physiological/psychological I—partly unconscious
4. communicative—socio-psychological—conscious
5. physical-acoustic

[3] Adorno 1958, p. 39.

The first, or lowest, fundamental layer of the musical process is physical-acoustic: Music as sound, sonorousness. This layer represents the "outer" world, the real object in relation to humans; but not a nature that is pre-human, or apart from humans. For music is always man-made. Music as language of sounds and the language of words share the acoustic medium, however instantly with the division of labor between predominantly emotive and cognitive, which goes beyond mere naturalness. The acoustic substrate forms the basis, which is "apart from humans", for music as "tonal art", as well as for music as noise. Horst-Peter Hesse has under the generic term "acoustic", being a science of "sound and its perception", pointedly, but factually, reconciled the outer and human (physical as well as psychological) nature:

> The physical acoustic is concerned with development, spreading, transmission and ranging of sound, whereas the physiological acoustic or physiology of hearing is dedicated to the biological process of hearing and those organic processes, tied to it, whereas psychological acoustic, or psychoacoustics turns to phenomena of hearing and their interpretation and ordering in the psyche.[4]

The physical layer of the musical process has insofar a special status, because it has, nearly exclusively to bear the total burden of sensuality and sense of music; acoustically as sound—and visually as view and ambience of playing music-contradictory. On the one hand, in the physical-acoustic event of music, anything that is non-acoustical, gets objectified in the resounding 'thing', or process,—inasmuch as music is just tonal art. But on the other hand, there exists much more that is not part of acoustics. These layers correspond in their outlines to one layer on the side of humans, who are producing or perceiving music: In the musical process the anthropological-social substrate is concretized, virtually incarnated, because humans are a mediated unity of different layers. Thus, firstly, the musical process is layered vertically, from the physical-acoustic to the non-material layer of the ideas.

2.1.4 Horizontal Segmentation I: Succession. The Arrow of Time

Secondly, the musical process proceeds in time, that is, directed and along the arrow of time. The term 'process' takes temporariness more seriously and more substantially. Because art in time is the first of the three basic definitions of music. The second is the definition of music as an executive art, that is, it needs for its performance a practical-mental realization. (Thus, against common prejudices, the latter, the mental, belongs also to the performance of music.) What is often ignored, is the fact, that basically music shares both definitions with theatres, or interpretative arts. Finally, thirdly, music is tonal art, with a priority of the acoustics.

---

[4]Detailed in Hesse 1995, col. 1104f.

2.1.5 Horizontal Segmentation II: Fanning Out in Various Sensorial Areas

Thirdly, the musical process is fanned out 'horizontally' in various sensorial areas, centered on the auditory. In this third dimension, here only briefly outlined, music appears, genetically as well as structurally, to be part of the 'Mimetic Ceremony'.[5] It comprises as a total artwork, in virtual-perspective or actually realized, all sensorial areas and arts. On the one hand, these are the different dimensions of senses of the musical process as such, obviously besides the acoustic and visual, also the tactile, i.e. touching and gripping musical instruments, sense of balance, sense of kinaesthesia etc. There is a tendency for the participation of all senses, while varying qualitatively and quantitatively in emphasis and blending. On the other hand, from the point of view of human nature, the mediated unity for the nature of instincts and reasoning, such dimensions of senses are differently configured in various arts: As for some types of the total artwork, which originates in the Mimetic Ceremony, there is a tendency for completeness, in specific parts of the respective separate arts.

2.1.6 The Three Main Phases: Production/Composition, Reproduction/Interpretation, Reception

As a chronological sequence the musical process is segmented in various phases. The polarity of production—reception is the most ordinary segmentation. Then under the precondition of composed and notated music, production is split up in production/composition for the realization of the (work) conception and in reproduction/interpretation for the tonal realization of what was composed/notated. In analogy to the case of the term musical process, the term art process is as well essentially, mutatis mutantis, valid for the other arts. Likewise, the concept of forms of existence. As in both other dimensions of the musical process, there are between and within such temporary phases, layers and sensorial areas, tensions and contradictions between ideas and realizations, conceptual and tonal, ideational and material, together with relative solutions. Again and again, the fundamental dialectic of sharpening and blurring turns up. The musical process, as a whole, as in its separated phases and components (layers and sensorial areas) is always moving oppositely: Between specification of the vague, contouring of the fuzzy, elaboration of concise figures on the one hand, while on the other hand blurring in the productive as well as the receptive realization of music. In a certain harmonizing exaggeration, one may call it complementarity, when two poles are joined and raise up to form a higher unity.

[5] To the explication and explanation of this term see detailed Heister 2007, *Mimetic Ceremony—total artwork and all senses. Aspects of a concept*, pp. 143–184.

2.1.7 Partial Advancement from the More Vague to the More Precise—Tendencies and Countertendencies

The main tendency within the layers is an ascendant tendency to a higher distinctness, and above all awareness from physiological to the intellectual, which behaves like this: It is more blurred in the bottom layers; it becomes sharpened in proportion to a deeper involvement of the intellect. Thus, the more cognition, the more sharpened becomes the expression, and likewise the perception of shapes. Seen in the light of Fuzzy Logic, the principle of sharpening acts as filtering and crystallization. Anyway, cognition and emotion are closely connected with each other. Here, affects, or emotions are only to be treated as marginal and basic conditions, above all in connection with the forms of existence after the form of notation, even as they naturally motivate while making a draft and a composition. They are one factor, which can lead to blurring as well as to sharpening. That depends on the concrete circumstances. For instance, someone who is particularly interested in a certain piece of music, will consequently pay much higher attention to his playing/singing or listening. Though to validate this empirically and experimentally here is impossible. It is at least complicated. For this would demand a great effort, but principally it can be achieved.

Within the phases and forms of existence there is no linear sequence from the elementary to the complexity, from fuzzy to precise. For after the ascendant tendency to the fair copy form, or score form, follows a mainly descending tendency in the further forms of existence, as will be explained in detail. For instance, in one's memory, or remembrance, there are superpositions of various reminiscences in temporal succession, like a kind of palimpsest, i.e. the superscription on a leaf, already filled with writing. There are attempts of a precise verbal transformation of musical phenomena, such as imaginations or impressions and drifting into vaguely impressionistic. Generally, within the different forms of memory, coexist various kinds, contents and degrees of accuracy of mental representations of musical processes, alternate and counterpoint. Thus, sharpening (in the double character of filtering and crystallization) and blurring (not-intended and intended) are actually interconnected.

Certainly, such a dialectic may also be found in other arts, even in similar dimensions. Essentially they differ in their specific material and their sensorial areas (acoustic, visual, in motion or steady), but already much less in techniques, which are art overlapping, i.e. 'composition', rhythm, montage etc. And in the various arts, the arrangement of the respective art process in phases between initial ideas and later aftereffects is similar, quite often even identical, depending on the extent of abstraction of an approach.

Between and within the phases there are various, heterogeneous interactions, even in the terminology of systems theory and cybernetics, feedbacks. At the latest, if it concerns circumstances of reflections in the broader sense, which may also be reciprocal, then the principles of Fuzzy Logic are gripping. For example, at the opera not only the setting and the scenic action refer to the scenario, the "drama", but also the acoustic dimension, i.e. words and music (whereby, words are already largely set in music). Here, only the respective sharpness or fuzziness of such references may be inquired, even if it is hard to quantify them.

2.2 Dialectic—Becoming and Being, Contradictions and Unity, Change and Constancy

These interactions etc. may be named, as proposed in the title of the book, as "dialectic". As such, there is no need for a closer specification. Nevertheless, some short quasi-defining explanations about it. Contradictions are generally viewed as necessary constituents of reality and as driving forces for developments. In particular and in regard to modal logic one may quote Ernst Bloch: "I am, but I do not own myself, therefore we are in the making."[6]

2.2.1 Sublation as "Processing Identity"—A New Category for the Entire Process and Result

Dialectic signifies: (1) Objective inner contradiction of matters of fact, (2) Its movement—of reality and therefore, if and insofar as suitable of thinking and art production and—reception. Such a "proceeding" aims at equalization. But the 'lifting' of contradictions into a new unity on a higher stage may also pass through a sharpening of the contradiction until it reaches an (antagonistic) opposition. (3) The mentioned movements take place through interactions between the respective poles, i.e. sharpening versus blurring. Between such poles oscillates, here, more concretely, especially the multilateral relation between concepts and imaginations about how a piece of music might be visualized: On the one hand as a notated composition or improvisation, and on the other hand, what emerges in the end on paper or on the podium as the real result, and later on, what reaches the percipient, and in turn, what might emerge as mental 'projection'. Interactions also do occur between the different phases of the musical process in its temporal proceeding. In regard to this manifold and segmented entirety, the artwork appears, as already mentioned, as 'processing identity'.

[6]Quoted by the theologian Klaus Schmidt, *Junge Welt*, Berlin 14./15.1.17, Beilage S. 2.

2.2.2 *Difference and Identity: Heraclite's River Bank, River and Transformations of Object and Subject*

Historically Heraclitus of Ephesus (about 550–475 B.C.) appears to be the first to have formulated the conception of a processing identity.[7]

> Fragment B12. „potamoisi toisin autoisin embainousin hetera kai hetera hudata epirrei. On those stepping into rivers staying the same other and other waters flow."(Cleanthes from Arius Didymus from Eusebius) [...] Three versions of this fragment exist: B12 | 49a [...] B91
>
> [...] B49a: "We step and do not step into the same rivers; we are and are not."
>
> B12 is a fragment about the unity/diversity inherit in reality. Reality is changing and in flux (like the different waters), yet it is all stable and unified (like a river). B49a assigns change and stability to the one stepping into the river as well as the river. B91 focuses just on the diversity and not on the unity. I think that B49a above is authentic (at least the first half). B12 also seems authentic, giving the reason for B49a: "different and different waters flow" upon those who step into a river. [...] On those who enter the same rivers, ever different waters flow.
>
> [...] The major theoretical connection in the fragment is that between '*same rivers*' and '*other waters.*' B12 is, among other things, a statement of the coincidence of opposites. But it specifies the rivers as the same. [...] We call a body of water a river precisely because it consists of changing waters; if the waters should cease to flow it would not be a river, but a lake or a dry streambed. There is a sense, then, in which a river is a remarkable kind of existent, one that remains what it is by changing what it contains [...] Another reading: On the same people stepping into rivers, other and other waters flow. With this reading it is people who remain the same in contrast to changing waters, as if the encounter with a flowing environment helped to constitute the perceiving subject as the same. [...] Yet if the rivers remain the same, one surely can step in twice–not into the same waters, to be sure, but into the same rivers. [...] Something much more subtle and profound. It is that some things stay the same only by changing. One kind of long-lasting material reality exists by virtue of constant turnover in its constituent matter. Here constancy and change are not opposed but inextricably connected. A human body could be understood in precisely the same way, as living and continuing by virtue of constant metabolism – as Aristotle for instance later understood it. On this reading, Heraclitus believes in flux, but not as destructive of constancy; rather it is, paradoxically, a necessary condition of constancy [...]. In general [...] high-level structures supervene on low-level material flux. [...] In general, what we see in Heraclitus is *not a conflation of opposites into an identity*, but a series of subtle analyses revealing the interconnectedness of contrary states in life and in the world. There is no need to impute to him a logical fallacy. Opposites are a reality, and their interconnections are real, but the correlative opposites are not identical to each other.[8]

The reason, why I cite these arguments in detail lies in the fact, that they indicate a great deal of the complex interrelations between identity and differences in the musical process, so that I must not refer to them each time in detail. Although the processing identity is one that is becoming and one that came into existence, it is nevertheless an identity. In becoming, instead of being, real, because lastly everything

[7]In the following according to http://www.heraclitusfragments.com/files/e.html, call 2016-12-02.

[8]http://www.heraclitusfragments.com/files/e.html, retrieved 12 Febr 2016. Emphasis H.-W.H.

is historical, even the universe, nature. There one can search after the—relatively!—invariableness. Parmenides, one of the younger generation of Ionic philosophers, or "Pre-Socratics", forces invariableness, as a counter-reaction, as "being". Even the work *will be* and *is*.

2.3 Horizontal Segmentation III. The Sequence of the 'Forms of Existence'—Progressive Elimination of Fuzziness up to the Score Form

The work is, it exists as a product of compositorial or improvisatorial and music making activity, through artistic working with technique and material. Although the whole musical process, like artworks in general, is already subjectively mediated, because it is a matter of art production, or consciousness production. But the work is not just constituted by some cognitive acts of perceptive subjects, and it does not exist as a "construct", but production and its product precede the subjective "reconstruction", temporally and technically, almost "ontologically" in the receptive appropriation, or perception. The work is object, constituted through the activity of the artistic subject, and objectively advancing. But it unfolds as a partial moment of a process—with, through and in this process.

2.3.1 The Concept 'Forms of Existence'

According to the present state of my deliberations, at least 13 forms of appearance of the musical work, or object, result from the temporarily succeeding phases of the musical process. In each case these forms of appearance are differing from each other and are also at variance in themselves. Fuzzy processes and fuzzy facts in a different musical context are reflected and quasi 'fractally' on higher levels reproduced in the complex of forms of appearance or forms of existence ('Existenzformen')—the sequence in time of different states, stadiums of conception, productive and receptive realization of a piece of music between idea and storing in memory. In all these forms of existence there are two opposed processes in regard to sharpening and blurring.

About the notion: In 2008, I decided, for the time being, to use the term 'form of existence' instead of 'form of appearance', which may also be possible. Both implicate, respectively, certain conceptions of the musical process and the position of the musical artwork therein. 'Form of appearance' would require something as its complementary opposite expression, that would be the essence character. The first candidate for such an honorary post would probably be the concrete unfolding and precision of the idea for a work in it's notation. But, on the one hand, this formation of a concept would unnecessarily narrow the orientation on the type of European-American art music. While, on the other hand, such an identification of a notated work and character of the work would be more than problematical, as there is no identity, at most a trend of congruence. The notion form of existence avoids such a hierarchy, though it allows additional differentiations of dignity if necessary.[9]

The systematization pertains not only to composed (classic) music, but also to folk and popular music. In the case of music without notation the sketched form and score form are naturally omitted—even though preparatory notes and even models for performances in the form of guitar tabs or chord cipher sequences may exist even in these cases. The European forms of notation started in Greek antiquity with a letter notation supplemented by prosodic signs. Interrupted and continued then in early feudalism tonal movements were indicated by gestural-iconic signs, the so-called 'neumes', leading in some steps to "our" usual notation. There are also notations of finger positions, specific for different instruments, the "tablature". And in modern music there is again a mainly pictorially prescribing 'graphic notation'. (See Chapters 6 and 7) There are on the other hand globally a lot of other systems of notation, like Indian, Chinese, Arabic etc., as well as a North American Indian graphic notation. But here as a rule, the notation is not a binding written fixation of a firmly defined work-character, thus from the outset blurring has to be taken into account.

2.3.2 List of the 'Forms of Existence'

In the following survey the ideational and only individual-subjective forms are given in *italics*, the others, the materially realized (and therefore available for collective appropriation), are not in italics.

1. Work as notion and idea. *Conceptual Form*
2. Work as outlined and notationally fixed conception of the idea of the work. Sketch Form
3. Work as completed idea. *Thought Form*
4. Work as fixed idea in written form. Score Form
5. Work as sound, anticipated in the conception. *Anticipation Form*
6. Work as idea realized in sound. Sound Form
7. Work as sensorily perceived idea. *Perception Form*
8. Work as technically stored and (re-)produced sound. Phonographic Form
9. Work as sensual idea preserved in memory. *Recollection Form*
10. Work as verbally expressed perception Form and recollection form. Word Form

[9]In linguistics the use of the term 'form of existence' is totally different; for instance, dialects are considered as a form of existence within a national language. See e.g. Serébrinnikow 1975.—Carl Dahlhaus likewise employed the word, following Roman Ingarden, also in another sense as it is used here, that is only as singular in order to differentiate the music from the other arts—*Über die Existenzformen der Musik* (about music's forms of existence), part of a chapter in Dahlhaus 1971, pp. 98–103. I personally used the term already back in the 1970s in a chapter of my doctoral thesis about the theory of the concert, without reflecting on the term itself, in a similar sense as here—*Erinnerung als Existenzform* (remembrance as form of existence).

11. Work as imaginary-real reflex in Visual Arts. Picture Form
12. Work as imagined idea according to Image- and/or Word form. *Imagination Form*
13. Work as intra-musically or intertextually sublated sound-idea-combination. Quotation Form

The phases and forms of existence relate not only on principle as a temporal sequence one by one (apart from the forms 11–13). They are in their contents and concretely retroactive with each other, even in various dimensions, grades and modes, as it was mentioned in the explanation for the musical process as a category. Such retroactions and interlinkages take in each case different shapes. And they are not in all forms of existence of equal intensity. But they form a continuous whole—the musical process is the whole of these forms of existence.

2.3.3 'Segmented Continuum': Overarching Holistic Process with Cesuras

The forms of existence constitute a ‚segmented continuum': A continuous, overarching total process, but with distinct breaks. These forms may occur in very diverse manifestations—short ideas on a paper napkin, or more elaborate, repeatedly revised sketches. In the case of Beethoven these fill entire series of notebooks. On the other hand, if music is not notated, then written sketches are omitted, at least to a large extent, mainly the basic form of the notation form, as well as the anticipation form, which derives from the former. In all these forms of existence there are, as said above, the two opposed processes of sharpening resp. filtering and blurring. Concepts, sketches, notes, or mere mental premeditation and reflection are means for a sharpening specific to art—the formation of concise forms in the process of making and producing, of poiesis. These activities can be found in all forms of existence. But they are especially prominent in the compositional procedure. But even more generally it can be applied to the relation of indistinct beginnings and continuations and the carving out of succinct thoughts and forms—thereby for the entire field from idea through sketches up to the completely notated musical art work.

1. The gradation of the existence forms is a temporal sequence, as a specific structuring of the musical process for each individual work: One after the other.
2. But on the other hand, the processes and activities within each form of existence intertwine through retroaction and anticipation. In this context the concept of enkapsis, the interlocking of various circumstances, could be applied productively. Composers control through the Score Form and later the Sound form, in how far they have succeeded in precisely and clearly formulating their idea from conception to notation.For instance, it is well known, that Anton Bruckner was constantly modifying and revising many of his symphonies, and that he gave recommendations to the conductor for improvements, which were qualified as well as less qualified. But also Mahler's improvements, who even knew how to

counsel Mahler, the composer. Moreover, here one may already see the repercussions of the practice of performance and reception on composing. In the case of Boulez such a retroaction lead to permanent work of retouching, paralyzing his productions. In the end this turns into acts of compensation for a lack of real compositional productivity: In order to play his old but touched up work as a new one for the first time, he called his approach work in progress.
3. Thirdly, beyond the individual work there exists a polyphonic texture of realizations, perceptions, recollections: At one with and next to many others.

All these interlinkages, intertwinings, interactions flow into the musical process. Thereby deliberate and unintentional alterations form a cycle of variations. This goes for productive as well as receptive realization, the realization of the work as sounds and the perception and resulting stages of reception, but also already for the formation and (notational) realization of the work idea. Altogether this sequence of existence forms from 1 to 6 with 7 to 9, 10 and 12 results in a sort of unintentional or directionless morphing—which does not only exist for pictures, but also for music, that is a changing of "Gestalten" but a conserving of some identical relations. (See below) The whole of the artwork as processing identity evolves and is established in and by the different forms of existence. The identity is, again following the configuration of core and contours, in itself necessarily fuzzy. But this does not mean the "anything goes", whateverism and mere subjectivism: The objective measure of the differences is qualitatively and quantitatively definable.

2.3.4 Processes of Composition and Decomposition, Sharpenings Versus Blurrings

At first a growing order is set up: As in a calculator, in life, in the brain and on paper (or on the screen) as well. Then the order thus grown is transformed. Within the gradation of the existence forms of music, the order will not be cut down in the strict sense, though similarities with the original character of the work become continually fewer—at any rate as a main tendency of the sequence. The work fades away; gradually its details, contours, distinctness will be blurred. But especially the sound form itself, as an intersecting point, will be refreshed through repeated performances of the same piece, although in this case variations within the repetition are required. Like it is similar in repeated perceptions, that is, in repeated lending an ear to the same piece or the same interpretation. In particular technical reproductions or the phonographic form offer suitable prerequisites for both. The fading reminiscence is filled with new life, and is kept alive—hopefully, somewhat different than the shades in the Nekyia of Homer's Ulysses: In Hades they press eagerly around Ulysses' blood sacrifice, and when satiated, inform him about past and future, then they relapse back into the realm of shades.

The 'sound writer', the "Phono-Graph" transforms the unique distance into a repeated proximity—and this for many people, in a certain difference to current prejudices about notation, as being at least somewhat hermetic. It creates a reification of music resp. the musical process exterior of the subject and the body, furthermore as an extension and enhancement of the musical notation, and is likewise the storage of what is perceived and therefore relieves the memory. So to say, it functions as a mental prosthesis, or "projection of organs" (Arnold Gehlen). It is a material perpetuation as a reflection of—unattainable—immortality.

Also in the sound form there are losses, as in each stage, here in particular, just to mention it in passing, losses of extensive sensorial realization, and blurs. Though the real acoustical noise was at least partially reduced in history, in ideology it heavily increased. Likewise, benefits also became apparent, as the conservation of long gone sound, the support for traditions of interpretation and much more. Separate passages, nice or difficult, may be singled out and repeated—the more advanced such a technical reproduction became, the easier it is, namely for masses of sound as well as democratically for mass use. (Anyway soloists or musicians playing in small ensembles were always capable of doing this.) For its part, the sound form starts to operate a whole process of circulation from 'performance', recorded performance, perception, recollection—always with the chance, to repeat a relatively identical realization of the work with itself (or an improvisation etc.). Changes in the objective context conditions, such as locality of play, from its acoustics to the humidity, as well as subjective context conditions, such as mental state and mood are then unavoidable blurrings of identity: For the work as for the perceiving subjects.

Here as well, a dialectic of chaos and cosmos unfolds. The contours are increasingly blurred, obscured, nevertheless still perceptible, even if all the details had faded or vanished long ago. These different forms of existence are connected with one another, they change and transform into each other: when the 'work' is transformed into sound, when the work listened to is re-transformed into sensations, imaginations, and ideas. In this context, it is always essential to realize, how shapes are build and dismantled, sharpened and blurred contours and to take in account unavoidable blurs. While cybernetics and systems theory on the whole grasp above all the general principles of regulation also here, Fuzzy Logic supplies revealing aspects, in context with cognition research and other branches, especially for the study of perception and memory. The modes of thought and procedures of Fuzzy Logic help to understand more accurately the important aspects of the relation of particular forms of existence in detail, as well as the development and the changes of the 'work' in its different forms within the sequence of phases of the musical process. In the musical process, first of all, the ideational and material formation of patterns is the problem and task. The production of patterns precedes the 'recognition of patterns' and the storage of recognized patterns, which are researched with priority in cognition science and also by Fuzzy Logic: Forms in art are not simply given, thus differing from nature. (See Chapters 4 and 5) Perception is not only mediated through storage and thus to the memory, but also by production. Processes of memory and perception are thereby interlocked, once again also in circular causality: What we appreciated of earlier listening is present, if we listen to something new. Furthermore, there are correlations between learning, remembrance and forgetting, not wholly symmetrically, but nevertheless quite extensively. For instance, different types, like "procedural" and "declarative" (semantic) memory, likewise for implicit and explicit learning, or learning of facts (episodic learning) and the practicing skills (procedural learning).[10]

Also the temporal phases of ultra-short-term, short-term and long-term memory, which is especially important for issues about music as a timely art, belong also to the complete framework of learning and remembrance, unlearning and forgetting and practicing what is learned. As already mentioned apropos the emotive layer, consciously and unconsciously, affectively directed attention, aesthetical interest and surely all kinds of emotive motivations for the production and perception of musical forms, function as basic conditions.[11]

In regard to the cognitive aspects this gradation means: At every stage, i.e. in every existence form of music, entropy is reduced, contrary to the universal laws of thermodynamics, since these are processes of living/animated systems. In this case the living systems are the participants of the musical process, foremost the composers, interpreters/artists/performers, recipients, critics and reflecting bodies in their contestations of their inner states and tasks/purposes/assignments as well as with the respective 'surroundings' including objects and systems. The musical art work at first seems a fully accomplished unity („in sich vollendetes Ganzes")—a classical definition by Karl Philipp Moritz, a contemporary of Wieland, Goethe, and Schiller—, even as a "world in its own" („Welt für sich selbst"), in a closed system. At the same time it is also an open system, since in the phases of the musical process, or rather in its various forms of existence, new aspects of the work may emerge as anti-entropic 'information'. For each productive and receptive activity, material and ideational, bears additional information for the entire system and the individual existence forms—including the jeopardy of practically or theoretically faulty interpretations; all of them are necessarily subjective, apart from the notation form—and even here individual idiosyncrasies such as notation mannerisms flow into the written materialization of the object. Thereby subjectivity generally gains some margin and leeway, in spite of all its linkage to the musical object. Altogether the music makers and listeners act as systems capable of learning.

However, after the precision reaches its real-imaginary peak in the fully elaborated idea and its realization in the score form, the fuzziness increases in every subsequent stage: Not everything in the notation is sung and played correctly and completely, not everything that has been realized is perceived accordingly fully to the realization/interpretation and to the work. And even the ideal listener cannot describe her/his entire music experience, the music critic cannot—or due to allotted space may not—express and write down all her/his impressions appropriately and so on. This discrepancy already comes into play, taking the point to an extreme, with the—real-material—notations in relation to the idea and the conceptions.

[10]Cf. e.g. Spitzer 1996, especially p. 315f.

[11]Here, this thematic can not be treated in detail. For instance, M. Spitzer 1996 presents thereto inspiring and instructive ideas along with charts, e.g. in connection with „Kohonen networks"and other computer controlled simulations of neural nets, especially p. 108–112.

2.3.5 Complete and Incomplete Cycle of 'Forms of Existence'. Word Form, Picture Form and Quotation Form as Exceptional Cases

Not always and in every case does a work go through all stages. For example, in spite of the fact that visual associations complement our memories only rarely is the picture form fully artistically articulated. But as a rule at least the first ten forms of existence are taken. In principle, nothing is changed, if works are not somehow at least in outlines objectified, not performed or unheard, and in similar cases. Although this happens not quite unexceptionally.

Even improvisations cannot neglect exercise and preparations. The latter precedes the actual process of realization. Although it shortens, reduces or condenses the first five forms of existence, namely, conceptual form, sketch form, thought form, score form, anticipation form. But even here their traces and outlines may be distinguished. In the setting of European-American "western" music a certain minimal rudimentary script was established. Therefore, even in jazz sketches of "sheet music" notes of compositions with melodies, occasionally along with lyrics, and harmonic changes as well as arrangements for several voices can be found. Compared to this, scriptures for Indian or Arabic-Islamic music are obviously fewer.

The system of existence forms could be refined even more by introducing more gradual stages.[12] In the case of hallucinations the anticipated form and the perception form take on a life of their own, independent of the specific occasion of the work: External sounds or tones, or internal ones as in a tinnitus may suffice to trigger an accordingly musical imagination. In principle these are well-known musical pieces, mostly more simple melodies ranging from nursery rhymes to chorales and pop songs, but depending on educational background and experience also more complex works[13] At the same time, just like in comparably rare appearances of complete musical works in dreams, the past, in its cognitively recollected form, is retroactively transformed into present perceptions, and thereby in a certain sense from potential into kinetic energy; but in a way that the perceptions, lacking an external point of reference, take place within the individual. The existence form "Work as complete idea" is the culmination and turning point in regard to precision. From here on the indistinctness increases, unconscious or conscious formation of variations and more—just to mention the problem of musical interpretation.

---

[12]For some instances, the terms too are certainly also capable of improvements.

[13]More fully on such and similar phenomena, see Sacks 2007/2009; see Sect. 6.1.

2.4 Core and Contours. The Variable and the Relatively Invariable

In every phase of the musical process, at every stage occur gains and losses in regard to precision of forms, to sharpness or fuzziness, and finally this also applies to information. These 'fuzzinesses' are probably specific for each form of existence, in the way they form variations and in their structure (and their content). Everything mutates, changes, and hardly anything remains the same. As in biological evolution the main problem is not so much the changing as instead the stability. But the stable one is often substantial, like the core motif embedded in thematic variations, or the contour of melodies specific for the respective work objectively and—that is not the same—the contours stored in the subjective, individual memory.[14] Similarities of all kinds are crucial. Altogether the Cartesian „clare et distincte"(clear and distinct) proves to be faulty—if it can at best be found in the final result of a process, or more concise mainly as specific intermediate result.

But even here out of chaos a cosmos comes into being, for instance through the development of musical themes from motivic germ cells. Altogether the whole process refers to the fact, that the constitution of the work could somehow be compared with the constitution of the self, or the subjects. Concepts, sketches, notelets or even simply mental pondering and anticipating are means of a "sharpening" specific to the arts, or like a crystallization for the creation of suggestive forms in the process of preparing and producing, in poiesis. Such activities can be found in all forms of existence. But they make themselves clearly conspicuous in those of the compositional process in a stricter sense, namely in the forms up to the fair copy form, or notations form (1 to 4). "The Frenchman says, l'appetit vient en mangeant, and this sentence of experience remains true, if one parodies it, saying, l'idée vient en parlant." So Kleist in his famous vignette *Über die allmähliche Verfertigung der Gedanken beim Reden* (1805; "On the gradual production of thoughts while talking"). Initially this goes for all kinds of improvisations, like talking and rhetoric (Kleist's principal illustrative model), acting in a theatre or playing music. But more generally this goes actually for the relation of blurred startings and continuities and working out distinct thoughts and forms. Therefore it goes for the whole range from the idea to sketches to the completely notated musical artwork.

2.4.1 Dialectic of Variance and "Constancy of Shape" Within and Between the 'Forms of Existence'

The dialectic of variance and constancy of forms concerns also the forms of existence itself. Their unreflected respective identity is taken for granted—a work is a work, "rose is a rose …", the tonal realization is once again the purely objectively interpreted

[14]To the significance of 'contour' cf. Quinn 1997.

work, the technical reproduction redoubles the tonal realization and so on. But all that is already an identity of constancy and changing within an in principle never ending process of changes and modification: Therefore, identity is coming from and as difference. Here it has to been taken for granted, that in the frequently referred to neural nets, also inside the brain, there is not only learning, but what is often forgotten, also unlearning. As already mentioned, even the growth of blurring appears to be as a predominant tendency, though in the course of the musical process it can never be the exclusive tendency. Usually this happens unintentionally, against the will for preservation and distinctness. But occasionally it may even have been already from the outset and during the production the artist's intention: For instance in visual arts, Gerhard Richter's technique of photo-paintings or Arnulf Rainer's "overpaintings"—blurring as art.

At first it may appear, that in the contradiction of production—reception lies the main source for blurs: Simply said, the public does not understand new ideas and sounds, or only vaguely and insufficiently. Quite often this may be the case. But if one looks closer and reflects deeper, this cannot be decisive. For the respective contradictions date back to the composition itself and its phases. Fuzziness up to impossibilities to fully realize compositionally of what was thought and deliberated existed even then, marking a dialectic in the process of creation of musical forms with the main tendency of growing precision. As already mentioned, by some deeper considerations the underlying tensions between idea and realization is revealed. Shades and facets of this relation, as the contradiction between theory and practice, between conception and realization, and more of such polarities, fan out this basic problem. Like already explained at the beginning, generally mathematical or mathematical-analogous perfection are in conflict with artistic and technical completeness in the realization.

This implies necessarily also blurs, in any case coarsening—but in return also sensualisation: If art is only imagined, that is (still) nothing. Therefore "concept art" is only seemingly a counter-example. If it should ever reach an inter-subjectively real existence, then at least concepts must have been written.[15]

2.4.2 System and Surrounding, Motivations and Functions

As already mentioned, after all, within the frame of forms of existence, the musical process and the separate partial processes as a 'system' along with subsystems, are inserted in an 'environment' of comprehensive lineary developments, such as culture, the system of 'art', along with the dimension of history, or history of culture and music. System's, or rather subsystem's relations with environments constitute numerous further starting points for the dialectic of blurrings and sharpenings—which can not be treated here in detail.

[15]See below *senza titulo* (untitled), by Dick Higgins), in 6.3.

The real psychical, social, historical and music-cultural causes and motives, the reasons and motives for the cycle of variations of the forms of existence pose a complex, once again newly problem, which can here not be dealt with in detail. Thus a whole bunch of areas, questions, and problems has to be excluded from the start. Above all, as mentioned before, individual and social emotions provide primarily causes and aims to set the musical process actually in motion. Ultimately this is a matter of catharsis, in other words removal of desires, excitement and subduing of individual and social affects. Likewise not extensively treated will be the material, energetic substrata of changes and the gradation, such as the transformation of chemically bound mechanic energy into heat, diaphoretically while playing music, or emotionally while listening or dancing to it. All this belongs to the 'lower' levels of the musical process. Nevertheless, there is definitely a dialectic connection between id- and egohood, physical and psychical, emotive and cognitive. Above all, here it is a matter of the ideational, or rather cognitive dimensions of the musical process,[16] and this also essentially in structurally-formal regard, but over and over again with a view on concrete associations, meanings and contents. It remains to be seen, whether immediately a new subdiscipline of musicology in form of a "cognitive musicology" must be established, one that belongs to the realm of "artificial intelligence" and which is especially preoccupied with computerized modeling of musical cognition processes, and on the other hand should be closely associated with "biomusicology", "gestalt psychology" and others.

References

Adorno, ThW: Philosophie der neuen Musik [1949]. Europäische Verlagsanstalt, Mannheim (1958)

Bellos, D.: Was macht der Fisch in meinem Ohr? Sprache, Übersetzen und die Bedeutung von allem. Eichborn-Verlag, Köln (English as: Is that a Fisch in Your Ear? Translation and the Meaning of Everything. Penguin Group, 2011) (2013)

Dahlhaus, C.: Musiktheorie. In: Ibid. (ed.) Einführung in die systematische Musikwissenschaft, pp. 93–132. Arno Volk Verlag, Köln (1971). Reprint in: Danuser, H. et al. (eds.) Carl Dahlhaus, Gesammelte Schriften in 10 Bänden, vol 2., pp. 209–232. Laaber Verlag, Laaber (2000–2008)

Heister, H.-W.: Natur, Geist, Gesellschaft. Musik-Anthropologie im Gefüge der Musikwissenschaft. In: Petersen, P., Rösing, H. (eds.) 50 Jahre Musikwissenschaftliches Institut in Hamburg. Bestandsaufnahme - aktuelle Forschung—Ausblick (Hamburger Jahrbuch für Musikwissenschaft, vol. 16), pp. 65–82. Peter Lang GmbH, Internationaler Verlag der Wissenschaften, Frankfurt a. M. (1999)

Heister, H.-W.: Geschlechterverhältnisse als Modell. Musik-Anthropologie: Gegenstände, Themen, Forschungsperspektiven. In: Stahnke. M. (ed.) Musik – nicht ohne Worte. Beiträge zu aktuellen Fragen aus Komposition, Musiktheorie und Musikwissenschaft (Musik und. Eine Schriftenreihe der Hochschule für Musik und Theater Hamburg, vol. 2), pp. 13–46. Von Bockel-Verlag, Hamburg (2000)

[16] New approaches such as the 'Neuro-Psychoanalysis' are trying to build bridges; so for instance Solms/Turnbull 2002, Hüther 2006.

Heister, H.-W.: Natura, spirito, società. Antropologia musicale e nuova musicologia. In: Musica/Realtà, vol. 22, issue 65, pp. 25–54 (Part I); Musica/Realtà vol. 22, issue 66, pp. 115–150 (Part II). Edizioni Unicopli, Milano (2001)

Heister, H.-W.: Mimetische Zeremonie—Gesamtkunstwerk und alle Sinne. Aspekte eines Konzepts. In: Ibid. (ed.) Mimetische Zeremonien—Musik als Spiel, Ritual, Kunst (Musik und. Eine Schriftenreihe der Hochschule für Musik und Theater Hamburg. Neue Folge, vol. 7), pp.143–185. Weidler Buchverlag, Berlin (2007)

Hesse, H.-P.: Gehör, IV. Psychoakustische und psychophysikalische Grundlagen. In: MGG^{2}...L. Finscher (ed.) (Die Musik in Geschichte und Gegenwart) 2nd, neubearbeitete Ausgabe, Sachteil vol. 3, col 1103–1118 (1995)

Hüther, G.: Die Macht der inneren Bilder. den Menschen und die Welt verändern. Vandenhoeck & Ruprecht, Göttingen, Wie Visionen das Gehirn (2006)

Quinn, I.: Fuzzy Extensions to the Theory of Contour. In: Music Theory Spectrum 19, https://doi.org/10.1525/mts.1997.19.2.02a00050, pp, 232–263 (1997)

Sacks, O.: Der einarmige Pianist. Über Musik und Gehirn. Rowohlt, Reinbek (2009) (English 2007)

Serébrinnikow, B.A.: Allgemeine Sprachwissenschaft, vol. 1: Existenzformen, Funktion und Geschichte der Sprache, 2nd, revised Auflage. Akademie-Verlag, Berlin (1975) (Russian 1970)

Siebert, C.: Qualia. Das Phänomenale als Problem philosophischer und empirischer Bewußtseinstheorien (Phil. Diss., Humboldt-Universität zu Berlin). Berlin (1998). http://edoc.hu-berlin.de/handle/18452/15088

Solms, M., Turnbull, O.: Das Gehirn und die innere Welt. Neurowissenschaft und Psychoanalyse. Walter-Verlag, Düsseldorf (2002) (English 2002)

Spitzer, M.: Geist im Netz. Spectrum Akademischer Verlag, Heidelberg (1996)

Chapter 3
Fundamental Fuzziness: The Imperfect Perfection

Abstract This chapter unfolds the central concept of blurs or fuzziness from the side of FL as well as from the side of art, especially music, and thus forms, continuing Chapter 2, a basis, also for the analyses in the following chapters. "Vagueness" is avoidable, fuzziness is principally not. But: In many processes, both in the compositional and the tonal realization of the idea of the work, fuzziness has a productive and formative effect, sometimes indirectly, sometimes purposefully. With reference to the 'forms of existence' it is shown precisely how the blurs unfold in different phases of the musical process and in different musical genres, and which mechanisms work and are used in order to master and make them productive for the musical composition and reception, such as 'leitmotiv', transcription, reinforcement of recollection by technical re-/production. Attention is also drawn to border experiences, such as 'forgetting', false memories, prejudices, hallucinations.

Keywords Avoidable "vagueness" and indelible fuzziness · Bias · Entropy and neg-entropy · 'Halo' and 'Devil' Effect · „High Fidelity" · Information Theory · Improvisation · 'false memory' · Memory: ultra-short-, short-, long-term · Musical notation · Phonography · Postmodern "constructivist" and "deconstructivist" nihilism · Perception · Productive fuzziness · Quotation · "Leitmotif" · Soft computional modeling · Technical re-/production of sound · Tuning systems · Uncertainty

Blurriness, the "fuzzy", is not only the specific difference for fuzzy logic, but also foundation and starting point of this book. They relativize realistically the absolute, here the musical artwork. Once created and accomplished it is the "a complete entirety in itself" („ein in sich vollendetes Ganzes", Karl Philipp Moritz, 1785) and as such consummate. This applies not only to artworks in the field of "classical" music, but more generally to all artworks; the long and rightfully famous chorus chain of tenor saxophonist Coleman Hawkins (1904–1969) on the Jazz standard *Body and Soul* from October 1939 has the qualities of an artwork, and is permanently conserved on LP record in a repeatable manifestation (Fig. 3.1).

The ballad was composed in 1930 by 1930 John W. Green, lyrics by Edward Heyman, Frank Eyton, and Robert Sour. The first line of the song. The artful variation by Hawkins is evident even without knowledge of musical notation (Fig. 3.2).

H. Heister, *Music and Fuzzy Logic*, Studies in Fuzziness and Soft Computing 406,
https://doi.org/10.1007/978-3-662-62907-9_3

Fig. 3.1 Coleman Hawkins: *Body and Soul*
The beginning of the improvisation
http://www.sokillingman.com/wp-content/uploads/2012/01/Body-And-Soul-Bb.pdf, retrieved 8 April 2020

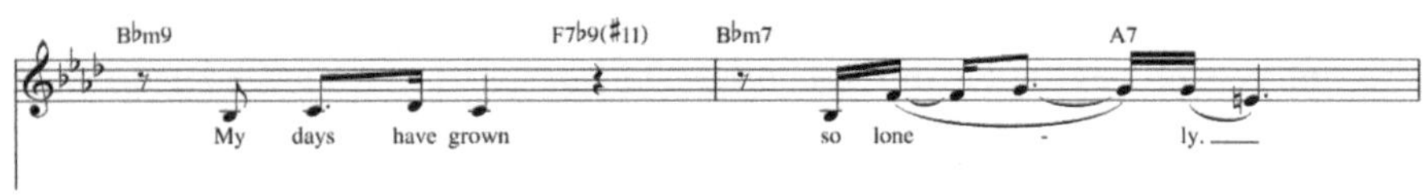

Fig. 3.2 *Body and soul*
The beginning of the song
https://www.virtualsheetmusic.com/score/HL-15084.html, retrieved 8 April 2020

3.1 Idea vs. Sound. Fuzziness in Composition and Interpretation

Even the consummate work of art always contain fuzziness: Never, as mentioned, does the execution of the idea completely conform with its sensual realization. "The work is;" stated Walter Benjamin, 1928 "the death mask of conception."[1] As we move from production in a close sense, the composition, to reproduction and reception, fuzziness increases and multiplies. In the abstract, blurriness runs analogue to the 'tolerances' (see Sect. 1.1) which are ever present in engineered and technical applications and processes, right up to 'quality management', which must allow for diffusion and some rate of error. This already applies in preliminary stages of the work to the musical material and its formation, not least by sharpening through filtering. Interesting with music in general and with artworks in particular are the quantities of deviation, beyond this, however, qualitative aspects of why and based on which factors these deviations from the notation as prescription, as 'blueprint' of the execution. This constitutes a kind of practical, unintended and involuntary blurring

[1] Benjamin, W. 1928: ANKLEBEN VERBOTEN! DIE TECHNIK DES SCHRIFTSTELLERS IN DREIZEHN THESEN, XIII, p. 33.

of shapes and contours. Therefore, in principle the opposite process is decisive in the production and origins of an artwork in the spread between concept, idea, notion, sketch and completed work: The emergence of concrete, distinct shapes from the amorphous, unformed, the elaboration of material chaos to a work cosmos.[2]

3.1.1 *Productive and Receptive Realization*

At first it may seem the main source of blurriness lies in the contradiction of production and reception. But it is not only the audience that cannot fully comprehend a new artwork immediately. Contradictions go back to the composition itself and its forms of existence. And they are likewise found in the various forms of tonal realization. Realization here is, as expounded in the foreword and the first chapter, meant more broadly and does not only include the transformation of notation into sound. For blurriness exists at all stages, in all forms of existence and phases. In 1998 Reicher-Marek developed a distinct differentiation primarily for literary works. Her 'ontology of artifacts' „is based on the distinction between works and their realizations“. „Works determine classes of realizations. “Realization is thereby, in analogue to the relation between score and performance, from notational form to tonal form, the transformation of scripture to sound – „works of literature may be realized in concrete readings, but not in copies of books etc.“–, and so the return from the written form to the older, primordial sound form. However, Reicher-Marek assumes that different works might have the same tonal realization: „There might be distinct literary works which may be realized in exactly the same readings.“[3] She thereby underestimates the difference between the varying tonal realizations of different works. Analogue comparisons such as comparisons of interpretations are another topic discussed by Fuzzy Logic. She therefore inserts a second level: 'world-types', as strata of literary works."[4] But even that is not enough for her; she senses too many identities: „But this is still not sufficient, because there might be distinct works of literature with both identical "form-strata" and identical "world-strata"." She assumes identities where I would rather assume differences. She therefore introduces yet another meta-level: „a third 'stratum' which may be realized in concrete mental occurrences.“

A revealing parallel to the relation of artwork and realization in music is provided in another art, gardening art, which by its procedural character is related to music. In the relation of garden plan and realization of the garden this analogy shows similarities and differences. Here the asymmetrical accent lies, in the literal sense of the word, inevitably on sharpening. For the plants only partly grow according to

[2]On preliminary stages see below Chapter 5 *Filtering*, on the development of shapes see below Chapter 6 *Crystallization*.

[3]Reicher-Marek 1998, Abstract, without pagination, http://www.philosophie.rwth-aachen.de/global/show_document, Abruf 2.4.19. Cf. Reicher-Marek 2018/1019, http://www.philosophie.rwth-aachen.de/global/show_document.asp?id=aaaaaaaaaabrgwm, retrieved 2 April 2019.

[4]Ibid.

the composition of the garden plan. They are, as "natural material" obstinate. In order to preserve the core identity of the garden as artwork in its procedural identity weeding, pruning, etc. are carried out. One extreme in this regulating sharp intervention in the organic, and from a human perspective often chaotic growth is reached in the so-called topiary cut, which was particularly common in baroque gardens: Here primarily evergreen trees or bushes (such as box tree and privet) are cut contrary to their natural forms to achieve sculptural effects in animal or human shapes or architecture, often from today's perspective with a slightly grotesque appearance. And these too become more and more blurry through the growth of the plants and have to be refined anew again and again.

For musical artworks a primary aim is to represent imaginations, concepts, thoughts and feelings. In a further sense it deals with the realization of ideals. Blurs are unavoidable in this process of realization. And they are, as will be shown, also indispensable. They already occur in the process of emerging musical shapes with the main tendency auf growing conciseness (see Sect. 2.2; see Chapter 6) up to the completed work in notation, which doesn't exist in the case of improvised or technically produced music. In these processes of sharpening we find a dialectic counterpoint in fuzziness/blurring and even impossibilities of fully realizing the preconceived and desired in a composition. Just like producers in other arts composers and interpreters often complain that they only approximate the ideal and consummate that they had envisaged.

Behind this is the tension between idea and realization, which is fundamental to the subject matter here in the book. Shades and facets of this relationship such as the contradiction between theory and practice, between conception and realization and similar polarities, fan out this basic problem. In general, as mentioned in the beginning, mathematical perfection stands again artistic or technical perfection in realization. And specifically this perfection itself exists in a charged relationship to the 'liveliness' of the musical interpretation as realization of a composition.[5] This realization necessarily also implies blurriness, or at least coarsening—but at the same time sensualization: art only conceived is not (yet) art (See Sects. 3.8 and 6.3). "Concept art" as art consisting only of an idea is only seemingly a counterexample. If at all it is meant to reach intersubjective real existence, the concepts at least need to be noted down: as a rule in language, sometimes also as schemes, but practically never without words. Transitional forms make concrete at least the outlines of a potential work and give some sporadic indications (See Sect. 7.7). But the musical work is only created in its sound form as a realization of a vague template.

3.1.2 Avoidable "Vagueness" and Indelible Fuzziness

In contrast to avoidable "vagueness" (Zadeh), as for example through careless notation, in singing or playing, fuzziness remains indelible even in spite of greatest

[5]Extensively Heister 1983, within the framework of an implicit theory of interpretation, passim.

efforts. Even precisely defined, crystallised shapes have a dimension of blurring. For one thing, the boundaries between motifs and themes are not always clearly defined. We already discussed the B-A-C-H leitmotif in this book, whose head B-A is also the "sigh motif", which is immediately repeated in the C-H. Secondly, their respective meanings in the language of music, but also in the seemingly more precise language of words, in accordance with the meaning of words and morphemes, are often blurred, sometimes even vague—in music the meaning is usually connotative, hardly denotative.

Thirdly there are blurred margins and overlaps even in the meanings, as for example with words and semantic fields. And here we are still talking about the notational form, the relatively most exact forms of existence of the work.

Necessarily there are blurs nearly everywhere in the music process, not only in general in the realization of ideas, but also in detail as e.g. the edges of motifs. Thought patterns and procedures of Fuzzy Logic serve as methodical instrument that expands and enhances conventional thought patterns of music analysis, beginning with the segmentation or filtering of relevant elements of the musical matter and emphasis on the crystallization of significant motifs from the score and its flow, but also regarding the concrete definition of "musical logic": for fundamental aspects like the interrelations of motives, not only within a single piece of work, the quotation and/or the allusion, the unaware quote, the reminiscence, in general the perception and remembering of motif complexes and more.

I must leave aside here the psycho-logic—causes, motivations, motives, the question of affects, emotions—, the social and sociological questions of interaction between musical process and public, the influence of situations on realization and reception etc. Inversely music (and similarly the fine arts) appear as part and concrete elaboration of the aesthetic and sensual perception, which form their own laws, logics, layers and types of cognition, and therewith nothing less than a paradigm and models for Fuzzy Logic. Fuzzy Logic acknowledges various degrees and even types of blurriness. I can neither list them in detail here, nor always provide concrete possible musical applications given the current state of research. Some pointers must suffice. Three representations in graphic form are most prominently used, shown and described (Fig. 3.3).

This graphic representation provides a characteristic line to describe groups of similar but differentiated elements. In the group of contemporary good singers of Norma their quality or other distinguishing features vary (style of interpretation, voice, acting skills and others) between 0 and 1. The threshold values bottom left (X) and top right (Y) can be determined quantitatively or qualitatively and all other variations placed accordingly on the characteristic line (Fig. 3.4).

This graphic representation is useful for depicting gradations of belonging to a group. One popular example for illustration is the question of a pleasant room temperature. Musically more relevant is the discussion of for example tempos in conducting. Of course one may choose any random musical piece for this. But numerous directions for tempos by Mahler are here particularly significant. For they are not only, as already with Beethoven, Schumann and so on, simultaneously poetic and loaded with aspects of expression, but also often acquiesce at the outset to the fuzzy logic of "a bit

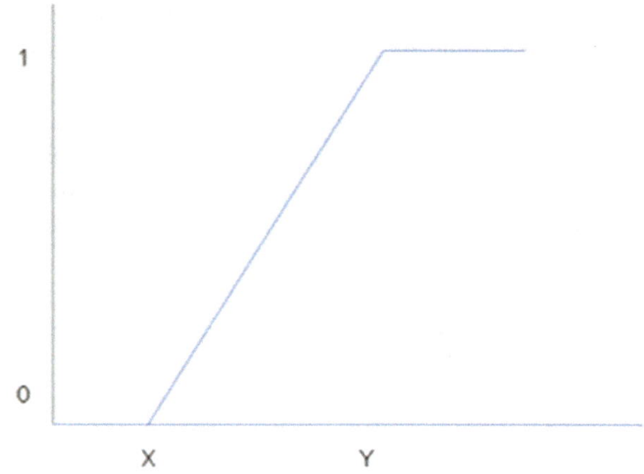

Fig. 3.3 Variant 1: direct graphic representation in the form of a characteristic line or element function

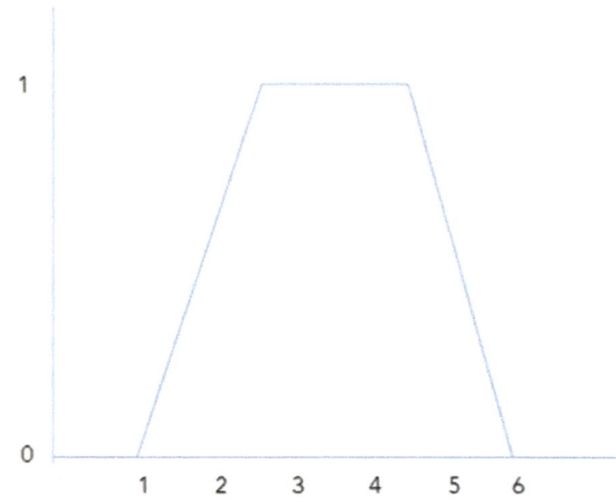

Fig. 3.4 Variant 2: the most often used graphic representation in trapeze form

more/a bit less". In the 1st symphony, for example, it says "slowly halting", "always very leisurely", "solemn and measured, without halting", "again a little livelier, as in the beginning". A musical interpretation, moving between 2 and 5 in the graph, could be considered appropriate and yet there remain many options/possibilities. Between 1 and 2 the tempo is still bearably, but lags behind somewhat, just as between 5 and 6, but altogether seems somewhat exaggerated. Anything that would be posited beyond this graph is not permissible (Fig. 3.5).

This graphic representation may on the one hand be employed when one is asking about the gradual belonging to a given group where there exists an optimum peculiarity that finds itself at the apex of the triangle—i.e. the "perfection" even if it always remains relative and never is absolute. The acute angle at the top is chosen

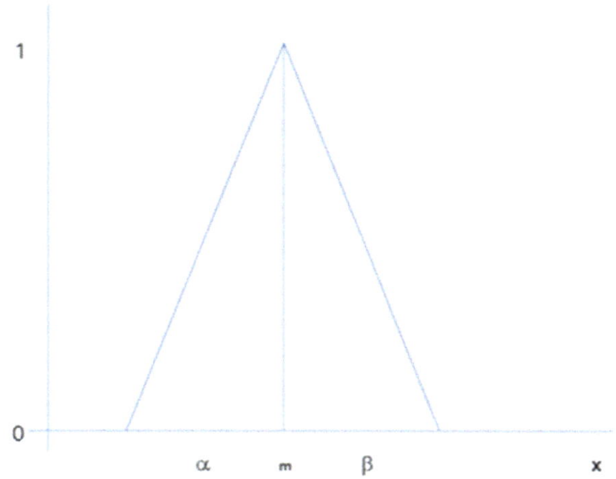

Fig. 3.5 Variant 3: the triangular fuzzy group, often depicted but rarely described, but mathematically easy to apply

randomly, because it is often depicted so, but it could also be more obtuse and the triangle more plane.[6]

One example is the successful realization of a particularly difficult tone by a human voice, as with the famously notorious "high C" of sopranos and tenors. Even more difficult is of course the high F (f^4) of the Queen of the Night in Mozart's *Magical Flute.* More difficult is the disjunction of parameters: conventionally this refers to combined elements (conjunction of parameters such as the combination of rising pitch and volume in approximation of the Olympian motto of "Citius Altius Fortius") are separated and in given cases reassembled in a new and different way. A well-known and famous example, which so far has not been fully appreciated in its aesthetic and systematic significance, is the ending of Verdi's "romanza" *Celeste Aida,* in its final phrase „ergerti un trono vicino al sol" (erecting you a throne near the sun) the voice in accordance with the text has to rise very high. For the first syllable ("er-") on a high b Verdi prescribes forte together with a fermata. Then the high and loud is separated and the first "vicino…" immediately is assigned no less than a fourfold pianissimo *pppp*, while the second "vicino…" is threefold with diminuendo, and the sequencing third "vicino…" *pp*, and for the concluding "sol!" again on b even a *pp morendo.* This disjunction of parameters contravenes the usual, the natural, and in this unnatural and particularly artful demand is accomplished only by few tenors. At point lies the value of optimal success in singing. In proximity one can describe values that are still acceptable. What lies outside of the triangle is unacceptable. Such graphs may be used for both qualitative analysis and quantitative calculations. Latter requires one to be able to identify mathematical values. This should be possible, but might not always be necessary. Further methods of representation are available as

[6]For elaborations and explanations of the graphs including the lead to the modes of presentation in Mahler and the accomplishments of vocalists I thank Inge Rubbert (March 2020). Graphic representations and more material in detail found in Gerhard Müller n.d.

complementary sets, intersectional sets or unions/composite sets and in addition a number of complex models.

> A Type-1 Fuzzy Logic System (T1 FLS) has membership functions which only represent a type of imprecision, or ambiguity.
>
> An Interval Type-2 Fuzzy Logic System (IT2 FLS) has membership functions which can represent varying degrees of uncertainty.[7]

A further innovation is "granular computing" with special modeling procedures.

> In Rubio et al. […] through the use of *Neutrosophic Logic*, which extends the concepts of Fuzzy Logic and uses a three-valued logic that uses an indeterminacy value, granular neural-fuzzy models are created. This leads to more meaningful and simpler granular models with better generalization performance when compared to other recent modeling techniques.[8]

The musical process involves a multi-layered system of contradictions. Sharpening operates within the general contradiction of blurriness and focus with various procedures. (1) Minimizing blurs and aiming for optimal precision. (2) Allowing blurs for the sake of pragmatic optimizing. (3) Blurs are produced in purpose to achieve certain artistic effects—the 'blurring'. (See Chapter 7). In regard to the tension of blurring and sharpening the artwork may be regarded through the cybernetic model of system-environment—the work as a system that has to prevail in an environment filled with fields of potentiality and conditions. Starting out from the notational form as center the musical process brings forth concentric circles with increasing blurriness, the further they move from the "text". The relationship with/context of existential forms adds precision to this yet too simple model.

The ontological and gnoseological primacy of fuzziness is a fact, but not the last word in the matter. For sharpening is a fundamental counter-principle. The goal of focus is in itself a moment of reality, especially of the reality of music. Sharpening sets limits and confines the blurriness, as much as possible and necessary. Musical performers should not realize a score with a precision like a clockwork. The sharpening of the blurriness asymptotically approximates the optimal focus immanent in the context. For this optimum there exist general rules such as that all sounds presented in the score must be realized in some form, but always together with what was intended but not explicitly stated in the notation. Specifically this has to be decided on a case-by-case basis for each artwork. Blurriness is inevitable. But it is also, as will be seen at the conclusion of this book, something that cannot by bypassed in the sense of what is necessary for artistic productivity and development (See Chapter 9).

[7] Castillo et al. 2016.

[8] Solis/Panoutsos 2013; see also Solis/Panoutsos 2017.

3.1.3 Information Theory and Improvisation

The objective uncertainty in view of the second law of thermodynamics and the irreversible "time arrow", not to mention "chaotic" developments, is reflected in the subjective un-predictability in the sense of Shannon's information theory, which also plays its part in any prognosis, without coinciding with it. This difference concurs with the difference of objective indeterminacy and subjective uncertainty, at the same time constituting a difference of ontology and gnoseology—which in interpretations of the quantum mechanical uncertainty principle is always carefully considered. Possibly, 'blurredness' is a common denominator of both fields. The partial unpredictability especially in the development of musical "thinking" (a term from Schönberg), of motives and themes, is to varying degrees of "fuzzy" an essential component of artistic productivity. Listening to music with minimal information content provides little pleasure, as one could predict yesterday how it will sound tomorrow. This partial unpredictability manifests itself in the realization of original ideas in the layered levels of existential forms. In a countermovement, indeterminacy and improbability in the production process as well as in the reception are linked back to actual or assumed expectations. As a rule it is more likely not the totally unpredictable, but rather the small differences between the probable and the expectable and the final realization that create particularly pleasing aesthetic tensions and relaxations, because they maintain a balance between novelty and recognizability.

The dialectic of the expectable, the expected, and the resounding is especially noticeable in improvisation. This musical "gradual production of thought in speech", to quote Heinrich von Kleist's famous formulation, is in the context of the existential form a special case of relations between idea and realization. The imagining of a whole, the ideational anticipatory finalization, at least in contours, may be on the mind of the improviser; but it will not proceed as far as in thought form. (See Sect. 6.3.3). For the relative uncertainty and unpredictability of the progression and even of the ending, even for the improviser, is an essential component of the principle of improvisation—for it is 'improviso'—unforeseen, unexpected. The jazz pianist Michael Wollny belongs to those musicians who can sit down at the piano without any preconceived plan and out of the moment play an entire concert. Stage fright in a milder form is still familiar to him. But:

> I would be far more nervous, if I knew exactly what I am to play. Imagining working with a template and detailed guidelines, even when the moment feels so very different feels much more terrible for me. I would be forced to constantly align my decisions with a plan or a score. The more open things are, the less afraid I feel.[9]

But to fully engage with the moment, the flow of inspirations and the not quite foreseeable process, the improvising musician must be able to rely on experience, practice, and preparation. All this pervades a "kind of basic trust in the moment" and makes possible a floating "unintentionality".

---

[9]Jazzpianist Michael Wollny in an interview with Gregor Dotzauer: „Das Klavier ist mein Avatar", https://www.tagesspiegel.de/kultur/jazzpianist-michael-wollny-im-interview-das-klavier-ist-mein-avatar/21104498.html, retrieved 1 March 2020.

> Without good handcrafted technical preparation it doesn't work. One has to do one's fluency exercises and analyze the works of others. From this I build up a stock of tones, which I then practice in various fingerings and voicings. The point is to surround myself with themes that might then spark off something. Sometimes I note down small etudes. When you do that, you become more and more free in the musical language.[10]

Even here freedom is linked to necessity, the use of the spontaneous and random together with the planned.

> At the same time one needs to remain earthbound while on stage, so that the endless numbers of possibilities become a necessity. I define improvisation pragmatically: a given model, in an oral tradition or written, as is the case with today's jazz standards, gets played with in the moment and in the process of a performance, gets developed, varied and so on. In "free" improvisation such models can also be absent. The execution follows rules. This may be rather loosely or stricter.[11]

Improvisation comes without notation, which could serve as yardstick for deviations and adequacies. Basically the objectified existential forms such as the conceptual form and foremost the sketch form may be drawn on for the relationship of idea and realization. But the source material for this is usually scanty, even more so if considering the subjective existential forms: the thought form as ideational anticipation of the whole cannot exist in any way, and the anticipation form as ideational anticipation of the sound can only be rudimentary at best. The Indian ragas, Arabic maqamat etc. or the "standards" in jazz as compositions with melody, harmonic 'changes', and maybe text have another function. Instead of a precise as possible reproduction it demands change that is as imaginative as possible, filigreed, ornamental, an extension of the given, which usually is not noted down (though it could be). A further dimension of the dialectic of fuzziness and sharpening lies in the fact that earlier sound forms may serve as measure of each actual realization via the recollection form. These deliberate blurs can bring forth a productive further development (See Chapter 7).

In jazz improvisation, in so far as choruses go, all choruses will resemble another based on the harmonic sequences, the changes, and on the melody, as free as this in turn may be—it follows rules of dissonance-consonance treatment and replacement possibilities. Not every tone fits to every chord. This begins already with the paraphrasing improvisation that embellishes the given melody. Here we get an area of study which could begin with pitch statistics, but then has to include in its considerations design qualities and series of motifs or motif-like series. A notational form is not necessary, as long as jazz largely remains folklore, as it was of course with the early blues, but also in the New-Orleans style. But they can exist; Jelly Roll Morton and others like him already recorded compositions in notational form at that phase. With the technical reproduction of jazz on record since 1917 it became emancipated from its folklore status. At about the same time and in the same way the expression of social and aesthetic ascendancy jazz compositions get noted down. A precursor was the rag-times composed for piano, like those of Scott Joplin. They now serve

[10]Ibid.

[11]Ibid.

as representation of the work in its existential notational form and will be used as orientation and starting point. This can already be seen in the yet rather folkloristic variations and formations of variations in the late New-Orleans jazz. The variations largely follow the principle of similarity. They also include in the form of repetitions dissimilarity as contrast or equivalence.

This falls mainly into the area of variation (See Chapter 8). Fuzzy logic could test, how and how far the rules of play with its implicit and explicit rules are followed, expanded, or broken—an arduous but not impossible endeavor. For example, in the chord Cmaj7 the notes c, e, and b may be used to form the melody; further notes will require special consideration of contextual circumstances, of the before and after of chord and melody sequence, which can be learned and trained. The boundaries between legitimate, musically meaningful and sensually attractive deviations from the rules and nonsense breaking of the rules are in themselves blurry.

Precondition and starting point is the sound form, usually in phonographic form, since it allows for the repeated listening necessary for writing down. On a larger scale and with some precision respective 'hands-on' analyses presuppose a transcription of what is played, transference from the sound form to the notational form. Graphic representations such as sonograms allow 'automated' comparative fuzzy logic analyses, comparisons of improvisational variations with the starting point, and comparisons of variations with each other, all this without the notational form. At the next step, beyond the singular 'recording', this also makes immediately comparable different interpretations of a standard, various realizations of a maqam, a melody model in Persian, Arab, or Turkish music. To a certain degree improvisation assume the character of works, substantial, clearly defined, objectified. In improvisation the upcoming and secondarily also latter is deliberately and in the imagin-real anticipated, as the negotiation of sound form and anticipation form in the process of musical performance. This is not immediately verifiable or empirically researchable; but maybe a methodological detour might help. Different takes of the same work, i.e. recordings including those that didn't make the final release, show traces of the tension between intention and realization. In this manner improvisation becomes accessible in the area of research outlined here for research employing fuzzy logic.

In the following sections of this chapter the diverse blurs, that necessarily result from the realization process of music, are analyzed and described in detail using the theory of existence forms expounded in Chapter 2 (see above, cf. also Heister 2008).

Of the thirteen forms of existence eight will serve in the order of their occurrence in the musical process: score form, sound form, perception form, phonographic form, recollection form, word form, picture form, imagination form. In the conception form, sketch form, thought form, and anticipation form the idea of a work becomes continually refined up to the score form; these forms will therefore be discussed in the Chapter 6. A special case is the quotation form. In it we only see an incomplete work as a rule. It will receive special attention in Chapter 8. All forms of existence discussed here show blurs that resemble analogous types or specific blurs in relation to the original idea of the at first planned and then realized work. These analyses demonstrate the art-theoretical use of the theory of the forms of existence and are thereby a proof of their usefulness for fuzzy logic research.

3.2 Artwork as Idea in Fixed Written Form. Sharpening of Blurs Through the Notation

In improvisation the chain of existence forms is shorter. Many of the relations of blurs between (subsequent) existence forms do not apply here. The 'idea' in many music styles is not materialized in a notation form is such a way, that this forms becomes an identity to which the various realizations relate—which would allow determining degrees of blurriness or of precision. In much folklore music and "traditional music" there is no notation at all, in a lot of modern pop music, including jazz, there exists mostly only a skeletal or sketched notation, that indicates a model of what is to be executed or variated. In 'classics' this is different, but not always so that the idea is completely or commensurately determined. In *basso continuo* music for example there is considerable room for what we call, in a broad sense, improvisation (See Sect. 6.6.1).

3.2.1 Enlarged and Restricted Frames of Determinations in the Notation

Every procedure of sharpening in notation tends to include *per negationem* starting points for fuzziness. When for example the intended pitch is exactly and unmistakably noted, this does not mean that the actual tone following this prescription will always be the same and the originally intended. In this case simple local, regional, and varying tuning systems according to musical styles result in differences and deviating blurs—only in 1885 an international standard concert pitch was introduced, which, as will be explained further, is subject to changes, though now at least on a global scale. Or even when, relatively late, basically since Monteverdi's opera *Orfeo* in 1607, it is indicated that for example one voice is to be played by a violin, in general the violins in their individuality significantly differ, even more so in the special case of 'baroque violin' and modern violin, and within that framework yet again can be distinguished as originating from either Italian or French violin makers and so on. Multi-dimensional comparisons, in regard to tone color, volume, etc. in correlation with measurements and proportions of the instrument could be a rewarding task for fuzzy logic.

Historically important development stages are most easily traced in the European tradition of notation. This begins with gestural indications of melodic movements, which are vaguer than blurred, called neumes, literally 'winks'. It continues to progress from bare notes to indication of instrumentation and thereby tone color, up to the excessive graphically notated and verbal expressions or duration proportions.

Even writing in word languages is far from exact, as is often naively assumed. Correlating phonemes and graphemes is rarely unambiguous, but mostly ambiguous. Often the notation of phonemes is already blurred. Especially English orthography seems often arbitrary, and is not explainable without information about the history of a language. By analogy this applies to other languages as well. And ultimately the prosodic dimension or the emotive coloring of speech is almost impossible to include in notation. The supra-segmental, the paralinguistic, and the prosodic, i.e. in speaking the self-evident coloring through modulation of sound, rhythm and tempo, emphasis and volume, accentuate, add nuance and precision to the meaning and sometimes even determine it as in the difference between statement and question. The not notated intonation of a sentence is at least indicated by a special punctuation, the question mark. All these mentioned components are precisely the domain of music. But even in the notation of music much is lost. However, rhythm is notated, historically with growing precision, as are the pitches which in European music even have the priority.

> the *score is merely a map* that helps us to explore, and to discover the buried treasure that we so eagerly seek: music.[12] Not all of the composer's wishes are contained within the printed score and the most important and meaningful of them simply cannot be, precisely because it is not possible to write them down. [...] A score is two things: the tools of communication, and the message itself. The first is visual, and the second is aural.[13]

3.2.2 Starting with Mere Indications of Pitches and Durations to Excessively Precise Notation of Technical and Expressive Execution

The score form has a hinge function from the viewpoint of antecedent existence forms, the conceptual form, the sketch form, and the thought form. Investigating the possibilities and impossibilities of notation has a historical dimension. Seemingly unalterable limitations are continuously transformed into exceedable and exceeded barriers. In musical notation there has been, especially in European music, tremendous progress since the humble beginnings in the 9th century. Unlined, 'adiastematic' neumes show only the text and approximate gestural images of melodic contours. Many intermediate stages lead from here to the notation system with several lines (at first 4) and various clefs around 1025, invented by Guido d'Arezzo, ['di' here von is not a title of nobility but a designation of origin]. Through tis the pitches, but not yet the tone durations, are precisely demarcated as points. This already made possible the notation of several voices in spatial superimposition, even though a more complex polyphony had not yet been realized in composition. A decisive transition—and an emancipation from the 'linear' script as notation of word language—was the exploration of this optical vertical pitch space, a productive analogy between the real optical, here two-dimensional space and the imaginary-real acoustic space. An exceptional case, as in the Codex Montpellier H159 from around 1280, with both neumes and letters representing the notes, even offers a twofold safeguard of sharpening, difficult to read but precise.

The development goes so far, that eventually not only every tone gets exactly determined in its dimensions of pitch, duration, volume, but also with nuances of

---

[12] To the difference of map and territory see below in Chapter 4.

[13] Frost 2007. Emphasis HWH.

Fig. 3.6 Artur Schnabel, sonata for piano (1923), 4th movement, m. 2
George A. McGuire (Hrsg.) Peer 3618. Mit freundlicher Genehmigung von Peermusic Classical, New York und Hamburg
Pitch, duration, volume and tone color or rather ways of pressing the keys are exactly noted. Almost every note additionally comes with directions for technical and/or expressive terms of performance

sound color and expression, which are rather connotative, and have to be because they can neither be really exactly notated nor executed and remain relatively blurred (Fig. 3.6).

In this long historical development line existing blurs in the relation of compositional realization and idea have been reduced step by step. And even though this produces ever-new blurs, altogether it has resulted in progressive sharpening and increasing precision of notation. Put differently: In part because of and in part despite such progress and growing sharpening of notation of composition blurs remain indelible in the realization. Notation as prescription emancipates itself and takes on a life of its own to a certain degree, as "eye music" in the Renaissance. Here the notation itself as optical system of signs obtains an additional semantic field. Following the keyword "Figure circles", the notation and not the music in Alban Berg's *Wozzeck* draws just such Figures—because of the horizontal time axis these notation-pictures are not actually in the shape of circle, but as a circle in motion ellipsoid, and in the sound as well as the perception are hardly realized as "circle" (See Fig. 4.13. Alban Berg: *Wozzeck* (1925), I/4, m. 561).

3.2.3 'Transcription' of Sounding Music as Re-Translation

A different, but analogous case is presented by the 'transcription' of sounding music as re-translation or reverse engineering from sound to notation. Hearing works considerably faster than writing. Only the extraordinarily gifted, with a tendency for eidetic memory, can note down from memory a piece of music—even an improvisation—after hearing it only once. Mozart demonstrated this with the polyphonic *Miserere* of Allegri, which was performed only in the Sistine Chapel and of which the score was kept secret. Transcription only became easier with the arrival of technical reproduction of music, which allows repeated hearing of recorded improvised music, or allows for special help such as slowed down replay. The digitalization and the subsequent almost automatic transcription present another leap. The general historical development therefore shows a growing refinement of the notation up to the score form, which does not equally apply to all dimensions of a work. This progress significantly foregrounds blurs in the realization, which become more distinct and noticeable (See above 2.2). Productive realization means transformation and realization of the score form in the sound form of a work.

> Musical scores are not yet music: They indicate to performers the points to reach. […] A classic Western score mainly contains information about pitch, onset [not an actual parameter], duration, loudness, and timbre. […] Other verbal indications, such as *con fuoco, con dolcezza, con espressione*, don't have any obvious mathematical or mechanical correspondence. They must be translated by the performer into gestural variations that imply sounding variations, and gestures (not only musical but any bodily gestures) *can be mathematically described.* […][14]

Gestures are—primarily iconic—signs. Here it means more than that is actually practical, not merely symbolic bodily movement, mostly of the voice or the hand and its treatment of the music instrument or vocal apparatus, which produce material-sensual vocal and instrumental sounds. Every intersection, every segment of the chain, necessarily produces blurs in this permutation of the ideal to the material.

3.2.4 Notation and Finalization of Uncompleted Works

The question of similarity, of blurs and sharpening, also comes into play in the finalization of uncompleted works. Of course, as in the case of Alban Berg's opera *Lulu*, there must be enough completed material available in order to make legitimate comparisons, reconstructions and constructions of what the composer intended.

This is surely not the case for the following.

> One simply puts out the headline 'Beethoven's Uncompleted Symphony now completed'. The rest will follow. There was never an 'Uncompleted Symphony' by Beethoven. Now every member of the educated class recognizes Schubert's symphony fragment in b minor […], but of a 10th symphony, let alone an 'Uncompleted Symphony', by Beethoven not even fragments are available […]. Now an 'international team of musicologists', composers and computer scientists is making an effort to train an 'algorithm in such a way, that it can re-compose the missing passages in the spirit of Beethoven'. That's how simple it is with Beethoven's spirit. One only needs to connect the right algorithm with the right people and one can complete what was never even begun […]. What is more ridiculous […]: The boundless overestimation of artificial intelligence in union with the contempt of human intelligence—or the insensitivity of the advertising department of the middle class degrading one of its greatest cultural achievers to a promotion clown?[15]

[14]Mazzola/Mannone/Pang 2016, p. 269. For more detail see chapter 33. Emphasis HWH. Cf. numerous further studies about the relationship of music and mathematics, cf. among others Mazzola 2002.

[15]Stefan Siegert: *Schimäre des Tages: Beethovens Unvollendete, Junge* Welt, Berlin, 9.12.19, Commentary of a dpa news report (Deutsche Presseagentur).

Even from the viewpoint of fuzzy logic the difference between the completion of a work recorded only in fragments and completion of a work never begun is fundamental. If, as in this case with Beethoven, a template for comparisons is almost non-existent, a comparison with more or less fictional material is not possible. Berg's case is different: Here we have a mostly completed full score draft, a condensed score ('Particell'), a kind of piano excerpt with indications of instrumentation. Thereby the available notation of the score draft allows a triangulation of the instrumentation through comparisons with analogous sections, for which Berg had already completed the instrumentation. The latter are the actual reference objects for the comparison. The composer Friedrich Cerha took on the task of completing the missing instrumentation (Berg died rather suddenly) in parts of the third act. The result is basically a success. In a methodologically and technically rather arduous and complicated, but manageable comparison the success (or failure) could be verified and quantified.

3.3 „Tolerances" in the Real Music. Work as Acoustically Realized Idea: Sound Form

Once the work has been put into score form with as much precision as possible, blurs remain and even various new ones occur with varying dimensions. The first main area for these is the sound form, which usually receives large attention in the discussion of 'inaccuracies'. The second main area is since 1877 the phonographic form, which 'automatically' reproduces the sound form. The third main area is the perception form and the subsequent existence forms. A remarkable special form of realization as paradoxically silent musical rendering shall be briefly introduced here. One of them is the reading of the score. It involves a more or less sharp imagination of the sound. Reading, thinking and anticipatory imagination interact.

The other type is practice on dampened or even noiseless musical instruments. Appropriation through reading and rote memorization precedes this or goes hand in hand with it. Best known example is the "silent claviature". Travelling virtuosos in the 19th century used them for practice. Here the motor skills for sound production are trained in anticipation, but without a controlling feedback loop of the resulting sound. Analogues are pianos for home use with a mechanism to silence the claviature or string instruments with fretboard but no resonating body. All this is training for the real world.

A guiding principle for realization and interpretation, so to say the 'set point', is the highest possible adequacy of the realization of the work. Since it is a form of reflection it makes little sense to speak of equity or even identity.

The issue is even more complicated. For the 'artwork' is on the hand, in improvisation, always already realized as sound. In that case the problematic divergence of work and realization is omitted, even though not completely. The divergence is deferred and of another kind, apparent in works existing in notation. For on the other hand, such a work is given only in its written form. Notation is the hinge that opens the door between work and realization. But work and composition and notation are not identical. Rather they refer to a common third value, which we could call the idea of the work. But now the idea cannot be grasped by itself, it becomes tangible only in the notated work—additional information such as formulated conception, sketches, commentaries, declarations of intent, explanations of the composer help to decode the idea or to reconstruct it: What did the master try to tell us or what did she mean, when she wrote this? And still the idea of a work is accessible in its precedent form only in a limited fashion. In principle it is, like a thing itself, that which appears *in* the work.

For sound realization this redoubles the problem of adequacy. It must realize the reconstructed imminent idea and the notated idea of a work, that means it must be true to the 'spirit' as well as the 'letter'—an old divergence comparable to that of spoken and written language.

The performance of a musical artwork, its realization in sound form appears simultaneously as the work brought to life, so to say as a return to the work in its notation form. This can also only be relatively perfect and requires the transformation into actual social and artistic aesthetic liveliness.[16] Essential for the realization in sound is not least the dialectic of perfection and liveliness. An additional complication lies in the demand that perfect art should compensate the imperfect life. But within certain limits the technical imperfection is a sign of artistic liveliness. For technical perfection, vocal or instrumental, is not the same as expressive and meaningful interpretation.[17]

The transformation of a work from score form to sound form is only 'reproduction' in regard to its temporal-logical sequence of composition as production. In essence this 'reproduction' is the production of a subject. It expresses and relinquishes itself in a work both objectively and personally. The appearance of quasi-compositional creativity, the reproduction as production is created through the by now for solo interpreters obligatory recital by memory—as if the work originated from the interpreter. This covers up the actual creativity of the interpretation.[18] Its subjectivity is essential. For sure the person should recede in favor of the work. The representation of the work, not of the self, should be in the foreground. But self-denial is not an ideal, since objectivity can transition to indifference.[19]

Even mistakes can be tolerated as inescapable component of liveliness.

> "I'm only human" […], what we say when we make a mistake […] is a rather odd expression. We seem to be implying that we'd rather be a computer, or at least that we aspire to the technical perfection a computer can achieve. We want *to play all the right notes*, in the right order, at the right time and tempo, and with perfect intonation and ensemble. We desire to be as true as possible to the score. This is the Holy Grail. […]

---

[16]The seeming liveliness of a mechanical-automatic musical realization can be uncanny. In the 19th century this was still largely science fiction, but often finds its expression in literature, as with E.T.A. Hoffmann. See Hirt 2010.

[17]In detail cf. Heister 1983, pp. 380ff.

[18]Cf. Heister 1983, p. 463.

[19]Cf. Heister 1983.

> What is this truth we seek, and does it really reside in the notes themselves or does it hide behind them, between them and, most importantly, in the fragile nature of the humanity we bring to them in a performance?[20]

Corresponding to the saying "too sharp makes for jagged" the blurs are not only unavoidable and to be tolerated, but even necessary. However, a high degree of technical capability is an indispensable precondition. Those who cannot even realize the letter of a work, cannot at all reach its spirit. The 'work' is relatively identical and as a stable orientation point constitutes the core of the many non-identical interpretations.[21] The differences in performances refer, in interplay with the identity of the work, to the collective context of the appropriation of music and its historical perspectives. Interpretation here is, through all blurs, unfolding and enrichment of the work.

Interpreters often claim that they want to bring to realization for themselves and for the audience a very definite, individual and the same time work-specific auditory imagination. This is assuming that they don't simply adopt what is being done anyway, but develop their own interpretation using the score, which represents the work. How far this interpretation will satisfy the criteria of refined adequacy can be assessed through analysis of each concrete interpretation and additionally through verbal records.

'Set points' etc. applies to every work as well as its realization – both productive and receptive. Dialectic-contradictory compositional guiding principles like those of complexity versus comprehensibility overarch productive and receptive realization, singing or playing and hearing. For logical reasons composers should appreciate the audience's reception of what is composed. This is why Anton Webern, especially in his highly complex compositions, made an effort for "Fasslichkeit" (*comprehensibility*) to avoid the aforementioned contradiction. The idea of a work has to be conferred in and through the realization of what is notated, including the audience as third geometrical point of the triangle work idea plus notated work, auditory realization and reception. A more radical version of this concept was formulated by music theoretician and historian Hugo Riemann in 1914, suggesting

> that not the actually resounding music, but rather the tonal phantasy of the creating artist preceding the notation of a score and the tonal phantasy evolving anew in the listener and the concurrent imagined tonal relationships constitute the alpha and omega of musical art. Both the stipulation of artistic sound creations in notation signs as the resounding performance of a work is the means to transplant the musical experiences from the imagination of the composer to that of the music listeners.[22]

A motto from engineering sciences may be adopted as guiding principle of realization and interpretation of works in the context of fuzzy logic: 'Not as exact as possible, but as exact as necessary!'[4] Imprecise, sloppy and slapdash singing and playing, initially mentioned as examples of 'vagueness', are obviously not justifiable as artistic

[20]Frost 2007.

[21]Cf. among others Heister 1983, esp. pp. 417–421.

[22]Riemann 1916, p. 2, quoted in Wicke 2011, p. 1063.

freedom of interpretation in the realization. Instead it requires the practical realization to be as accurate as possible both technically and artistically, in order to realize the work in an optimally, both in regard to the composition and in regard to the reception.[23] From this there always results a larger or smaller gap between intention and what is actually achieved—yet another example of fuzzy logic at work. An extreme case is described by Franz Grillparzer in his novella *Der arme Spielmann* (*The Poor Minstrel*). The minstrel plays, objectively seen, highly inadequately. But he has the illusion that he is the true interpreter. For he hears what he thinks he is playing, not what he is actually playing. In terms of categorization one could say that here the anticipation form and the imagination form become strangely mingled and result in a self and object perception that we might call a *hallucinatory form*. But for it to be introduced as an independent existence from it is not frequent and relevant enough.

Another type of deviation from adequate sound realization or interpretation deserves to be briefly mentioned: Mistakes, drastically wrong singing or playing, do nt fall in the category of blurriness and are therefore not part of the field of fuzzy logic. Under the aspect of flagrant non-perfection and the keyword "all-too-human" it has become a topic on You tube in anecdotes or short movies from musicians. In the technical reproduction of a musical process such mistakes become permanent and can thereby maybe be analyzed with fuzzy logic as blurriness when compared to the actually intended shape of the work in its sound realization.

> (Gregor Dotzauer) Miles Davis claimed: "Do not fear mistakes—there are none." Do you share his view that ninety percent of all mistakes can smoothed out by an intelligent trick? (Michael Wollny) Ninety percent, that's a very optimistic assumption. But it is insofar right, that every process is cramped if it is thought from the goal. And first of all we would have to define what a mistake is. A wrong note is in view of musical tension not per se a wrong note.[24]

3.3.1 "Miserable Violin" (Beethoven). the Composable and the Composed, the Feasible—and then the Audible

One main historical tendency is the striving to close the gap between the real and the optimal and ideal in an asymptotic approachment. Just as with notation in connection to composing, the abilities and skills of those, who realize the music develops on a historical scale. The *Mannheimer Hoforchester* (*court orchestra*) after 1740 had developed such unprecedented 'collective virtuosity' (Heister 1983), following the model of contemporary military and manufacture drill, in the main identical

[23] Questions of interpretation are among the most discussed in music journalism and musicology. The literature is innumerable; an anthology may be mentioned as an example, Danuser (ed.) 1992 and Heister 1983.

[24] Jazzpianist Michael Wollny im Interview mit Gregor Dotzauer: *„Das Klavier ist mein Avatar"*, https://www.tagesspiegel.de/kultur/jazzpianist-michael-wollny-im-interview-das-klavier-ist-mein-avatar/21104498.html, Abruf 1.3.2020.

bowing technique and homogenous increase of volume throughout the entire apparatus—the 'Mannheimer Crescendo'—that they became like an "army of generals, equally deft at devising a battle plan as at fighting it" (in the words of music historian Charles Burney). Burnley had heard the *Mannheimer Hoforchester* in 1772 at an opera performance at a summer residence in Schwetzingen. And Burnley adds regarding the combination of interpretational virtuosity and compositional productivity or competence: "There are really more soloists and good composers in this orchestra than in any other in Europe."[25]

Yet, in spite of all technical progress in execution and performance: the tension between production and reproduction or interpretation is exacerbated by the emergence of autonomous music, i.e. music as music for a concert, without functional integration or demarcation as in types of music like dances, marches, etc. The demands posed by compositions have greatly increased since the *Wiener Klassik* (Vienna classical period). In the context of this development the personal union of composers and performers has become rare; it had put certain limitations on technical requirements due to the two-fold and thereby less intensive specialization. But the personal union remains as an option: Mozart, Beethoven, Liszt, Brahms, Busoni, Béla Bartók and others played their own works as pianists, Louis Spohr or Paganini in the first half of the nineteenth century did the same on the violin.

Basically the demands are made from the side of the composers. To the objection, a passage in the string quartet op. 59,1 was too hard and unplayable, Beethoven replied to his friend Ignaz Schuppanzigh, leader of the string quartet, with the characteristic avantgarde sentence: "Do you believe that I think about your miserable fiddle when the muse strikes me?" This typical collision is a topos. The relation between mental (read intellectual) and physical (read menial) work comes into play—making music is not only spiritual-intellectual but spiritual-physical work realized through a body, through voice and hands. Conceptually more can be composed and notated than can be realized as material sound. "Narrow is the world, and the mind is vast. / easily do thoughts cohabitate, / but in space things roughly collide." (*Wallensteins Tod,* II, 2) Through a feedback loop of growing demands and increased quality of music education and training, the capabilities and skills of interpreters are improving everywhere on the world—at least from a technical perspective. It is the path from execution to interpretation, from the simply more or less vague execution to a deeper reading of what is in the score and of the idea of the work. Summarily this is a process over historically long periods of time of perfecting as generalization of highly advanced models and standards.[26] And so we find works of the avant-garde, considered nearly unplayable in the beginning of the twentieth century, in the repertoire of students' final exams at the beginning of the twentyfirst century. Methodical work, learning, and practice increases, via feedback loops, not only the capacity for productive realization and representation of complex structures, but also trains perceptive faculties. In this context the full scope of senses and its modalities may be applied, as in types of learning centered on kinaesthetic, optical, auditory faculties. Training and

[25] Burney 1773/2003, p. 73.

[26] Cf. e.g. Heister 1983, pp. 412f.

practice nowadays comprise of almost Taylorian processes of decomposition into ever smaller individual steps, but at the same time the psycho-physical interplay receives more attention as part of holistic learning. In this application of the general dialectic of labor division and cooperation lies a decisive progress of the late twentieth century—mechanical-manufacture practicing of scales, short phrases finger and grip techniques and so on were already known during the early industrialization. Thus by improved training and intensive practice previously insurmountable boundaries become increasingly transgressed limitations through individual and finally also historical-collective efforts.

Actual natural limitations are given by the human body and its performance capacity, which includes the psychological capacities. However, as these heightened demands on a performance increase, so do the performers' aspirations and the by now internalized perfectionism, which lastly tolerates no mistakes, which results in increased fear and hypertension when playing music, potentially inducing inhibitions and blockades.[27]

That professional playing of music is work and not playing, and that this equally applies to the 'playing' of musical instruments, is given by the definition. Yet the progress, which characteristically aims for steadily increasing sharpening, comes at a high price. Its dialectic involves at least two phenomena. Not only does the technical perfection overshadow the artistic perfection, the soulful and spiritual and sentimental interpretation, which brings to sensual expression the contents of a work. Furthermore the joy, which should be essential to music, suffers losses. In this regard the realistic insights into the inevitability of imperfections, as provided by fuzzy logic, may be helpful.

As in the reproduction of music, so too in the perception: The human senses and bodies have a historical dimension. But all the practice in the world will resolve the fundamental contradiction of composition/notation and reproduction/interpretation instead it can only progressively mediate it. Only the technical reproduction removed boundaries (but not limits) of the body and enabled its emancipation as is the case with Conlon Nancarrows electrical piano, which can polyphonic music at a much faster speed than a human pianist could. Analogous effects had in the pre-digital era been produced primarily with the technique of overdubbing. Starting points is still—or again—the individual playing of a human and not mechanical performer, as with Nancarrow and his music, which was played by soloists but on a polyphonous piano. The recording of several sound tracks on magnetic tape (multi-tracking, overdubbing) occurs sequentially, and is then arranged vertically as extreme polyphony and simultaneity. The jazz pianist Lennie Tristano made such experiments as early as the 1950s in experimental works such as *Requiem* and *Turkish Mambo* as well as *Line Up and East 32*nd, where he increased the speed beyond human capacity through altered tape speed. At the time he was much criticized for it, as the procedure was felt to be a deception.

Digital technical re-/production further expands the creative space, since it does away with several material-technical limitations and makes the realization of sound

[27] Cf. e.g. Möller 2018.

easier than it had been in analogue techniques that involved the cutting and pasting of magnetic tape. Now more objectively unrealizable music can be played and realised, as for example the absurd and formally complex work Maximiliano Maximilianos (pseudonym) *Faerie's Aire And Death Waltz.*[28]

The tension between idea and realization, relevant for fuzzy logic, and the interplay of fuzziness and sharpening are largely suspended here. In electro-acoustical music the compositional idea becomes completely independent from its realization as interpretation—naturally at the price of a very reduced and artificial (meant in a negative way) sound (See Sect. 6.8). However, it has its own and unique qualities. As can be seen especially in pop music since its electrification and lately in its partial digitalization, these "mechanically" produced sounds can find preference over traditional sounds. The electro-acoustical music production in both classical and pop music often reaches gigantic proportion in regard to performance-technical impossibilities. But even that has to pass through the bottleneck of our senses. And those who don't want to listen or can't, must feel it: Vague tactile sensations, preferably at unreasonably high volume bordering painfulness, so to say "good vibrations" of art versus "bad vibrations" in life, can surrogate more specific audiovisual sensations.

3.3.2 Compositional Guiding Principles Complexity Versus Comprehensibility

But those involved, for the sake of art, resort to deliberate rules concerning naturally occurring contradictions. These get resolved largely in a kind of homeostasis. For this end those guiding principles serve best that are already conceived as ambivalent and polar, thereby involving aspects of fuzzy logic. Guiding principles are comparable to cybernetic categories such as 'set points' in control circuits for both the work and its performance—this even applies to the conduct rules for ceremonies of the type concert or rituals of the type god or idolatry and other social norms. The capabilities of the interpreter as well as the perceptive faculties of the recipients constitute limiting circumstances. These are again layered. Of course voices can be stacked nearly without limits except for pergament size or paper format, as it has been done in some pomp works of the Renaissance, as in those of Ockeghem, Tallis, and Striggio, with up to 64 voices.

Such works may even be performed, within the economic limits, but already at a low threshold of around 5 to 6 voices maximum the individual voices are longer discernible in the polyphonic sense. Here the extended contradiction between the main phase production and reception come into play as one of feasible and audible. It is resolved in the process in the form of a relatively flexible dynamic equilibrium. More specifically in regard to the relation of composition and reproduction/interpretation this contradiction takes the shape of polarity of perfection versus

[28]Video e.g. https://www.youtube.com/watch?v=RbT4lIrdnYk and https://www.youtube.com/watch?v=Da911wDQHpA, accessed 2.4.2019.

liveliness, which has to be reconciled in the execution. Such guiding principles are ambivalent and bi-polar, which almost appears to be a contradiction between cybernetic and fuzzy logic and is resolved at least partially by the characteristic "a bit more – a bit less" of fuzzy logic in the mediation of the two poles.

For the production as the decisive phase of the entire musical process, it is mainly about the dialectical unity of the guiding ideas complexity versus comprehensibility. This contradictory unity has effects up to the level of melody formation, of compositional organization and style, and appears in numerous variations, as in Eisler's work understood as musical-political education as the contradiction of "high art" and "popularity". Complexity is of course more than the multiplication of voices or relations of motives. However it manifests itself: It is not a musical-aesthetic end in itself or a quality feature, as some avant-garde circles often like to assume. Primitivism and reductionism as in minimal music is certainly a popular, but not necessarily a good alternative.

Guiding principles serve as 'education of the educators'. Principally they are the result of social and individual decisions and therefore deliberate and set. They significantly determine the internal regulations of systems and sub-systems as well as procedural details. They interact with given circumstances, be it tonal or rhythm systems or liturgical practice, star cult or Orisha cult. Through the course of historical processes they are transformed into prerequisites, thereby often becoming half conscious or subconscious. This multi-layered context has systemic character itself. And it opens the perspective of the societal and historical dimensions of the existence forms. All this constitutes a generative deep structure and helps to better understand and explain usually unconnected phenomena, perceived individually and scattered, as a framework that is determined as exactly as possible. An example could be the emergence and inflation of interpretation styles, ranging from interpretation trends and interpretation history up to the performance of Beethoven on historical musical instruments or the constant battle of the 'critical musicus' (title of a work of enlightened music theoretician Johann Mattheson) over how to express in an intersubjective communicable and adequate manner the personal recollections and experiences (See Sects. 1.3.3 and 8.3).

A series of experiments is conceivable that compares various versions of the realization of a work. This is commonly known as qualitative comparison of interpretations in the field of music criticism. Further objectifying, quantifying and specifying with the tools of fuzzy logic would be rather innovative. With the works of 'classics' this becomes comparatively easier. Orientation point for the adequacy of 'work' and 'realization' would be the hidden center of aforementioned triangle of work idea, notated composed work as well as its auditory realization. Here the primary orientation point is neither the later covered primary "original" nor the phonographic or sound form of a work, but rather its score form. Herewith the orientation point itself is not immediately material and simply given, but rather must be constructed in each case. At least the score provides some orientation. This can reveal how through the indelible subjectivity of interpretation, the capabilities of the personalities of the performers, not only abstractly new and different music can emerge but sometimes even concretely qualitatively better music. In some ways this would be not a blurriness of realization but a sharpening of what the work expresses and shows. The conventional wisdom that realization must be blurry compared to the idea and the work is here not proven wrong, but should itself be reflected on as a problem. In spite of the methodical challenges outlined here such inquiry could be helpful for making subjective, mostly untenable moral and aesthetic judgements more objective and to a certain degree even quantifiable.

3.3.3 Fuzzy Edges. Framework Institution, Tradition, Historical Changes of Interpretation

As a rule, norms and rules change over history. Take the ideal of 'faithfulness to the original', an attempt to escape from crude blurs as caused for example by the subjective caprice of the interpreter or the ad hoc arbitrariness of orchestration in accordance with easy availability as opposed to what is required by the work. Faithfulness aims to approximate as closely as possible the letter and spirit of a work in a performance. It is also a manifestation of the principle similarity. It developed historically with the growing precision of notation of works on the one hand, and of execution and performance on the other hand. The place in society for this process was primarily the concert, that after the epochal watershed of 1848/1849 up to the beginning of the twentieth century had developed into an institution in which the realization of this ideal was possible, necessary and desired.[29]

The next stage was, especially after 1945, the historically accurate of for short 'historical performance practice'. Through the use of instruments and styles of playing from the respective time period of the work its—historical—spirit and letter should be realized in a more faithful and 'similar' fashion. The accompanying aporias shall only be mentioned briefly here, such as the impossibility of reconstructing today the exact contemporary historical performance situation, be it the spaces or the mentalities, experiences and tastes of the audience of those days. This reflection of historical developments, of the temporal difference between the respective 'then' and 'now' together with the changes in practice, constitutes a legitimate approach that since the dawn of the twenty first century has been more moderately termed 'historically informed performance practice'. Just as the faithfulness to a work it in principle resembles a sharpening. Yet the constitutional blurred fringes of the institutional framework, such as the imponderableness of audience behavior (particularly in their acoustic dimension), the acoustics of the space related to air temperatures and humidity etc. These can only be mentioned here, not analyzed in detail, as they constitute the 'environment' of the system 'work-realization' and are hardly accessible for fuzzy logic.

[29]In detail cf e.g. Blaukopf 1968, Heister 1983 etc.

> The score is not the music itself but merely a set of instructions – and a very vague set at that – formulated in order to make the music happen. Surely, when we say we wish to be true to the score, it is not the computerised mirror-imagery.[30]

The blurs within the 'system' work/idea/realization can be comprehended qualitatively and tendentially be conceived quantitatively at least in a number of its components, such as the four dimensions of tone pitch, duration, volume, and color, with the methodology of fuzzy logic.[31] Mazzola/Mannone/Pang (2016) aspired to capture in mathematical-quantifiable terms even the expressive and connotative components of a realization. At first they presume a very simplified definition of realization, which is then differentiated and expanded.

> Performance is understood to be the transformation of a symbolic musical object of notes [...] to a physical object composed of sound events.[...] This description is a huge abstraction since performance includes the performing artist, gestures, and real instruments, all being realized in a concert hall in front of an audience. But it is the approach that is traditionally taken when addressing "expressive performance research."[32]

However, the issue is much more complicated. For music does not only resound in concert halls, but also in the opera house. In music-theatrical works, which are still part of the domain of music, a scenic—dramatic dimension is added to the vocal and instrumental music. This is one the one hand the materially manifested set design, including costumes and props, on the other hand the personal-procedural dramatic activities, tied to the body similar to playing an instrument, or, as in ballet, the art of dancing. Mazzola's mathematical models take this layering and expansion into account:

> the gestural aspect of performance, adding another aspect to a most complex total image. However, the... abstraction to notes and sound events can be helpful as an introduction when trying to piece together the whole picture. It will turn out that gestural performance can be built upon this note-oriented abstraction [...]. Expressive performance is a semiotic action, a rhetorical endeavor of transmitting messages (of gestural, symbolic, or emotional nature) via instrumental interfaces to a more or less interested audience.[33]

3.3.4 **Desafinado.** *Tuning Systems and Standard Concert Pitch and Historical Change*

Further fundamental blurs, constitutional to the system, occur both in the numerous different tuning systems for pitches between nature and culture, as well as in the

---

[30] Frost 2007.

[31] In full detail and addressing the full picture Lerch 2008. (See below Chapter 5) Cf. also the attempt of evaluating sound qualities quantitatively, or vice versa, using quantification to make qualified judgements: Wrzeciono 2008. Yankovskii 1965 focuses on one component in his still analogous and pre-digital approach.

[32] Mazzola/Mannone/Pang 2016, p. 266.

[33] Ibid., p. 267.

problems that result from the historical changes of the concert pitch that serves as standard and benchmark (See Sect. 5.1.1). In music there exists a difference between tone realized as sound and the compositional tone intended in the tone system; similar to the difference between phonetics and phonology in language. (1) The individual articulation/a/can be expressed in different ways, but within limits (2) Within these limits—articulatory etc.—it will 'count' as phonetic realization of the intended phoneme. This again, just like the phone, stands in structural system contexts. In the respective context of an individual language the phonemes are filtered from many phonetic varieties. Regarding musical tone qualities, commonly referred to as pitch, the notation is more precise and more distinct than the sound articulation. Once again we come across the relation between idea and realization. The musical a' can be—in new notation since around 1600—distinctly determined as intended. How it sounds, that is on what pitch, is defined by no means. In old tunings up to the nineteenth century the same tone in notation can sound quite different. Considering all the identity in another medium this difference is considerable. And even in standardized modern tuning since 1885 there have still been historical changes. The gap between notation and realization remains. In a typical fuzzy logic framework around 440 Hertz the standard tuning has been driven upwards in the last three to four decades from 435 Hertz to 441 even 443 Hertz. This leads to somehow more brilliant orchestras, but also to big despair of singers—for they cannot extend their vocal chords higher and higher. And so the 'a' from today is the b flat from yesterday. Whoever now says A, is saying B from a historical perspective. Listening attuned to the musical system context will gloss over these differences.[34] (See Chapter 1).

3.3.5 The Work as Unspecific Stimulus. Postmodern "Constructivist" and "Deconstructive" Nihilism

Still a whole historical phase before the development and assertion of the ideal of 'faithfulness of the original' a countertendency emerged among the artistic avant-garde of the last turn of century, which expressed criticism of the 'work', both in its definition and its matter.

The traditional works of art were in this context considered as dead weight to be disposed of—maybe also in order to get rid of annoying competition and to place the own products on the 'market'. Especially the Italian futurism stands out with its efforts.

Until the 1970s, with the beginnings of the structural crisis of the 'market economy', this tendency remained mostly confined to avant-garde circles. After this, these tendencies became part of the mainstream. The 'director's theatre' is probably the best-known and most pervasive phenomenon of this conviction. Here the existing (total) art work is not realized in accordance with its own shape and content, but rather

[34] Cf. e.g. Lerch 2006.

serves as template for the realization of ideas of the directors. From productive realization such tendencies also encroach on receptive realization. According conceptions and procedures can rely on decade-old traditions, which have attributed primacy in the art process to the reception; and a century-old tradition of subjective idealism, which was condensed in the formula of Bishop Berkeley in the eighteenth century: "esse est percipi", meaning being is being-perceived, by which ontology is replaced by gnoseology. The newest variation is postmodern 'constructivism'. In matters of art this ideology leads to a subjective arbitrariness in dealing with art works, where lastly fuzzy logic has no more say. A particularly drastic example is provided by the comments of a music archivist, in his own words the 'director of an internationally significant collection of musicalia.'[35] According to this concept the work does not actually exist, and if it does, then it is 'meaningless'. As far as it does exist, it only stimulates sensual execution rather than being a meaningful prescription.

> An artifact presents itself to us as a foil, behind which we suspect a truth that only becomes an affective object in the moment of perception. We find back to ourselves and our emotions, feelings, passions, associations and projections, *the object itself* as instrument of reflection has no life of its own, is *meaningless*. The achievement of the sound creator lies in forming and shaping the initial disposition, the musical idea which has not yet become sound, in such a way that it enables the interpreter to evoke a resonance in us and thereby enabling us to imagine the illusion of teleology of the sound event.[36]

With another turn of the caleidoscope the work is replaced by 'the emotional impression' of the recipient, which in a keen turn is said to be a 'quality of the recipient'. He or she are full of themselves and don't actually need any art.

> It is not necessarily the structure prescribed by the creator that affects us. In music and the fine arts it is the emotional impression, which alone by motivating our feelings or associations or our thinking is experienced as meaningful and causal. This is – and this is the central point – *a quality of the recipient*, not a quality of the artifact to be objectified.[37]

To ontologize an 'impression' of a recipient as a 'quality', as one might do with your biometrical data, is twisted thinking. But it may serve once more as support of the anti-art credo of the time shortly following Duchamps—a mantra preached with particular pleasure and impact by concept artist J. Cage with passion and severe consequences:

> Art is, what the viewer, reader or listener perceives as art or identifies as such.[38]

This is nothing else than an aggravation of the claim that beauty lies in the eye of the beholder—in part justified with natural beauty, not however with the beauty of art. And this again reflects the wide-spread subjective-idealistic dictum of Bishop Berkeley in the eighteenth century of 'esse est percipi'. In adaptation to the populist ideology of musical illiteracy the negation of the objective gives no credit to the 'read text', but commands easy listening for the reception of music, which will only be

---

[35] Nägele 2019, p. 47.

[36] Ibid.

[37] Ibid., p. 48.

[38] Ibid., p. 47.

> a resounding phenomenon and not a text to be read -, and this experience communicates itself without requiring any necessary knowledge of the structures prescribed by the sound creator. This applies equally to a baroque four-part fugue as to a serial piece of music. Both cannot structurally be listened to in their entirety or when only with considerable intellectual effort (which at the moment of listening would not constitute a *surplus value* of pleasure).[39]

When dealing with 'surplus value', one that can be realized in the market, everything else becomes secondary. This irrational basic tendency leads to claims without any proof or even the attempt to provide supporting arguments for the opinion.

> But music is not a language, not a concrete semiotic system, in spite of all attempts by music science to prove just that. It is comparable to the equally affective glossolalia: not a monument that has been recorded in writing and has to be interpreted as such, but an event beyond language. Music in this sense does not carry meaning, but is an indifferent stimulus, which at the moment of reception evokes a pattern of arousal.[40]

Even using constructivist criteria this makes little sense. If the stimulus in this behavioristic understanding of humans would be totally 'indifferent', then one single piece of music or a computer game would suffice to induce the various subjective reactions. And humans are indeed diverse. This would save a lot, but also diminish the mentioned 'surplus value', and of which more could be realized through more works than just one. This may suffice. Such thinking, that exacerbates common patterns of practice, thinking, and arguing, does away with all intellectual and practical exertion, which would attempt to resolve the problems of adequate realization of creative ideas. Work-based performance practice, object-related and adequate perception, science and with it fuzzy logic are forced to abdicate here.

3.4 'Uncertainty'. Work as Sensually Received and Mentally Reflected Idea: Perception Form

Perception is itself a theme of fuzziness. This involves both subjective and situational aspects and in many cases also the perspective of the layman regarding music as a profession. But also in professional contexts of music production the perception of the executing subjects blurs and mutates the experienced and reproduced shape of the musical work. Human perception, with its own limitations, therefore constitutes a specific existence form of the work in the world of musical experience. This results in new tensions between subject and object, which can be understood and described with the analogy of 'fuzziness relation'. At the same time these processes spawn new leverage for sharpening; for every perceiving subject tries to change experienced fuzziness into a 'precise picture'. The deviation from a given musical work can become quite significant.

---

[39] Ibid., p. 48.

[40] Ibid.

3.4.1 Unconscious Amending of Hearing, Playing, and Singing

The transition field from sound form to perception form includes phenomena such as foremost 'amending hearing' in the musical system context, and in the production context, as term rather uncommon, 'amending playing' and 'amending singing': Tone pitch, duration, volume and color are adapted to the own conception and the situation or 'negotiated' and jointly approximated by the interpreters in the intersubjective context of a performance situation. On all sides fuzziness is allowed for and remains as such; the perceiving assimilate it by overhearing an imagined, virtual precision. A variational manifestation of aforementioned 'amending hearing' is the not fully accurate listening. 'It'll get lost in transmission' (original: „Das versendet sich") was a common phrase in the recording room for broadcasting in the 1970s, particularly in radio broadcasting. What was meant was that in the process of listening, such mistakes are not fully perceived and thereby hardly ever remembered. A not precisely articulated word, a not quite precise cut would then not be again refined, but left as it was—hoping that in the temporally bound hearing process not every detail could be fully perceived. This belongs to the general questions of semantics. In an active social context this can be adjusted and corrected by mutual feedback. In the unilateral sender-receiver model of a performance this is only possible in a very limited way, even though it is well known that performers absorb the resonance of an auditorium and probably let it influence and adapt their playing. In all dialogue forms the intended meaning can be inferred only from what has been audibly perceived. What is maybe incompletely or imperfectly stated is amended according to context and transposed into a fitting structure of meaning. In an active social context this can be adjusted or corrected by mutual feedback. In the unidirectional sender-receiver model this is possible only in a very limited manner, though it is well known that performers absorb the resonance in the auditorium and probably incorporate it into their performance. In the course of 'live' performances', such as a concert or an opera, small mistakes are easily amended. Generally they are not noted by the listeners, since they concentrate more on the whole than on details. This applies also to the perceiving of wrongly tuned musical instruments or the deviation of the right pitch in singing, "detuning". Such blurs are thematized textually and slightly ironically in a standard of bossa nova by Antonio Carlos Jobim, who texted and composed *Desafinado* ('out of key') (1959). (see above Tuning Systems and Concert Pitch)[41] Conversely such blurs or even mistakes can naturally occur just as well in the receptive realization, not lastly in the somehow mistuned and mistuning perception of pitches.

> Musically trained and untrained participants provided magnitude estimates of the size of melodic intervals. Each interval was formed by a sequence of two pitches that differed by between 50 cents (one half of a semitone) and 2,400 cents (two octaves) and was presented in a high or a low pitch register and in an ascending or a descending direction. Estimates were larger for intervals in the high pitch register than for those in the low pitch register and for descending intervals than for ascending intervals. Ascending intervals were perceived

[41] Cf. e.g. Weihs/Berghoff/Hasse-Becker/Ligges 2001.

> as larger than descending intervals when presented in a high pitch register, but descending intervals were perceived as larger than ascending intervals when presented in a low pitch register. For intervals up to an octave in size, differentiation of intervals was greater for trained listeners than for untrained listeners.[42]

Perception could be defined as a special form of refraction. Refraction here also means: The perceptive apparatus of a subject is proactive and does not remain passive in regard to the object. This does not mean that the perceptive apparatus constructs what is perceived.[43] It already exists as a work of art and does not need to be constructed yet again. It suffices that the perceiving subject reconstructs, as precisely as possible, the artwork for him—or herself. Since we are dealing with a refraction relationship, and the work ontologically and gnoseologically antecedes the perception, a perfect conformity in the perception of a work that itself is only relatively perfect is impossible (see Chapter 4). Fanning out types and degrees of adequacy regarding the hearing and understanding of music with the methods of fuzzy logic constitute a difficult but worthwhile task. Existing hearer typologies would themselves constitute a sub-object of such an investigation. Th.W.Adorno for example in his rather unsystematic typology postulated an ideal that can only be attained by those professionally involved with music. The musicologist Harry Goldschmidt carved out as ideal and goal "explicit musical understanding", which as such is quite demanding and, in contrast to Adorno, includes explicitly semantics, which further complicates the matter. But in the end Goldschmidt's ideal is more attainable.[44]

The initially vague, then fuzzy contour of the whole serves as guideline almost literally by drafting an outline that continuously becomes clearer.[45] Frame and orientation—points are also provided by the respective idiom—of the work, the composers, the genre and the style. Due to the musical logic of musical language there results an immanent stringency. The consistency of an object—which surely

---

[42]Russo, Frank & Thompson, William 2006.

[43]Cf. e.g. *How the Brain Constructs the World,* Neuroscience News, February 9, 2018. *Source: SISSA,* http://neurosciencenews.com/sensory-signaling-8468/, retrieved Feb. 10, 2018.

[44]Adorno 1962/1968; Goldschmidt 1974.

[45]The "extrapolation" of incomplete and fuzzy information and the evolutionarily evolved evaluation of such under the aspect of the necessity survival occur with frenzied speed in areas of the brain preceding the processing in the cerebrum. "(T)he brain can begin processing value just 80 minutes after seeing something. That's less than a tenth of a second—and means the brain is basically figuring out if something is quality or junk at the same time it recognizes what it is. [...] (T)hat value representation is strongly associated with later responses in the prefrontal cortex. The new findings show that value processing can begin in the visual cortex, before any value signals appear in the prefrontal cortex. The visual cortex is well equipped to discriminate the fine details in appearance that underlie value judgments about natural objects [...] for evolutionary purposes [...] being able to evaluate objects can be just as important as recognizing them [...] to form fast, appropriate behavioral responses" (Connor 2018). However, all of this at first relates to visual perception. For us it would be important firstly how this relates to the auditory, whether it is the same, similar, or different, and secondly the question of accuracy or conversely of fuzziness, difficult because the evaluation has a subjective component. One difference is already apparent: As essential as works of art may be for life—their perception is not a matter of life and death.

is manifested in a variety of ways—makes it easier for the subject to still perceive what it has not heard or even objectively was not able to hear. By going back and forth between the unique and the moment on the one hand, and the outlook onto the ideally-sensually anticipated whole at the end of the work process, the subject reconstructs the work that never appears consummately and without fuzziness, and which ideally itself represents the fully accomplished unity (as stated by the writer and aestheticist of the late enlightenment, Karl Philipp Moritz) (See Sect. 2.3.4). Through amending hearing the immediate perception of works and sonic realizations in the present can change into to fragmentary memory. This then approximates a hallucination.

3.4.2 Perception in the Trinity of Text, Sound, and Context

In all perception processes fuzziness are unavoidable as a form of special refraction in varying degrees. This applies equally in the context of other existence forms, such as the perception of music only in its notation form. Issues and investigation become particularly complex when considerations of semantics and of word language are included.[46] The perception of a work process in a 'performance', regardless whether live or technically recorded, moves within the triangle of (1) the text that precedes the score form (2) the respectively present actual sound form and (3) the given, objective preconditions such as quality of space, behavior of the audience and so on as well as subjective situational preconditions such as preexisting knowledge and respectively updated 'horizon of expectations'. Sensually realized object and mentally reflected idea go hand in hand in this.[47]

One way that information becomes incrementally more fuzzy in a channel prone to disturbances, as can be seen in the game 'Chinese whispers', a so-to-say global play using different names: politically incorrect named *téléphone arabe* in French, or more correctly as *téléphone sans fil* (wireless telephone), in German correctly and similarly referred to also as *Schnurloses Telefon* or, more common, *Stille Post* (silent mail).

In Britain, the local refer to it as "Chinese Whispers" because historically, the British believed the Chinese language to be incomprehensible, whereas in politically-correct America, the game is referred to as "The Broken Telephone Game."

The idea of Chinese Whispers is for several people to form a line and for one person to think up a phrase. He then whispers it to the player who is next [...]. This player should now whisper what they think was said to the next player, and so on, until the last player shouts out the phrase so that everybody can hear [...].

- I took my dog for a walk today and then I gave him some food
- I took my dog for a walk today and then I gave him something good

[46]Cf. among others Koelsch 2002; Koelsch et al. 2004.

[47]Cf. e.g. Heister 1983, Reicher-Marek 1998.

- I took Michael for a walk today and then I shaved him something good
- I mistook Michael for an ork today and I shaved him something good.[48]

The growing fuzziness is evident in the example and has comical features (see below the less comical growing of 'white noise' in the phonographic form) the pedagogic reasons for playing this game are expressed as follows:

> as example of why accurate communication is important; to illustrate why gossip should be ignored; as a fun game to see how language can evolve.[49]

Less funny are these circumstances, when we relate them by analogy to the transmission history of works. Here the oral transmission is by far closer to the 'Chinese Whispers' than the written transmission.[50] The human auditory perception is highly efficient. One step further and we would be able to hear the friction noise of molecular movements, which would make no sense from an evolutionary perspective and would be catastrophic for our appropriation of reality. The volume threshold therefore is very low. Also, the range from there to the zone, where auditory perceptions transitions to tactile sensations and reaches the acoustic threshold of pain, is considerable and lies in the order of magnitude between—defined by logarithms—90 and 120 dbs. The perceived volume depends a lot on the distance from the source of sound and can therefore not be determined in absolute numbers. Generally our hearing is in principle strongly differentiated physiologically and thereby has little fuzziness—except of course in the fringe zones of acoustic perception, where our capacity for differentiation and discrimination is diminished. The limits of the perception of volume and frequencies are known: frequencies above 22.000 Hertz are not even audible even for younger people, but only for example for bats and dogs. Blurs before the actual esthetic-musical perception move in this area of the auditory range, but naturally intensively affect this. Further details from auditory physiology and psychology cannot and must not be further explicated here, as important as they are for an in-depth and expanded investigation of this topic. The hearing sense is therefore only the last instance of impact, when in the hearing of music and the receptive realization of a work of art, difficulties and blurs occur.[51] However, in the further processing of sensual impressions these appear as contradiction of mathematical precision and biological-psychological fuzziness.

---

[48] http://www.chinese-whispers.com/, retrieved 15 April 2020.

[49] Ibid.

[50] This also applies generally to memory processes and many historical processes.

[51] Hearing disorders are of course a different matter. About the special case of deafness alleviated by prosthesis cf. e.g. Wright/Uchanski 2012.

3.4.3 Reception as Processual Sensory-Mental Perception Between Anticipation and Recollection

Sensation, perception and neuronal processing constitute the main stages of the reception process. At each stage there is yet again a new chance for fuzziness. Immediate sensation and mediated anticipation, recollection and expectation, categorical cognitive preformation and context interact with each other.

Recollection processes are an indispensable component of the reception process even during perception, at first as retention in the ultra-short/short-term memory or working memory, and then following the immediate perception process of a work in the long-term memory. In analogy to the thresholds and limits of pitch and volume there exists a temporal perception limit and a threshold and limit for the short-term memory of about 3 seconds—a timespan in which quite a lot of music and possibly additional word language can be accommodated. In the present there are therefore, as stated, both past and future present. And at the same time, as we know, the 'present' is itself an ever-shifting point in time moving unstoppably along the timeline. Because such diverse memory forms are intertwined and meshed together, new and inevitable blurs emerge at the intersections and points of contact.

> the brain can begin processing value just 80 milliseconds after seeing something. That's less than a tenth of a second—and means the brain is basically figuring out if something is quality or junk at the same time it recognizes what it is. [...] (T)hat value representation is strongly associated with later responses in the prefrontal cortex. The new findings show that value processing can begin in the visual cortex, before any value signals appear in the prefrontal cortex. The visual cortex is well equipped to discriminate the fine details in appearance that underlie value judgments about natural objects [...]
>
> for evolutionary purposes [...] being able to evaluate objects can be just as important as recognizing them [...] to form fast, appropriate behavioral responses.[52]

3.4.4 Conditions, Situations, Distractions

At every stage of perception there are, as mentioned, relatively autonomously originating blurs of both hearing and of neuronal processing. Additionally blurs or rather shortcomings regarding complete precision occur through always possible and often actual disturbances, distractions, or shorter or longer phases of inattentiveness. These can come from outside or inside: The neighbor in a concert coughs or rustles with a candy wrapper; one thinks, instead of listening attentively, of the neighbor or, by association and more distant of the soon due tax declaration, landscapes or whatever.

All this is not easily accessible for self-observation, especially in retrospect, and even harder to verbalize and to objectivize. By empirical-experimental means this is in principle possible to investigate and therefore accessible for fuzzy logic, by measuring brain activity during the process of perception in combination with retrospective commentaries of participants. Sensual aesthetic perception is never only

[52]Connor 2018.

"immediate" but always embedded in a web of preconditions, preconceptions of cognition. But this does not mean that the mind or rather the perception "constructs" the perceived. It is rather an interplay and interaction of subject and object. Said preconditions might slightly blur the object perception, since they do not refer exclusively to the given object. But on the other hand they make such perception possible in the first place. Kant contended, perception without concepts is blind, but concepts without perception are empty.

Crucial reasons for fuzziness of perception are found in the manifold objective and subjective general preconditions, which precede each current perception process and provide a frame and embedding for it. They range from the acoustics of an opera house including the respective current room temperature and many more parameters up to the psychological condition of the listeners such as alertness, sleepiness, attentiveness including their musical, aesthetical and historical prior knowledge and experiences.

Moreover, all these preconditions also have a historical component.[53] The intertwining of the diachronic and synchronic will add complexity to the matter.[54] To contrive an even only approximately complete list of conditions and factors is beyond the scope of the work presented here. In the following I shall at least sketch out some important conditions. From the viewpoint of fuzzy logic these must always be seen as preconditions. These co-determine the respective actual perception, which is in part an already unreflecting, unconscious result, in part and in the best case resulting from preconditions and the given special situation. The preconditions themselves are not object of fuzzy logic analysis. They can hardly be that, since what could be called 'fuzziness' here would be too complex to determine appropriately. In view of the methodological difficulties there are understandably relatively few empirical investigations and even less experimental ones. But despite the methodological difficulties, there are some empirical studies in this direction. (For a sketchy overview, see 5.2.2 below).

3.4.5 Preconditions Activated and Updated: Expectation

General and special preconditions interlock within expectation. The 'expectation horizon' shapes through a variety of information what the perceiving party will imagine and conceive of as what is heard, and this already before the perception process has even begun.[55] This anticipation continues immediately with the beginning of the actual hearing process. In the perception process the neuronal processing proceeds by trial and error and reaches ahead through probabilities. Between this and

[53]Regarding historical cultural and musical background cf. e.g. Gratzer 1997.

[54]It is not very helpful for fuzzy logic investigations when the famous "reduction of complexity" (Luhmann) descends into simplifications. As for example: Zivic/Shifres/Cecchi 2013. This investigation about reception in regard to historical styles assumes schematically and in an unreflecting way pseudo-epochs as 'baroque' and so on, as if different styles never co-existed at the same time (Cf. Knepler 1977).

[55]Cf. e.g. Pearce/Wiggins 2006.

the actual perception it is most likely that contradictions emerge, which find their expression as fuzziness.

> Human perception is based on mathematically optimal principles, but the brain implements those principles imperfectly [...]. (H)uman perception is blueprinted on optimal strategies, even though the brain's execution of these strategies seems to be imperfect [...]. (T)here are indications that *neural computation is inherently imprecise*, which makes it implausible that humans would perform optimally on any non-trivial task. [...][56]

Studies about visual perception are however not directly transferrable to auditory perception. In a large-scale study N. Dupuis-Désormeaux expands the components of music perception into a system, that in a communicative 'dialogue' explicitly includes composition, and not only modelled or digitalised types of composition but genuine ones. Herein expectation plays a key part.[57] To comprehend the real complexity of perception methodically, it is often simplified through modeling or simulations via computers.[58] Closer to fuzzy logic are probability-based procedures.[59] As part of perception expectation has already been discussed extensively in music phenomenology in succession to Husserl and especially by Roman Ingarden.

> The role of *expectation* in listening and composing music has drawn much attention in music cognition since about half a century ago. [...] (W)e provide a first attempt to model some aspects of musical expectation specifically pertained to short-time and working memories, in an anticipatory framework. In our proposition *anticipation* is the mental realization of possible predicted actions and their effect on the perception of the world at an instant in time. [...] (C)omplex musical behavior such as long-term planning and formal shapes [... is] a result of the anticipatory architecture.[60]

Anticipation and reenactment of momentary sounds interact continuously in the subjectively reproduced formation of shape. The moment progresses perpetually together with the temporal process. Perception is not only receptive or even passive. But nonetheless: The object is given and must be reconstructed. At the same time recollection processes are involved. They become effective through non-conscious 'implicit' mechanisms.[61] Only by this automatization is the evolutionarily required tempo of perception warranted.

> a biological purpose of memory is to prepare organisms for the future, so it can usefully be construed as a kind of "readiness circuit." At least three memory systems are thought to be involved in expectations: episodic, semantic, and short-term memory. Each system leads to a distinct form of expectation.[62]

[56] Stengård/van den Berg 2019. Emphasis HWH.

[57] In detail Dupuis-Désormeaux 2017.

[58] Cf. e.g. Schellenberg 1997.

[59] Cf. e.g. Temperley 2006 and Temperley 2007.

[60] Cont/Dubnov/Assayag 2007.

[61] Expectancies can be conceived of as a form of implicit memory. Thompson/Balkwill/Vernescu 2000.

[62] Huron 2006, quoted in the recension of William Forde Thompson, in: Empirical Musicology Review Vol. 2, No. 2, 2007, pp. 67–70, quotation p. 69, https://kb.osu.edu/dspace/bitstream/handle/1811/24826/EMR000024a_.?sequence=1.

3.4.6 Pre-preconditions. 'Halo' and 'Devil' Effect and Other Blurs—Prejudices as Pre-formations and Deformations of Perception

History continues in the present, even in the individual. The biography particularly preceding experiences of art, impact the respective current perception. Memory imbues, even shapes, as previously noted, the expectation, and thereby the immediate perception. Therefore within perception all three times are stacked together: Past, present, and future. Furthermore this reflects the general relationship of experience and perception, in some aspects even that of theory and empiricism. 'Theory' in a broad sense here refers to what enters into sensual perception as knowledge, education, and skills.

A further-reaching methodological approach shows how preconceived judgements of an issue influence not only the judgement but even the perception. Preconfiguration of perception and prejudice simultaneously guide perception through expectations in such a manner, that as anticipation they are also its fulfillment, a kind of self-fulfilling prophecy – as has been generally noted, a strange merging of times. Really new aesthetic experiences are thereby made more difficult or even impossible.

A special, but far-reaching fuzziness of perception is due to the 'halo' effect. The metaphor of the halo already refers to the two-fold and thereby blurred contour of the matter—normally referring to people. The halo effect refers to the tendency of wrongly thinking together qualities of people or things that are factually independent or only loosely related.[63] This issue borders on typical perception and thinking mistakes, as the confusion of (statistical) correlation and causality or of temporal and causal sequence—latter the classical 'post hoc ergo propter hoc'. Single characteristics of a person (such as looks, voice, attractiveness, social status, renown generate a positive or negative impression, which wrongly generalized is transferred to the whole, thereby 'overshadowing' further perceptions of the person and influencing the impression of the whole person disproportionately.[64] This wrong or rather inductive generalization has its inversion, which strangely enough does not feature in the discussion that is the deduction of single qualities from the affective and evaluated whole. And this not only applies to people, but also to issues and objects. For the individual the effect remains sub-conscious. It is intentionally used primarily in advertising and marketing—and thereby also for works of art. This involves prejudices and preconceptions—such as descriptions that guide and determine reception, be it written or oral, or the components of the work itself, which function as triggers. This can then be transferred from people to issues or more particularly works of art.

[63]„Edward Thorndike, known for his contributions to educational psychology, coined the phrase "halo effect" and was the first to support it with empirical research. He gave the phenomenon its name in his 1920 article "The Constant Error in Psychological Ratings". He had noted in a previous study made in 1915 that estimates of traits in the same person were very highly and evenly correlated." In detail https://en.wikipedia.org/wiki/Halo_effect 17.1.14, retrieved 23 Jan 2014.

[64]Cf. primacy and recency effect—http://www.soft-skills.com/sozialkompetenz/menschenkenntnis/halo/effekt.php, retrieved 23.1.2014.

Because of a beautiful passage, which is especially impressive, the entire work is considered beautiful. And vice versa: If a work pleases us as a whole, every single detail, even if it is misconceived, will be considered well-made. By analogy this also applies to work realizations. And finally this pre-judgement can be transferred to the interpreter: If we like them in principle, or they are generally regarded as good, their interpretation of a work will be more likely rated as good.[65] The 'halo'-effect has its opposite, the 'devil' effect that is a radiating generalized negative bias, methodically the same only with inverted values. Newer research has given a new spin to the halo effect. Murphy, Jako and Anhalt claim

> that the assumption that halo effects are always detrimental is incorrect, going so far as to say that *some halo effects in fact increase the accuracy of the rating.* Additionally, they discuss the idea of 'true halo' – the actual correlation between, for example attractiveness and performance as an instructor, and 'illusory halo', which would be cognitive distortions, errors in observation and judgement, and rating tendencies of the individual rater.[66]

The analogy to ambivalent effects of fuzziness especially in the arts is evident.[67] It is challenging to rehash all this concretely for a fuzzy logic investigation, but not impossible. The approach would be a two-stage experiment. Just making people aware of the sub-conscious prejudice already changes the perception, in principle towards increased correctness and accuracy. A specific research design or an empirical study will bear the same or similar problems (such as verbalization), tasks and procedures, as they are discussed at the end of this chapter.

3.4.7 *Soft Computational Modeling. Computerized Entanglement of Perception and Production*

Often targeted is the entanglement of procedures and particularities of reception, especially through perception, for purposes of production, as

> *applications to automatic improvisation and style imitation.* The proposed model, based on cognitive foundations of musical expectation, is an active model using reinforcement learning techniques with multiple agents that learn competitively and in collaboration. We show that compared to similar models this anticipatory framework needs little training data [...].[68]

As interesting as this might be methodically, as questionable it is in the nature of things. We may question whether it is relevant to automate improvisation or even

[65] https://en.wikipedia.org/wiki/Halo_effect, see above.

[66] Murphy/Jako/Anhalt 1993. Emphasis HWH.

[67] We do not have to concur with the slightly agnostic tendencies: „They claim that any true differentiation between true and illusory halos is impossible in *a real-world setting*, because the different ratings are strongly influenced by the specific behaviors of the person observed by the raters." Ibidem.

[68] Cont/Dubnov/Assayag. Emphasis HWH.

imitation of style, which itself is rather dubious, except for learning purposes—which should not result in an automated process for the subject. After all we are here considering a human-machine interaction:

> (A) set of stochastic rules [...] allows for the creation of new musical improvisations by means of intuitive interaction between human musicians or music scores and a machine.[69]

With the alternative option mentioned above of interaction between "music *scores* and a machine" this approach by far exceeds the interaction model that emerged as early as the late 1950s as well as the almost traditional "live-electronic". Precursors of such mechanical improvisation or music production guided by algorithms are Lejaren A. Hiller's early experiments with computer music in the late 1950s.[70] In another way and without scores such appliances with algorithms become semi-autonomous co-players and opposing players of humans in the application to jazz.

> GenJam is an interactive *genetic algorithm (GA)* that *models a human jazz improviser and performs regularly as the author's sideman on jazz gigs.* GenJam learns to improvise full-chorus solos under the guidance of a human mentor and "trades fours" [the quick dialogic interplay between two soloists with 4 bars each, HWH] in real time with a human performer in "chase" choruses.[71]

Investigations of this type aim to combine results from reception studies with an automatization of production. The result is generated by statistically guided pseudo-randomness, or rather pre-formed pseudo-randomness steered by algorithms.[72] As far as I can tell, this actually implies no previously intended concept of a work, which could serve as scale for fuzziness or sharpening in the context of fuzzy logic. The "machine production" of music the automatisation of personal playing of music seems not to be highly desirable for societies these days. It is not a social need at most it constitutes an economic demand, in the hope that it might save costs for the workforce—a well-known general desire of "market economy". When the word "creativity" appears somewhere, it does not immediately imply that it actually comes into play in the matter at hand.[73] The musical products themselves, not to mention the restricted codes of the resulting sound forms, can lastly not compete in quality with what is musically produced by humans instead of machines.[74] Humans generally

---

[69] See Cont/Dubnov/Assayag.

[70] See e.g. Hiller/Isaacson 1959.

[71] Biles 2003, emphasis HWH.

[72] E.g. Conklin 2003; more general e.g. Pearce/Wiggins 2005.

[73] https://www.researchgate.net/profile/Geraint_Wiggins/publication/253385688_Computational_Modeling_of_Music_Cognition_and_Musical_Creativity/links/544782900cf2d62c30507c6a/.

[74] The following example might raise suspicions: "We examine the use of a recurrent neural network called Long Short-Term Memory (LSTM) with a prediction algorithm called temporal difference (TD) to predict the outcome of a music pitch sequence, while the sequence is being played. This is part of a larger system that will use this prediction in order *to choose pitches to play.* We describe our previous results using the LSTM network for musical tasks and then show its ability to predict a positive or negative outcome for a short musical task of *chromatic lead-into a chord tone.*"

have larger capacities of permutations and variations. Highly qualified musicians can be found all over the world. And they are certainly more capable of learning and possess more imagination than even well-trained algorithms. To altogether automate spontaneous musical improvisation is contradictory.[75]

More sensible, though limited by economic constrictions, is the application of respective procedures for 'filtering', as in the automated internet search for specific works (See Chapter 5). Such filtering could be conceived as a specific type of reception—it selects quantities of works instead of individual works, and does not involve human perception but instead digitalized or automated methods. Many of the investigations to be mentioned here may have dual-use and could be used to both ends.[76]

As with production and filtering, there is room for further theoretical processing of perception or rather reception.[77]

3.4.8 New and Negentropy—The Unexpected

The recipients are looking for an—individually different—homeostasis between the comfortingly familiar and the excitingly different, between the known and the unknown. Too much uncertainty about what to expect next in the flow of a musical work is irritating and frightening, too much predictability is boring. In music, too, there are pleasant and unpleasant surprises. But surprises, technically speaking contrasts instead of repetitions or even modification of given things are necessary to convey musical delight. The musical flow of Heraclitus should be neither too sluggish nor too wild. The steady or unsteady flow of information requires continuous re-balancing. An overarching object-subject-system with interdependencies in the perception process is methodically envisioned by Dubnov (2005/2006),

---

Franklin, J. A.: Predicting reinforcement of pitch sequences via lstm and td. In: Proc. of International Computer Music Conference, Miami, Florida (2004) Google Scholar JA Franklin—ICMC, 2004—cs.smith.edu; see further http://www.cs.smith.edu/~jfrankli/papers/icmc04JudyFranklin.pdf.

[75]See e.g. „We describe variable Markov models we have used for statistical learning of musical sequences, then we present the factor oracle, a data structure proposed by Crochemore & al for synchronization of data chains. We show [...] how it can be adapted for learning musical sequences and *generating improvisations* in a real-time context." Assayag, G., Dubnov, S.: *Using factor oracles for machine improvisation.* Soft. Computing 8–9, 604–610 (2004), https://hal.archives-ouvertes.fr/hal-01161221/file/index.pdf.

[76]See e.g. Dubnov/Assayag/Lartillot/Bejerano 2003; Dubnov/El-Yaniv/Assayag 1998; Dubnov/Assayag/Lartillot/Bejerano 2003.

[77]Explicit e.g. Conklin/Witten 1995: „This paper examines the prediction and *generation of music* using a multiple viewpoint system, a collection of independent views of the musical surface each of which models a specific type of musical phenomena. Both the general style and a particular piece are modeled using dual short-term and long-term theories, and the model is created using machine learning techniques on a corpus of musical examples. The models are *used for analysis and prediction* [...]." (Emphasis HWH); see further Cross 1998.

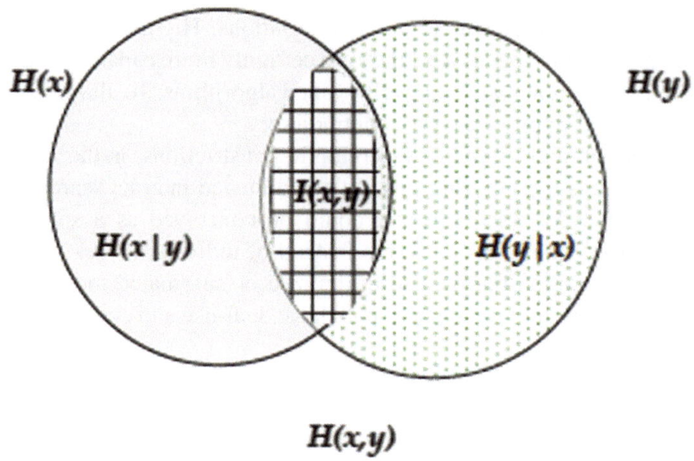

Fig. 3.7 The intersection shows the resulting neg-entropic information
Dubnov 2006, p. 2

> a novel approach to (automatic) analysis of music based on an "anticipation profile". This approach considers dynamical properties of signals in terms of anticipation property, which is shown to be significant for discrimination of noise versus structure, characterization and analysis of complex signals. Mathematically, this is formalized in terms of measuring the reduction in the uncertainty about the signal that is achieved when anticipations are formed by the listener. Considering anticipation as a characteristic of a signal involves a significant paradigm shift in approach to analysis of audio and musical signals. In our approach the signal is no more considered to be characterized according to its features or descriptors alone, but its characterization takes into account also an observer that operates intelligently on the signal, so that both the information source (the signal) and a listener (information sink) are included as parts of one model. (Fig. 3.7)[78]

Beyond the problem of wrong prejudices and judgements as with the halo effect, perception is ambivalent. On the one hand is refines the noticing and temporary remembering, so to say 'just in time'. On the other hand it pushes the redundant, somehow already familiar or seemingly familiar, to the foreground, instead of directing attention also, or even primarily, to the unexpected, new, the negentropic. But finally the object-centered perception altogether does not only confirm what we already know. Whoever listens closely enough can, in spite of fuzziness and *through it*, experience the music-aesthetically new. The pleasure of art is owed on the one hand to the repetition of the pleasant, the renewed encounter with the known, familiar, associated with positive experiences. This would be, related to the categories of bonding theory by Bion, an artistic reflection of a "secure bond". This specifically enables an 'exploratory attitude', which reaches further and further in the course of its ontogenetic development, literally in a spatial sense. This corresponds with the pleasure of the new, the unknown, of discoveries on the other hand. Here curiosity is a strong affective driving force. As in the works themselves in such

[78]Dubnov 2006, p. 2.

encounters contrast and variation are preferred over repetition of what is already known, even if it was and is pleasing.[79] (See below in Chapter 8. the triad of variation: Repetition, transformation, contrast). In particular the appropriation of the new necessarily begins with a lot of fuzziness, initially even with vagueness as proposed by Zadeh. In the course of the perception process, provided there is a secure scaffolding and basis of attention, there results step-by-step sharpening. The latter can be improved and expedited by the efforts of the perceiving parties of preparing ahead, both through more general experience and practicing as well as through preparations for the specific work.

Whoever encounters a work for the first time, regardless of whether in the concert hall or through technical media, and has not perused the score beforehand—which probably is the exception—cannot but perceive fuzzily at first, since she or he will not know how it continues, what constitutes an important motif or an embellishment. Of course this passive acceptance of fuzziness cannot be the end of the story and isn't. Piece for piece the work unfolds for the listener, and its shape emerges more and more clearly and with more and more focus. The perception process itself is analogous to the work as process, and in some regards also to the multi-stage process of production of a work. The listener does not 'construct' it—which is unnecessary, since it is already made and available—but reconstructs it, conveyed subjectively and objectively. The perceiving must themselves refine, attentively and with some awareness, filtering from the objectively provided shapes the relevant, the particular, or new (See Chapters 5 and 6). In perceiving they must crystalize in a subjectively adequate manner the work as pre-existing object, ideally not only as a sequence of beautiful or meaningful parts of the whole.[80] Naturally the object side, the work of art, must provide something fitting for the subject: Something new to it, enough incentives for precise perception and for following the work process, variety, in short variation (see below chapter on variation).[81] In the 'logically' abstracted temporal process of music, the existence forms following the perception form allow the subjective side to become more and more foregrounded. The objectified and objective work thins, evaporates to a singular occasion and starting point. The reasons for this are found in the matter itself, not in the ideological predisposition as in the mentioned (de-)constructivist opinion about the music process. By the necessities of nature memories fade, which were already involved in the perception, sinking from a conscious state to the pre-conscious and the sub-conscious, are superimposed on or displaced by new perceptions.

---

[79] Especially the unexpected, surprising, unpredictable, which counters expectations and derails them, has been assumed by 'Russian formalism' of Tynjanov, Mukarovsky and other to be a general essential characteristic of art.

[80] Perception processes, especially in regard to individual components of music, can be modeled and simulated and serve for automated filtering; cf. e.g. Tzanetakis/Essl/Cook 2002.

[81] This of course also applies to other arts, especially to those that inherently have a procedural sequence. Cf. e.g. Lehne/Engel/Rohrmeier/ Menninghaus/Jacobs/Koelsch 2015.

3.5 Remembering and Forgetting. Work as Sensual Idea Preserved in Memory: Recollection Form

The recollection form presupposes the reception form. Both essentially refer to the sound form of the work—or the sound of a technically reproduced work, the phonographic form (See Sect. 3.5). Already when directly perceiving a work the memory processes play a vital and indispensable part. The expectation during perception then impacts indirectly also the memory—and vice versa. The precondition transforms steadily into a positing, which itself is resolved in the course of the musical process.

3.5.1 Ultra-Short-Term Memory, Short-Term Memory and Long-Term Memory. Cognitive Frames

Not completely symmetrically, but mostly so, there are positive and negative correlations between learning, remembering, and forgetting. The various types of memory such as "procedural" and "declarative" (semantic memory) correspond ceteris paribus to implicit and explicit learning or learning of facts and the practicing of skills (procedural learning). Part of the entire constellation of learning and remembering, unlearning and forgetting, and the application of what was learnt is still the temporal gradation of ultra-short, short-term and long-term memory, essential especially in questions pertaining to the temporally bound art of music (Fig. 3.8).

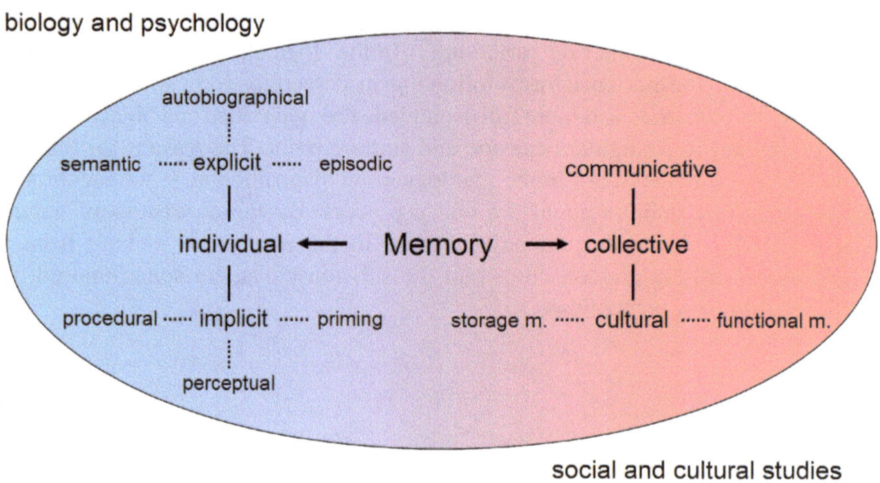

Fig. 3.8 'Memory'
Bernhard Wenzl: OVERVIEW OF THE TERM "MEMORY" AS USED IN VARIOUS BRANCHES OF ACADEMIA, https://en.wikipedia.org/wiki/Memory, 11.4.20, retrieved 21.4.20

Parameters are conscious and sub-conscious, affectively guided attention, aesthetic interest and of course all kinds of possible emotive motivations for the production and fabrication of musical shapes. As already noted in the context of ultra-short, short-term and long-term memory the overarching cognitive framework moves back and forth between immediate and mediated perception, recollection or technically speaking "recording". The third step in a media, the 'rendition' is rather secondary in the musical process on the side of the subject, as opposed to the technical re/production. It becomes relevant again in the derived word, picture, and imagination form. Recollection is the making present of what is past.[82] But it is also, as mentioned, part of perception, in interplay with expectation as anticipation of what is to come. Learning, the collecting and inculcating of experiences and insights, is the decisive mediation for this.[83] How the recollection form is stored in the brain of the subject seems not to have been investigated thoroughly. One limitation is the objectification of the subjective, which already occurs in the verbal expression of impressions (see below word form 8.3). Another limitation is the correspondence of subjective, mental perceptions with their neuronal storing in the brain, difficult to objectify in any research. This leaves much to do for fuzzy logic, as with 'artificial intelligence', the digitalized comparison and the classification of big amounts of data. A core area of memory in regard to music is the recognizing of works, usually researched by using melodies—melody meaning the unit of pitch and duration movement.

> The aim […] was to find structural descriptions of melodies that influence recognition memory for melodies. 24 melodies were played twice to 42 test persons. In the second turn, some of the melodies were changed, and the subjects were asked whether they think that the melody has been exactly the same as in the first turn or not. The variables used to predict the subject judgments comprise data about the subjects' musical experience, features of the original melody and its position in the music piece, and information about the change between the first and the second turn. Classification and regression methods have been carried out and tested on a subsample.[84]

Some challenges emerge in view of the complexity of the processes even with a relatively simple, but effective experimental design:

> In general, the results are much worse than expected and demonstrate that the involved regression methods extract only slightly more information from the data than trivial predictors.
>
> We suspected that this tendency is due to the fact that between-subjects differences dominate the judgments in a more complex manner than captured by the variables on musical background or the additive random effect of the mixed model. Therefore we repeated the comparison (without classification methods) on a partition of the data set where the same subjects have been present in all data subsets. The regression forest and the ordinal regression were the best methods in this setup […].[85]

---

[82] See e.g. Snyder 2000.

[83] See e.g. Saffran/Johnson/Aslin/Newport 1999.

[84] For more detail see Müllensiefen/Hennig 2006, p. 732.

[85] Ibidem, p. 737.

Apart from the refinement of mathematical methods there seems to be a need for an elaboration of methods that further differentiate the component analysis of music and grades it, as well as methods that include in their reasoning varying subjective preconditions.

3.5.2 The Lady Vanishes. *Fading, Sharpening, Forgetting*

That memories become blurred is almost literally natural. In correlation to the growing temporal distance from the perceived musical event the 'white noise' increases, the accuracy of the remembered decreases. Fading is virtually the 'default mode' of the memory. The escalation of fading is forgetting, which itself comes in varying degrees.

An illustrative example is offered by the movie *The Lady Vanishes*, a British thriller by Alfred Hitchcock from 1938. Music plays an important part in this movie, as it sometimes does in Hitchcocks work, and one melody is even a central leitmotif of the plot. A secret message is encoded in a folklore melody. Hitchcock hereby draws on the widespread function of primarily songs as unity of text and melody.

> The function of song embodying and encapsulating memory [...], has been part and parcel of artistic representations in literature and music for hundreds of years. Its earliest manifestations arguably are "songs" in the Pentateuch, where particularly poignant moments in the historical narrative are captured by a more poetic, song-like rhetorical style.[86]

The usual function is that of retaining and refining a memory, often as a return of something suppressed, that rises from the sub-conscious, because the song worked as a trigger.[87] In contrast to this Hitchcock makes a theme of forgetting. He does not have to explain, how and what is actually encoded in the ominous melody. It remains a black box and an unresolved dramaturgical prop and is insofar a 'MacGuffin'.[88]

---

[86]Thym 2012, p. 263.

[87]„The close connection of memory and song [...] can be traced in other musical genres. Wagner's *Götterdämmerung* [...]: When Siegfried, suffering from drug-induced amnesia, remembers his past life with Brunhilde and thus recaptures his wholeness and identity, he does so in the form of a Lied—he recapitulates a long forgotten song the forest bird sang to him in the previous opera *Siegfried*. In Weill's *Lady in the Dark* (1941; its title at the German premiere ten years later was *Das verlorene Lied*—the lost song), the protagonist is haunted by a childhood song, the melody and lyrics of which she has forgotten; with the help of a psychoanalyst, she remembers the song, alluded to throughout the play in fragments and paraphrases, in its entirety, and this process of rediscovery restores her wholeness and identity—it is the beginning of the healing process. The link of memory and song is also at work in Schoenberg's cantata *A Survivor from Warsaw*, whose central mo ment is the entrance of a half-forgotten Jewish prayer, the *Sh'ma Yisroel*: Humanity in the form of ethnic and religious identity is emphatically asserted here in the face of barbarism by invoking memory—memory as song " (Thym 2012, p. 272f).

[88]„(U)sually a mysterious package/artifact [here: melody] / superweapon that everyone in the story is chasing." (http://wordsmith.org/words/mcguffin.html) „A *MacGuffin* is an action goal that is void,

> (T)he person who has memorised the MacGuffin is kidnapped by enemy agents […]. Much of the plot is then taken up with the protagonists' attempts to track down and rescue this character, thereby ensuring that the MacGuffin gets back to England […].In *The Lady Vanishes*, Gilbert and Irene work together to find and rescue Miss Froy, and the film ends with her joining hands with them as if to unite them. Again the ending emphasises the formation of the couple, here mediated through the benevolent mother figure whose adventures have served to bring them together.[89]

The three essential characters: Miss Froy, music pedagogue and British agent, Iris, well-off future bride, Gilbert, ethnomusicologist. The story opens with an exposition of the melody[90]:

English tourist Iris Henderson […] arrive(s) at the "Gasthof Petrus" inn in the country of Bandrika, "one of Europe's few undiscovered corners". Iris is returning to Britain to marry a "blue-blooded cheque chaser", but an avalanche has blocked the railway line. The stranded passengers are forced to stay the night at the inn […] Miss Froy tells […] of her experience being governess and music teacher for the past six years and now returning home to England. Upon *hearing a folk singer in the distance.*

Of interest are also distinctness and accuracy of perception. The comment of Froy constitutes a blurring—what is heard is not folk music but a coded message. 14'10:

> Do you hear that music? Everyone sings here. The people are just like happy children … with laughter on their lips and music in their hearts.[91]

As the friends say goodbye to Iris, the same song can be heard, quietly, and so perceivable only in a fuzzy form, then in the following ten seconds from 14'57 to 15'07 it becomes more distinct again. Thus there is yet another fuzzy perception and a straight out misunderstanding of the work. Short silence. Generally Hitchcock used music very intelligently: not as a sound-goo, and thereby quite against the trend of typical contemporary and later use of music in Hollywood, but intentionally, enduring long passages completely without music or only complemented by real noises and sounds. And even these can have a dramaturgical function when they drown out the

---

most often an object, sometimes a secret message. But the MacGuffin is never completely void: It has to meet functional requirements that the context provides it with. For example is has to be rare, dangerous, expensive oder valuable in any other way." (HJW, https://filmlexikon.uni-kiel.de/index.php?action=lexikon&tag=det&id=472). Term discussed e.g. in http://www.openculture.com/2013/07/alfred-hitchcock-explains-the-plot-device-he-called-the-macguffin.html. „The World-War-I-era actress Pearl White used the term "weenie" to identify whatever object (a roll of film, a rare coin, expensive diamonds, etc.) impelled the heroes, and often the villains as well […]. The name MacGuffin was coined by the English screenwriter Angus MacPhail […] and was popularized by Alfred Hitchcock in the 1930s." (Extensively discussed with many examples https://en.wikipedia.org/wiki/MacGuffin, 3.5.2020, retrieved 10.5.2020. See also Walker 2005, pp. 296–306.

[89]Ibid., p. 304.

[90]The plot in the following according to https://en.wikipedia.org/wiki/The_Lady_Vanishes, 7.4.20, retrieved 10.4.20, with own time markers added to the relevant music passages and added commentary. Emphasis always HWH.

[91]Quotations of dialogues here and always following SergeiK, http://www.script-o-rama.com/movie_scripts/l/the-lady-vanishes-script-transcript.html, retrieved 17.3.2017.

dialogue and thereby blur it. Back in her room, Froy opens the window to listen more intently to *the folk singer, who sings on the street below her window.* At 15'25 the song is once again heard distinctly. It contains short, rather monotonous phrases; at 15'43 the singer appears on scene with a guitar, becomes visible and not only audible. Mrs. Froy deftly taps her hand, as if it were Morse signals, aiming for a sharpening. But then noise from above disturbs the scene, a clarinet and stomping on the floor above can be heard—the latter even becomes visible through the shaking of the lamp in the room below. Music 2 blurs, put differently, music 1. Froy meets Iris.

IRIS—What's happening? An earthquake?

FROY—That wouldn't account for the music.

IRIS—What a noise. What can they be doing?

FROY—I don't know but I'll soon find out.

IRIS Hello. Musical country this.

FROY I feel quite sorry for that poor singer outside having to compete with this. [...] Someone upstairs is playing musical chairs with an elephant.

On screen: Gilbert is playing the clarinet and is simultaneously transcribing indigenous folk music of Banderika while three natives dance. In the nature of things this is slightly inconsistent: Simultaneously playing music and transcribing it is nearly impossible. Apart from that Gilbert seems to already know the folklore melody and has mastered it.

In the consequent dispute between Gilbert and the hotel manager a new perspective is adopted regarding the question of music versus noise. Music and dance are repeatedly interrupted by the dialogue, but continue all the way to 19'16.

As Miss Froy listens, *the folk singer is*—[the murder unseen by her].

Later: 23'14 Froy at the window, singer in the picture, sings briefly, is then strangled by an unseen murderer—for the viewers represented by a shadow, invisible for Froy. 23'39: Froy throws down coins—we see that they remain lying on the ground. She takes off her jewelry, and shortly memorizes the melody, up to 24'. It has only 8 tones, a bit sparse for a longer message, the formulation of a 'secret pact' (see below). After all, it is a MacGuffin. The next morning, before catching the train, Iris is hit on the head by a flower pot that was apparently aimed at Miss Froy, who then helps Iris onto the train. Their sleepiness during the train ride is dramaturgically important as are their later insecurities regarding their perception, both of which are well motivated by the previous scene. Next to the acoustical the optical also plays an important role as sign for perception and memory. In a conversation with Iris, Miss Froy introduces herself by name. Since remembering—the sub-conscious, forgetting, stifling—is a main theme, psychoanalysis may not be amiss, at least in the funny form of mishearing,[92] the equivalent to the slip of the tongue and the prototypical Freudian slip. 29'16:

[92]Probably prepared and provoced by the choice of the name, which seems a little bit strange.

- I'm a governess. My name's Froy.
- Did you say Freud?
- No. O-Y, not E-U-D. Froy.
- L'm sorry l can't hear.
- Froy. lt rhymes with joy.

Now it is not music as noise but simply noise that affects the scene. Train noise drowns out the dialogue. Therefore Froy writes her name on the dusty window with her finger. Then they return to their compartment. 30‘49–30‘54: Froy briefly memorizes her melody, only the 4 initial notes—the text is not relevant. Iris falls asleep. When Iris awakens, Miss Froy, the Lady, has vanished. She has been abducted by agents of the country Bandrika and, as will be seen later, has been bandaged like a mummy as supposed patient to make her unrecognizable. This is a quite substantial form of blurring. Iris searches for Froy. The people on the train all deny, for various reasons, having seen the vanished lady. It is insinuated that Iris is hallucinating. During the search, nothing less than a quest, she encounters Gilbert again. He is slowly convinced by her and tells her, among other things, the reasons for the racket in the evening: ‘I'm writing a book on folk dancing.’ A masqueraded and costumed double of the lady accounts for additional fuzziness, as does an Italian magician—again raising the question of illusion and deception—on the side of Bandrika, the dictatorial state, from which they want to get out. The magician has a speaking name. It alludes to ‘doppio’, the double, and he is famous among other things for a trick that we see on an advertising poster, that relates to the movie plot and its title:

> “The Great Doppo. Magician, illusionist, mind reader… he will appear in all the towns and cities. See his fascinating act…—The Vanishing Lady.”

However, together with Gilbert Iris finds the vanished lady, 1.10’, and they now wrap up her double. The question of remembering, of fading memories, of false memories and intentional deception of perception and memory is therefore a main theme; the issue of the melody supplements this as a secondary theme and a sensually perceivable motif.

The motif of the ‘double’ is a topos of myths and of art across the globe. Similar to the ‘shadow’ these would be worthy objects of investigation with methods of fuzzy logic, to clarify differences and blurs between depiction and originary picture, as far as the latter still exists. The swapping of the vanished lady and the bandaged double now taking her place, the correction of the relation of original and imitation, is discovered. A gunfight ensues between the military of Bandrika and the train passengers. During the gunfight, Miss Froy reveals to Gilbert and Iris that she is a British agent who must deliver a message to the Foreign Office in London. The message is encoded in the tune that the folk singer sang. 1.22‘:

> If I'm unlucky and you get through, take back a message […] to Mr. Callendar at the Foreign Office.
>
> – Then you are a spy.
>
> – I think that is such a grim word.

– What is the message?

– It's a tune. It contains, in code […] the vital clause of a secret pact between two European countries.

I want you to memorise it.

The first part goes like this… 1.22'12

Again the 8 tones resound. They are, as mentioned, not really sufficient to include 'the vital clause of a secret pact'. 1.22'21 bis 1.22'31:

I'd better write it down. Let me get a piece of paper.

GILBERT I was brought up on music. *I can memorise anything.*

Now there are two phrases with the structure of question and answer. With Gilbert's help, Miss Froy slips through the window away into the forest. 1.22.37: In order to memorize the melody, Gilbert sharpens the memory by repeating it again and again.

– You're sure you'll remember it?

– I won't stop whistling it.

1.28'19—1.28'34: Gilbert whistles the melody, interrupted by dialogue. After escaping Bandrika the signal whistle of the locomotive is intended to announce the victory.

GILBERT to helping passenger: Take your hand off that thing. [Steam whistle] I've got to remember a tune. Remember … […]

1.28'50: Gilbert continues humming, even at the arrival in London.

We're home, Gilbert. Stop humming that awful tune. You must know it backwards.

[…] I've got to deliver this theme song to Miss Froy […].

1.29'45: He sings a few notes from the beginning. Gilbert recognizes that Iris would rather marry him than her betrothed, and kisses her. 1.30'15:

Yes, you heartless, callous, selfish, swollen-headed beast…

– Are you going anywhere?

– Foreign Office.

– Where are we going on our honeymoon?

– Somewhere quiet. Somewhere where there are no trains.

Having arrived at the Foreign Office, they are seated in a waiting room.

1.30'33 Mr. *Callendar*[93] will see you now.

---

[93]The name is ostentatious. Though an H.L. Callendar exists, who was involved with the elaboration of thermodynamics, mentionend at the opening of this chapter. More likely Hitchcock or rather the two scriptwriters Sidney Gilliat and Frank Launder were alluding to 'calendar' and thereby time, of which the time arrow—realized not least in thermodynamics—provides a essential frame for remembering, fading away, and forgetting.

Suddenly Gilbert is unable to remember the vital tune.

– Wait a minute. It's gone!

– What's gone?

– The tune. I've forgotten it!

– No! No!

Gilbert had been retaining the melody truthfully in his memory during the turbulent and life-threatening events. Having been saved and arriving, he forgets it. An additional joke: He had claimed that he never forgets a melody, wherefore he had disallowed the notation form as safety and sharpening. Fateful irony: His forgetting is probably motivated by the abrupt positive twist of his infatuation, as if love's kiss had erased his memory like in a mythical fairy tale.

Wait a minute. Let me concentrate.

1.30'51: He sings.

IRIS No, that's the Wedding March.

She—or rather the production team—is possibly mistaken. It is not the usual suspects, meaning neither Mendelssohn nor Grieg, but it is actually, even though blurred, the beginning of the secret-bearing melody. Viewed from within the work this—supposed—fuzziness probably constitutes a slip on her side: Iris wants to marry Gilbert, and already anticipates the wedding ceremony.

It's awful. I've done nothing but sing it since that day. Now I've forgotten it completely.

The message has arrived nonetheless. Gilbert and Iris hear the tune on the piano—first off screen, then on screen: The doors of the office open to reveal Miss Froy playing the tune on the piano surrounded by ministers listening intently.[94] According to information theory, and technically calculated prudently through the securing of a second channel, the information was saved, the uncertainty and fuzziness are eliminated.

3.5.3 „O, Du Lieber Augustin, Alles Ist Hin" (O My Dear Augustin, All Is Gone) by Arnold Schönberg. Fading Away, Composed

Fading away was composed as grand gesture in Schönberg's 2nd string quartet with soprano solo op. 10 $f^{\#}$ minor. In this work from 1907/08 he completed the transition

[94] If we take the film as a work of art, there could be further extensions for fuzzy logic investigations: (1) the relation of Hitchcocks 1938 film adaptation to the 1936 original novel *The Wheel Spins* by Ethel Lina White; republished after the film release 1938 as *The Lady Vanishes*. However, there might not be enough identity to determine difference s and blurs, since the plot was adopted only in outlines by Hitchcok and his scriptwriters. (2) the relation to two remakes of the movie in 1979 and 2013. (3) *The Lady Vanishes* (2018 stage play) by Derek Webb. (4) the relation of the remakes and the drama to the original novel.

Fig. 3.9 Arnold Schönberg: 2nd String quartet with soprano solo op. 10, 4th movement *Entrückung*, cipher 30
Universal Edition, Wien, revised 1921. Plate U.E. 2993/W.Ph.V.229. IMSLP29725-PMLP66179-Schoenberg_-_SQ_No._2_score

from d minor tonality to free atonality without reference to a key. This departure from d minor tonality is programmatic and verbalized explicitly through the choice of text. Schönberg already uses a poem by Stefan George (from the collection *Der Siebente Ring* [*The Seventh Ring*] from 1907) in the third movement. In the final fourth movement *Entrückung (Rapture)* the pertinent formulation is found:

Mir *blassen* durch das dunkel die gesichter
Die freundlich eben noch sich zu mir drehten.
Und bäum und wege die ich liebte *fahlen*
Dass ich sie kaum mehr kenne. (Ziffer 30)

I feel air from another planet.
The faces that once turned to me in friendship
Pale in the darkness before me.
And trees and paths I loved grow wan
So that I hardly know them (Fig. 3.9).[95]

The growing fuzziness coming with growing distance is thought by George spatially and temporally as well as psychically. Schönberg elevates this on a meta-level. Already in the third of the four movements he departs more and more from the traditional d minor tonality. There is no longer any indication of key. This transition is finally completed in the fourth movement, *Rapture.* All four instruments strive with strongly reduced volume and dissolved phrases to a high range that one could call ethereal—"I feel air from another planet", as it says programmatically at the beginning of the soprano solo. Schönberg himself conceived this process on the one hand as symbolic, as distancing from the tonality that was grounded in the earth as

[95]Translation Richard Stokes, https://www.oxfordlieder.co.uk/song/2222, retrieved 24.4.2020.

Fig. 3.10 Arnold Schönberg: 2nd String quartet with soprano solo, fourth movement *Entrückung,* beginning
Universal Edition, Wien, revised 1921. Plate U.E. 2993/W.Ph.V.229. IMSLP29725-PMLP66179-Schoenberg_-_SQ_No._2_score

nature, on the other hand naively as a sort of space travel and science fiction—surely not intended by George.

> The fourth movement 'Rapture' begins with an introduction describing the departure from earth to another planet. The visionary poet relates phenomena, which might soon be confirmed as true. This introduction attempted to depict the liberation from gravitation—the passing through the clouds to thinner and thinner air, forgetting all earthly worries (Fig. 3.10).[96]

The process of fading is realized as sound charged symbolically, music-historically and music-aesthetically. arlier in the work the return of the past is represented as shadow, ghost. In the trio of the second movement, a scherzo, the second violin quotes the Viennese folk song *O du lieber Augustin, alles ist hin.*[97] Using the citation, which at first focusses the memory, the growing fuzziness is elaborated in the composition. Schönberg allows the citation to disintegrate and dissolve.

He composes the fading, forgetting and leaving-behind more drastically here. The phrase "All is gone" refers both to the d minor tonality as well as Schönberg's biographical crisis. His marriage was also 'all gone', since his wife left him for a painter (Fig. 3.11).

The cited melody in the second violin accompanies the first violin playing 'jumping octaves' used in Viennese music, especially in waltzes, but also found

---

[96]Arnold Schönberg 1976: »Bemerkungen zu den vier Streichquartetten«, in Schönberg 1976, p. 421.
[97]In detail e.g. Waeltner 1978, pp. 246–262.

Fig. 3.11 Arnold Schönberg: 2nd String quartet with soprano solo, 2nd movement, after Zi. 160, 2. violin
Quotation of the melody of the first line: „O, du lieber Augustin, Augustin, Augustin"
Universal Edition, Wien, revised 1921. Plate U.E. 2993/W.Ph.V.229. IMSLP29725-PMLP66179-Schoenberg_-_SQ_No._2_score

Fig. 3.12 Arnold Schönberg: 2nd String quartet, 2nd movement
„O, du lieber Augustin", fading
Universal Edition, Wien, revised 1921. Plate U.E. 2993/W.Ph.V.229. IMSLP29725-PMLP66179-Schoenberg_-_SQ_No._2_score

Fig. 3.13 Arnold Schönberg: 2nd String quartet, 2nd movement
The viola, continuing the song, and ending it
Universal Edition, Wien, revised 1921. Plate U.E. 2993/W.Ph.V.229. IMSLP29725-PMLP66179-Schoenberg_-_SQ_No._2_score

in the yodelling of the Alps.[98] Here it jubilates only highly ironically. This accompaniment one the one hand blurs the melody and reduces its recognisability, but overall it focusses the reference to the cited past and tradition (Figs. 3.12 and 3.13).

Schönberg shortens the original text somewhat. The rest could be sung along to the melody: „O, du lieber Augustin, Augustin, Augustin / […] Alles ist hin."Then,

[98] According to Schönberg it is appears even earlier in the second string quartet as theme with a 'more lyrical character' than the third theme. Schönberg 1976, p. 417.

Fig. 3.14 Arnold Schönberg: 2nd String quartet, 2nd movement, cipher 170
Universal Edition, Wien, revised 1921. Plate U.E. 2993/W.Ph.V.229. IMSLP29725-PMLP66179-Schoenberg_-_SQ_No._2_score

after the viola has ended the citation from the folk song, its dismantling already begins (Fig. 3.14).

1. Cipher 170, m. 1: First violin and violoncello lead upwards with a three-syllable chromatic figure, like a kind of musical double dot. If we liked, we could recognize the transposed BACH motif, incomplete and complete in the E# F#G/G# A A# B. The two voices run in widely spaced sixth parallels, ending with as fifth. This is a slightly corny reference to folk music of the Viennese tradition, but it violates the rules of counterpoint, which prohibit fifth-parallels, even concealed ones like here. The blurring of the principles of traditional counterpoint sharpens the distinctness of the transition to new musical principles, to atonality.
2. Measure 2ff. The first violin picks up the jumping figures. The first figure still remains meaningful harmonically. It marks the transition from the original major to minor through a change from the major third to the minor third ($f^{\#}$-f over d).
3. The second violin realizes this transition to minor with the leitmotif „O, du lieber Au-". Instead of the b adequate to the citation the dismantling b flat appears now. Like a stuck barrel organ, in Viennese 'Werkel', the more and more fuzzy destruction of the melody continues, always in the same rhythm, each descending—to fading and forgetting—with chromatic connections (g-$f^{\#}$, e-$d^{\#}$, $c^{\#}$-c).
4. The violoncello proceeds with the drop by a fifth of the first half line with a different melody fragment, and almost obtusely repeats the first 'Augustin' without the melodically variational continuation of the original.

After this the *Augustin* melody dissolves more and more into denaturalized fragment and finally disappears by lack of contours. Schönberg himself speaks of 'liquidation' and refers to another memory immanent to the work and here incorporated as memory that transcends the work, namely that 'remnants' *remind* of motives and phrases from the first movement.[99] Altogether Schönbergs distorted citation appears slightly mocking. It corresponds with his biographical and music-historical situation from

[99] Schönberg 1976, p. 418. Emphasis HWH.

his perspective. It therefore constitutes more a conscious blurring than a spontaneous fading of memories. But: By *composing* the fading and blurring, he makes it conscious together with what has become blurry, and thereby again refines this—a real paradoxy (See Sect. 8.2).

3.5.4 Refreshing and Reinforcing

Thus there are alternatives to fading: Memories can be refreshed, and even without ambivalences as in Schönberg. The most obvious procedure is repetition, the repeated listening to the same work. Refreshing and reinforcing are themselves learning processes. This interplay of fading and reinforcing, becoming blurred and sharpening, again provides a polarity, analogous to fuzziness and sharpening. And evidently even more than merely an analogy. For in the brain processes that follow the principles of fuzzy logic take place during sensory sensations, perception and its further processing.

A functioning memory is indispensable for receptive and productive realization. This existence form plays a role not only in reception, but already during production or composition, sometimes even a crucial role. In any case this applies, as mentioned, in principle to all music types without notation: That which has already been achieved, flows consciously or sub-consciously, into each realization. This offers opportunities to examine the fuzziness as well as the sharpening in the context of fuzzy logic, at least up to a certain gradation, in particular through the comparison of various figurations of a work, such as an Indian raga or a jazz improvisation on standards. If they are recorded as technical reproductions, the already sketched out automated methods could be applied.

3.5.5 Reconstructing from Memory as Special Case of Reinforcement and Sharpening

Making music from memory (instead of from a score) is common in traditional and folk music, in some regions of the world, styles, and genres even the only access.

Differently weighted than this playing primarily using memory as a source, the function of memory serves as secondary reconstruction, if access to the usual sources is denied. One extreme case of this was the playing of music in concentration camps. Here the conductor and, in the case of the women's orchestra, a female conductor, had to reconstruct the works from their memories and at the same time rearrange them according to the various incidentally available options of musicians and musical instruments. Reconstruction of what is no longer available in externalized notation or as effort against forgetting also occurs in written music. For not uncommonly composers lose parts of scores of the already composed work and are forced to *reconstruct* the respective passages from memory. Regarding specifically the fuzziness this might be hard to examine, unless the original version was recovered. A special case of this type is the *new-construction*, which only loosely follows the lost version. In the first week of November 1923 Leoš Janáček composed

> his first string quartet with almost volcanic force [...]. Only eighteen minutes of playtime for four movements – and what abundance of phantasy, expression, richness of musical shapes [...]. » I had the poor, tortured, beaten, battered woman in mind, as [...] portrayed by Tolstoj in his work *Kreutzersonate* « , relates Janàcek in the year of its first release 1924 to [his lover] Kamila Stösslovà. He had engaged through music dramaturgy with a Tolstoyan adulteress, Anna Karenina, thirteen years earlier, another, Katja Kabanova from a drama by Ostrovsky, he had only few years ago glorified as his most warm-blooded opera character. The 'poor, tortured' murder victim from the *Kreutzer Sonata* had already inspired him in 1909 to a sublimation in a piano trio, the score of which got lost and seemingly was already missing when Janàcek composed his first string quartet in October 1923. The memory of what had already been formulated must have lived on in him. Otherwise it is hardly conceivable, that he, as he claims, wrote down the quartet within a period of one week.[100]

3.5.6 Recalled Memories—Quotation, Leitmotif. Composed Memory as Reinforcement

Composers may also refresh memories within the work process, as a sharpening of what is fuzzy. A method transcending all ages and cultures, the simple repetition within a work, is the easiest option. This is structurally entrenched in one universal of music that also transcends cultures, in the relationship of cantor and choir, to which the relationship of varying stanzas and unchanging refrain corresponds. Repetitions, even when unchanging, are never identical; they also can't escape the fundamental fuzziness.

The intentional citation, as in the outlined example from Schönberg, as well as the self-citation, more restricted in scope and time, are more sophisticated procedures for sharpening memories and bringing to the present what musically occurred in the past. A variation of citation is the allusion, or paraphrase. In that case what is adopted from the work of others or one's own work has already been processed, changed, and thereby made fuzzy. In line with the dialectics of fuzzy logic the sharpening results in new fuzziness. In both cases what is newly composed frames or superimposes the old, sometimes in more than one layer, as in a palimpsest, leaving a faded impression of the old shimmering in the back-/underground.

Repetition and variation connect the basically epic leitmotif, common especially in the music theatre of the late eighteenth century. It keeps alive in the memory what has already been said and sung in the past by bringing it into the present, which sometimes serves also as indicator of what is to come and thereby generates something like a temporal polyphony.[101] The *Götterdämmerung* (*Twilight of the Gods*) as finale of Richard Wagner's ring tetralogy summarizes tendentially nearly

[100] Honolka 1982, p. 220f.

[101] Cf. e.g. Heister 1991.

all leitmotifs of the three previous musical dramas.[102] Bernd Alois Zimmermann has extended this simultaneity of different temporal layers in the 1960s as "Kugelgestalt der Zeit" ("*Time shaped as a sphere*") (for more detailed examples see Chapter 6 *Kristallisation* and 8 *Variation*).

3.5.7 Uncalled Memories as Hallucination Form

A very ambivalent intersection of fading and renewed sharpening can be observed in a special type of musical hallucination.

> *Neurology – 'nice variety'* / Neurologist Eva Schielke, 38, from Berlin about musical hallucinations.
>
> SPIEGEL: Mrs. [Ms.] Schielke, at the Charité Berlin you just examined a 57-year-old patient who suffers from a rare form of musical hallucination. What did the man hear?
>
> Schielke: He heard folk songs sung by male and boy [children] choirs, sometimes several at once. The cause was a suppurative focus at a certain place of the brain stem, the so-called dorsal pons. After being treated for eleven weeks with antibiotics, the abscess receded and the hallucinations disappeared.
>
> SPIEGEL: What is special about this type of hallucination?
>
> Schielke: Musical hallucinations occasionally occur in older people who have lost their hearing. With our patient the faculty of hearing was only slightly impaired. At the same time probably some neural pathways were damaged, which normally dampen a part of what is heard. This evidently *released musical memories stored* in the frontal lobe. [...][103]

It is always memories, not new material: what is past is recalled in the present.

> SPIEGEL: Did other patients also hear folk music?
>
> Schielke: *All would perceive music that was familiar to them.*
>
> A French patient heard Verdi and Mozart, but also chansons.
>
> A Japanese patient, who fell sick during a trip through China, heard Chinese folk music.[104]

The generating of an imaginary-real object by the subject as such seems to be hardly irritating, more so those cases where the hallucination becomes noticeably fuzzy.

> SPIEGEL Are the hallucinations uncomfortable for the patients?
>
> Schielke Some feel *bothered, because they hear several melodies at the same time*
>
> Others enjoy it as a welcome diversion. Ten years ago I had a patient, who at first heard a tenor from an Italian opera *singing very loud and very wrong*. But after a few days it changed to arias from 'Madame Butterfly', which he found far more pleasant."[105]

---

[102] Cf. e.g. Jeßulat 2013.

[103] *Der Spiegel*, 34/2000, S. 169.

[104] Ibid.

[105] Ibid.

This matter is intricate and ambivalent. On the one hand it is a sharpening, from the view of musical perception. For faded, no longer conscious memories of melodies are activated and sensually perceived. On the other hand, from the view reality perception and the proper functioning of the brain, this sharpening is just an illusion. For the melodies do not actually resound and are not 'objectively' heard but rather hallucinated by the subject. In this tendentially pathological case the constructivism is right, since the subject constructs the object. A comparable phenomenon is the earworm (See Sect. 6.9).

3.6 Blurring in the Technical Reproduction of Music. Work as Technically Recorded Sound: Phonography Form (I)

One important means against blurring and fuzziness of memory is the technical reproduction. It retains and conserves the past in sound form, can be replayed anytime, and thus makes the singular repeatable. This makes it possible ceteris paribus to replay a work as often as desired and to thereby commit it to memory more and more precisely. As in art generally, the main aim of technical (re-)production is the imaginary-real immortalization of what is otherwise transient [ephemeral/fleeting/evanescent].

A pragmatic gradation of this function of the phonographic form in counteracting the fading of the memory of an impression made by a work works specifically by reinforcement by learning through repeated listening or more generally by storing and (re-)playing. The repetition of the singular becomes particularly poignant and sustainable, when it is not a studio production, comparably timeless and unspecific in space, that is recorded in the phonographic form, but instead a specific performance, so that this again can be recalled through recollection as a firstly singular event.

3.6.1 Media Mediated Power and Deception of Memory—'False Memory'

The potency of memory and of what is remembered has already been mentioned in the context of perception.[106] Especially the fausse mémoire, false memory, the objectively wrong memory of something that never happened in that way intricately

[106]For general information on uncertainties and errors, compare for example the classification approach: "memory's imperfections can be classified into seven basic categories or "sins." Three of the sins concern different types of forgetting (*transience, absent-mindedness*, and *blocking*), three concern different types of distortion (*misattribution, suggestibility*, and *bias*), and one concerns intrusive memories (*persistence*)"). (Schacter/Chiao/Mitchell 2002).

combines and interlaces present and past, both sharpening and blurring.[107] The mediation through technical reproduction, the medially ostensible confirmation can serve as an important vehicle, which itself promotes the self-deception.

> Thomas Mann visiting Weimar was a deeply moving occasion, which I remember *as if I was there*. And I was there, but only through the radio. […][108]

In this case the acoustic live radio experience was induced by the mere *reading* of the spoken text.

> When I today read the script I have to summon all my sobriety to not *confuse reality and imagination*. The words and their meaning, the recordings of the *Wochenschau* [weekly news reports on German TV] or pictures from newspapers, the children singing a welcome, the theatre hall, in which I would later often sit and sometimes also work, and the voice, strangely familiar right from the beginning and congruent with the way in which I had read his texts, all this intermingled to an inner picture so convincing, that I occasionally seriously doubted, *whether I might not have after all experienced that hour face to face*. But in fact it was only the radio.[109]

All this is not a pathological auditory hallucination, but yet an irritating autosuggestion.[110]

3.6.2 A New Type of Sound Form from Technical (Re-)Production. Blurs and Permutations in the Transformation to Another Medium

The phonographic form is a new type of sound form. It results from the technical reproduction of music, more precisely not of music absolutely and generally, but more specifically of the sound form of music.[111] The main area of concern is the recording medium. Yet another section of the phonographic form is broadcasting, which developed somewhat later in the mid-1920s. Though it has entirely different technical-media foundations—the electromagnetic transference instead of

[107] Cf. e.g. Schacter 2001; Draaisma 2016; Draaisma 2013; Draaisma 2006.

[108] Hans-Dieter Mäde: *Weimarer Vorfälle. Begegnungen mit Thomas Mann,* Vorabdruck aus: *Nachricht aus Troja—Fragmente einer Motivation.* Edition Schwarzdruck, Gransee 2012, zit. n. *Junge* Welt, Berlin, 14.03.2012. S. 10. Emphasis HWH.

[109] Ibidem. Emphasis HWH.

[110] Related to this in many regards are dreams. Usually the optical has primacy in them. But acoustical components (particularly spoken language, also music and sounds) as well as kinaesthetic and other sensual perceptions play a part—always merely imaginary-real, since during dreams the motor functions are dormant. Dreams are purely subjective, barring eventual external stimulation disturbing the sleep and potentially being triggers. Along with classics such as Sigmund Freund, ample, also historical material in Draaisma 2015.

[111] Comprehensively on the here discussed complex of themes see Sterne (ed.) 2012; in particular the contributions of John Mowitt, Rick Altman, James Lastra, Louise Meintjes, Lisa Gitelman, and Jason Stanyek/Benjamin Piekut.

the mechanical recording, storing, and playback of the sound form. But in regard to this it underlies similar conditions and limitations. Along with performances, that were transmitted simultaneously—the actual 'live'—broadcasting soon employed sound carriers. In regard to the level of static noise and the accompanying fuzziness broadcasting was altogether lagging behind the sound carriers. Here problems and tensions, that already occurred in the transition from notation form to sound form, are reproduced or are repeating themselves at a new level, and thus also the problems of fuzziness. Therefore we can keep this rather short.[112] Fundamentally opposing the identity illusion in the context of technical re/production of music and acoustics, the trenchant formulation goes:

> Recordings do not *reproduce* sound, they *represent* sound.[113]

This emphasizes that it is a specific reflection, as in perception though in a different way. And this implies, as always, as such necessarily differences and thereby fuzziness.

The closest thing to an identity of depiction and the depicted is the more or less 'automatic' impression[114] as a mechanical symmetrical reflection, be it a fossil in slate or a record from the negative matrix. And though this act of reproduction of the original unicum 'record' is extremely important socially, it is medially not a central but rather a derived process. Nevertheless, there are differences here too, as a result of material or technical blurring, as in the multi-stage 'printing' process of records, in casting and molding processes from hollow molds, etc.

Since the automatic reproduction only records and stores the print or imprint objectively, nature-like, any subjectively caused fuzziness can basically be ruled out. In a second instance however it does reappear. For as soon as basic technical problems of the static noise drowning out everything else where largely solved, there emerged with more sophisticated recordings aesthetic-technical questions such as the positioning of the receiving hoppers, positioning of the orchestra and so on—electrical microphones only came into use with broadcasting technology in the mid-1920s. Such questions become more and more complex and differentiated in the course of media history and shall not be discussed extensively here.[115] Basically the problems occurring here on the side of subjectivity are analogous to the problems of the fuzziness of the sound form itself. Essentially these are aesthetical considerations. But they are intermeshed with technical aspects.[116]

> In the recording studios, right from the beginning, the transformation of sonic waves as the acoustical material of audibility and sound as the sensual medium of music-making had at first to be laboriously restored, since the given cultural context would not reconstitute itself [set in/appear/ensue] considering the limitations of the recording apparatus. From this restoration procedure at the beginnings of the developments there evolved under the

[112]Cf. e.g. Hermann/Hunt/Neuhoff (eds.) 2011.

[113]Demers 2010, 170. Cf. e.g. Krause 1998.

[114]Information and numerous aspects, Didi-Huberman 1999.

[115]For comprehensive detail on aspects of sound see Morat/Ziemer 2018.

[116]Cf. e.g. Corey/Benson 2017.

> label „High Fidelity"a reconstruction process and this again later developed into a design process—the art of *phonography*.[117]

The composer Stephen Frost provides some remarkable observations and thoughts about classical music editing. A central consideration is the dialectics of perfection and liveliness, which we already encountered in the sound form. An important factor is the relationship of continuous recording and discontinuous montage of single takes and patches.

> Once a first "take" is recorded, the conductor and producer will listen back to it and pinpoint the things that need attending to – specific aspects like intonation and ensemble, as well as those more ephemeral elements like energy and passion. Sometimes the whole piece or movement is played again, sometimes not, but either way we soon cross the Rubicon and enter that strange, bizarre world known in the business as "patching". Technically, a "patch" is anything less than the whole piece, or movement. It might be as much as three-quarters of it, or sometimes as little as eight bars. I have occasionally been asked to work with patches of only one or two bars in length.[118]

Editing dates back to the beginnings of sound carriers, at first simply as repetition of the recording. The actual montage only became possible with the introduction of the magnetic tape in the late 1940s.

> (W)e have been editing since the dawn of recording. Even with wax cylinders we could give ourselves the luxury of two or three "goes" and then choose the best one – which is a simple, but nevertheless very powerful, form of editing. However, the kind of cutting and splicing that we are now familiar with wasn't really possible until the advent of magnetic tape. With razor blade in hand, a whole new vista of possibilities opened up before us: we could choose from different performances of the same piece and select and join together not only different movements but also different sections, passages, bars and even individual notes![119]

The digital cut due to its simplicity has again increased the unscrupulousness of interventions.[120]

> With computerised digital editing almost anything technically (though not musically) is now possible. At least with a razor blade there was a sense that we were all potential Sweeney Todds and that without great care before every editing decision we might slip up and cut our own throats. Today, because all edits can be made and unmade in any order, and at great speed, there is the danger of being so cavalier with our virtual razor blade that we end up building ourselves a Frankenstein's Monster.[121]

To the demand for adequate realization of the idea of the composer, incarnated in the score form of the work, is added a further demand for adequate realization of the idea of the interpreter, incarnated in the sound form. We come

---

[117] Wicke 2011, p. 1062. Emphasis HWH.

[118] Frost 2007. Emphasis HWH.

[119] Ibid. Emphasis HWH.

[120] The „record in short sections, at least in the case of orchestral sessions [...] saves time and money " (Ibid.).

[121] Frost 2007.

> closer to the composer's or performer's vision […] if we edit truthfully, honestly and with integrity, then the artist's vision will not only survive but be magnified.[122]

Especially this selection of "beautiful passages", made possible by the phonographic form, augments the contradiction of technical perfection and artistic lively perfection in comparison to the sound form. Especially with predictable playing, and therefore less lively playing, a "patchwork" of small, separately recorded sections becomes more feasible. But:

> It is a highly efficient way of re-creating the *score*, but almost totally ineffective at revealing the *music within it*.[123]

Artistic liveliness is not even guaranteed in „Live Recordings".

> […] before the advent of digital editing (c.1979) and every live concert ever given.
>
> Some artists are troubled by the moral issues raised by editing so they turn instead to "live" recordings in the belief that they represent a true and honest account of a real performance. Disenamoured by the recording process, they gravitate towards recording live concerts in a kind of back-to-basics approach, in the hope of retaining the integrity and honesty of their performance and also regaining some control over the end result. Record companies are increasingly interested in live recordings too, because they are, on the face of it, cheaper to produce (though that does not always turn out to be the case).
>
> […] Some live recordings are, indeed, genuinely "live": one performance, one recording, and no edits. But these are very much in the minority. Most live recordings are edited, and the more high-profile the project is, the more heavily edited it is likely to be. How is this possible? Well, if the concert is repeated we have two, and sometimes even three, recorded versions to work with. The rehearsal is often recorded too. In theory, this means that you can edit back and forth between each version as much as you like—every bar or note if necessary—and some live recordings are, indeed, very heavily edited. Furthermore, there will often be a "patch" session, recorded after, or between (or even before!) the live concerts. Sometimes patch sessions are so extensive that on the finished CD the entire movement of a symphony (say) may not actually be live at all.

For the sake of liveliness even obvious mistakes can be tolerated, as they are nearly impossible to completely avoid in a live performance:[124]

> Less obviously, a passage is often exciting despite some questionable ensemble and intonation. So, can we dare to ask the question: is it exciting because of those "questionable" qualities? (I suppose I am using the word "exciting" in the all-encompassing scientific sense of the word—in the way that temperature is transmitted through an object by one atom exciting the next—and not just to describe the obviously "big" moments in music; in other words I use it in the sense of having the power to change, move, and communicate with the listener on all levels).[125]

---

[122]Ibid.

[123]Ibid. Emphasis HWH.

[124]See above the catchwords unconscious amending of broadcasting and of hearing.

[125]Frost 2007. "For example, Alfred Cortot's recordings are somewhat notorious for their technical inaccuracies (wrong notes), yet he is considered by many to be the greatest interpreter ever on record of Chopin and Schumann. His white-hot interpretations are just as compelling now as they were then" (Ibid.).

3.6.3 "Noise" vs. „High Fidelity". Technical-Medial Limits of Accuracy

The technical considerations present a slightly different case. For here occurs new, additional fuzziness. The blurs of the phonographic form result from the nature of technical reproduction as reflection in another, new medium through the medium itself. The existing old tones appear new and different and yet remain similar: one the one hand expansion of temporal and spatial range and social availability, on the other a multi-dimensional reduction of the sound spectrum as opposed to the original sound form and especially its accompaniment, interference, superimposition of static noise.[126]

A psycho-social equivalent of the purely technical noise can be found in concert halls and operas in the form of whispering, rustling candy wrappers, coughing foremost and more. The unnatural technical (re-)production abstracts from this fuzziness of the lively performance its aim of "naturalness". The sound ideal is "just like in concert". Norms such as „High Fidelity" (Hifi) and so on are guided by this. The general aim is a similarity relation to the sound form of a lively performance, something like 'faithfulness to nature', meaning the precise as possible imitation through the means of technical reproduction, 'faithful to the work' on a meta level in regard to the sound form, of the technically not processed and un-reproduced sound as in the opera or the concert.[127] Surely:

> The result was not really a resounding depiction of a performance, but much rather a simulation of it in the temporal shape of sound.[128]

These positive aims are supplemented and supported by the eradication of the negative, of blurs, the battle against fraying fringes and non-musical sounds of the phonographic form. In analogy to the fuzziness of the sound form and the psychological fuzziness of the fading of recollections and the superimposition of other, not actually related impressions in the recollection form, the phonography form brings about technical blurs, such as the rustling, clicking, static and suchlike of types like the Edison phonograph, the gramophone record and magnetic tape.

'Noise' becomes a metaphor for the informational-theoretical increase of fuzziness, of entropy and the loss of information, before the historical-technical background of technical (re-)production of the acoustical world; on all 'channels' of perception 'channel noise' causes this deficit in the transmission of information. The term 'noise' can also be used as metaphor in the context of optics. In sound noise occurs non-metaphorically materially as acoustical blend of frequencies that doesn't allow any distinct sound to be heard. As white noise with a blend of all frequencies or colorful noise with a more limited spectrum of frequencies it can become the starting point for electro-acoustical music and filtering and processing of musical

---

[126] Cf. e.g. Bates 2004.

[127] On the critique of this concept under aspects of ontology of media and art in detail see Wicke 2011, pp. 1068–1074.

[128] In detail Wicke 2011, p. 1066.

material. But in the phonographic form it is nothing but bothersome fuzziness and imperfection.

This form shows gradations of fuzziness and possibilities of sharpening, from the phonographic cylinder, where the noise is louder than the music, to the continuously improving digit technologies such as ‘Bluray’ and so on, which maximize pitch and volume while minimizing noise until it can practically not be heard anymore.

Digitalization, and as sound carrier the ‘compact disc’, were a leap in the development for eliminating the disturbing noise. This progress also came at a price. For during „pulse code modulation“, the limitations of the sampling rate resulted in limitations of the frequency spectrum; this being the price of increased precision. Analogue coding contains broader formants. Especially during the early years following the introduction of digital technology, when comparing was still rather simple, many felt that analogue recordings sounded ‘warmer’,[129] and by analogy more lively. At this new, higher level of the media the dialectics of technical perfection and aesthetic liveliness reappear. It runs parallel to the contradiction of liveliness with its imperfections versus technical perfection of the sound form, just as historically a new and more sophisticated general state and standard of instrumental playing capability was achieved (see Sect. 3.2). Just as then no considerable longing for the lower standards of the past came about, so today the nostalgic romanticization of clicking and noise seems hardly appropriate.

3.6.4 The Audiovisual Form as a Variant of the Phonographic Form

With sound film (talkies) since 1929 a new stage of a more comprehensive, grittier technical reflection than the sound form became available. For nor the performance of a specific, individual work can be represented in technical media, in both its acoustical and its optical dimension. Furthermore: The entirety of the singular performance can be rearranged slightly through montage, while remaining a temporal continuum in the technical (re-)production. The blurs or even mistakes of ‘live’, inevitable in the realization as sound, give the whole a more lively appearance compared to a studio production, where slips and mistakes can be repeated and corrected.

Another change to a new quality results from the digitalization of technical (re)production of sound. In audio-vision a new form of storage becomes possible, namely of music or rather the work of art as graphical representation, and in principle automatically, be it as sonogram or as transcription of an improvisation, independent without sound or as addition to the soundtrack as a moving score track. This way the original notation returns, such as the scores of classical works, only now directly *coordinated with their sound form* (See Sect. 6.8). Conversely the notation form can be transformed directly, almost automatically, into sound form through input into keyboards and special programs and interfaces. Admittedly this comes with two

[129]See for example Hans Krauss in a communication from November 2018.

major deficits, which it would be all too polite to call blurs: Firstly, the voice is excluded; secondly the sound of the realization of the work is more than meagre and confined to the limited capabilities of the apparatus, as a rule electro-acoustically reduced and limited synthesizer sound.

> Benjamin Britten described [in 1964] the magic of music as being "[that] which is not in the score." To find out what that magic is, perhaps we first need to ask the question: what is in the score?
>
> With software like "Sibelius" or "Finale" this is easy, and we soon find out what Britten meant. Input the dots of a Chopin Ballade, for example, and out of the sausage-machine comes a lifeless deluge of sound that has little more to do with Chopin than Morse Code. The computer can give us rhythm, certainly, but only the kind of meaningless ensemble of a marching army, the soulless accuracy of a Swiss clock mechanism. The pulse, such as it is, comes not from the heart but from a life-support machine. There is none of the subconscious rubato and variation in color and agogics that defines the musical outpouring of a human being. After all, the *only thing consistent about a human performance is its inconsistency.* (It is true that the software mentioned above can use algorithms capable of introducing a simple form of "espressivo", but it is fake, unconvincing and, worst of all, utterly consistent.)[130] (See Chapter 7)

A work 'represented' in such way and not 'reproduced'—to make use of this distinction—is similar to what is achieved by a piano excerpt of an expansively orchestrated chamber music or orchestra piece: It gives an impression of the work, oscillating between fuzzy and focused.

The black-and-white sound form of the work resulting from this compares more to a silhouette than a painting. The subjective mediation complements and augments the fundamental fuzziness. It must be decided on the one hand, which voices or motifs of the score are important, on the other hand what is feasible for the piano.

Parallels to the reduced orchestration or even the piano excerpt can be found in the fine arts. Apart from independent functions such as flyers, pictorial broadsheet etc., various engraving methods are used for monochrome, black-and-white reproductions of colored paintings.[131] The engraving, as the shellac record, can be mechanically reproduced. Only photography, i.e. color photography, has reached the stage of mechanical, in part automated technical reproduction that corresponds with phonography: refraction without, or rather with principally small subjective input as fuzziness of its own kind, in addition to the technical fuzziness.

3.6.5 Phonographic Form as Primary Form of Existence

The phonographic form can also not be a derived existence form, but constitutes a primary existence form of the work. For the largest part of newer pop music since about the 1960s it is that. At the same time it is the primary medium for music reception for the vast majority of humankind.

---

[130]Frost 2007. Emphasis HWH.

[131]Cf. e.g. Blübaum et al. 2007.

> Since about half a century, music is no longer *performed music but produced music*. That means music that has received its definitive sound shape in a studio assisted by technical apparatus in a step-by-step process. Live on stage follows a reconstruction of the sound shape produced in the studio, which is heavily dependent on technical support.[132]

In the production in the former 'recording' studios

> not only the playing of music, but also the material, in which it takes place, the sonic materiality of music, has undergone dramatic changes. This refers not only to the now technically possible interventions regarding the acoustic identity—changes of frequency composition through equalizers, changes of the formant structure through aural exciter or the technical synthesis of sounds. What is meant here in a more fundamental sense is how the physical parameters of sound are connected [related] to the aesthetic parameters of music, the conceptualization of sound as material for making music.[133]

Digitalization brought about also a technically qualitative leap. The origin of sounds', their spatial-temporal situation and their semantics are blurred to unrecognizability.

> In the binary representation sound has detached itself from the modalities of its generation. It has become a calculated real-time simulation of itself, whereby the simulation procedure, i.e. the making audible again the values gained through digitalization using D/A (digital/analogue) converters, has not only made oblivious the differences between bodily, mechanical, or signal-based, electrical forms of sound generation, but has made it disappear altogether. The rhythmic patterns and sound streams of techno or hip hop tracks can be generated by machines and played manually on a MIDI-Keyboard or be a mix, as in the technically fabricated endless loop of a figure played by hand or taken from a sample. From the resulting sound it is no longer possible to accurately identify the methods of sound generation—traditional instruments, technical effect devices, later forms of sound editing and transformation, technical synthesis of sounds. Sound has thereby undergone a comprehensive de-referentialisation. It used to be the symbolic medium *par excellence*—every audible sound referred as a vector of meaning more or less distinctively to the spatial-temporal mode of its production and its producer, and was thereby inextricably connected to its meaning that made it significant. In digitalized form sound has become completely detached, without cohesion, without origin, without traces and thereby at first with meaning in context.[134]

In this almost autonomous manifestation of the phonographic form there exists no scale in regard to aspects of the work on which fuzziness could be determined according to fuzzy logic. Apart from that the already discussed factors apply, which produce fuzziness such as noise. As at other points of the matters discussed in this book, this is a point where fuzzy logic in the perspective of work realization reaches its limits. And still it can be applied to the already mentioned covering. The starting point of various cover versions ex-post assumes the character of a work of art, which serves as point of reference. What was not a work of art paradoxically becomes one through its reproduction, which is a variational new production and makes no claim to similarity between adequacy and quasi-identity.

Covers or covering self-evidently refers implicitly only to the phonographic form or rather its media transmission and realization of a preexisting musical work in

[132]Wicke 2011, p. 1062. Emphasis HWH.

[133]Ibid.

[134]Wicke 2011, p. 1079.

this form. Calling it a cover when „Meier“ or „Harteros“ sing the same aria would be an absurd use of the term. A cover version is in the most general sense a new interpretation of a ‘work’ in phonographic form by other musicians (or producers, sound technicians etc.).[135] (More specific usage as remix etc. falls under the term quotation form; see Sect. 8.2). A special from a cover versions, that intentionally and drastically variate the original and reduce it, so that it is produced purely instrumental, without text or singing—comparable to the piano excerpts of operas, that contain no vocals, only sometimes some text noted between the piano voices.[136]

> There are both cover versions that in their arrangement and sound strictly adhere to the original, as well as versions that are hardly even recognizable. Chubby Checker’s version of the *The Twist* is of the first type, its original having been composed by Hank Ballard and recorded on 11 November 1958, was so similar to the original that Hank Ballard on hearing it in the radio, thought it was his piece.[…] Checker’s version was identical in pitch, rhythm, and vocals.[137]

Alongside such almost already traditional procedures of covering, which date back to the beginnings of the phonographic industry, there are also newer ones. They expose the covering and to some extent give it a life of its own. Of the three commonly employed types only the first does covering in a strict sense.

> Tribute acts or bands are performers who make a living by recreating the music of one particular artist or band. Bands such as Björn Again, Led Zepagain, The Fab Four, Australian Pink Floyd Show and The Iron Maidens are dedicated to playing the music of ABBA, Led Zeppelin, The Beatles, Pink Floyd and Iron Maiden respectively. […] Many tribute bands attempt t*o recreate another band’s music as faithfully as possible,* but some such bands introduce a twist.[138]

The second type can imitate the sound of the original, but doesn’t have to instead it can relate to the underlying work *before* its phonographic form and follow its own path.

> Cover acts or bands are entertainers who perform a broad variety of crowd-pleasing cover songs for audiences who enjoy the familiarity of hit songs. […] Since the advent of inexpensive computers, some cover bands use a computerized catalog of songs, so that the singer can have the lyrics to a song displayed on a computer screen. The use of a screen for lyrics as a memory aid can dramatically increase the number of songs a singer can perform.

The third type finally doesn’t cover individual works at all in phonographic or notational form instead it broadly covers styles and genres.

> Revivalist artists or bands are performers who are inspired by an entire genre of music and dedicate themselves to curating and recreating the genre and introducing it to younger

[135]“The term “cover” goes back decades when cover version originally described a rival version of a tune recorded to compete with the recently released (original) version. The *Chicago Tribune* described the term in 1952: ‘trade jargon meaning to record a tune that looks like a potential hit on someone else’s label’”. (https://en.wikipedia.org/wiki/Cover_version, 26.4.20, retrieved 1.5.2020).

[136]Cf. https://de.wikipedia.org/wiki/Coverversion, 29.4.20, retrieved 1.5.2020.

[137]Ibid., Emphasis HWH.

[138]In detail https://en.wikipedia.org/wiki/Cover_version, 26.4.20, retrieved 1.5.2020.

audiences who have not experienced that music first hand. Unlike tribute bands and cover bands who rely primarily on audiences seeking a nostalgic experience, revivalist bands usually seek new young audiences for whom the music is fresh and has no nostalgic value.

This could be described as of the late Alexandrinian era or as an ever-growing lack of the frequently invoked 'creativity' or even as 'parasitic', all of which is not uncharacteristic of the late 'market economy'. Here we can imagine experimental designs, which could compare with each other the various versions of evergreens—including the cover versions of pop music—with the means of fuzzy logic. For *La Paloma* by Yradier, for example, one of the most edited works of newer popular music, as well as for many 'standards' of jazz there exist hundreds of different versions, which should provide enough statistically significant material. Complications of methodical aspects arise already before commencing with the actual investigation. What exactly is a 'version', or rather of what is it a version? For a 'cover' this remains rather simple: Orientation point and yardstick is what is covered. And while jazz standards on the other hand are indeed compositions, they are not explicated in enough detail to serve as orientation point. Obviously the first realization through the respective composer could be a possible point of reference. More prudent would be to filter out a median value from all realizations. But this again, as the center of the triad work idea, notation, interpretation, is nothing material and immediately given.

3.6.6 Guiding Principle Immersive Reception as Overwhelming—Focus on Effect Instead of Work

Technical procedures such as two-track stereophonics and four-track quadrophonics were still following the mentioned ideal of 'naturalness': Recording and replay of a work was intended to come as close as possible to its 'live' performance, here especially through simulation of a more or less real space. A considerably different approach was followed by cinemas in the adoption of sound systems such as „Dolby Surround" and others, which are also used in high-end homes, and which aim to reproduce an acoustical space or spatial acoustics as acoustical three-dimensionality. They reproduce and realize what has already been reproduced on record, magnetic tape, CD, and as film, only now customized for perception in the spatial reality of the cinema. The aim is no longer, as it once was for the 'high fidelity' of records, a similarity relation to the sound form of a lively performance. This new procedure is geared towards an immersive reception as overpowering of the senses. They are geared towards effect, not oriented by the 'work'. The fuzzy logic question of fuzziness and sharpening is no longer applicable.

3.7 Work and Effects. the Multiplication of Fuzziness Possibilities

With the effect conveyed by the sensual-mental perception, the musical process reaches its end. Its goal, however, is only reached when those who perceive the work not only allow it to have an effect on them, but also when it has consequences in their everyday lives. The effect, which at first appears as purpose, is thus itself again a means. When music, like art in general, becomes effective, it fulfils a social function, an objective purpose. The purposes themselves, like the effects, cover a broad spectrum. They are music-unspecific or music-specific, intended or unintended. Music serves as an invitation to dance or to work, to escape the world or to know the world. It is used by ethos as well as by Eros, by the upper as well as by the lower classes. It beautifies humanity or barbarism, enlightenment or religion.[139]

3.7.1 Difference Between End and Aim of the Music Process

Music as such does not 'do' anything. For its effectiveness to unfold and for the musical process to fulfil its purpose, it must be appropriated and implemented by human subjects and their practice. This requires additional activity beyond those activities that are constitutive of the music process and music culture. This return of art to the practice from which it originated takes place outside the musical process, but is its logical continuation. Needs in human activities are the driving motive and determining purpose for the music process to be set in motion at all. Even in the case of catharsis specific to art, in the broader sense as 'effect' the final link in the actual musical process, the question of the adequacy, the appropriateness of perception and effects on the object that produces the effect is legitimate. It is thus once again the question of similarity (see Chapter 4). To investigate the question of similarities in the sense of an appropriateness of the effects, purposes and functions of the works, for example as 'analogies', and thus at the same time the range of deviations from this appropriateness, i.e. from corresponding fuzziness, is difficult, but not a completely hopeless undertaking.

3.7.2 Considerations and Suggestions for Fuzzy Logic Experiments for Objectify Subjective Experiences

The congruence of effect as a result of reception and production in at least one elementary parameter is a special case. When listening to a music recording, a conductor

[139]See among others Heister 1984.

had the same profile of his pulse frequency as when conducting.[140] Usually the relationship between fuzziness, sharpening and blurring can also be found in the effect. When an adult falls asleep at the *Marseillaise* with its double fourth call to the "Allons enfants de la patrie", he did not listen properly or did not understand something correctly. When a toddler does not fall asleep at the lullaby "*Schlaf Kindlein, dein Vater hüt die Schaf*," he/she has not listened properly or does not want to hear the message. *Or*, however, he/she may have listened too closely, when, as a palimpsest, he listens to the rhythmically-prosodic and rhythmically identical, but very different textual sequence: "Maikäfer, flieg! /Dein Vater ist im Krieg. /Deine Mutter ist in Pommerland, /Pommerland ist abgebrannt. / Maikäfer, flieg!" ("May bug, fly! /Your father is at war /Your mother is in Pommerland, / Pommerland has burned down /May bug, fly!").

The baby, who continues to cry in George Gershwin's opera *Porgy and Bess* despite the beautiful lullaby '*Summertime*' behaves appropriately and cannot be calmed by the father: The father dies later. The idyll was endangered and destroyed. Such inappropriate reactions cannot be quantified, however. In fact, they are analogous to the accidental mistakes in the productive realization of the works (See Fig. 3.15).

However, the situation is different with questions about the spread of effects. They can be explored and even quantified by means of verbally articulated surveys such as the polarity profile and self-observations of the test subjects on the one hand, experiments and tests below and outside the conscious mind up to brain scans on the other hand. After all, the goal is to compare something that is essentially already similar: The reference to the work is the same, the common third of the differences. The work forms measure and benchmark. An evaluation of the differences is difficult, necessarily also subjective, but not completely impossible. First of all, statistical distributions and their edges should be taken as indicators of the spread.

Suggestions could be given e.g. by the differently directed, methodically-statistically careful investigations of Eun-Hee Jeong. She was mainly interested in effects and especially in the interaction of the already mentioned imagery, the anticipation of different kinds (see above imagination form) with the actually reached state of the dancers. A descriptive table for this:

In order to transfer results, methods and experiments from the effects to the works, some considerations and some effort are required. Some of this will be outlined below from the point of view of music.

As has already been shown several times, perception and effects are not impossible to access with Fuzzy Logic, however difficult it is to measure subjective and not exteriorized elements. It could, for example, start with typologies and classifications of hearing and the hearing person. It is precisely here that the 'definitions' in the literal sense, the boundaries, but also the transitions, are often indeterminate and fuzzy.[141]

---

[140] See Damasio 1997, p. 265f.

[141] Cf. e.g. Pampalk et al. 2003 and other titles in 5.2.1 and 5.2.2.

Raw Data Themes	1ST Order Themes	General Dimensions
Total immersion in music, feeling, and dance	Total concentration	Total concentration 17(85%)
No worldly thoughts; just concentrate to dance at that moment		
Automatically absorbed in dance		
Total concentration		
Feel it is real life; Forget stage, audiences, and that it is performance		
Uniting everything such as music, lighting, movements, and story of performance	Uniting with music, feelings and movements	Merging of action and awareness 10(50%)
Uniting with music and feel like playing music through body movement		
Breath, dancing, and music becoming one and uniting		
Filled with positive thoughts	Experience positive thoughts	Positive feelings 10(50%)
Energy or catalyst for life	Energy for life	
Self–contentment (satisfaction)	Self-satisfaction	
Removal of stress and feeling relieved	Feel relieved or refreshed	
Feeling refreshed		
Just happiness or enjoyment because moving	Autotelic experience	Autotelic experience 10(50%)
Happiness because dancing		
Delight and enjoy		
Feeling "up" and happy		
Feel spine-tingling	Remarkable experience	
Feel eerie		
Feel like taking drug		
Experience an ecstatic state	Ecstatic state	
Feeling like reaching performance climax		
Everything going well	Everything going well	Sense of control 8(40%)
Freedom of movement, music, feeling, and body	Freedom of music, feeling, and movement	
Expressing one's feelings well through dance	Expressing one's feelings	
Everything is correct while dancing	Kinesthetic awareness	Unambiguous feedback 5(25%)
Positive response from audience	Positive response from audience, teacher, peers	
Positive clear feedback from teacher		
Positive feedback from peer		
No consciousness of others	No consciousness of others	Loss of self-consciousness 4(20%)
Freedom from negative judgment of self		
The absence of time realization (losing track of time)	The absence of time realization	Time transformation 2(10%)

Fig. 3.15 Characteristics of flow
Jeong 2012, p. 127

As always with such questions, cooperation with the neurosciences is necessary. These, in turn, require further development in order to measure subjective aspects, excitation patterns, etc. directly on the brain with sufficient precision and, moreover, which is much more difficult, to determine more precisely the quality, the what and how of the effects and the mental correspondences of brain patterns. But all this is not impossible in principle up to a certain penetration depth. Especially the question of sharpness/fuzziness, which is specific to Fuzzy Logic, is even more difficult if we relate it to the extent to which the effect is adequate to the work. It is always only a gradual adequacy. Answers can hardly be obtained without verbal cognitive mediation, such as the well-known polarity profiles, which are always comparatively general. Here, too, refinements are necessary and possible. In order to make progress in the difficult question of objectifying subjective perceptions, meta-analyses of existing studies in the light of fuzzy logic are another generally applicable approach. They can be used to make fuzziness conscious, and to clarify, quantify and thus sharpen differences between individual perceptions according to type and degree. It must suffice to mention this here as a general suggestion. On the other hand, a complicated, multi-sectional experiment for the investigation of perception should be developed in more detail. This would be complex, but it is possible. Ideally, the neurophysiological and neuropsychological approach should be supplemented and expanded—imaging procedures that can, or will be able to, capture similarities, differences and fuzziness at a deeper level as technology progresses. (See below in 4). The question is: Which perception has which degree of sharpness? The same question applies to memory, to the imagination that the test subjects have of a work, and finally to the subjectively experienced and objectively produced effect of a work. The steps of an experimental setup can be outlined as follows.

1a. Play a piece, live and/or technical reproduction. The latter is much easier to handle.

1b. thereby examine all possible parameters in the test subjects, not only EEG, but also including the usual suspects such as skin resistance, pulse rate, etc.—and the changes over time in the music process *and* afterwards, which are important from the point of view of perception and effect.
    Whichever method is used in detail—it would provide objectified data that can give information about degrees of sharpness of perception and even allow scaling and quantification in further investigations—and the changes that are important in terms of perception and effect. However, more advanced procedures such as CT or brain scans cannot be applied in the case of live music. In general, the differences between the laboratory situation of the experiment and the usual reception situation, the "artificial" demarcation—the blood pressure-increasing "white coat effect" e.g. due to the sheer presence of test supervisors is well known and must be considered as one of the disturbing factors. The objective, necessarily unconscious physiological reflection of the subjective is now to be supplemented by the conscious to semi-conscious subjective.

2. The test persons record what they have heard.

These can be their own free formulations alone or they can be combined with specified grids of the questionnaire type, such as the polarity profile, which, despite all justified criticism, still gives a fuzzy approximation.
3. The result is now their individual, personal word form.

II

Here are forks, branches and different paths possible.

1. The different word forms are compared with each other and could be checked for common features, similarities and resemblances. This step is a kind of text analysis. One hypothesis would be that the common core corresponds most closely to the work. This may not always be true: It is a question of what relevance extremes, averages and median values have.
2. The currently obtained word forms are supplemented with those of pre-existing other experimental groups and then compared. In addition, a comparison with other "types of texts", like music reviews, concert or opera guides, introductions to works in program booklets and the like can be made.

III

The following stage takes the step from the word form of the effect to the imagination form. This detour tells us something about perception in general.

1. The word forms, whether pure or mixed are brought to the attention of *other* experimental groups. It remains to be seen whether reading aloud or individual self-reading is more suitable. These groups have *not* heard the work, at least not currently.
   In order to shorten the experiment, it probably makes sense to produce a condensed version of the various word forms, in which, as mentioned, the essential and common elements are already filtered out by a kind of superimposition—as in a kind of palimpsest.
2. The test persons must now reconstruct or guess which work is meant by the word forms (or the optimized one word form). For this purpose 2+ n works are played—similar to identity parades.
3. As a test of the recollection form, the same could be done with the original group after a proper interval. There are many difficulties, but the problems are in principle solvable, when considering the effort for the experiments as possible.

A major difficulty may be the selection of suitable pieces. They should be rather unknown, but still correspond to fairly familiar idioms. Extreme, peripheral works are therefore out of the question, whether in pop or classics. They should be as short as possible in order to keep the experiments within a reasonable time frame. A programmatic support, however, might make it easier, so Debussy's orchestral piece *Prélude à l'après-midi d'un faune* or the flute solo piece *Syrinx* would be suitable. The usual social contexts, i.e. social classes, educational levels, musical and artistic experiences and the like would have to be taken into account, but are rather secondary here, apart from the question of the choice of works. The systematic, empirical-experimental application of fuzzy logic could in any case make a significant contribution to the questions of the degree and nature of a reception appropriate to a work that are often discussed at length and with little result.

All the following principles are now under the sign of fuzziness. These accept the fuzziness partly as a necessary evil, partly they try to minimize it by different methods of sharpening or at least to grasp it quantitatively, but partly they also take it along as a 'method', as a "detour" and stimulus on the way to discovery and invention of something new. The latter is one of the leitmotifs of the following chapters and concentrates in the final chapter.

References

Adorno, Th. W.: Einleitung in die Musiksoziologie—Zwölf theoretische Vorlesungen Suhrkamp, Frankfurt a.M. (1962) (extended as: Einleitung in die Musiksoziologie. Rowohlt Verlag, Reinbek 1968)

Assayag, G., Dubnov, S.: Using factor oracles for machine improvisation. Soft. Comput. **8–9**, 604–610 (2004)

Bates, Eliot, Glitches, Bugs, and Hisses: The Degeneration of Musical Recordings and the Contemporary Musical Work. In: Washburne, C., Derno, M. (eds.): Bad Music. The Music We Love to Hate, pp. 212–225. Routledge, New York and London (2004)

Benjamin, W.: Einbahnstraße. Ernst Rowohlt Verlag, Berlin (1928)

Biles, J.A.: Genjam in perspective: a tentative taxonomy for genetic algorithm music and art systems. Leonardo **36**(1), 43–45 (2003)

Blaukopf, K.: Werktreue und Bearbeitung. Zur Soziologie der Integrität des musikalischen Kunstwerks (Musik und Gesellschaft, vol. 3). Braun-Verlag, Karlsruhe (1968)

Blübaum, D., et al. (eds.): Gestochen Scharf! Die Kunst zu reproduzieren, Catalogue. Kehrer, Heidelberg (2007)

Burney, C.: Carl Burney's der Musik Doctors Tagebuch einer Musikalischen Reise, vol. 2: Durch Flandern, die Niederlande und am Rhein bis Wien Hamburg (1773). http://www.zeno.org/nid/20007748108, retrieved February 23, 2020

Castillo, O., Kacprzyk, J., Melin, P. (eds.): Nature-Inspired DESIGN of Hybrid Intelligent Systems (Studies in Computational Intelligence vol. 667). Heidelberg etc., Springer (2017)

Connor, E.: Separating trash from treasure: how the brain assigns value to objects. In: Neuroscience News February 9, 2018, http://neurosciencenews.com/object-value-8469/, retrieved February 10, 2018

Conklin, D.: Music generation from statistical models. In: Proceedings of Symposium on AI and Creativity in the Arts and Sciences, pp. 30–35 (2003)

Conklin, D., Witten, I.: Multiple viewpoint systems for music prediction. J. New Music. Res. **24**, 51–73 (1995)

Cont, A., Shlomo, D., Assayag, G.: Anticipatory model of musical style imitation using collaborative and competitive reinforcement learning, http://link.springer.com/chapter/10.1007/978-3-540-74262-3_16, retrieved 2 February 2018

Correia, J., Ciesielski, V., Liapis, A. (eds.): EvoMUSART: international conference on evolutionary and biologically inspired music and art. Computational intelligence in music, sound, art and design. 6th International Conference, EvoMUSART 2017, Amsterdam, The Netherlands, April 19–21, 2017, Proceedings. https://link.springer.com/book/10.1007%2F978-3-319-55750-2

Corey, J., Benson, D.H.: Audio production and critical listening. technical ear training. technical ear training software, 2nd ed. Routledge, New York and London (2017)

Cross, I.: Music analysis and music perception. Music. Anal. **17**(1), 3–20 (1998)

Danuser, H. (ed.): Musikalische interpretation. In: Dahlhaus, C., continued by Danuser H. (eds.) Neues Handbuch der Musikwissenschaft, vol. 11, 12 vols. Laaber-Verlag, Laaber (1992)
Demers, J.: Listening Through the Noise: The Aesthetics of Experimental Music. Oxford University Press, New York and Oxford (2010)
Didi-Huberman, G.: Ähnlichkeit und Berührung. Archäologie, Anachronismus und Modernität des Abdrucks. Dumont, Köln (1999)
Douwe, D.: Wie wir träumen. Galiani, Berlin (2015)
Draaisma, D.: Warum das Leben schneller vergeht, wenn man älter wird. Von den Rätseln der Erinnerung. Piper, München (2006) (Dutch 2001)
Draaisma, D.: Das Buch des Vergessens. Warum unsere Träume so schnell verloren gehen und sich unsere Erinnerungen ständig verändern. Galiani, Berlin (2013) (Dutch 2010)
Draaisma, D.: Halbe Wahrheiten: Vom seltsamen Eigenleben unserer Erinnerung. Galiani, Berlin (2016)
Dubnov, S.: Spectral anticipations. Comput. Music. J., 1–39 (2006)
Dubnov, S., El-Yaniv, R., Assayag, G.: Universal classification applied to musical sequences. In: Proc. of ICMC, pp. 322–340. Michigan (1998)
Dubnov, S., Assayag, G., Lartillot, O., Bejerano, G.: Using machine-learning methods for musical style modeling. IEEE Computer Society **36**(10), 73–80 (2003)
Dupuis-Désormeaux, N.: Tutti!—music composition as dialogue (Phil. Diss., York University Toronto, Ontario) (2017)
Franklin, J. A.: Predicting reinforcement of pitch sequences via lstm and td. In: Proc. of International Computer Music Conference, Miami, Florida (2004)
Frost, S.: Striking the wrong note. J. Art Rec. Prod. (01), Industry Perspective (2007)
Goldschmidt, H.: Musikverstehen als Postulat. In: Faltin, P., Reinecke, H.-P. (eds.), Musik und Verstehen. Aufsätze zur semiotischen Theorie, Ästhetik und Soziologie der musikalischen Rezeption, pp. 67–86. Arno Volk, Köln (1974)
Gratzer, W. (ed.): Perspektiven einer Geschichte abendländischen Musikhörens (Schriften zur musikalischen Hermeneutik vol. 7). Laaber, Laaber (1997)
Heister, H.-W.: Das Konzert. Theorie einer Kulturform, 2 volumes. Noetzel-Verlag, Wilhelmshaven (1983)
Heister, H.-W.: Zeitliche und szenische Polyphonie, dargestellt an Ausschnitten aus "Götterdämmerung" (III/2, "Siegfrieds Todesgesang" und "Trauermarsch beim Tode Siegfrieds") und "Falstaff" (II/2, 2. Teil). In: Marinelli Roscioni, C. (ed.): Quaderni dell' I.R.TE.M. (Istituto di Ricerca per il Teatro Musicale) 11, Serie 3: Atti, N. 5: Tempo e spazio: Problemi di un rapporto tra opera e televisione, pp. 115–136. Rom (1991)
Hermann, T., Hunt, A., Neuhoff, J.G. (eds.): The Sonification Handbook. Logos Verlag, Berlin (2011)
Hiller, L.A., Isaacson, L.M.: Experimental Music: Composition with an Electronic Computer. McGraw-Hill Book Company, New York (1959)
Hirt, K.M.: When Machines Play Chopin. Musical Spirit and Automation in Nineteenth-Century German Literature (Interdisciplinary German Cultural Studies, vol. 8). De Gruyter, Berlin and New York (2010)
Honolka, K.: Leoš Janáček. Sein Leben—sein Werk—seine Zeit. Belser Verlag, Stuttgart und Zürich (1982)
Huron, D.: Sweet Anticipation: Music and the Psychology of Expectation. MIT Press, Cambridge (2006)
Jeßulat, A.: Erinnerte Musik. Der Ring des Nibelungen als musikalisches Gedächtnistheater (Wagner in der Diskussion, vol. 8). Königshausen & Neumann, Würzburg (2013)
Kluge, R.: Skalen und Stimmungen des bronzezeitlichen Glockenspiels aus dem Grabe des Zeng Hou Yi. Zur Deutung eines tonometrischen Befundes. In: Ereignis und Exegese. Musikalische Interpretation—Interpretation der Musik. Festschrift für Hermann Danuser zum 65. Geburtstag, pp. 231–242. Argus Verlag: Schliengen (2011)

Knepler, G.: Geschichte als Weg zum Musikverständnis. Zu Theorie, Methode und Geschichte der Musikgeschichtsschreibung. Reclam Universalbibliothek, Leipzig (1977) (2. erw. Aufl. 1982)

Koelsch, S., et al.: Bach speaks: a cortical language-network serves the processing of music. Neuroimage **17**, 956–966 (2002)

Koelsch, S., et al.: Music, language and meaning: brain signatures of semantic processing. Nat. Neurosci. **7**, 302–307 (2004)

Krause, M.: Original oder Illusion: Ziel der technischen Reproduktion von Musik? In: Kopiez, R., Barthelmes B. et al. (eds.): Musikwissenschaft zwischen Kunst, Ästhetik und Experiment. Festschrift Helga de la Motte-Haber zum 60. Geburtstag, pp. 319–330. Königshausen und Neumann, Würzburg (1998)

Lehne, M., Engel, P., Rohrmeier, M., Menninghaus, W., Jacobs, A.M., Koelsch, S.: Reading a suspenseful literary text activates brain areas related to social cognition and predictive inference, PLOS ONE May 6, 2015, pp. 1–18. https://journals.plos.org/plosone/article?id=10.1371/journal.pone.0124550

Lerch, A.: On the requirement of automatic tuning frequency estimation, urn:nbn:de:kobv:83-opus-20822 https://doi.org/10.14279/depositonce-2037 (2006), retrieved July 4, 2011

Lerch, A.: Software-based extraction of objective parameters from music performances. Phil. Diss. TU Berlin 2008. urn:nbn:de:kobv:83-opus-20677 https://doi.org/10.14279/depositonce-2025, retrieved July 4, 2011

Mazzola, G., Mannone, M., Pang, Y.: Cool Math for Hot Music. A First Introduction to Mathematics for Music Theorists. In: Mazzola, G., Andreatta, M. (eds.) Computational Music Science. Springer International Publishing, Switzerland (2016)

Möller, H.: Angst, Entmutigung und Perfektionismus bei Musikern. Mit Fallbeispielen aus der Praxis. Self-publishing, Berlin (2018)

Morat, D., Ziemer, H. (with the collaboration of R. Rutz) (eds.): Handbuch Sound. Geschichte—Begriffe—Ansätze. J.B. Metzler Verlag, Stuttgart (2018)

Müllensiefen, D., Hennig, C.: Modeling Memory for Melodies. In: Spiliopoulou, M., Kruse, R., Borgelt, C., Nürnberger, A., Gaul, W. (eds.) From Data and Information Analysis to Knowledge Engineering, pp. 732–739. Springer, Berlin (2006)

Müller, G.: *2. Mathematik der Fuzzy-Mengen*, n.d., http://ejb-ressourcen.de/docs/FuzzyLogic/node5.html, retrieved February 27, 2020

Murphy, K.R., Jako, R.A., Anhalt, R.L.: Nature and consequences of halo error. a critical analysis. J. Appl. Psychol. **78**(2), 218–225 (1993)

Nägele, R.: Musikwissenschaftsdämmerung, Anmerkungen zu einem unzeitgemäßen Fach. In: Merkur, issue (842), *July 2019*, https://www.merkur-zeitschrift.de/2019/06/24/musikwissenschaftsdaemmerung-anmerkungen-zu-einem-unzeitgemaessen-fach/, retrieved 20 April 2020

Pampalk, E., Dixon, S., Widmer, G.: On the Evaluation of Perceptual Similarity Measures for Music. In: Proceedings of the international conference on digital audio effects, pp. 6–12 (2003)

Pearce, M., Conklin, D., Wiggins, G.: Methods for Combining Statistical Models of Music. In: Wiil, U.K. (ed.) CMMR 2004. LNCS **33**(10), 295–312. Springer, Heidelberg (2005)

Pearce, M.T., Wiggins, G.A.: Expectation in melody: the influence of context and learning. Music. Percept.: Interdiscip. J. **23**(5) (June 2006), 377–405. University of California Press (2006)

Reicher-Marek, M.E.: Works and Realizations. In: Nicola Guarino (ed.): Formal Ontology in Information Systems. Proceedings of the First International Conference (FOIS 98, June 6–8), IOS-Press. Amsterdam et al. (1998), pp. 121–132

Reicher-Marek, M.E.: Werk und Autorschaft. Eine Ontologie der Kunst. mentis. s.l. (2018/2019)

Riemann, H.: Ideen zu einer "Lehre von den Tonvorstellungen". In: Jahrbuch der Musikbibliothek, vol. 21/22, pp. 1–26. Peters, Leipzig (1914/15), C.F. Peters, Leipzig (published 1916)

Rodriguez Zivic, P.H., Shifres, F., Cecchi, G.A.: Perceptual basis of evolving Western musical styles. In: Sejnowski, T.J. (ed.) PNAS **110**(24), 10034–10038, s.a.

Russo, F., Thompson, W.: The subjective size of melodic intervals over a two-octave range. Psychon. Bull. & Rev. **12**, 1068–1075 (2005)

Saffran, J.R., Johnson, E.K., Aslin, R.N., Newport, E.L.: Statistical learning of tone sequences by human infants and adults. Cognition **70**(1), 27–52 (1999)

Schacter, D.L.: Wir sind Erinnerung. Gedächtnis und Persönlichkeit. Rowohlt, Reinbek (2001)

Schacter, D.L., Chiao, J.Y., Mitchell, J.P.: The seven sins of memory. Implications for self, retrieved April 17, 2020

Schellenberg, E.G.: Simplifying the implication-realization model of melodic expectancy. Music. Percept. **14**, 295–318 (1997)

Schönberg, A.: Stil und Gedanke. In: Vojtěch, I. (ed.): Aufsätze zur Musik, pp. 409–436. Fischer-Verlag, Frankfurt a.M. (1976)

Snyder, B.: Music and Memory: An Introduction. MIT Press, New York (2000)

Solis, A., und Panoutsos, G.: Granular computing neural-fuzzy modelling: A neutrosophic approach. https://doi.org/10.1016/j.asoc.2012.09.002 (2013)

Solis, A., und Panoutsos, G.: An ensemble data-driven fuzzy network for laser welding quality prediction. IEEE Int. Conf. Fuzzy Syst. (FUZZ-IEEE). https://doi.org/10.1109/FUZZ-IEEE.2017.8015496 (2017)

Stengård, E., van den Berg, R.: Imperfect Bayesian inference in visual perception. PLOS Comput. Biol. **15**(4), (2019), https://doi.org/10.1371/journal.pcbi.1006465

Sterne, J. (ed.): The Sound Studies Reader. Routledge, Abingdon, Oxon and New York (2012)

Streich, S., Herrera P.: Toward describing perceived complexity of songs: computational methods and implementation. In: Proceedings of the 25th international AES conference (2004)

Temperley 2007: Music and probability. MIT, Cambridge. http://www.emis.de/MATH-item?1136.00303. http://ismir2006.ismir.net/PAPERS/ISMIR0630_Paper.pdf (2007)

Temperley, D.: A probabilistic model of melody perception. Cogn. Science. Multidiscip. J. **32**(2), 418–444 (2008). https://doi.org/10.1080/03640210701864089

Thompson, W.F., Balkwill, L.L., Vernescu, R.: Expectancies generated by recent exposure to melodic sequences. Mem. & Cogn. **28**(4), 547–555 (2000)

Thym, J.: Song as memory, memory as song. *AfMw* Archiv für Musikwissenschaft **69**(3), 263–273. Franz Steiner Verlag. Stuttgart (2012)

Tzanetakis, G., Essl, G., Cook, P.: Human perception and computer extraction of beat strength. In: Proceedings of the international conference on digital audio effects (DAFx-02), pp. 257–261. (2002), https://www.researchgate.net/publication/228769561_Human_perception_and_computer_extraction_of_musical_beat_strength

Waeltner, E.L.: "O du lieber Augustin". Der Scherzo-Satz im II. Streichquartett von Arnold Schönberg. In: Bericht über den 1. Kongreß der Internationalen Schönberg-Gesellschaft, edited by R. Stephan (Publikationen der Internationalen Schönberg-Gesellschaft, vol. 1), pp. 246–262. Lafite Wien (1978)

Walker, M.: Hitchcock's Motifs. Amsterdam University Press, Amsterdam (2005)

Weihs, C., Berghoff, S., Hasse-Becker, P., Ligges, U.: Assessment of Purity of Intonation in Singing Presentations by Discriminant Analysis. In: Kunert, J., Trenkler, G. (eds.): Mathematical Statistics and Biometrical Applications, pp. 395–410. Josef Eul, Bergisch-Gladbach (2001)

Wicke, P.: Ästhetische Dimensionen technisch produzierter Klanggestalten. In: Phleps, T., Reich, W. (eds.): Musik-Kontexte. Festschrift für Hanns-Werner Heister, vol. 2, pp. 1062–1085. Monsenstein und Vannerdat, Münster (2011)

Wright, R., Uchanski, R.M.: Music perception and appraisal: cochlear implant users and simulated CI listening. J. Am. Acad.Audiol. **23**(5), 350–379 (2012)

Wrzeciono, P.: Statistical evaluation of violinist's performance based on recordings from the AMATI multimedia database. In: 55th Open Seminar on Acoustics, Wrocław-Piechowice, Poland (2008)

Yankovskii, B.A.: Method for the objective appraisal of violin tone quality. Sov. Phys.-Acoust. **11**(269) (1965)

Chapter 4
The Principle of Similarity: The Resultant of Identity and Difference

Abstract Similarity Music, as a reflection of the world, is like this and different from it: It is similar. Reality is depicted in specific material, and thus reproduced, reshaped and transformed into an artistic form. We have to mind the gap between signifier and signified, as between the realm of matter and the realm of ideas – the map is not the territory. Fuzziness is the gap, the difference between reflection and the reflected. This chapter uses the classical aesthetic terms mimesis and poiesis to analyze the artistic approach to the world. To this end, the working with iconic and symbolic signs, with material such as real sounds, ready-mades, numbers, geometric elements and with intertextuality (music, texts and pictures) is analyzed together with the interactions between different components. On the basis of this, and again including the B-A-C-H motif, I discuss how fuzzy logic and musical analysis complete each other in recognizing the various phenomena and dimensions of similarity in the artwork.

Keywords Abstraction · Aesthetics: mimesis and poesis · Analogy · (Asymptotic) Approximation instead of identity · B-A-C-H-Motif: variations and "constancy of shape" („Gestaltkonstanz“) · Difference and identity · Mimicry · Mirror-neurons · Musical onomatopoeia and soundpainting · 'Neuronal iconic' · Palimpsest, ‚transtextuality‘, ‚hypertextuality‘ · 'Readymade' vs. handmade · Real tones · Reflection ('Widerspiegelung') · Signifier and signified · Signs: index, icon, symbol · Synaesthesia · Typology and classification · Total art work ("Gesamtkunstwerk")

4.1 Such as and Different from: Approximations, Convergencies—Difference Instead of Identity

Blurring means basically, in the context we are concerned with, Fuzzy Logic and music, as well as realization of the idealistic. But it is likewise fundamental for the relation of the idealistic with reality, that is, the mimesis of reality in a work of art (see Sect. 3.2). Here, in both dimensions the utmost that can be achieved, is similarity. The work of art is like and different from something, that is represented in it. Similarity results in a contradictory-unity of sameness and difference. And, from its process

H. Heister, *Music and Fuzzy Logic*, Studies in Fuzziness and Soft Computing 406,
https://doi.org/10.1007/978-3-662-62907-9_4

of creation or production it is a resultant of the movement in opposite direction for Fuzzy Logic's specific circumstances of blurring and sharpening.

4.1.1 Different from and Such as: Expression Versus Imprint, Misleading Identification of Signifier and Signified

Similarity means for an objective circumstance, a—virtual or even real—object-object-relationship. Criterions and characteristics of this similarity may be very different. The periodic table of elements may serve as a borderline case of what still may be called similarity. Here, proximity and distance, „affinities", are directly quantifiable according to their respective number of electrons, but not sensually perceivable or visible. Other typical problems, like a greater part of criterions are indicated by the Linnaean system of classifying plants and animals, and, later in history, speeded up through the Theory of Evolution. But, on the other hand, and in this context, similarity is eminently important, a circumstance inside the object-subject relations. Here, we may talk about representation, expression, and even imitation or reflection.

A special case is the biologic mimicry: here, a living being resembles another species, or adepts his *appearance* to the environment, just like the proverbial chameleon, yet remains objectively different—the pretended identity is the effect, which is ultimately aimed for, leaf grasshopper as a leaf, the butterfly as a tree's bark etc. Mimicry is, in a strict sense, a three-sided configuration: 1. An imitating animal (rarely a plant), which in the course of the evolutionary development assumed a poisonous animal, because this gives better protection against its predators, so to say, the „after-image"; 2. An imitated animal as model, 3. An animal for whom the model as well as the after-image is a potential prey.[1]

Similarity plays also an important role, in context with the quotation, and especially in matters of plagiarism, the deceptive similarity. By analogy, this applies to the non-deceptive similarity, in the complex stylistic copy and in similar paraphrasing, in the context of the form of quotation, which may be meant critically or honorably.[2]

Similarity, instead of equality or even identity, is generally determining and important for image, re-presentation, reflection—of reality in consciousness in the object-subject-relationship, but also inside the matter itself, the object-object-relationship. An identity of image, and what is represented in it, is in the best way achieved by the more or less ‚automatic' print[3] as a mechanic symmetrical reflection, whether as a fossil in slate, or as a record disc from the master negative (See above 3.5.2). But even then, differences of blurrings, which have material or technical causes occur, like in the various stages of disc's ‚printing' process, or in procedures of casting

[1] For differentiations and complications, see Wickler 1973.

[2] See below in detail 8.2 New-configuration of already existing shapes quotation form; compare also 5.2 Filtering—the search for plagiarism.

[3] In detail and with numerous aspects, see Didi-Hubermann 1999.

and hollow mould casting. Otherwise, in conscious intentional reflection there is a subjectively determined choice of criterions, which depends on the motivation and aim of the comparison—and it is not only arbitrary, but also pertinent.

Now, this has amazingly turned out to be in exact conformity with Fuzzy Logic's basics and approach. Obviously its pre-requistion is not an identity,—which is stated from the outset and maintained—, but it is—only to be achieved asymptotically—the procedure's aim of sharpening and 'precisiation'. Because, as logic of non-identity Fuzzy Logic requires, as outlined, principally as well the difference between idea and realization, of what was thought and what is done, resulting, in what was aimed at, and what is finally achieved.

Quite often, image and reality, art and life get with each other identified and therefore mistaken. While in Paris, a Nazi official of the occupation authorities questioned Picasso faced with *Guernica*: „Have you done it?", whereupon Picasso gave him the answer: „No, you've done it." Picasso's brilliant bon mot plays with the ambiguity between cause and effect, art and life. For marking a point, it mistakes artistic production of an image, with the non-artistic production of what appears in this image, to be its real material prototype. René Magritte's picture-word joke, emphasized the difference of art and reality, strictly the other way round. The writing, inserted in a picture of a tobacco pipe, explains „C'est ne pas un pipe" (It is not a pipe), as it is only showing the picture of a pipe.

There can be no homogenousness, or even identity of image and of what is represented in it, for ontological, as well as for gnoseological reasons. Because reflection generally, and likewise the particular type of reflection in the arts, imply a difference between sign and signified.

A paradoxical, or even preposterous persisting on identity, instead of allowing differences, is a map on a scale of 1:1. The concept was invented by Lewis Caroll (with Harry Furniss) in *Sylvie and Bruno concluded* (New York 1894). A dialogue in the chapter *The man in the moon* (p. 169):

> "What do you consider the largest map that would be really useful?"About six inches to the mile."Only six inches", exclaimed Mein Herr. "We very soon got to six yards to the mile. Then we tried a hundred yards to the mile. And then came the grandest idea of all! We actually made a map of the country, on the scale of a mile to the mile!" "Have you used it much?" I enquired. "It has never been spread out, yet," said Mein Herr: "the farmers objected: they said it would cover the whole country, and shut out the sunlight! So *we now use the country itself, as its own map*, and I assure you it *does nearly as well*."[4]

In the case of Carroll, a steadily minimized difference of image and what is presented in an image, finally turned out, to be categorially identical with the image, and functionally paradoxical, as design of itself. Jorge Luis Borges continued with the short story "*Del rigor en la ciencia*" (1946, "*On Exactitude in Science*" *or* "*On Rigor in Science*"):

> In that empire, the art of cartography attained such perfection that the map of a single province occupied the entirety of a city, and the map of the empire the entirety of a province. In time,

[4]Text quoted from https://ia802609.us.archive.org/6/items/conclsylviebruno00carrrich/conclsylviebruno00carrrich.pdf, retrieved 3.3.2019. Emphasis by H.-W.H.

> those unconscionable maps no longer satisfied, and the cartographers guilds struck a Map of the Empire whose size was that of the empire. The following generations, who were not so fond of the study of cartography as their forebears had been, saw that that that vast map was useless.[5]

Borges, at variance with Carroll, clings to the categorial difference of image, and of what represents an image. Thereby, he takes only the formal and quantitative difference in account. Such a category of thought may be found in a number of art concepts, for instance, in the context of ready made.

The artistic idea of Carroll, being an author and mathematician,[6] has something to tell about reflection of reality, in general, about the relation between signifier to signified, and particularly about scientific formation of model. The misleading identification of image and what is presented of it, is doublesided. Seen from one or the other perspective, it concerns different matters: „The map is not the territory" and „the word is not the thing" (Alfred Korzybski), or, more wittily „The menu is not the meal"[7] (Alfred Alan Watts).

In the area of the arts, the problem is always of some relevance, when mixing, mistaking or identifying art and life. There is a common observation, that actors get mistaken for their parts. But even the reverse may happen. In art, both of them can be *performed*. Kurt Weill demonstrated in his one-act play *The Protagonist*, after Georg Kaiser, how the named protagonist fuses maniacally theatre with reality. In a courtyard, a company of actors rehearsed a mimed comedy. After the rehearsal, „confesses the sister of the protagonist that she has a lover, and the Protagonist insists on meeting him, not convinced she is telling the truth." Then, the miming company's director demands the performance of a tragedy instead of the comedy.

> In this version, the Protagonist plays a jealous cuckold. The rehearsal nears its climax, and he prepares to enact the killing of his wife. At that moment the Sister enters to tell him that her lover is here. In a fit of madness and rage, only partly induced by his role, he kills her. As the other actors call for help, the Protagonist recovers himself, only to ask the Duke's servant to protect him from arrest until after that night's performance, so the Duke can admire his masterly simulation of madness.[8]

The actor, who committed the murder, wants to give a great performance. His part will be „most successful", there shall be „no difference between real and acted insanity". In their barbaric aestheticism, Weill and Kaiser critically present such a crossing of

[5] See https://de.wikipedia.org/wiki/Karte_im_Ma%C3%9Fstab_1:1, 19.12.16, retrieved 7.8.18.

[6] See e.g. Seneta 1984 and 1993: Of "a few key problems and their solutions from the 13 probability problems" "some are badly posed with imaginative but incorrect solutions; other are difficult, interesting and with correct solution." See e.g. Seneta 1984—Further see: Vakhania 2009, http://science.org.ge/old/moambe/3-3/Vakhania.pdf; with relation to Carroll's "Wonderland" sub "Scienceland".

[7] See e.g. Bonini's paradox, simplified: „simpler maps are less accurate though more useful representations of the territory." https://en.wikipedia.org/wiki/Map%E2%80%93territory_relation, 1 July 2018, retrieved 7.8.2018. More generally Korzybski [1933], e.g. Book I: *A General Survey of Non-aristotelian Factors,* Book II: *A General Introduction to Non-Aristotelian Systems and General Semantics*; against exaggerations and with connections to a central problem of Fuzzy Logic e.g. Wit 2012.

[8] https://www.kwf.org/pages/ww-der-protagonist.html, retrieved 8.3.2019.

the seeming and the real, of deception and self-deception. In a well known episode, from *Don Quixote* by Cervantes: Maese Pedro, a puppeteer, performs *Historia de la libertad de Melisendra* (*The Tale of the rescue of Melisendra*), a romance of chivalry, from the circle of legends featuring Orlando/Roland (II. Book, chapter 26). As Don Quixote suffers basically from the problem to devide between his imagination and reality, he identifies himself totally with the performed acting. He mistakes the puppets of the story, for heroes in need of his help. Thinking to secure the escape of Melisendra and her husband from the pursuing Moors, he intervenes, to destroy the enemies—and wrecks the puppets and the scene. The identification of the map with the territory, as of art with life, can lead to catastrophic results. .

Looking at it structurally, there is always a confusion of similarity with identity. In face of such a misleading identity, similarity offers the mentioned compromise between identity and difference: The first 'something' is like and at the same time different than the second 'something'. Here, the second 'something' is primarily the reality, respectively concrete by extracts. In the following, the first 'something' is not solely, yet preferably, the 'work' in its various forms of existence and motifs or themes within a work. Therefore similarity is a leitmotif-category for the relation of various forms of existence to one another. To make matter a bit more complicated, in toto, there are four dimensions of similarity in music process: Relations with reality, relations between a work's components, relations with other works' components, and relations between its forms of existence (see Sect. 4.3). In the end, similarity itself is once again dialectically relative, stacked and boxed, according to spheres of reality and particular circumstances, which put in relation, comparatively to each other.

> About a real object, or an event, even mutually contradictory determinations can be said. A real thing may be at the same time part of the idea of similarity, as well as of the idea of *dissimilarity*. (A dog is similar to a wolfe, but dissimilar to a tiger. A dog is similar to a tiger, but dissimilar to a stone etc.)[9]

In reflection, for instance, H. H. Holz, a philosopher, comprehends still more widely and differentiated, the dialectics of identity and non-identity.

> Something remains itself, while at the same time, it becomes something different, it is identical in itself, and at the same time non-identical. We experience it in our life. The bureaucratic expression is, the identity card; it guaranties, that I am the same, but my portray needs to be renewed after a couple of years, in order to be assigned to me.
>
> With the question of identity, the dialectic starts. The unity of entity and development must be perceived.
>
> (Heraclitus). Beyond every border expands a new immeasurable reality.[10]

This points towards a—polymodal—‚Philosophy of Fuzzy Logic' (see Sects. 1.3 and Chapter 9). This is not the matter of blurred or even contradictory statements, it is rather the

[9]Plato, as reported by Peter Möller, Berlin. http://www.philolex.de/platon.htm (call 13.1.14).

[10]Holz, H. H.:*Einheit und Widerspruch. Über Theorie und Erscheinung der Dialektik*. Vorabdruck. In:*junge welt*, Berlin, 18 January 2017, p. 12f.; quotation p. 12. In detail see: Holz 2017.

> contradiction in the matter itself, which is covered under the seemingly logical identity of the concept. Hegel's example of ,master and servant' explained it. A master can only be master, if there is a servant, whose master he is. [...] By comprehending this concept, the circumstance will be perceived as being contradictory in itself and the form of its abolition imaginable.[11]

The variation is an interplay of identity and difference, of sameness and otherness (see Chapter 8). Their limits are formed, in each case, by recurrence of sameness, on the one side, while, on the other side, a wholly otherness, novelty, as decided dissimilarity. It is an universal of reality; for instance, its central role in evolution is known, whereby essentially accidental mutations produce variations, as substance of selection. Variation is likewise one of art's universals, both in its historical, so to say, phylogenetic development, as also in the ontogenetic development of a single work of art. Variation has also two sides, as universal of music, and in the special case of motif-thematic work in modern European art music. One the side, is its production by composers and also by improvisors. On the other side, such a genetic process is furthermore complemented by a receptive and especially analytical one: Herein, problems, which exist in a ,work', how it actually proceeds, for instance, in case of motific similarities, are detected and re-constructed. The intercession is the sound form: Thereby, facts, fixed in writing, get re-produced. Knowledge and comprehension/understanding of similarity-relations improve the realization-quality, for example, through appropriate articulation and phrasing; but, on a lower level, it can be achieved without some deeper understanding. In analogy, this is valid for awareness and recalling/visualization of the receptive realization.

4.1.2 Comparing—Production and Determination of Similarities: Concrete and Abstract Analogy, Typology, and Classification

One of the great lacunae, a gap to be bridged by common efforts, and of common relevance for different spheres of reality as of disciplinary matters, is, what I would call a '*general theory of similarity*'. Of course, similarities of styles (by the way, in music,[12] as in literature and other arts) are important in different fields and methods of research and intensively done.[13] Here, Fuzzy Logic would be in its primal element. But (as far as I see,) there is no very clear distinction, differentiation, hierarchical and reflected system of terms (and meanings), systematisation in general/special, relations and interrelations etc. I think of categories as reflection, representation, image, isomorphy, homomorphy, analogy, mirroring etc. Here, on the whole, all

---

[11] Ibidem.

[12] See e.g. Dubnov et al. 2002.

[13] For searching stylistic similarities and patterns in arts and architecture, cf. e.g. Sondereguer 2010; for filtering similarities of words, in different dimensions („Wortähnlichkeiten"), from a corpus for scopes of psychological-linguistical research, e.g. Heister/Polk 2010, p. 13.

this can not be discussed extensively. Simply, a few remarks about analogy, as an essential procedure, may suffice.[14] Analogy takes a double position: It is both subject and procedure of examinations. On the one hand, it is a method of reflection, an important path to it, which results in objective similarities; on the other hand, it is a method for knowledge, in search for objectively existing similarities, and indeed, in reality as in its reflecting. It

> is in philosophy and numerous particular sciences a common notion to design a certain result, or procedure, of the activity of knowledge. In distinction to equivalence (conformity in a single respect) and equality (conformity in every respect), analogy is the result of a comparison, on the basis of which may be concluded, between features and conditions of objects or systems, the relation of similarity, a ‚correct' proportion/relation or correspondence/equivalent. Analogy is a *similarity of different degrees,* in connection to certain aspects and characteristics between different objects or systems, which can turn into *identity in a borderline case* […].[15]

Then ‚similarity' may become quite abstract. Thus it designates not only

> any knowledge by comparison and findings of similarity, but in a relational-theoretically way – in contrast to standards of art and generic – the unit of relation, or conformity of relation […], of mutually structural or functional similar features/characteristics or conditions, and thus, it means correspondence of structural and functional properties of varying entities. This in mind, even the relation between a model and an object or a system, which is isomorphic (being of the same shape) to it, may be designated as similarity.[16]

Just in music, such an analogy is often relevant and made use of. Indeed, active parallelisms between different spheres of reality, together with parallels between various departments of science, also adhere to analogy. Here, for instance, the tendency to describe neuronal processes of perception and memory development, according to the model of evolution with mutation, selection and „survival of the fittest" is of relevance.[17] Perhaps, the comparison of the solar system, with the atomic model of Bohr, as analogy of macrocosm and microcosm, is a formal, spatial-optical analogy. The parallelism of operations with computers and the brain is a functional analogy.[18] In a tendency of the history of art, both types are blended: Thus, especially since 1905, one can detect in all arts an increased ambition to dissect the material of the arts in always tinier, elementary units—tones into microtones, the words into sounds etc. Obviously, the model of ‚nuclear fission', the dismantlement of the „indivisible" atom, helped to bring this into being. Then, in Fuzzy Logic's perspective, in each case, the degree of blurredness, or sharpening, of such analogies needs to

---

[14]Analogy, in opposition to ‚homology', refers in biology to adaptive similarities of *shape* and equality of functions of organs, which are of different evolutionary origin, like for instance, gills and lungs, and the wings of insects and bats.

[15]Sandkühler 1990, p. 101f.

[16]Sandkühler 1990, p. 102. For Kant (*Prolegomena*,1783), such a kind of analogy is not an „incomplete similarity of two objects, but a totally completed similarity of two relations between totally dissimilar objects" (quoted after ibidem, p. 104).

[17]However, the application of natural processes, directly on the development of life in society, by social Darwinists is questionable.

[18]For utilization in the context of Fuzzy Logic, compare Kacprzyk et al. 2017.

be explained. For the origin and development of verbal language, analogies and metaphors, which are based on comparisons and similarities, play an essential role. Metaphorical expressions can start from a human body anthropomorphically and serve in pictorial-iconic transference as designations of landscape elements. Above all, they serve under the principle of pars pro toto, and through metonymies of developing abstract expressions from concrete words—such as, for instance, the word ‚crasp(ing)‘, which derives from manual intervention. In modern electro-acoustic music, the use of metaphoric procedures takes a peculiar turn.

> Electronica often propagates metaphors linking different types of sounds with larger concepts. For instance, the sounds of synthesizers, especially vintage instruments of the 1960s and 1970s, frequently evoke science fiction and futurism.[19]

The retrieval on the past is then associated with a formerly and presently future-to-be. In another variant, noises from former media, in this case the record disc, serve as some kind of music‘s ennoblement, due to its historic dignity. Yet, mostly the former turns out to be merely similar, as the reproduction is a construction, and identity an illusion.

> Phonographic noise such as the pops and scratches of a record player's needle can serve as a metaphor for the age of the underlying music, even if those noises were artificially constructed rather than innate to the material.[20]

The significance of analogies in language, and thus also for cogitation is, for instance, emphasized by Deutscher, as well as by Mary LeCron Foster:

> „Language is an *analogical* system for classification on multiple levels. Language systems build upon semantic analogies and analogies in phonological, morphological, and syntactic distributions (positional analogies). New meanings are created through the process of *metaphorical extension*. The direction of language change is determined in large part by this process and by *analogical systematization* – hierarchical congruence of classes."[21]

Starting spheres, which are chosen for the formation of metaphorical expressions, both the emotive and also the cognitive semantic groups, are tinged by words. In an array of experiments, the synaesthetic associations, together with their references on two different separate languages, was probed. The arrangement for the test

> presented Dutch- and Farsi-speaking people — who use different musical pitch metaphors — with musical tones of different pitches. The tones were accompanied by images on a computer screen: horizontal lines that varied in how high or low they were placed on the screen, or vertical lines in the middle of the screen that varied in thickness. After each trial, the participants were asked to sing the tone they heard. Farsi speakers, who talk about *pitch in terms of "thin" and "thick,"* were affected by the relative thickness of the lines. Thicker lines made them sing slightly more flat, and thinner lines made them to sing slightly more sharp. The researchers found that the association between the metaphors and musical pitches holds even if the participants are prevented from using the words themselves. Dutch participants could be trained to use the Farsi metaphor, and they produced pitches as if they

---

[19]Demers 2010, p. 171.

[20]Ibid.

[21]http://www.massey.ac.nz/~alock/hbook/foster.htm, retrieved 01 Febr. 2019. Emphasis H.-W.H.

> had been using the thick/thin metaphor all their lives. But […] *the metaphor couldn't be reversed*. Dutch participants who were taught to associate high-pitched sounds with "thick" and low-pitched sounds with "thin" were unaffected by thick and thin lines presented on the screen.[22]

Metaphors for musical pitch vary, but the basic principles are similar. The decisive conveyance is the intermodal perception, where one sense may function as substitute for the others.

The authors of the study conclude:

> that the low = thick and high = thin associations *might be innate* in humans, *or could be learned* from physical experiences very early in life: Another study revealed that even before children can talk they respond to associations between pitch and height and between pitch and thickness.[23]

Except for the fact, that the meaning of „innate" and early „learned", also nature or nurture, are approximately opposites, their interpretation is hardly conclusive. Most likely, the associations are probably specific to culture and language. But, what seems to be correct, is the relational language-comprehensive significance of analogy. In this respect, synaesthetic associations, like thick and low, thin and high, which refer to gravitation, may assume to be an universal by tendency. In Greek, „low-high" has the meaning of ‚heavy-light'—probably in line with accent patterns and dancing steps—whereas in Latin, ‚gravis' and ‚acutus', means ‚heavy and pointed'. Thus, analogies are twice taken from the sphere of vestibular and kinesthetic, as well as the haptic sensory perception, in contrast to the spatially-optical like in modern European music, where analogies are created through the musical notation: The „low" tones are positioned on notation lines „below", while „high" tones are placed „above". The designation „pitch", is in line with this paradigm. It is even strengthened: In notation lines, the violin clef is in charge of higher pitch levels/register, while the bass clef is for the lower ones. Exactly in between, c^1 is situated, which again is placed in the middle of the piano keyboard. From respectively analogous associations, hardly any producing and reproducing performer could escape. As an example, chosen from a multitude, states: „Lower orchestral instruments, plus the organ pedal at 16′ and 32′ pitch, suggest the subterranean nature of the catacombs"[24] in Ottorino Resphigis *I pini presso una catacomba* (*The Pines Near a Catacomb*), 2nd movement of *I Pini di Roma.* Although, in a religious context, heaven is more important, there is nowadays a tendency for esteeming high-pitched tones as celestial sounds. Fuzzy Logics could be used to prove, where, in adaption to the mode of notation, or by deviating from it, the transitional spheres of pitch-levels for heaven, hell and earth, might be subjectively placed.

---

[22] Dolscheid et al. 2013.

[23] Ibidem.

[24] https://en.wikipedia.org/wiki/Pines_of_Rome, 8.3.19, retrieved 14.3.19.

Lastly, what is here of relevance, the—various—procedures of classification and typology, as they are important procedures for determination and analysis of similarities. Aside from the linearly-unidimensional, there are numerous complex, hierarchical and encaptic-boxed kinds, often depicted as tree (trunk) structures, up to more complex, but once again, not any longer ‚hierarchical' nets.

> The designations 'typology' and 'classification' [...] are clearly different from each other. A typology places types in a mass of individuals, these are masses of shaping principles, whose concentration in some of these – called typical – individuals is exceptionally shaped, while it is not so dominant in others. The first lot constitutes a type; while the latter belongs only marginally to it, and in some respects, it even belongs to another type. Thus, there are individuals, who belong to more than one type; and possibly, some individuals [...] will not enter the typology.[25]

A (complete) classification is based on a number of criterions, which may either be true or do not apply for each element of the classified mass. Therefore, it includes all elements of a mass, and orders each of them exactly to one class. Hereby, no element has a particular status.[26]

A well-known example is the Linnaean binomial biological taxonomy; sometimes it has to be enlarged by a third term, for instance: homo sapiens sapiens, to differentiate the Cro-Magnon- or modern man, from homo sapiens neanderthalensis. The latter is a classification, which is not arbitrarily or contingent, but it derives „from the nature of the being" and serves for „its further finding".[27] Occasionally, there is no interest for any classification, to delimit elements clearly from each other by individualization and special attributes. Then, a kind of vague, unsharpened or intentional blurred totality is sufficient.

> In Crete, in spite of abundant religious pictorial themes at the time of the new palace, it seems that, in the Aegean early epoch, no standardized, consistent iconography of single deities was ever formed, or at least has not in the same way succeeded [... and], that again, in the in era of the Mycenaean palace, no efforts can be observed, which – according to the testimony of Linear B-Texts – would have clearly defined an abundance of numerous gods and goddesses in iconography, [... so that] the contradictions and ambivalencies in the iconography of Minoan and Mycenaean images of deities /idols [...] suggest the conclusion of a 'Pantheon without attributes'.[28]

Especially here, the reason could be in the nature of power-technique.

> In face of the omnipresent religious iconography, in neo-palatial Crete of 17.-16. centuries BC, one might well imagine, that in the palace of Knossos, the political 'elite', confronted, with manipulative religious-political strategies and for socio-political binding and integration, the existing regional diversity, by initiating a new, systematized religion for the creation of a 'new Minoan world'.[29]

---

[25] A standard example is "bird-likeness". From the human point of view, in particular, the robin—is a «prototype» (Eleanor Rosch)—considerably more obvious, than a penguin, who is unfit to fly, although both are classificatorily birds. Subsequent graduations are again standard opportunities for the application of Fuzzy Logic; see e.g. Drösser 1994, p. 16f.

[26] On this, extensively, see Chr. Lehmann: *Klassifikation*, undated.

[27] So Chr. Lehmann: *Klassifikation*, undated.

[28] Blakolmer 2010, n.p.

[29] Ibidem.

For such innovations, the blurring of traditional arrangements is quite appropriate. Then they were filtered, and on such a manner being sharpened, were ‚overwritten' by some new classifications, in conformity with existing conditions. In the arts, these facts are also structurally-methodically of relevance for the development of new types, by a fraying out, or mixing of old types, quite often, in line with the development of new materials and new techniques, like, for instance, in case of the „invention" of „symphonic poetry" by Ferenc Liszt.

Fuzzy Logic, on the one hand, is apt for awareness of similarities and differences, while on the other hand, for the quantification of degrees of similarity *and* similarities—a second step, from the point of view of music-analysis. Fuzzy Logic analyses and measures, above all, not types and kinds, but degrees of similarity. Again, it is the quantitative dimension that matters.[30]

The potentially vast area of cases of similarity is limited by the question of their relevance with Fuzzy Logic, respectively, and inversely, according to how and where Fuzzy Logic, here, for the music-analysis, has any relevance. The exemplary cases, which are treated here, show *kinds* of similarities and different expressions. But, there are no complete systematics, just a few exemplary model cases. They form the material and the objects of research, for which, with the help of Fuzzy Logic, the *degrees* of similarity can be researched, in an exact and quantitative manner.

4.1.3 From Known to Unknown—Heuristic, Mirror-Neurons, and Empathy

An essential factor of heuristics is comparison, just like the search for analogies.[31] And, the „comparative method" was raised to—relatively—autonomous branches in philology, study of literature, ethnology, and in some other scientific departments, respectively. For its part, heuristics, already in its method of research, takes uncertainty and the ‚blurredness' of circumstances, into consideration, and it achieves often results, among others, via mental shortcuts, not only faster, but also new ones, which were not previously included in a question and a procedure. By Nicolaus Cusanus (1404–1464), a particular epistemologic-methodologic variant of similarity, which includes elements of heuristics, was developed: „Any research about something unknown is based on similarities and differences, in relation to what is known"; and thus these must be positioned to one another „in comparativa proportione", that is comparable, or comparative proportion.[32] Already, a quantification and mathematical formalization were involved.

---

[30]Possibly, it could be more deeply clarified, to what extent the methods of Fuzzy Logic might deal with such questions better than conventional statistical methods. Eventually, this has to be dealt with by Fuzzy Logic itself.

[31]In this context, analogy describes „a *cognitive procedure*, which draws analogous conclusions as a methodological aid for discovering the unknown from the known." Sandkühler 1990, p. 102.

[32]As reported by Kanitscheider 2002, p. 96.

In various areas and in different stages of life, people use an analogous heuristics, starting from things familiar, which they set in relation with something unfamiliar. To relate anything what is novel to already acquainted patterns, marks already in early childhood the continuous classification of faces in familiar-unfamiliar—as the relation similar-dissimilar of patterns stored in one's memory, at the same time, one reflects well-known/nearer and further.[33] For the recognition of similarities and learning, such a recognition is essential, even necessary for survival, and it is genetically-anthropologically embodied. Instead of a time-consuming statistical filtering out of a mass of senses-data, there exist shortcuts: Namely, the selection of significant patterns. Humans start early to recognize patterns, even as intrauterine learning: In case of patterns of excitement, neither identities nor something totally new, may be used to search for similarities, even if they are absolutely vague, because both will not lead to learning.[34]

Such a kind of filtering, on the basis of a guiding line of already familiar patterns as models, sets at the same time a valuation. In case of experiences in art, it works not only in a positive way: The known is higher valued than the unknown, things familiar appear more beautiful, up to downright anxiety of anything new, by which new aesthetic experiences are usually blocked.

It has to be noted, that here, it concerns altogether phenomena, which affect the surface or the outside. Actually, and also finally, this is a feature of aesthetic perception. Phenomena *are* not essence. But normally, they reflect something of it. (Even here, too, relations are involved, which may be understood as similarities.) For instance, in a split second, somebody recognizes, whether a face signals a friendly mood or not. This can be performed in pantomine or pictorially in arts, and even in music, as analog motional contours of emotions. Because, such phenomena are closely connected with affects. And these, again, also possess a cognitive component themselves. About all these similarities, not least empathy for others takes place, empathy for what is fundamentally similar for humans, but even for others, like strangers. These circumstances are reflected in art as well. Fritz Breithaupt's *Kulturen der Empathie* (Cultures of Empathy) (2009)

> combines the research of mirror neurons with the literary analysis of narrative. Breithaupt provides the toolkit for looking at literature and its relation to empathy with the aid of neurosciences. Beginning with a fundamental assessment of empathy as an evolutionary pattern, that in its simplest form implies a narrative character through the *necessity of the predictability* of future actions. In a fashion similar to Costello's statement on the unlimited potential of the sympathetic imagination, Breithaupt states: […] ,We seem to possess the ability to empathize ourselves into anything without limitations, in order to relate it to familiar schemes and thereby erect imaginary bridges between us and others, without having to think the actually dissimilar as dissimilar (Breithaupt 2009, p. 20; translation Hilmar Heister).

[33]Extensively, for instance, in Kandel 2012. Both as a result of the common, evolutionarily embodied inclination to perceive relations of similarity, and also particularly, as a result of the importance of faces, we often see faces as projections in clouds, being a central phenomenon of ,shapes' as such, or also some other features, mainly animals or plants, thus something vivid.

[34]Hüther 2011, 92f. Compare the dialectics of new information, redundancy, and entropy in information theory. By the way, already here, the configuration ,Innovation through Non-Identity' turns up.—Extensively treated in Chapters 7 and 9.

> Breithaupt considers how easy it is to see similarities and assumes that the production of dissimilarity might strongly contribute to the process of empathy. [...] A more complex cognitive empathy [...] relies on the possibility of insight into the motivations and intentions of the parties, whereby they become the ‚soundbox of experience‘ (*Resonanzkörper des Erlebens*; Breithaupt 2009, p. 145f., 170). Breithaupt formulates a *narrative theory of empathy* and assumes a dualistic setting based on the three-fold structure of two opposing parties and an observer. The setting must provide the observer with the opportunity to engage with the protagonist and his actions by implicating him in the outlook and predictability of the future course of events. However, in a second step the observer must be able to beware himself from a loss of self in this process, wherefore the empathetic process necessarily includes blockades of identification [...], allowing the observer to stay at a safe distance – otherwise he would risk losing himself in the process of identification."[35]

The realization of similarity also means to recognize oneself in others, and thus to accept the equality of the whole mankind as humans, irrespective of any individual distinction.

4.2 Art-Specific Iconicity and Materiality of Reflection, Mimesis, Poiesis, Aisthesis, and Catharsis—Representation, Production, Perception, and Achievement

Thus, similarity is the basic precondition for reflection in general, as well as for artistic reflection of reality in particular. At the same time, similarity, under the sign of Fuzzy Logic, differentiates itself, according to tendency in an infinity of graduations and shadings between equal and unequal. Art, together with science and religion, is one of the basic forms of reflection of reality. ‚Mimesis‘ is suitable as the differentiating expression for the art-specific type of representing reality.

4.2.1 Art ≠ Science. The Artistic Reflection: Mimesis

'Mimesis' is specific for arts, one of the four essential and basic principles of art (with poiesis, aisthesis and catharsis). Art is mirroring, but as a special 'mirror' it is *as* reality and *not as* reality: It is similar. But perhaps, we should emphasize: During the interplay between ‚such as‘ and ‚not as‘, being different may possess a trifle predominance. Art is transforming the world. 'Mimesis' is not meant as eidetic or pictorial. It is totally independent from any stylistic or idiomatic options, be it naturalistic or „abstract“-nonfigurative. It may also be a matter of structural

[35]Hilmar Heister 2015, p. 38f.—In a psycho-analytical setting this corresponds to the analyst‘s problem to admit and contain the transferences from the analysand, while he admits and also reflects his own transferences, and for all empathy takes care, not to be inundated by respective affects, but to keep a specific coordinated balancing detachment.

relations, of isomorphs, of metaphoric or synaesthetic transferences and analogies, for instance, fierce melodic movements, which reproduce the storm of the outer nature, or as stormy passions of the inner feelings.

Then, under the sign of similarity, reflection achieves, in relation to its respective level of development, in each instance, objective, capturing reality, relative truths. Unfortunately, in each instance we obtain only them, but at least something. Because, in fact, it is an illusion, to expect that „absolute truth" could really be achieved completely, and in a completed form. Only, as an aim in perspective, could this be possible. Under the precondition of the relations of illustrations or reflections this statics are dissolved in a progressive process. Nevertheless, we obtain this kind of truth, however, in an approximate, asymptotical, endless process of—progressive—approach. Under the aspect of Fuzzy Logic, this is a progressing process of interaction between blurrings and sharpening as well. We might also state: Relative truth will gradually resemble, more and more, an absolute truth. But, never will it become the same nor identical. This also marks a partial moment of Fuzzy Logic's way of thinking; it is in conformity with a dialectic epistemology.

Firstly, different from sciences, in arts there are no epochs overlapping with all their turning points and even retrogressive, yet continuous, processes of progressing cognition. Truth, as guiding principle for science, serves also as an important aim for the arts. Still, the essential principle is beauty. Arts, or their expression by works possess, in each case, an epochal- and cultural-specific intrinsic value. Such an intrinsic value may even be epoch-outlasting; for this, a prominent example is the art of Greek antiquity, as it is still today relatively directly important and effective. (While on the other hand, nowadays, Greek sciences are predominantly of historical importance.) Arts are not exhausted to be a relativated and outdated station in an forever renewed endless progress. Thereby, the epistemological character of arts shall not be denied (see Chapter 9). Especially, if it concerns naturalistic art in a wider sense, there are admittedly certain analogies, namely, tendencies for an increased precision and sharpening. In this respect, here, mimesis resembles both ordinary and scientific-epistemological reflections. But here, too, there is no lineary further development. Since millennia, late-paleolithic cave paintings, or portraits in late antiquity, represent a similar standard of „imitating nature". The development, in this direction, went even further: Then, with photography, a qualitative leap in „similarity of nature" occured. There are analogies—as a lagging behind and catching up development—in music. Above all, with Richard Strauss the technical means of naturalistic visual art were attained. But, the naturalistic trend marks only one side. Its contrast is not just a ‚principle of dissimilarity'. But abstraction from the sensual surface, recourse to structural analogies and stylizations are likewise always present and effective throughout history.

Secondly, different from sciences, in the arts mimesis possesses a concrete, sensual materiality and shape of a higher, indispensable intrinsic value.

In fact, artistic ideas, just like ordinary, scientific, and philosophical ideas, also belong to the spiritual sphere. But, right from the start, they are rooted in the sensual-empirical. Firstly, because materiality in general, and the specific material are fundamental for the arts—if the term and the word ‚horse' is written in white or in black, in Arial or in Gothic, does in no way alter the facts, while considerable differences evolve, if in the picture the horses are in black, white or brown, or in blue, as in the painting by Franz Marc, the expressionist. Secondly, arts can never attain the same stage of abstraction like in sciences, and it even should not attempt to. For, it is bound to the sensual-material by the basic *aesthetic* relations of similarity of the art-processes, or art-objects with aspects of reality. However, such similarities are transmitted in a highly complex way, and may, though lastly mimetic, as was mentioned before, be even fully abstract; for instance, The Black Square by Kazimir Malevich, or the squares, rectangles, lozenges and lines, grids, and primary colors by Piet Mondrian, or the musical constructions by Iannis Xenakis, which are based on statistical-stochastic processes.

Affective, aesthetic and cognitive components were explicitly combined in Early Enlightenment‘s musical theory of affects. Musical shapes and motifs, which are formed according to the principle of similarity, operate with various styles and degrees of similarity between music and world, by methods like analogy, synaesthetic association, metaphor, pars pro toto, and much more.

Just in the arts, besides non-intentional blurredness, the intentional, materially-oriented blurring belong, in part, to their characteristics: Components of beauty are metaphors, all kinds of ornaments, anything that is in the best sense superfluous, or „decoration“. Arnold Schönberg‘s dictum, after the fin de siècle, art should not decorate, but has to be true, is not in contradiction to this; for here, it fulfills the social function to „face-lift“ unpleasant real-relations. Such an intrinsic value of sensual material is different, truely in opposition to the „principle of frugality“ in sciences, which after „Ockham‘s razor-blade“ demands to present facts as well as possible, without frills, in brief and simple ways. For, in mathematics, simplicity of proofs, not really serve as a measure for their validity, yet for elegance and beauty.

In the setting of ‚mimesis‘, in whichever way it was and is thought, formulated and discussed in detail, if in part of „theories of reproduction“, „representation“, „copying“ (here, the space for the discussion of such a theme is lacking). Some kind of relation between a work of art, in this case of music, and reality does exist, and similarity in the wider sense, as it is here presented, is deemed to be useful as an expression for such a relation. Mimesis is component of a configuration.

1. Mimesis, a classic expression in Aristotle‘s aesthetics, means not only „copying (of nature)“, but much wider conceived as the specific artistic ‚presentation‘ of reality in particular arts with given specific materials and techniques.
2. Such a work and elaboration of similarity demands for poiesis, specific methods of production, of handling with sensual material, together with a perceptible presentation of the innate characteristics of a presented object for somebody else. May be, ‚metamorphosis‘ is an expression to define the art-specific such as and not as, the conversion of appearance *and* characteristics of reality into imaginary-real genuine, sensually-tangible shapes.
3. With aisthesis, the perception, as well,—whatever may be its particular form of—a relation of similarity between the object, in this case, a work of art, which

necessarily implies its subjective appropriation, and for sharing the given qualities of an object, an utmost sharpened adequacy is advantageous.

4. Finally, the art process results in catharsis, the shock and „purification" from what strains, mentally as well as socially, with a reverberating effect still after the art process. For a respective work of art, there are inappropriate or appropriate effects and reactions. In case of the latter, sympathetic reasoning is required, lastly even a form of empathy as object-subject-relation. It is generally quite difficult for exactly researching the degrees of adequacy, hence also in particular for Fuzzy Logic. The same applies for the forms of existence in the subject without exact exteriorization or only as a verbal expression (see Chapter 5).

4.2.2 Four Dimensions of Similarity in Music Process: Relations with Reality, Relations Between Components of a Work, Between Its Forms of Existence, and to Components of Other Works

Similarity, instead of identity, is the realistic ideal of adequacy in connection with relations, which are external to music, and just as well as to those, which are internal to music. On one side reality—in excerpts, which are represented in a given work of art—serves as point of reference and standard, while on the other side for the work itself. In art's context, one of the important questions for Fuzzy Logic is to trace relationships in the setting of a ‚theory of similarity', whereupon here only to some extent can be referred to. Thus, what matters is, on the one hand, the qualitative kinds and types of relations between particular results of the music process, as well as altogether in the context of the forms of existence of the art process, right to the after-effect in forms of memory and imagination. For Fuzzy Logic this is no immediate affair. But it may give a keen eye, if in case of musical works and components, they are already examined closely for a possible fuzzy-logical quantification. On the other hand, what matters is quantification of analysis and practical treatment of facts, as the degrees of similarity are quantifiably graded, especially in the light of the differenciating Fuzzy Logic.

If this constitutes a practical form of fuzziness, then in the production, the creation of an art work in the scope of concept, idea, sketch and finalized work, the converse process is decisive: The development of concrete, distinct forms from the amorphous, the formless, the evolving of the chaos of the material into the cosmos of artwork (see Chapter 6). In the context of a 'theory of similarity' it's about degrees of relatedness in their quantification, and thereby at least indirectly also qualitative forms of relatedness between individual results of this production process. This, as well as the overall context of forms of existence of the art process up to the resonances in forms of memory and imagination, are in the arts one of the most important duties of Fuzzy Logic.

In the context of music process the question of ‚similarities‘ concerns at least four different dimensions. There is no direct connection between them. Thus, it has to deal with:

1. The relation artwork—reality—reflection of the world in the work;
2. relations of components to one another *within* the respective work, from the level of forms and shaped parts, along motif-theme relations to the level of idiomatics, intonations, colors and similar effects, which are work-specific;[36]
3. relations of works or components of a work with those of other works, either from a given composer or from someone else; (see verbal-, image-, quotation forms)
4. relations of various forms of existence to one another: In each case they reflect the work, or rather the work through various other forms of existence.

The dimensions 2–4 are more directly accessible for methods of Fuzzy Logic, however, as usual, the problems with non-objectivated forms of existence have to be taken into account. They will be discussed in a part of Chapter 3.3, and later extensively in this chapter and 5–8.

Subsequently, dimension 1 is of concern.

As already explained, for ontological as well as for gnoseological reasons, homogenesis, or even identity of image and what is represented in it, is impossible. Because reflection in general, and likewise the particular type of reflection in art, imply a difference between sign and signified. But, exactly, in connection with art, like a chimaera, the setting up of identity steadily reappears in the shape of the previously mentioned equation of „art and life“. Analogously to the map on a scale of 1:1 (see Sect. 3.1.1), Hugo Ball, in 1916, designed his relationship with vibrations and words, at the inauguration of a Dadaists-Café:

> I want my own nonsense, and with added vocals and consonants, which correlate with it. If a vibration has a length of seven ells, then I want justifiably words of a length of seven ells.[37]

Here, not only due to vagueness, the crossing over of identities and differences is complex. On the one hand, Ball minimally concedes for his utterances a character of reflections, being his „own nonsense“. But, on the other hand, elements of words should fit to such nonsense, but in a way, that he defines the length of words, geometric-spatially and of identical length, with the lengths of „vibrations“—he keeps unsettled, whether such „vibrations“ are facts of reality or reflections of this „nonsense“. Certainly, on the whole and in spite of all complexity—which is consciously produced by Ball—lastly, it turns out to be solipsistical nonsense and no proper critical anti-sense of dadaism. However, it is an abstract and unproductive

---

[36] Verdi called a work‘s specific style and ‚colour‘ its „tinta musicale“, like a palette of colours for the work as such. It originates through a net-like combination of various factors, from text to instrumentation and manner of performance. For instance, in *Macbeth* of 1848, Lady Macbeth should sing in belcanto-voice, but in a „rough“, „choked“ voice, in order to express her character as well as her behaviour, along with the prevailing gloomy mood of the whole opera.

[37] *Inauguration-Manifesto*, 1. Dada-Evening Zurich, 14. July 1916, in: Ball 1991, p. 133.

alternative to deny completely any relations of images agnostically or even constructivistically as a countermove to such absurd and arbitrary equivalents, or classifications of images, and of what is represented in it. There are many ways and procedures for the asymptotic approach of the similarity of an image with a perfection—which does not mean identity, but an optimal sharpening.

4.2.3 Analogies, Synaesthesias, Subjects, Sources and Material of Musical Mimesis in Nature and Society

In the setting of the relation ‚work of art-reality', the particular material of a music culture, an epoch, a style, a genre, and a male or female composer for a single work has each time to be chosen, adapted and arranged in a specific manner. In case of material, there exist general procedures and characteristics, which are overlapping times and cultures. These are essential for the creation of music, but in the concrete situation of a work's creation—which may also be an improvised song—each time they are actualized and newly applied. Basically, what for mimesis matters, is similarity of various kinds in a range between indexical, iconic and ‚symbolic' signs. For the process of such relations, a tripartite structure might be useful. What is represented in particular, are objects. They become sources if their musically relevant characteristics are chosen as starting points for mimesis, normally, as pars pro toto, tonal or even motion-like. After all, the material that is used, as presentation on part of music for the mimetic poiesis, the carrier or body of signs. Again, as a rule, they are tonal. Such objects exist in different spheres of reality, in the (exterior as well as in the) internal human nature, moreover, also in the interaction with spoken language, both for the musical dimension of chronological order, rhythm, and for the sound, as noises and tones, with timbres and volumes. Noises are found in a large range among thunderclaps, bursting of toppling trees and sounds of animals. Sources for tones with unfaltering pitch levels are howling and whistling stormy winds, the buzzing of some insect species, the melodiously croaking red-belly toad, howling wolves, trumpeting elephants, whistling marmots, mooing cows, bleating sheep, and above all, naturally, the songbirds „singing".[38] Reality offers optical and acoustic models for the rhythm.[39] Acoustics are preferably of relevance for the percussive, the sound effects, while the optical for the dance, which is by its rhythm as well as by gestures and miming intimately connected with music. Rhythm is mimesis of interior and exterior nature.

1. Models of human rhythmic are without any explicit acoustic dimension, walking, running, jumping, and the ceremonial walking.

[38] See Sect. 3.2.4.

[39] More extensively, above all, see Knepler 1977.

2. In chronologic spatial distance and only indirectly perceivable biological rhythms, for instance, the circadian (sleep-wake cycle), and the toneless model; added to this are cyclical, periodical major rhythms in the exterior nature.
3. However, acoustically relevant models are events in nature, like sea waves, or generally wind and waves, and plenty of such kinds of noise.
4. Likewise, acoustic models are processes of the ‚inner‘ human nature, above all heartbeat and breathing. They are physiological events in the body itself, which are acoustically noisy, and in their chronological routine periodic-rhythmical. Thus, already a prenatal disposition comes into being.[40] Just as in the first months after one‘s birth, here, listening takes definitely priority over sight—for inside the uterus hardly anything can be seen, but something can be heard; other senses, in particular, a sense of taste, or, more significant for music, a sense for balance, are likewise already in a rudimentary form involved.
5. Another rhythmical model, extra-uterine and a speeding-up of the physiological normal condition, is the coitus, noisily-percussive by way of body movement; pre- or proto-musical vocal utterances are merely accompanying it, mainly in form of interjections, but if necessary also with words.[41]
6. Finally, the sphere of work with its numerous, mainly percussive processes supplies models of rhythmic, starting from hammering with pebble stones of early humans, then sledge hammers in big industry to the in these days ubiquitous, wide-ranging traffic noises.[42] Thus, by aesthetically relevant elegance productive efficiency is enhanced, for which regularity and rhythmicality are means and standards.

4.3 Nature and Arts, Mimesis and Poiesis. Real Tones as Readymade and Handmade, Musical Onomatopoeia and Painting

The use of signs is indispensable for communication in general, and arts in particular. They refer in various ways to the signifier, with similar, or dissimilar to the signified. The following sections concern the relation sign/artwork-reality in various dimensions and facets. There are several stages and developmental lines within the relations of signs, carrier of signs (material of signs) and what is signified, in regard

[40]For a pioneering work, see Tomatis 1990.

[41]Well-known examples are the prelude to the I. Act of *The Knight of the Rose* (*Rosenkavalier*) by Richard Strauss or Ravel‘s *Bolero*. Something more recent: “A collaboration album between Breakcore artists Venetian Snare and Hecate entitled *Nymphomatriarch* was composed entirely from sounds of the two performing various sexual activities together, which were distorted and time-stretched, to resemble typical Breakcore samples” (https://en.wikipedia.org/wiki/Biomusic 12.12.2013, retrieved 15.3.19).

[42]In the history of mankind, noises at work are a constant stimulation. Since the 1910s, futurism recorded the noises of workplaces, traffic and war in the industrial world directly, and integrated such noises within music—being up to the present a continuous developmental line.

to the relationship of tension between mimesis and poiesis. The following eight lines are part of a larger mass, but as examples they must be sufficient. They are intricate and hard to disentangle. Although I bear their historical order in mind, yet primarily, they are set in a systematic order. Vagueness between components and efforts to minimize those in an artistic way and to sharpen their similarity. Thus, everywhere there are attempts for the use of Fuzzy Logic too, even if one encounters difficulties in case of qualitative problems.

4.3.1 Things as Signs

In communication, the most obvious thing is the simple presentation, or show-piece: The thing, the object—which also could be a human being—itself is an optic-acoustical presentation—the yelping dog, whose yelping is shown and thereby according to tendency shall be pictured (not only itself). If it will be shown and introduced to the recipient of communication, then possibly it will be incited to start yelping again. In some ways, this is the oldest, already pre-verbal layer of communication. Yet, such a presentation of an object is still neither a re-presentation, nor a reflection. And, instead of similarity, it is about identity: The dog is the dog.

Thing-signs are an early form of communication by means of signs as an elementary form. Somewhat different than the map on a scale of 1:1, which cannot be a sign anymore, the thing-sign is certainly a genuine sign. Because, it is not identical with the signified, and, it also does not pretend, but normally, it refers to, pars pro toto, as metonym. Thing-signs are with regard to the similarity of sign and what is signified with their partial identity of sign and what is signified, seemingly more precise, closer outlined, more unambiguous than iconic-visual or even abstract, symbolic. This also includes verbal signs, which are hardly similar with what is signified. (Thing-signs can be understood as a special variety of indexical signs, namely, causally dependent indications from what is signified, like smoke for fire.) However, things talk rather inarticulately, especially, in view of an objectified meaning and subjective adopted sense.

In the myth of Pyramus and Thisbe (as told in Ovid's *Metamorphoses*), a piece of clothing is mentioned, Thisbe's blood-stained veil as a thing-sign for her death. It is a relic, which she left-behind, when she fled from a lion. Yet Pyramus take it as pars pro toto, deductively as being what of her remained and as a sign for her death. But in fact, it was only a part. A tragic fallacy, which led Pyramus to commit suicide. We apprehend: Even seemingly unambiguous, reliable nonverbal signs may deceive somebody. The interpretation by the recipient is the essential element of meaning. Yet once again, it is according to the given reified sign by no means arbitrary, but there is a core, as well as—no clear-cut—borders. In Sophocles' *Electra* an analogy is the urn, containing allegedly Orestes' ashes, which is delivered by Orestes himself and by Pylades, his companion. Yet, in *Elektra,* by Hofmannsthal/Strauss, these signs are mentioned, but not presented. Hofmannsthal/Strauss assume that Sophocles' version

Fig. 4.1 Ottorino Respighi: Pini del Gianicolo, Fading-in, nightingale singing from a record First edition Verlagsinfo Milan: G. Ricordi & C., 1925. Plate P.R. 439. IMSLP05603-Respighi_-_Pini_di_Roma_(full_score)

and its content are known. Thus, they reduce this thing-sign already into a—verbal—form of report. It is a blurredness, that the urn's content remains objectively uncertain, and depends only in bona fide. But here, these seemingly partly-identical thing-signs are deceptive, as Orestes' thing-sign is likewise false, just like the fabricated verbal reference by Electra about it, and it only serves to keep Aegisth, the patricide, in a delusive sense of security.[43] From this we may draw a conclusion: By recognition of blurring, Fuzzy Logic contributes to sharpen the sense for the mere appearance of identity.

4.3.2 Real Tones as Readymade and Thing Signs

„Real tones" are most directly entered by music into the thing-world. In the language of movies and broadcast sounds or noises are called „real tones", as they are part of acoustic reality, and then they can be integrated into a movie, a radiophonic work, and later into a television-work. Music, which is actually heard at a location of shooting, adds to this, while music, which is to be heard during the takes as „on", is a borderline case. Real tones are semiotically index-linked signs, which causatively derive from what is signified. Thereby, they must be differenciated from iconic signs, which are allocated to what is signified by analogies and abstractions, and also from symbolic signs, which are not similar, and whose allocation is arbitrary. For instance, the countless birds, which start to „sing" in music, probably in any time and place, these are no songbirds, but artificial and artistic imitations of this special communication among animals, with their calls for mating, courtship, or warning cries. An exception happens in Ottorino Respighi's program-musical work *I Pini di Roma* (*The Pines of Rome*, 1924): In the 3rd movement, *Pini del Gianicolo,* a real nightingale appears on the concert podium, however, by no means as a living creature, but in a technically reproduced form: Its singing in the concert hall is played from a disc (Fig. 4.1).

Respighi wanted to present such real tones as „truely" as possible. The "genuineness" is emphasized by naming a probably „authentic" scene for recording, which is identical with an area that is described in these words: The nightingale was recorded in the yard of the McKim Building of the *American Academy* in Rome, situated on

[43] While deceiving by camouflage or mimicry—whether an animal or a plant—the thing-like material acts as sign or signal as such, though in case of animals it is a complimented and reinforced behavioural pseudo-characteristic of their species.

Janiculum hill.[44] Just like in the fitting saying: „Program-music is something for the inner eye“,[45] so Resphigi‘s music itself was used in movies as film music for the exterior eye, for instance, to pep up nature films with melodramatic effect. In the animated cartoon film Fantasia 2000, an abridged version of his work, without the gloomy 2nd set, was chosen for incidental music to th movements of hump-back whales. What is typical for the soundtrack in documentaries and particularly in nature films, is a fading out of the acoustic communication of these whales, which is attractive and highly esteemed in itself, and their replacement by some meaningless sound-sauce, instead of the real noise.

With all these sounds of nature, and also what will be discussed later, with the treatment as onomatopoeia in a given musical idom, under Fuzzy Logic‘s aspect of measurements of blurrings/sharpenings the following procedure may be used for the relation artwork-reality. For real natural sounds, such as birdcalls, and also in their musical conversions, a reduction to sonograms or other similar technical ‚notations‘ is carried out. With animal sounds in polyphonic compositions, then if necessary, particular voices have to be filtered out. On this basis, then they are more easily accessible for a computerized quantifying analysis, than with the musical notation. At the same time, there are methods available for this, such as the ones which, in search for plagiarism, were developed by Fuzzy Logic.[46]

Since the middle of the 1920s, the procedure to add real tones to artworks, or—initially more often—musical-breaks, by means of a sound storage medium, from record discs or later from tape recordings, and still much later compact discs and other media. Up to the present, this is especially a common practice for works of the theatre staging plays and music theatre. Respighi‘s pioneering feat, to install into musical works animal sounds as technically reproduced „real tones“, was applied by many emulators.[47] In particular, the musically traditional and plainly arranged Missa Gaia/Earth Mass of 1982, by the soprano-saxophonist Paul Winter, with Paul Halley, Jim Scott, Oscar Castro-Neves, and Kim Oler, gained popularity.[48] As original tones „the voices of wolf, whale, eagle, harp seal, Amazonian musical wren, Russian loon“ are to be heard.[49] But above all, the composers transformed the sounds of the animals into musical material and recorded them within it: „The singing of whales describe

[44]https://en.wikipedia.org/wiki/Pines_of_Rome, 8.3.19, retrieved 15.3.19. Such a procedure was in ethnology, music-ethnology and zoology, soon after the development of Edison‘s phonograph, in 1877, in popular use, as field recording, audio footage taken on location. Field recordings were originally used in sciences, but, already with Respighi, they became part of art, and, finally in post-modernism, even an independent genre of art, „a popular form of sound art“ Demers (2010, p. 170) is critical of the indentity-illusion, which is in connection with it: “Although any act of recording necessarily involves some editorial decisions on the part of the recordist, many field recordings also aspire toward an unmediated representation of reality“ (see Sect. 4.5).

[45]https://www.swr.de/swr2/musik/musikstueck/ottorino-respighi-pini-di-roma-sinfonische-dichtung-fuer-orchester/-/id=2937886/did=16340214/nid=2937886/hskkpm/, retrieved 14.3.19.

[46]See Chapter 5. Filtering, and the literature, which is indicated there.

[47]There are many examples in https://en.wikipedia.org/wiki/Biomusic 12.12.2013, retrieved 12.8.14.

[48]https://en.wikipedia.org/wiki/Missa_Gaia/Earth_Mass, 1.11.18, retrieved 15.3.19.

[49]http://www.paulwinter.com/projects/missa-gaia-earth-mass/, retrieved 15.3.19.

in Sanctus the expanse of heaven and earth […]. The wolf's howl becomes a Kyrie-melody."[50] After this primal invocation of Kyrie in alternation between saxophon, realtone-wolf and choir, the compact-disc is gradually fading out (Score p. 24). If anything, then anything, a conductor thought.

> The demand of recording a compact disc for the compositionally essential sounds of animals made me realize, I might also add some film sequences.[51]

The sound of an imaginary movie for the „inner eye" is thus completed by a real movie: Two heads are better than one. A second type of musical utilization for real tones made by animals are "sounds made by humans in a directly biological way […] created by the brain waves of the composer […] by neuro-feedback […] An electro-encephalo-phone […] was first designed by Erkki Kurenniemi in the 1960s." David Rosenboom, the Brazilian composer Eduardo Reck Miranda[52] or Alvin Lucier (see Sect. 9.2) followed suit. By means of modern digital technique one may even draw music from a plant, a bonsai-tree as a musical instrument, as if nature is directly transferred into art.

> To determine the key I used the lowest note I could play and recorded the rest around it. Besides playing the leaves, I used bows of different sizes, a piano hammer and a paint brush. As far as microphones I used my Røde NT6, a customized stethoscope and tiny MEAS piezo transducers. I played all the sounds and rhythms only with the bonsai, I didn't use any synthesizer or samplers to create or modify the sounds.[53]

But here, it has to be considered, that this kind of „bio" requires an utmost extent of extremely refined „techno", and moreover for the composing/realization, by means of neuro-feedback, musical instruments, normally percussive, in order to realize what is „thought" tonally.

A third type of utilization of real animal sounds as „biomusic" belongs presumably mainly to the sphere of musical comic effects: „The use of animal notation: music scores created by animals, often in the form of paw prints. Biomusic can also take the form of animals trained to perform specific behaviors as part of a musical performance".[54] Sadistic versions of such utilizations of animals as musical instruments, including Monty Python and the Muppet Show, exist at least since the—fictitious—„cat organ", as is was described by Athanasius Kircher, in his Musurgia universalis (1650):

> A bear that played the organ; instead of pipes, there were sixteen cat heads each with its body confined; the tails were sticking out and were held to be played as the strings on a piano, if a key was pressed on the keyboard, the corresponding tail would be pulled hard,

---

[50] https://www.evangelisch-beuel.de/wordpress/?p=1130, retrieved 15.3.19.

[51] Ibidem.

[52] https://en.wikipedia.org/wiki/Biomusic 12.12.2013, retrieved 12.8.14.

[53] Self-comment by Diego Stocco, *Música con bonsai*, http://www.youtube.com/watch?v=qvyHHX6hNkY&feature=relmfu, retrieved 25.3.14.

[54] https://en.wikipedia.org/wiki/Biomusic 12.12.2013, retrieved 12.8.14.

> and it would produce each time a lamentable meow. The historian Juan Christoval Calvete, noted the cats were arranged properly to produce a succession of notes from the octave.[55]

It might well make some sense, to limit such an expression like „biomusic“ terminological to these three types of utilization of real tones. In fact, Fuzzy Logic and the search for similarities have hardly anything to with these almost identities. But, it is a different matter with mimesis of numerous *composed* sounds of animals, which were set into the given musical material, starting in Early Renaissance Florence, across Janequin up to the birdcries by Messiaen or the singing whales by George Crump and Witwulf Malik. This type belongs to onomatopoeia, but not to „biomusic“.[56]

By real tones, a piece of unmediated nature is dubbed into music. They are something like the parallel case in fine arts called readymade or ‚objet trouvé‘ (a found object). But readymade is primarily borrowed from the industrial world. And, being an isolated thing from reality, it is quite unlike the real tones, which were, so far dealt with, embedded in a work of art. The term ‚readymade‘was invented by Marcel Duchamp.[57] What is new, that is the isolation of a pre-existing thing, but not its artistic use. Already in 1912, the Cubists (Braque and Picasso) pasted collages with paper. Then, in 1916, during WWI, Dadaism started with the montage, which is artistic-technically somewhat higher developed.[58]

> Duchamp‘s earliest readymades included *Bicycle Wheel* of 1913, a wheel mounted on a wooden stool, and *In Advance of the Broken Arm* of 1915, a snow shovel inscribed with that title. In 1917 in New York, Duchamp made his most notorious readymade, *Fountain*, a men's urinal signed by the artist with a false name and exhibited placed on its back.

Here in an U-turn, Duchamp uses an universal counterpointing artistic device, an inversion, the reflection of a motif, or theme, along an imagined horizontal central axis. Thus, he does not make nothing. He does not leave the real thing as it was, instead, firstly, he eliminates its use function, while secondly, transforming it into a work by his signature „Mutt“, even if shamed, but defined as art by his existence as an artist, then thirdly, transposing such a work into art-contexture by an exhibition. Hence, for mimesis, a minimum of poiesis is indispensable—even concept artists have at least to write down their concepts, which are not performed. From its origins and its materials, readymade is similar to real tone. Yet, it differs fundamentally in its aesthetic functionality. Because, in fine arts, an isolated real thing is a work. In music, this does not occur, for the difference between things of art and usage is not existing—no piece of music could be used as an urinal. Solely, Respighi‘s record

---

[55]Extensively, with some further variants like the “pig‘s piano” (Piganino), see: https://en.wikipedia.org/wiki/Cat_organ, 27.10.15, retrieved 17.12.15.

[56]With extensive materials see e.g. Doolittle 2012 and 2014, and recently Gredig 2018.

[57]It refers to *industrial* contexts: “By the end of the nineteenth century the term ‘ready-made’ was being used to describe objects that were manufactured as opposed to being handmade. By the end of the nineteenth century the term ‘ready-made’ was being used to describe objects that were manufactured as opposed to being handmade” (https://www.tate.org.uk/art/art-terms/r/readymade, retrieved 9.9.18).

[58]For some contexts see: *Sound art*, http://www.oxfordmusiconline.com/public/page/sound_art, retrieved 12.12.16.

presentation of nightingale‘s singing, and as such, it could at best be, institutionally and aesthetically, imagined in exhibitions or museums, but never on the concert rostrum. Moreover, animal sounds, like the nightingale‘s singing, are—for humans—in reality not functional, like an urinal in the Fountain, and already art-analogous. Readymade in music is a thing inside the work.

4.3.3 Real Tones as Handmade—Mimesis by Means of Poiesis

For Respighi real tone meant the technically reproduced nightingale. It appears as a natural sound, coming from the outside, to enter the symphonic poem, and it is also medially dissociated—technically reproduced, versus cheerfully played music. The tones possess a function of reference, but without any real mimesis. Like in music, the real tones possess also in film, or in a radio transmission, as part of the soundtrack, components of an overlapping whole. Here, the noises take centre stage. They are tokens for environment, atmosphere, time and place. Nevertheless, even the use of real ‚things‘, which seems to promise identity, shows a considerable blurredness. In fact, the remake/make, it also bears the pioneer‘s name „Foley“, who started with the „making of noises“, is placed historically before the mechanical-automatic recording of real tones, which only began in 1924, when the electric microphone was invented.[59] The counterpart to readymade in fine arts was only about a decade later introduced for acoustics. Because of difficulties in recording soundtracks for movies there is made use of artificially produced additional, or instead of, real or „original tones“.„[…] The sound in Hitchcock's The Birds, singing birds, flapping of wings, originates from the trautonium only […]“,[60] an electro-acoustic instrument of the late 1920s. Such artificial real tones often appear to be more authentic than the real noise. In one of Hans Christian Andersen‘s fairy tales, „The Nightingale“, a stupid princess prefers a nightingale‘s mechanical replica in place of the genuine nightingale. Thus, similarity by means of something similar, like imitating the clatter of horse hooves by the clapping of coconut halves,[61] is turning into a similarity by means of difference, almost dissimilarity. Thus, rifle shots may be represented by

---

[59]Extensively, among others, in: https://www.videomaker.com/article/7220-real-time-sound-effects-the-foley-way; http://web.archive.org/web/19970516041845/http://citd.scar.utoronto.ca/VPAB93/course/readings/prenderg.html; http://www.filmsound.org/game-audio/ An endless link-list is given in: http://www.filmsound.org/theory/, retrieved 10.3.19. An extensive alphabetical list with hints for the production of sounds is given in: David Filskov: The Guide To Sound Effects. Tips and ideas for making your own sound effects, https://www.epicsound.com/sfx/, retrieved 10.3.19; for instance, to horse hooves, besides the obligatory coconut halves "by beating two potatoes in a 3-beat rhythm in a pan, filled with sand, rice, and crushed crackers. Pitch adjusted to suit" or simply "cup your hands and clap them against your thighs to the running horse rhythm." Product placement is a bonus extra: "Works best when you're wearing denim pants."

[60]klangmaschinen.ima.or.at, retrieved 18.9.2013.

[61]Compare with the Monty Python‘s parody.

pattering beans on a saucepan's lid.[62] Thereby, an analogy in the sense of biology is created. What used to be indexical, turns into iconical.

Quantifying the degrees of similarities in the sign of Fuzzy Logic appears difficult, but viable; though it remains questionable, to what extent it will make sense. Considering the degrees of similarity, in the sense of Fuzzy Logic, the appropriate common basis for comparisons could be the reduction to sonograms, both of real tones, like birdcalls, and also of their mimesis in a soundtrack or a musical work. They could be facilitated by optical fixing, as it eliminates the problem of chronological processes. In contrast to film-procedures, in Dieter Schnebel's five-movements work ‚Symphony-pieces'(Sinfonie-Stücke) from 1984/1985, a sort of real tones play an important, semantically an almost crucial part.

> It suits for each piece [...] to express a significant sound of an almost semantic nature – as in all the Symphony-pieces, the matter of sound demands to be expressed. In the [IInd movement, a] funeral march is grounded by a stony noise of grinding.[63]

Such real tones were not readymade, they were handmade, namely, self-created symphonic-orchestral noises. In this respect, Schnebel's procedure was similar, to both Foley's, and to the one used by futurism as well. For in futurism the production of noise was due to the state of the art, still dependent on imitation and faking. The „intonarumori", or noisily-intoned, are musical-*instruments*. They do not produce real tones, but noises similar to them, which are more or less unsharpened, and subdivided in different classes. Schnebel makes self-imagined noise-intonation. The score-instruction for the two percussionist players indicates:

> In a box, filled with gravel, (in part I fine, in part II coarse) create crunching and grinding noise with the help of a roll or something similar; not loudly, but always clearly audible.[64]

In the A-part of „Marcia funebre", this naturalistic-symbolic tonic-ground is distinctly perceptible, underneath gruff and scanty chord steps. Further on, a hidden sound layer follows, which was at the beginning of the A-parts already noticable: The noise of stirred water. For the instrumentation the the instruction from the beginning „container with water, in it splashing noises" was enlarged (score cipher E):

> With containers (bottles, jugs) scooping water from a tub and pouring it back again.[65]

Like any other elementary, *Dripping Water* is also popular with happenings. There as here, the elementary is, in a quite traditional sense, reintegrated into art as a piece of life. The motif and semantic group „water", including its symbolic dimension, even shows its marks in the instructions of the performance in their particular form parts: „Flowing", with various additional directions, changing from „quiet" to „agitato":

---

[62] So during the early 1930s; Henckels 1966, p. 114.

[63] Schnebel, *Extrem komprimiert. Anmerkungen zu den Sinfonie-Stücken für Orchester*, in: Programmheft des Philharmonischen Staatsorchesters Hamburg (Extremely compressed. Notes to the symphony-pieces for an orchestra. Program sheet of the Philharmonic State Orchestra), 23. 6. 1985, n.p. Emphasis H.-W.H.

[64] Ibid.

[65] Ibid.

slowly running (A) – slightly stirring (B) – very quiet (A') – accelerando (B') – quite running (A") – ritardando (B") – again slowly running (A"').

In spite of similarities in the procedure, Schnebel, for his purpose, passes beyond it. In a Foley, the real tones should not really represent reality, but have to „be" it; to imitate them, is for making reality in movies, broadcasts and television more visible, in plasticity and acoustic. In futurism the noises are a—naturalistic—mimesis of reality. For Schnebel they are the same, yet, they symbolize something. Of course, lastly, it concerns more and something much higher than merely the „matter of sound", that is something metaphysical and not solely physical. In his interpretation for both elements of natural sounds, the grinding noise in the funeral march and the noises of water in the Adagio, Schnebel himself, is reaching too high.

[This may] „widely point beyond the two aspects of being, which symbolize time, indeed one [might] say: The procedure, in which we are positioned, and the eternity, in which we are living."[66]

4.3.4 Real Tones and Existing Pieces of Music as Readymade Material of a Work

As mentioned above, only with a highly developed technique, electricity and electro-acoustics being provided, then mechanical, quasi-automatic recording of real tones is possible, from nature or culture. This recording is primarily mediated with the aid of technical equipment, without any art-analogical human activity: *Mimesis without poiesis*. Since the mid-twenties the appropriate state of the art was for the film available, by means of a microphone-technique, which derived from the radio, that was ‚electrified' already from the start. Nevertheless, the ‚artificial' production of natural sounds survived, as be it was technically easier and often „more similar", than a „live soundtrack".

In fact, the opportunity to compose with real tones, further required the magnetophone-technique, with its possibilities of cutting and montage. Shortly before 1945, this standard was attained by musique concrète. In fact, since the mid-1940s, musique concrète made use of readymade real tones as material, whereas in futurism they had to be instrumentally produced, still with a lot of effort. In this case, it did not concern natural sounds, like a nightingale's singing, but the noise from industrial technique, above all from railways. So Pierre Schaeffer, one of the protagonists of musique concrète, entitled a piece of his works accordingly *Chemin de fer*. The real tones it uses are taken mainly from the industrial reality—found audio footage—. But these are used as such. These tones are not ‚displayed' or performed. But, in a larger context, they are musical material, to be further processed, assembled, and composed, or as a work of its own. However, the ideal is not to achieve similarity.

Schaeffer's earliest concrete works used recognizable sounds, although Schaeffer advocated "reduced listening" as a way of bypassing external associations to focus on

[66]Ibid.

a sound's inherent qualities. [...] A point of contention in post-Schaefferian electro-acoustic music is whether materials can be mimetic or should adhere to the modernist ideal of abstraction.[67] Already Schaeffer made the decision against mimesis, and his students readily adapted to it. He wanted real tones, without any mimetic relations with reality. Whenever possible, he wished eagerly to discard sigmatics and semantics.

The sound object (*object sonore*) is sound isolated from its means of production or notation as well as the state of mind of the listener. Schaeffer's conception of the sound object is a reduction, a bracketing out of external information in order to arrive at a pure phenomenological experience. The concept of the sound object has held wide sway among post-Schaefferian electro-acoustic musicians.[68]

Contrary to a listener‘s spontaneous experience, the reception of sound sources should consciously be fading out, thus creating a listening, which is accordingly named "reduced listening"—

> Schaeffer's term for listening that ignores the source and origins of sound. Schaeffer advocated reduced listening for his musique concrète, a genre that often relied on recognizable footage of everyday sounds such as trains. Several composers, including many of Schaeffer's own students, have criticized or rejected reduced listening for placing unrealistic expectations on listeners.[69]

Art critic Michael Fried called the emphasis of material as such, without any further meaning, "objecthood". It occurs in minimalistic overpowering-aesthetics up to the "glitch micro-sound" and other forms of recent electronic music as "a way of avoiding musical signification."[70] Then, a new turn is achieved by the fact, that pre-existing music and not only real noises function as readymade. Here, Duchamp‘s art-concept set a precedent. In the neo-liberal post-modernism, since the late 1970s, its spreading received once more a powerful push. Sampling,[71] "Appropriation Art“[72] and similar trends and genres take something as material, that is already art, not anymore as a reality apart of art.[73] In this respect, realtone and readymade are here only an analogy. Because it is a matter of something, which was already artistically formed, that is henceforth transferred as material and language into a new context. By „Plunderphonics“, being rather the composer‘s John Oswald trade name than a term, sampling obtains a provoking kind of cultural criticism.

> Musical instruments produce sounds. Composers produce music. Musical instruments reproduce music. Tape recorders, radios, disc players, etc., reproduce sound. A device such as a wind-up music box produces sound and reproduces music. A phonograph in the hands

[67]Demers 2010, p. 173, and p. 172.

[68]Demers 2010, p. 174.

[69]Ibidem.

[70]Ibidem.

[71]An instructive typology was attempted by Ratcliffe 2014. It could be differentiated and further developed, not at least with the help of Fuzzy Logic. For sampling in Hip-Hop, see Kautny 2010.

[72]Generally, but especially focussed on trends during the 1980s in the USA, extensively. https://www.tate.org.uk/art/art-terms/a/appropriation, retrieved 9.9.18.

[73]See e.g. Döhl/Wöhrer 2014.

> of a hip hop/scratch artist who plays a record like an electronic washboard with a phonographic needle as a plectrum, produces sounds which are unique and not reproduced – the record player becomes a musical instrument. A sampler, in essence a recording, transforming instrument, is simultaneously a documenting device and a creative device, in effect reducing a distinction manifested by copyright.[74]

Historically, in music, the open, explicit collage and montage spread rather later than in fine arts. Yet in fact, its beginnings started earlier. Apart from Charles Ives, who started it in 1900, that was even before cubism, here, historicity looked like a sheet lightning, long time before modernism. In Brahms, in conformity with his classicism, appropriations of Vienna Classic are to be found, up to invoking the structure of its forms and themes, instrumentation, idiomatic. Already before 1900, there were many, who composed "in the old style", like Max Reger, or Pëtr Ilich Tchaikovsky, in his *Variations on a Rococo-theme* for violoncello and orchestra, op. 33 (1876). Then, Stravinsky, by means of stylistic citations and collages of artistic "Readymades", especially emphasized the making and what is made. In avant-garde, between Alois Zimmermann and Luciano Berio, such procedures are quite common, since the 1960s. Particularly, in today's popular music at least some minimal re-arrangement, of what was found, is accepted, and eve required, not at least to avoid accusations of plagiarism for reasons of copyright infringements[75] (see Sects. 5.2 and 8.2).

To put one's hands into the pockets of somebody else, such infringements are generalized by people, who are only secondary reproducers, like DJs, as they practice a kind of post-compositional and post-interpretive 'secondary sampling'[76] and by a quasi-compositional interference of the recipient, for example, in form of ‚mash-up'.[77]

4.3.5 *Real Tone Transformed into Onomatopoeia*

The simplest method for the transport of real tones into a work is to have them technically reproduced, as with Resphighi, or later, in musique concrète. But this implies complex technical devices. From the outset, the method, that was applied for music was technically simple, but artistically more complex. In a wider sense, it means onomatopoeic procedures, choosing real tones, as pars pro toto, for the thing or the factual situation, and to transform them into a given musical material and

---

[74]Oswald 1985: cf. e.g. Collins 2007, 2011, Kornauth 2011.

[75]See e.g. *The Man Who Stole Michael Jackson's Face*, https://www.wired.com/1995/02/oswald/, retrieved 16.3.19.

[76]„In many genres of electronic music, the DJ enjoys the status of an auteur and constructs elaborate collages of sound" (Demers 2010, p. 167). One manifestation is the "turntable", the "practice of playing record turntables as musical instruments" in Hip-hop of the mid-1970s (Demers 2010, p. 177).

[77]See e.g. Boone 2013, Döhl/Wöhrer 2014, Mundhenke et al. 2014 and Döhl 2016.

Fig. 4.2 Charles-Camille Saint-Saëns: *Le carnaval des animaux*
II. movement *Poules et coqs* (Hens and Roosters), Excerpt
First edition, Paris: Durand, 1922. Plate D. & F. 10,154. IMSLP576538-PMLP6099-Le_carnaval_des_animaux_-...-Saint-Sae-ns

idiom. Besides, for the similarity, universals exist, yet above all, it is determined by music-culture and epoch-specifics.

The sound imitative stylization does not musically go beyond this, as it takes another direction: What is noise-like is translated into pitches, the natural sound, as a whole, is fitted to the respective idiom. Generally, the sound imitative mimesis in verbal languages is reaching to its limits by the principle of phonemes in general, and especially, by its respective development of individual languages. Blurrings in music-language are comparable. Likewise, onomatopoeia suffer fractures and filtering by styles, idioms and in similar ways through the respective historic state of the material and its system, above all, precisely, in its systems of pitches and rhythms. Here, the two differentiated and yet similar examples for the cock's crow are sufficient. Charles-Camille Saint-Saëns invoked in the suite *Le carnaval des animaux* (The Carnival of the Animals) of 1886, in the II. movement *Poules et coqs* (Hens and Roosters) in an oversimplified way the piece for harpsichord *La poule* ("The Hen") from Rameau's Suite in G major: The hens peck in the 1. violin and cackle in the 2., while the rooster crows in the 1. and 2. piano (Fig. 4.2).

In Rimski-Korsakov's satirical fairy-tale opera *The Golden Cockerel* (1909), after Washington Irving and Pushkin, the title character, who plays the hero, appears on stage in manifold variations. Befitting to his rank, he should serve as a warning watchman, and to save the incompetent rulers from troubles. His crow has two variants. The first is a call for reassurance (Fig. 4.3).

By the orchestra the three-tone kernel of the crow is nonverbally incorporated like an echoed reply. The second variant is the warning cry. It is similar, only transposed into other pitches (Fig. 4.4).

Here, we have to keep it comparatively brief, although many of the appropriate works are quite attractive. In the context of Fuzzy Logic, the basic issue is always as to how far the real model and the musical imitation are similar, and to what extent this similarity may be quantifiable. This is still comparatively easy with directly used, cited and installed real tones. The issues become considerably more complex

Fig. 4.3 Rimski-Korsakov, *The Golden Cockerel* (1909)
Rimsky-Korsakow, dergoldnehahnein00rims2 KA dt. engl., Piano Score Rimsky-Korsakow, C.F.Peters 1907
Reassurance

Fig. 4.4 Rimski-Korsakov, *The Golden Cockerel* (1909) Rimsky-Korsakow, dergoldnehahnein00rims2 KA dt. engl., Piano Score Rimsky-Korsakow, C.F.Peters 1907
Warning

with a compositorial subsequent treatment of the given material and idiom of time, composer, work. And it becomes even more complex if the analyses include intercultural variants. This is also worthwhile as a search for mimetic-musical universals, because it brings up both a—probable—invariant core and its various manifestations, i.e. another facet of the dialectic of identity and difference. Nevertheless, according to tendency, the problems may be solvable, even if it demands a considerable expenditure.

4.3.6 *Painting as Mimesis and Poiesis. Hypernaturalism, Trompe D'Oeil/"Optical Illusion", „Organized Sound" (Edgard Varèse)*

For readymade, the object's sign-character is of no interest. It is simply neglected. But, actively denying the character of sign and of reference for mimetic creations, the illusion of an identity of signifier and signified is something different. It may acquire extreme forms, as was already hinted at, when Don Quixote involuntarily confuses theatrical appearances and reality of existence.

What might be in analogy to an optical illusion could be more difficult to realize in music, as sounds are more likely to be procedural than objectively-real, and hardly tangible in a haptic sense. After 1877, with the epochal invention of Edison's phonograph, it became principally possible to create the *illusion* of real tones and acoustic reality with recorded real tones; this was considerably more intensive than with photography, or in the next decade even more impressive with films. The interest for musical paintings was in Viennese classic already in decline, as noted by Beethoven in a programmatic statement for his, actually still picturesque 6th symphony, the *Pastorale*, "more the impression of sensation than picture". Nevertheless, imitations of sounds and musical paintings in program music were to remain an essential skein. The picturesque excellence and excesses of Strauss are dated in the days after phonograph and film. Here, the spectrum of possibilities was drastically enlarged by the electro-acoustic music. Although being most primarily interested in abstraction and material as such, it promised, once again, more of an identity of representation and what is being presented, in any case, approximative-asymptotic, instead of just similarity.

> Before the advent of electronic music, musical imitations were necessarily constrained by the limitations of musical instruments, whose timbres did not necessarily approximate nonmusical sounds with any precision. Thanks to sampling and synthesis, however, electronic music can produce mimetic representations that are *virtually indistinguishable from their real-world* counterparts.[78]

Edgar Varèse, being one of the pioneers of electro-acoustic music adopted the term „organized sound"

> for what he predicted as the future of music, when electronic instruments would be liberated from the dictates of conventional music to produce any type of sound.[79]

But, to a large extent, the possible approximative identity with minor blurrings remains unrealized. In so far as music really was turned into pictures, then these were mainly of an "abstract", non-representational "non-figurative" kind of painting—once again, like in the mentioned limitation of the technical reproduction of real tones. On the other hand, the acoustic "optical illusions" are still realized, though at the periphery, in film soundtracks or in the morphing.

4.3.7 Mimesis of Matter

The dialectics of mimesis and poiesis, of readymade and handmade, in fine arts were by photography already changed earlier than in music. Reacting to photography's almost automatic portrayal with its seemingly absolute exact imitation of

[78] Demers 2010, p. 172.
[79] Demers 2010, p. 174.

optical reality. At the latest since the 1840s, the avant-garde turned against a most similar representation of reality. That was a conscious, but yet also a rather unconscious reflex, and then with industrialization, a conscious reflection of the seemingly possible total feasibility and creation of new worlds. And moreover, already in impressionism, the aisthesis as a theme was further taken on, as a perception in the interplay of picture-object-contemplators. Nevertheless, the naturalist line of tradition was maintained. Not only the "photorealism", with all its conscious blurring serves as an example for this. But, the object is shifting, according to tendency. Particularly, in the age of post-modernism, material and formal aspects often took priority over those of representational-mimetic and with regards to content. This may include a reflection of the character of representation and art.

As usual, here too, Fuzzy Logic should have the task to measure degrees, as well as even measuring types of similarity quantitatively too—for instance, as a simple mode of procedure, by the comparison of photos of *real* surfaces with those artistically reproduced. Thus, we returned again to the initial point of departure, at the thing and at the real tone, but in a spiral movement, as it now takes place on the level of artistic mimesis.

Enno Poppe (born in 1969), for a large part of his works has taken as a programmatic theme real materials and objects, among other things, such as:

salt	*Salz* (2005)
sweat for violoncello and keyboard or for four instruments	*Schweiß* (2010)
felt for viola solo, 4 bass clarinets and strings	*Filz* (2014) etc.

In Poppe's cycle *Wood* for clarinet and small ensemble (1999/2000)—*Bones* for an ensemble (1999/2000)—*Oil* (2001–2004) the natural materials, as he alludes to in the title, become "Ciphers of the Organic" (Poppe). Yet, paradoxically, he uses even for the bones the principle of becoming and growth.

> *Holz – Knochen – Öl* (Wood – Bones – Oil) is a poetic and dramatic cycle, which may not only be listened to, as a study of material constitution – of the pliable stability of a fibrous blending of voices in the wood over the hard "martellatissimo" and dry "secco" in bones, up to the viscous, but energetic stream of lines in oil, running into one another. Poppe also tells the story of becoming and decay. Wood becomes tangible as a living, growing material. Bones promise growth, and stability, hardness and – as a symbol for vanitas – death. Oil, on the other hand, as residues of past lifes, reminds of transitoriness and a decay, lasting millions of years, while as a source of energy, it creates at the same time life anew.[80]

Moreover in *Knochen* (bones) splintered, hard, even aggressive sounds render associatively, and with the help of analogies, thoroughly iconic relations with the object audible.[81] Already Charles-Camille Saint-Saëns took a path, specific to music, in the use of bones as a substance for the presentation with the help of analogies, metaphors, icones and symbolism, in his suite of 1886 *Le carnaval des animaux* (*The Carnival*

[80] Björn Gottstein, http://www3.musikrat.de/index.php?id=2178, retrieved 30.3.19.

[81] Cf. https://www.youtube.com/watch?v=RcdRIC_gxXI&list=RDRcdRIC_gxXI&start_radio=1&t=21.

of the Animals), as it was mentioned before. The XII. movement, *Fossiles* (*Fossils*) is set for strings, two pianos, clarinet, and xylophone. The—humorous—representation of fossils is multilayered and multidimensional. Saint-Saëns had with his tone poem *Danse Macabre* (1876) introduced the xylophon as a new orchestral instrument in the „classic" orchestral literature, which was unusual in those days. In the first place, by its material it is similar to a fossil. Because, „the wooden claves, made of hardwood, well dried up and for a long time seasoned, are as such already ‚petrified' fossils".[82] Secondly, the instrument's seasoned, sharp sound reminds of the rattle of bones. As soon as he sets those actually in themselves inert dormant fossils, being like relics and readymades, in motion, then drastically. Thirdly, Saint-Saëns alludes to the danse macabre—also with the swift tempo, humorously designed by him as *allegro ridicolo*, a „ridiculous" allegro. He „mimics his own composition, the *Danse Macabre*, which makes heavy use of the xylophone to evoke the image of skeletons playing card games, the bones clacking together to the beat."[83] Fourthly, by playing the solo-violin, simultaneously appears the traditional image of death as he plays dance music with his fiddle. Finally, in the fifth place, Saint-Saëns with ironic side cuts in every direction moves to the symbolic plane of signs. He quotes, after the self-citation, the popular *J'ai du bon tabac*, and the in the 18th century already well-known nursery rhyme *Ah! vous dirai-je maman* (better known in the English-speaking world as *Twinkle Twinkle Little Star*), whose opening bars inspired Mozart to the theme of his twelve variations for piano (KV 265), the French nursery rhymes *Au clair de la lune, Partant pour la Syrie*, and finally the aria "Una voce poco fa" from Rossini's *The Barber of Seville*—altogether musical pieces, which according to Saint-Saëns, could be heard as fossils.[84] In particular by classifying *Partant pour la Syrie* among the "fossils", he even risked a political statement. For the song was attributed to Bonapartism.

Originally written as a love song [ca. 1807], the melody was attributed to Hortense de Beauharnais, [...] mother of Emperor Napoleon III [...], but musicologist Arthur Pougin determined the true author of the melody was Louis-François-Philippe Drouet, [...] a favourite of Queen Hortense. The song was a favourite of the Bonapartists during the times before the Second Empire (when this was used as the anthem), and was also popular during the First Empire. [...] During the Second Empire *Partant pour la Syrie* was the unofficial national anthem, while *La Marseillaise* was forbidden but for the very end. With the collapse of Napoleon III's rule [1871, 15 years before *Carnaval*], the popularity of the song waned. [...] It remains part of the repertoire of French military music.[85]

[82] See https://de.wikipedia.org/wiki/Karneval_der_Tiere, 11.2.19, retrieved 27.3.19.

[83] https://en.wikipedia.org/wiki/The_Carnival_of_the_Animals, 26.3.19, retrieved 27.3.19.

[84] Description as a mix of own, https://en.wikipedia.org/wiki/The_Carnival_of_the_Animals, 26.3.19, retrieved 27.3.19. and https://de.wikipedia.org/wiki/Karneval_der_Tiere, 11.2.19, both retrieved 27.3.19.

[85] http://www.nationalanthems.info/fr-70.htm, retrieved 29.3.19.

4.3.8 Music as Painting

Musical pictures of sound is concentrated on the acoustic side of the representational. Besides, in picturesque music non-acoustic elements are mimetically presented. Here, 'picturesque' is the term for the naturalistic line in musical mimesis, with the aim of an iconic-picturesque presentation. Until the days of the Vienna classic the European music followed an ideal, as it was formulated for literature by Horace: "ut pictura poesis erit", literature had to be painting. The transition between both types of pictures of sound and of musical painting are without any clear-cut dividing line. Important vehicles of analogies are movements and rhythms, within the real space as well as in the imaginary-real space of sound, between music and reality. The musical devices, materials and modes of procedures as well, stretch along the cognitive, across the emotive, and further into the biologic-physiologic layer of the music process.

A well-known 'figure' is the "sighing motif" as an expression of grief, the descending semitone with a downbeat intonation, for instance: *B A*. This most narrow, shortest step within the twelve tonal, chromatic division of the octave is pointing iconically and indexically at a choked breath and voice. So, the motif is similar to psychical emotion and bodily motion, in which it is expressed. Being extended as a descending chromatic pace in the fourth-space, such a motif becomes a phrase of lament, or the "passus duriusculus", "a somewhat forced pace", for instance: C H *B A* As G, as it belongs to the standards and topoi of European music. Such topoi systematize the musical doctrine of affects, in combination with the musical rhetoric in shape of 'rhetoric figures'. They form a system of signs with sigmatic, semantic and pragmatic valence and an iconism, which includes by the usual procedures of formation of analogies and metaphors even abstract symbol-signs with meanings of a virtually notional-denotative nature. Let us stick to the example of the motif a sighing. An appropriate German texture is "Wehe!". This lament became the leit-motif in the tetralogy *The Ring of the Nibelungs (Der Ring des Nibelungen)* by Richard Wagner. In *The Valkyrie (Die Walküre)*, second part, it appears frequently, in particular, expressively in this form, in the III. Act, as wailing of the valkyries (Fig. 4.5).

Without difficulty one may recognize, that the four tonal treble contour of the repeated "Wehe" wail, is nothing else than the—transposed—BACH-Motif. Wagner repeated it immediately in the next two bars with the pitches G F# A G#. The designations of the leitmotifs did not go back to Wagner himself. Here, they are blurred, neither plain and nor coherent. The present piano score designates the motif of wailing as "corvée-motif" ("Fron-Motiv"), a sign indicating forced labour, as Sieglinde had to endure after being seized by Hunding, who then took her as his spouse. (However, in the prelude of *The Twilight of the Gods (Götterdämmerung)*, the motif, differently harmonized, passes as "wailing motif" with the treble-contour Ab G.) (Fig. 4.6).

Here, Wagner, who was due to his instruction well acquainted with Bach's music, had probably consciously alluded to the BACH-Motif. The horizon of meaning with

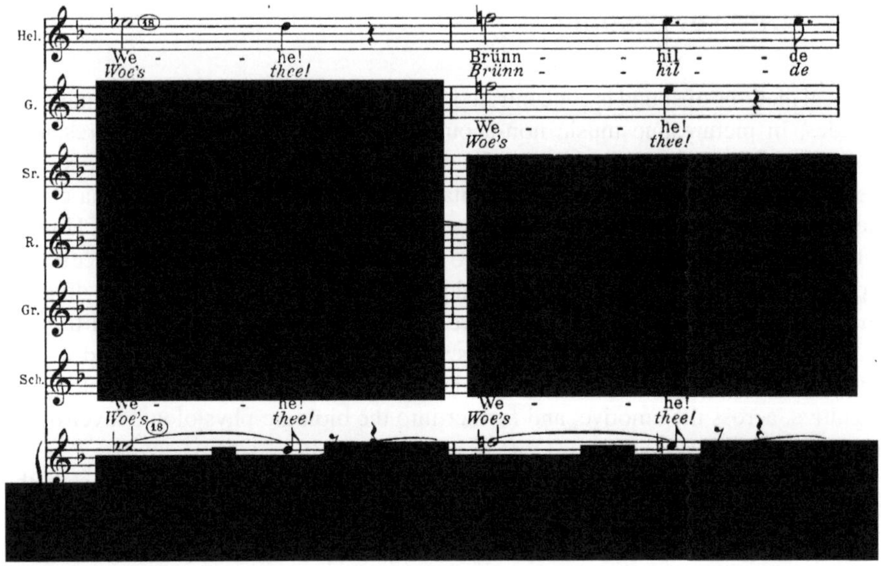

Fig. 4.5 Richard Wagner: *The Valkyrie*, III. act
Wailing
Piano Score, Karl Klindworth; Frederick Jameson, English text. Mainz: B. Schott's Söhne, 1900. Plate 26590. Nachdruck New York: G. Schirmer, 1904. Plate 37218. IMSLP16815-Wagner_-_Die_Walküre_(vocal_score)

Fig. 4.6 Wagner, *The Valkyrie*
"Corvée-motif" ("Fron-Motiv")
Vocal Score, Karl Klindworth. B. Schott's Söhne, Mainz (1900). Plate 26590. Nachdruck New York: G. Schirmer, 1904. Plate 37218. IMSLP16815-Wagner_-_Die_Walküre_(vocal_score)

a cross, secularizing grief, which Sieglinde, Siegmund="Wehwalt", Brunhild and Odin (Wotan/Woten) had to bear, suggests it. Wagner himself, at least twice passes a hint: The "corvée-motif" has the treble-contour BA. With the first appearance of "Weh!", as an explicit cry in the II. Act, it appears anew with the tones BA. (Brunhild tried to persuade Odin to revoke the death warrant for Siegmund: "Weh'! Take back (the word) remorsefully!") (Fig. 4.7).

Already this short selection may indicate the richness of available material for a comparative analysis with the help of Fuzzy Logic. The range of facts and the scope of search can be extended. In his great cycle *The lied of the earth* (1907), Gustay Mahler enlarged the descending short second at the closing of the VI. and final set *farewell* (*Abschied*) with a descending whole tone on "forever" ("ewig"). The lament

Fig. 4.7 Wagner: *The Twilight of the Gods*, at the beginning of the III. act
The phrase of lament by the Rhine daughters
Richard Kleinmichel: piano reduction (simplified). Schott, Mainz (n.d.). Plate 23263. IMSLP225682-SIBLEY1802.17368.4e8c-39087011392265score

is to such a degree neutralized and entwined with consolation, just like some kind of "die and become".[86]

Die liebe Erde allüber*all*
Blüht auf im Lenz und grünt
Auf's neu! Allüber*all* und *ewig*
Blauen licht die Fernen!
Ewig … ewig …

Everywhere, the beloved earth
blooms in the spring and
is newly green! Everywhere and forever
the distances are blue and bright!
Forever … forever …[87]

As a keyword Mahler choose 'all over *all'* (Allüberall), as if it contained accidentally the pitches with the wailing-motif BA (Fig. 4.8).

Likewise, for the next keyword "and forever", initially the short second of the wailing-motif is his choice, but ascending (Fig. 4.9).

All the more emphatically is the effect of the final descending long second on "eternally", which is afterwards next to endlessly often repeated (Fig. 4.10).

Only sweepingly can be referred to relations concerning biophonetics and phonosemantics. As the scope of mimicry in nature, so the effect of similarity in the arts may as well be intensified and changed by something, like the already mentioned „super-normal dummies"in ethology. As in fine arts, and likewise in music as well, caricatures are musical exaggerations, which produce similarity with a participation of dissimilarity. In music, they are less common than in fine arts or literature,

[86]Incidentally: "forever/ewig" is to be heard repeatedly later on in the tones E-D. They constitute also the central phonemes of *Er-De* (‚earth') in German. Thus Mahler condensed heavily and combined two words into a single statement – the earth will exist forever – musically as a unit, in addition to the variation of the wailing-motif.

[87]Translation: Larry Rothe, http://www.keepingscore.org/content/farewell-das-lied-von-der-erde-song-earth-symphony-tenor-and-contralto-or-baritone-and-orc, retrieved 25.3.19.

Fig. 4.8 Gustav Mahler: *The Lied of the Earth* (1907), VI. *Farewell*
Excerpt of the final "(all over-)all"
Josef Venantius Wöss: Vocal score. Universal Edition, Vienna (1912). Plate U.E. 3391. IMSLP112602-SIBLEY1802.14766.d11c-39087017290620score.pdf

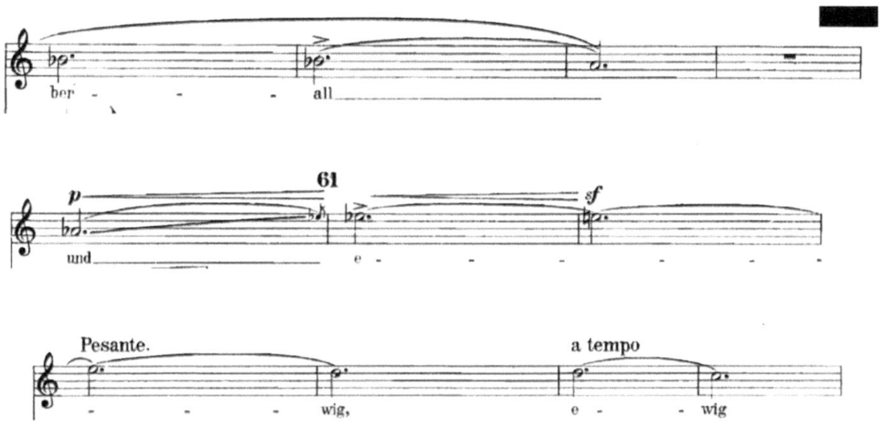

Fig. 4.9 Gustav Mahler: *The Lied of the Earth* (1907), VI. *Farewell*
"and forever"
Wöss (1912), ibidem

Fig. 4.10 Gustav Mahler: *Das Lied von der Erde* (1907), VI. *Farewell*
„Forever … forever …"
Wöss (1912), ibidem

although they occur as deliberately unpleasant, and therefore as music, which characterizes something in a negative way. It improves expressive quality and intensive cathartic effect.[88] We may also apprehend this as some special, purposeful type of conscious blurring, but *at the same time* as a sharpening, so to speak, excessively sharp. Likewise Bertolt Brecht distorted with the help of alienation effect, amongst the use of other elements for stage acts, the range of facts or characters „in a discernibleness".[89] Instead of upright imitation, or an even a less honest plagiarism, which does not want to be discerned as imitation, but pretends to appear as far as possible to be like the original, the difference of critical or comical parody of

[88]Extensively to this, see Kandel 2012.

[89]See Sect. 3.2. Foley and producing noises: Similarity through dissimilarity.

style must be conspicuous (see Sect. 8.2). Wagner, in *The Meistersingers of Nuremberg,* ingeniously composed an unintentional distortion of self-exposure, and at the same time a theatre-figure's characterization. When Beckmesser during the singers contest, in the finale, performs Walther's *Lied of praise* (*Preislied*) as an—involuntary—caricature, Wagner compositionally imitates self-deceiving and deceiving, thereby imitating plagiarism (Extensively see Sect. 7.7.1).

4.4 Art with Numbers. Musical Mimesis with Proportions as Paradoxical Abstract Iconic

The real in music may also be relations, quantitatively measurable and describable proportions between a range of facts, or in shape of things or persons. There are still other ways and methods of mimesis than those, which were outlined in the previous section. Sometimes, the relations of similarity, of reality in art work follow strange, winding paths. In vocal music, the usual mimesis produces essentially iconic, or, in a wider sense picturesque similarities: Real ranges of facts, as referred to in a text, are musically reflected in such a way, that in a plain case, for instance, „upward", will be transposed into a melodic upward movement. It is a matter of a kind of metaphor: Movement in real space is translated into a movement in a imaginary-real space of sound. Already in Gregorian chants, can traces of such an analogy be detected, for example, in the turn of the *Credo* „ascendit ad coelum" (*he ascended to heaven*). Thus, it is a matter of a common, so to speak, doubled sigmatism, the relation of sign and object: The fact of ascension is reflected, both, in the text and the music as well. Even a modern example shows both directions of movement, obviously at the same time (Fig. 4.11).

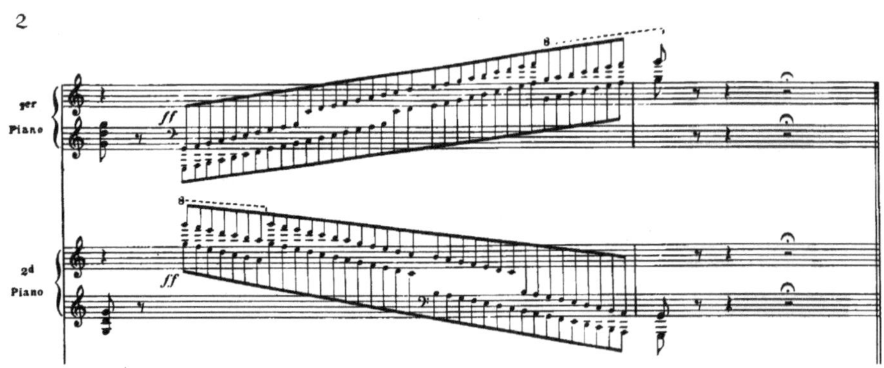

Fig. 4.11 Camille Saint-Saëns: *The carnival of Animals* (1886)
III. Set, *Hémiones (Animaux veloces)*—Wild Donkeys (Swift Animals), *Presto furioso*, final bars
First edition. Durand, Paris (1922). Plate D. & F. 10,154. IMSLP576538-PMLP6099-Le_carnaval_des_animaux_-...-Saint-Sae-ns

4.4.1 *Mimesis with Visual Abstractions and Geometric Proportions—Alvin Lucier and Alban Berg*

But, in music, there are also spatial analogies, which „methodically" take the detour of notation. And quite often, they are more vivid and comprehensible than in their sound form, as in both times the matter is primarily visual. For instance, Stockhausen‘s diagram of basic spectra along seven basic durations derives from the contours of a landscape, a mountaineous panorama of a health-resort, once visited by Stockhausen. This could have even passed into the conception of work as hidden motivation and representational reference.

Alvin Lucier reproduced both geometric objects as well as concrete real facts, namely, special objects and preferably fruits and vegetables in their geometrized contours. It is not a question of a musical illustration of abstract pictures and shapes, but of real facts, which are not at least, rather ordinary things. Lucier draws them more or less meticulously. The material of his special mimesis are, in these cases, rather abstract geometric shapes, sometimes objectivized in the work or the sound, sometimes in the notation. The decisive arrangement is a synaesthetic optic association, but primarily more notational-optic than tonal—the glissando as diagonal (see Sect. 7.4.2.3).

Whereas Lucier‘s other procedures do not show any visible relations of similarity: Since the 1960s generated percussive sounds through the deduction of his own brain waves (see Chapter 9).

Lucier requires the spatial concept of Western notation.

> In compliance with the convention of musical notation, Lucier interprets the pitch as ordinate, the time as abscissa. […] Hereby, no frequencies are graphically depicted, but tone locations derived from intervals.[90]

Glissandi as diagonal motions in imaginary-real tonal space, remain Lucier‘s preferred means for the presentation of geometric figures.

> *Kettles* (for five kettledrums and two sinus generators) were probably the first instrumental composition, in which Alvin Lucier explicitly specified the illustrative character of slow glissandi. In the preface to the music score it is stated about the sinus tones: ››The vibrations are moving in opposite directions in slightly different speeds, whereby they form asymmetric mirror images. « The emerging parallelogram will be graphically manifested in the score-scheme, which shows the glissandi of the oscillators and the resultant beat frequency. […] *Music for Piano* [pick up] in their sinus tone glissandi the parallelogram, which was in *Kettles* (1987) already established. Both oscillators begin in unisono, and the first requires 7' 12", in order to rise two octaves. Then, it returns within 8'48" back to the original sound. The second oscillator in reverse requires 8'48", in order to descend two octaves. Then, it returns within 7'12" back to the original sound.[91]

For Lucier, too, the expression of angels is only an approximately solved question (Fig. 4.12).

---

[90] Straebel 2018, pp. 54f. and 56.

[91] Ibid., p. 54.

Fig. 4.12 *Music for Soprano* II. Set *Triangle* Quoted after Straebel 2018, p. 55

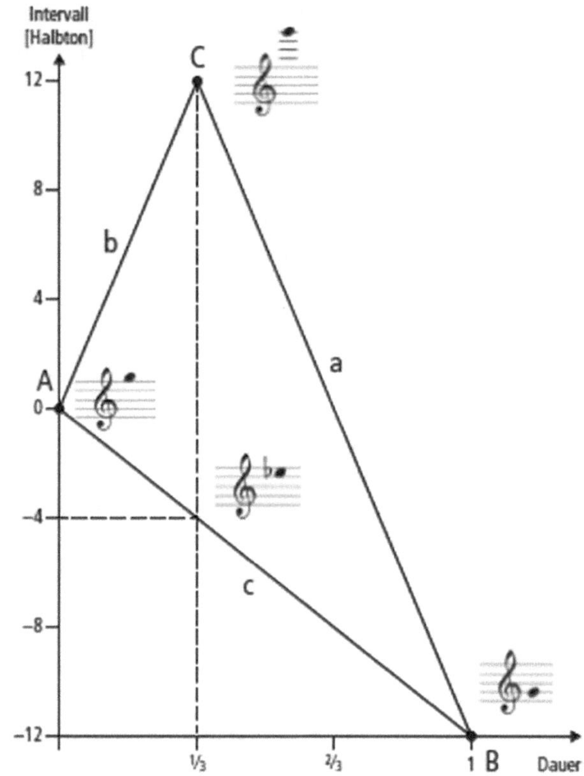

Abb. 2: Alvin Lucier, *Music for Soprano* und Oszillatoren, II – *Triangle*, grafische Analyse

> The first two sets of *Music for Soprano* […] form simple geometric figures, for their presentation the proportion of scaling of the basic note and time axis is of importance. In the ››triangle « of the second set the lower voice glides with a steady tempo in an octave below. (The starting and end points being marked A and B, the side which they form is C). The upper voice glides in a third of the time, from the same starting point A, one octave upwards (towards C), then, falling two octaves below, and in two thirds of the total duration of the set, reaching the end point B.[92]

In such works abstract geometric forms are matter of mimesis. But, Lucier takes, in another group of works, almost exceeding concrete things as matters. The optic-acoustic geometric mathematization remains, material and technique are in both cases the same.

> In *Fruits and Vegetables* for baritone, piano and oscillators (2000) in 13 short sets, inter alia peppers, cucumbers, spring onions, figs and papayas are depicted. Lucier made a very precise measurement of his presentation and formed surface lines partly with individual sinus tone sequences, so that in case of cellary it reached to no less than twelve oscillator

[92]Straebel 2018, p. 59.

voices. Finally, in *December 12th* for flute, violin, cello, piano and oscillators (2013) four sinus tone glissandi follow in the shape of a grapefruit.[93]

This mimesis'procedure is not simply abstract and arbitrary, as in a wider sense it is concerned about similarity, which are arranged by means of numerical proportions and geometric shapes. These are represented, however, once again rather abstract—scarcely in sound, but above all through the note picture—the concrete before the abstract.

The objects are not „drawn" by sound, on the contrary, two-dimensional graphic representations (outline sketches) of objects will be aligned with two-dimensional representations of sound processes, which are codified in musical notation.[94]

Lucier's mimesis is rather abstract, but it *is* mimesis, and as such it is intended with the aim for a similarity of musical reflection and reality. Even though it concerns mostly rather trivial, everyday objects. But, in our case, this remains a secondary problem. Straebel points in his meticulous analysis again and again to inaccuracies, conscious-arbitrary variances of Lucier's self-imposed proportions and blurrings. Furthermore, with the help of Fuzzy Logic it could be probed, how acute or blurred is the mimesis of real matters within the given abundant repertoire, also in comparison of single works to *each other*.

Like angles, circles as well, cause a special problem for presentation.

Only since *Fruits and Vegetables* for baritone, piano and oscillators (2000) Lucier has more precisely considered the curvatures of the depicted objects and realized in *Two Circle*: for flute, clarinet, violin, cello, piano and oscillators (2012) and *Double Rainbow* for female voice and oscillators (2016) circular curves and parabolic lines with a continuously changing gradient.[95]

In European history of music the musical representation of the circle was frequently attempted, starting with Guillaume de Machault in the fourteenth century. The form is solely representable as an imaginary-real motion, and it was commonly combined with the comparable temporal motion of a crab, or a palindrome, as for instance, in the circle canon (canon perpetuus). Thus, the circling would be slightly more extensively defined: Its beginning and end merge, the ending returns to the starting point. For Fuzzy Logic it would be a difficult, but attractice task, to measure the various types and degrees of approximate approaches, especially across extended historic fields with their huge differences in material, stylistic and idiomatic ways. The circle, like its three-dimensional extension, the ball, with its philosophical, and even in everyday consciousness with its emphasized form can only rudimentarily be represented in sound; it would be easier, to combine the metaphoric, or pictural, notation with the optic dimension.

Even more pronounced than Lucier, Alban Berg refers to the tradition of „music of the eyes" (Augenmusik) in a prominent passage of his opera *Wozzeck*. During

[93]Ibid., p. 62.

[94]Ibid., p. 70.

[95]Ibid., p. 54f.

Fig. 4.13 Alban Berg: *Wozzeck* (1925), I/4, m. 561 Score U.E. 7379/12100, p. 133. Arranged, detail

Fig. 4.14 Circulatio 2009_Bach_MP_Duellmann_Figurentabelle (figures' table), retrieved 26.3.19

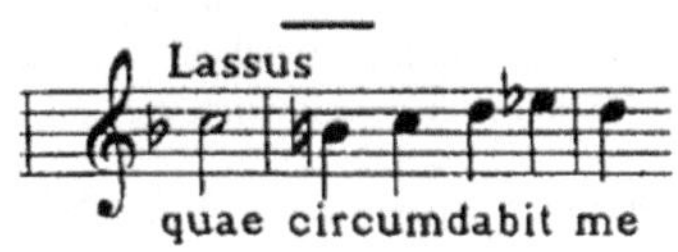

the dialogue with the medical doctor, I/4, T. 561 (a single 7/4 bar) Wozzeck gives an account of his hallucinations: „Lines and circles – strange figures – who could be able to read them!“ We may read such ominous signs, that is seeing them, and allusively even hear them, in so far the two-dimensional spatiality is transferred and associated by an intermodal-synaesthetic analogy into the imaginary-real spatiality of music (Fig. 4.13).

Most remarkable is the circular form, as far as possible, with its each time contrapuntally-opposed, ascending and descending lines in the 1st and 2nd solo violin-cello with the whole tone hexachord d-e … or c-B … etc. The circle figure is thus formed in a slightly shifted diastematic symmetry with an each time arc-shaped, 11-tone melody.[96] Berg integrated in his melodic lines the traditional rhetoric figure of ‚circulatio‘ or ‚kylosis‘ (revolving movement, ‚encirclement‘). The passage in Orlando di Lasso‘s (Orlande de Lassus) text has ‚the one who encircled me‘ (Fig. 4.14).

Mostly it performs a wavy—and therefore actually no circular—motion. Berg has, at least notationally, sharpened the circular form as far as possible.

4.4.2 Lengths of Structural Parts and Numerical Proportions as an Expression of Gender-Relations in Case of Artur Schnabel and Johannes Brahms

Primarily tonally and in temporal proportions the spatial proportions are depicted by the pianist and composer Artur Schnabel. In his Piano Sonata (1923) he draws a musical self-portrait, being at the the same time a double portrait with his wife,

[96]Here it must be done without further details on the formation of symmetries and on semantic numbering. The fact that Berg's choice of 561 as the first of the strange Carmichael numbers for this remarkable passage is possibly not accidental, shall not be further commented here. This Carmichael-number—$3 \times 11 \times 17$—includes numbers, which were of importance to Berg.

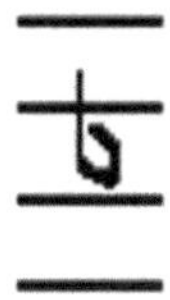

Fig. 4.15 B-molle
Membeth: B molle in der Quadratnotation. 14 July 2007. https://de.wikipedia.org/wiki/B_molle#/media/Datei:B.molle.gif. Retrieved: 14 July 2019

the singer Therese, née Behr. The features reflected in the real facts, here, for both individuals, are expressed by Schnabel in and with music in different ways and with different means in line with the principle of similarity. They are mostly acoustic, and not optical.

1. Firstly, the similarity appears rather common-sketchy, and non-iconic: The excessive indications in the score, trending towards one per tone, are a symptom of Schnabel's hyper-expressionism. While at the same time they fit with her as well: The singer Behr was frequently attested a special expressiveness.
   More over Therese Behr is in many ways represented.
2. Initially, not yet with icon, but with symbol signs: The complete name in tone letters with forenames and last names T = D[97] ErESE BEHr is in arDur scHnaBEl fully embodied, proto musically and thus music-capable.

.CHnAB.l
eS E
A – eS
(D) H E r E eS E
B E H r

The German „B" is in English the „B flat", can therefore be used also in English. It does not well with the German letter „H" = „B sharp". The English terminology of tone letters maintains with the ABC … the historic origins in method and notation more accurately than in German. Already in times of „Gregorian Chants" the tones of the seven-step scale (diatonic) were denoted alphabetically as A B C D E F G. With the coming up of chromatism the B was divided in B molle or rotundum (B♭ = B flat) and B durum or quadratum (B# = B sharp); thus, already around 1025, by Guido of Arezzo within the scope of his Hexachord-teaching (Figs. 4.15 and 4.16).

The shape of the durum sign is *similar* to the letter h. In German, as well as in Scandinavian and Western Slavonic, it was referred to as H—so likewise the names for the major and minor modes derived from here, in difference to the English major and minor, which refer to the constitutive major and minor third above the keynote. Moreover, the signs for low and high alterations derived from♭und #.

---

[97]The substitution of T with the music-capable D is not uncomon. In an album sheet by Ignaz Moscheles for Louis Spohr, dated 4.3.1815, Moscheles characterized his substitution of t with d (with the expression „aecht" (genuine)) as „lizentia poetica" (poetic licences).

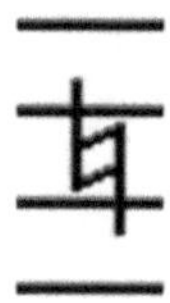

Fig. 4.16 B-durum
Membeth: B durum in der Quadratnotation. 14 July 2007. https://de.wikipedia.org/wiki/B_durum#/media/Datei:B.durum.gif. Retrieved: 14 July 2019

3. The cipher Therese Behr is connected through the semitones E-eS and B-H with the semitone steps BA CH.
4. Therese Behr has not only a further big B, but the B featured prominently as the initials of her maiden name, which she kept on in her double-barreled name.[98] It also implied the tritone, B-E, this was certainly a desired parallel to the tritone A-eS (E♭), (the acronym or initials of *A*rtur *S*chnabel.
   The almost triumphant triple B in the completion of the sonata is ambiguous. Primarily meant is, aside from Bach, the greatest of all B's, namely Beethoven, as whose congenial interpreter Schnabel was famous. But just as well, this B includes Behr. Immediately, before the final chord the tone pitches of the letters AS-Es-H-C appear, followed by B B B / A.
5. Therese Schnabel-Behr enters the work through the prosody, or the number of syllables of her Christian name jointly with Artur's prosody, through the number 5 (3 + 2). This happens both with preferred lengths of phrases and also with the large-scale arrangement, here, in the guise of a joke. Then, from the perspective of „double portrait" another part of the title and aspect of the work becomes then relevant: Sonata for piano. Piano piece in five parts of A.S. (1923). With *five parts*, Schnabel deliberately differs from standards—this is once again a manifestation of his principle „completely free". The violin sonata *Sonata for violin solo* of summer 1919 and the piano suite have also five sets. The freedom from standards is not the sole motif. Moreover, Schnabel may still have linked such a freedom with himself and his personality otherwise. As by coincidence, his first name ARTUR has five components, or letters. (A critic censured Schnabel's dropping of the ‚h' in Artur as mannered.) The number five itself reveals still another motivation: Artur has 2 *syllables*, while Therese has 3; together this makes 5. In respect to the *number of letters* the proportional result is 5:7. The number five and the number seven hint to irregular lengths of phrases and motifs. Probably, for him they became a kind of second nature. Because, the deviation from the „squaring of the composition", the composition of themes in ranking or encaptic 2-bars-clustering units of music, which became standard since the music of absolutism, is as „prose" a feature of avant-garde composing. However, if Schnabel forms explicitely „quadratic" lengths of phrases with a „hidden eight bars mode",[99]

[98] Artur Schnabel did not bear a joint name, but he led in respect to his marriage a double life. This is certainly under patriarchal conditions somehow irregular, but regrettably not very unusual.
[99] Detailed in Dümling 1987, p. 15.

then, this is anew regularity as contrast in second potency.[100] Thus, for the sake of diversity, Schnabel thwarted his prevailing deviance from the norm of powers of two. From the avant-garde's viewpoint, such a recourse is a blurring in the sense of Fuzzy Logic, while from the view of tradition it is metric freedom and irregularity. Numerical-mathematical motivations and motifs are here, and as a rule, yielding, especially in regard of relations between work and life.[101]

6. After all, in another dimension of composition numerical proportions play a semantic role too. A contrasting relation between life and work results from *proportions of duration of sets.* They are of rather unequal lengths, in fact to a degree which attracts attention: 150-235-195 / 555-555", hence approximately 1-1,6-1,3-/3,7-3,7. The three sets I, II, and III together are only sligthly longer than each of the two sets IV and V on its own (580 versus 555). There hardly exists any internal musical motifs or reasons, in difference to the lengths of phrases. To ask for whys in order to explain this situation, the answer lies in an aspect of the relation between the couple Artur + Therese, as in the case for the motifs for the five-sets. The stature of his body was fairly smaller than hers. A visual and significant deviation from the normality of gender relations. Moreover, it seemed unconventional, that Therese Behr was older than her husband. Therefore, she is not only spatially longer, but temporally too, insofar, as she had already experienced a longer lifetime when both met. Time and again, either was discretely, or to a lesser extent discretely addressed, so both of them and the public were conscious of it. At a joint concert at Berlin's Singakademie, even before their marriage (9th of July, 1905), Therese Behr, accompanied by Artur Schnabel, sang Hugo Wolf's *My Darling is so small* (*Mein Liebster ist so klein*), from the *Italian Songbook* (*Italienisches Liederbuch*), and twinkled in direction of her companion. The audience applauded with roaring laughter about such a dark joke.[102] Now, in the imaginary-real space of the piano sonata Artur turns the real relations upside down. If the ratio of numbers for the set periods in the specific number of syllables (2 + 3 = 5 syllables, accordingly in 5 sets) are related to Artur + Therese as well, then Schnabel turns the relation of lengths as relations of durations simply in reverse: The 2 for Artur is almost double in size, in relation to the 3 for Therese.[103] Politely he concedes precedence for the lady.

---

[100] »Like in his piano playing and during his instructions, Schnabel very strictly considered in his compositions as well [...] the deviations from the regular periodicity of groupings with two-, four- or eight bars. Groups of three-, five- or six bars are always used as an expressive device and in the score of the 1st string quartet they are continuously indicated with Roman numbers.« (Dümling 1987, p. 14).

[101] Compare the relevant studies of Thomas Phleps, among others, Phleps 2001a und 2001b, 2002.

[102] Compare Searchinger 1957, p. 96.

[103] Here. I follow a hint by Hanjo Polk (given on *3rd of October.2015*.—Schnabel's joke, in the structural iconic, widely surpasses the subsequent joke. Question: How large in the difference in length between the *Nile* and the *Amazon*? Answer: 2 letters. In substance, the effect of the joke is rooted in the fact, that in this case the lengths are not taken and compared, but instead a totally arbitrary reflection of word language is made, and this in the written but not in phonetic form. See: https://schlechtewitze.com/buchstabe, call 3.8.18.

> Therefore one may not say, that the musical mimesis is simply a depiction. In fact, the reflection holds a mirror up to the real relations, which virtually turns them around—depending on one's viewpoint, it can be critique or a distorting mirror. The psychic gains of pleasure and power form such a musical compensation is evident, and needs no further discussion nor reasoning.

Johannes Brahms, composer and pianist, enciphered also a gender relationship with reference to relations of lengths and ages in his *Variations for Piano* op. 9, also by numbers, but in a somewhat different way, non-iconic and tonal, more implicitly than Schnabel. Actually he signed the particular variations either with „Br.(ahms)"or „Kr.(eisler)", modelled on the odd music director of E.T.A. Hoffmann. „The number of ‚Br.' and ‚Kr.'-variations points to a significant proportion of 8:5 (Johannes : Clara)."[104] She was born in 1819, he in 1833, thus 14 years older, thus in terms of time and thereby numerically ‚longer'. Here Brahms turns the real proportions musically around, likewise for his benefit. Actually he was of extremely small stature, compared to Clara's impressive size, may be in their biographics this problem is not adequately assessed, whereas under patriarchal circumstances it is relevant. Thus Brahms, like Schnabel afterwards, takes the *lengths* of bodies as his point of departure, he does not present them as time relations, but only as the respective letters of names. Hence Clara is losing out. That moreover, quasi as pseudonym, the quirky and even crazy music director is assigned to her, while Brahms remains himself, shows a certain added malice.

Here, Brahms abstracts and analogizes, in order to swap the levels of signifier and signified, what may not be acceptable from a scientific point of view, as in the mentioned comparison of lengths of the rivers Amazon and Nile. But for an artist his conduct is justified and correct, like in the following joke. Question: The ABC has how many letters? Answer: 3, namely abc.[105] But for Brahms' joke it is essential, that there is a secondary iconic, and thus a ‚real' similarity. Because the respective size in the type*face* of Clara versus Johannes shows beyond the arithmetic, once again geometric measures of *length.*

4.5 Art and Art—Intertextuality: Reflections and Interactions of Music with Texts, Pictures and Other Arts

Subject of musical mimesis may be texts or pictures, works of poetry or fine arts, anything that was already formed in an artistic way, as medial layer between music

[104]More detailed, see: Phleps 1997. It is a coincidence, but one that is musically suitable, that 5 and 8 plus their addition 13, are numbers from the start of the Fibonacci sequence, to which we already frequently referred to,—1, 1, 2, 3, 5, 8, 13 etc. .. Musically, this is especially highly relevant in the context of asymmetry, which is fitting to Fuzzy Logic. If we further add, that both had met in 1853, or permuted in sequence 1-3-5-8 (without 2), so it is rather to much of a coincidence.

[105]https://schlechtewitze.com/buchstabe (bad jokes, letter) called 3.8.2018.

and reality. This includes relations between various arts, with their mutual reflections on each other. The relation between poetry and musical art is very intense: In vocal music both are intertwined with one another to become a new entity. Through rhythm, as their common element, even dance becomes a component of music. On the other hand, dance and texts, just like pictures, may become subject of musical reflection. For the dance a predetermined choreography is necessary, for instance, for Salvatore Viganó's ballet *The Creatures of Prometheus* Beethoven wrote the action-related and scene-related music. This falls under the category of ,program music'. Frequently duplications and multiple reflections occur. Thus architecture, which is (partially) also a genre of fine arts, is subject of pictures, which in turn may be arranged in program music themselves, like the finishing touch in the 15th piece *The Great Gate of Kiev* (Kyiv) of Modest Moussorgsky's *Pictures at an Exhibition.*

4.5.1 Medium Change and Total Art Work ("Gesamtkunstwerk")

From the outset, identity is excluded as an ideal for all these ,medium changes', because it involves the various spheres of senses. medias and artistic materials. Objective-subjective contexts provide analogies and special synaesthesias. Similarities between different spheres of senses through such associations are imaginary-real. They are often entirely subjective, though they may be compulsive and not intentionally guided, or, on the other hand, arbitrary, free, but high-handed. Admittedly not wholly, because there are still certain intersubjective correlations, and such limits of arbitrariness are particularly interesting. For instance, for the characterization of keys in European music the associative color coding of C major will certainly never be in black, or hardly in dark at all, nor does it from assignments of affects express a tonality of grief.

Intertextuality is a central aspect for the total art work's concept. Therein, the various arts are indeed fusing, but never fully, as they remain individually distinguishable. For instance, under the aspects of the total work of art and the complex of ,ready-made', the ,hip-hop' of the 1970s combined various arts and even non-artistic cultural areas. "Hiphop includes music as well as graffiti, breakdancing, and fashion. It is an art of (often illicit) appropriation: hip-hop music samples or borrows from preexisting recordings, graffiti marks public transportation, breakdancing incorporates robotic and mechanistic movements, and hip-hop fashion literally rips off car emblems and fashion labels to make new collages."[106]

In other types of total art works, as part of the reception process, the intermodal, multi-sensory sense of touch, and even the „lower" nearer sense of smell, in combination with conscious, active physical movements are integrated. "Installations often require spectators to interact physically with the work through actions such as walking, *touching,* sitting, *speaking*, and *smelling*. Installation art arose in

[106]Demers 2010, p. 171.

tandem with other attempts (minimalist, site-specific, environmental, and conceptual art) to dispose of the frame and the museum as preferred vehicles for separating the artwork from the spectator and the outside world. Sound and/or music often play a role in installation art, usually underscoring the contingency of perception on bodily placement."[107]

This results in a three-digit configuration of relations: 1. musical work of art—literary or picture-artistic work of art; 2. literary or picture-artistic work—reality; 3. musical work as their neutralization—reality. „Reality" as such, is layered and differentiated—in nature, society, worldviews, and mentalities, on which works of art refer to. Similarities, blurrings and sharpness in the sense of Fuzzy Logic are to be analyzed in a more sophisticated way. A possible alleviation could be, if one reduces both of the respective works, for instance, The Great Gate of Kiev, as a painting by Hartmann, and as a piece by Moussorgsky, to find a common third, which could consist likewise of preferably reduced elements. The reduction can be multifarious, like in the sequence of situations, or in the structuring of a given whole.

4.5.2 Word Art and Musical Art, Setting to Music and Textualization (Tropatura)

For the composition of texts there are two universally complementary procedures in the ratio of text and music: Setting to music and what I call „textualization"—a traditional, often slightly broader expression for it is troping. Normally the texts, which are suitable for music, are not prose, but poetry, such as lyrics, poems as clearly defined structures. The rule does tolerates no exceptions, like, for instance, in a „Gregorian" chant.

Setting to music: 1 text, n (many) melodies and compositions

Textualization: 1 melody, n texts

A setting to music of lyrics, which operates with abstract-visual elements, is *Six Geometries*, the choral work of several parts, by Alvin Lucier, processing poems by William Carlos Williams. He paid special attention to insignificant everyday objects. Like in the VI. Set, *Plums*, „plums are mentioned, which the protagonist had tasted in *This is just to say*, though they were probably to be kept for the breakfast (from *Poems*, 1934)."[108] The II. set, *The Letter X*, relates to the central letter in the word „saxifrage", in the well-known poem *A Sort of Song* (from *The Wedge*, 1944). "Lucier designed the »letter X« in the second set [...] in a graphic formal-scheme of the composition with two octaves pitch and 2′24″ wide. Both of the sinus-tone glissandi are sliding within in a speed of six seconds per semitone towards each other, crossing in the middle of the set, and move on in the same speed (Fig. 4.17)."[109]

---

[107] Ibidem. Emphasis H.-W.H.

[108] Straebel 2018, p. 59.

[109] Ibid., p. 59. In 2002, Lucier revived the X in *Almost New York* for five flutes (one player) and oscillators. „Here the X is four octaves up and 24′ wide. Thereby each of the semitones, which

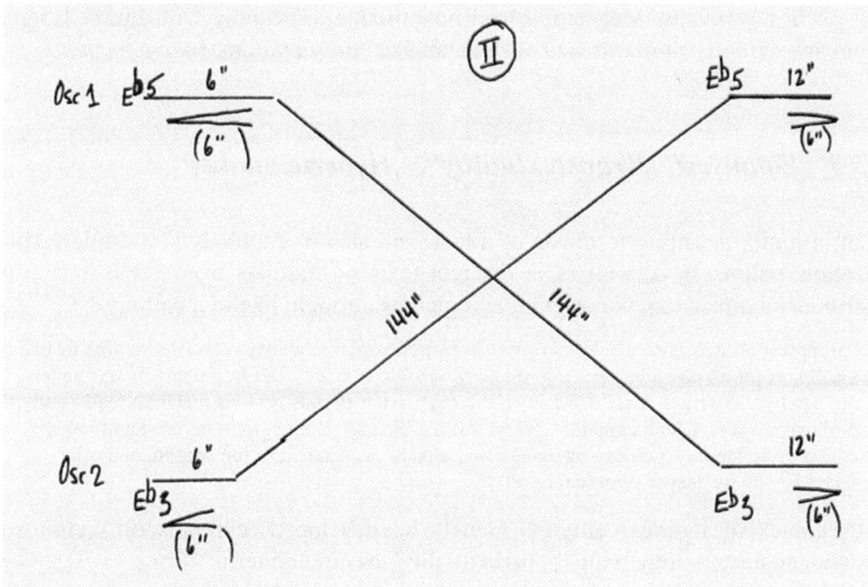

Fig. 4.17 Alvin Lucier: *Six Geometries*, II – *The Letter X*, scheme of the composer
Cited after Straebel 2018, Fig. 4, p. 59

For Fuzzy Logic, a particularly rewarding starting point are multiple settings in music, as well as multiple textualizations. Specially the contrafactum, the multiple-textualization is probably almost an intercultural universal—generally infolk and in popular music there are more texts than melodies available. In European music contrafact is common in art music and in folk music, documented since the troping in (Gregorian) chants in the high-medieval period, or high feudalism. Here, this special far-flung issue cannot be persued further. An objectifiable yardstick for the different musical presentations is the principally consistent text—along with the reversal, music as yardstick for each of the different textualizations. Here, this cannot be further persued. There are certainly on both sides sharpenings and blurrings, for instance, to the effect, that the reference to reality, or to the object of the text and its meanings, are presented in detail, either more profiled or less noticeable, such as in tonal painting (onomatopoeia), being the simplest procedure. The similarity of the text‘s mimesis in music, being a specific form of mimesis, with only indirect allusion to reality, is as such to a certain extent measurable. Because, in this medium change music and lyrics are basically closely related. There are certainly common roots for both, that is genetic relations, and structural as well in prosody, or rhythm

---

are reached by the sineus-tone glissandi, corresponds with a sustained note of the flutes, whereby their performances can be varied, so the tremolos at the beginning and the ending of flute notes may differ“ (Ibid.). Even more elaborate is the crossing of lines in Conlon Nancarrow‘s *Study 21 – Canon X* for mechanic piano (Extensively, see: Thomas Phleps 2000).

and pitch-movements. Measurable and quantifiable, especially with Fuzzy Logic, are in any case the similarities of various musical presentations *among each other.*

4.5.3 Palimsest, „Transtextuality", „Hypertextuality"

Both in multiple setting to music of a text, and also in a melody's multiple textualization, palimsest, as well as in discreet texts or citations from vocal music in instrumental music too, the original text gleams through, like in a palimpsest.

> Palimpsest is first of all a rewritten or overwritten papyrus, whose original text was technically, as far as possible, erased and is only visible in traces and fragments in and under the newly overwritten text. Nowadays, the expression is mostly used metaphorically and designates a text, which contains „below"the legible text another, marginalized and largely erased primal text, which can be understood as sub- or metatext of the second, and offers a secret key on its deeper significance.[110]

In the context of ‚intertextuality', G. Genette has developed a complex, differentiated system and terminology with "hypertextuality" as guiding term.

> Hypertextuality concerns the most complex type of transtextuality. Genette defines as hypertextuality all connections between a previous text *(the hypotext)* and the actually available text *(the hypertext),* as well as each of the derivatives, which were caused by various kinds of transformation processes. There are two basic types of hypertextual procedures: the *transformation* and the *imitation.* Transformation means, to transpose a style or a subject into another text – a deformation of the hypotext. Imitation means, to form a hypertext precisely after the model of the hypotext.[111]

4.5.4 'Music Form' of Texts: Program-Music I

‚Text' in music, or to put it in another way, subject and matter, can both be taken from literary works as well as visual arts. Music, corresponding to images, or to poems, dramas, or even to concepts, which the composer himself has formulated, in poetic or prosaic form, is program-music. It is principally instrumental. In contrast to vocal music, with its manifest word-linguistic text, here, the ‚text' is latent and in another way suspended, such as by sound painting or musical painting. From the point of view of literature, this is the *music form* in the literary process. Mainly it refers to epic poetry and dramatic art, though not exclusively. Normally, in vocal music, lyric functions immediately as text, which is set to music.[112] In vocal music, an exceptional case of lyrics set in music is Claude Debussy's straight instrumental orchestral piece

[110] Wulff, Hans Jürgen: Palimpsest, http://filmlexikon.uni-kiel.de/index.php?action=lexikon&tag=det&id=3566, 8.2.2012, retrieved 21.2.19.

[111] Gérard Genette 1993. Here, following his lecture, in http://de.wikipedia.org/wiki/Palimpseste._Die_Literatur_auf_zweiter_Stufe, 11.5.18, retrieved 8.8.18.

[112] See below, 8.3 Verbal form, and 8.2 Quotation form. See also Genette 2001.

L'après-midi d'un Faune (*Prelude to the Afternoon of a Faun*, 1892/1894), after a poem by Stéphane Mallarmé, an „églogue" in the tradition of the Bucolics and Idylis poetry of late antiquity, in particular by Theocritus. But Mallarmé breaks this line, by creating his eclogue with 110 alexandrines, that is in the rigid verse metre of French classicism, not ideal for setting in tones. As usual, he then breaks again with the regularity of verse lines. Debussy doubted the possibility of a merging for this poem.

> It [the music] aspires in no way for a synthesis with it [the poem]. On the contrary, they are successive depictions of the mood, through which are moving the faun's cravings and dreams in the heat of an afternoon. Then, he becomes tired of chasing the frightened fleeing nymphs and naiads, succumbs into a deep slumber, finally, when dreams came true he is saturated, by a complete rule in the all-embracing nature. (Debussy).[113] (See Sects. 4.6 and 8.2)

Already the intertextual disposition of the source is peculiar. Because, in 1876, the poem with illustrations was published by Édouard Manet. Little wonder, that afterwards Debussy's work itself became later the source for a ballet—Mallarmé's poem as a program for the choreography was at hand. In 1912, under the influence of Sergei Diaghilev, Vaslav Nijinsky danced and directed himself the ballet with the same title *L'après-midi d'un Faune.* Then, in 1953, Jerome Robbins choreographed his version of *Afternoon of a Faun.* In this sequence of intertextualizations, one link is missing, namely, the picture after the ballet.

> Rodin was deeply impressed when he attended the performance, thate intended to portray Nijinsky; his project was doomed to fail, because the dancer became extremely restless to stand as a model for him.[114]

Both ballets, by Robbins and by Nijinsky, to the same music allow a comparative examination between two kinds of arts in regard to blurring and sharpening, as hard as such a methodic-technical procedure could turn out to be. As a hint:

> Nijinsky's is set in the ancient world, with designs by Leon Bakst; the movement is two-dimensional, like a bas-relief in motion. Robbins's occurs in a ballet studio, designed and lit by Jean Rosenthal: the dancers not only move three-dimensionally but also refer to the unseen fourth wall as if it were the studio mirror.[115]

Debussy had chosen solo flute for Faun's monologue. Historically, in the literary as well as in the visual arts tradition, it is a symbol, being associated with the arcadian world of herdsmen, nymphs and satyrs, which appears in two shapes, as aulos, a kind of oboe, or as pan flute—naturally Debussy selected for it the modern transverse flute, like in *Syrinx,* the piece that is discussed below (Fig. 4.18).

---

[113]Cited after https://www.mdr.de/konzerte/werkeinfuehrung-debussy-prelude-a-lapres-midi-dun-faune100.html, retrieved 28.3.19.

[114]Por 2002.

[115]Kathleen Leary: *Connected Choreography? Nijinsky's "Faune" & Robbins's "Faun"*, https://www.nypl.org/blog/2018/01/05/robbins-faun-nijinsky-faune, January 5, 2018, retrieved 28.3.19.

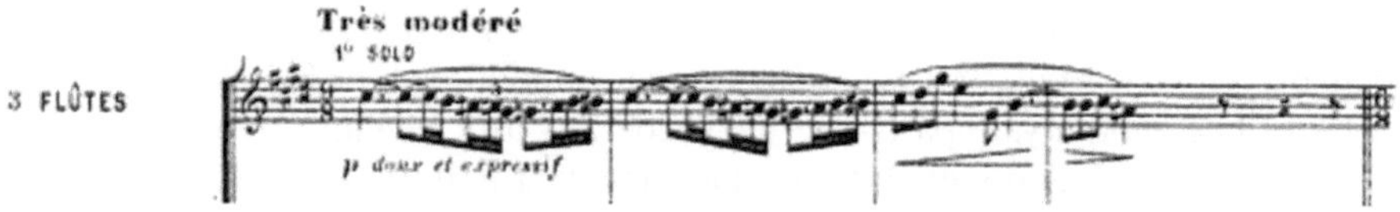

Fig. 4.18 Claude Debussy: *L'après-midi d'un Faune*
Opening, Main theme in the Flutes, m. 1-4

4.5.5 'Music Form' of Pictures: Program-Music II

Analogous to musical mimesis from texts, there exists also the music form from pictures as an ingredient of program-music.[116] Here, in addition, is a further analogy of correlation to the forms of existence in music process. Both of the ‚music forms' are from the view-point of music the counterpart and the reversal of musical forms of existence, ‚word form' and 'picture form'[117] (see Sects. 4.6 and 4.7). These again possess ultimately variants in the ‚form of conceptions', which may be an eidetic draft or a text (see Sect. 7.1), and the musical graphic as picture *before* the music (see Sect. 7.2). Here too, the principally quantifying access of Fuzzy Logic on music is most easily possible and reasonable, if it is a matter of comparison between different arrangements for the same picture. Then, degrees of similarity, or blurring/sharpness, may be determined.

Indeed, Alvin Lucier always relates in several of his works on fine arts, to landscapes. In *Ever Present* (2002) he emulates the layout of a watercourse with a walk in Getty Garden.[118] In its realization of the concept of garden art, it already contains visual art. *Panorama* (1993), is an intertextual dually mediated presentation of a panorama in the Swiss Alps, „which was prepared by Fritz Morach, after a landscape photography by Hermann Vögeli."[119] In doing so, Lucier composes explicitly with the on-looker's viewpoint in mind. The complexity is further enhenced in *Copied Lines* (2011), originally called *Panorama II*. At first, a mountain in Colorado was meant, which Lucier featured by using internet-photos; in a second step, contours of landscape are changed into the wall drawing *Copied Lines* by Sol Lewitt, an outline, which is then, while on an exhibition, when traced by other artists in parallel—necessarily blurred.

[116]To the multiple examples of music set to pictures, see among others: Christensen/Fink 2011, and also the data base *Musik nach Bildern*, where compositions, related to pictures, are listed and indexed with meta data. [http://www.musiknachbildern.at] respectively http://webapp.uibk.ac.at/musiknachbildern/mnb.xsql?&&id=442.

[117]For this context, see among others, the standard reference by Maur 1985.

[118]Extensively in Straebel 2018, p. 67ff.

[119]Extensively in Straebel 2018, pp. 64–67.

4.5.6 *Real and Imagined Pictures as Objects—Respighi's* Vetrate di Chiesa *(1926)*

Mimesis as a reflection with the to and fro between picture and music, like in a hall of mirrors, is already reflected in Lucier's works *Panorama* and *Copied Lines*. No less than an at least fourfold refraction appears in O. Respighi's series *Vetrate di Chiesa. quattro impressioni per orchestra* (‚Church Windows'. Four impressions for orchestra) of 1926:

1. *La Fuga In Egitto*—The Flight Into Egypt
2. *S. Michele Arcangelo*—St. Michael Archangel
3. *Il Mattutino Di Santa Chiara*—The Morning Mass of Holy Light
4. *S. Gregorio Magno*—St. Gregory The Great.

The transformation from pieces for piano into pieces for orchestras is to be added—the first three sets being the instrumentation of *Tre preludi sopra melodie gregoriane* (1919).

> For *Vetrate di Chiesa* ('Church Windows') 1926 Respighi returns to the twin inspirations of early music and the visual arts […] – though the stained-glass panels depicted in each of the four sections existed purely in the composer's imagination, and indeed the descriptive titles were only *dreamt up after he'd orchestrated* the piano originals several years after composition. The most arresting […] is the tumultuous 'San Michele Arcangelo', depicting an apocalyptic battle between St Michael and a dragon embodying the forces of darkness: the Liège brass breathe fire and brimstone aplenty, rasping and thundering in the cod-Wagnerian opening passages and revelling in the cinematic, Spaghetti-Westernish writing for the aftermath of the conflict […].The final 'picture' is almost as remarkable, representing the sixth-century pope St Gregory the Great in a bells-and-smells blaze of glory that carries more than a whiff of the Coronation Scene from *Boris Godunov* and the *Te Deum* from *Tosca* […].[120]

The relation of the work on his wife can be viewed as a fifth level. In 1919, Elsa Olivieri Sangiacomo, one of his composition-students, became his spouse, and she contributed to raise his curiosity for the Georgian chant and the early modes of church music. Then, these became sources of inspiration for many other works, such as the *Concerto gregoriano per violino e orchestra* (1921), the *Concerto in modo misolidio per pianoforte e orchestra* (1925) or the *Quartetto dorico* (1924). A sixth level is the—subsequent—integration of sacral texts as commentary for the respective imagined pictures. In doing so, Claudio Guastalla, his friend and librettist, assisted him. The *Flight into Egypt,* is related in the Gospel of Matthew 2, 14. In turn, the geographic location motivated Respighi for an orientalising instrumentation in 5/4 time, which reminds of Rimsky-Korsakov's *Shéhérazade*. The final IV. movement is a great imagination about the Kyrie Gloria of the *Missa VII* "de Angelis", and a phrase from the *Graduale Romanum* (Commune Sanctorum, 33): «Ecco il Pontefice Massimo!… Benedite il Signore… intonate l'inno a Dio. Alleluia!» as its motto.

[120]https://www.prestomusic.com/classical/articles/2052—recording-of-the-week-respighi-from-john-neschling-and-the-orchestre-philharmonique-de-liege, retrieved 15.3.19.

(See, the Pontifex Maximus! Praise the Lord … sing a hymn to God.)[121] Thus, we have intertextual interrelations in the configuration Early Music—(piano pieces)—Orchestration and New Composition—imagined pictorial models, as source for the titles of the movement.

4.6 Similarities, Affiliations, Structural and Processional Relations: Fuzzy Logic and Music-Analysis

The preceding subchapters 3.3–3.5 were about relations between work and reality, or how these, on their part, were already presented in figurative works in other arts. Now, however, on the one hand, it is about the components *within* the respective work, while on the other hand it is about the relations of works, or components of a work in relation to other works, performed by the respective composer himself or by a stranger. In the latter case, there is a range between accidental parallel, deliberate allusion, and citation. For the types of similarity there are in each case different points (of comparison), on which ‚similarity' refers to. Thus, different shapes may be queried as a whole, but also for their various specific characteristics and components. In case of motifs, certainly the works'most important component in search for similarities, on the one hand, these are overall melodic processes, on the other hand, reduced on elements, stock of pitches, or rhythms.[122] This applies both for work immanent and for work all-embracing, that is to say, for allusions, borrowings, adoptions, and even essentially identical citations. For instance, BACH and his permutations of the quarter-notes store may be taken as a figure, but also possibly a single note of them, the interval structures, the contour of short-melodies, and many more. As repeatedly mentioned, the issue of types of similarity is a quantitative issue as well, and it complementarily includes degrees of similarities, which belong to the topics to be discussed by Fuzzy Logic. Here, Fuzzy Logic has an important task, namely, in the first place, to check the motivic-thematic shapes for consistency or inconsistency according to a basic model, which is hypothetically assumed to be significant, as well as for—statistically registered—frequencies of significant tone pitches and rhythms. It has to be considered, that much of folk music and „traditional music" does not use any explicitly realized basic model, but only (improvisatory) variants with regard to generative models.[123] An example from North African folklore (Fig. 4.19).

In folk and popular music particular similar coincidences and figures are normally already objectified, that means, they are phonetically implemented. In art music, they must still be objectified. In both cases, different kinds and degrees of blurrings occur. On the one hand, in the receptive realization. On the other hand, both in the receptive

[121] For further information see Giovanni Panella: Ottorino Respighi, *Vetrate di Chiesa – 4 impressioni per orchestra*, http://interlitq.org/blog/2018/03/14/ottorino-respighi-vetrate-di-chiesa-4-impressioni-per-orchestra-scritto-da-giovanni-panella/, retrieved 16.3.19.

[122] Ibidem.

[123] So, for instance, Oesch 1987, particularily pp. 124–175, 275–288; see also Elsner 2011.

Fig. 4.19 Excerpt from the nūba Ġrīb after a recording by the well-known Algerian singer Mohamed Khaznadji
Elsner 2011, p. 206, Fig. 7

realization, as well as already at the productive, acoustic realization, which is only objectified in notated form.

4.6.1 The Range of Possibility Between Identity, Equality and Blurred Sophistication

Debussy tried in his piece for solo flute, *Syrinx* (1913), to emulate improvisational practices in a crossing of exotism and folklorism. This short piece was written as incidental music to the uncompleted play *Psyché* by Gabriel Mourey, in1913, and was originally called *Flûte de Pan.* However, since one of Debussy's *Chansons de Bilitis* had already been given that title, it received its final name in reference to the myth of the amorous pursuit of the nymph Syrinx. Syrinx is French flute of Pan.[124] Debussy deployed the traverse flute for this purpose. It is tightened in a network of relationships. Firstly, corresponding to the pan flute, it refers to antiquity, and secondly, on nature as Pan is the god of nature, and thirdly, the correlation of nature and its mental mimesis is thereby represented. Because, Debussy has

> elevated the transverse flute as the instrument of the mythic nature. From *L'après-midi d'un faune* across the *Chansons de Bilitis* up to the late *Sonata for Flute, Viola and Harp* they are

[124] See Georg Predota: *Claude Debussy, La Flûte de Pan/ Syrinx. Melodrama and Flute Solo*, http://www.interlude.hk/front/claude-debussy-la-flute-de-pan-syrinx-melodrama-flute-solo/, January 5th, 2017, retrieved 27.3.19.

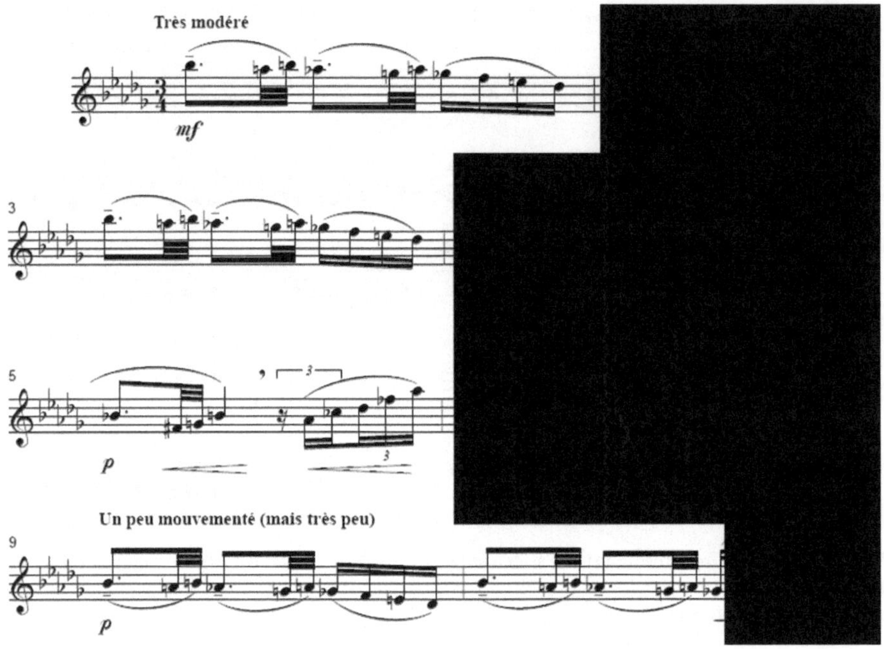

Fig. 4.20 Claude Debussy: *Syrinx*
Measures 1–10, arranged
J. Jobert, Paris (1927) Plate J.J. 344. IMSLP175411-WIMA.14bc-Syrinx.pdf

> carriers of an aestheticism, which the composer called "correspondence mystérieuse de la nature et de l'imagination", mysterious correspondence between nature and imagination.[125]

Based on the mood of the work, the issue of coherence and of appropriateness plays also a role for the music's piece embedding in the theater play (Fig. 4.20).

> Inserting the story of the death of Pan according to Plutarch's version, Mourey supposedly requested "the last melody Pan plays before his death" to be performed from the wings of the stage. However, it is highly unlikely, that Debussy's music was played on the occasion of Pan's death. Instead, it properly belongs to the first scene of the third act, a moment of intense lyricism in which the atmosphere is full of sensuousness and eroticism.[126]

The variant in T. 10 has the same header, but a different end: It moves upward instead of downward. The point of departure remains the same, Gb; Debussy mirrors the phrase on this tone not in an identic-symmetric way, but rather contrasts excerpts of two exotic musical scales.

−1−1 −*3* becomes +2 +3 +*3*. Thus the concluding minor third remains the same (Fig. 4.21).

---

[125]https://www.kammermusikfuehrer.de/werke/496, retrieved 28.3.19.

[126]See George Predota 2017. *Claude Debussy, La Flûte de Pan/ Syrinx. Melodrama and Flute Solo*, http://www.interlude.hk/front/claude-debussy-la-flute-de-pan-syrinx-melodrama-flute-solo/, January 5th, 2017, retrieved 27.3.19.

Fig. 4.21 Claude Debussy: *Syrinx* m. 9f.
J. Jobert, Paris (1927), ibidem

Fig. 4.22 Claude Debussy: Syrinx m. 3 and m. 5.
J. Jobert, Paris (1927), ibidem

In particular, the variant in m. 5 varies greatly from the basic figure (Fig. 4.22).

Yet, the tones 1 to 4 of the principal theme are rhythmically identic. And, the store of pitches rests also within a changed, permuted sequence almost entirely preserved:

Bb A B Ab G A Gb F E Db

1 2 3 4 5 6 7 8 9 10 becomes

Bb Gb G B A B Db E Ab[127]

1 7 5 3 6 3 10 9 4

Thus, only the 2. tone A is missing.

Towards the end Debussy variegates the theme rhythmically,[128] and completes it in a melodic-diastematic way, by converting the descending line in an exotistic six-tone whole tone scale.

Bb A Ab G Gb F E Db becomes

Bb A G F Eb Db (Fig. 4.23).

A deliberate motivic work is created through mixing genres with „Classics" in the Cool Jazz of the MJQ, the „Modern Jazz Quartet". Thereby, the piece *Fontessa*, of 1956, has noteworthy intertextual connections. Lewis relates the four mouvements (Prelude and Part I-III) on characters of the Commedia dell'Arte. In so far, it is program-music (see Sect. 3.5). The sequence: Colombina *Colombine*, Arlecchino *Harlequin*, Pierrot, Pantaleone *Pantaloon*. However, Lewis remains vague about the assignments, as he writes in the liner notes stereotypically—in verbal variants

---

127 Notated, regardless of the correct orthography.

128 **See Jochen Scheytt: *Syrinx*,** http://www.jochenscheytt.de/debussy/debussywerke/syrinx.htm 12016, **retrieved 27.2.19.**

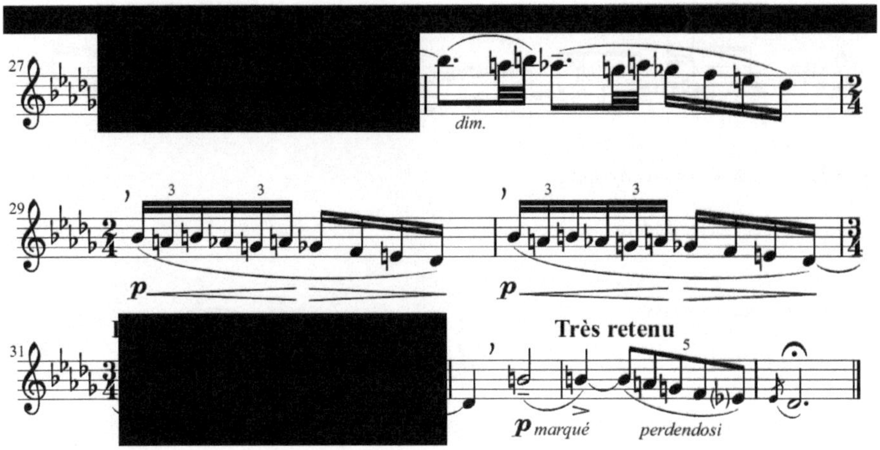

Fig. 4.23 Claude Debussy: *Syrinx*
End, arranged
J. Jobert, Paris (1927), ibidem

„This piece could perhaps be the character of".[129] Therefore, a sharpening in the interpretation is difficult.

> Lewis in the liner notes: *Fontessa* is a little suite inspired by the Renaissance Commedia dell'Arte. I had particularly in mind their plays which consisted of a very sketchy plot and in which the details, the lines, etc. were improvised. This suite consists first of a short Prelude to raise the curtain and provide the theme. The first piece after the Prelude has the character of older jazz and improvised parts are by the vibraphone. This piece could perhaps be the character of Harlequin. The second piece has the character of less older jazz and the improvised parts are played by the piano. The character here could perhaps bePierrot. The third piece is of a still later jazz character and develops the main motif. The improvised parts are by the drums. This character could perhaps be Pantaloon. The opening Prelude closes the suite. Fontessa is the three-note main motif of the suite and is perhaps a substitute for the character of Colombine.[130]

An analogy between Commedia dell'Arte („plays which consisted of a very sketchy plot and in which the details, the lines, etc. were improvised") and *Fontessa* is formed by the combination of composed, or set out in writing and improvised parts on the basis of a '*canovaccio*' (Canavas).[131]

The note example shows in each case the head theme in the prelude, as well as in theme I, II and III. Lewis himself states: „Fontessa is the three-note main motif of the suite" (Fig. 4.24).[132]

Thereby, the themes themselves are similar to each other, a transposition included. The slight reduction marks it (Fig. 4.25).

---

[129] Quoted after https://en.wikipedia.org/wiki/Fontessa, 14.12.18, retrieved 23.3.19.

[130] https://en.wikipedia.org/wiki/Fontessa, 14.12.18, retrieved 23.3.19.

[131] Baldizzone 2007.

[132] Quoted after https://en.wikipedia.org/wiki/Fontessa.

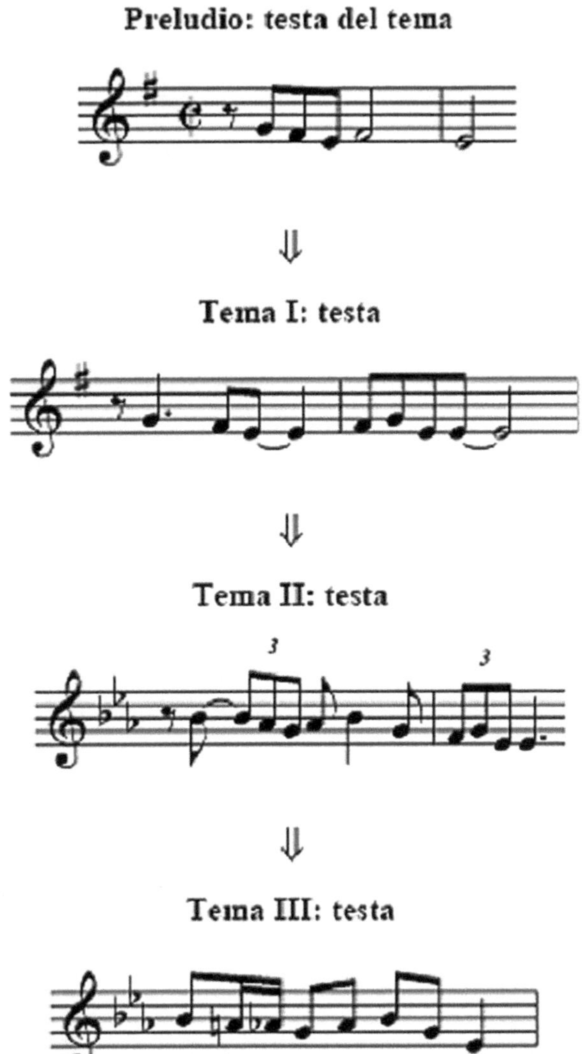

Fig. 4.24 John Lewis: *Fontessa*
Themes table, arranged after Baldizzone 2007

While Lewis chromatized, permuted, and transposed the initial theme, he obtains for the theme of the III. mouvement the head-motif BACH (Fig. 4.26).

This could have been his intention, setting an example of combining jazz with European art music, as for John Lewis, J. S. Bach and his polyphony were highly significant.

In regard to their similarity, figures may be classified according to their store of pitches (and then, their divergences or correlations), for—significant, dominant—interval-structures, for formal-schemata and form-processes, if composers emulate *special, single* formal-processes of another work, of their own, or by anyone else.

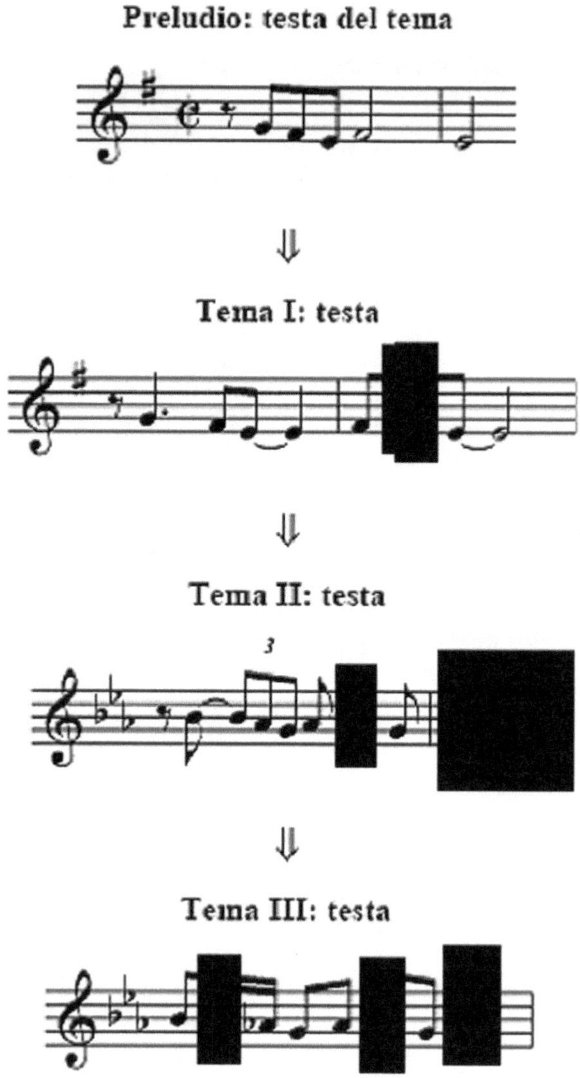

Fig. 4.25 John Lewis: *Fontessa* Themes table, reduced

The musical forms themselves are as form-types, or -schemata per se similar to each other—the „sonata-form" consists in general of the parts introduction (optional) + exposition (main theme + secondary theme) + development + recapitulation + coda (optional). Viewed differently, it is pattern of the same, though not of an identical topology. Now, the question is the following, how such a form-scheme finds its expression in a given concrete work. Then, by comparison, further classes may be formed, according to differentiated individual characteristics, such as the amount of themes in the exposition, length of themes, implementation of new themes or motifs in the development, proportions of keys, and much more. An example from

Fig. 4.26 John Lewis: *Fontessa* Theme III

the realm of the work encompassing relations shows a number of similarities in the use of motifs, as well as in the structure between Beethoven's 5. and 6. symphony. Both were

> together conceptualized, outlined, and premiered at the same time […]. Even though they are in their character as diverse as possible – the V. is focussing on something, which was described by „overcoming", „endeavour", while the VI. on something, which was called „remembrance", „contemplation" –, they present a […] long series of common features […]. Both start with a „motto", which is marked-off by fermatas. Both introductions provide material, with which the movement has to handle, as clearly as possible: the V. shows by its unisono, that we will have to deal with a compositional technique, in which the concentration

on the progression of voices will play a prominent role in any case; the VI. sets in its beginning the Bordun-fifth under the central theme. It is noteworthy, that both main themes start in the same time signature with the same rhythmic figure:[133]

But the seemingly same is not the same.

The themes are diametrically opposed; nevertheless, this concurrency can be brought into play, because the motifs on the basis of the grapheme are to each other immediately equal – and as […] Beethoven constantly and essentially works with the written form of a musical unit, it appears to me, that the comparison […] turns out productive for the appreciation. Although these introductions are completely related to the individuality of the movements – so to speak, they call them by name, but in toto they link both compositions with each other.[134]

The aim of the analysis is synthesis. It reconstructs the work as a composed unity, consisting of individual parts. But at the same time, according to the aim, resulting in a general unit of various things.

In the meantime, science passed from the age of reductionism into an age of emergence, an age, where the search for the final causes of things shifts from attitude of particles to the attitude of the collective. The antagonism between both of these conceptions of the ultimate – the law of particles and the law of the collective – is not only very old, but nevertheless, still relevant. […] The general statements are also valid for the technical sciences, and thus for technology as well. Only with the development of computers was it made possible to solve successfully the more complex technological tasks *quantitatively*. It started with the decomposition of complex technological lines in unit operations, which happened in particular machines (such as centrifuges) and apparatuses (such as dryers, filters, grinders). For this purpose the methods of the *theory of similarity* for scaling-up were developed and applied. […] With the development of computer technique the more complex processes of unit operations were further advanced for solving (reducing) and calculating the so-called micro- and sub-processes. At the same time, the synthesis of unit operations into cybernetic systems was made.[135]

Quantities, whose elements bear something in common with each other, are classes. In our leitmotif, the standard example being BACH, this would consist of a class of motifs of halftones, which are pairwisely combined with each other, therefore as well F-E-As-A. Still further refined, they would consist solely of downward half-tone intervals in pairs (like in the model)—a typical one-dimensional classification. Already multidimensional, as with more than just one criterion, or feature, are the characteristics downward versus solely upward, or if both are combined and sequentially interlaced; if necessary, with further intervals, for instance B-A-C#-H etc. The degree of similarity will then apparently decrease. There is little to be said against the disposition, to see potentially more similarities than there were perhaps intended, even in motivational features. For it happens frequently, that subconscious impulses of composers are more important than the deliberate intentions, which likewise attempt to form between the motifs a narrow network of relationships, or of references, to the utmost. Here and there, a main reason for this is the never ending and unavoidable search for patterns of human sensation.

---

[133] Weber 2014, p. 283.

[134] Extensively, ibidem.

[135] Banse/Reher 2011, pp. 39f.

Fig. 4.27 Gerhard Rühm: *eintonstück für piano / onetonepiece for piano* (1952), first line. Quoted after: Patrycja de Bieberstein Ilgner 2014, fig. 3, without pagination

4.6.2 One-, Two-, Three-Tone-Motifs and Four-Tone-Motif B-A-C-H – Modifications, Transformations, Permutations

Motif can be defined as the smallest unit bearing meaning in music, semantically and syntactically.[136] It is comparable to the morpheme in language. Motives are filtered from the total set of the music material, the store of pitches and rhythms, specifically for a piece, for a composer, or, as in the case of B-A-C-H in a manner transcending individual circumstances.

That is done during the more or less long process of composition, in different, even temporarily disrupted phases, pre- and retro-thinking, combining, cutting, amending etc., with or without conscious decisions for the respective selection or choice. And it can be done during an improvisation, with more pressure from the uninterrupted stream of time and (therefore) less possibilities for choices. A motif mostly contains about 2–5 tones. The upper limit should lie around six tones—but this would need to be ascertained empirically by analyzing historical styles, genres, art works and more. As border case, not very known and reflected, there exist 1-tone-motives. A famous one of these relatively rare cases of 1-tone-motives is to be found in Alban Berg's opera *Wozzeck*—the H (engl. B) as symbol of death, played two times by the entire orchestra in an immense crescendo, formally-dramaturgically functioning as interlude Act III, 2nd Scene, to Act III, 3rd Scene (measures 109–120). By contrast, Gerhard Rühm's e*intonstück für Klavier* (1952) points rather in direction of concept art. This "onetonepiece for piano" implies the identy of the octaves—"the 8 lines indicate the 8 possible As on the piano" (Fig. 4.27).

More popular might be A. C. Jobim's *One Note Samba (Samba de Uma Nota Sò*, 1959): only 1 tone is enough in each of the both first sections—in the melody; but the harmony changes of course, and therefore in a strict sense it is not a real one-tone-motif (Fig. 4.28).

[136]See extensively e.g. Heister/Polk 2017.

Fig. 4.28 A. C. Jobim: *One Note Samba, mm. 1-5*

Fig. 4.29 Duke Ellington: *C Jam Blues* (1942)
http://folkhogskolan.com/wp-content/uploads/2017/02/c_jam_blues.pdf, retrieved 26 March 2019

Two melody tones, G and C, needs Duke Ellington for his *C Jam Blues* (1942),[137] even with more than one chord in the changes, a simple 12-bars blues scheme and a good base for improvised chorusses (Fig. 4.29).[138]

[137]„New Orleans-born clarinetist Barney Bigard is likely the originator of this tune [...]. Since Bigard was a veteran member of Duke Ellington's Orchestra in 1941, Duke had a slice of the pie, too, and undoubtedly arranged the piece for the orchestra" (James Lincoln Collier, http://www.jazzstandards.com/compositions-1/cjamblues.htm, retrieved 26.3.19).

[138]"Basically a vehicle for jazz instrumentalists to display their improvisational skills, it is one of those pieces that is far more enjoyable for the player than the listener" (K. J. McElrat, quoted in http://www.jazzstandards.com/compositions-1/cjamblues.htm, retrieved 26.3.19).

4.6.3 BA, BAC, BACH—Variations and "Constancy of Shape" („Gestaltkonstanz")

A motif is, according to historical style and genre, more or less a distinct and independent element of the complete work. But it is also a building unit, element of a superordinate "set". The theme represents this larger unit. In the process of the work (and/or a movement, a chorus, a folklore improvisation) the motif undergoes changes—in language respectively phonetic-phonematic—as well as semantic, within and through the given context. I mentioned morpheme in correlation to language. However, the musical motif seems to be rather more easily and extensively variable, as by permutations of sequence, augmentation or reduction, abbreviation or extension, of pitch or duration, and so on.

At first, the BACH-motif as a whole.[139]

BACH
BA
CH
BAC
HCA

Already a part, the BA, is a motif: for one thing as a descending half tone step forming a topos of lament, specifically, depending on rhythm, a "sigh motif", when accentuated on the first tone (see Sect. 3.3.8). In a similar way the transposition to a higher pitch, CH. Even the BAC as an extension is in itself again a motif—as for example a "question motif" in Wagner's *Götterdämmerung (Twilight of the Gods).* In the orchestra the three-tonal kernel of the question-motif, A♭ G B♭, forms the completion of a chromatically descending line in the fifth room (Quintraum)—a prolongation of the mentioned lamento-phrase (Fig. 4.30).

In a variant of the question, the sung question follows in a modulated melody G A B♭- being the orchestra's kernel-motif; the instrumental composition repeats twice like an interlaced echo, the words (When?/How will it be? /Wann? Wie das wird?)—and thereby the different verbal variants of the question are condensed, and they are contracted in a melodic-harmonic formulation. The leitmotif is also named as „lament for the dead"(„Todesklage"). It is no pure coincidence, that Wagner turns to the BACH-formula: It will be rather dreadful than joyous (Fig. 4.31).

A question sentence in German (and in other languages) has an ascending intonation of the voice as marker. An answer to this question could sound H-C-A, thus rounding the melodic line. Also C-A descending resp. falling is the so-called "Kuckucks"- or "Rufterz", a minor third often used with texts for calling someone, in the kernel a two-syllable name as e.g. "Ru-di". And so on. Important: The motifs are similar in structure, mainly 'logical' and evolutionary, but very different in meaning.

---

[139] As mentioned, B in English = B flat, H in English = B. For musical reasons I will stick to the German denotation of B as H.

Fig. 4.30 Wagner, *Götterdämmerung,* Vorspiel (*Twilight of the Gods, prelude*)
2. Norn Frage Schicksalmotiv (Norn, question: motif of fate)
„Wann das wird?" („When will it be?")
Karl Klindworth, piano reduction; Übersetzer Frederick Jameson. B. Schott's Söhne, Mainz (1900). Plate 26809. IMSLP16827 vocal_score_.pdf

Fig. 4.31 Wagner, *Götterdämmerung,* Vorspiel (*Twilight of the Gods,* prelude) 1. Norn
„Wie das wird?/„How will it be?"
wie das ward?/„How it became?"
1. Norn: Weisst du, wie das wird?/Do you know, how it will be?
Klindworth (1900), ibidem

As can be seen here, the dissection or assembling, division, permutation of sequence (anagram), all without changing the intervals of the tones, and other methods lead to semantic and syntactic variations. This variety is multiplied through combinations with rhythmic variations—not to be further discussed at this point. With the pronouncement of death for Siegmund, in the IInd act of *Walküre,* the question becomes an order, with syncopated rhythm (Fig. 4.32).

Like the „Wehe!"-motif, which is related through the descending half-tone step, a nearly infinite abundance of similarities is achieved, and thus a perfect material for comparative studies with Fuzzy Logic. The kernel should be the two-toned „sighing motif"(„Seufzermotiv"), for instance, the tone pitches B♭ A. The identic or equal motif-kernels, which are further developed in variations to motives and themes, are objectively present, and may be compared and questioned for similarity. That is also a question about the invariant kernel and variant contours: How far may a concrete form differ from the kernel, in order to be still recognizable as a figure? Yet, to perceive it, depends on prior knowledge, just like the subjacent similarities in word-language.

For English ears words like Islam, Muslim, Salâm, which have hardly any vowels in common, may sound quite dissimilar, but for speakers of Semitic languags, such

Fig. 4.32 *Walküre*, IInd act
Brünnhildes Todesverkündigung/pronouncement of death
Karl Klindworth, Vocal Score; Frederick Jameson, English text . B. Schott's Söhne, Mainz (1900). Plate 26590. Nachdruck New York: G. Schirmer, 1904. Plate 37218. IMSLP16815-Wagner_-_Die_Walküre_(vocal_score)

words, as well as names like Salman, Suliman, Salim, Solomon, (Ab-)salom, are all perceived as closely related *variations on a theme*: the root *s-l-m*.[140]

4.6.4 *BACH: Counterpoint as Procedure of Variation and Transformation*

Basically B-A-C-H is—only—a motif. But historically it would best be named a "soggetto", an older expression for "theme". After J. S. Bach it becomes subject to numerous variations, be it by Robert Schumann, Max Reger or Arnold Schönberg or Hanns Eisler or many others.[141] Variations of the BACH-motif are traditionally created by the method of counterpoint: basic figure, inversion, retrograde, retrograde inversion. Only retrograde leaves the pitches unaltered. The second type alters the pitches: transposition and variation of the interval structure. The third type, permutation, alters the sequence of pitches. Further methods are rhythmical alterations and variations.

Inversion: Intervals, spaces accordingly running in (vertical) opposite direction in the tonal space—what went up, now goes down, and vice versa.

Retrograde (palindrome): order of tones inverted horizontally in time, back to front.

[140]Deutscher 2005, p. 38. Emphasis H.-W.H.

[141]Meanwhile more than 350 composers in more than 400 works (https://www.bachueberbach.de/johann-sebastian-bach/b-a-c-h/, retrieved 29.3.19).

1	BACH	Grundgestalt	basic shape/form
2	HCAB	Krebs	retrograde
3	BHAsA	Umkehrung oder (transponiert) Umkehrung:	inversion or transposed inversion
4	AsAHB	Krebsumkehrung	retrograde inversion

Basic form and retrograde in this case remain at an identical pitch. The non-transposed inversion (3) introduces new pitches (As).

Transposition: 11 further tonal steps within the 12-tone scale of the well tempered octave (last transposition = 1), altogether 12 steps of transposition

01 BACH
02 HBDesC
03 CHDCis
04 Cis …
05 D
06 Dis
07 E
08 F
09 Fis
10 G
11 Gis
12 A

Here, we can see a chart of possible variations. Transposition means in the shifting of the motif to different pitches. In this case, the similarity includes the differences. In isolation, the transposed forms are identical relative to pitch. Only with perfect pitch can the differences be detected within a framework of pitch classes. But even without this, in this context, the forms are noticeably different, if the motif is repeated directly and seriatim at these different pitches.

The symmetries or rather reflections (or inflection) over a horizontal or vertical axis, as well as the shift of transposition, parallel to the axis, all preserve the "constancy of shape" (Gestaltkonstanz) to various degrees. Boundaries of this constancy in the reflections in tonal space are rotations by 90 degrees or such. As remarked, the form ("Gestalt") of a motif or melody etc. is relatively stable, at least in its kernel; on the other side, it's one of Fuzzy Logic's tasks to analyse also quantitatively the grades and degrees of the "more/less" similarities of variants and starting point.

Permutation

01 BACH
02 ABCH
03 ABHC etc. …

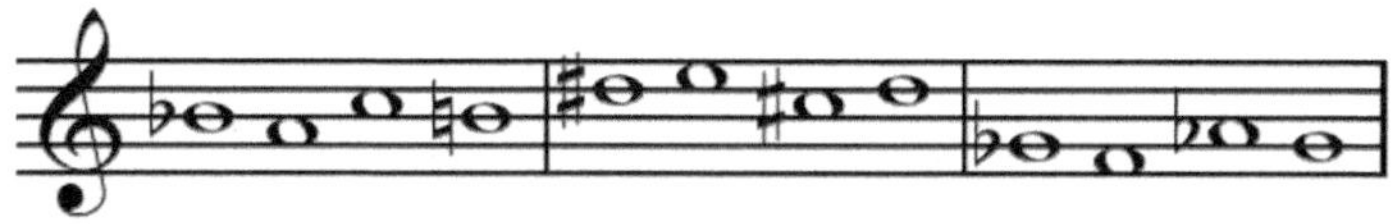

Fig. 4.33 Anton Webern
String Quartet op. 28, tone row https://commons.wikimedia.org/wiki/File:Webern_String_Quartet_tone_row.png
Hyacinth, 7 December 2010, retrieved 19 February 2019

In the thinking of the dodecaphony, or the twelve-tone music, these methods of counterpoints are reintroduced, basically in the construction of the tone row (Fig. 4.33).[142]

In addition, the procedure of permutation, the rearrangement of sequence constitutes another type of similarity, with internal degrees of similarity: In general the similarity is mathematically greater than musically, in particular varying—ABHC, for instance, is a simple chromatic sequence, with relatively little significance as shape. What remains unchanged here, is the ambitus of the motif, the border or edges/contour as upper and lower pitch demarcation. This ambitus changes in the following variations. The intervals, the spaces between tones are changed: augmented or extended. (In principle, a reduction or compression is possible—but not with half-tones, being here the smallest interval). Nevertheless, contours remain visible. Here, we have variations of the basic shape, with more and more augmented intervals.

Formula:

B A C H
−1 +3 −1
1 semi-tone
– descending
+ ascending

Variation of intervals

B A C H
−1 +3 −1
B As C H
−2 +4 −1
B As C B
−2 +4 −2
B As Des C
B G C H
B G C B
B G C A

[142]The whole Viola-Concert of Karl Amadeus Hartmann, the [concert for viola with piano, accompanied by wind/brass players and percussion] *Konzert für Bratsche mit Klavier, begleitet von Bläsern und Schlagzeug* (1954/1955) is a study in and about BACH—not dodecaphonic, but influenced by Webern.

B G Des C
… etc.

Contour and kernel pattern remain as a constant shape, at least to a certain degree, but still intact and visible. However, similarity is steadily reduced.

The sequence and cross structure as an essential aspect (see Sect. 5.2) is maintained, even when the tone intervals are augmented. The borders are once again fluid, the edges fringed—a typical marker of Fuzzy Logic. Furthermore, the borderlines are fringed; there is no clearly defined set of variations—a frequent occurrence in music. These borders can be empirically verified by comparably simple experiments. As common in Fuzzy Logic, the degrees of fuzziness could and should be precisely quantified. Thereby we have an opulent quantity of variations. On the one hand, they are a precisely defined set, created through various procedures following exact rules or algorithms. In so far, the variations result from a simply binary logic: *EITHER* B before A, *OR* A before B etc., as mentioned above. *But*: The shapes differ from the basic shape according to the degree of their similarity. Insofar, the variations follow a Fuzzy Logic with fringed borders, transitions: once more, a structured or segmented continuum.

Precisely this potential variety is employed by J. S. Bach himself; in countless instances it is to be found in his work with just as many variations. In the 14th and last "Contrapunctus", in his final opus, *Kunst der Fuge* (*Art of the Fugue*), he quotes himself and his name, in a way, leaving a signature (bar 194f.). Furthermore, in this huge cycle, the BACH-signature was implied and intended, right from the start.

4.6.5 Cross Motif and Chromatics—Wagner's **Ring**, *Strauss'* **Arabella**

The cross-sequence structure of BACH appears very frequently, even without an allusion in form of a quotation about the motif, or the name of Bach, is necessarily meant (see Fig. 5.3, St. Andrew's cross). By means of various keys and an accidental, such a motto-motif may actually be written with 4 keys by a single note—a procedure, that was employed a long time before, in renaissance-music, but in this particular case, not by J. S. Ba ch himself.

In Richard Strauss'operetta-like opera *Arabella* (1929), Zdenka, younger sister to the protagonist Arabella, secretly loves Matteo, who himself is openly in love with Arabella, and proposes to her. She has to bear her cross, that her elder sister first has to get married, and she has to disguise herself as a man.

Her friend Matteo's leitmotif turns out to be particularly apparent as a concealed variant of BACH. G A A#=Bb H=B is a permuted and transposed figure. Here, most likely, Strauss was neither willing to cite nor to allude to it. But, the compact chromaticism operates by its own logic, which also asserts its claims. In a wider sense, it symbolizes, what is dreadful, gloomy. The iconic even adopts indexical characteristics: The BACH-motif is in so far a doubled sigh-motif (see Sect. 3.3.8).

Virtually the motif is sung with a choked voice; it does not, in joyfulness, stride along extensively.

> Matteo's poverty receives already an expression, as he is the only important figure to receive just a single motif. This motif [act I, cipher 12], in frail E-minor, is not only extremely short, but also short of breath, therefore it is not qualified for a melodic continuation, but solely for the sequence; it appears mostly in isolation.[143]

A second feature is a partial mimesis of Matteo's character and behavior.

> Although the motif is characterised by an impulsive upward tendency, which alludes on Matteo's desire, it seems to be frantic and feeble, because it almost immediately comes to a halt, and hence, the melodic energy of the upward mouvement is wasted. At its first appearance, only the third link of the sequence possessed enough energy to reach the high tone, which then tellingly plunges in the glissando [act I, cipher 6].[144]

The sequencing characterizes Zdenka's motif too.
K. Hottmann interprets this symbolically and per analogiam as similarity:

> Zdenkas mental misery sounds as a self-tormenting upward chromaticism: She is as a figure, in her room for manoeuvre most restricted, what is reflected in the motif's narrow ambitus [Act 1, 6 T., after cipher 23].[145]

But the motif is also a BACH-variant: BbABb-Cb = H, and still more articulated in the second phrase C H=B Bb A— the reversal of the chromatic four-tone sub-motif of Matteo's leitmotif.

Anyway, once again Fuzzy Logic might turn into action. With its support, BACH-topics in corresponding or similar chromatic tone sequences, motives, topics of 2 till n tones (5 is a reasonable, but no absolute limit) may systematically after degrees of similarity be investigated, and then quantified. (On relevant filterings, see Chapter 5) The permutations (modifications of the running order, and thus, of the interval-structure), such as transpositions (the motif's re-assignment on different tone pitches) produce an enormous amount of test material, last not least, with a diachronic, historic dimension. This becomes further extended, if we add rhythmical variants and moreover—last not least, emotionally controlled—variants in the sound volume and other parameters of music. „Manually" and by merely more or less intuitive comparison, such material will never be managed. The hazy limits will further be widened, if we include ornamentation. The turn consists of 4 tones. Mahler frequently gave it the rank of a motif. If it starts with H, it receives at once multiple times the half-tone step h c h, and thus a two-tone motif, referring on BACH.

> For the turn the melody note is at first played around the upper auxiliary note, then through the lower auxiliary note. In doing so, it starts with the upper auxiliary note, followed by the main note, then, the lower auxiliary note rings, and finally again the main note.[146]

With the alteration of tone pitches the four-tone BACH-motif can be produced (Figs. 4.34 and 4.35).

---

143 The following very shortened and condensed according to Hottmann 2010.

144 Ibidem.

145 Ibidem.

146 Extensively http://www.musikzeit.de/theorie/verzierungen.php, 19.08.2007, retrieved 21.2.19.

Fig. 4.34 Turn Motif http://www.musikzeit.de/theorie/verzierungen.php. 19.08.2007. Accessed: 21 February 2019

Fig. 4.35 Turn Motif http://www.musikzeit.de/theorie/verzierungen.php. 19.08.2007. Accessed: 21 February 2019

> The auxiliary notes can be chromatically changed by accidentals. An accidental above the double strike sign stands therefore for the upper auxiliary note, while below the double strike sign, it stands logically for the lower auxiliary note. Both auxiliary notes may be changed at the same time, also mixed with # and *b*. Besides # and *b* a natural sign is also possible.[147]

The three-tone figures Es-D-Cis and Ais-Gis-Fis should at first be transposed, permuted, and combined, and will result in D-Cis- E-Dis. Then, we take a fourth step of transformation, another transposition, to obtain BACH. (To the turn-ornament as semantic motif see below 7.4.2.1 *Ornamentation*)*

4.7 Neuronal Iconic and Fuzzy Logic

Dissimilarity—on the level of appearance and the iconic-pictorial—as contrast, as well as complement are inherent to the principle of similarity. We encountered this, as an abstraction in various kinds and degrees, as a steady counter- and parallel-tendency of the naturalistic mimesis. And, we encountered similar phenomena when producing noise for the soundtrack of movies, key word: Foley. Therefore, corresponding to 0-1-principle of the digital, digital representations of real analogous matters are not similar to what is depicted, whether they are images taken from nature or letters from texts. Also the relation between matters in the world and neuronal representation in the brain are generally considered as abstract. For such a kind of reflection, neither about ‚depiction/Abbildung‘, nor about similarity shall be said anything. Admittedly there are contradictory findings. The first example comes from the AI-research. We find processing procedures in neuronal networks, analogue to the technic of retrieval, in the living brain as well as in its simulations via Artificial Intelligence. M. Spitzer describes a Kohonen-Network learning patterns, here letters—which has a certain similarity to BACH.

[147]Ibidem.

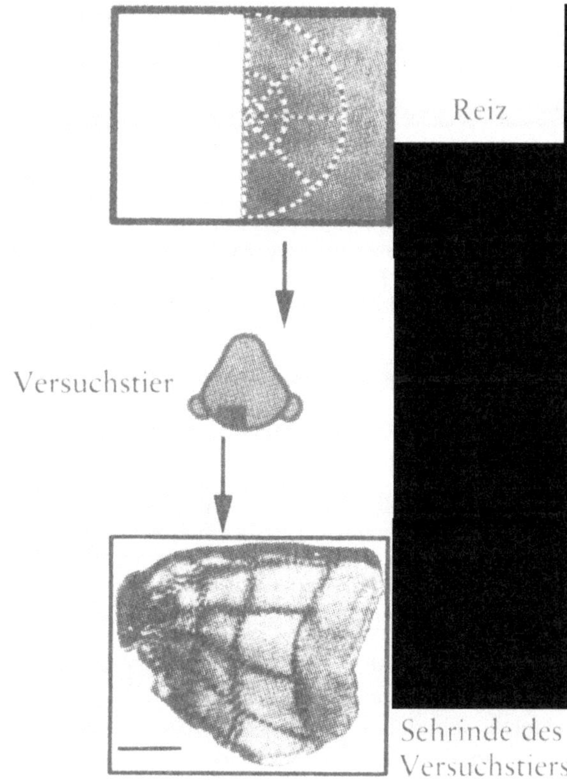

Fig. 4.36 Damasio, Antonio R.: *Descartes' Error: Emotion, Reason, and the Human Brainc*, 1994, p. 104. The illustration is edited and shortened. It shows a geometric stimulus, represented as geometric pattern, even transformed. in the brain of an animal

> In the beginning, the neurons are not configured on letters, this means, it depends on casual values of the weight of synapses, which of the neurons will win after the presentation of an input signal. The „learning steps" correspond to the [...] multiple filtering. [...] Already after 200 learning steps [...] is a reasonable arrangement of the neurons, which represent the input signals, foreseeable, whereas still an insufficient differentiation is apparent. After 700 learning steps in total, [...] the letters are well distributed within the network, in conformity with their graphic properties.[148]

The second example concerns the following finding, which was attained from a living brain. I believe, this is sensational. Aside from abstract representations, there are obviously iconical neuronal representations too, similar to those represented. In the illustration below, the two-dimensional pattern of a three-dimensional brain is slightly blurred, but still clearly recognizable (Fig. 4.36).

At least, on the level of elementary signs, such results contradict the widespread disposition to expand the range of dissimilar and arbitrary matters, last but not least in arts, as far as possible (see Sects. 6.8.5 and 8.5.3).

But even if such a specific iconic could neither be any further proven nor improved by experiments:

[148] Extensively and with illustrations, Spitzer 1996, p. 106.

The representations inside the brain may appropriately serve as patterns of „firing" to be compared with each other, like neurons, which enter into relationship with each other.

> Brain–computer interfaces enable active communication and execution of a pre-defined set of commands, such as typing a letter or moving a cursor. However, they have thus far not been able to infer more complex intentions or adapt more complex output based on brain signals. Here, we present neuroadaptive generative modelling, which uses a participant's brain signals as feedback to adapt a boundless generative model and generate new information matching the participant's intentions. We report an experiment validating the paradigm in *generating images of human faces*. In the experiment, participants were asked to specifically focus on perceptual categories, such as old or young people, while being presented with computer-generated, photorealistic faces with varying visual features. *Their EEG signals associated with the images were then used as a feedback signal* to update a model of *the user's intentions*, from which new images were generated using a generative adversarial network.[149]

The objective, quasi-automatic results in the brain are then evaluated on the basis of subjective evaluations of the test participants.

> A double-blind follow-up with the participant evaluating the output shows that neuroadaptive modelling can be utilised to produce images matching the perceptual category features. The approach demonstrates brain-based creative augmentation between computers and humans for producing new information matching the human operator's perceptual categories.[150]

This naturally generates—still—double fuzziness: 1. between the image of the face as represented and its representation in the brain, and 2. between the latter, which is interiorized, and the exteriorization in the verbal utterance. Nevertheless, the methodological significance of these experiments is presumably extremely high.

The fineness of neuro-physiological analysis, the 'sieve' will some day be sufficient, in order to depict the subtle differences, such as the ones between BACH and BAC#H—not only the signifiers but the signified, not the letters but the sound they represent—and more; if not for today, then certainly tomorrow.[151] Perhaps the just quoted experiments reveal something of this 'tomorrow'.

The transcription of note-names into letters, with their distinct contours is undoubtedly a great big step forward, and said experiments probably even a leap. And from such a data basis, any degree of fuzziness will, be fairly easy be detected and presented, if a cue for Fuzzy Logic is set in motion.

References

Baldizzone, R.: Spontaneità organizzata ad 'arte'. Il metodo compositivo di John Lewis. In: Philomusica on-line. Rivista del Dipartimento di Musicologia e Beni Culturali. Università degli Studi di Pavia, vol. 6, N° 1 (2007). http://riviste.paviauniversitypress.it/index.php/phi/article/view/06-01-LAU01/101

[149] Kangassalo et al. 2020, n.p.

[150] Ibid.

[151] Cf. e.g. Calvin/Ojemann 2000, p. 154.

Ball, H.: Erklügelte Nervenkultur. Pendo Verlag, Zürich (1991)

Banse, G., Reher, E.-O.: Ambivalenzen von Technologien – Chancen erhöhen, Gefahren mindern, Missbräuche verhindern. In: Sitzungsberichte der Leibniz-Sozietät der Wissenschaften zu Berlin, vol. 112, pp. 19–41. Trafo Verlag, Berlin (2011)

Banse, G.: „Nicht so exakt wie möglich, sondern so genau wie nötig!" – Das Einfachheitsprinzip in den Technikwissenschaften. In: Sommerfeldt, E., Hörz, H., Krause, W. (eds.) Einfachheit als Wirk-, Erkenntnis- und Gestaltungsprinzip (Sitzungsberichte der Leibniz-Sozietät der Wissenschaften), vol. 108, pp. 93–104. Trafo Verlag, Berlin (2010)

Blakolmer, F.: Wie Bilder lügen. Die Frühägäer und ihre Götter. In: Forum Archaeologiae 54/III/2010. n..p. http://farch.net (2010)

Boone, Chr.: Mashing: toward a typology of recycled music. Music Theory Online **19**(3), 1–14 (2013) 2011

Calvin, W.H., Ojemann, G.A.: Einsicht ins Gehirn. Wie Denken und Sprache entstehen (English 1994). dtv, München (2000)

Collins, N.: The Cambridge Companion to Electronic Music. University Press, Cambridge, MA (2007)

Collins, N.: "Beyond notation: communicating music." Leonardo Music. J. **21**, 5–6. MIT Press, Cambridge, MA (2011)

Demers, J.: Listening Through the Noise: The Aesthetics of Experimental Music. Oxford University Press, New York and Oxford (2010)

Deutscher, G.: The Unfolding of Language. The Evolution of Mankinds Greatest Invention (1st edn., 1988). Random House, London (2005)

Didi-Huberman, G.: Ähnlichkeit und Berührung. Archäologie, Anachronismus und Modernität des Abdrucks. Dumont, Köln (1999) (französisch 1997)

Döhl, F.: Mashup in der Musik. Fremdreferenzielles Komponieren, Sound Sampling und Urheberrecht. Transcript, Bielefeld (2016)

Döhl, F., Wöhrer, R. (eds.): Zitieren, appropriieren, sampeln. Referenzielle Verfahren in den Gegenwartskünsten. Transcript, Bielefeld (2014)

Doolittle, E.L., Brumm, H.: O Canto do Uirapuru: Consonant intervals and patterns in the song of the musician wren. J. Interdiscip. Music. Stud. **6**(1), art. # 12060103, 55–85, Spring (2012). Doi:https://doi.org/10.4407/jims.2013.10.003

Doolittle, E.L., Gingrasb, B., Dominik, M.E., Fitch, W.T.: Overtone-based pitch selection in hermit thrush song: unexpected convergence with scale construction in human music. PNAS **111**(46), 16616–16621, 18 Nov 2014. https://www.pnas.org/content/111/46/16616

Dolscheid, S., Shayan, Sh., Majid, A., Casasanto, D.: The thickness of musical pitch. Psychol. Sci. **24**(5), Mar 2013. https://www.researchgate.net/publication/236090333_The_Thickness_of_Musical_Pitch/link/542956870cf26120b7b5bdfa/download

Drösser, C.: Fuzzy Logic. Methodische Einführung in krauses Denken. Rowohlt, Reinbek (1994)

Dubnov, Sh., Bar-Joseph, Z., El-Yaniv, R., Lischinski, D., Werman, M.: Synthesis of Sound Textures by Learning and Resampling of Wavelet Trees, Oct 2002. https://www.researchgate.net/publication/2906148_Synthesis_of_Sound_Textures_by_Learning_and_Resampling_of_Wavelet_Trees

Dümling, A.: Der Komponist Artur Schnabel. In: Dissonanz/Dissonance 1987/12, pp. 10–20. Zürich

Elsner, J.: Maqām – abstrakte Einheit – konkrete Vielfalt. In: Phleps, Th., Reich, W. (eds.) Musik-Kontexte. Festschrift für Hanns-Werner Heister, 2 vols., vol. 1, pp. 198–214. Verlagshaus Monsenstein und Vannerdat, Münster (2011)

Genette, G.: Palimpseste. Die Literatur auf zweiter Stufe (French 1982). Suhrkamp, Frankfurt am Main (1993)

Genette, G.: Mimologiken. Reise nach Kratylien (French 1976). Suhrkamp, Frankfurt am Main (2001)

Gredig, M.: Tiermusik. Zur Geschichte der skeptischen Zoomusikologie. Königshausen & Neumann, Würzburg (2018)

Heister, H.: The sympathetic imagination in the novels of J.M. Coetzee. Empathy and Mirror Neurons. Phil. Diss. Humboldt-Universität zu Berlin. Kultur-, Sozial- und Bildungswissenschaftliche Fakultät (2015). https://edoc.hu-berlin.de/bitstream/handle/18452/17838/heister.pdf

Heister, H.-W., Polk, H. (eds.): Bewegtes und Bewegendes. Der Motiv-Begriff in Künsten und Wissenschaften (Musik/Gesellschaft/Geschichte, vol. 7). Weidler Buchverlag (2017)

Henckels, P.: Ich war kein Musterknabe. Eines Lebenskünstlers lachende Weisheit, 3rd edn. Lothar Blanvalet, Berlin (1966)

Holz, H.H.: Speculum mundi. Schriften zur Theorie der Metapher, spekulativen Dialektik und Sprachphilosophie. From the legacy edited by Jörg Zimmer. Aisthesis-Verlag, Bielefeld (2017)

Hottmann, K.: Männlichkeitskonstruktion in der Oper am Beispiel von Hugo von Hofmannsthals und Richard Strauss' *Arabella.* In: Grotjahn, R. et al. (eds.) Musik und Gender. Grundlagen – Methoden – Perspektiven (Kompendien Musik 5), pp. 183–194. Laaber Verlag, Laaber (2010)

Hüther, G.: Das Geheimnis der ersten neun Monate. Unsere frühesten Prägungen. Beltz Verlag, Weinheim and Basel (2011)

Kacprzyk, J., Castillo, O., Melin, P. (eds.): Nature-Inspired Design of Hybrid Intelligent Systems. Springer, Heidelberg etc. (2017). http://www.springer.com/978-3-319-47053-5. Retrieved 5 Feb 2017

Kandel, E.: Das Zeitalter der Erkenntnis. Die Erforschung des Unbewussten in Kunst, Geist und Gehirn von der Wiener Moderne bis heute. Siedler Verlag, München (2012)

Kanitscheider, B.: Kosmologie: Geschichte und Systematik in philosophischer Perspektive. Reclam, Stuttgart (1984) (2., extended ed. 2002)

Kangassalo, L., Spapé, M., Tuukka Ruotsalo, T.: Neuroadaptive modelling for generating images matching perceptual categories. Scientific Reports. Nature Research (2020)

Kautny, O.: Talkin' All that Jazz — ein Plädoyer für die Analyse des Sampling im Hiphop. In: Appen, R.von, Doehring,A., Dietrich Helms, D., Phleps, Th. (eds.) Online-Publikationen des Arbeitskreis Studium Populärer Musik e.V. (ASPM). www.aspm-samples.de/Samples9/Kautny.pdf. Jahrgang 9 (2010) — Version vom 28.10.2010

Knepler, G.: Geschichte als Weg zum Musikverständnis. Zu Theorie, Methode und Geschichte der Musikgeschichtsschreibung. Reclam Universalbibliothek, Leipzig (1977) (2. extended ed. 1982)

Kornauth, G.: Plunderphonics – Plagiarismus in der Musik. https://plagiatsgutachten.de/wp-content/uploads/2011/06/Kornauth_Gaar_Past_Plunderphonics.pdf, https://plagiatsgutachten.de/wp-content/uploads/2011/06/Kornauth_Gaar_Past_Plunderphonics.pdf (2011)

Korzybski, A.: Science & Sanity. European Society for General Semantics. http://esgs.free.fr/uk/art/sands.htm (1933)

Lehmann, Chr.: Klassifikation. https://www.christianlehmann.eu/ling/typ/typ_vs_klasse.php, not dated. Retrieved 13 Nov 2013

Mundhenke, F., Arenas, F.R., Wilke, Th. (eds.): Mashups. Neue Praktiken und Ästhetiken in populären Medienkulturen. Springer VS, Wiesbaden (2014)

Oesch, H., in collaboration with Haas, M., Haller, P. (eds.): Außereuropäische Musik (Part 2), Neues Handbuch der Musikwissenschaft, vol. 9. Laaber-Verlag, Laaber (1987)

Oswald, J.: **Plunderphonics, or Audio Piracy as a Compositional Prerogative**—as presented by to the Wired Society Electro-Acoustic Conference in Toronto in 1985. http://www.plunderphonics.com/xhtml/xplunder.html (1985)

Phleps, Th.:"In meinen Tönen spreche ich" – Biographische Chiffren in Kompositionen von Brahms. In: Johannes Brahms oder Die Relativierung der "absoluten" Musik. Hg. v. Hanns-Werner Heister (Zwischen/Töne 5). Hamburg: von Bockel 1997, S. 147–198

Phleps, Th.: Zwischen Adorno und Zappa. Semantische und funktionale Inszenierungen in der Musik des 20. Jahrhunderts (Zwischen/Töne. Neue Folge 1). Weidler Buchverlag, Berlin (2001)

Phleps, Th.: Das Schaffen des Künstlers ist triebhaft. Über das Bewusstsein von Zahlen in Schönbergs kleinem Klavierstück op. 19,2. In: Neue Zeitschrift für Musik 162. 1/2001, pp. 17–23. Schott International, Mainz (2001)

Phleps, Th.: Affectionen einer lebhaft begehrenden Sinnlichkeit. Verschlüsselte Botschaften in Schubert-Liedern. In: Heister, H.-W. (ed.) Kunstwerk und Biographie. Gedenkschrift Harry Goldschmidt (Zwischen/Töne. Neue Folge, vol. 1), pp. 335–360. Weidler Buchverlag, Berlin (2002)

Por, P.: Die Entstehung des Fragments (Mallarmé: L'après-Midi d'un Faune) und die Entstehung des Fragments aus dem Geist des Fragments (Hofmannsthal: Der Tod des Tizian). In: TRANS. Internet-Zeitschrift für Kulturwissenschaften. No. 14/2002. http://www.inst.at/trans/14Nr/dejong14.htm, Abruf 27 Mar 2019

Ratcliffe, R.: A proposed typology of sampled material within electronic dance music. Danc.: J. Electron. Danc. Music. Cult. **6**(1), 97–122 (2014). https://www.semanticscholar.org/paper/A-Proposed-Typology-of-Sampled-Material-within-Ratcliffe/77a3470b049e48875f097413cfee5c36dd17bf5e

Searchinger, C.: Artur Schnabel. A Biography, with a Tribute by Clifford Curzon, Dodd, Mead & Co, New York (1957)

Sandkühler, H.J.: Analogie. In: Sandkühler, H.J. et al. (eds.) Europäische Enzyklopädie zu Philosophie und Wissenschaften, vol. 1, pp. 101–108. Felix Meiner Verlag, Hamburg (1990)

Seneta, E.: Lewis Caroll as a probabilist and a mathematician. Math. Sci. **9**, 79–94 (1984)

Seneta, E.: Lewis Carroll's "pillow problems": on the 1993 centenary. Stat. Sci. **8**(2), 180–186. https://projecteuclid.org/euclid.ss/1177011011 (1993)

Sondereguer, C.: Sistemas compositivos Amerindos. Morfoproporcionalidad / El concepto arquitectonico-escultorico en Amerindia. Praxis de un pensmiento morfoespacial, urbanistico y simbolico. Corregidor, Buenos Aires (2010)

Spitzer, M.: Geist im Netz. Spectrum Akademischer Verlag, Heidelberg (1996)

Straebel, V.: Aspects of conceptual composing. The Open Space Magazine (10), fall 2008, pp. 69–78

Straebel, V.: Musikalische Repräsentation geometrischer Objekte in Alvin Luciers Kammermusik. In: Tadday, U. (ed.) Alvin Lucier (Musik-Konzepte 180/81), pp. 49–71. Edition text + kritik, München (2018)

Tomatis, A.A.: Introduction and editing S. Manassi: Der Klang des Lebens: vorgeburtliche Kommunikation – die Anfänge der seelischen Entwicklung. Rowohlt, Reinbek (1990)

Vakhania, N.: On a probability problem of Lewis Carroll. Bull. Georg. Natl. Acad. Sci., **3**(3), 8–11. http://science.org.ge/old/moambe/3-3/Vakhania.pdf (2009)

von Maur, K. (ed.): Vom Klang der Bilder. Die Musik in der Kunst des 20. Jahrhunderts. Prestel, München (1985)

Weber, P.: Zyklusbildung in Beethovens letzten Sinfonien. In: Archiv für Musikwissenschaft, vol. 71(4), pp. 282–290. Franz Steiner-Verlag, Stuttgart (2014)

Wickler, W.: Mimikry. Nachahmung und Täuschung in der Natur. Fischer-Verlag, Frankfurt am Main (1973)

Wit, E.: 'All models are wrong…': an introduction to model uncertainty. Statistica Neerlandica (2012). Doi:https://doi.org/10.1111/j.1467-9574.2012.00530.x1, quoted after http://www.philos.rug.nl/~romeyn/paper/2012_wit_et_al_-_all_models_are_wrong.pdf. Retrieved 1 Mar 2019

Chapter 5
The Principle of Sharpening (I): Filtering. Cosmos Out of Chaos—Aspects and Elements of the Musical Materials

Abstract Filtering' is the differentiation of the essential and the secondary – depending on context and goal. With music it is the basal stage of sharpening. Thus natural material is processed, transformed into tones, sounds, noises, and systematized. This then serves as material for the second stage, the 'crystallization' (see Chapter 6). These filtering processes repeat themselves, now already in the field of music, when chords are filtered out of heterophonic lines, and when in electroacoustic musical tones are worked out of noises or sine tones. The automated filtering by soft computing and FL-oriented methods is analysis of similarities. Our 'leitmotif' BACH serves as a test for internet research, which shows the productivity of Fuzzy Logic. Filtering as highlighting essential information via the Phonographic Form supports the re-sharpening of the memory. The compositional use of the partial-tone-spectrum as in Wagner's Prelude to Rheingold is already a transition to crystallization.

Keywords Audio- and data-compression · Automatic data collection for marketing, advertising, and and internet control · BACH-'leitmotif' as internet-test · Chaos to cosmos · Culture/'nurture' versus nature · Electro-acoustic music · Information · Natural and musical material · Tone versus noise · Pitch-, rhythm-, and tuning-systems · Marketing and Werbung Reklame · Medieval 'solmization' (syllables for pitches) by Guido d'Arezzo (ca. 1033) as 'sogetto cavato' ('carved out theme') · Phonography for sharpening the memory · „Precisiation" (L. Zadeh) · Segmentation of continua · Similarity · Soft computing · „Spectralism" (a musical trend since the 1970s)

The paired terms of the dialectic blur/fuzziness-sharpening, in German, *Unschärfe-Schärfung*, unfortunately dissipate in English. The polarity is just fuzziness-sharpening or making precise. L. Zadeh himself has proposed, especially for the question of meanings, the artificial word „precisiation",[1] a portemanteau-word of precision plus clarification or specification; the latter is more in the direction of detailing. But even more appropriate might be precisi-cation or precisi-fication, in analogy to 'fuzzification'. 'Precision' would be an equivalent to sharpening in German.

[1]Zadeh 1984, Zadeh 2004, and Seising / Zadeh 2013.

H. Heister, *Music and Fuzzy Logic*, Studies in Fuzziness and Soft Computing 406,
https://doi.org/10.1007/978-3-662-62907-9_5

5.1 The Development of Musical Materials Under the Aspect of Sharpening. Segmentation: Discrete Parts Instead of Continuum

Filtering is a phenotype of sharpening. It filters out something already contained in the filtered material without changing this itself. The result is not really something different—but just something new: Because what is filtered out there, was already there, in the case of music as a natural material, but not in and of itself, that is, usable on the first level for artistic purposes as material of art, and on the second level of filtering suitable for further processing of various kinds. In contrast to the crystallization and the production of the work of art, the filtering is probably mainly done without deliberate intention. Their results, especially as far as the fundamentals of the musical material are concerned, are the result of collective processes. But they are not actually emergent, i.e. do not develop by self-organization in the object itself, rather by human intervention. I will introduce some cases, which are even about individual access.

5.1.1 Sharpening. Filtering of Components, Crystallization of Shapes

Filtering creates on a first elementary level, 'atomic' to 'molecular' shapes, musically above all sound and rhythm systems, in spoken language about the level of the phone, phonemes and syllables. On the other hand, crystallization generates new things on the one hand, and more complex structures on the other, such as musical motifs, words or phrases and phrases in spoken language. This happens primarily consciously, as intentional shape generation; Emergences also come about admittedly in the second instance (See above Sect. 1.4). However, the delimitation is in turn blurred and needs clarification in turn. On a second level of filtering, it is about filtering out already shaped, thus crystallized forms, of special patterns, patterns of larger contexts. Similarities do not play a significant role in the first stage: The tone A is its raw material in sound, not similar to nature, at most in the very general sense that it could be considered as a type of—selective—reflection and insofar would be 'similar'. Completely for the relations between the tones A, B and C, etc., the question of similarities is irrelevant.

It is different on the second level. Because this is about the filtering of more complex shapes rather than simple elements, similarity becomes an important criterion of filtering. Classifications and typologies of forms play a role as „pre-order", as intermediary, so to speak, upstream or intermediate sieves. The fineness of the screens is a default—a control variable for controlling the filtering—also with regard to an expected or desired result. The degrees of freedom of choice are great. But this subjective side must be fed back with the objective nature of the characters to be filtered out; if not, instead of the bandwidths, the fuzzy logic, the „precise fuzziness“, arbitrariness arises. The filtering can be repeated, not least in consideration of increasing precision. For this purpose, e.g. also the learning Kohonen networks, which simulate and reproduce neural processes. An example on a historical scale is the emergence of pitch systems. It is already tangible in the Late Palaeolithic (for example, a bone flute from Geißenklösterle, a cave near Blaubeuren, Swabian Alb, Southern Germany, around 37,000 BP with so many holes that at least 5- to 7-note scales result).

5.1.2 Selection. Filtering in Arts and Nature

Filtering is a fundamental process in the realm of the production of art, in respect to the evolution of single signs as to the combination of these signs. It could be discussed if perhaps s*election* would be a better term, with its relation to historical processes, especially the ‚evolution‘—certainly without implications of biologism.[2] Anyway, 'filtering' as term may fit for historical reasons in the context of discovering Fuzzy Logic; and it covers a wide span from natural (water through gravel aggregate bed filter), industrial (mechanical, chemical) filters, trivial as coffee filter, mathematically abstract metaphorical („Peano-Sieve“) to social „infiltration" by „foreigners" from outside the own ethnos, nation and so on—and their filtering out as potential so called „terrorists" or similar. Filtering has further implications that show also correlations to music-external processes. One essential aspect is the reduction of entropy. Chaos becomes cosmos, but as a „moving, fluid order" (as nature for Goethe), especially in music as fluid equilibrium and dynamic, dialectical homeostasis of stability and variability. To a certain degree it is possible to imitate or reproduce the dialectics of mutation aka variation and (environmental) selection in respect to the 'evolution' of music: slices of music as reproducing and mutating population, a (selected, probably not statistically representative sample of public as 'environment'—a „Darwinistic" experiment, perhaps not without a certain irony:

> Music evolves as composers, performers, and consumers favor some musical variants over others. To investigate the role of consumer selection, we constructed a Darwinian music engine consisting of a population of short audio loops that sexually reproduce and mutate. This population evolved for 2,513 generations under the selective influence of 6,931 consumers who rated the loops' aesthetic qualities. We found that the loops quickly evolved into music attributable, in part, to the evolution of aesthetically pleasing chords and rhythms. Later, however, evolution slowed. Applying the Price equation, a general description of evolutionary processes, we found that this stasis was mostly attributable to a decrease in the fidelity of transmission. Our experiment shows how cultural dynamics can be explained in terms of competing evolutionary forces.[3]

---

[2]See e.g. Koza 1992 „on the programming of computers by means of natural selection“.

[3]Imitation and reproduction of the evolution: *Evolution of Music by Public Choice*, 18 June 2012, http://schall-und-mauch.de/artificialmusicality/2012/06/evolution-of-music-by-public-choice/, retrieved Jan. 30, 2014. .

Filtering is also a basic category for music and its evolution. Even in a structural and historical perspective the constantly renewed filtering of (especially acoustic) elements of reality continues in producing music. This process creates general music material—sounds, sequences, pitch systems or rhythmic systems, the combination of sequences of pitches and rhythms to melodies and more. Not only the motif, but already the tone is filtered from the fullness of the real clangors, noises, sounds, and is embedded in a system of at least partially 'encaptic' structure. In general we must note that musically not the physical-acoustical fact and technical signal but the musical-aesthetical and psycho-social sign is essential and decisive. That counts already on the elementary level of pitches. Even an acoustical fuzzily realized tone which deviates more or less from the exact pitch, a nearly wrong tone, can in the framework of the system be judged and perceived as a right tone.[4] This aspect and component too is a specific form of filtering. In music, being communication and art, consciousness and intentionality play a great role.

This filtering as a basic process can be found, as in the case of the material of signs, as well in its connections and combinations in and through artistic production in general. The material as and in arts has special aspects and qualities. Musical materials, meaning sounds, colors, forms, have an aesthetically higher intrinsic value as material in scientific reasoning, deviating from everyday experiences. Apart from that the filtering runs analogue to processes and methods of abstraction and terminology as used for scientific reasoning. *Here* one speaks of reality as the total set, of which subsets are extracted through filtering, *there* one sees the musical world of sound with all its cognitive components divided, segmented and gridded in various manners. The aforesaid sign material, that is to say sounds, colors, forms, has a higher intrinsic value in the aesthetic and the arts than in the gnoseological, in the cognition, namely, as something beautiful, deviating from the usual and everyday things. Otherwise, however, the filtering is analogous to the processes and methods of abstraction, generalization and conceptualization in scientific knowledge. As here the reality appears as a total quantity, out of which structured subsets are filtered out—which can often overlap—the musical world, together with its cognitive components, is screened, structured, segmented in various ways.

5.1.3 Filtering of Elements: Musical Tone, Pitches- and Rhythm Systems, Tuning Systems

The world of sounds is systematized for music, as music and in music, historically at first purely pragmatically, then with increasingly complex theoretical considerations.

[4]See León / Liern Carrión 2010 and 2011—Many very interesting types of acoustical illusions http://deutsch.ucsd.edu/ with further links (retrieval 29.6.2009). See also for the special branch „Cognitive Musicology" with relations to cognitive sci(ence in general and psychology of music Deutsch 1999; cf. Risset 2002; De Poli / Rocchesso 2002.

In turn, a generalizing symbolic composition builds on this particular musical, theoretically mediated basis. It reaches as far as cosmological speculative universism (see Chapter 9), which seeks to encompass microcosm and macrocosm, human beings and the world as cosmos in holistic connections and interactions. It unfolded mainly in Old India, in China,[5] then in Greek antiquity—keyword Pythagoreanism—and was taken up again in the feudal Middle Ages. Continuations and resumptions still survive today.

Systems of pitch orders have mostly been developed primarily for the dimension of pitch. Rarer are systems of durations and rhythms such as the Indian talas as part of the raga concept. In some cultures, tone color systems are emerging, especially in East Asia. In China and Korea, they start from a classification of the musical instrument groups according to the production material, specifically the magic eight: metal, stone, earth, leather, silk, wood, pumpkin, bamboo. The least systematic seem volumes. Because with them there is no scaling as the pitches are the frequencies as natural bases of the partial tones and the octave identity. In the case of the volumes, a scaling has been sought since the age of absolutism and in the name of rationalism, that is shortly before 1600, and is realized as precisely as possible by progressive refinement: in the 20th century, for example, from *pppppp* to *ffffff*. The problem of scaling volumes is reinforced in the object-subject relation: the boundaries e.g. between simply „loud" and „noise" have broad transitional zones that are subject-specific and situational-related, among other things. It is only with the help of electrical engineering and electro-acoustic methods that a real objectification becomes possible. Units of measure such as phon or the logarithmic dB allow precise scaling. And it is necessary, too, given the growing problems of noise, for a humanitarian sound ecology.

5.1.4 Tones from Noises and Sounds: Organizing and Structuring of Rhythmical-Temporal Acoustic and Non-Acoustic Processes—Culture Versus Nature. Historical and Logical-Systematic Dimensions

From the abundance and diversity of the acoustic in the natural and social reality elements are isolated, selected, filtered, prepared, processed further: in principle and in general universal, in detail both depending on the particular environment of the sound systems (not everywhere there is cuckoo, nightingale and donkey) as well as culture-specific goals and values. The aim is even more special: filter out musical material. However, even in pre- and extra-musical objective reality, the continuum of sounds and sounds is not complete. It is already objectively an 'articulated continuum', structured and differentiated by causers, places and times,

[5] See e.g. Zoltai 1970; Chen 2000.

addressees and more. Indexical and iconic characters form the starting point of the musical material, and are often further developed into symbolic signs.[6]

We remember the layers of the music process. (See Sect. 2.1) Material supplies, firstly, the inner human nature, namely through biological-physiological, sonic processes such as breathing, coughing, etc., walking or, already rhythmic, striding; especially important is the heartbeat, which also affects the music as a pulse—as in the case of breathing and walking at the same time there are already rhythmic models. A second source of the material is the external nature, so wind and waves, thunder, etc., on the one hand, animal noises from the chirping of the cicadas and croaking of the frogs to the bird „song" on the other hand. A third source is the „second nature" of human technology, working sounds from hammering with pebble stones or hand axes to forging or typewriting noises, etc. Here comes from the working a certain higher regularity and order—before stylization as a principle of art. For an aesthetic dimension already has the work. A leitmotif is here the combination of efficiency and elegance. A fourth source is a layer anchored mainly in the emotional and social psychology: components of language as material of music, the intonational and emotional in word language.[7] As with the natural sound of bird communication, the aspect of the tones and the specific pitches here comes to the fore in contrast to sounds and noises as well as rhythms. The „speech melody" thus becomes a music melody as a unit of pitch and duration movement.

Fuzziness always plays a role in one with sharpening as filtering: what is transformed as material into music is not identical with the original sounds, but only similar to them.

Both structurally and historically, an ever-renewed filtering of (especially acoustic) elements of reality is constantly taking place. This generally forms musical material—tones, sequences of tones, systems of pitches or of durations and rhythms systems, tones from noises and sounds, the combination of courses of pitches and rhythms to tunes and more.[8] Thus, rhythms, as found in numerous phenomena of reality, become *musical* rhythms, not least through filtering processes.

5.1.5 Development of Chords as Structured Vertical from Heterophonic Lines

The development of pitch systems historically goes back to the depths of the times. It requires the development of musical instruments. Because only they offer the concrete vision for thinking, forming, and practicing systems, the „intuition" the

[6]The terms index, icon, symbol in the sense of Peirce and Morris.

[7]In detail to the spectrum of this starting material especially Georg Knepler 1977.

[8]„I hypothesize three components as necessary for ensemble playing and dancing: pulse *extraction*, beat entrainment, and motor pattern generation. The first involves auditory processing only, the second cross-modal (auditory to motor) integration, and the last sensorily-guided motor behaviour." (Fitch 2011, p. 2 [of 17], Emphasis H.-W.H.).

„theory“: they are objectified-exteriorized, representational, visible and tangible and correspondingly materially modifiable. The mainly invisible voice inside in the body does not offer these possibilities. Sound systems date under these conditions at the latest to the time of Homo sapiens. The oldest datable discovery so far is a late Paleolithic bone flute from the cave Geißenklösterle near Blaubeuren, Swabian Alb, about 37,000 years old. The flute is a total of 18.7 inches long. She has at least three holes. It was „made from two carved halves that were tied together and glued to the accurately designed, airtight seam.“[9] (See above p. 201.)

> Other flutes at the same site were made of hollow bird bones. This is much easier than carving hard ivory. The adhesive technology is therefore an already more sophisticated manufacturing technique.[10] Three holes is not just only three tones. We get more by the clogging of the tube at one end, probably at least five. Through diversified blowing techniques and gripping combinations even „almost all tones of our modern scale within a certain tonal ambitus"[11]

However, we do not know if all the *possible* techniques were *actually* used then. So once again, Fuzzy Logic's relationship in terms of modal logic between possibility and reality is important. But we can almost certainly assume that at least one five-tone pentatonic system was created in connection with these flutes. This can also be read off from other flutes of this time. Surprisingly, it seems that there was a hole for corrections of the tuning. Cultural comparisons such as natural principles speak for the probability of five-tone or even seven-tone scales.

> „the widespread cross-cultural fact of pentatonic and 7-note diatonic scales in our own history (and the acoustic basis for these scales) [are a] justification for the probabilities being higher regarding the Neanderthal bone matching a diatonic rather than matching a more obscure or hitherto unknown scale.“[12]

Even with later archaeological discoveries it is possible, to reconstruct the pitch system, be based on the object instrument by means of physical, objective measurements and technical musical experiments.[13] It is easier if there are written sources for the pitch and tuning systems. Then precision and blurs can be more easily recognized and calculated.

> Despite the difference between Chinese and Western music development, the basis of the equal-temperament scale (twelve equal semitone intervals per octave, we call this the 12-tempered […]) was first mathematically calculated by the Chinese mathematician and musician Zaiyu Zhu (1536-1611) in 1584 […]. But this was never widely used in composing

[9]http://www.knochenfloeten.de/pdf/HEIN%20Wulf%20knochenfloeten_blaubeuren.pdf, retrieved Oct. 19, 2017. These paleolithic flutes are evidently earlier than the neolithic flutes presented by Zhang et al. 2007.

[10]To other flutes of the Aurignacien art in detail among others Käfer 1998.

[11]http://www.knochenfloeten.de/pdf/HEIN%20Wulf%20knochenfloeten_blaubeuren.pdf, retrieved Oct. 19, 2017 can be generated. The ambitus is defined by the overall length of the flute.

[12]*Correspondence III: Neanderthal Flute,* Kent Nickersson. Jul 08 1997, Updated Mar 2003, exhaustive discussion, https://web.archive.org/web/20060527135823/http://www.greenwych.ca/fl3debat.htmWhat, retrieved Jan 15, 2018.

[13]See e.g. Falkenhausen 1993, p. 50f.; Kluge 2011.

> the indigenous music of China until 1685. Before Zhu's tuning system was conceived there wasn't a standard tuning system between different instrument types such as strings, winds, and keyboards. The formula he discovered allowed pitch-pipes in a equal-tempered scale of twelve equal-ratio semitones per octave, which led to a revolution of music and physics. […] Zaiyu Zhu calculated the equal-tempered system in 1584 […]. However, the earliest Chinese tuning system […] has several correspondences with the Pythagorean system. It is called the thirdsdivide subtract-add (三分损益) method. Like the Pythagorean system, it is based on the evaluation of frequencies using the length of a portion of vibrating string.[14]

5.1.6 Tuning Systems—Sharpening, Blurring and Fuzziness

There are multiple connections from fuzzy logic to music in this area, too. Blurs of all kinds are almost everywhere in their context: already elemental on the level of sound, not to mention the system relations between music, culture, society. These are more complex and difficult to handle or to grasp than any elaborate set of sounds. Thus scatterings in pitch and timbre are characteristic of transient phenomena. They are blurs, which are essential for the impression of a „living" sound. The different tuning each show limits of the precision and sharpness. Although these are needed as a scale and a benchmark, they are historically and culturally widely different realized. What seems numerically and mathematically unequivocal, in the sonic reality proves to be only seemingly accurate: So are the various comma of just intonations, but also the equal temperament. Just the highly rational equal temperament, which has been treated several times by the Fuzzy Logic,[15] uses the 12th radical of 2, an irrational number, for its completely even division of the octave. And with the equal temperament, actually all intervals except the octave are a bit detuned, because the Pythagorean comma is distributed to all fifths in the same way.[16] In equal temperament the octave is divided so that there are twelve semitones with completely equal interval intervals. This is a significant departure from the acoustical-physical nature of the sub-tone structures and the traditional modes or scales. It serves the cultural-artistic progress. For all the keys are equally accessible and one can compose with them, and all tend to be connected to all.[17] This progress, in turn, has its price. The key differences together with the hitherto objective key characters through qualitatively different interval structures of the individual scales as well as through various roughness and bulkiness, that means graduated dissonance degrees of chords, disappear. Since all scales on all 12 levels have the same interval structure, instead of the previous variety of modes or scales, there are actually only two more, namely

---

[14] Mazzola et al. 2016, p. 247.

[15] See e.g. León / Liern Carrión 2010 and 2012.

[16] As a kind of 'archaeology' in sound records one can discover and select different tuning systems of the respectively used instruments—and the detuned and fuzzy variants of the intentioned systems. See Dixon et al. 2012.

[17] On the historical dimension of tuning systems for composing and thinking music see e.g. *Pythagorean Tuning and Medieval Polyphony,* http://www.medieval.org/emfaq/harmony/pyth5.html, retrieved Jan. 14, 2018.

major and minor. If we take the distinction between „melodic“ and „harmonic“ minor seriously, then there are three.

Interval structure major (1 = half tone)
$0 + 2 + 2 + 1 + 2 + 2 + 2 + 1$
Interval structure „harmonic“minor
$0 + 2 + 1 + 2 + 2 + 1 + 3 + 1$
„melodic” minor
$0 + 2 + 1 + 2 + 2 + 2 + 2 + 1$

The slight deviations from the just intonation, however, are barely perceptible in musical practice, because the objective acoustic reality is adjusted to the intersubjective generative system of pitch order as an ideal and conceptual background through subjective „correcting listening“. This background is historically, socially, culturally mediated. In terms of perception, as a further example, I refer only to the broad field of acoustic illusions even at the very elementary level of intervals or short motives or short melodies. There are even more differentiated microtonal divisions of the octave, such as in Indian and Chinese music, and in Europe theoretically since the late Renaissance, practically since the beginning of the 20th century, in the latter case mainly on the basis of equal temperament as a further „nuclear fission“ of the 12 semitone of the octave. The US-American composer Harry Partch (1901–1976), on the other hand, built new musical instruments or adapted existing ones to the just intonation, dividing the octave into 43 partials instead of the twelve semitones like in the equal temperature. He found successors who in turn developed variants.[18]

Culture is based on (here physical-acoustic) nature and is at the same time filtering and mastery of nature in deviation from the natural, physical overtone series: namely with regular and regular blurring, as mentioned. These are, however, blurs only in the frame of reference nature /partial tones versus art /equal temperament. These themselves are a rationalizing sharpening in the reference system of pitch system—composition, a clarification that makes wider-spanned harmonic contexts possible. On the other hand, in the practice of music realization there are again multiple blurrings. The mood of the pianos is deliberately not exactly equal-tempered. And in the orchestra, the musical instruments necessarily are adjusted with compromises to the now fundamental tuning, again different according to instrumental genres, such as the base of pure fifths in the stringed instruments or, to put it a bit blankly, the construction on the overtone series in wind instruments. The ratio of blurring and sharpening /filtering is generally reconfigured differently in terms of the pitch and duration systems in the concrete compositional design and above all in the productive as well as the receptive realization. For example, the difference between g♯ and a ♭ is in non-equal tempered systems an objective-real difference of the frequencies. And in the equal tempered system there still remains a difference in the intentional context according to the musical logic—g♯ is leading-tone up to d, a ♭ leading-tone downwards to g. The difference influences and colors productive realization insofar

[18]See in detail http://www.swr.de/swr2/programm/sendungen/jetztmusik/glossar/-/id=4979696/1m5qq8g/index.html, June 24, 2009. (Link now decommissioned).

that string instruments intone g♯ a little bit higher, and a ♭ lower. Even in the receptive realization the right, in the context of system, syntax and musical logic meant tone is filtered out.

5.1.7 Filtering: Technical Production of Tones and Sounds from Noises in Electro-Acoustic Music

In the process of history and music history, new material from the world of sound is always filtered out. Thus the Futurists celebrated their discovery of the sounds as musical-compositional material from 1912—sensational actually only because in the „classical" music in the 19th century the acoustical noise and the percussive had been pushed back more and more. With the electroacoustic processes since the 1940s, a new wave of material innovations and extensions came. It was followed by further expansion in the course of digitization. The term filtering is metaphorical—that too is a piece of blurring. How often, here scientific terminology is transmitted from everyday life and in terms of the acoustical, especially from the sound engineering. Because there is filtering technically literally, namely in electro-acoustic music. As part of the technical reproduction of the sound, it recapitulates in this frame historical developments in no time. On the one hand, the technique here goes even further into the elementary, decomposing the so-called „atomic" tone towards musical „elementary particles". Secondly, it supplements the sound analysis by sound synthesis, the recombination of the elements and their processing. Central are different types of filters such as sawtooth, rectangle, etc. They make of the most elemental raw material, the sinusoidal tone, really musical tones, sounds, noises. Conversely, at least theoretically, even if practically nonsensical, the absolutely pure sinusoidal tone can be filtered out of the so to say „absolute" random noise or „snow", which contains all the audible frequencies, and is the nearly absolute vagueness and blur in terms of pitch. Usually, „colored noise", a certain section of frequencies, serves as raw material. And thirdly, real sounds from technique and nature are used and processed as musical material—as in the *musique concrète* since the 1940s. Thus the filtering, by which the sound systems are historically formed in a dialectic of partially planned processes and mostly primordial and unplanned results, is raised to a higher level: Technical reproduction /production complements traditional filtering as a conscious, rational positing—„thesei" not substituting but supplementing „physei", nature nurtured by technique.

5.1.8 *Solmization as Soggetto Cavato (Carved Out Theme): Filtered Syllables as Pitch Names. System Building Within an Already Established Historical System of the Musical Material*

The emergence and development of the musical material are complemented and extended by formations of additional systems within an already established historical sound and pitch system. An immediate musical-practical model of great historical reach is solmization. Guido d'Arezzo (991/992—1050 at the most) developed it using pedagogical functional lyric.

In his writing of about 1033 *Epistola de ignoto cantu* Guido, an impressive music theorist and music educator, develops a system for singing from the sheet, for a relieved reading and a more relaxed singing. He introduces two things: the ingenious staff notation and the use of the ut–re–mi–fa–so–la (solmization) as mnemonic and re-cognition technology. These ut–re–mi-fa-so-la syllables extracted from the initial syllables of each of the first six half-lines of the first stanza of the hymn *Ut queant laxis* in honour of St. John the Baptist. The text traditionally is attributed to the North-Italian monk and historiographer Paulus Diaconus in the 8th century.

Ut queant laxis *re*sonare fibris
*mi*ra gestorum *fa*muli tuorum
*sol*ve polluti *la*bii reatum
Sancte Iohannes.

A paraphrase by Cecile Gertken, OSB (1902–2001) in 1981 preserves the key syllables and loosely evokes the original meter:

Do let our voices
resonate most purely,
miracles telling,
far greater than many;
so let our tongues be
lavish in your praises,
Saint **J**ohn the Baptist.[19]

The melody possibly was composed by Guido, but most likely he used an existing melody of an—very similar—eleventh-century musical setting of Horace's poem Ode to Phyllis (*Odes* 4.11), recorded in the French Montpellier manuscript H425,[20] obviously a pretty secular song. So this is also a transformation and a double filtering process, from secular to sacred, and then from words to syllables (Fig. 5.1).

The melody starts at c and ends at a, so it contains the six notes of a hexachord. It acts as a gamut resp. musical scale. Because the first syllable of each half-verse is

[19] https://en.wikipedia.org/wiki/Ut_queant_laxis, retrieved Jan. 2018, 14.
[20] See ibidem.

Fig. 5.1 Guido d'Arezzo: Ut queant laxis. https://www.pinterest.de/pin/188447565635805455/

sung higher by one tone (or semitone) than the previous one. The hexachord forms in the musical language of (Gregorian) chorale, an organizing principle of the tone system from which then later in different form scales were. Guido combines and composes with a special application of the acrostic method. This is at the same time a special case of the soggetto cavato respectively of the soggetto e vocalis (from the vowels). It is a pitch sequence deriving from a word or phrase, scrapped out in the form of tone syllables or tone letters. Guido's filtering procedure appears ex-post arbitrary compared to the alphabetical order of the ABCD, etc., the pattern for scales and modes. But the syllables are originally motivated, as semantically and syntactically coherent text. With an extended notion of the soggetto cavato, acronym and acrostic can be taken as its special cases: in these, it is always the first letter of the word sequence.[21] In the soggetto cavato as a *musical* method, however, the selection is directed to syllables or letters suitable to music. (See below examples in 6.1.1) Thus, from higher-order figures, namely, poetic-religious text and melody, lower-order shapes are filtered out, namely, abbreviated terms for the tone stages within the hexachord (6-tone sequence) in the form of syllables, correlated with the tones of those stages. In the light of Fuzzy Logic, this is a sharper detour, truly methodical and ingenious: poetry and meaning and the melodic context of the anthem are not only blurred, but wiped away. But in return Guido wins a precise correlation of pitch-name within a small system that captures all sounds of the Western European sound

[21] Cf. e.g. to Animal Alphabet: A is for Armadillo, B is for Bat, C is for Camel, etc.—The method is also found in the word language, in the word game. „Ditloid" is about guessing from the first letter a text—a saying, a book title, an everyday wisdom, etc.—to guess. the form in which a number is declared, with the filler words being written out. For example 4 = H on a H (hooves on a horse). (See: Die Presse, Spectrum, Vienna, 25.6.2011).

system in his time. This solmization method has a great educational value for the vocal didactics and practice. The exact name for relative tone levels facilitates the change between the three hexachords naturale, durum, molle.

naturale: c d e f g a
molle: f g a b flat c
durum: g a b natural c d

Later in the major-minor tonality solmization then allows the transposition of a melody into different keys. The system still lives on today, adapted to the major-minor tonality, which developed since early absolutism from about 1600. In some languages, these relative pitch designations are also used as absolute, e.g. French ut majeur or Italian do maggiore = C major.

The Florentine humanist Giovanni Battista Doni (c.1593–1647) brought about an extension of the subject and a change in nomenclature. In 1635 and 1640 he changed the name of note „Ut"(C), renaming it „Do". He named two main reasons. The first, continuing Guido's practical motivation: „Do" is easier to pronounce than „Ut". The second is a reversal and revision of Guido's arbitrary selection in naming the pitches:

> „'Do' is an abbreviation for 'Dominus', the Latin word for The Lord, who is the tonic and root of the world."[22] The third, not openly stated reason could be: Probably Doni „also wanted to imprint himself into musical canon in perpetuity because „Do" is also ulteriorly an abbreviation for his family name."[23]

To adapt the nomenclature to the new state of the development of music, namely the diatonic seven-tone system, and to complete the diatonic scale, soon after Doni a seventh note was added: „Si"(from the initials „Sancte Iohannes. In the early nineteenth century Sarah Glover changed „Si" to „Ti" so that every syllable might begin with a different letter. (This also freed up Si for later use as Sol-sharp).[24]

5.2 Filtering as a Search for Similarities Using Fuzzy Logic

This dimension of filtering is nothing more than the practical, computerized application of what we have already discussed in terms of 'similarity' in relation to Fuzzy Logic. (See Chapter 3)

5.2.1 Detection, Storage and Retrieval of Patterns or Shapes

Because only the search for similarities is filtering. The filtered self is pretend. As with the filtering out of sound and rhythm systems, everything is already there, but

[22]See https://en.wikipedia.org/wiki/Giovanni_Battista_Doni, 21.7.17, retrieved Jan. 15, 2018.
[23]Ibidem.
[24]Ibidem. Cf. https://en.wikipedia.org/wiki/Sarah_Ann_Glover, 18.4.16, retrieved Feb. 20, 2018.

on a higher level of complexity, that of the already artistically formed figures. While the what, the object of the filtering is objectively present, as works or works-parts, in the same way as the filtering in the broad sense comes subjective. Formally, e.g. the fineness of the sieves selectable. And in terms of content, the search for similarities in melodies, pitch gradients, rhythms, timbres, etc., may be essentially preconditioned for classification and typing, for the purpose of order and retrieval of digitally stored music or, for legal purposes, clarification of plagiarism. Reproaches or motivic-thematic contexts. Just as it is in the nature of the principle of similarity, the filtering of the subjects of the possible spreads widely between identity, equality and only rudimentary, blurred or blurred similarity. There is already a large amount of research and publications on this topic. Especially elementary questions and special practical problems are extremely well and competently researched. On the fan of functions:

> Such work has applications in the areas of classification (e.g. organisation and navigation of music collections), recommendation (e.g. discovery and marketing of new music) and annotation (e.g. automatic transcription for education, musicological research, and music practice), and it complements traditional research methods in musicology, enabling more quantitative and larger scale analyses to be performed.[25]

As I said in the beginning, my researches are complementary, and therefore it is my program to bridge gaps, to fill lacunae, and to open up new fields of research, topics, questions and aspects for the main theme music and fuzzy logic. (See above 0.2 and 1.) It may and must therefore suffice here to relate to certain aspects, along with supporting titles, to mention some systematically relevant aspects and to give hints for further readings. There are nearly endless lists of literature to this subject, with hundreds of titles—e.g. not less than 194 titles already solely already in Adak (1998). It is, summa summarum, not necessary here to treat exhaustively this field of retrieval, especially as I am a non-expert in mathematics and hard or soft computing.

5.2.2 Melodic and Other Similarities—Computer-Based Analysis

In the analysis of and search for similarities with the help of fuzzy logic the melody, the unity of movements of pitch and duration, has according to the everyday consciousness a certain primacy. But there are also numerous retrieval methods that focus on other parameters and dimensions between chord, key, and even cultural contexts. Wojcik /Kostek 2010 add:

> main issues related to music information retrieval (MIR) domain. Within this domain, there exists a variety of approaches to musical instrument recognition, musical phrase classification, melody classification (e.g. query-by-humming systems), rhythm retrieval, retrieval of high-level-musical features such as looking for emotions in music or differences in expressiveness, music search based on listeners' preferences, etc.

[25]Dixon et al. 2012, p. 1.

In the following I give some surveying hints to the broad spectrum of themes and dimensions of the researches. Similarities are the subject of some early pioneering work, even before the actual „digital age“, such as *Mathematical analysis* of *formal structure of music*—Fucks 1962; *Struktur und Ähnlichkeit: Methoden automatisierter Melodieanalyse* (Structure and similarity: methods of automated melody analysis) – Steinbeck 1982; Müllensiefen /Frieler 2004; further *Computational models* of *music similarity* – Pampalk 2006b; *Ähnlichkeit, Nähe, Distanz: zur Anwendung multidimensionaler Skalierung in musikwissenschaftlichen Untersuchungen* (Similarity, proximity, distance: on the application of multidimensional scaling in musicological studies) – Markuse /Schneider 1996; Gardiner et al. 1996.

I „Parameters", Dimensions

Pitch order, tuning

Estimation of Harpsichord Inharmonicity and Temperament – Dixon et al. 2012; *multipitch estimation of piano music* – Lee et al. 2012; combining spectral and temporal representations for multipitch estimation of polyphonic music – Su /Yang 2015.

Melody

Measures of melodic similarity – Müllensiefen /Frieler 2004b.

One can also combine two dimensions: *modeling meter and harmony* – Temperley /Sleater 1999; *register classification by timbre* – Weihs et al. 2008; Temperley /Sleator 1999.

Rhythm, metrics

hypermetric manipulation in Canadian blues – Ethier 2001; *rhythmic similarity* – Foote 2002; *beat tracking* – Goto 2002; *rhythm description, tempo induction* – Gouyon 2005; Gouyon /Dixon 2005; *pulse salience and metrical accent* – Parncutt 1994; *forming and ranking rhythmic hypotheses* – Wojcik 2006; Wojcik /Kostek 2010.

Tempo, pulse, beat

Tracking musical beats in real time – Allen /Dannenberg 1990; Allan et al. 2007; *tempo tracking algorithms* – Alonso et al. 2003; *beat spectrum* – Foote /Uchihashi 2001; Foote 2002; *cyclic-beat spectrum* – Kurth et al. 2006; *tempo and beat analysis* – Scheirer 1998; *dance style into tempo detection* – Schuller et al. 2008; *rhythm and tempo* – Takeda et al. 2007.

Volume

Filtering of dynamics – Arenas-Garca 2006.

Harmonics, chords, keys

Harmonic content – Grekow 2007.

Instruments

Representing musical instrument sounds for their automatic classification – Kostek /Czyzewski 2001; *Automatic recognition of musical instruments* – Bruderer 2003; *percussion-related semantic descriptors* – Herrera et al. 2004; *musical instrument sounds – automatic classification* – Kostek /Czyzewski 2001; *identification of musical instruments* – Röver et al. 2005; *automatic indexing of musical instruments*

– Wieczorkowska /Raś 2005; *multi-way hierarchic classification of musical instrument sounds* – Pearce et al. 2010; Wieczorkowska et al. 2005; Wieczorkowska et al. 2007; *instrument recognition by timbre and pitch* – Hung /Yang 2018; *Instrument Role Classification* – Ching /Ramires /Yang 2020.

Timbre, 'sound'

Spectral similarity of songs – Flexer et al. 2005; *timbral similarity* – Levy /Sandler 2006; *spectral similarity* – Pampalk 2006a; *frame-level instrument recognition by timbre and pitch* – Hung /Yang 2018.

Structures and structural elements of the composition

"Due to its apparent relevance to music listeners and in commercial applications the automatic detection of vocals in particular has received considerable attention in the recent Music Information Retrieval literature".[26] As a counterpoint to this prevalent filtering of vocals Mauch et al. 2011 have developed methods for „detecting *instrumental solos in polyphonic music recordings*".[27] – Mauch et al. 2012.

Compositional structure, style

Automated methods for the analysis of style – Jürgensen /Knopke 2004; *style recognition* – Kranenburg /Backer 2004.

Multimodal structure segmentation – Cheng /Yang /Lin /Chen 2009; *confounds of accompaniments in singer identification* – Hsieh /Cheng /Fan /Yang /Yang; *comparative study of Western and Chinese classical music* – Fan /Yang /Dong /Pasquier 2020.

Artist

Artist classification – Knees et al. 2004; Pampalk et al. 2005.

Genre

genre classification systems – Ahrendt 2006; Li et al. 2003b; Lidy /Rauber 2005; Lin et al. 2009; *genre for music emotion classification* McKay /dictionary learning for music genre classification – Yeh /Yang 2012.

II More complex components with a distinctive dialectic of the objective and the subjective

The following topics include subjective aspects and are composed of relationships between different forms of existence and of relationships between the music process and the cultural environment. The research refers to different aspects.

Productive realization, interpretation, expression

Detecting emotion in music Li et al. 2003a; *objective parameters from music performances* – Lerch 2008; *automatic detection of vibrato* – Pang /Yoon 2005; Rossignol et al. 1999; Widmer et al. 2003; Goebl et al. 2004; *subjective intensity* – Sandvold /Herrera 2005; *embedded feature selection for multi-label classification of music emotions* – You et al. 2012; *performance expressiveness in intercultural view* – Fabian et al. 2014: *cross-cultural and cross-dataset generalizability of music mood regression models* – Hu /Yang 2014; magnitude and phase-derived spectra for playing technique classification – Su et al. 2014; common mistakes in novice violin

---

[26] Mauch et al. 2011, n.p.

[27] Ibid. (Emphasis H.-W.H.).

playing – Luo et al. 2015; *expressive musical terms in violin using score-informed and expression-based audio features* – Li et al. 2015.

Receptive realization—perception, tastes, value judgements

Human perception of musical similarity – Herre et al. 2003; *Subjective music-similarity measures* – Berenzweig et al. 2004; *cognitive adequacy in the measurement of melodic similarity* – Müllensiefen /Frieler 2004a; *rhythmic salience of sound with association rules and neural network* – Kostek et al. 2005a; *learning system for estimation rhythmic salience of sounds* – Kostek et al. 2005b; *perception of accents in pop music melodies* – Müllensiefen et al. 2009; listening behavior and user mood – Liu 2014.

‚Emotion'

Emotion is a complex object of investigation that is frequently addressed here, as in music aesthetics in general. Before modelling and quantifying, the fundamental, at least two-digit relation must be considered: emotion in the music itself and the emotion of the recipients. Thirdly, there is the emotion of the composer—as a motivation for objectification, at least in minimal form, relatively abstractly necessary, but not necessary in concrete and detailed ways. The fourth is the emotion of the interpreters. C.Ph.E. Bach in the Age of Sensitivity wrote that only those who are moved themselves can move others with their music making. In fact, this identification and connection is only possible, but not really necessary. The situation is somewhat different in jazz and in many areas of popular music: here, the emotional empathy and participation of the performers are important or even essential. Following the four Aristotelian causae, the musical representation of emotions in material and technique, an aspect of the polyvalent 'expression', would be the combination of causa materialis and formalis, the composers the causa efficiens. With a certain re-interpretation of the Aristotelian categories, what has been effected, the work, becomes in turn a causa efficiens, an effective agent, namely in relation to both the interpreters and the recipients.

Detecting emotion – Li /Ogihara 2003a; *perceptual similarity measures* – Pampalk et al. 2003; *classifying emotion in popular music* – Liu et al. 2006; *mood detection* – Liu et al. 2003; *user-adaptive music emotion recognition* – Wang et al. 2004; *Modeling Emotional Content of Music Using System Identification* – Korhonen et al. 2005; *classification of emotions in music* – Wieczorkowska et al. 2006; *emotion classification: fuzzy approach* – Yang et al. 2006; *emotion in popular music* – Liu 2006; *emotion recognition: the role of individuality* – Yang et al. 2007a; *emotion classification* – Yang et al. 2007b; *emotion recognition: role of individuality* – Yang et al. 2007c; *continuous measurement of emotions in music* – Nagel et al. 2007; *multi-modal music emotion classification* – Yang et al. 2008; *emotion ranking* – Yang /Chen 2009; *neural network model for the prediction of musical emotions* – Coutinho et al. 2010; *emotion recognition* – Kim et al. 2010; *music emotion recognition* – Yang /Chen 2011; *automatic highlights extraction for drama video using music emotion and human face features* – Lin et al. 2011; *exploiting online tags for music emotion classification* – Lin et al. 2011; *emotions: predicting second-by-second subjective feelings of emotion from low-level psychoacoustic features and physiological measurements* – Coutinho et al. 2011; *relationship between categorical and dimensional emotion semantics of music* – Wang et al. 2012; *cross-cultural music mood classification: a comparison of English and Chinese songs* – Yang /Hu 2012; *machine recognition of music emotion* – Yang /Chen 2012; *music emotion recognition: from content- to context-based models* – Barthet et al. 2013; *music listening behavior in a social and affective context* – Yang /Lin 2013; *role of time in music emotion recognition* Caetano et al. 2013; general survey e.g. Aramaki et al. 2013; *fusion of electroencephalogram dynamics and musical contents for estimating emotional responses in music listening* – Lin et al. 2014; *fusion of electroencephalogram dynamics and musical contents for estimating emotional responses in music listening* – Lin et al. 2014; *principal component analysis to alleviate day-to-day variability in eeg based emotion classification* – Jao et al. 2015; *affective analysis in multimedia* – Soleymani et al. 2015; *cross-dataset and cross-cultural music mood prediction: a case on Western and Chinese pop songs* – Hu /Yang 2017; *component tying for mixture model adaptation in personalization of music emotion recognition* – Chen et al. 2017; *mood of Chinese pop music:* – Hu /Xang 2017; *predicting the probability density function of music emotion using emotion space mapping* – Chin et al. 2018; *pop music highlighter: marking the emotion keypoints* – Huang et al. 2018; *cross-cultural music emotion recognition* .– Chen et al. 2018.

Emotion-oriented retrieval for trivial commercial purposes

Automatic mood detection and tracking – Lu /Zhang 2006; *iMR: Interactive music recommendation* – Yang /Chen 2009; Personalized music emotion retrieval – Yang et al. 2009; *prediction of the distribution of perceived music emotions using discrete samples* – Yang /Chen 2011; recommendation based on multiple contextual similarity information – Chen 2013; *context-aware music recommendation on smartphones* – Teng et al. 2013; *linear regression-based adaptation of music emotion recognition models for personalization* – Chen et al. 2014; music emotion recognition models for personalization – Chen et al. 2014; *interactive music delivery system for private listening as wherever you go* – Lin et al. 2014; *Exploiting latent social listening representations for music recommendations* – Chen et al. 2015; *addressing cold start for next-song recommendation* – Chou et al. 2016; *query-based music recommendations* via *preference embedding* – Chen et al. 2016.

Even other, practical-motional methods of reception or appropriation are taken into account: *Danceability estimation* – Streich /Herrera 2005.

III Methods of special filtering, particularily for retrieval

As the topics, the search, analysis and processing methods are both expanded and refined, as e.g. *rhythm finding systems* – Kostek et al. 2005; *hierarchical organization and description of music collections* – Pampalk et al. 2005;, *audio-based music similarity and retrieval* – Pampalk 2006a; *computational models of music similarity and their application in music information retrieval* – Pampalk 2006b; Ranking-based emotion recognition – Yang /Chen 2011; *machine recognition of music emotion* – Yang et al. 2012; bag-of-frames representation for music information retrieval – Su et al. 2014; annotation and retrieval using unlabeled exemplars – Jao /Yang 2015; *marking the emotion keypoints* – Huang et al. 2018; *singing voice detection based on convolutional neural networks* – Huang et al. 2018; *singing voice detection based on deep neural networks* – You et al. 2018.

Without notes such as *query by humming /voice* – Lesaffre et al.2003; ***new query interfaces* – Tzanetakis et al. 2002a.**

In addition, the search with notes by entering more or less sharply remembered „key notes" in the musical sense, further simple input of text for the search for title, artist, composer, genre, and dating etc.; beyond that even transformation of melodies into texts: e.g. *melodic n-grams as text* – Downie 1999. The correction of failures is also ensured: *Solving failing queries* – Ras /Dardzinska 2006.

Automatic 'transcription' (transformation of the sound form into a notation form)

*Automatic rhythm transcription from multiphonic MIDI signal*s – Takeda /Nishimodo 2003; *Non-negative matrix factorization for polyphonic music transcription* – Smaragdis /Brown 2003; Ligges 2006; *automatic music transcription* – Takeda et al. 2007; Building Polyphonic Datasets for Automatic Music Transcription – Su /Yang 2015; Lead sheet generation and arrangement – Liu /Yang 2018; Score-to-audio music generation with multi-band convolutional residual network – Wang /Yang 2019.

Visualization, self-organizing neural networks

Grekow 2006; Lehwark et al. 2007; *visualizing timbre distances of music as topographical maps* – Mörchen et al. 2005; v*isual mining in music collections* – Risi et al. 2007; *visualizing music genre*s – Schedl et al. 2006; *pitch histograms* – Tzanetakis et al. 2002b; s*elf-organizing neural networks for visualization and classification* – Ultsch 1993; Ultsch 1996.

Automatic recognition, retrieval, composition

The methods of filtering for retrieval as a „reception" in the broad sense can even be used for the production of music. The claim was about a decade earlier still the quite modest—„automatic drum accompaniment to a given melody".

> The objective of this study is to propose a method for retrieval of *hypermetric* rhythm on the basis of melody. A stream of sounds in MIDI format is introduced at the system input. On the basis of a musical content the method retrieves a *hypermetric* structure of rhythm of a musical piece consisting of rhythmic motives, phrases, and sentences. On the basis of the *hypermetric* structure retrieved, a system capable of *creating automatic drum accompaniment to a given melody* supporting the composition is proposed. A method does not use any information about rhythm (time signature), which is often included in MIDI information. Neither rhythmic tracks nor harmonic information are used in this method. The only information analyzed is a *melody*, which may be monophonic as well as polyphonic. The analysis starts after the entire piece has been played. Recurrence of melodic and rhythmic patterns and the rhythmic salience of sounds are combined to create an algorithm that finds the metric structure of rhythm in a given melody.[28]

But the appetite grew while eating.

Acousticvisual emotion Gaussians model for automatic generation of music video – Wang et al. 2012; *Automatic melody harmonization with triad chords* – Yeh et al.; *adversarial networks with binary neurons for polyphonic music generation* – Dong

[28] Wojcik / Kostek 2010, Abstract.

/Yang 2018; *Multi-track sequential generative adversarial networks for symbolic music generation and accompaniment#.* – Dong /Yang 2018; *Jazz and Pop piano music from audio* via *MIR techniques* – Yeh et al. 2019; *Improving automatic Jazz melody generation* – Hung et al. 2019; *Drum fills detection and generation* – Tamagnan /Yang 2019; *Beat-based modeling and generation of expressive Pop piano compositions* – Huang /Yang 2020; a*utomatic composition of guitar tabs* – Chen et al. 2020; *Jazz Transformer on the front line: Exploring the shortcomings of AI-composed music* – Wu /Yang 2020.

Even the mediator and performer per se, the DJ, does not remain untouchable: Dream for making an automatic DJ – Huang et al. 2017.

However, the—relative—overproduction of musicians is part of the expanding „precarious" conditions of labor: "relative", since it is about an overproduction in relation not to needs and demand at all but to the solvent demand. It is an open question whether, in view of this situation, it is a particularly urgent social need to replace these human workers and to automatize their work. It may be, but it is uncertain, whether it will pay off economically. Culturally, it certainly contributes to a further, damaging flooding of the social landscape as a „soundscape" with mediocre to bad music. Questions, procedures and results of such research are also otherwise not always completely coherent and convincing from the standpoint of the music. A short case study must suffice. The research briefly discussed below is centered around the question of timbre. This is treated here as „global timbre quality". A possibly more appropriate complex category for this would probably be 'sound'. The starting point, however, are in principle plausible considerations:

> listeners are sensitive to timbre, at least in a global manner.[29]

From a pragmatic point of view, the search based on the sound is probably more robust towards blurs than that based on melodies:

> Indeed, the notion of melodic similarity is problematic, as a change in a single note in a melody can dramatically impact the way it is perceived (e.g. change from major to minor). Conversely, small variations in timbre will not affect the timbre quality of a music title, considered in its globality.[30]

A first, not insignificant blur, however, comes in that the authors correlate sound with taste, timbre with key; but this is not fundamentally, but conceded only „often" the case:

> (A)lthough it is difficult to define precisely music taste, it is quite obvious that music taste is *often* correlated with timbre. Some sounds are pleasing to listeners, other are not.[31]

Furthermore, the somehow historically meant term „music periods" is vague.

> Some timbres are specific to music periods (e.g. the sound of Chick Corea playing on an electric piano), others to musical configurations (e.g. the sound of a symphonic orchestra).[32]

---

[29] Aucouturier / Pachet 2004, p. 1.

[30] l.c., p. 10.

[31] l.c., p. 1.

[32] Ibidem.

Corea's „playing on an electric piano" only insufficiently fills the term „period" with content. As implied, „similarity" is conceived as „natural", abstract and contextless, referring only to sound and without any regard for differences in epoch, style, genre or idiom:

> (T)imbre similarity is a very *natural way* to build relations between music titles. The very notion of two music titles that „sound the same" seems to make *more sense than, for instance, query by humming*. [...] Typical examples of timbre similarity as we define it are: a Schumann sonata („Classical") and a Bill Evans piece („Jazz") are similar because they both are romantic piano pieces. A Nick Drake tune („Folk"), an acoustic tune by the Smashing Pumpkins („Rock"), a bossa nova piece by Joao Gilberto („World") are similar because they all consist of a simple acoustic guitar and a gentle male voice, etc.[33]

From the point of view of music this is crude and a little bit anti-musical, not just fuzzy but vague. Trivial and sweeping categories like „romantic" appear as objective standards. However, people do not listen so roughly, and tastes are generally more different as well—whoever wants Schumann does not want Evans, and vice versa. (Of course, the two do not exclude each other.) Noticeably, in a roundabout way, a prominent role of the „melody" comes into play over the—in principle true—special position of the vocal part:

> Some timbres (i.e. here sets of frames) are more salient than others: for instance, the first thing than one may notice while listening to a Phil Collins song is his voice, independently of the instrumental background (guitar, synthesizer, etc.). This saliency may depend on the context or the knowledge of the listener and is obviously involved in the assessment of similarity.[34]

As here, certain non-coincidences, indeed collisions between musical, psychological, cultural facts and their treatment under the isolated aspect of convenient retrieval, are not altogether rare. Because of such limitations, not only items and topics are out of focus of researches. Also, results themselves are not as precise as they should be—and could be—in accordance with the matter. Also in this area obviously Fuzzy Logic still could learn a lot or at least a little from musicology.

5.2.3 Call Wolfgang *via* Voice Over IP *(Johannes Kreidler). Automatic Phone-Tapping, Internet Censorship and Fabrication of "Terrorists"*

On another level new, different discrepancies may arise, with regard to the filtering under the aspect of signifiant-signifié or of the filtering out of keywords from texts. If the methodological segmentation is inadequate and the context is not sufficiently taken into account, there are merely mechanical-automatic, abstract and therefore semantic and pragmatic nonsensical similarities—an only apparent sharpening. The

[33] Aucouturier / Pachet 2004, p. 1.

[34] Ibid., p. 10.

filtering techniques are used also for purposes other than retrieval of music from databases on the Internet. From the point of view of Fuzzy Logic and its principles, as discussed here, it is secondary if filtering no longer refers only to elementary, pre-artistic raw material but to already existing, crystallized shapes. An often only seemingly exact filtering provides the Internet censorship (or even the simple search program within a digital text, spam filters, etc.). In the two following cases, figures are filtered out of a whole, and thereby another shape is created or re-created. But this shape, originally integrated into an overarching complex, gets disintegrated and reconstituted as an isolated form.

> Johannes Kreidler's (born 1980) *Call Wolfgang. terrorist generated content* (2008) is a VoIP-generated conversation between two computers interspersed with „suspicious"words such as „frame of reference,",„precarization,",„reproduction,"or „Marxist-Leninist"; that the telephonic use of these words the harmless, formerly blameless Berlin sociologist. Andrej Holm in 2007 at the BKA suspected terrorist and three weeks in custody, raised the question: Is every phone call a telephone conversation with Interior Minister Wolfgang Schäuble? Kreidler's piece provides the practical test.[35]

An example taken from my own experience in electronic communication with the administration of my university at that time, after 2001, complements these discoveries. Mails referring in the subject line „Staatsexamen" (States' Examine) never reached its destination. Because: Staats-Examen can be segmented in: Staat SEX amen. This is even imitable in English: State Sex Amine. „Amine" with a little bit of blurring could be pronounced similar to „Amen". The combination of sexuality and Christianity is spicy in both languages.

Instead of just regarding single isolated words, the context has to be taken into account—a general problem, and in detail a complicated issue. The cases are also reminiscent of the already sketched procedure of soggetto cavato, which uses a grammatical order of magnitude higher, namely words from sentences instead of clay letters /sounds from words. The filtering is mechanical or automatic, precise and „sharp". However, because of the automatic nature and the isolation of individual keywords context conditions are not taken into account, semantic blurring or errors arise. It is a mechanical, as automatically generated 'soggetto cavato', without regard to meaning as syntax, here in contrast to Kreidler's composition even without regard to word and morpheme boundaries. In neither case is there any semantic-syntactic 'cohesion' of the filtered-out text, even in the sense of text linguistics, nor a pragmatic 'coherence' in relation to reality. It is the further question to what extent and how such problems are solvable with Fuzzy Logic. Contextual conditions, especially societal, political, and in many cases diversified references, including multiple meanings, elude the question of blurs, at least for the time being. However, answering them is difficult, but not impossible, because fuzzy logic can proceed in multi-factorial and multi-dimensional ways and can accommodate many variables. Anyway, computer programs are now able to successfully calculate—and win—even poker games with their elements of objective indeterminacy and subjective uncertainty and moods

[35] Kunkel 2011, p. 480f. Works of Kreidler are documented on www.kreidler-net.de, retrieved July 7, 2010.

including bluffing. Like the game of chess for about two decades, at the turn of the millennium, they surpass now (often) the best human players in this game. In the case of Kreidler as in the case of „state-sex-amen" it is about similarity—an actual or supposed -, thus a basic problem for Fuzzy Logic. In order to grasp the full reality, that is, actual collusion of terrorists or pornographic contents, however, the complementary consideration of the dissimilar is necessary. There is no identity without difference. That, too, may be considered dialectic.

5.2.4 *BACH-Searching as a Test on the Internet. Sequence, Cross, Fate: Expected and Unexpected Results—Flexibility and Productivity of Fuzzy Logic*

However, the automatic search with its „mechanical" results can produce surprising and productive results for the analysis of music instead of comic, stupid and destructive results. The search for relatively simple motives similar to the BACH-motif in the internet was in 2009 nearly an experiment looking for mechanisms of pattern perception, implying learning processes. One will once more encounter analogies to the physiological „neuronal" assimilation of motives, artificial intelligence and networks. Search engines and algorithms reflect in a technical manner some of the problems of *intramusica*l similarities, relations and correlations (See above our 'leitmotif' B-A-C-H in Sect. 4.6.5, and below in Sect. 6.2.1) (Fig. 5.2).

The internet search for melodies runs mainly according to key motives. The algorithms in the background are probably often inaccurate. And they are restricted: A four-tone motif as BACH is excluded from the contour search. This requires at least five tones—feasible, but not very sensible. The musically essential qualitative evaluation of the significance, especially of the four-part cross structure, is not possible within those parameters (Fig. 5.3).

Searching with the various interfaces mostly leads to the same results. The vital, essential aspects, the chromatic and sequenced or chiastic order, is not really taken

Fig. 5.2 BACH. https://upload.wikimedia.org/wikipedia/commons/thumb/f/f6/B-a-c-h.svg/2000px-B-a-c-h.svg.png. 22 Jan 2006, retrieved 3 April 2017

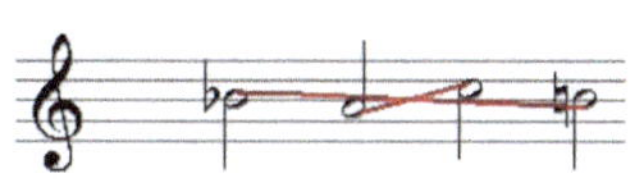

Fig. 5.3 BACH as St.-Andreas-Cross. Boeks~commonswiki, 3. April 2006. https://de.wikipedia.org/wiki/B-A-C-H#/media/Datei:Musik.motiv.b-a-c-h2.jpg, retrieved 3 April 2017

into consideration by the search filters, but is taken along as contour and marginal phenomenon. As often the search engine delivers a vast amount of results, but mostly only seemingly valid or helpful information. They in turn must be filtered. Probably the algorithms behind this are not precise—or fuzzy—enough. The program recognizes half tones and chromatic patterns, but takes them into account quite randomly (as far as I could see). The sequence of results reflects dimly some degree of similarity, but not very precisely or logically. In total I would dare say, this is in part more a fuzzy logic than a truly multi-dimensional Fuzzy Logic.

Yet some of the results I got by this special type of fuzzy logic at least were interesting. They pointed at similarities starting with BACH which with conscious and classical research methods nearly never would have been found: César Franck, *Les Djinns* /Grieg, *Peer Gynt* (Hall of the King of the Mountains) /*El Garrotín*—a dance of Asturian origin and belonging to the group of the Flamenco. If these similarities make sense—that is another story.

There is most probably no explicit, even intended musical connection here, neither genetically or diachronically nor synchronously. The common denominator is, relatively abstract, the core of the motive, including some extensions and variations. The motif follows the already mentioned configuration 'expression becomes material'. Nevertheless, it will not be insignificant. As a widespread topos it proves its generative and interconnecting power.

Yet there are not only structural motivic but also some semantic similarities. *The Hall of the King of the Mountains* Edvard Grieg wrote in 1875 as incidental music for the sixth scene of act 2 in Henrik Ibsen's 1867 play *Peer Gynt*. The music announces and describes the "Trolls", supernatural, non human and very negative figures. Grieg, as Ibsen, associated them by the way with negative national characteristics of Norwegians (Fig. 5.4).

After the first five tones of the diatonic scale in the traditionally sinister key b minor, B C# D E F#, the core, the decisive chromatic motivic-intervallic structure—a sequence of major thirds with 3-tone-segments:

F# D /E# = F /E C#

The conspicuous and not accidental three times three has something magical, musically discretely implemented. Grieg composes a jagged line of sequences, in which as a hidden counterpoint two chromatic descending lines intertwine:

F# E# E

Fig. 5.4 Edvard Grieg: Music to H. Ibsen, Peer Gynt, *In the Hall of the Mountain King*
Peer Gynt Suite No.1, Op. 46, m. 1-5. http://imslp.org/wiki/Peer_Gynt_Suite_No.1,_Op.46_(Grieg,_Edvard). 29 November 2009. Retrieved: 8 March 2013

D C# C
Transposed to the *BACH* we get the reduced and permutated – 431 – core:
C H B.
Theme transposed, m. 10f.
With the scale of b minor from the dominant F# G# A# B C# D, the main motif echoes transposed and with a permutation at the end, 2143 instead of 1234 BACH, also intertwining the chromatic intervals H (English B) B = A# D C#—obviously an intervallic expanded version of BACH:
−1 +4 −1 instead of
−1 +3 −1

César Franck wrote *Les Djinns* in 1884 for piano and orchestra as program music inspired by Victor Hugo's poem *Les Orientales.* "Djinns" are ghosts, supernatural forces as the Trolls, and eerie. As similar the subjects—fuzzily or vaguely, but sensible—, as similar the motifs.

Right at the beginning, Franck exposes a motif characterizing the gradual approach of the Djinns in a bleak, nocturnal landscape, in line with the poem by Victor Hugo. In an interplay of long-held, rhythmically pent-up chords in clarinets, bassoon, and double basses with a brief, sharp reverberation in the other strings, an increase in volume and tempo develops. The interval structure of the upper voice is chromaticized throughout, as is the entire four-part setting, and is unmistakably a derivative of BACH, in the simplified and reduced form of four ascending semitones A A# B C (= German A B H C), that is, a simple permutation. (On this and other contrapuntal procedures, see section 4.6.) Beginning in measure 11, Franck gradually expands the tonal space until a wide-ranging motif structured by chordal refractions is added to complement the counterpoint. The work, which rises from the depths in F-sharp minor, ends in high registers and with a ghostly, transfiguring F-sharp major.

Here also the cross structure as well as the motive of the *Dies irae* may play a role. Nonetheless, yet another result in the context of this research was also productive for music analysis, even if at the first glance seemingly estranging (Fig. 5.6).

The retrieval varied and transformed the search for a pitch motif, BACH, in the result of a rhythmic motif. The main motif of Beethoven's 5th symphony added another model, where the rhythmic configuration was more essential. I would not have expected this. But not without reason the search engine chose this as an example.

Fig. 5.5 Grieg: *In the Hall of the Mountain King.* Ibidem

Fig. 5.6 Beethoven, *5th symphony*, 1st movement, main theme

Fig. 5.7 *Dies irae*. Benedikt Emmanuel Unger: Dies Irae Choralnotation.https://de.wikipedia.org/wiki/Dies_irae#/media/Datei:Dies_irae.gif.25. Juni 1006 (sic!). Retrieved 25 June 2016

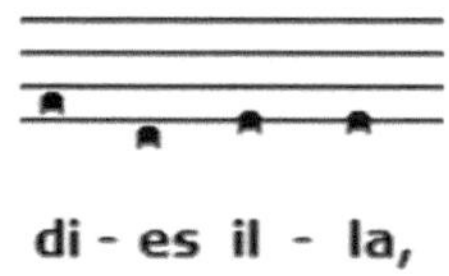

Fig. 5.8 *dies illa*—sequence. Benedikt Emmanuel Unger: *Dies Irae* Choralnotation. https://de.wikipedia.org/wiki/Dies_irae#/media/Datei:Dies_irae.gif.25. Juni 1006 (sic!). Retrieved 25 June 2016

For the sequence and cross figure plays a vital role in the Beethoven motif. Therefore it is a variation of BACH with very augmented intervals. A general contour is retained by the identical diastematic movement – + – (descending /ascending /descending). Acoustically the cross-structure, articulated as a sequence, refers to further traditions of religiously influenced music, just as the dense chromaticism, completely within the minimal space of ABHC (English: A B♭ B C).

The sequencing cross-motif is related to an old and as BACH countlessly reproduced motif: the "Dies irae" (Day of Wrath) of the Requiem. The famous Dies irae, the *Sequence* of the requiem, the mass on Judgement Day, has a similar chiastic structure—the cross again. The procedure of sequencing creates it. Sequence means: the same interval or mini-motif is repeated at a different pitch, higher or lower. The *Dies-irae*-motif remains within the frame of a diatonic, primarily 7-tone system. Bach, however, already applied the "well tempered" 12-tone system, tendentially in the above mentioned equal temperament (Figs. 5.7 and 5.8). (See below sect. 7.4.5, p. 477f.)

The formula (transposed from F to C):
di-es /i-rae … (dies illa)
−1 +1−3
CH CA (H-G-A-A)
The varied, but very similar shapes as survey in letters:
BACH
−1 +3−1
Dies irae
−1 +1−3
Beethoven, 5th symphony
−4 +2−3

Fuzzy Logic allows for this technical and aesthetical "tolerance". Surprising is how the detour of a negatively fuzzy filtering process can lead to a highly relevant connection. However, the motif undergoes considerable changes in character. Instead of

Christian fear and submission to fate Beethoven articulates with his "fate motif" the uprising against fate: the self-confident and active citizen. In the Second World War this motif became the acoustical signal of the BBC broadcasts to Germany. Primarily it had been found as "V" = Victory as common abbreviation both for the Flemish and Wallonian resistance in Belgium ("vrijheid"/"victoire"). Secondarily the rhythmic similarity, nearly identity of this "V" as Morse character and of Beethoven's motif was discovered. It was indeed adequate not only structurally but also semantically as symbol of resistance against Nazism.

Beethoven Motif = Signal Morse ... – = V

BBC-Signal V · · · — = V from Victory

5.3 Sharpening Without Proper Filtering and Transitions Towards Crystallization

There are some border cases in the area or around the filtering. With them, in the sense of Fuzzy Logic, it is a bit indefinite whether it is specifically about filtering or more generally cases of sharpening.

5.3.1 Precisioning of the Recollection Form Through Objectivation, Exteriorization and Repeatability—Phonography Form (II)

A special form of filtering is the perception of—in the best case essential—components of the composition: sometimes it is only the contour, the leading voice, often the vocal part—in the broad sense "the melody". This is a certain analogy to the contours seeing in the realm of the optical; this is evolutionarily so important for orientation in space that special brain areas have developed for it. What from the viewpoint of polyphonic composition is an 'edge', becomes a 'core'. The whole of the composition becomes a kind of marginal condition and becomes blurred or vague. However, this whole is in this reducing listening also listened to, partly consciously, partly unconsciously. And especially in this form the perception is remembered. One thing correlates with this receptive fact on the side of the production of music: an individual, even and especially out of memory, can (re) produce "by heart" only a single voice, however much it from the original whole, no matter how much of the whole she/he may have in mind. The other type of reproduction on an instrument capable of polyphony is rather the exception. The rule here is singing—possibly with lyrics—or humming or whistling without the text, which is often remembered only fragmentarily anyway, in a certain contrast to the melody. Perception and memory are therefore in the filtering and yet blurred reduction. Technically, it should already be possible today to correlate the reduced and the remnants of the unreduced by means of brain scans. Fuzzy Logic would thus have another field for more precise and quantifying research.

This applies analogously and expands for relationships between sound, phonography, perception and recollection form. Sound and phonographic form are objectified forms of existence and therefore appear more accessible to scientific access. However, the phonographic form preserves something in the present and at the same time for the future that has passed away irrevocably and irreversibly, hence history and no longer directly accessible. The sound form must be reconstructed from the phonography form essentially by inference. For the comparison of two inherently objectified forms of existence, however, subjective forms of existence can and must be used. They complement the objective findings. In order to clarify differences and fuzziness within the framework of the Fuzzy Logic, statements of the type "I had that differently" (or equally) in memory, but now …" are inevitable. Corresponding questions and evaluations are difficult and methodically a squaring of the circle, or reminiscent of Baron Münchhausen, who pulls himself out of the swamp on his own thatch. On the other hand, the phonography form itself helps the memory along. It works against unlearning and forgetting, the other side of learning. For the phonographic form as such sharpens something that exists, but over time becomes fuzzy and unintentionally blurred.

This sharpening has at least three dimensions. First, it refers to the past perception of the phonographic form, which is triggered by the current reproduction of perception. Specifically, so the memory of the sound of a record, MusiCassette or CD is reactivated. Secondarily, the sharpening relates indirectly to the sound form that has already been technically reproduced and thus preserved in the phonographic form, and to the still present (rather fuzzy) memory of it at the time of the former sound form. Third, if necessary, the verbal resp. report form of both the original sound form of a work *and* its phonographic form, so encaptic-nested, now in turn mediated by the imagination form, can reactivate the former perception forms to a certain extent and recrystallize them.

Another aspect to be mentioned is the objective and subjective filtering 'away' of the noise already mentioned. For with analogue records as broadcast, noise as background is pushed away by ear and brain to a certain degree. For that, they filter, sharpening shapes, out the music. In this way they are emphasizing something that is already contained in the object, without changing it in itself, but profile the intended shape. On the other hand when cleaning, the usual process is reversed: instead of filtering *out* something, it filters *away* what obscures the original shape, blurs its contours and profile. 'Cleaning' simply means wiping dust off records, but also using noise filters, restoring damaged records—or later files, and finally transferring them to new formats—from the phonograph cylinder to shellac then to polyvinyl records, then into various digital formats. As is characteristic of filtering, it involves a kind of negation: the background, as mentioned, is minimized to make the figure appear as clear and precise as possible. It is not about making something new, but restoring something already existing, something old. Of course, in the digital age, not a few are longing for the patina of analogue noise as a reminder of a "good old time" that has never existed before, here too. On the other hand, filtering out the shapes, filtering away blurred edges and components, is really an improvement in sound quality on the part of the object and therefore also for the subject.

5.3.2 *Audio- and Data-Compression as Ambivalent Filtering*

The situation is different with data compression. It rather causes the opposite. In the context of the sound form and the phonography form that objectifies it, data compression can be regarded as—somewhat ambivalent—sharpening. Because filtered out and removed is everything that does not belong to the core. However, what is to be heard or should be heard, while not blurred in the strict sense, is reduced in its full conciseness and beauty. As usual with digitalization, it rushes less, but for that it sounds less. The noise-free sound has a narrower ambitus and a less differentiated spectrum. The same applies analogously to data compression for pictures and films. Already with the digitization of analog recordings around 1980 at that time for the compact disk by means of the "Pulse Code Modulation" frequencies disappeared. Apart from the "digitization noise", however, the noise of analogue records (or radio and television broadcasts) has also disappeared. This increases the sound and perception quality and is more noticeable than the loss of data. Somewhat different is the slightly later explicitly intended and realized data compression for the Mp3 player and various following procedures. Here the sound spectrum is reduced significantly. As in the case of retrieval, data compression (source coding, bit-rate reduction) is always about comfortable handling of the data, above all about the transmission speed, but not about the quality of the transmitted data (Fig. 5.9).[36]

Manowar is, by the way, the group which broke with 129.5 dB in 1994 the concert loudness record 1976 of the English rock group "The Who", 126 dB at 32 meters: the renaissance of barbarism in the cloths of modern high tech. "Perhaps in the interest of public health, the Guinness Book of World Records no longer keeps a record for loudest concert."[37]

5.3.3 *Compositional Use of the Filtered Nature for Crystallization: Shapes and Motifs from the Partial-Tone-Spectrum—Wagner's Prelude to* **Rheingold**, *„Spectralism" Since the 1970s*

As far as the musical material in the sense of sound and rhythm systems is concerned, what is said here may suffice. Because it is, as already mentioned in the introduction, on the one hand already often examined from the Fuzzy Logic, e.g. the tuning systems.

---

[36]Of course, with these short explanations, the entire problem is barely sufficently discussed. In our context, however, it must be temporarily enough.

[37]See https://en.wikipedia.org/wiki/Manowar, 23 January 2018, retrieved Feb. 13, 2018.

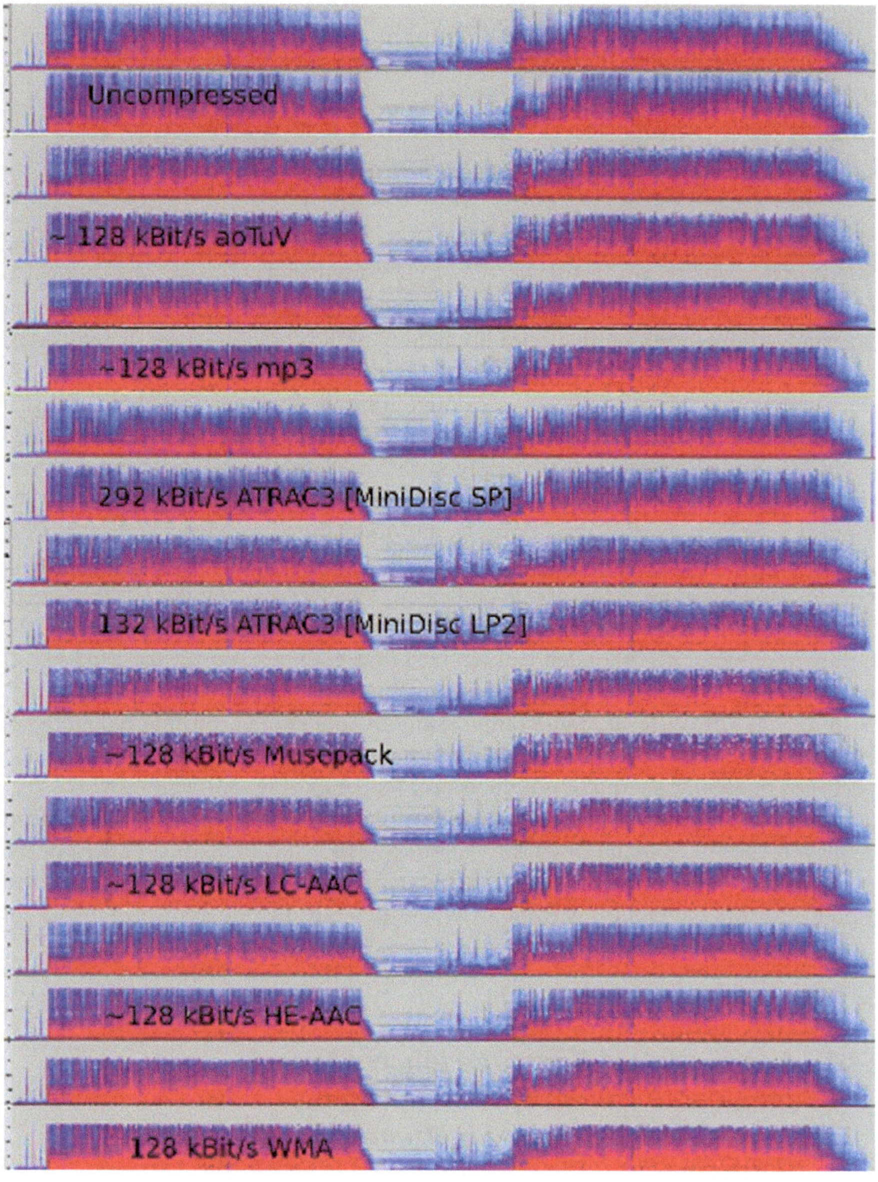

Fig. 5.9 Data-Compression: Manowar, *The Power of Thy Sword* (1992) of the Album *The Triumph of Steel*. Frequencies of the uncompressed, 128 aoTuV, mp3 and ATRAC3 Files. 9 April 2011 https://upload.wikimedia.org/wikipedia/commons/3/3a/AudiodatenkompressionManowarThePowerOfThySword.jpg, retrieved Aug. 15, 2017

On the other hand, this filtering in principle precedes the respective artwork and only indirectly enters the process of forms of existence and the dialectic of sharpening and blurring. Exceptions are possible. By resorting to the filtering of the pitch material, which has long been historically accomplished, it can in turn be recycled as part of crystallization and motivic design.

The opening of Haydn's oratorio *Die Schöpfung* (*The Creation*) is prototypical for musical evolvement of cosmos out of chaos: an unshapely, harmonically diffuse sound arrangement without melodic contours gradually crystallizes to clear shape and forms. Wagner radicalizes this topos in the opening of his musical drama *Rheingold*, when for several minutes the orchestra reverberates only with a single E-flat-major chord. The elementary constitution of tonal systems by filtering, as elaborating a pre-compositional raw material, here is itself as crystallization of motives and themes part of the composition and the artwork as process.

This prelude to *Rheingold* is a historically far-reaching model. (see Sect. 9.2) It is no longer about actual filtering. The process is primarily crystallization, also because it is art-immanent as part of the music process. For out of the unformed of the pure tone gradually emerges the chord E flat major (Fig. 5.10).

As second stage after intervals and chord follows first an elementary, simplest possible melodic movement that rises to the top. Melody and harmony unfold as an orchestral fan (Fig. 5.11).

A finesse: at the second appearance of the melody, Wagner completes the fifth by a fourth-note opening, so it has the complete octave. All this happens in the swaying 6/8 time. This alludes to the 12/8 time of the Siciliano, which was popular in Christmas and pastoral music. Only after he has fully tasted the delights of the waters, Wagner begins with an actual melody, always tender, in cellos and bassoons. However, it is still an elementary diatonic, the tones of the scale of E flat major: E♭ F G A♭ B♭ e♭ g, then offset by a third up G A♭ B♭ d e♭. The scale of E flat major is complete.

Fig. 5.10 Richard Wagner: Prelude to *Rheingold*, m. 1-10. Beginning: Fundamental tone and fifth, E♭ and B♭. B. Schott's Söhne, Mainz. n.d.[1873]. Plate 20800. IMSLP33405-PMLP21241-Wagner_-_Das_Rheingold__Scene_1_

Fig. 5.11 Prelude to *Rheingold,* orchestral fan, m. 29-41. B. Schott's Söhne, Mainz. n.d.[1873]. Plate 20800. IMSLP33405-PMLP21241-Wagner_-_Das_Rheingold__Scene_1_

The crystallization also has aspects of filtering. For harmony as well as melody have as a common set of sound extracts from the overtone series of E♭. For the purpose of the mimesis of a world formation that is the theme here, Wagner therefore resorts to the filtering behind the crystallization as motif formation. This recourse to filtering, which historically precedes crystallization, here helps a modern and more precise representation and expression.

Not least following Wagner—or in opposition to "bourgeois": romantic" music—many trends in the arts returned to the elemental. And they tried, as mentioned, at the same time to further disassemble the elements of music. One of these trends is the *Musique spectrale* or the "spectralism" that developed around the group *L'Itinéraire* in Paris in the 1970s. They transform extracts and single tones of the partial tone spectrum into new sound experiences.

The electro-acoustic procedures are a 'natural' prerequisite in the background. Gérard Grisey (1946–1998), Tristan Murail (* 1947) and the US-American James Tenney (1934–2006) belong to this group. The sound is fanned out into its frequency spectra and becomes the starting point of whole works: "A sound becomes a timbre, a chord a spectral complex, and a rhythm a wave of unpredictable durations." (Gérard Grisey)[38] Spectralism is thus particularly close to the prelude to *Rheingol*d in terms of technique and material, but is aesthetically and ideologically remote from it. "It

---

[38]Quoted from http://www.swr.de/swr2/programm/sendungen/jetztmusik/glossar/-/id=4979696/1m5qq8g/index.html, 24.6.2009 (Link now decommissioned).

is no longer possible for me to regard the notes as fixed and mutually permutable objects. They seem to me more like bundles of temporal forces.[39]

The sound, the tone that otherwise appears as an almost atomic whole, are thus segmented and filtered. The resulting changes in the partial-tone structures become audible in a kind of blurring as changes in timbre. The filtering recourse to elementals and—relatively—simple basic can thus serve to create something complex and new.

References

Adak, S.: Time-dependent spectral analysis of nonstationary time series. J. Am. Stat. Assoc. **93**, 1488–1501 (1998)

Ahrendt, P.: Music genre classification systems—a computational approach. PhD thesis, Technical University of Denmark, DTU (2006)

Allan, H., Müllensiefen, D., Wiggins, G.A.: Methodical considerations in studies of musical similarities. In: Proceedings of the Eighth International Conference on Music Information Retrieval, pp. 473–478. Vienna (2007)

Allen, P.E., Dannenberg, R.B.: Tracking musical beats in real time. In: Int. Comp. Music Conf., Glasgow, Scotland, pp. 140–143 (1990)

Alonso, M., David, B., Richard, G.: A study of tempo tracking algorithms from polyphonic music signals. In: Proceedings of the 4th COST 276 workshop, information and knowledge management for integrated media communication, Bordeaux, France, pp. 1–5 (2003)

Aramaki, M., Barthet, M., Kronland-Martinet, R., Ystad, S. (eds.): From sounds to music and emotions. 9th International Symposium CMMR 2012, London, UK, June 19–22, 2012, Revised Selected Papers. Information Systems and Applications, incl. Internet/Web, and HCI, vol. 7900. Springer-Verlag, Berlin and Heidelberg (2013)

Arenas-Garca, J., Larsen, J., Hansen, L.K., Meng, A.: Optimal filtering of dynamics in short-time features for music organization. In: Proceedings of the 7th international conference on music information retrieval, Victoria, Canada, pp. 290–295 (2006)

Assayag, G., Feichtinger, H. G., Rodrigues, J.F. (eds.): Mathematics and music. A diderot mathematical forum. Springer-Verlag, Berlin, Heidelberg, New York, Barcelona, Hong Kong, London, Milan, Paris, Tokyo (2002)

Aucouturier, J.-J., Pachet, F.: Improving timbre similarity: how high is the sky. J. Neg. Results Speech Audio Sci. **1**(1), 1–13 (2004)

Aucouturier, J.J., Pachet, F.: Representing musical genre: A state of the art. J. New Music. Res. **32**(1), 83–93 (2003)

Barthet, M., Fazekas, G., Sandler, M.: Music emotion recognition: From content- to context-based models. In: Aramaki, M., Barthet, M., Kronland-Martinet, R., Ystad, S. (eds.) From Sounds to Music and Emotions. CMMR 2012. Lecture Notes in Computer Science, vol 7900. Springer, Berlin, Heidelberg. https://doi.org/10.1007 (2013)

Berenzweig, A., Logan, B., Ellis, D., Whitman, B.: A large-scale evaluation of acoustic and subjective music-similarity measures. Comput. Music. J. **28**(2), 63–76 (2004)

Bruderer, M.: Automatic recognition of musical instruments. Master thesis, Ecole Polytechnique Fédérale de Lausanne (2003)

Chen, C.-M., Chien, P.-C., Tsai, M.-F., Yang, Y.-H., Lin, Y.-C.: Exploiting latent social listening representations for music recommendations. In: ACM International Conference on Recommender Systems (RecSys), September 2015

[39]Gérard Grisey in his lecture *On the Origin of Sound …* at the "Darmstadt Summer Courses for New Music" in 1978, quoted from ibidem.

Chen, C.-M., Tsai, M.-F., Lin, Y.-C., Yang, Y.-H.: Query-based music recommendations via preference embedding. In: Proc. ACM Recommender Systems (RecSys), pp. 79–82, September 2016

Chen, C.-Y.: Early Chinese work in natural science. A re-examination of the physics of motion, acoustics, astronomy and scientific thoughts. Hong Kong University Press, Hong Kong. https://www.questia.com/read/120363874/early-chinese-work-in-natural-science-a-re-examination (2000)

Chen, Y.-A., Wang, J.-Ch., Yang, Y.-Hs., Chen, H.: Component tying for mixture model adaptation in personalization of music emotion recognition. IEEE/ACM Trans. Audio, Speech, Lang. Process., vol. **25**(7), 1409–1420 (2017)

Chen, Y.-A., Wang, J.-Ch., Yang. Y.-Hs., Che, H.: Linear regression-based adaptation of music emotion recognition models for personalization. In: IEEE International Conference on Acoustics, Speech and Signal Processing (ICASSP), May 2014

Chen, Y.-H., Huang, Y.-S., Hsiao, W.-Y., Yang, Y.-Hs.: Automatic composition of guitar tabs by Transformers and groove modeling. In: Proc. Int. Society for Music Information Retrieval Conf. (ISMIR), October 2020

Chen, Y.-W., Yang, Y.-Hs., Chen, H. H.: Cross-cultural music emotion recognition by adversarial discriminative domain adaptation. In: Proc. IEEE Int. Conf. Machine Learning and Applications (ICMLA), December 2018

Chin, Y.-H., Wang, J.-Ch., Wang, J.-Ch., ang, Y.-Hs.: Predicting the probability density function of music emotion using emotion space mapping. IEEE Trans. Affect. Comput. **9**(4), 541–549, October 2018

Chou, S.-Y., Yang, Y.-H., Jang, J.-S., Lin, Y.-C.: Addressing cold start for next-song recommendation. In: Proc. ACM Recommender Systems (RecSys), pp. 115–118, September 2016

Coutinho, E., Cangelosi, A.: A neural network model for the prediction of musical emotions. In: Nefti-Meziani, S., Grey, J.G. (eds.) Advances in Cognitive Systems, pp. 331–368. IET Publisher, London (2010)

Coutinho, E., Cangelosi, A.: Musical emotions: predicting second-by-second subjective feelings of emotion from low-level psychoacoustic features and physiological measurements. Emot. **11**(4), 921–937 (2011)

De Poli, G., Rocchesso. D.: Computational models for musical sound sources. In: Assayag, G., Feichtinger, H. G., Rodrigues, J.F. (eds.) Mathematics and Music. A Diderot Mathematical Forum, pp. 257–285. Springer-Verlag, Berlin, Heidelberg, New York, Barcelona, Hong Kong, London, Milan, Paris, Tokyo (2002)

Deutsch, D. (ed.): The Psychology of Music, 2nd edn. Academic Press, Boston (1999)

Dixon, S., Mauch, M., Tidhar, D.: Estimation of harpsichord inharmonicity and temperament from musical recording. J. Acoust. Soc. Am. **131**(1), 878–887 (2012)

Dong. H.-W., Yang, Y.-Hs.: Convolutional generative adversarial networks with binary neurons for polyphonic music generation. In: Proc. Int. Society for Music Information Retrieval Conf. (ISMIR), September 2018.

Downie, J.S.: Evaluating a simple approach to music information retrieval: Conceiving melodic n-grams as text. PhD thesis, Faculty of Information and Media Studies, University of Western Ontario, London (Ontario), Canada. http://people.lis.uiuc.edu.jdownie/mir_papers/thesis_missing_some_music_figs.pdf (1999)

Ethier, G.: Techniques of hypermetric manipulation in Canadian blues. Can. J. Tradit. Music., pp. 33–53 (2001)

Fabian, D., Timmers, R., Emery Schubert, E. (eds.): Expressiveness in Music Performance. Empirical Approaches Across Styles and Cultures. Oxford University Press, Oxford (2014)

von Falkenhausen, L.: Suspended Music: Chime-Bells in the Culture of Bronze Age China. University of California Press, Berkeley (1993)

Fitch, W.T.: The Biology and evolution of rhythm: Unravelling a paradox. In: Rebuschat, P., Rohmeier, M., Hawkins, J.A., Cross, I. (eds.) Language and Music as Cognitive Systems. Oxford University Press, Oxford. https://people.finearts.uvic.ca/~aschloss/course_mat/MU320/

Global%20Rhtyhm%20and%20Human%20Consciousness/ARTICLES%20AND%20REFS%20FOR%20320/FitchParadoxOfRhythm.pdf (2012)
Flexer, A., Pampalk, E., Widmer, G.: Hidden Markov models for spectral similarity of songs. In: Proceedings of the 8th International Conference on Digital Audio Effects, Madrid, Spain (2005)
Foote, J., Uchihashi, S.: The beat spectrum: a new approach to rhythm analysis. In: Proceedings of the IEEE International Conference on Multimedia and Expo, Tokyo, Japan, pp. 224–228 (2001)
Foote, J.: Audio retrieval by rhythmic similarity. In: Proceedings of the 3rd International Conference on Music Information Retrieval (2002)
Fucks, W.: Mathematical analysis of formal structure of music. IEEE Trans Inf. Theory **8**(5), 225–228 (1962)
Gardiner, J.M., Kaminska, Z., Dixon, M., Java, R.I.: Repetition of previously novel melodies sometimes increases both remember and know responses in recognition memory. Psychon. Bull. & Rev. **3**(3), 366–371 (1996)
Goebl, W., Pampalk, E., Widmer, G.: Exploring expressive performance trajectories: Six famous pianists play six Chopin pieces. In: Lipscomp, S.D. et al. (eds.): Proceedings of the 8th International Conference on Music Perception and Cognition. Evanston, IL Illinois, 2004 (ICMPC8), pp. 505-509. Causal Productions, Adelaide, Australia (2004)
Goto. M.: An audio-based real-time beat tracking system for music with or without drum-sounds. J. New Music. Res. **30**(2), 159–171 (2002)
Gouyon, F., Dixon, S.: A review of automatic rhythm description systems. Comput. Music. J. **29**, 34–54 (2005)
Gouyon, F.: A computational approach to rhythm description: Audio features for the computation of rhythm periodicity functions and their use in tempo induction and music content processing. PhD thesis, Universitat Pompeu Fabra, Departament de Tecnologia, Barcelona, Spain (2005)
Grekow, J., Raś, Z.W.: Detecting emotions in classical music from MIDI files. International Symposium on Methodologies for Intelligent Systems ISMIS 2009: Foundations of Intelligent Systems, pp 261–270. Springer Verlag, Heidelberg. https://link.springer.com/chapter/10.1007/978-3-642-04125-9_29 (2009)
Grekow, J.: An analysis of the harmonic content—main parameters in the AKWET method. In: Proceedings of II Conference on Technologies of Knowledge Exploration and Representation, TERW. (2007)
Grekow, J.: Broadening musical perception by AKWETS technique visualization. In: Proceedings of the 9th International Conference on Music Perception and Cognition, ICMPC9. (2006)
Heister, J. et al.: dlexDB – eine lexikalische Datenbank für die psychologische und linguistische Forschung, In: Psychologische Rundschau, 62 (1), pp. 10–20 Hogrefe Verlag, Göttingen 2011
Herre, J., Allamanche, E., Ertel, C.: How similar do songs sound? Towards modeling human perception of musical similarity. In: Proceedings of the IEEE workshop on applications of signal processing to audio and acoustics, pp. 83–86 (2003)
Herrera, P., Sandvold, V., Gouyon, F.: Percussion-related semantic descriptors of music audio files. In: Proceedings of the 25th international AES conference, London, United Kingdom. https://ieeexplore.ieee.org/xpl/conhome/8891805/proceeding (2004)
Hu, X., Yang, Y.-H: The mood of Chinese pop music: Representation and recognition. J. Assoc. Inf. Sci. Technol. **68**(8), 1899–1910, June 2017
Hu, X., Yang, Y.-Hs.: Cross-dataset and cross-cultural music mood prediction: A case on Western and Chinese pop songs. IEEE Trans. Affect. Comput. **8**(2), 228–240, April 2017
Huang, H. Chen, W., Liu, C., You, S. D.: Singing voice detection based on convolutional neural networks. In: 7th International Symposium on Next Generation Electronics (ISNE), Taipei, 2018, pp. 1–4. https://doi.org/10.1109/isne.2018.8394727 (2018)
Huang, Y.-S., Chou, S.-Y., Yang, Y.-H.: Pop music highlighter: Marking the emotion keypoints. Trans. Int. Soc. Music. Inf. Retr. **1**(1), 68–78 (September 2018)
Huang, Y.-S., Yang, Y.-Hs.: Pop music transformer: Beat-based modeling and generation of expressive Pop piano compositions. In: Proc. ACM Multimedia, October 2020

Huang, Y.-S., Chou, S.z.-Y., Yang, Y.H.: Pop music highlighter: Marking the emotion keypoints. Trans. Int. Soc. Music. Inf. Retr. **1**(1), 68–78. Doi: https://doi.org/10.5334/tismir.14 (2018)

Hung, Hs.-Tz., Wang, Ch.-Y., Yang, Y.-Hs., Wang, Hs.-M.: Improving automatic Jazz melody generation by transfer learning techniques. In: Proc. Asia Pacific Signal and Information Processing Association Annual Summit and Conf. (APSIPA ASC), November 2019

Jao, K., Lin, Y.-P., Yang, Y.-H., Jung, T.-P.: Using robust principal component analysis to alleviate day-to-day variability in eeg based emotion classification. In: Proc. Int. Conf. IEEE Engineering in Medicine and Biology (EMBC), August 2015

Jürgensen, F., Knopke, I.: A comparison of automated methods for the analysis of style in fifteenth-century song intabulations. In: Parncutt, R., Kessler, A., Zimmer, F. (eds.) Proceedings of the Conference on Interdisciplinary Musicology (CIM04), pp. 98f. http://www-gewi.uni-graz.at/staff/parncutt/cim04/CIM04_paper_pdf/JurgensenKnopke.pdf (2004)

Käfer, B.: Paläolithische Knochenflöten im Ostalpenraum und dem nordöstlich vorgelagerten Lößgebiet (Magisterarbeit). Hochschule für Musik und darstellende Kunst Wien (1998)

Kim, Y., Schmidt, E., Migneco, R., Morton, B., Richardson, P., Scott, J., Speck, J., Turnbull, D.: Music emotion recognition: A state of the art review. In: Proc. ISMIR (2010)

Kluge, R.: Skalen und Stimmungen des bronzezeitlichen Glockenspiels aus dem Grabe des Zeng Hou Yi. Zur Deutung eines tonometrischen Befundes. In: Ereignis und Exegese. Musikalische Interpretation—Interpretation der Musik. Festschrift für Hermann Danuser zum 65. Geburtstag, pp. 231–242. Argus Verlag, Schliengen (2011)

Klüsener, S.: Qualitative Heuristik. Strukturierendes Entdecken. Ruprecht-Karls-Universität Heidelberg, Geographisches Institut. http://www.kluesenernet.de/Deutsch/Pdf/Qualitative_Heuristik.pdf. Retrieved 7 August 2017. dead link (1998)

Knees, P., Pampalk, E., Widmer, G.: Artist classification with web-based data. In: Proceedings of the 5th International Conference on Music Information Retrieval, pp. 517–524. Barcelona, Spain (2004)

Knepler, G.: Geschichte als Weg zum Musikverständnis. Zu Theorie, Methode und Geschichte der Musikgeschichtsschreibung. Reclam Universalbibliothek, Leipzig (1977) (2nd enhanced ed. 1982)

Korhonen, M.D., Clausi, D.A., Jernigan, M.E.: Modeling emotional content of music using system identification. IEEE Trans. Syst., Man, Cybern., Part B, Cybern. **36**(3), 588–599 (2005)

Kostek, B., Czyzewski, A.: Representing musical instrument sounds for their automatic classification. J. Audio Eng. Soc. **49**, 768–785 (2001)

Kostek, B., Wojcik, J., Holonowicz, P.: Estimation the rhythmic salience of sound with association rules and neural network. In: Intelligent Information Systems, Gdansk, Poland. Advances in Soft Computing (AINSC, volume 31) (2005a)

Kostek, B., Wojcik, J.: Machine learning system for estimation rhythmic salience of sounds. Int. J. Knowl.-Based Intell. Eng. Syst. **9**(4), 275–284 (2005b)

Koza, J.R.: Genetic Programming: On the Programming of Computers by Means of Natural Selection. MIT, Cambridge (1992)

Kunkel, M.: Gibt es neue Typen des Engagements in der neuesten Musik? Positionen und Negationen. In: Phleps, T., Reich, W. (eds.) Musik-Kontexte. Festschrift für Hanns-Werner Heister, pp. 472–487. Verlagshaus Monsenstein und Vannerdat, Münster (2011)

Kurth, F., Gehrmann, T., Müller, M.: The cyclic-beat spectrum: Tempo-related audio features for time-scale invariant audio identification. In: Proceedings of the 7th International Conference on Music Information Retrieval, pp. 35–40 (2006)

Lee, C.-T., Yang, Y.-H., Chen, H.-H.: Multipitch estimation of piano music by exemplar-based sparse representation. In: IEEE Trans. Multimed. **14**(3), 608–618, June 2012 (第二屆美律電聲論文獎 – 銅質獎)

Lehwark, P., Risi, S., Ultsch, A.: Visualization and clustering of tagged music data. In: Proceedings GfKl 2007, Freiburg i. Br. Conference Paper, January 2007. https://www.researchgate.net/publication/221649327_Visualization_and_Clustering_of_Tagged_Music_Data/link/56f4daab08ae81582bf0a868/download (2007)

León, T., Liern Carrión, V.: Music and similarity based reasoning. In: Seising, R., und Sanz, V. (eds.) Computing in Humanities and Social Sciences (Studies in Fuzziness and Soft Computing, vol. 273), pp. 451–468. Springer, Berlin and Heidelberg (2012)
León, T., Liern Carrión, V.: Obtaining the compatibility between musicians using soft computing. In: Hüllermeier, E., Kruse, R., Hoffmann, F. (eds.) International Conference on Information Processing and Management of Uncertainty in Knowledge-Based Systems, IPMU 2010 Part II, CCIS 81, Dortmund, Germany, June 28–July 2, 2010, pp. 75–84. Springer-Verlag, Berlin and Heidelberg (2010)
Lerch, A.: Software-based extraction of objective parameters from Music performances. Phil. Diss., TU Berlin (2008) urn:nbn:de:kobv:83-opus-20677 https://doi.org/10.14279/depositonce-2025. Retrieved July 4, 2011
Lesaffre, M., Tanghe, K., Martens, G., Moelants, D., Leman, M., De Baets, B., De Meyer, H., Martens, J.P.: The MAMI query-by-voice experiment: collecting and annotating vocal queries for music information retrieval. In: Proceedings of the 4th International Conference on Music Information Retrieval, Baltimore, Maryland, and Library of Congress, Washington, DC, pp. 65–71 (2003)
Levy, M., Sandler, M.: Lightweight measures for timbral similarity of musical audio. In: Proceedings of the First ACM Workshop on Audio and Music Computing Multimedia (AMCMM), pp. 27–36. ACM, New York (2006)
Li, T., Ogihara, M., Li, Q.: A comparative study on content-based music genre classification. In: Proceedings of the 26th International ACM SIGIR Conference on Research and Development in Information Retrieval, pp. 282–289. ACM, New York (2003b)
Li, T., Ogihara, M.: Detecting emotion in music. In: Proceedings of the Fifth International Symposium on Music Information Retrieval, pp. 239–240 (2003a)
Lidy, T., Rauber, A.: Evaluation of feature extractors and psycho-acoustic transformations for music genre classification. In: Proceedings of the 6th International Conference on Music Information Retrieval, pp. 34–41 (2005)
Ligges, U.: Transkription monophoner Gesangszeitreihen. Diss., Fachbereich Statistik, Universität Dortmund. Dortmund, Germany (2006) http://hdl.handle.net/2003/22521 https://eldorado.tu-dortmund.de/bitstream/2003/22521/1/Diss_Internet.pdf #
Lin, K.-S.., Yang, Y.-H., Lee, C.-T., Chen, H.-H.: Automatic highlights extraction for drama video using music emotion and human face features. Neurocomputing **119**, 11–17, November 2013
Lin, Y.-C., Yang, Y.-H., Chen, H.-H.: Exploiting online tags for music emotion classification. In: ACM Trans. Multimedia Computing, Communications, and Applications, volume 7S, number 1. Special Issue on Social Media October 2011,
Lin, Y.-P., Yang, Y.-Hs., Jung. Tz.-P.: Fusion of electroencephalogram dynamics and musical contents for estimating emotional responses in music listening. Front. Neurosci. **8**(94), 1–14, May 2014
Liu, C., Yang, Y., Wu, P., Chen, H.: Detecting and classifying emotion in popular music. In: Proceedings of the 9th Joint Conference on Information Sciences (JCIS)/CVPRIP (2006)
Liu, D., Lu, L., Zhang, N.: Automatic mood detection from acoustic music data. In: Proceedings of the Fifth International Symposium on Music Information Retrieval (2003)
Lu, L., Liu, D., Zhang, H.J.: Automatic Mood Detection and Tracking of Music Audio Signals. IEEE Trans. Audio, Speech, Lang. Proc. **14**(1) (2006)
Markuse, B., Schneider, A.: Ähnlichkeit, Nähe, Distanz: zur Anwendung multidimensionaler Skalierung in # musikwissenschaftlichen Untersuchungen. Systematische Musikwissenschaft / Systematic Musicology / Musicologie systematique **4**, 53–89 (1996)
Mauch, M., Fujihara, M. Yoshii, K. Goto, M.: Timbre and melody features for the recognition of vocal activity and instrumental solos in polyphonic music. International Society for Music Information Retrieval Conference: Proceedings of the 12th International Society for Music Information Retrieval Conference, ISMIR 2011, Miami, Florida, USA, October 24–28, 2011. http://citeseerx.ist.psu.edu/viewdoc/summary?doi=10.1.1.656.2322 (2011)

Mazzola, G., Mannone, M., Pang, Y.: Cool math for hot music. A first introduction to Mathematics for music theorists. Springer International, Heidelberg (2016)

McKay, C., Fujinaga, I.: Automatic genre classification using large high-level musical feature sets. In: Proceedings of the International Conference on Music Information Retrieval, pp. 525–530 (2004)

Mörchen, F., Ultsch, A., Thies, M., Löhken, I., Nöcker, M., Stamm, C., Efthymiou, N., Kümmerer, M.: MusicMiner: visualizing timbre distances of music as topographical maps. Tech. rep., Department of Mathematics and Computer Science, University of Marburg, Germany (2005)

Müllensiefen, D., Frieler, K.: Cognitive adequacy in the measurement of melodic similarity. Algorithmic vs. Human judgments. In: Comput. Music. **13**, 147–176 (2004a)

Müllensiefen, D., Frieler, K.: Optimizing measures of melodic similarity for the exploration of a large folk song database. In: 5th international conference on music information retrieval, Audiovisual Institute, Universitat Pompeu Fabra, Barcelona, Spain, pp. 274–280 (2004b)

Müllensiefen, D.; Pfleiderer, M., Frieler, K.: The perception of accents in pop music melodies. J. New Music. Res. **38**(1), 19–44. http://dx.doi.org/10.1080/09298210903085857. Retrieved Sept. 11, 2012 (2009)

Nagel, F., Kopiez, R., Grewe, O., Altenmüller, E.: EMuJoy. software for the continuous measurement of emotions in music. Behav. Res. Methods **39**(2), 283–290 (2007)

Pampalk, E., Dixon, S., Widmer, G.: On the evaluation of perceptual similarity measures for music. In: Proceedings of the International Conference on Digital Audio Effects, pp. 6–12 (2003)

Pampalk, E., Flexer, A., Widmer, G.: Hierarchical organization and description of music collections at the artist level. In: Proceedings of the 9th European Conference on Research and Advanced Technology for Digital Libraries, pp. 37–48 (2005)

Pampalk, E.: Audio-based music similarity and retrieval: Combining a spectral similarity model with information extracted from fluctuation patterns. In: 3rd Annual Music Information Retrieval eXchange (MIREX'06), n.p. (2006a) http://pampalk.at/publications/

Pampalk, E.: Computational models of music similarity and their application in music information retrieval. PhD thesis, Computer Science Department, Technical University Vienna, Austria (2006b)

Parncutt, R.: A perceptual model of pulse salience and metrical accent in musical rhythms. Music Perception **11**(4), July 1994, pp. 409–464. https://doi.org/10.2307/40285633 (1994)

Pearce, M.T., Müllensiefen, D., Wiggins, G.A.: Melodic grouping in music information retrieval: New methods and applications. In: Zbigniew, R., Wieczorkowska, A. (eds.) Advances in Music Information Retrieval (Studies in Computational Intelligence, vol. 274), pp. 365–390. Springer, Berlin (2010)

Pianists play six Chopin pieces. In: Lipscomp, S.D. et al. (eds.): Proceedings of the 8th International Conference on Music Perception and Cognition. Evanston, IL Illinois, 2004 (ICMPC8), pp. 505–509. Causal Productions, Adelaide, Australia (2004)

Predicting the probability density function of music emotion using emotion space mapping – Chin et al. 2018;

Ras, Z.W., Dardzinska, A.: Solving failing queries through cooperation and collaboration. Special Issue on Web Resources Access. World Wide Web J. **9**(2), pp. 173–186 (2006)

Risi, S., Mörchen, F., Ultsch, A., Lewark, P.: Visual mining in music collections with emergent SOM. In: Proceedings workshop on self-organizing maps (WSOM). https://pure.itu.dk/portal/en/journals/proceedings-workshop-on-selforganizing-maps-wsom07(db3fda23-ee17-4aff-be9f-7e2b0871dfb0).html, pp. 1–6. (2007)

Risset, J.-C.: Computing musical sound. In: Assayag et al. 2002, pp. 215–231

Rossignol, S., Depalle, P., Soumagne, J., Rodet, X., Collette, J.L.: Vibrato: detection, estimation, extraction, modification. In: Proceedings of the COST-G6 workshop on digital audio effects (DAFx-99) http://citeseerx.ist.psu.edu/viewdoc/summary?doi=10.1.1.25.6408 (1999)

Röver, C., Klefenz, F., Weihs, C.: Identification of musical instruments by means of the Hough-transformation. In: Weihs, C., Gaul, W. (eds.) Classification—the ubiquitous challenge, pp. 608–615. Springer, Berlin (2005)

Sandvold, V., Herrera, P.: Towards a semantic descriptor of subjective intensity in music. In: Proceedings of the international computer music conference. http://mtg.upf.edu/files/publications/Sandvold-ICMC-2005.pdf (2005)

Schedl, M., Pohle, T.P., Knees, P., Widmer, G.: Assigning and visualizing music genres by web-based co-occurance analysis. In: Proceedings of the 7th International Conference on Music Information Retrieval, pp. 260–265 (2006)

Scheirer, E.D.: Tempo and beat analysis of acoustic musical signals. J. Acoust. Soc. Am. **103**(1), 588–601 (1998)

Schuller, B., Eyben, F., Rigoll, G.: Tango or Waltz?: Putting Ballroom Dance Style into Tempo Detection. In: EURASIP Journal on Audio, Speech, and Music Processing, volume 2008, Article ID 846135 2008, 12 pages. https://link.springer.com/article/10.1155/2008/846135 (2008)

Seising, R., Zadeh, L. A.: On fuzzy sets and the precisiation of meaning—An interview with Prof. Dr. Lotfi A. Zadeh. APHSC 1:2013 – DOI tbp – http://www.aphsc.org, retrieval 12.11.2013

Smaragdis, P., Brown, J.: Non-negative matrix factorization for polyphonic music transcription. In: IEEE workshop on applications of signal processing to audio and acoustics, pp. 177–180 (2003)

Soleymani, M., Yang, Y.-Hs., Irie, G., Hanjalic, A.: Challenges and Perspectives for Affective Analysis in Multimedia. IEEE Trans. Affect. Comput. **6**(3), pp. 206–208, July 2015

Steinbeck, W.: Struktur und Ähnlichkeit. Methoden automatisierter Melodieanalyse (Kieler Schriften zur Musikwissenschaft XXV). Bärenreiter, Kassel (1982)

Streich, S., Herrera, P.: Detrended fluctuation analysis of music signals: Danceability estimation and further semantic characterization. In: Proceedings of the 118th AES convention. https://www.researchgate.net/publication/238677665_Detrended_Fluctuation_Analysis_of_Music_Signals_Danceability_Estimation_and_further_Semantic_Characterization/link/0046352d3b8a4599cf000000/download (2005)

Su, L., Yang, Y.-Hs.: Combining spectral and temporal representations for multipitch estimation of polyphonic music. IEEE/ACM Trans. Audio, Speech, Lang. Process. **23**(10), 1600–1612, October 2015

Syst. Technol. 3, 3, Article 40 (May 2012). https://doi.org/10.1145/2168752.2168754 http://doi.acm.org/10.1145/2168752.2168754 (2012)

Takeda, H., Nishimoto, T., Sagayama, S.: Rhythm and tempo analysis toward automatic music transcription. In: IEEE International Conference on Acoustics, Speech and Signal Processing, ICASSP 2007 Volume: 4. https://doi.org/10.1109/icassp.2007.367320 (2007)

Takeda, H., Nishimoto, T.: Automatic rhythm transcription from multiphonic MIDI signals. https://www.researchgate.net/publication/2891340_Automatic_Rhythm_Transcription_from_Multiphonic_MIDI_Signals/link/02e7e53481c238cd54000000/download (2003)

Tamagnan, F., Yang, Y.-Hs.: Drum fills detection and generation. In: Proc. Int. Symp. Computer Music Multidisciplinary Research (CMMR), October 2019.

Temperley, D., Sleator, D.: Modeling meter and harmony: A preference-rule approach. Comp. Music J. **15**(1), 10–27 (1999)

Teng, Y.-Ch., Kuo, Y.-Sh., Yang, Y.-Hs.: A large in-situ dataset for context-aware music recommendation on smartphones. In: IEEE Int. Conf. Multimedia and Expo., July 2013

Tzanetakis, G., Cook, P.: Musical genre classification of audio signals. IEEE Trans Speech Audio Process. **10**(5), 293–302 (2002)

Tzanetakis, G., Ermolinskyi, A., Cook, P.: Beyond the query-by-example paradigm: New query interfaces for music. In: Proceedings of the International Computer Music Conference, pp. 177–183 (2002a)

Tzanetakis, G., Ermolinskyi, A., Cook, P.: Pitch histograms in audio and symbolic music information retrieval. In: Proceedings of the 3rd International Conference on Music Information Retrieval, pp. 31–38 (2002b)

Ultsch, A.: Self organizing neural networks perform different from statistical k-means clustering. In: BMBF Statusseminar Künstliche Intelligenz, Neuroinformatik und Intelligente Systeme, München. Bielefeld: transcript, pp. 55–74 (1996)

Ultsch, A.: Self-organizing neural networks for visualization and classification. In: Opitz, O., Lausen, B., Klar, R. (eds.): Information and classification—concepts, methods, and applications, pp. 307–313. Springer, Berlin (1993)
Wang, J.-CH., Yang, J.-Hs., Jhuo, I.-H., Lin, Y.-Y., Wang. Hs.-M.: The acousticvisual emotion Gaussians model for automatic generation of music video. In: ACM Int. Conf. Multimedia (MM), Grand Challenge solution paper (First Prize), November 2012
Wang, M., Zhang, N., Zhu, H.: User-adaptive music emotion recognition. In: 7th Inter-national Conference on Signal Processing, ICSP (2004)
Wang, Y.-Ch., Yang, Y.-Hs., Chang, K., Wang, Hs.-M., Jeng, Sh.-K.: Exploring the relationship between categorical and dimensional emotion semantics of music. In: Int. Workshop on Music Information Retrieval with User-Centered and Multimodal Strategies (MIRUM), in conjunction with ACM Multimedia (MM), November 2012
Weihs, C., Ligges, U., Mörchen, F., Müllensiefen, D.: Classification in Music research. Adv. Data Anal. Cl.Ification **1**(3), 255–291 (2008)
Widmer, G., Dixon, S., Goebl, W., Pampalk, E., Tobudic, A.: In search of the Horowitz factor. AI Magazine **24**(3), 111–130 (2003)
Wieczorkowska, A., Ras, Z.W., Zhang, X., Lewis, R.: Multi-way Hierarchic Classification of Musical Instrument Sounds., In: IEEE CS International Conference on Multimedia and Ubiquitous Engineering (MUE 2007), Seoul, Korea, April 26–28 (2007)
Wieczorkowska, A., Raś, Z.W.: Do We Need Automatic Indexing of Musical Instruments? In: Bolc, L., Michalewicz, Z., Nishida, T. (eds.): IMTCI 2004. LNCS (LNAI), vol. 3490, pp. 239–245. Springer, Heidelberg (2005)
Wieczorkowska, A., Synak, P., Ras, Z.: Multi-label classification of emotions in music. In: Intelligent Information Processing and Web Mining, Advances in Soft Computing, Proceedings of IIS 2006 Symposium, Ustron, Poland, vol. 35, pp. 307–315 (2006)
Wojcik, J., Kostek, B.: Intelligent Technologies for Inconsistent Processing. In: Nguyen, N.T. (ed.) Intelligent Methods for Musical Rhythm Finding Systems (Int. Series on Advanced Intelligence, vol. 10), pp. 187–202 (2004)
Wojcik, J., Kostek, B.: Representations of music in ranking rhythmic hypotheses. Advances in music information retrieval. Stud. Comput. Intell. **274**, 39–64. Springer Nature, Switzerland (2010)
Wojcik, J.: Methods of forming and ranking rhythmic hypotheses in musical pieces, Ph.D. Thesis, Faculty of Electronics, Telecommunications and Informatics, Gdansk University of Technology, Gdansk (2006)
Wu, Sh.-L., Yang, Y.-Hs.: The Jazz Transformer on the front line: Exploring the shortcomings of AI-composed music through quantitative measures. In: Proc. Int. Society for Music Information Retrieval Conf. (ISMIR), October 2020
Yang, Y.-H., Chen, H.H.: Machine recognition of music emotion: A review. ACM Trans, Intell (2012)
Yang, Y.-H. Chen, H.-H.: Music emotion ranking. In: IEEE International Conference on Acoustics, Speech and Signal Processing (ICASSP), April 2009
Yang, Y.-H., Chen, H.-H.: iMR: Interactive music recommendation via active interactive genetic algorithm. In: Workshop on Computer Music and Audio Technology (WOCMAT), 2009
Yang, Y.-H., Chen, H.-H.: Machine recognition of music emotion: A review. ACM Trans. Intell. Syst. Technol. **3**(3), Article No. 40, May 2012
Yang, Y.-H., Chen, H.-H.: Prediction of the distribution of perceived music emotions using discrete samples. IEEE Trans. Audio, Speech, Lang. Process. **19**(7), 2184–2196 (September 2011)
Yang, Y.-H., Liu J.-Y.: Quantitative study of music listening behavior in a social and affective context. IEEE Trans. Multimed. **15**(6), 1304–1315, October 2013, Special Issue on Social Media as Sensors (2013)
Yang, Y.-Hs., Chen, H. H.: Music Emotion Recognition. CRC Press Book Taylor & Francis / Routledge, London (2011)

Yang, Y.-Hs., Hu, X.: Cross-cultural music mood classification: A comparison of English and Chinese songs. In: International Society for Music Information Retrieval Conference (ISMIR), October 2012

Yang, Y., Su, Y., Lin, Y., Chen, H.: Music emotion recognition: The role of individuality. In: Proceedings of the international workshop on Human-centered multimedia. HCM 2007. https://www.citi.sinica.edu.tw/pages/yang/publications_en.html (2007)

Yeh, Y.-Ch, Hsiao, W.-Y.,Fukayama, S., Kitahara, T., Genchel, B., Min Liu, H.-M., Dong, H.-W., Chen, Y., Leong, T., Yang, Y.-Hs.: Automatic melody harmonization with triad chords: A comparative study. In: ArXiv e-prints. abs/2001.02360, January 2020

Yeh, Y.-Ch., Yu Liu, J.-Y., Hsiao, W.-Hs., Huang, Y.-S.,Yang, Y.-Hs.: Learning to generate Jazz and Pop piano music from audio via MIR techniques. In: ISMIR demo paper, November 2019

You, M., Liu, J., Guo-Zheng Li, G.-Zh., Chen, J.: Embedded Feature Selection for Multi-label Classification of Music Emotions, pp. 668–678. https://www.tandfonline.com/doi/abs/10.1080/18756891.2012.718113 (2012)

You, Sh. D., Chien–Hung Liu, Ch.-H., Chen, W.-K.: Comparative study of singing voice detection based on deep neural networks and ensemble learning. Hum. Cent. Comput. Inf. Sci. (2018) 8: 34. 18 pp. https://link.springer.com/article/10.1186/s13673-018-0158-1 (2018)

Zadeh, L.A. 1983: Precisication of Meaning via Translation into PRUF. In: Vaina, L., Hintikka, J. (eds.) Cognitive Constraints on Communication, pp. 373–402. Kluwer Academic Publishers B.V., Alphen an den Rijn (1984)

Zadeh, L.A.: Precisiated Natural Language (PNL). In: AI-Magazine (AI MAG) vol. 25/3, pp. 74–91. Publisher: Association for the Advancement of Artificial Intelligence (2004)

Zhang, J., Harbottle, G., Wang, C., Kong, Z.: Oldest Playable Musical Instrument Found at Jiahu Early Neolithic Site In China. https://web.archive.org/web/20071008043743/http://www.shakuhachi.com/K-9KChineseFlutes-Nature.html (2007)

Zoltai, D.: Ethos und Affekt. Geschichte der philosophischen Musikästhetik von den Anfängen bis zu Hegel. Akademie-Verlag, Berlin, Budapest (1970)

Chapter 6
The Principle of Sharpening (II): Crystallization. Development and Advancement of Musical Shapes

Abstract In the processes of crystallization the material is used, which was formed through the filtering (Chapter 5). The idea of the work develops into a richer, more and more concrete shape in the various stages of the work process (see 'The 13 forms of existence' Chapters 2, 4 and 5). The musical motif, e.g. our 'leitmotif' BACH, is a germ cell of the formation of form, in composing as well as in improvising. Tools for the production of the hinge 'Abstract composition' are partimento and thorough bass, then complemented by ornaments, instructions for the execution, instrumentation unto the Score Form. Blurs are bypassed by direct composition of the Sound Form without notation and human interpreters via automatophones, "Musique concrète", and "Electronic Music". The perception is also sharpened. The dialectic of sharpening through blurring is manifested in the development of the European melodic polyphony including poly-textuality, with Fuzzy Logic as a partial logic of compositional history.

Keywords 'Abstract Score' ("Tonsatz") · Animation film · Audiovisualization · Basso Continuo · Covering · Digitalization · Earworm · 'Forms of Existence' · History of European melodic polyphony · Hook · Karaoke · Notation · Perception · Player and Reproducing Piano · Recollection/Memory · Sing along · 'Soggetto cavato' (sculptured theme) · Synesthesia · Synchronization of movements · Synthesizer

In the long-winded historic system of music process, in an according to a potentially infinite net of relations, fuzziness may appear in every place, and usually it does so. (This was extensively explained in Chapter 3.) But, at the same time, producers or recipients interfere at each of these junctions in an attempt to get rid of such blurs, or to minimize them. Yet here, the already repeatedly registered peculiar dialectics comes into being, that—nearly—every effort to create order and sharpening, produces further chaos and fuzziness. In Chapter 3, this was demonstrated partially with the model of notation. However, this leads not to a simple to and fro, like an oscillation, but, in the light of history, a spiral of a gradual higher development—out of the realm of coincidences, arbitrariness, somehow obvious in something intentional, deliberate, consciously supposed. In such a spiral movement the differences between ideal and reality become more precise and the extent of fuzziness is diminishing.

H. Heister, *Music and Fuzzy Logic*, Studies in Fuzziness and Soft Computing 406,
https://doi.org/10.1007/978-3-662-62907-9_6

Then, moreover points of attraction for the sharpening are crystallized, in multiple approaches, varying according to music culture, as, for instance, the notation, which does not exist in every music cultures. The compositional subject as well, which originated earliest in the beginning of advanced civilizations, but it unfolded only with the rise of individualization in High Feudalism, after A.D 1000. In its turn, it produces new consequences in regard of the consciousness, and it operates on the basis of a planning, which anticipates and prudently includes results, and these are implemented purposefully. All this belongs to sharpening, in historical perspective and development as well.

Despite being methodically an analogy to conscious sharpening, the conscious blurring is substantially the opposite (see Chapter 7). In this case, what was is or could be clear, precise, and sharp, will intentionally be made unclear. However, on a higher level, such a reverse is at the same time a component of a communion. Even blurring supports a work's variety of figures. It concerns the improvement of expressiveness, and thus a work's worldliness. The improvement of complexity, including the formal, means of expression serve to this end. Content and figure are interlocking and reciprocally improve their quality—in a positive case. Complexity is double-sided: more components, thus more differentiation, but at the same time, more integration, coherence. This is analogous to facts in nature and in society (see Chapter 9).

6.1 From Filtering to Crystallization: Extraction, Concentration, Core, Contour

It stands to reason: crystallization is a metaphor, just like filtering. Crystallization is based systematically **on** what was formed by filtering: some material, though systematized, but yet hardly configured, arranged and form-like, i.e. above all, systems of sound and rhythm, as well as, organization of orchestration for the timbre. In European-American music the dimension of sound volumes is subject to a system, or scaling, which is only bluntly and vaguely patterned by steps, from p to f, piano softly till forte loudly, especially since the Viennese classicists it became ever more refined in nuances, for instance between pppppp and ffffff. But this remains relative without a standard like dB, which likewise includes in the measurement subjective and situational factors, such as, above all, distance from the source of sound.

After filtering, the material became in terms of the Gestalt-psychology a „figure", now, on a higher level it becomes a „ground". Developing figures means above all the production of motifs, themes, melodies, thus of work's components, which musicians and composers in principle handle and form consciously. Then, the singular and special, just in form of the „movable" and „moving" motif serves to bring about coherence. At first, this is a connection between the particular components, but then, it will as a—becoming—whole, in its time, be an all-embracing structure, reacting on the components in a formative way.

Here, the relation between entropy and information gets involved. Once again in contradiction to the second principle of thermodynamics, in a section of a system, life as a higher dimension of order is created in the neg-entropy. Aside of consciously creating, there is the emergence too, here as configurations, who arise out of structures, just like from themselves. Even, something not figured, chaotic and also by coincidence, which may appear playfully in the work process or even as hitting a wrong note, during the work's productive realization, can be made fitting as part of the whole (see Sects. 3.1 and 3.2). Similar to such scraps in production, there are also analogous results accidentally created through emergence. And examples of how a new, higher order may evolve from relative disorder are offered by patterns resulting from an ensemble playing without any conscious purpose, as in the Amadinda-music for xylophone at the former royal court of Buganda, nowadays a province of Uganda, or the style brisée in European music during Absolutism, for instance, in the prelude in C major of J. S. Bach's *Wohltemperiertem Klavier*, I, No. 1, where the implicit melody evolves from the peak tones of the arpeggios, which Gounod then used as *Ave Maria.* This is at the same time sharpening and blurring (extensively, Sect. 7.4.4 and 7.4.5).

6.1.1 'Soggetto Cavato'

A first type of figure formation is a transition from the filtering to crystallization: the extraction of figures from already existing as being arranged and its transformation, new configuration and concentration in a new figure. The environment, from which this figure was gathered as a kernel, will then become an almost irrelevant hull—again an inversion of the figure-ground relation. What was already somehow finished, a kind of work, is being reduced to be a semi-finished product, just a material-system for potential formations of figures. This is illustrated by a filtering technique, called „soggetto cavato", "excavated" or "carved out theme" theme, a term used in the Renaissance for such a semantically coined motif of notes expressed in syllables or letters (see Sect. 5.1.5). It may also be employed para-musically.

6.1.1.1 *BTHVN*

For instance, Beethoven occasionally used signature with an abbreviation of his name.

> *Extremely modern: Beethoven as a brand Marke / occasionally* Beethoven signed his compositions with "BTHVN". He was not realizing, how trendy such contractions would become

> in times of Twitter & Co. Actually it was chosen to be the official logo for the Beethoven jubilee.[1]

In this case, Beethoven used decidedly 3 of 5 letters of his name, which are *not* suitable for naming notes these are T, V, N. How far, he used the letters, which are compositionally suited, remains to be analyzed. These are by no means only a few: luDwiG vAn BEEtHovEn. The fifth-row G-D-A-E-H as well as the triton B-E is certainly included.

It might be appropriate to ask for something, which is obviously ultimately impossible, about the very relevant relation of motives, themes, melodies, with invariance and variance, of kernel and hull, of detailed line, something of a melody, and roughly sketched contour—the latter is also a problem of the fractal, for instance, for measuring coast-lines and their length exactly. But, Fuzzy Logic already knows it.

6.1.1.2 Josquin's Mass *Hercules Dux Ferrariae* and in Conan Doyle's *The Hound of the Baskervilles*

Whereas, what appears in the case of BTHVN to be pragmatic and rather jocular, is for „Soggetto cavato" emphatic and musical seriousness. Thus, what was called in Renaissance-music an „excavated" theme, is a semantically coined motif, or theme, of tonal letters, which were part of a name. A well-known example is the Feast-Mass *Hercules Dux Ferrariae* of Josquin Desprez, Missa *Hercules Dux Ferrariae* of Josquin Desprez, originated in 1480, or in the early 1490s. Here, the vowels of name and title of Duke Ercole of Ferrara by means of solmization syllables form the tones re - ut - re - ut - re - fa - mi - re, which Josquin takes as Cantus firmus—in the Kyrie I in absolute pitches d c d c d f e d—thus, here and there, cryptography. Josquin composed and combined this figure in many of the possible variants.

> Josquin uses the figure in each of the three hexachords available to Renaissance music theory: the natural hexachord beginning on C (giving A-G-F-D-E for the figure), the „hard" hexachord beginning on G (giving E-D-C-A-B natural for the figure), and the „soft" hexachord beginning on F (giving D-C-B flat-G-A for the figure).[2]

A soggetto—in which motif and theme overlap—appears moreover downward in a crab figure, by extracts or even with the change of the hexachord-type in the middle of the Soggetto. Thus, Josquin receives an immense stock of tones and produces with the theme a very compact net of motivic relations. Josquin's work, being the most famous European composer in his days, was due to his method so famous, that many princes also demanded to have it dedicated as their Mass. With an identic composition the work's new dedication became certainly an unintentional joke. For the composition is by its central musical theme closely intertwined with the name

[1] Ibid.

[2] Extensively. https://en.wikipedia.org/wiki/Missa_La_sol_fa_re_mi, 24 June 2017, retrieved 6 September 2017.

Hercules, so that another name would have been inappropriate and ridiculous. But it seems that the princes have not noticed it. Such an extraction method is not dependent on syllables, as it also works with letters of tones and writing, or with complete words. For instance, in one of the stories about Sherlock Holmes, *The Hound of the Baskervilles*, an anonymous warning note in the mail turns up. The extraction of the words is by their permutation restored to rearrange their original sequence.

> Out of the envelope he [Sherlock Holmes] took a half-sheet of foolscap paper folded into four. This he opened and spread flat upon the table. Across the middle of it a single sentence had been formed by the expedient of pasting printed words upon it. It ran: *"As you value your life or your reason keep away from the moor."* The word "moor" only was printed in ink. […] "Might I trouble you for it — the inside page, please, with the leading articles?" He glanced swiftly over it, running his eyes up and down the columns. "Capital article this on Free Trade. Permit me to give you an *extra*ct from it. '*You* may be cajoled into imagining that *your* own special trade *or* your own industry will be encouraged by a protective tariff, but it stands to *reaso*n that such legislation must in the long run *keep away* wealth *from the* country, diminish the *value* of our imports, and lower the general conditions of *life* in this island.' […] "And yet, my dear Watson, there is so very close a connection that the one is extracted out of the other. *'You,' 'your,' 'your,' 'life,' 'reason,' 'value,' 'keep away,' 'from the.'* Don't you see now whence these words have been taken?" "By thunder, you're right! Well, if that isn't smart!" cried Sir Henry. "If any possible doubt remained it is settled by the fact that '*keep away*' and '*from the*' are cut out in one piece.""Well, now—so it is!".[3]

Entirely brilliant and sharp-witted—the extraction has a tiny flaw, for in the news item the opening „As“ was not included, at least not in the passage found by Mr. Holmes. Obviously, in the dialogue Doyle ignores this blurring, while he expressively remarks, that „moor“ is hand-written. Undoubtedly, in the news item‘s core, from which the message was extracted, none of these words appeared. Only from the point of view as a warning, which in this context is the essential message. On the other hand, concerning the Missa *Hercules Dux Ferrariae* it would be a mistake, to see the essential in the vowels, or tonal syllables, the core of name and function of the Duke of Ferrara. On the other hand, they are for the *work*, which is related with him, the relevant pitches-substance. Another matter concerns acronyms. The initials of a name are often musically transformed and may, as they are often used as an abbreviation for the whole, be taken as a kind of core.

6.1.1.3 Verdi's "parola Scenica": "Sculpt and Render the Situation"

A less mechanical and cryptographic variant of „excavating“ an important message, consisting of one or several words and sentences from a given text, is represented by Verdis‘ notion and implication of „parola scenica“, i.e. the scenic word. He speaks of „scolpisce“ (sculpturing), using a sculptor‘s metaphor, just as Josquin used a wood-carver‘s metaphor, which are similar to excavating. Verdi writes to *Aida*'s librettist Antonio Ghislanzoni (14 August 1870)].

---

[3]https://en.wikisource.org/wiki/The_Hound_of_the_Baskervilles/, retrieved 13 October 13; cf. Doyle 1993.

> I do not know how to explain what I mean when I say parola scenica; but I mean to say that the word should *sculpt* and render the situation cleanly and obviously. [„la parola che *scolpisce* e rende netta ed evidente la situazione."].[4]

Obviously, for Verdi, the theatre, the drama has priority among the components of the „Gesamtkunstwerk".

> I know you will say to me, "what about the verse, the rhyme, the strophe?..." I do not know what to say... except that when the action demands it, one should abandon rhythm, rhyme, strophe; I could abandon verse, if it meant I am saying clearly all that the action demands.[5]

So, in the categories of Fuzzy Logic, he accepts blurriness in details, especially in the text, but also even in bel canto, in order to sharpen the whole and make it more precise.

6.1.1.4 "Agathe!" Brahms' String Sextets Op. 18 and Op. 36

Similar to Josquin, the extract-filtering of tonal-letters from names is an universal operation in recent European music. Above all, it is used for dedications or declarations of love, whereas the latter were often secretly confessed. In succession of Robert Schumann this appears regularly with Johannes Brahms: op. 18 and 36,

The pair of both of his string sextets, op. 18 of 1859/1860 and op. 36 of 1864/1865 are marked by the common sign of a short, nevertheless ardent love-affair with Agathe von Siebold in Göttingen, from summer 1858 until about February 1859. Brahms encodes her name within the interplay of revealing and hiding. Both to conceal it as well as to expose it musically, he utilized on the one hand the letters-tone assignation for the pitches (so A = A, and so on), on the other hand, prosodic equivalents in the metric-rhythmical (»*Agáthe*« appears thus in three tones as unstressed-stressed-unstressed). In op. 36, as Brahms himself confessed, in the second main theme of the first set in a high, exposed position (violin I, T. 162–171), the letters of her name ring out thrice in a row in the sequence of notes *A–G–A–H–E*, plus *D* for »t«, thus at the same time a „Ade!" (farewell) in remembrance of the goodbye.[6]

6.1.1.5 J. O. Grimm: *Zukunfts-Brahmanen-Polka* (1854)

A musician's joke, markedly male, is crossing in a dedication, just like *Hercules Dux Ferrariae*, honoring and mockery, sharpened filtering with musical blurring. The joke not only leaves little to be desired in verbal bluntness and also little for many similar gags in Happenings, Fluxus, et cetera. Thus, as Brahms' friend, Julius Otto Grimm, musical director in Göttingen, composed for Brahms' 21st birthday, in 1854,

[4]English translation Du Plessis 2007, p. 32. Emphasis HWH. The same term and ideas in a letter to his publisher Giulio Ricordi, 10th of July 1870.

[5]Du Plessis, ibid.

[6]Extensively in Heister 2013.

a parodistic *Zukunfts-Brahmanen-Polka* (Future-Brahmans-Polka) with the theme *B*-rest-*A*-*H*-rest-*eS*, hence with letters to be used for tonal letters of his BrAHmEs'.[7] The sequence of notes is.

> thirteen times [!] till sixteenth time accelerated […]; the second part sets for the Soggetto's tones straight fifth-parallels, which were forbidden in a harmonic set.[8]

So far so good as far as music-and arts are concerned. But, for the performance of the Trio the following directions were given:

> „One has to sit quite squarely and ppp […] 8 times on the keyboard alternately in the deepest and the highest octaves, and in addition throws up from the front." At least, the last instruction would probably hinder a practical revival.[9]

In blurring, crude actions of anti-arts included, simultaneously various sharpening of composition are concealed. Grimm even connects two forms of encoding: the letters-tonal letters-assignment and the gematric coding of letters by numbers. For here, the years of a life to be celebrated appear as well, matching in a quite simple addition of numbers: $13 + 8 = 21$. Probably the 16tenth possess some gestic-semantic value too. Because, if we take the „quite squarely", of how the performer has to sit with his bottom (4 letters) on the keyboard, not literally as words, but as a factual corporal statement, so by adding casually „backside" with „front" $2 \times 8 = 16$. Frequently one number is several times determined, thus in a way more than sharpened. A likewise rift, or doubling, 2×8 is achieved by the number of letters of the name Johannes = 8. Thus, Grimm even integrates the Christian name, which, as far as apparent, does not crop up explicitly, into his work. Grimm could have been capable, that he alluded moreover to the six letters in the family name of Brahms with the six-tenth. Finally, the fifth-parallels, besides a musical-caricatural, have possibly also a number-semantical meaning: they indicate whereas number 5 to the year of birth 54—while number 4 appears in rifts and doublings often enough.

6.1.1.6 HF and AB. A. Berg's *Lyric Suite*

A shrunken form of musical filtering out of names is the application of an acronym for pitch letters, like the already mentioned A.S. = A Eb = eS for A.rnold S.chönberg (see Chapter 3), as well as for Alban Berg and—multiple times—by Hanns Eisler; the latter used his own name cipher H-E quite often in his works. Dmitri Shostakovich acted likewise with the cipher DSCH as D.eS, after the German tone name. Alban Berg in turn dedicated for his clandestine great love, Hanna Fuchs-Robettin, the married sister of Franz Werfel, his string quartet *Lyric Suite* (*Lyrische Suite*), which he began in 1925. Hanna's initials HF and his AB, combined with gematric tecniques, the initials of her son and some significant quotations are decisive motives of the whole work. He gave to her an annotated copy of the score with the comment:

---

[7]B in German B flat, H B sharp.

[8]Together with samples of music Kross 1997, vol. I, p. 226.

[9]Cited after Kross 1997, and Kross himself, ibidem.

> It has also, my Hanna, allowed me other freedoms! For example, that of secretly inserting our initials, HF and AB, into the music, and relating every movement and every section of every movement to our numbers, 10 and 23. I have written these and much that has other meanings, into the score for you. ... May it be a small monument to a great love.[10]

Almost naturally, Berg quotes the beginning of *Tristan*. Exactly these notes H-F are also the two lowest notes of the Tristan chord, in reversed order F-H-d#-g#. And even further: In his Violin Concerto, his last finished work, in 1935, there are echoes of H.F.

> The tone row of the violin concerto ends with the same four notes that begin the aforementioned Bach Chorale. The first of these notes is B. B in German is written as H. The last is F. HF. Hanna Fuchs.[11]

These ciphers, or mini-motifs, are not limited to both of these works. Already in summer of 1925, after the encounter with Hanna Fuchs-Robettin, the H-F is found in Berg‘s arrangement of one of his earlier lieder.

> "Schliesse mir die Augen beide" (Close both my eyes / with your beloved hands!), a poem by Theodor Storm, was twice set to music by Berg. He composed his first setting in 1907, dedicating it officially to his future wife. The second version he composed in 1925 to celebrate the 25th anniversary of the Universal Edition music publisher. This second setting was Berg's first full-fledged 12 tone composition, and he re-used its tone row, or rather its secondary set (Perle, George 1996: *Twelve-Tone Tonality*, p. 20) in the first movement of the Lyric Suite.[12]

It is undeniably a sharpening of perception and of understanding, to know about the meaning of motives. Only then, the full content of a work as lyric-tragic love-story can be revealed. Through additional citations, as well as through the musical arrangement as a whole, the work represents within the particular at the same time the general. As another facet, the twelve-tone row which Alban Berg took as a basis for the first movement refers to the general, and at the same time to the individual. The 12th and last note is a B, slightly masked as the enharmonic Cb, the H of Hanna, directly after Berg's Bb as the 11th note, and the 1st note is her F (Fig. 6.1).

The row is very personal, and very general—in fact an all-interval twelve-tone row. It contains each interval within the octave. The first known all-interval row was named the „Mutterakkord" (mother chord) by Fritz Heinrich Klein, who created it in 1921 for his chamber-orchestra composition *Die Maschine* and published his invention in 1924.[13] The acronym of Fritz Heinrich, *F.H.* was a welcomed hint for Berg. At first, this special twelve-tone row appeared to be outstanding. Naturally it suited Berg on account of his love affair. Only much later numerous others were

[10]Cross 2003, quoted after https://en.wikipedia.org/wiki/Hanna_Fuchs-Robettin, 6 July 2019, retrieved 11 June 2020.

[11]Christopher Purdy and admin: *Columbus Symphony Presents Alban Berg's Violin Concerto*, 2 May 2013, https://radio.wosu.org/post/columbus-symphony-presents-alban-bergs-violin-concerto#stream/0, retrieved 11 June 2020.

[12]See, 30 June 2017, retrieved 12 June 2020; cf. Redlich 1957, pp. 177–182. https://en.wikipedia.org/wiki/Schliesse_mir_die_Augen_beide.

[13]In 1938 Nicolas Slonimsky invented the „Grandmother chord". See https://en.wikipedia.org/wiki/All-interval_twelve-tone_row, 24 September 2019, retrieved 11 June 2020.

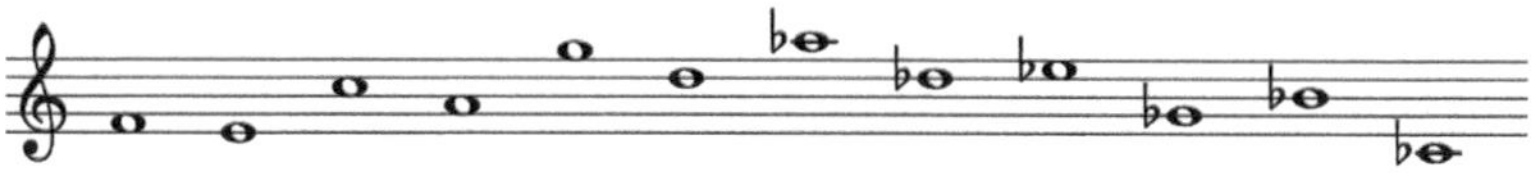

Fig. 6.1 The tone row from the first movement of Alban Berg's Lyric Suite, according to George Perle.
Benjamin D. Esham (bdesham), 1 September 2007, https://upload.wikimedia.org/wikipedia/commons/thumb/c/cb/Lyric_Suite_Movement_I_tone_row.svg/1280px-Lyric_Suite_Movement_I_tone_row.svg.png, retrieved 11 June 2020

found[14] and the mathematic regularities by which they are operated. To these belongs among others:

> Since the sum of numbers 1 through 11 equals 66, an all-interval row must contain a triton between its first and last notes,[6] as well as in their middle.[15]

Berg could not know that yet. In any case, with the initial note F he automatically received the final note H (here enharmonic as Cb). Moreover, the song has exactly 20 bars, i.e. twice Hanna's fate number 10.[16] Berg doubles in the song her, the number, and makes her, the beloved, thus once more his own. Finally, also on the meta-level of the musical-technical methods, the song refers per analogiam to the love of the two.

> Consciously or not, Berg had sought to give this new technique his own personal signature, to find a way to integrate it with the drama. And when Hanna's appearance had caused new and high emotional complications in his life, the application of the new method of composing had similar results in his music.[17]

Nearly naturally, Berg quotes the beginning of *Tristan.* Exactly these notes H-F are also the two lowest notes of the Tristan chord, reversed F-H-d#-g#. Numerous variants of the coding procedures are interesting, but do not bring any additional facets here. All in all, it is a specific sharpening of shape and content with a compacted crystallization core. Instead of an arbitrarily elaborated motif, a semantically charged motif is used from the beginning.

[14]„There are 1,928 distinct all-interval twelve-tone rows. […] These sets may be ordered in time or in register. ‚Distinct' in this context means in transpositionally and rotationally normal form (yielding 3856 such series), and disregarding inversionally related forms" (ibid.).

[15]Slonimsky 1975, p. 185 and Slonimsky 1975, p. iv.

[16]See Monson 1982. Cf. e.g. Floros 1992, Gratzer 1997.

[17]Ibidem. Own translation. „Coscientemente oppur no, Berg aveva cercato di dare a questa nuova tecnica una sua impronta personale, per trovare il modo di fonderla col dramma. E quando l'apparizione di Hanna era stata motivo di nuove ed elevate complicazioni sentimentali nella sua vita, l'adozione del nuovo metodo di composizione ebbe risultati simili anche nella sua musica."

6.1.2 *Pre- and Para-Compositional Form of Hallucination: Arnold Schönberg's* War-Clouds Diary—*Anthropomorphic Projections of Socially Significant Shapes into Natural Phenomena*

A complementary variation of this type of crystallization and extraction is a mere mental filtering. The projective viewing of subjective figures into objectively different forms, into clouds as an almost paradigmatic "fuzziness", remains compositionally without consequences in Schönberg's case, in contrast to Debussy, for example, who in his orchestral piece *Nuages* (*Clouds*) gives acoustic-musical expression to the optical-mental impression of clouds. Clouds, symptoms and symbols of weather, wind, water and processes of heat circulation and more, have from the beginning been significant and fascinating phenomena of nature, even before the Neolithic revolution with the ensuing development of agriculture that depended on climate and weather. In regard to Fuzzy Logic the clouds' two-fold character as form and amorphous object is significant. The groundwork for the classification of clouds was established by Luke Howard as early as 1802 in accordance with the example set by Carl Linné's model for the classification of animals and plants. Today's settings are based on this. The specification of vague forms in detail, the more precise classification of transitions and caesuras in the real ‚structured continuum' is one of the classical tasks of Fuzzy Logic, which can contribute quite a bit to this. The moving, flowing, and ever-changing character of clouds, again and again solidifying into forms with distinct contours, invites anthropomorphic projections: All people imagine seeing faces, or animals, etc. The anthropomorphic projection thereby seemingly clarifies the actual fuzziness of the amorphous appearance of these natural phenomena. Arnold Schönberg formulates nearly delusional cloud representations in his "*War-clouds diary*" of 1914. For a while he notates (verbally and graphically) meteorological conditions. He attempts to read and interpret the "Book of Nature"—from a kind of universal perspective that establishes correspondences between everything, lastly in aesthetically mediated analogies.

> [14/5] suddenly all brightened up again, but only for a in total [sic] very short time; then slightly cloudy; again partial brightening up and around half past five surprisingly fast clouds gather (storm!) and bring heavy rains, that soon abate. The storm continues. Around 1 h 48 A tone point the sun appears. Storm; dense clouds remain. 27/6.1915 8 h 50 in the evening after sunset the following sword can be seen in the sky (form of a cloud) the longer part (the sword points South-West [)] is shorter than it should be in relation to the hilt. But it appears to extend beyond what is clearly visible, as if it were stuck in something.[18]

Schönberg amalgamates "Faith", "God", and "War" with nationalism and occultism (Fig. 6.2).[19]

[18]Illustration and transcription: https://www.usc.edu/libraries/archives/schoenberg/painting/exteriorhtms/noritter27.htm. 12 January 2012. Arnold Schoenberg: "War-Clouds Diary". *Journal of the Arnold Schoenberg Institute* 9/1 (June 1986), p. 55, plate no. 5 (color). Owner and Location: Arnold Schönberg Center, Vienna, Austria.

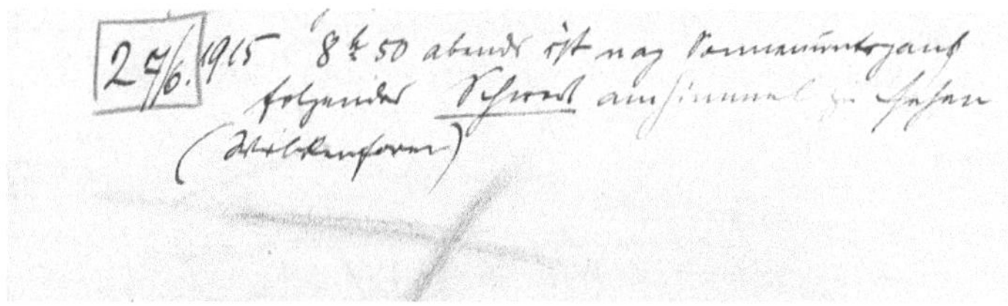

Fig. 6.2 Arnold Schönberg: *War-clouds diary* 19 May 1915, excerpt https://www.usc.edu/libraries/archives/schoenberg/painting/exteriorhtms/noritter27.htm. 12 January 2012. Arnold Schoenberg: *War-Clouds Diary. Journal of the Arnold Schoenberg Institute* 9/1 (June 1986): p. 55, plate no. 5 (color). Owner and Location: Arnold Schönberg Center, Vienna, Austria

> War clouds diary; commenced on 24 September 1914 / Many people will have tried, like me, to read the events of the war in the skies today, now that finally the faith in higher powers and in God returns. […] I […] hope to find correspondences, once more exact reports are available, since a number of the war events so far have only been foreseeable by the "atmosphere/mood" of the sky. Repeatedly the "golden glow", the "wind of victory", a "dark blue sky", "bloody clouds" (at sunset) have caught my attention, which have unfailingly preceded victorious German events. Similarly worsening weather conditions with storm and rain, deep black clouds preceded all negative turns of events in the Austrian-Russian war zone. [...] I must note that not always does a cloudy sky give me the impression of unfortunate events.[20]

Schönberg believes the natural object to be a socially significant sign. In his irrationality he remains sufficiently realistic in order to relativize the trivial attribution and also the dissonance of the non-beautiful nature in regard to usual perceptions of good weather as possibly positive quality of the bellicose. Surely his unreflecting chauvinist assumption that the skies are on the side of the Austro-German troops is naive. There is even a glimpse of skepticism or at least reservation when Schönberg hopes for empirical verification through lengthy empirical surveys. His master student Alban Berg attempts a rational explanation within the discourse of natural sciences in his letter to Schönberg from 8 October 1914 (Fig. 6.3):

> Your suggestion to Webern to observe the clouds and the sky has also moved me strongly. Already this summer—admittedly without connecting it with the war—I have noted incredible appearances. […] The consistently good weather of this summer […] was striking to me, too; only that I saw a more material connection to the war. For I believed, after often having had the opportunity to observe the strong effects of weather shooting ["Wetterschießen"], that the millions of shots over a period of weeks – including almost the entire area of central Europe – across battle lines that stretch over several hundreds of kilometers might have influenced the weather. Surely, what I now see in the sky seems to have other, deeper reasons.

[19]With this 'anthropomorphic projection' Schönberg alludes to magical-religious divination, such as the Roman auspice or also the omina from the hetascopy from ancient Greece and ancient Babylonia and many other instances. Gavin Pretor-Pinney, author of Wolkengucken and founder of the "Cloud Appreciation Society", rightly comments, "Clouds are the Rorschach inkblots of the sky, so to say abstract forms, into which we project our desired images." https://www.sueddeutsche.de/panorama/guck-in-die-luft-club-gemaelde-am-himmel-1.921434. 16 September 2011.

[20]In detail see Arnold Schönberg 1986, l.c.

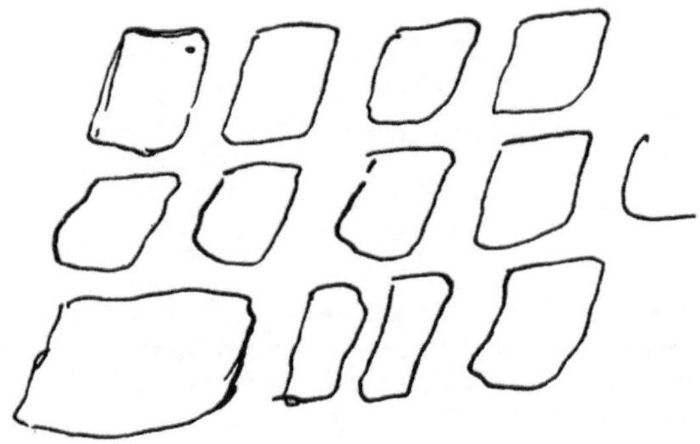

Fig. 6.3 Alban Berg: Letter to Schönberg on 8 October 1914
Regular pattern of 12 clouds
Quoted from Nono-Schoenberg 1992, p. 135

> Very particular was for example the night sky some days ago. The clouds were juxtaposed in regular rhombus shapes; about so.[21]

Alban Berg depicts the clouds in very regular shapes, almost in a constructivist manner. On top of this, the sum total of the cloud forms, if we exclude the vaguely indicated scribble to the right of the second row, as chance would have it, equals 12, which is a highly significant number not only for the Schönberg school and its twelve-tone technique. From Berg's addition to this observation of clouds thus leads a rather direct path to compositional, twelve-tone chromatic material.

Schönberg's projection of shapes is guided by interests: He wants to see in the clouds that his people will triumph. In Hamlet the courtier Polonius flexibly sees many things that the prince suggests him to see.

LORD POLONIUS

My lord, the queen would speak with you, and presently.

HAMLET

Do you see yonder cloud that's almost in shape of a camel?

LORD POLONIUS

By the mass, and 'tis like a camel, indeed.

HAMLET

Methinks it is like a weasel.

LORD POLONIUS

It is backed like a weasel.

HAMLET

Or like a whale?

LORD POLONIUS

Very like a whale.

HAMLET

Then I will come to my mother by and by.[22]

---

[21] Letter and sketch quoted from Nono-Schoenberg 1992, p. 135.

[22] Hamlet—Act III, Scene 2.

However, an eye with interest in art-production on indefinite or blurred figures stimulates the active imaginative faculty in a productive way and not for mere adaptions. Thus, already Leonardo proposed to look at walls et cetera, in order to activate through seed crystals associations and fantasy and to obtain at the same time already concrete models. The direct mimesis of such surfaces, instead as image, was much later materially realized as ‚impression' by Max Ernst with the „Frottages", the rubbings of irregular structures of wooden boards. Whereas Inge Rubbert since 2019 makes rubbings from different materials with a regular structure like corrugated boards, wrappers and something like that. Max Ernst mounted these structures that have been recovered in this way as components horizontally in a picture. While Inge Rubbert mounted vertically and lays one or two further laminates above, in geometric figures like squares, trapeziums or at the corners opened squares, which were cut out of cardboard or painted. She calls it „Doppios", doublings. Blurring in detail leads to sharpening of figure and content of the whole.

6.2 ‚Motif' as Nucleus for Formation of Structures in Improvisation and Composition

‚Motif' as a unity of content and figure is an elementary and essential component of music in improvisation and in composition. It is a nucleus for formation of structures and also a central category for the development of larger contexts within a work and even between several works. The category is to be found in numerous spheres of sciences and arts. The core of meaning is motif as motive (force). Thus a common material kernel for the various fields of ‚motif' is ‚movement' to the cause or the causes for a process and for further movements, and till an objective reason and subjective motive of an action, an activity. The double character of the motif, namely it being potentially at the same time motive and moved, bears implications and consequences. Thus, a musical motif is in the context of the compositional subject and his situation both psychic as well as music-linguistically based. Once formed, its music-linguistic processes start in turn. But this does not happen automatically. For there are always some reasons behind it and they are deeply-seated. Nearly almost the momentum of certain deliberateness resonates, notwithstanding the primary subconscious psychic parts and basics.[23] In US-American English *»motif«* and *»motive«* coexist. Ernst Toch, being exiled, formed in 1948 a play of words: »*It, the little motif, becomes the motive, the motive power, the motor*«.[24] Just the aspect of form, as a ‚shape', gets accentuated by the English equivalent »*figure*«. Therefore a motif, and not only if it is a matter of the dominant theme (leitmotif) is first an agent, central for musical logic, context, content, and figure in the work. Then, secondly in this determination it functions both as main idea and crystallization nucleus for the perception and for the memory as well. Probably, it acts frequently as a trigger, wherein the motif

[23] In detail cf. e.g. Heister 2017, pp. 13–26.

[24] *The Shaping Forces in Music*, New York 1948, p. 201, quoted after Wehnert 1966, col. 289.

stands, as pars pro toto, for the whole of a composition or ,melody‘. So, 3rd it may function as the crystallization nucleus on the level of semi-automation of „memory“ and recollection in computerized storage and retrieval.

Here, for our purpose, we have to supplement, by something organic, the chemical-mechanic metaphor of ,crystallization‘, which is the nucleus. A joint third of nucleus and its development, and of formation of crystals is ,growth‘. The musical-motivational nucleus originates in the composer's brain or the brain of an improviser, and therefore it is a component of the conceptual form, in some vaguely or in an already firmly outlined impression. It contains already the draft in principle for anything that may follow, just like in nature. This unfolds in dialectical reciprocity with the environment, for instance, in experiences or reactivated recollections of music performers, and in cooperation with others by cell division, splitting up and off, and in development of cell growth. The further sharpening of motives takes place partly in the brain, partly in the form of sketches on paper or in another external medium. The operation with motivational ,crystals‘ and ,cells‘ is an universal of playing music and utilizes both similarities and contrasts. We may consider this as an expression and special case of the universal principle of variation (see Chapter 8).

As mentioned, as a compositional consciously handled procedure, it reaches back, at least to the times of early Renaissance. Scientifically the cell was only discovered later. In 1658, Jan Swammerdam discovered the red blood cells. Since 1665 Robert Hook discovered cells in cork, and in living plant tissue using an early compound microscope, followed since 1668 by Antoni van Leuwwenhook—1677 the sperm cells. In sciences the problem of technical tools for work arises in the first place, whereas for the arts and minds the problem lies in the ability to think in abstract terms for analysis and for synthesis, and for artistic production. Here, what is needed are mental microscopes, the capabilities of anticipation, of considering something in advance and in foresight, in an emergency also taking a look into something and listen into something, like Schönberg and his clouds of war. An example of respective growth processes, taken from Franco Donatoni's *Lumen*, for Piccolo, Bass Clarinet, Celesta, Vibraphone, Viola and Cello, in memory of Luigi Dallapiccola, 1975 (Fig. 6.4).

If one believes in chance this belief may be untouched and shall remain undisputed. If not, one should see some connection between this tribute-work and the person, resp. his name, Luigi Dallapiccola. The title of the work begins like the name, *Lu*igi and/as

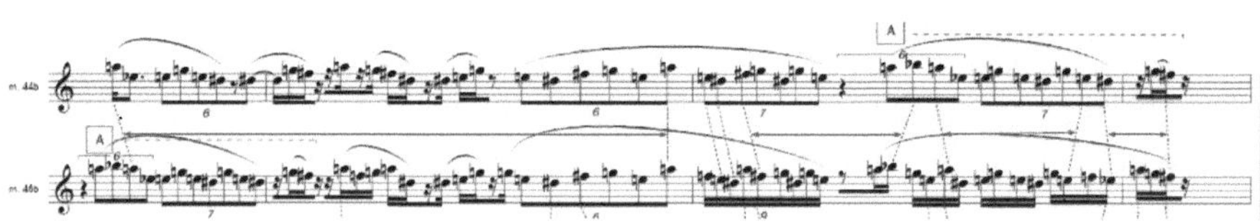

Fig. 6.4 Franco Donatoni: *Lumen* Paradigmatic scheme of the Piccolo part, mm. 44b and 46b
Two accumulation phases: additive and subtractive filtering in pitches and rhythms
Quotation from De Sanctis De Benedictis 2013, p. 43, Fig. 7, excerpt, beginning (see also an earlier version of this essay: De Sanctis De Benedictis 2011)

*Lu*men (Light, as Enligthenment). The scoring of the work was freely selectable and insofar arbitrary. However, it is not arbitrary but motivated: The *piccolo* refers to Dalla*piccola*. The first note of the piccolo's melody is an A, in Italian La. This La is not only to be found in the first name, but also in the surname, and there three times, if we count the retrograde version al-la: *L*uigi D*al-la*picco*la*.The crystallization is undergoing structural and historical developments.

1. A first level of crystallization is the formation of figures, still as extraction, almost filtering. A second level is the formation of figures out of something not yet configured. That means, above all, the creation of motives, or analogous patterns, and their further improvement into themes, melodies, and greater horizontal and vertical composition structures. In their kernel are times, styles and genres overlapping figures, which possess the character of topics. Again, to demonstrate it with the *BACH*-cipher that has already been cited repeatedly. After all it is concise *and* flexible and as letter-cipher often even comprehensible without exemplary notation (see above to *BACH* 4.6 and 5.2).
2. Such topics will be continued by composers, but also in non-notated music, revised and newly worked out, and in doing so often drastically varied. Much of these motivational and thematic work is done subconsciously or pre-consciously, yet some things fully consciously too. Thereby topics as motives can produce connections between several works, both between of one's own, as well as in relation to other works. In the latter case, crossings into borrowing and citation are fluid. Such relations, in turn, may be more likely hidden and blurred, or explicitly emphasized. In the case of BACH, the status of this pattern changes drastically with J. S. Bach. Because now, it is no more only a motif with various variants and meanings, but a cipher, which exactly points to J. S. Bach, even if it might have slipped without such an intent into a purely chromatic composition, as a third of the 12-tone chromatic total. ABHC has to be complemented by the transpositions C# D D# E and F F# G G#, and instantly we get all of the 12 semi-tones within the octave.
3. In analogy to the founding of relations stands what happens during compositional work with one's own, self-created new motives and themes. As new they may be as figures out of the musical material, yet they are still in line with the historic, cultural, and social impressions of such material of tradition and greater currents. On the one hand, they serve the formation of contexts just in a single work, or on the other hand between various works of a composer. For both together, it was already demonstrated by the example of Alban Berg with the name-ciphers AB and HF.

Fig. 6.5 One-Note BACH https://www.bachueberbach.de/johann-sebastian-bach/b-a-c-h/, retrieved 28 February 2019

6.2.1 *Crystallization in the Historical Process. BACH as an Emblem Before, In, and After J. S. Bach—The Cross as Generating Deep Structure Since Guillaume Dufay*

With BACH[25] the filtering of the core motif is unproblematic. However, even here we encounter problems at the fringes in the transitions, degrees of similarity—and even types and sorts of similarity, as in a subset of cyclic replacements within the permutations. On a meta-level the BACH itself is a self-structured „set" of—perhaps: „fuzzy sets"; and someone more qualified in math than me could in a second step of analysis determine the exact degree of relatedness (see Chapter 4).

But there is a further, rather hidden generating and regulating agent of the BACH-cipher. The pious Christian J. S. Bach consciously made use of it. Historically these traces would require empirical scrutiny to be securely confirmed. The structure is that of the cross. With the help of different clefs and the accidental ♭ the four-tone motif can even be represented by a single note (Fig. 6.5).

On the one hand J. S. Bach inscribes his name into the tonal structure by using the BACH-motif, motivated biographically. A coincidence readily made use of by

[25]BACH in German; the motif is not translatable: English B♭ A C B#.

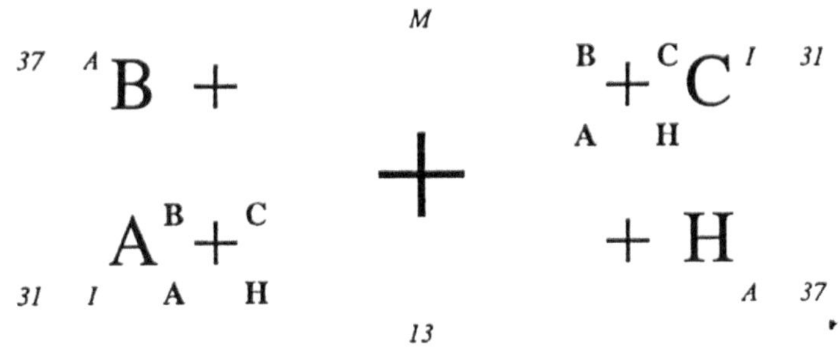

Fig. 6.6 J. S. Bach: Duett F-Dur BWV 803, in III. Part of „Clavier-Übung" (*Piano-Exercise*; printed in 1739)
Shape of the cross in fractions on various levels of the piece
Scheme by Ulrich Siegele, quoted and extensively discussed in Hartmann 1996, pp. 906f

Bach, and due to his extraordinary position in music history and beyond, and due to its structure this motif has lived through a spectacular career in music history and composition. On the other hand this motif via its cross structure makes a reference to religion, mentality and history. Bach delivers a confession. This rhetorical aspect, together with some wordings found in canons or cantatas, implies even an emblematic metaphor. The shape of the cross is primarily conveyed by the notation, rather for the optic than the acoustic sense. And it is the St. Andrew's cross, a diagonal (not the orthogonal "normal" cross), used as railway crossing sign—and additionally laid down with a rotation of 90° (see Sect. 5.2) (Fig. 6.6).

In the duet BWV 803 Bach probably inflected the Bach-cross-shape on various levels of his work. This figure or configuration is in no way to be heard. It is possible, but not confirmed, that Bach himself arranged this configuration. But geometric arrangements generally play an important role in music,[26] at least, African metrical and rhythmical patterns referring to wickerwork included, from Guillaume Machault in the fourteenth century till Iannis Xenakis in the twentieth and afterwords. Therefore we can readily assume Bach's intentionality as a pious composer, giving the BACH-motif such a prominent position. Even if this question is always discussed, mainly with the not very honorable intention to negate semantics in music, it is principally mostly of secondary importance, if motivic metamorphoses, inventions, variations and developments are made consciously or unconsciously. For it might be one of the characteristics of creativity that both psychic spheres are permeable to each other, more fluid and moveable than commonly and in non-creative persons or states of mind. Acoustically the cross-structure, articulated as a sequence, refers to further traditions of religiously influenced music, just as the dense chromaticism, completely within the minimal space of AB♭HC.

First on *sequence*: another line of tradition comes into play. The sequencing cross-motif is related to an old and also countlessly reproduced motif: the "Dies irae" (Day of Wrath) of the Requiem. The famous *Dies irae*, the sequence of the

[26]Cf. e.g. Mazzola/Muzzulini/Hofmann 1990, Mazzola 2002.

requiem, the mass on Judgement Day, has a similar chiastic structure—the cross again (see Sect. 5.2). The procedure of sequencing creates it. Sequence means: the same interval or mini-motif is repeated at a different pitch, higher or lower.—The formula (transposed from F to C):

di-es / i-rae … (dies illa)
−1 +1 −3
CH CA (H-G-A-A)
di-es / i-rae …
−2 +2 −3
CBCA

This *Dies-irae*-motif remains within the frame of a diatonic, primarily 7-tone system. Bach, however, already applied the 12-tone system. The dense chromaticism BACH can be extended to a 6-tone motif, which as a lament-phrase entirely fills the tetrachord, the frame of the fourth: D D♭ CHBA or ABHC C# D. I illustrate the motif with musical examples from the Finale of Gustav Mahler's 2nd Symphony. It begins with a scene of the Last Judgement—which, as later in the last parts of the movements is revealed, is adjourned in favour of a redemption and reconciliation.

Mahler variates the *Dies-irae*-motif in extenso: with intervallic formulas as −1 + 1 −5 or −1 +1 −6 (instead of the original −1 +1 −3).[27] We see, that the identical initial core is retained, the cauda (or tail) varied. And then, too, Mahler makes it similar or transforms it into his semantic counterpoint, the "theme of resurrection" ("Auferstehungsthema")—formulas −2 +2 −4, … −2 +2 +5, partial inversion the of the initial core +1 −1 −3; +4 +2 −2. The similarities are as evident as fuzzy; once again a challenge for Fuzzy Logic. Retained is, as contour, the return to the beginning tone. The descending diatonic line *Dies irae*, bar 3s., and *Resurrection*, bar 5s. as sequence complete the metamorphosis of the motifs (d♭-C-B♭… A♭ / A♭-G-F-E♭; and inverted and expanded to a 6-tone-scale, bar 6 f.., A♭b-B♭b-c-d♭ b-e♭-f) (Fig. 6.7).[28]

The *Dies-irae*-motif remains within the frame of a diatonic, primarily 7-tone system. Bach, however, already applied the 12-tone system. The dense chromaticism BACH can be extended to a 6-tone motif, which as a lament-phrase entirely fills the tetrachord, the frame of the fourth: D d minor C H B A or ABHC C# D.

Characteristic for the BACH-motif is the easily reached combination with text of the lament of Mi-se-re–re, equivalent to Kyrie eleison (Lord, have mercy). The semantics of tone and word are in conjunction with the doubled tone of lament, the descending minor second (of course there are also other *Miserere*-motives) (Fig. 6.8).

Again this can be deduced from the descending half tone, the submotif BA or CH, as mentioned earlier. Josquin des Prez in the *Agnus Dei* from the Mass *Missa L'homme armé super voces musicales* uses the doubled Lamento-semitonestep as ABHC (a permutation von BACH) *ascending*, adapted to the keyword "(tol-)lis"

[27]Musical examples in Floros 1985, vol. III, p. 71.

[28]See also Mahler, 2nd Symphony, Floros 1985, vol. III, p. 69.

Fig. 6.7 Mahler, 2nd Symphony
Dies-irae-motif, then "theme of resurrection" ("Auferstehungsthema")
Quoted from Floros 1985, vol. III, p. 69

Fig. 6.8 Guillaume Dufay: *Ave Regina celorum*
Quoted from Hartmann 1996, p. 322

(you *bear*). In all three voices except the upper superius we find furthermore massed the semantic semitones, as BC, B♭A etc. (Fig. 6.9).

The BACH-motif thus results historically as well as aesthetic-'logically' in its structure and in its semantic field from manifold, crossing and self-reinforcing reasons and motives—a multiple, complex logic.—Quasi as a palimpsest the *Miserere*-Lament gets an ambiguous, doubled sense in Schubert's late mass in E♭major *Agnus Dei* and the late Lied *Der Doppelgänger* (Heinrich Heine; 1828). In both cases the same 4-tone resp. 4-chord figure (in The Double/*Doppelgänger* in

Fig. 6.9 Josquin des Prez: *Agnus Dei* from the mass *Missa L'homme armé super voces musicales* Quoted from Hartmann 1996, p. 323

the sinister key B minor) is functioning as an ostinato throughout the whole movement. The chronology is uncertain; therefore Schubert probably connects sacred and secular spheres not blasphemously but under a sign of sighing as a nation and a person crucified by repressive social-political circumstances of the Habsburgian restauration (Fig. 6.10).

Intervallic formula
-1 + 4 -1
=
B♭ A C# C
B A# d c#

Here, amongst others, the task for Fuzzy Logic would be, besides the clarification and specification of natures and above all of the quantitative grades of affinities, or similarities, to define the limits of fixity, or identity of the respective motif. It would mark

Fig. 6.10 Franz Schubert: *Der Doppelgänger* (Heinrich Heine, 1828)
Eusebius Mandyczewski (ed.): Schubert's Werke, Serie XX: Sämtliche einstimmige Lieder und Gesänge. Band 9. 1827–28 b.z. 'Schwanengesang', No. 554–567 (pp.134–85). Breitkopf & Härtel, Leipzig (1895). Plate F.S. 920–933. IMSLP60828-PMLP02204-Schubert_Werke_Breitkopf_Serie_XX_Band_9_F.S.920–933

the first step, to establish this on part of the musical object. The second step: the empirical countercheck on the part of the listening subjects. In this direction most likely one or a number of research results should already be available. Thereby questions like „is it the same motif?" or similar ones are indispensable. By them a subjective awareness is obtained. But, in any case, it can to a certain degree once again be objectivized: by neurological measuring, which indicates conformities in the respective active areas of the brain—for instance, by comparing the motif's previous perceptions. As suggested by the mere heuristic sketch, hence follow complex research arrangements and comparatively protracted series of experiments. At first, this is just a scientific, music-analytical statement of the problem. But, it could have some collateral benefit for the already practiced employment of Fuzzy Logic in search-processes for the retrieval of down-loaded music. Here, the melody is in the focus, as a somewhat larger unit of music, which often consists of motifs (see Sect. 8.1).

There are other types of variation and development of variants in some styles and genres of music. In these the dialectic of core und contour is not fully brought to bear, because there is no stable, fixed kernel as identical starting point and reference point for the differences. This is nearly the rule in many fields of music outside of the European art music and is often in folk music—but not in Popular Music at least in the present.

> The chain of variants [...] is also not stamped by forms of memory, by any form-building restatement of something already performed. Following the principle of generation and formation of variative "Gestalten" (shapes) the variation takes place successively in the melody and in case of heterophony simultaneously in different parts."[29]

The immediate forms of existence of a work as well as the historical formation of tradition follows a sort of Fuzzy Logic.

> Passed on from generation to generation are the basic shapes ("Grundgestalten") on one side and on the other the methods to vary them."[30]

It is a different matter in case of Persian-Arabic maqamat. On the one hand, they have a non-variable melodic kernel and on the other hand they take shape as works in the sense which is used here. In detail Elsner 2002, 2011 among others (see Sect. 2.1). Here and there in itself variative and flexible Fuzzy Logic is necessary to realize identity including differences.

6.2.2 *Crystallization Within the Work. Motif, Theme, Movement, Work*

The formation of larger units is a kind of natural process in the work's production. In numerous details it reminds of the emergence, of spontaneous „self-organization

[29] Oesch 1987, p. 473.

[30] Ibid.

of matter" (Prigogine). With the difference, that this production in principle is intentional controlled and finished by a subject. This is valid, even if undeniably much of the unconscious is involved, and not every decision is consciously made, for instance „bifurcations", where it might turn in this or that direction. But in the end, the accomplished work appears, in case of a successful one, as if it could hardly have been different.

In the 1960s the musical „aleatory", like in the past the praxis and theory of basso continuo (see Sect. 6.6), left consciously some decisions about the final shape of the composition open and thus created zones of vagueness, fuzziness and blurring within a work. In fact, it does not dispense with the concluding determination of the whole instead its interpreter is charged with it, so that somebody else has the final say in the realization of the work. Vagueness and fuzziness in the score form never have the final say. By the further work in the sound form they get neutralized. Though, how far this is a sharpening, may depend on some kind of uncertainty (see Sect. 7.3). In non-notated as in notated music there are literally countless practicable examples for the creation of new figures, for variants of a starting form, for motivational-thematic work, for varying, contrasting, repetition, for complete new creation and recombination of what is already in existence. Some of it was already presented some of it will follow later always bearing in mind that it refers to extracts. In regard to Fuzzy Logic, the methodical and to what is of relevance in its own light should become clear enough, in spite of such a material limitation.

To the development of configurations within a work belongs in the setting of twelve-tone technique (dodecaphony) also the derivation of a „ground figure" and its three variants out of the chromatic total of 12 tones. This is set in a fixed, prethematic order by a row. Usually the row-technique is employed as a twelve-tone series. It contains all the 12 tone of the octave within the equally tempered tonic system. Since the early eighteenth century it was developed within the music of European Absolutism, and during the nineteenth century it prevailed in the music of the ascending middle classes. But there are also scales of less than 12 tones. For instance, Schönberg initially was experimenting with them. But principally the chromatic total as pitch-dimension of the material-system serves for crystallizing from it, in each case, a special scale for the work. Then, motifs and themes are configured from it. Quite often they might have been the first inspiration, thus origin and beginning, but compositionally are treated as a first result, which afterwards serves as a starting point for subsequent formations of structures.

In the process of work this belongs usually to phases before the completed notation form that is mainly to the sketch form.

Alban Berg's opera *Lulu* (1928/1935) is based on a twelve-tone row, from which further configuration rows for the characters of the drama derive (Fig. 6.11).

Such a basic scale itself may derive from a still deeper lying basic cell. Its structure: $+5 -1 +5$ ($1 =$ semitone). It is also found with Bartók.

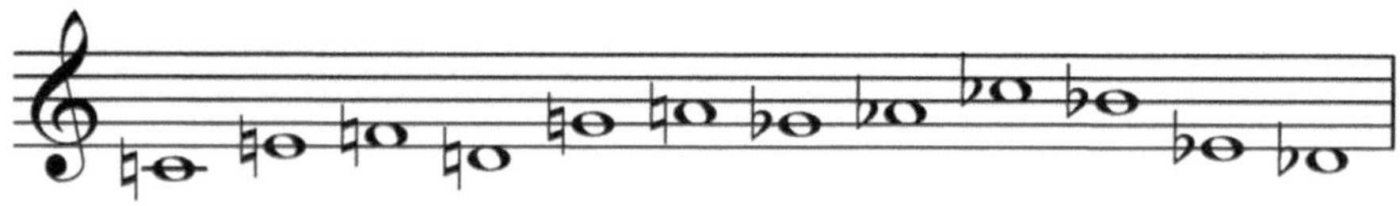

Fig. 6.11 Alban Berg: *Lulu.* Basic row (Redlich 1957, p. 228; figure in https://en.wikipedia.org/wiki/Lulu_(opera), 27 May 2020, retrieved 2 June 2020. The following musical examples and some quotations are also quoted from this article. In details see also Redlich 1957, pp. 210–282; Perle 1989)

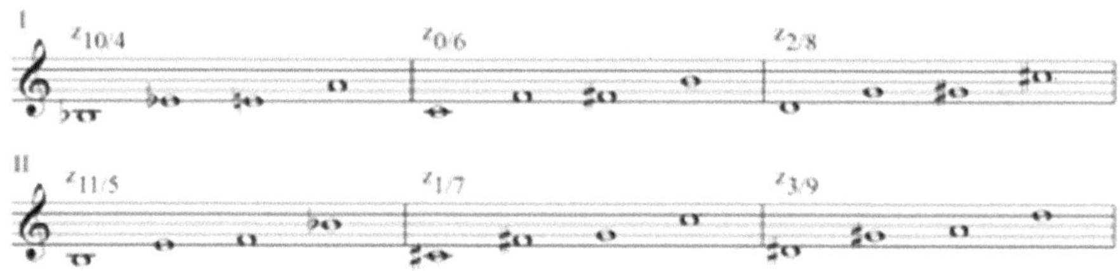

Fig. 6.12 *Lulu* Trope I. „The two non-identical forms of trope I, each generated from all even or odd forms of cell z". (https://en.wikipedia.org/wiki/Lulu_(opera), ibidem)

Cell z is, connecting this motive to other works, is also one of the basic cells in Béla Bartók's String Quartet No. 4) is the basic cell of Lulu and generates trope I (Fig. 6.12).[31]

In the tradition of Wagner, Berg orders by analogy the derivates of the basic row to keynotes (leitmotifs) and thus to particular characters of the drama. Essentially, he obtains them by a simple filtering count and subsequently by appropriate selection.[32]

> The tone row associated with Lulu herself is: F, G, A♭, B♭, C, D, F♯, D♯, E, A, B, C♯. This row is constructed by extracting one note (F) from the basic row's first trichord, then taking the next note (G) from the basic row's second trichord, then taking the third note (A♭) from the basic row's third trichord, and so on, cycling through the basic row three times. […] The tone row associated with Alwa is arrived at by repeating the basic tone row over and over and taking every seventh note […]: B♭, F♯, E♭, G♯, F, B, E, D, A, C, C♯, G. […] Similarly, the tone row associated with Dr. Schön is arrived at by repeating the basic tone row […] and taking the first note, missing one note, taking the next, missing two, taking the next, missing three, taking the next, missing three, taking the next, missing two, taking the next, missing one, taking the next, missing one, taking the next, missing two, taking the next, and so on; this results in the following tone row: B♭, E♭, G, G♯, D, F, E, A, B, C, F♯, C♯.

In summary and easy to survey, the proportion of the various expressions of the *Lulu*-row, all of them transposed to C as basic tone (Fig. 6.13).

Berg's principles of derivates are remarkably mechanic-mathematical. On this level of the formation of forms no blurs in the sense of Fuzzy Logic exist. These enter the scene only on the next stage, when the personal rows within the composition

[31] Perle 1977, p. 12.

[32] In detail already Redlich 1957, pp. 228–231.

Fig. 6.13 *Lulu*-row, with phenotypes and variants
JoaDon, 3 März 2020, https://de.wikipedia.org/wiki/Datei:Lulu-Reihen.tif, retrieved 12 June 2020

are further developed as concrete figures, into melodies, harmonies, guiding rhythms. The Lulu-Love-Theme is combined with the fundamental-fate-rhythm-motif,[33] and the Alwa-Theme has an obvious and intended minor character.[34] Apparently here the question about identity and differences becomes instantly again relevant.

The example of *Lulu* marks once again the relevance of relations between kernel, contour and hull of figures. Under the aspect of Fuzzy Logic, within this relation, which is at the same time a process as a component of the work-process, the important question to ask for is, what remains as it was and what will change, about identity

[33]See Redlich 1957, p. 231 (with fig.).

[34]See Redlich 1957, p. 229 (with fig.).

and difference—and blurring as well as sharpening. Here, the matter of surface- and „deep structure" is particularly complex. The latter is the work-specific twelve-tone row on which all is based. Frequently Janus-faced, as shown in the example of the *Lulu*-series, or better rows: on the one hand a characteristic figure with a particular intrinsic value, while on the other hand—already structured—material for the development of additional new figures.

Quite another kind of nucleus and an already highly developed, improved, encaptic complex imagination form of the work-defining motif is the „formula" of Karlheinz Stockhausen. He used the expression „formula-composition" for his procedure since the piece *Mantra* for two pianists (1970). It bears the concept, that by use of traditional procedures, like transposition, „motif"-separation, rhythmic augmentation and diminution, tending towards a deduction for all structural features of the work from a single original melody, namely the „formula". It is either a single melody formula, or a two- or three-voice contrapuntal construction.[35] Frequently precise indications for the performance can be found.

> Thus the largest augmentation of the formula represents usually the form-scheme of the piece. A formula with more than one layer is called "super-formula".[36]

On the one hand, Stockhausen thus continues the motivational-thematic work with the development of figures within a work from a single nucleus. While on the other hand he continues procedures of the twelve-tone technique, where the entire pitch-series is determined by a single row, and by the serial technique, by which this determination tries to expand even for other „parameters", as above all length of sound, tone color and volume. Yet, the continuation is at the same time a rupture. Because Stockhausen's „formula" is already a stamped, clear-cut figure. As indicated, in twelve-tone and row-technique such things will only be produced as a second step from the rows.

> In contrast to serial music, where the structuring features are more or less abstract and remain largely inaccessible to the listener's ear, in formula composition the musical specifications of pitch, dynamics, duration, timbre, and tempo are always directly evident in the sound, through the use of a concisely articulated melodic tone succession, the formula, which defines the large-scale form as well as all the internal musical details of the composition (Blumröder 1982, 184f.).[37]

6.3 Work as Inspiration and Idea. Conceptual Form

The overlapping work-idea is like in a process made concrete and crystallized. It takes steadily a clearer, sharper form, at first, starting more vaguely, in the conceptual form. Then, in central motives and themes it is intensified. This intensity will turn into the

[35] https://en.wikipedia.org/wiki/Formula_composition, 25 December 2019, with further reading, retrieved 5 June 2020.

[36] https://de.wikipedia.org/wiki/Formelkomposition, 3 October 2015, retrieved 5 June 2020.

[37] Ibidem.

point of departure for the development of other figures within the work. And in reverse:

> The musical shape is not only formed in the progress of a piece. The overlapping artistic idea is anticipated as central shape by the theme.[38]

6.3.1 *Questions of Documentation and Translation. Verbal, Notational, and Pictorial Partial Concretization of the Idea*

The conceptual form is the first concretion of this „overlapping artistic idea“. It plays as guiding idea both as terminus a quo as well as terminus ad quem, as point of departure and aim in the work-process a considerable rôle, which is certainly hard to apprehend and to analyze. The conceptual form contains both ideas about details as well as anticipating visions and imaginations of the whole. Both components develop themselves, frequently in interaction with each other. And aspects of the conceptual form even survive in the subsequent existence form. Then, with the imagination form, the interface between composition and interpretation, notation and tonal realization it to a certain extent raised on a new level. Because such ideas exist only mentally and as such may neither be hypostatized nor exteriorized, the conceptual form is mainly to be grasped indirectly, in verbal or graphic expressions. Here, blurring is therefore in particular large. In-so-far, mostly too little substantial material is available for Fuzzy Logic. About Richard Wagner e.g. there are some records, which sketch hints of inspirations and initial ideas.

6.3.1.1 Sudden Revelation. Dream, Somnolescence, Unconsciousness

During the early 1850s Wagner wrote some musical sketches for parts of the Ring and noted down various motifs of the future work. The 5 September 1853 Wagner claimed in his autobiography *Mein Leben* as the date of the musical idea for the orchestral prelude of *Das Rheingold.* The idea came to him while he was half asleep in a hotel in La Spezia in Italy during the day after the night 3 on 4 September 1853. The reality of this dreamlike inspiration has been disputed, but it could have been happened, even if being only an "acoustic hallucination" (Nikolaus Bacht).[39]

> After a night spent in fever and sleeplessness, I forced myself to take a long tramp the next day through the hilly country, which was covered with pine woods. It all looked dreary and desolate, and I could not think what I should do there. Returning in the afternoon, I stretched myself, dead tired, on a hard couch, awaiting the long-desired hour of sleep. It did not come; but I fell into a kind of *somnolent state*, in which I suddenly felt as though I were sinking in swiftly flowing water. The rushing sound formed itself in my brain into a musical sound,

[38]Goldschmidt 1970, pp. 281f.

[39]See https://en.wikipedia.org/wiki/Das_Rheingold, 5 December 17, retrieved 12 December 2017.

> the chord of E flat major, which continually reechoed in broken forms; these broken chords seemed to be melodic passages of increasing motion, yet the pure triad of E flat major never changed, but seemed by its continuance to impart infinite significance to the element in which I was sinking. I awoke in sudden terror from my doze, feeling as though the waves were rushing high above my head. I at once recognized that the orchestral overture to the *Rheingold*, which must long have lain latent within me, though it had been unable to find definite form, had at last been revealed to me. I then quickly realized my own nature; the stream of life was not to flow to me from without, but from within. I decided to return to Zurich immediately, and begin the composition of my great poem.[40]

6.3.1.2 Intention and Non-intention

The blurring up to impossibilities, in fully realizing in composition what is thought and wanted, was paradoxically pointed at by Ferruccio Busoni. Because every realization of a compositional idea was in fact just a transcription, it would lastly not matter, if it concerns an original work or an arrangement. But art producers may even from the start abstain from intentions, concrete ideas for a later realization and thereby of somehow more clear-cut and outlined work-concepts.

In this case it seems, what is meant by intention is the intent for statement, effect, or forming:

> the purposeful realization of a certain plan. But, this is exactly what should be excluded in such works! Preferences and aversions, artistic designs and other intentions should not play any rôle. ,How difficult it might ever turn out to be, it shall be playfully without any aim or intention. Such a play is an aleatory game. However, the expression aleatoric does not mean completely coincidental or even arbitrary works. Aleatory describes a certain determined, not intentionally designed imagination form of the work. The term ,aleatoric' was for the first time used in 1954, being a crucial year for the later avant-gardists, by Werner Meyer-Eppler (1913-60) in a communication seminar. He borrowed the term from a, in the previous year published [...], comprehensive theory of the ,fonctions aleatoires' by Andre Blanc-Lapierre and Robert Fortet – a work, which tried to mediate between users and theorists of random functions.[41]

But generally, with composing the aim is to create a most possible „completed entirety" (K. Ph. Moritz), in which ideas are given shape. The conception takes, as already noticed, something of such a whole in advance, being still non-material, but already with components for a realization. The importance of anticipation was already the thematic subject in case of perception. As it is a component in the process of perception, it is also one in the process of production, or composition:

> (T)he principal emotional content of music arises through the composer's *choreographing* of expectation [...].[42]

The conceptual form anticipates like afterwards on a new level of the work's process the anticipation form, what is to come in future. Moreover, it may be a pattern

---

[40] Wagner 1865/1880, p. 603.

[41] Schulze 2000, p. 17. About another employment of the term, see Sect. 7.3.

[42] So Meyer 1956, quoted after V. G. Red'ko et al. 2007. Emphasis HWH.

for a presentation form. In this way, such forms of conceptions achieve a relative independence in comparison with what will later be actually realized—or as well what was not realized.

6.3.1.3 A. Lucier's *Still Lives*. Pictures and Sketches of Ordinary Things as Inspiration

Alvin Lucier's cycle *Still Lives* (1995) copies and transforms quite ordinary, daily things in an artistic way[43] (see Sect. 3.5). In a certain contradiction to the object and theme it demanded from Lucier a highly complicated production process in multiple phases: selection and perception of the objects, copying and measuring, computer processing, documentation of such a processing, copying and pitch-determination in a musical notation.

> The suite in eight sets for piano and two or three sinus-generators *Still Lives was written* in the summer of 1995. The title is an invented word, being a homophone to still life. About the compositional procedure gave Lucier two differing accounts: „For the forms of each set, I simply looked around in my house and choose pictures and objects, which were before my eyes, among them a hammock, tightened between two trees behind the house, a diamond of sunlight on the floor in the living-room, a pair of chopsticks, lying on the kitchen-board. I drew the shapes on paper with *concise* dates and pitches, and sent them to Bob Bielecki, who programmed them on a computer and recorded the waves on DAT. I copied the shapes on notation paper and then wrote down the pitches for piano, which would create audible tremulants" (Liner Notes 2001).[44]

In this version of the tale on the origins Lucier puts emphasis on sharpening, whereas in the later version on accepting vagueness and even blurring.

> Two years before he declared: »I simply made drawings of objects, which I saw in my home, among them a lampshade, a bread knife and a hammock behind the house. I made sketches of these objects and copied them freely on music-paper. I made no effort to draw them *true to scale*. The shapes and their musical manifestations are *more impressionistic than true to nature*.« ››[…] I sketched those items and freely outlined them on staff paper. I didn't try to draw them to scale. The shapes and their musical manifestations are more impressionistic than accurate.«[45]

Lucier changed the usual sequence of the existence form.

> The drawings, which served for Alvin Lucier as basis for the composition of *Still Lives*, were in 1999 exhibited in the gallery yellow MUSIK in Berlin. With the signature on each of the drawings and their exhibition, he contextualized them in the spheres of visual art and established *Still Lives* as *intermedia project*.[46]

Here, a sketch form already precedes the score form and perception form. The work could be understood as an extraordinary variant of program-music, or musical

[43]Cf. in detail Straebel 2008.

[44]Quoted after Straebel 2018 and Straebel himself, p. 62.

[45]Straebel 2018, pp. 62f. Emphasis H.-W.H.

[46]Straebel 2018, p. 63.

ekphrasis—unusual, because composers generally take already existent pictures from somebody else others as models (see Chapter 4).

> Moreover, the knowledge of drawings changes the reception of the piano-pieces, as the fixed form of the score together with the expressive title is no longer reserved for the imagination, but with the help of the corresponding contour drawings, it may be related, both to the real objects and as well on their graphic representations.[47]

Thus, the recipients may, like normally with program-music, use Fuzzy Logic themselves, by comparing the real object, a „readymade", with its graphic artistic-optic and acoustic and perhaps with the notational-optic picture form.

> The sunspot on the floor of the living-room (1 – *Sunlight Diamond*) has the shape of a lozenge, if one assumes the equivalence of the pitch of three octaves with the distance of both of its lateral corners of 2′24″. [...] The edge of the bread knife (7 – *Bread Knife*) has in the music like in the drawing six points, whereas the distance between cutting edge and knife-back is oscillating to and fro between a minor third and a fifth, and each point has a length of approximately 26 seconds.[48]

Straebel finds fault with some obscurities:

> The three grasses in front of Lucier's window (6 – *Fens*) form in both medias varying twist and multifarious overlapping, only their distances on the left and the right side of the drawing are not wholly harmonized. The drawing of the chopsticks (8 – *Chopsticks*) seems to have found its way into the composition like being mirror-inverted, in any case, its crossing point lies in the sinus-tone glissandi markedly in the first half of the set at approximately 20:30 of the [...] notes, while in the drawing on the right half, at approximately 5: 3. Therefore it is fitting, that the relation of the distances for the straight line lies from start to end at 2:3 semitones, while in the drawing in reverse from left to right at 3: 2.[49]

6.3.2 *Emulation of Processes of Creation and Crystallization in the Work Itself*

It is not seldom the case that musical works present processes of arising, beginning, becoming, between the genesis of the world, as in Haydn's *Creation*, or the genesis—including rise and fall—of the world of the "market economy" in Wagner's *Ring* tetralogy, or, on a smaller scale, the spontaneous and then consciously commissioned genesis of illusions of freedom under the conditions of the said "market economy" in Brecht/Weill's *Rise and Fall of the City of Mahagonny.* There is always an initially growing order, a continuous sharpening, which can then in turn collapse gradually or rapidly, according to the dialectical configuration of chaos and cosmos. The quiet, gentle beginnings sometimes correspond, especially in new music, in an arched process such as in Alban Berg's *Three Pieces for Orchestra*, to a slow fading and dying away, a *morendo* (see Sect. 7.1.1).

---

[47]Ibidem.

[48]Ibidem.

[49]Ibidem.

6.3.2.1 Haydn's *The Creation*

Haydn's oratorio *The Creation* of 1796/1798 begins with the depiction of the coming into being of the cosmos out of chaos. He presents this as the formation of concrete, sharply defined musical shapes from a blurred or even vague material (see Sect. 9.2). *Prelude. Die Vorstellung des Chaos* (The Representation of Chaos) is an overture in C minor in very slow tempo, *Largo*. Haydn, the genius of brilliant order and flexible form, here exposes the polarity of nothing and being and its dissolution in becoming as a tension which lifts the work above the pious idyll. An inflexibly kept unison C by the entire orchestra symbolizes the infinite void. Shadowy thematic rudiments appear in aimlessly wandering harmonies without clear functional meaning so that the key remains uncertain. Dissonant tension and resolution are interlaced with an audacity that is without example in classical music. The ambivalence is heightened by chromaticism, the formation of suspended notes and changes of key into distant harmonic areas. The tonic triad of c minor, c-e♭-g only appears in unadorned completeness and in the fundamental position as a final chord in the 58th and 59th bars.[50]

However, the artful presentation of chaos still lets even the three-part scheme of the sonata form shine through. The first part, which consists of two thematic groups, has its cadence in the parallel key of E♭ major and leads into a development of the secondary topic which leads back into the strongly abbreviated reprise of the main part; a chromatically gliding coda which fades in pianissimo concludes the movement in c minor. The formless has gained form. Haydn uses the same C key, moving from minor to major, to introduce the origin respectively the creation of light. The rise of the sun, the symbol of reason and enlightenment, is then extended in No. 12 as a great scene.

No. 1. *Im Anfange schuf Gott Himmel und Erde* (In the beginning God created Heaven and Earth; cf. *Genesis* 1:1–4). A recitative for bass solo in C minor is followed by a choral presentation of the creation of light, ending with an outburst, a sudden surprise fortissimo C major chord on the word *Licht* (Light). The latter is depicted first with a soft pizzicato note from the strings, followed by a sudden surprise fortissimo C major chord on the word *Licht* (Light).[51] (See below section 9.2.)

In Schönberg's Variations for Orchestra op. 31, 1926/1928, the "creation" is not a transcendent but an immanent process, the "genesis" of variations for the building of shapes based on a twelve tone row. The series is a kind of a priori prerequisite for organizing the pitch dimension of the work. However, it is presented in the work as a resulting consequence a posteriori. It becomes a theme in the literal sense of the word, as "thesis", setting, which only develops in a long introduction as a prelude.

---

[50] See https://en.wikipedia.org/wiki/The_Creation_(Haydn), 10 December 2017, retrieved 12 December 2017.

[51] See ibidem.

Fig. 6.14 A. Schönberg: Variationen für Orchester op. 31
Basic Series
Quoted from Dahlhaus 1968, p. 8

6.3.2.2 Arnold Schönberg's 'BACH Variations'

In a certain sense, the work demonstrates how a theme evolves in a work. In the large-scale variations for orchestra Schönberg uses the twelve tone technique (dodecaphony) for the first time in connection with a large, classic genre and performance apparatus. Starting point and foundation of pitch organization is a specifically organized twelve tone row resp. series. The different forms or tropes of this 12-tone-series constitute an already in itself variating generative deep structure.[52] From this substructure the concrete shapes of the orchestral score as further variations are developed. Through the use of complex techniques of composition, which incorporate classical procedures, by use of detailed setting of the score, using counterpoint, Schönberg attempts to link his radical innovation back to tradition as a means of justification—more clearly and timidly this can be found with his student Anton Webern. One of these means is the partial identification with Bach. The „shape"—as defined in "Gestalt psychology"—being a primer in our perception, here the concrete formation of motif and theme and the variation of the underlying sequential forms, becomes the "ground" or rather background. From this evolves the decisive part. As pertaining to sound Schönberg foregrounds the emblem as a quote. He almost chisels the BACH out of various sequence forms as the mentioned "soggetto cavato". The BACH is hidden resp. embedded in the series: Tones 1 and 6 of the first hexachord of the "Grundgestalt", and 12 and 11 of the second hexachord. They are therefore in a prominent, outstanding position, but are not immediately conspicuous. But gradually, as a development within the immanent work process, Arnold Schönberg becomes BACH—a metamorphosis by filtering or extraction, densifying and permutation—resulting in a transformation, nearly a 'morphing'. The tones resp. characters of the name and the semitone intervals Eb-D and E-Eb reproduce almost fractally the motif head BA (Figs. 6.14 and 6.15).

[52]Cf. e.g. Dahlhaus 1968, schedule with all row variants, appendix I.

Fig. 6.15 Schönberg: Variations for Orchestra op. 31
BACH as emblematic rebus-figure inside the variants of the series
Quoted from Dahlhaus 1968, p. 9

The *musical* metamorphosis into Bach is to a degree prefigurated in the characters

ArnolD eSCHönBErG
A .... D eSCH .. BE.G
A CH B

of Schönberg's own name. ACHB
2 3 4 2
1 2 3 4
BACH

In analogy to the beginning and the construction processes in the progress of a work, like here in op. 31, there exists the ending with decomposition processes as well. A popular and at the same time programmatic-biographically motivated model for this exists already, as the finale of Haydn's „Farewell Symphony", when during a lonesome duet 2 violins are left and these last two snuff out the candle. But a historic and factually much nearer analogy is the mounting and dismounting of „additive rhythms" by Olivier Messiaen, that is for example the series of quarter note + 1 sixteenth note, + 2 sixteenth notes, and so on, and symmetrically the dismounting of such a rhythmic row. And, comparable in structure, but without exotic relations, are the „variable metres" by Boris Blacher, who arranged bars of different measures—for instance 2/4, ¾, 4/4 measures and so on; also among others, this procedure was employed by Karl Amadeus Hartmann. All this may be a musical implementation, what Goethe called to „die and become" („Stirb und werde"), the eternal process of growth and decay—and then a new creation, which happens in all spheres of reality, with the variant of the entanglement of fuzziness, sharpening, fading and slurring. The chain of processes does not form a simple circle, but a spiral: nothing is repeated completely unchanged.

6.3.3 Conceptual Art as Discontinued Conceptual Form

Unlike the transitory conceptual form in art process, conceptual art is in various arts a conscious decision, to consider the idea for a project of art as the end, and as of priority in comparison with realization. Thus within this form of art exist works, which become independent from the draft and are being realized in the form of artifacts, which are neither visually nor acoustically perceptible. In other cases the concept-idea is noted down and its implementation is left for others. But, in some cases there is a draft of ideas as well as a real implementation. In this form of art written records are important, otherwise the idea will vanish swiftly. But, in any case, the concept, the inspiration, the idea has priority over the implementation. Thus, since the 1960s, a new kind of blurring exists in this art movement. Therefore the work of art is at least for the recipient often only hardly imaginable, or permits manifold associations. Thus, in many cases even sublation in a dialectic sense is impossible. On the contrary, only steps of the music process and the forms of existence, which usually follow after the form of representation, are either cut off, or not implemented. The dialectic contradiction between idea and realization can thereby be dismissed and denied, but not eliminated.

The idea of a work as concept, or idea is in the conceptual form still relatively abstract. To begin with, it develops with insular concretizations, still vaguely and lacking shape, but already with particular clearly defined shapes. At first this is rudimentary, but soon it becomes a still blurred, anticipating picture of the whole in contours. In the sketch form everything becomes one step more concrete, formed and designed, in short, sharpened, and is already closer to the end of a work, which rests „completed in itself". Thereby, the conception serves as regulator and motivation. Without such a dark, but mighty total idea that precedes all technical, no poetic work can be achieved.[53]

6.4 Work as an Outlined Written Conception. Sketch Form

With un-notated, improvised music sharpening and crystallization of concise shapes is achieved by testing and practicing in the sound form. At first, the conceptual form exists only in the composer's mind. For others, it becomes tangible only indirectly, through some kind of written stipulation, whether verbal or in note form, or through oral communications, which in the best case are later recorded in writing. It also becomes tangible for the composers by activating their memory. But for the composition process to continue, it must be exteriorized at some point. This is then the stage of the sketch form.

[53] Schiller to Goethe, letter from 27th March 1801, cited in Benary 1998, col. 1507.

6.4.1 Exteriorizations of the Conceptual Form

Inspiration is followed by working out. Paul Hindemith defined this as „materialization" of the „vision".[54] While an inspiration is often a „coincidence" (so Ernst Krenek 1959),[55] the working out of jottings represents already in the conceptual form all along the sketch form to the score form the part of necessity. Here in reverse of a popular German saying: At first enjoy the amusement, then the work. Richard Strauss emphasized, around 1900, that „normally his inspiration included a melodic idea of altogether four bars in length, which demanded for hard work afterwards.[56]

Mozart, contrary to the stereotyped image of him, often worked quite laboriously and with written jottings.[57] His inspiration and even a part of the working out was quite often to a lesser extent improved by rest and concentration as rather compensated through diversion, amusement and restlessness. Thus his work was directly embedded in entertainment. So, for instance, his brother-in-law Joseph Lange:

> If he was just engaged with an important work […], he was not only talking in confusion, but at times he made some kind of jokes, which one would not have expected from him […]. He either deliberately concealed […] his inner strain by exterior frivolities, or he had pleasure, to bring the divine ideas of his music in a sharp contrast with imaginations of dull ordinariness […].[58]

At times, something from the interior penetrates to the outside in a distorted and blurred way. Then, the ideas and inspirations are realized in the mode of vocals, which were simply at hand. At a game of billiard, so again Mozart's brother-in-law Lange, he did.

> not answer, but steadily droned something to himself. When he was questioned, he answered, that he just composed the melody for Papageno, which was to be performed with a locked mouth.[59]

Besides alcohol, tobacco, coffee (as Puccini consumed the latter two) or Schiller's famous rotten apples as assistance for inspiration, even working may serve as a midwife to let inspirations flourish. Like, for instance, Bellini or Donizetti, so Carl Maria von Weber adapted the text, he had to set to music, by learning it by heart.

> During lonely walks in Hosterwitz, and later in the same year in Dresden, one could meet Weber on the terrace or in the great garden […], where he usually got small, written leaves

[54] Danuser/Katzenberger (eds.) 1993, p. 14.

[55] Cf. Danuser/Katzenberger (eds.) 1993, p. 17.

[56] So after Danuser/Katzenberger 1993, p. 13.

[57] Cf. above all Konrad 1992.

[58] Quoted in Schmitz/Ure 2019, p. 1042. Analogous could also happen with the less sanguine Brahms. Max Kalbeck „surprised […] him *while writing a score* and instantly turn to leave. But calmly he continued with his writing and remarked: »You may stay here, Your presence does not disturb me at all. On the contrary: if I attend a pleasant conversation, et cetera …. […] I just finish this page […]" (Kalbeck 1915, pp. 247f.) Brahms' calm inspite of the interruption is certainly to be explained, that he was already preparing a fair copy, which is above all a mechanical task.

[59] Schmitz/Ure 2019, p. 1042.

> out of his pocket, reading silently and then, like in quiet monologue, continued his way. He memorized by heart, *the text he had copied himself* of »Euryanthe«, to the utmost fluency, until it became embedded in his brain, like a product of his own, for which he himself had made up the words.[60]

Then, the productive unconsciousness has its sense of duty. According to the jottings of his student Benedikt

> along a whole row of recapitulations of the text quite often his genius was silenced, and then again suddenly the idea of a whole number flashed up, like out of the night, and was formed, without leaving the master again, for days incessantly inside of him fermenting and *crystallizing*, in its totally completed form in his mind. Only then he took it down on paper. Here, his explanation is found, that Weber nearly almost worked out in the Opera numbers, which were quite apart, totally outside of their order.[61]

It is not unusual that inspiration is increased and further developed by transpiration. For instance, from Beethoven long walks in the green are well known. The same is valid for Brahms. His friend and biographer Max Kalbeck reports:

> In Ischl, I later had a few times the unexpected opportunity to observe Brahms while he worked. On a warm morning in July, as I went very early out into the open, [...] when I suddenly saw a man coming from the wood and crossing the meadow in my direction, [...I] was pleased to realize that it was Brahms. [...] Bareheaded and shirt-sleeved, without waistcoat and collar, in one hand he swung the hat, while the stripped off jacket with the other dragging behind in the grass and hurried so fast onwards, as if he were chased by an invisible pursuer. Already, from a distance I heard him panting and groaning. As he came nearer, I could see how sweat ran from his hair that was hanging in his face, in streams down his heated cheeks. His eyes were staring straight on into space and glowing like those of a beast of pray – he gave the impression of someone possessed.[62]

Verbal notes, which are in a wider sense also ‚sketches‘, exist in un-noted music too, presumably mainly in oral form. For sketches in notation form the written is by definition a prerequisite. Bruckner „realized the opening theme of the Seventh Symphony while dreaming, and he immediately *noted it down* after he woke up“ (Fig. 6.16).[63]

The strictly diatonic theme starts with a leap of a fourth downwards and then ascends steeply upwards over no less than two octaves. The rise form the subconscious is, like the shape of the theme itself, no coincidence, as both are deeply entwined with Bruckner‘s psychic peculiarities. Because this inspiration can be associated with Bruckner‘s known compulsion for climbing stairs and counting steps. The rear set forms a compensatory contrast with the narrowly stepped chromaticism.

One of the presumably most pleasant stories on vision and realization, including their imperfection, which gives the composer pain, is the one about the inspiration and the creation of Giuseppe Tartini’s „Teufelstrillersonate“, c, Sonata for Violin and

---

60 Weber, 1866, pp. 458f. Emphasis HWH.

61 Ibid., p. 459.

62 Kalbeck 1915, pp. 248f.

63 https://www.problemkreis-psychosen.de/psychose/psychosen_anton_bruckner_herzinsuffizienz_lungenentz%Fcndung.asp, retrieved 12 December 2019.

Fig. 6.16 Anton Bruckner: Seventh Symphony, 1st movement, *Allegro moderato*, main theme, m. 1–9, score excerpt
Hans Ferdinand Redlich (ed.). Ernst Eulenburg, Leipzig (n.d., ca.1925). Co-issue Universal Edition, Wien (n.d.) later W.Ph.V. 211 (Philharmonia series). IMSLP30703-PMLP07929-Bruckner-WAB107FSeu.pdf

Basso continuo G-minor, B:G5, "Il trillo del from about 1713 or, more likely, after 1740.[64] The story is told by Tartini himself to the French astronomer Jérôme Lalande and reported in Lalande's *Voyage d'un François en Italie*:

> One night, in the year 1713 *I dreamed* I had made a pact with the devil for my soul. Everything went as I wished: my new servant anticipated my every desire. Among other things, I gave him my violin to see if he could play. How great was my astonishment on hearing a sonata so wonderful and so beautiful, played with such great art and intelligence, as I had never even conceived in my boldest flights of fantasy. I felt enraptured, transported, enchanted: my breath failed me, and I awoke.[65]

Tartini tried to reproduce the inspiration from the dream immediately with his instrument, instead on paper.

> I immediately grasped my violin in order to retain, in part at least, the impression of my dream. In vain! The music which I at this time composed is indeed the best that I ever wrote, and I still call it the "Devil's Trill", but the difference between it and that which so moved me is so great that I would have destroyed my instrument and have said farewell to music forever if it had been possible for me to live without the enjoyment it affords me.[66]

The core of the sonata's nickname is a passage where the violinist trills while simultaneously playing arpeggiated triads,[67] with both elements ascending chromatically—a new effect in the 1730s.[68] As a subconscious component, an upward chromatic line is emerging in the fourth frame, the famous "passus duriusculus", the 'somewhat hard passage'; first E F in the first voice, continued in the second voice as F# G G# A (Fig. 6.17).

---

[64]See Betsy Schwarm, https://www.britannica.com/topic/The-Devils-Trill, retrieved 4 August 2020; cf. https://en.wikipedia.org/wiki/Violin_Sonata_in_G_minor_(Tartini), 7 July 2020, retrieved 14 July 2020.—The sonata story was the sujet for Cesare Pugni's 1849 ballet *Le Violon du diable.*

[65]Holmen, Peter. "*Sonata in G minor 'Il trillo del Diavolo', Bg5*". Hyperion Records. Retrieved 21 February 2016, quoted in https://en.wikipedia.org/wiki/Violin_Sonata_in_G_minor_(Tartini), 7 July 2020, retrieved 14 July 2020.

[66]Ibid.

[67]https://en.wikipedia.org/wiki/Violin_Sonata_in_G_minor_(Tartini), 7 July 2020, retrieved 14 July 2020.

[68]Cf. https://www.kammermusikfuehrer.de/werke/1799, retrieved 14 July 2020.

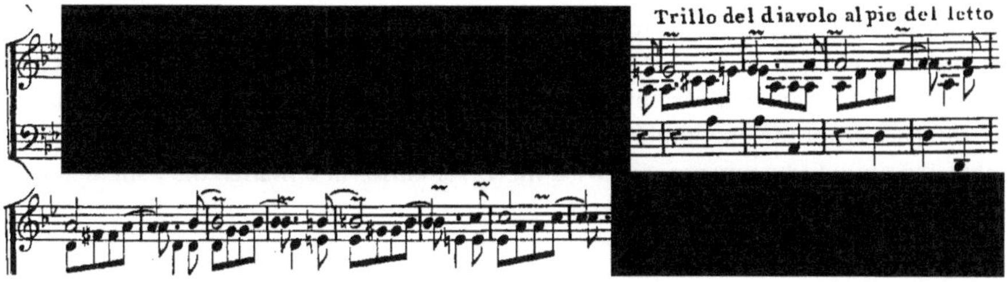

Fig. 6.17 Giuseppe Tartini: "Devil's Trill Sonata", excerpt
Giuseppe Tartini: *Sonata in G minor 'Il trillo del Diavolo'*, Bg5
First staff, right: "Trill of the devil at the foot of the bed"
https://imslp.org/wiki/File:PMLP346890-Tartini_vn_son_g_Devis_Trill.pdf

Not just in a dream, as in the case of Tartini, but as a nightmare and feverish hallucinations, something in analogy occurred to Chopin, ten years before with death, with an unfinished piece, intended as the 14th Prelude of the collection of 24 (Op. 28). In the published versions the Prelude, in E♭ minor, has 19 bars.

> It is just 33 measures long, shorter than the "Minute" Waltz, but it reveals a world about Chopin the innovator and Chopin the bedeviled. [...] "Here he was, in 1839, at age 29, half-crazed with sickness and dementia, with these *nightmarish visions of specters rising up out of the piano,* and trying out something completely wild," said Jeffrey Kallberg, [...] who reconstructed the piece. "For the first time, we have evidence of probably his most experimental work of all."[69]

Chopin's fragment as something unfinished is appropriate in this context. And it is a reflection of Chopin's improvisations[70]; these were partly, as already noticed, in Chopin's process of work an equivalent of sketches, partly also an independent play, like for George Sand or Delacroix.

Tartini's Sonata appears—in word form (see Sect. 8.4)—notably in a totally different context, in a comic strip series, in an Italian *fumetto.*

> The *Devil's Trill Sonata* is the only musical composition that Dylan Dog, the main character of the eponymously named Italian comic, can play on the clarinet. The sonata also has an important role in the plot of "Sonata macabre" (English: "The Macabre Sonata"), the issue #235 of the comic.[71]

A young and beautiful Russian violinist plays the piece under supervision of an aged composer. She is willing to achieve, wherein Tartini himself was unable to succeed: He wants to improve her play to such perfection, that she can perform this sonata the

[69] Rip Rense 2002. „There exist three further Préludes outside of the cycle; besides a piece in cis-minor op. 45 (1841) and another one in As-Dur (*Presto con leggierezza*, 1834) there is an unfinished one in es-minor, which was given the epithet *Devil's Trill* by Jeffrey Kallberg in analogy with the „Devil's 'Trill Sonata" of Giuseppe Tartini" (https://de.wikipedia.org/wiki/PrC3%A9ludes_(Chopin), 16 March 2020, retrieved 10 August 2020).

[70] See Rense 2002, ibid.

[71] https://en.wikipedia.org/wiki/Violin_Sonata_in_G_minor_(Tartini), 7 July 2020, retrieved 14 July 2020. Cf. also https://en.wikipedia.org/wiki/Dylan_Dog, 22 June 2020.

way the devil realized it in the legendary dream of Tartini.[72] But, the identity between dreamt and realized work becomes a nightmare. The tones of the new melody cause among the audience inexplicable reactions and bring them into a direct contact with the inferno. After the encounter with Tartini, the devil himself had for more than two centuries time, to become music-wise more versatile. In a manga the sonata is used to arrange the contact with him.

> The *Devil's Trill* also features prominently in a story arc of the supernatural horror manga *Descendants of Darkness*. A gifted music student accidentally contacts a demon while playing the Sonata.[73]

6.4.2 *Sketches and Music Process*

The developmental process is a to and fro between written exteriorization and mental thinking ahead. On the whole, many compositional work processes are *only* happening within the mind. Therefore many gaps exist in the records of jottings and sketches—a problem, even for Fuzzy Logic, if it concerns fuzziness, sharpening and blurring during the gradual completion of the work. Nevertheless, it is remarkable, how many written sketches, designs and notes are still preserved. On the other hand, sketches of a work are rarely fully handed down. And, anyway, much was never written in notes. As already mentioned, some stuff is prepared within the mind in such a way, that it may be transferred to a fair copy instantly.

Instead of saving the concept on paper or alike, it is, as mentioned in Sect. 6.2, also possible to save the concept in the mind. In the following case such storage within the mind is in reality already the reconstruction of what was written before. Mikis Theodorakis wrote in his autobiography in retrospect about the stay in Makronisos penal camp. During almost one year, in 1948/1949, he was interned there for participating during the civil war on the left side.

> All my musical sketches and even the works completed, which I had composed on Ikaria were lost. But I kept the *notes in my mind*. Under the influence of [...] new [noise- and sound-] experiences, I wrote the *Elegy and Lament for Vassilis Zannos* anew and projected for a second time the rudiments for the 1st Symphony.[74]

Exteriorized sketching belongs to the composition of complex compositions, especially to polyphonic music. Already in Renaissance existed a medium for it, conceived after the model of Roman wax tablet and slate, to be reused endlessly, this was the *tabula* as

---

[72] My description paraphrases https://it.wikipedia.org/wiki/Il_trillo_del_diavolo, 10 June 2020, retrieved 4 August 2020.

[73] See also https://en.wikipedia.org/wiki/Descendants_of_Darkness, 5 June 2020, retrieved 14 July 2020.

[74] Theodorakis, Mikis: *Mein Leben für die Freiheit (my life for freedom)*, Bern, München, Wien 1972, quoted in Folkerts, Gerhard: *Mikis Theodorakis. Seine musikalische Poetik*, Neumünster 2015. von Bockel-Verlag, p. 69. Emphasis HWH.

a device for drafting polyphonic sets, like it is described by [...] Lampadius [...], in 1537, as a tablet, made of wood or stone with a scratched system of ten lines [...], on which – *erasable* – polyphonic passages could be sketched and worked out. Lampadius [...] exemplified as an alternative for it also a tablet, on which the (for the production of voice-material necessary) separation of individual parts [...] in separate five-line-systems was already provided [...].[75]

Such a kind of ‚composing tablet' is documented till the nineteenth century. Sketches demonstrate quite clearly, how a work was created in a process. Once created and completed in the notation form, it exists further in the process of subsequent various appropriations in time. With music it is evident, even if there are no sketches at hand: The tonal realization is something new in comparison to the notation form. In the visual art the work process will be only evident with the receptive appropriation.

Unfortunately is this process in the completed picture, how the Marxist-Hegelian informed would say, "sublated" in the threefold sense: firstly, in the sense of "kept", secondly, in the sense of "uplifted" and thirdly, also unfortunately in the sense of "disappeared". [...] Because the tragedy of the visual art is manifested, that its essential process character in the completed work comes to a standstill and disappears. No completed picture – and it may even be highly accomplished – informs us in any way about the drama of its creation. Painting and sculpture are in the first place living exercises of art, *performances without an audience*, performative acts, which on the one hand cannot exist without their envisaged end, but on the other hand, they themselves are no longer present in the result. Indeed, this is a tragic situation in the classic sense: as evidently for him there does not exist any other possibility, the painter has to follow the path of painting incessantly, though he is conscious of the fact, that in the end he has to face death in the shape of the completed picture [...]. That the history of art, if told as a history of works of art, will always have something in common with a visit of a cemetery, is the inevitable consequence of this.[76]

Photos shot during the various phases while painting a work clarify the progress from vagueness to the sharpened.

(A) series of photographs [show ...] the great painting [...] in the various phases of its making [...]. With the help of such records of state it may be understood, how his way of painting actually advanced from a tonally transparent and constructively indeterminate *Vagueness* to an extensively opaque, representational *determination* [...]. In the meantime contours were shifted incessantly, chromatic values changed constantly. Up to the point where the painter judged to be sufficiently perfect.

Originally the expression ‚sketch' derives from the visual art.

sketch (n.) "a rough drawing intended to serve as the basis for a finished picture," 1660s, from Dutch schets or Low German skizze, both apparently 17c. artists' borrowings from Italian

[75] In extenso Jaschinski 2001, pp. 173f, Fig. p. 174.

[76] Rumold, Lothar: *I./15. Vom Ton zum Gegenstand – von der Linie zum Körper: Bemerkungen zum Werk von Michael Schneider und Doris Eilers. Rede anlässlich der Eröffnung einer Ausstellung mit Werken der beiden Künstler im Karlsruher Künstlerhaus* (25 October 2009) (From sound to object – from the line to the body: remarks about the work of Michael Schneider and Doris Eilers. Speech held at the opening of an exhibition with works of both artists at the artist's house in Karlsruhe), in: Rumold: *Dossier. Über Kunst. Einführungsreden, Artikel in Zeitungen und Zeitschriften, Beiträge in Büchern und Katalogen, zwei Vorträge und eine Art Interview 2003 – 2016*, https://lotharrumold.de/wp-content/uploads/2018/02/Rumold_Lothar_Dossier.pdf, p. 45, retrieved 1 May 2020. Emphasis HWH.

schizzo "sketch, drawing," which is commonly said to be from Latin *schedius ([...] schedia "raft," schedium "an extemporaneous poem"), from or related to Greek skhedios "temporary, extemporaneous, done or made off-hand," related to skhema "form, shape, appearance" ...). But [...] Italian schizzo is a special use of schizzo "a splash, squirt," from schizzare "to splash or squirt" of [...].[77]

In visual art the sketch bears another aesthetic place value than in music, namely often a high artistic intrinsic value, which is even recognizable for the layman.[78] P. Benary reconciles the categories sketch, design and fragment. Here, the fragment may be left aside as far as possible. It is the result of an unfinished process of work. It may be done deliberately like in Michelangelo's „non finito", or in some rather inflationary „Fragment"-titles in the new music since the 1970s. This may be coerced or accidental, through exterior condition like the „death of the author" (the real one, not the ideological of Postmodernism) or inner reasons in Schubert's „Unfinished" („Unvollendeter"). The terms ‚sketch' and ‚design' are not very clearly distinguished. The sketch is shorter, the design is longer.

In common they have incompleteness, provisional nature and supplementation or a need to completion.[79]

Similar to the term ‚design' are

expressions like concept of work, planning sketch, exposé or project.[80]

Thus it oscillates between that, what is here explained as conceptual-, sketch-, and thought form, and even contains some aspects of the imagination form.[81] Here, it can only be hinted at the multiplicity and diversity of variants of composition processes, the types, steps and procedures in the developmental process. Sources, also for what is not directly noted and jotted, are biographies and analyzes, and even anecdotes about, as well as autobiographies and working diaries by a composer.[82] Benary distinguishes the following types of sketches:

1. the thematic-motivic sketch, mostly noted for one voice, with or without (provable) relation to a certain work;

2. the combinatory sketch, with contrapuntal or other combinations of tonal sequences [...];

---

77 https://www.etymonline.com/search?q=sketch, retrieved 4 June 2020.

78 Cf. e.g. Benary 1998, col. 1506f.

79 Benary 1998, col. 1507. Manuscripts of workshops were called, for instance by Beethoven a „barely first concept [...] whereby, how it sometimes happens, a few things are still incomplete and not correctly indicated" (Letter to Adolph Martin Schlesinger in Berlin, 20th February 1822, *Beethoven Briefe* Bd. 4, edited by Sieghard Brandenburg, München 1996, No. 1458, p. 474, quoted in William Kinderman: *Zum Selbstverständnis der genetischen Textkritik bzw. textgenetischer Projekte*, https://beethovens-werkstatt.de/prototyp/expertenkolloquium/zum-selbstverstaendnis-der-genetischen-textkritik-bzw-textgenetischer-projekte/, retrieved 5 June 2020).

80 Ibid.

81 See ibid., col. 1509f.

82 See e.g. Bahle 1936 (*The musical creative process: Psychology of creative forms of experience and drive*); Stürzbecher 1973; Henze 1983, 1984, 1996, 2007; Danuser/ Katzenberger 1993; Gratzer 2003; Herwig et al. 2007; Gratzer/Neumaier 2013.

3. the trimmed sketch with or without any recognizable relation to the work [...];
4. the [...] sequential sketch, it cannot be distinguished from a design.[83]

In this sequential sketch, for instance Bruckner left often „eight or 16 bars for a future ‚filling in'.[84] Thus, he conceded, in the whole process of sharpening as its part, open spots of emptiness and indefiniteness even below the level of vagueness. But they are through the context, by their before and after, not totally empty. An important method for researching the creative processes is the sketch research, being a part of music philology. It was initiated mainly by Beethoven research, since around the 1860s, because from him, there are in particular many and even systematically collected sketches available. Present tendencies shift the accent from work towards the process, namely in the „genetic text critique",

> which on one side originates [...] in musical sketch research and on the other side incorporated methodic and theoretic impulses from the predominantly orientated *critique génétique*, which in the 1970ies was founded in France. The aim of genetic text critique is the research of composing work and thought processes – the genesis of compositions. The starting hypothesis states, that compositional writing is identical to compositional thought.[85]

Unfortunately this is simply misleading.[86] Some of the previous examples demonstrate that such thoughts may be even produced without writing, and in fact fully valid.

Again, to put it right:

> compositional writing is not only the conserving writing down of musical inspirations, but takes place as a dialogue between writing and (re-) reading of what was written.[87]

Such an idea of ‚dialogue' carries forward, but has to include the mentioned non-written procedures. It takes place within a communicative quadrilateral of notes (doubled as notes or verbal jottings), ‚sketching' on an instrument, sketching in the mind and the mentioned re-reading. Inside such a frame there are loops or zigzag movements.

> But compositional processes of thought and work are not instantly recognizable by a comparison of workshop documents (sketches, designs, work manuscripts, copies, galley proofs, authorized print editions et cetera), because their scripture or graphic manifestation form reveal only the ones that were compositionally achieved, quasi conserved as static textual (intermediate-) products, but not the respective dynamic process, which directly leads to them. Genetic text critique gains only insights through the re-dynamism of diverging textual,

---

[83] Benary 1998, col. 1508.

[84] Ibid.

[85] https://beethovens-werkstatt.de/glossary/genetische-textkritik/, retrieved 5 June 2020; cf. https://beethovens-werkstatt.de/, retrieved 5 June 2020.

[86] According to W. Kindermann one „may perceive compositional writing and thought only as ‚identical', if the context and the sense of writing is deeply researched and exposed. But this was not the case in the first prototype of the research project *Beethovens Werkstatt*" (https://beethovens-werkstatt.de/prototyp/expertenkolloquium/diskussion-teil-5/, retrieved 5 June 2020).

[87] Ibid.

„static" intermediate steps about the processes of writing and creating, on which they were actually based.[88]

This research attempts, with an extensive set of partly tried and tested traditional, partly tried and tested new methods, to reconstruct the developmental process from the conceptual till the score form as far as complete.

Handwritten workshop documents contain on the one hand texts of notations in different conditions of maturity and on the other hand scriptural signs, which are inseparably connected with the proceedings of writing. In genetic text critique these both manifestation forms of composition (product and process) are separated and at the same time they are set in relation to each other: different, successive textual stages (text stories, text stages) are reconstructed and set in a micro-chronological sequence, and finally the existing differences between such temporary textual intermediate products are interpreted in regard to writing processes on which they are founded and connected. The pair of working steps, which are here separately mentioned, by means of which an only rarely complete possible reconstruction of the processes of writing and working is attempted, are in text-genetic practice circularly interlinked with each other: the reconstruction and micro-chronologic sequence of text layers and stages depends on the interpretation of explicit and implicit meta-texts, and the reading of implicit meta-texts demands vice versa the reconstructive separation of text layers and stages.[89]

Further, subtle techniques are instrumental to close gaps and thereby—lastly to diminish unavoidable –fuzziness. For this are not at least the already mentioned „secondary references (e.g. letters, diary notes, contemporary credentials)"[90] of importance, as they assist in completing sketches not firsthand fixed by writing. For this reconstruction and construction of a ‚segmented continuum' in the scope of the work's sketch form, actually in general a discontinuum, affinities to Fuzzy Logic are evident. These would probably contribute to a methodical sharpening. The question of a re-accentuation remains.

In difference to philology of editions, which is directed towards the constitution of a performable text of a work, the genetic text critique is fixed on the process of production, on the gradual manufacture of musical thoughts while writing […].[91]

In the end, it is the question of „textual movements" as such, instead as movement towards a binding text, towards on, that is an „in itself completed whole" in notation form, as ‚score' in a wider sense. Admittedly, it is firstly somehow undialectical,

---

[88] Ibid.

[89] Ibid. The multiple repeated hints are convincing, that the textual genesis with a—simulated—digital three-dimensionality are easier feasible than on a two-dimensionality of the paper. The incorporation of coloris a further ‚dimension '—in print expensive, but digitally no matter. Cf. e.g. Almuth Grésillon: *Critique génétique und génétique musicale*, https://beethovens-werkstatt.de/prototyp/expertenkolloquium/critique-genetique-und-genetique-musicale/, retrieved 5 June 2020. But also convincing is, that the preconditions of such a presentation have to be correct: "A digital presentation of a sequence of variants has little value, if the transmissions are not secure and the selected sources are not researched in the context with other manuscripts" (William Kinderman: *Zum Selbstverständnis der genetischen Textkritik bzw. textgenetischer Projekte*, quoted ibid.).

[90] Ibid.

[91] Ibid.

to play the becoming and what is to become off against what became.[92] Secondly, as important might be the knowledge of the genesis to understand the structure, nevertheless, the work—or if possible its various—versions too, is right from the beginning envisioned end of the evolutionary process. Without such an end, being already conceived in the conceptual form, it would never happen at all. As such a „telos“ cannot be detected in every detail, is for now of secondary importance, and likewise, whether it may be sharp or fuzzy. The score form realized in „classics“, for what is here of concern, the reach of the end—certainly, if blurring of the asymptotic approximation between ideal and reality in realization of the work idea are taken into account. (3) Finally, the negation of the work as a result of the becoming, in favour of the becoming as such is one-sided. Because even a work realized in score form is actually a step of metamorphosis in the work process, just in its forms of existence. Even before the transition into the next form of existence, the sound form, intermediate steps within the notation form are inserted. Because even with completion of the ‚score‘, together with the training for what will be necessary for the performance the process of evolution has not yet arrived its final stage. It continues, on the basis of writing, with various steps of corrections, until later setting and printing, and moreover at an already completed work (see Sects. 6.6 and 6.7).

The claim, that no binding text would exist, is close to the assumption, that truth does not exist. Indeed, not an absolute, but a relative one. Here, obviously Fuzzy Logic may have something to say. Especially, if a series of sketches is at hand, the approximation process towards the completed shape may be researched by Fuzzy Logic even quantitatively. The comparison of different textual stages, the search for similarities and differences, the incorporation of a likewise re-constructional option for a work idea, together with the conceptional, thought, and anticipation forms, indeed are opening up the possibility, to find an authentic, suitable with the composer's intentions and ideas, work-shape—at least in asymptotic approximation. If there are really a number of versions of a work on the same level, there would be no reason, to deny the existence of the work,[93] as also in the multitude of versions there is truth. As far as written sources exist, there is plenty of matter for comparisons, both for between sketches and a completed work, as well as between the various authorized final versions of the work—as an example for a work‘s multiple truth.[94]

[92]See e.g.: „For *critique génétique* it is important, that the text or the work is not placed in the centre, that means something static, in principle completed, but a dynamic process, which does not run in straight lines-linearly, but partly volatile, chaotic, unpredictable“ (Grésillon, ibid.).

[93]See Almuth Grésillon: *Critique génétique und génétique musicale*, https://beethovens-werkstatt.de/prototyp/expertenkolloquium/critique-genetique-und-genetique-musicale/, retrieved 5 June 2020.

[94]Exemplary Gail 2009.

6.4.3 *"Writing" as Mere Dictation of Thought and Spontaneous Transcription*

In general, just fast writing often demand slowly and long preparations and work in advance, such tasks are either realized in written form or only mentally. So, for instance, Shostakovich:

> I compose very easily, but before I start, I have to think about it quite a while.[95]

Max Reger „wrote fair copies instantly",[96] well, on the basis of intensive preparations together with much thinking ahead. Once, in 1903, his student Alexander Bersche came too early into the class. Reger remarked:

> „So, now you have to wait, […] but the few bars, I will never skip." […] A huge score lay in front of him, and clear and bold strokes of his handwriting flew over the lines with an astonishing speed. Then murmuring partly to himself, partly towards me: ‚Well, here, the clarinets may seemingly follow up, but a particular counterpoint you will not get" […]. In this way he continued, quite calm and with the confidence of a *playful child*.[97]

Here, instead of jottings and sketches are given ad hoc verbal commentaries, which serve for a final sharpening of the idea's realization. These, together with the main features of the implementation, Reger had developed in his mind, as well as other composers, stored them in the mind and afterwards wrote them down.

> […Then] suddenly I realized, that there were no sketches … in front of him. „Well, Mr. Reger … You are here without a concept?" „Concept? That is stored in my memory." He really wrote it down in fair copy.[98]

Such not written down and for this reason at last after the producer's demise no longer tangible mental components of the evolutionary process complicate its reconstruction. Similar effects turn up, although motivated by other circumstances, in case of gaps and blurring through neglectful dealings with sketches. But Brahms was the opposite. He used blurrings up to the point of destruction and burnt all of his sketches, including uncompleted or fragmentary works, which were in his hands. A good deal is, like in Reger's case, at least indirectly, and as if through a peephole visible. Likewise only indirectly through verbal reports tangible is the improvised, groping composing at the piano. And, it may be undramatic or dramatic. Dmitri Shostakovich:

> I may compose without a piano, and even […] prefer […] this method. I try, to write music with my head […] and not with my hands.[99]

Stravinsky executed *and* sublimated the composing at the piano. His *Piano Rag Music*, 1919, is

---

[95] Schmitz/Ure 2019, p. 1044.

[96] His student Alexander Bersche, 1942, quoted in Schmitz/Ure 2019, p. 1046.

[97] Ibid. Emphasis HWH.

[98] Ibid.

[99] Schostakowitsch, quoted in Schmitz/Ure 2019, p. 1044.

> based on the «percussion possibilities of the piano» [Strawinsky: *Chroniques de ma vie*], pointing also to the importance of the physical-motoric aspect (*fingerwork) as primary inspiration source* of the musical thinking and composition.[100]

The work is dedicated to the pianist Arthur Rubinstein,

> whose „strong, agile, clever fingers“ came to mind when writing the piece.[101]

Brahms played *and* sang while making non-written sketches at the piano, as Kalbeck reported. He was „auricular-witness“ of Brahms' „inspirations“,

> which he, in all probability *before the first writing down*, confided to his discreet walls. Even there, the demonic touched the artistic in a strange way. One morning on a visit [...], as I was about to step in through the widely opened backdoor, I realized, that also the door to the music room was unshut. At the same time a marvelous piano-piece was played, by which I was held at the doorstep. It sounded like free fantasias, but by the frequent changed repetition of certain parts I became conscious, that was Brahms going through the work of a new composition, he had already completed in his memory, in order to improve and add polish to it. H repeated the piece many times in separate parts until he played it finally without a break. The pleasure could have been overwhelming, and might have surpassed the interest for the progress of his works, if the solo-part was not transformed into a most strangely duet. As the work became richer in its character, the presentation increased emotionally, accompanied by a constantly heavier strangely growling, whining, and moaning, that on the climax of musical intensification degenerated into a load howling. [...] After about half an hour the play together with the howling stopped, the armchair at the piano was moved, and I entered the room. [...] Brahms showed himself quite embarrassed [...] – he must have heavily cried, for the shining droplets still hung in his beard, and his voice sounded soft and shaky. I pretended, as if I had just arrived and had nothing noticed. Soon he became cheerful again and made a few jokes, and he played for me fugue by Bach.[102]

For Chopin a mixture of composing at the piano and with sketches is reported, likewise with dramatic, but even with epical aspects. Inspirations came especially in relaxed condition, so when walking.[103]

> His creative activity was spontaneous [...]. He found thoughts, without seeking them or to expect them. At the piano suddenly an inspiration occurred to him [...] or while on a walk it sung inside him, then he was in a *hurry, to play his thoughts on the piano.* But afterwards the most painful work started [...].[104]

The „whole“ is already contained within the inspiration. But it takes the strenuous course through the work on details.

---

[100] Angela Ida De Benedictis: *The "dual" form of Igor Stravinsky's Piano-Rag-Music: Musical structure as result of a rearrangement process*, CIM08 Abstract Proceedings, Thessaloniki, 3–6 July 2008, p. 47.

[101] Strawinsky: *Chroniques de ma vie*, quotation in David Truslove: 8.570377 – *STRAVINSKY: Music for Piano Solo*, https://www.naxos.com/mainsite/blurbs_reviews.asp?item_code=8.570377&catNum=570377&filetype=About%20this%20Recording&language=English, retrieved 28 November 2019.

[102] Kalbeck 1915, ibidem, pp. 247f.

[103] Georges Sand 1865, quoted in Schmitz/Ure 2019, p. 1042. Emphasis HWH.

[104] Ibid.

> There was no end of impatient indecisive attempts, to seize certain *particulars* of the theme, as he had them *mentally heard*. What he had *conceived as a whole*, he analyzed deeply when writing down [...]. He [...] broke the quills, repeated, changed the beat a hundred times and cross it out again just as often [...]. For six weeks working on one page, in order to write it down, *finally in such a manner,* as he had *sketched in the first draft*.[105]

In the end, the composer returns back to the first version. But, it went through a number of arrangements, and has proven its worth against alternative versions. Thus it will be something different than the first version—an almost Hegelian spiral movement.

Here again, within the triangle of mental shape, sketch and completion a good memory is necessary, in order to keep in mind, what was worked out through improvisations and experimentations. Generally the complexity increases in the course from sketches till the completed work. There are no exceptions possible. There are also simplifications, withdrawals of complexity.

> Ives composed most of the *Three Places* during the years circa 1908-1914. Ives completed the final score for full orchestra in the years 1913-14. However, *the final score was considerably simpler than his original sketches,* most notably in "The Housatonic at Stockbridge."[106]

Ives has for the work's II. Set some already existing works assembled—a kind of collage and montage, which is apt for Ives' general proceedings *within* his works.

> The second piece, *Putnam's Camp, Redding, Connecticut* was created from two short theater orchestra pieces composed by Ives in 1903. These pieces, *"Country Band" March* and *Overture & March: "1776",* were completed in 1904. Lyman Brewster, Ives' uncle, had asked him to compose the pieces for his play *Major John Andre* which was never performed due to Brewster's untimely death. In the early fall of 1912, Ives began tinkering with these compositions again.[107]

Although collage in principle means complexity, here Ives has simplified.

> By October, Ives had completed an ink score-sketch of Putnam's Camp. The final version of the piece clearly resembles its source materials, but many of the complex musical jokes that littered the originals had been replaced with simpler alternatives.[108]

Nevertheless, simplification was not Ives' last word in this case.

> In 1929, Ives rearranged the work for chamber orchestra for Nicolas Slonimsky's premiere performance. Ives *restored many of the complexities from the original sketches* that were not included in the version for full orchestra of 1913-14.[109]

---

[105] Ibid., pp. 242f.

[106] Scott Mortensen: *Orchestral Set No. 1:* Three Places in New England, 2002, https://www.musicweb-international.com/Ives/WK_OS_1_Three_Places.htm, retrieved 2 May 2020. Emphasis HWH.

[107] https://en.wikipedia.org/wiki/Three_Places_in_New_England, 28 February 2020, retrieved 8 May 2020.

[108] Mortensen 2002, ibidem. Scott Mortensen: *Orchestral Set No. 1: Three Places in New England*, 2002, https://www.musicweb-international.com/Ives/WK_OS_1_Three_Places.htm, retrieved 2 May 2020.

[109] Ibid.

Fig. 6.18 Fago: *Kyrie* (from *Fughe*), excerpt
Quoted from Sanguinetti 2012, p. 62

6.4.4 Accelerations on the Way to the Score. Partimento and Particell

Short-cuts of all kinds are characteristic on the long march from inspiration to the completed work, as thinking, musically and compositional included, is done essentially faster than writing (see above: Stenography). Thus, typically for sketches is a „hurried or shortened notation" down to „abbreviations", which are only individually used, and hard to decipher.[110] ‚Writing' means at least something threefold, as it was stated before, words and notes, sometimes also pictures and graphic representations. Such a trinity appears practically in all types of notation. The principle of short-cuts is concurrently mostly a sharpening intensification of the crystallization processes, but it generally incomplete and an abridgement. Here, the multiple and repeated interplay of mentally and externalized is typical, of processes inside the mind and on the piano (or with any other music instrument) and on paper or nowadays on the screen. In such an interplay the oral is bound up versus the written, and thereby also the intertwining of ‚implicit' and ‚explicit' knowledge. High-spot and last-stop of the sketches as the externalized start of the course, is the running through, the essentially complete design, the ‚continuous draft'. This is a possible, but no necessary and realized form of sketching. Further main stages of notation in this process after the sketches, in a narrow sense, represent partimento and particell, which are here to be presented more closely.[111] Now, the route is forked. In the work process both forms are alternative, but not successive, and they have a different historical age. Both forms have in common, that they present already the work constantly as a whole, but nevertheless still incomplete. Therefore they belong to the sketch form. But, particell stands considerably nearer to the completed score form and may already be understood as—a not fully executed—abbreviation of the score. *Partimento* means formally: One single staff, different clefs[112] (Fig. 6.18):

[110]Benary 1998, col. 1508.

[111]Cf. e.g. Jaschinski 2001.

[112]See above Fig. 6.5 the working with different clefs as playing with them in the case of the One-Note-BACH.

> partimenti, a pedagogical device [...] developed in Italy, and particularly in Naples, and shaped the musical mind of innumerable composers during the eighteenth century and part of the nineteenth all over Europe. [...] Characteristic of partimento notation is the *usage of different clefs*. Obviously, the main reason for that is that partiment is *written on a single staff*; but there are other reasons. Through a clef change the author may signal a thematic entry, a textural modification, a tutti-solo alternation, and many other events; in other words, clef changes are an important medium for conveying instruction concerning the realization.[113]

The usage of different keys (clefs) and the clef change goes back to the tradition of the mensural notation. Among other things, it saves space and here also time. Additionally figures are used for the chords, like the thorough-bass, or Basso continuo.

> Partimento became an *alternative notational system, as opposed to today's more familiar two-stave*, fully notated score; this latter system was called in Italy *intavolatura* (as the antonym of partimento; the term intavolatura has nothing to do with tablature). Obviously, partimento notation leaves ample space for improvisation. Indeed, unlike intavolatura, it *needs improvisation* in order to become music.[114]

The procedures are effective against one-sidedness and narrow specializations and it does not save the interpreting performer to work spontaneously in an imaginative way and in analogy to composing. What is blurredness from the viewpoint of the work and the score is sharpening from the viewpoint of tonal realization.

> It must have become evident that, besides being an efficient notational system — though only for highly trained professionals — partimenti also possessed remarkable potential as teaching tools. Through the practice of partimenti students could gain proficiency in a variety of fields: continuo playing, improvisation, unfigured bass, counterpoint, diminution, and fugue. Some of these fields were related to performance, but others — unfigured bass, fugue — were clearly of great interest for students of composition. Besides, all these *skills* entered the students' heads, so to speak, *through their hands and not through their eyes*. So, exactly as a performing student must develop a (quasi) instinctive response to the notational stimuli in order to obtain real proficiency on her or his instrument, likewise a composition student, using partimenti as a tool, could develop a (quasi) instinctive response to compositional stimuli. The faculty of developing *a kind of automatic composition* was the greatest advantage of partimenti as a teaching tool. Another benefit was that a composer trained in partimenti could gain unparalleled fluency in composition, and astounding rapidity.[115]

Thus the partimento-technique greatly accelerates composing.

> Every professionally trained Italian composer was able to compose an opera in five weeks, or less if necessary. This was the result of training, and not of genius. Partimenti were perhaps the most efficient composition exercises ever devised; for this reason generations of teachers and students used them for more than two centuries.[116]

Admittedly as most of the times, the increase in speed demands for the application an extension of learning, converted into the whole span of life, it accordingly means a slowing down of the actual work process.

---

[113]Sanguinetti 2012, p. VIII. To the partimento detailed see also Rabinovich/Slonimski 2015.

[114]See Sanguinetti 2012, p. 5.

[115]Ibid., p. 6. Emphasis HWH.

[116]Sanguinetti 2012, p. 7.

> The flip side was that partimento training was long and weary, and it needed a teacher — a very good one.[117]

Additionally a strange archaism becomes evident in the procedure, which corresponds to the time spent for the acceleration of production through manufacture and mechanized industrial system. The training demanded a personal, oral instruction.

> The extreme flexibility of partimenti, especially in the advanced stages, and the subtleties of their realization, made it impossible to commit to paper every aspect of the practice. Instead, partimento practice could flourish only in a context of a continuous oral tradition.[118]

Thus, the written rules were on the meta-level of the composition- and notation method not sharply enough specifiable—a dialectics of the written, how it is generally valid for the notation. Yet, even the writings can only partially emancipate from personal contexts.

> During the era of Bach and Mozart, Italian composers, theorists, and teachers also failed to write treatises. They did not even write essays, pamphlets, or — least of all — articles. They did, however, publish a number of treatises of a very peculiar kind — *almost wordless*.[119]

Like in practiced writing, in teaching materials as well, the word is stepping back. The treatises abstain from abstracted, generalized and generalizing rules and offer instead concrete case studies as models and examples. The combination in each notation of word and note (concurrently representation of the ‚picture') is shifting unquestionably in the direction of notes.

> Instead of presenting their ideas verbally, Italian musicians assembled large compilations of especially composed examples, with only a few intermittent verbal remarks here and there. Moreover, these Italian masters preferred to circulate their teaching material in manuscripts, rather than in published form [– …] an esoteric doctrine; it was for insiders only […].[120]

On the whole, the acceleration and swiftness of composing suggested a serial production, where the detailed, filigreed elaboration of the individual work did not stand in the centre of interests. A work of art as „a completed whole in itself" was therefore probably more often created inspite of this method. There are besides the common procedures, such as preemptive piano score, partimento or particell individual aids and proceedings, to accelerate the process of production, like grammalogues or kinds of notation as in the case of stenography. A decisive factor could be, on the one hand, fantasy and powers of imagination, above all from the future whole, on the other hand an outstanding memory above all in case of remembrance for details in sketch form. Just like C. M. v. Weber. He seems to have used an individual variant of the partimento-technique.

> He was in the habit early in the morning, after an extremely frugal breakfast, standing at his desk for the first writing down of his ideas, while he dedicated the hours of the evening for

[117] Sanguinetti 2012, p. 6.

[118] Ibid., p. 10. Emphasis HWH.

[119] Ibid., p. 10. Emphasis HWH.

[120] Ibid.

> the work of instrumentation and complementing the connecting elements. Mostly sketching at the first writing down *only the whole of the vocal parts, – at times without bass* – and only here and there hinted at a harmonic sequence, or the entrance of wind instruments, often in the most fragmentary form, well at times he himself gave, *only by signs known to him*, the marvellous, for him so characteristic effects of the orchestra. By this form of the mental function for the composition can be explained the enormous quantity of work, which at times he was able to produce. As further on will be more clearly revealed, the whole, complete, first act of »Euryanthe« was in 12 days, while Weber from Hosterwitz coming in between had to fulfill his entire duty in Dresden, the second in 13 days, the third in 15 days instrumented. Days, when he worked out 12 and 13, even 14 pages of this copious score, are not unusual in his diary notes. The whole elaboration of the entire opera was according to the sketches in no more than 60 working days done.[121]

An entirely different form than the partimento is the notation in form of two systems staves, known above all for piano music. The piano score was since about 1849 with the maturing of the concert as an institution generally *afterwards* as abbreviated reproduction of the score edited, for the laymen to play or read, the 'reduced score' or 'piano score' There are different forms. The basis was the reduction of the set for the orchestra (rarely for chamber music) on the two piano-systems staves. The types represented, on their part, different grades of blurring of the completed whole: with the singing voice or even the entire choral setting, a piano set without vocal part with written text only between the pair of staves, or purely instrumental.[122]

The piano score may be seen as mode of notation with two or three systems but also preliminary and intermediate stage on the way towards a fully completed score.[123] For itself it can count in this function once more as an earlier form of the *particell* = condensed score.[124] Particell notation has the form of a piano score, but with remarks and entries to the instrumentation. In the *particell* the structure of the score is already determined. However, the individual parts have not been stratified and dispersed to single staves as in the score. In principle, it shows the composition in chords or polyphonic, but still without instrumentation, so to speak, a body, whose dress is only fitted by pinning, but not yet entirely tailored. Such a completion is only delivered by the score. Debussy's symphonic poem *Prélude à l'après-midi d'un faune* (*Prelude to the Afternoon of a Faun*) after the poem with the same title by Stephane Mallarmé was his first important and influential work for orchestra, composed at the same time as his first important work for the stage, the opera *Pelléas et Mélisande*. For the *Faun*-particell Debussy uses mainly 4-staves (Figs. 6.19 and 6.20).

[121]Weber 1866, pp. 459f.

[122]Cf. Jaschinki 2001, pp. 182f.; Benary 1998, col. 1509f.

[123]Historically early, already in Renaissance, cf. e.g. Apel 1962.

[124]In detail Jaschinski 2001, pp. 182ff.

Fig. 6.19 Claude Debussy: *Prélude à l'après-midi d'un faune*. Particell. Holograph ca. 1891–1894, m. 10–12
https://imslp.org/wiki/Special:ImagefromIndex/411013/qruu

Fig. 6.20 Claude Debussy: *Prélude à l'après-midi d'un faune*. Full score, Autograph, m. 10–12
IMSLP

6.5 Work as an Idea Completed in the Composers' Mind. Thought Form

The work with and on sketches means above all working at details, gradual development of musical thoughts while reflecting and thinking ahead and then writing it down. But this is valid both for the previous conceptual form as well as for an anticipating imagination of the entire work, which is often already contained and encapsulated within the conception. The thought form is an independent existence form of the work. It mediates within the work process, on the one hand, between the stages of conceptual form a sketch form on the one hand and on the other hand between score form and imagination form. As only a subjective form of existence it is hard to grasp. But at least its contours are made visible[125] through sketches, above all in written notations.[126]

The actual production of the whole happens within the poles of a „writing down" as just a dictation of what is thought and a spontaneous writing down or in a conscious, strenuous, multi-staged production process with the work-whole as its final result. Mostly both are interconnected. As soon as Wagner had finished his copious and deviating preparations, he wrote his works in fair copy so swiftly, as he would just copy them, Schönberg, under favourable psychic states and general set-ups with the drive of inspiration and conception could very fast compose; contemporaries, like Hindemith, were even faster. The quick call and „copying", so to speak, from the mind is not bound to the medium of writing. It is also possible within the medium of the instrumental; as it was the case with C. M. v. Weber.

> Despite the fact that these sketches were utterly incomplete, [...] he played frequently [...] large sections from works, which he was preparing and had sketched, exactly in the form in which they were later published. Once a music piece was ripened, it was in his mind stereotyped [printed, like a photography] as it were of indestructible streaks, so that the *instrumentation, so to say, was made by him, instead of being the work of a copyist*, and out of such extremely insufficient sketches, he created as if pearls were running down the quill, without any intermediate work whatsoever, the flawless scores, from the flute tones till the bass were fully complete, with all signs, pauses, pianos, fortes, just like engraved in copper, perfectly written.[127]

6.5.1 „Always Having an Eye for the Whole" (Beethoven). „Gradual Generation" of Thought and Anticipatory Completion of the Work

Beethoven emphasized, he kept while working on the seventh symphony "always the whole in view", even when composing instrumental music. This he wrote, in spring 1814 to Georg Friedrich Treitschke, the collaborator when he was revising Fidelio. For him, such a "whole"[128] surpassed the whole of a set or work, and lastly was even related on his social surroundings. Beethoven bases his thinking ahead of the future whole on a material foundation, in the sense of the word. He created for himself, what is called 'writing space' by the genetic text critique:

> Physically present, with a single look one recognizes a space for writing and reading, where the composer actually writes and consequently is thinking. It may be just a single page or a sheet of notation paper (double-sided) or a number of codicological details (for instance, on the writing-desk there are several opened up sheets, sketchbooks single leaves). This directly in view focused *writing space represents a place for thinking*, in which musical contexts, being of relevance for the actual compositional work, may be perceived and worked out

[125] See among others, cf. Benary 1998.

[126] See ibid.

[127] Weber 1866, pp. 459f.

[128] Extensively on this, e.g. Gülke 2000.

> together. The conditions of the respective opened writing space are occasionally materialized in a special topography of the drawing-up, whereas, for instance, certain sections of form (like exposition, realization etc.) are always to be found at the same position of the writing space [...]. Casually, Beethoven in sketchbooks presents an enlarged writing space: then, one of two opposite laying leaves is folded, in order to take a view at the same time on three different pages and read any of the notes they contain [...].[129]

The writing space, in turn, is embedded in a working room. Its' furnishing circles between the poles of chaos and order. Both may advance the developmental process of the work. This was the case for Beethoven; Carl Maria von Weber, his student Benedikt and the publisher Haslinger visited him in 1823:

> (the) desolate, almost shabby room [...] was in utmost disorder. Music, money, pieces of clothing werescattered on the floor, on the dirty bed were clothes heaped up, the opened grand piano coatedwith lots of dust, broken coffee things on the table.[130]

Out of such a chaotic, collage-like working and writing space evolves, as is well-known, absolutely a harmonic, homogenous whole. The conceptual form as terminus a quo, as point of departure, and the thought form as terminus ad quem, as end point—but in the entire work process an intermediate stage for the notation form or sound form (for not-notated music)—are together steering a structuring the configuration of parts to become a whole. Charles Ives as well, produced his works as puzzle, patchwork and palimpsest and arrived finally to produce a convincing *whole*, not at last, guided by programmatic and by within the material objectified ideas. Ives has downrightly experimented compositionally, long before the „experiment" as a statement of the music cultural Up-to-date became fashionable after 1949. He created with combinations, permutations, reductions, amplification and so on of motives and melodies a kaleidoskopic cosmos of crossings and interconnections. In its occasionally chaotic gracefulness one may dissect the lastly well-structured, certainly at times hypercomplex order. One of the probate deep-structural means for connections between motives or ›citations‹ of different types and grades, for the communal in diversity, are »ordinary motives«. Their plasticity and capability to extend make them structurally to a kind of ›joint‹ or hinge, and they function, on the other hand, also semantically like musical ‚key words'. Just like the trichord e–f#-d: expanded with the deeper laying a, with a turn to f#-d-e-A, results in the Westminster Chimes, which itself, if complimented by an h (b) converges with the *Nearer, my God to Thee* f# e d d B B.[131]

---

[129] In detail https://beethovens-werkstatt.de/glossary/schreibraum/, retrieved 20 August 2020. In the case of J. S. Bach was the writing space less convenient. For preparation Bach still had to draw the 5 lines of notation at first, even by using the mechanical aid, called rastal, to draw them together in a single stroke. Then he had to use the paper very economically—in principle, like Beethoven too. That he took down his notes on the remaining open areas for writing, for instance, also for the sketches of various works, may have been the usual way in those days. But for researchers of sketches this economical behavior increases the lad of work.

[130] Weber 1866, p. 510.

[131] See e.g. Gail 2009.

But it may also happen faster. With a lightning as enlightenment, a usual topos for idea and inspiration, Hindemith, in 1959, compared the sudden appearance of the whole in thought form, as a

> vast landscape, not alone by its general contours, but with every detail. […] Musical compositions must be seen in the same way. Someone may not be called a real composer, to whom in a sudden flash of a creative moment a piece of music will not appear *in its entire wholesomeness*, with each of its structural links at the right place.[132]

Construction and reconstruction of an idea are often interlinked. In the following case something like lightning which is often ascribed to inspiration in the context of the conceptual form, already existent in the thought form. Because considerable parts of the works already existed, but the notation and sketches were lost.

> I had decided, to follow the way of the all-encompassing harmony, which music revealed for me to the end. But this music rested hidden behind hills and a heavy fog, which covered my soul entirely. I had to clear a path. But studies and my own thoughts were not insufficient. The music was not only contained in notes on paper and in the sounds of an orchestra […]. It was my other self, which I had to discover. A self that had knowledge of the secrets of harmony. Perhaps, self-destruction could lead to my true self. Perhaps. In any case, such a temptation produced inside me numerous sounds, which I did not hear at that moment. They were stuck somewhere inside of my soul, waiting to be discovered and blossom. […] So was that night for me filled with music. […] But my heart at the centre of the cosmos rested. Such a human heart, tortured with fright, pounding constantly louder, constantly faster, till the moment, when the brain suddenly, in a fraction of a second, grasped the incomprehensible: the all-encompassing work of music, whose melodic material escaped from the nightmarish landscape, the harmonics of the night and the rhythm from my heart […]. Thus the 1. Symphony was *created in a single moment*. […] It was *in the speed of light with its final shape* memorized in the fourteen billion cells of my brain. […] The mechanism of thinking is quite strange. When, in 1948, the storm raged over our camp on Makronisos, all of my musical sketches got lost. Including the material for the symphony. But *since then the main themes went through my head and appeared in my dreams.* And now, [like in] a moment before death, when, as people say, the whole life runs down like in a movie, I feel directly from the galaxies a rain of thousand sounds pouring down on me. […] Those musical themes, which were snapped by the wind during the storm in January recovered their shape, were combined crossed into clever counterpoint developments. As if they had been expecting this very moment to blossom.[133]

6.5.2 Disappointment. The Reduction of the Infinite Possibilities to the Only One Realized Possibility

In the end, actually it looks quite different again. The completed work is not perfect. There is a wide difference between what is possible and what is real, what was wanted and what was realized.[134] The potentially unlimited possibilities are with the

[132]Quoted in Danuser/Katzenberger (eds.) 1993, p. 14.

[133]Theodorakis, Mikis: *Mein Leben für die Freiheit*, Bern, München, Wien 1972, quoted in Folkerts 2015, p. 69. Emphasis HWH.

[134]Cf. e.g. Benary 1998, col. 1510f.

completion of the work necessarily reduced to a single realization (see Sect. 1.3.2). The relation or disproportion of grandiose idea and limited realization leads to disenchantment, disillusionment, and a certain frustration, which can also be explained as „narcissistic insult". The plunge into reality of realization is related to the height of fall more or less painful. As in the case of Tartini's „Devil's Trill Sonata".

> He later stated that his sonata was but a shadow of what he had witnessed in the dream, for he was unable to capture on the page the Devil's full intensity.[135]

In this case, the abyss of fuzziness between dream and reality gets so deep, that Tartini lastly for pragmatic reasons stuck to his job as breadwinner.

> Tartini later wrote that it was "so inferior to what I had heard, that if I could have subsisted on other means, I would have broken my violin and abandoned music forever."[136]

This seems to be true, and it is just a piece of the myth about artists. Under sober conditions D. Shostakovich complained:

> Towards the end of work it give me great pleasure, but it will be soon lost again, if the work is actually finished. Then it gets somehow boring.[137]

Schiller mentioned, the artist may

> esteem himself to be happy, if he reaches by the clear consciousness of his operations a point, where he can recollect the first dim total idea of his work in the completed work unweakened. Without such a dim but nevertheless mighty total idea, which precedes anything in technics, no poetic work can be achieved.[138]

The thought form is an intermediate stage and an aid. Thus, it may after the writing down, within the memory as a fair copy not only be blurred, but even totally wiped out.

> Well, in those days my head was in turmoil and worked so incessantly in musical ideas, that I personally on my ways to my students, as well as on walks, constantly composed and thereby soon found experience, to elaborate long periods, even whole *pieces of music completely in my mind*, which then written down without any further aid. Once this was done, they were deleted from the memory, and, I had regained space for new musical combinations.[139]

---

[135] Betsy Schwarm, https://www.britannica.com/topic/The-Devils-Trill, retrieved 4 August 2020.

[136] Gómez, Mario: "*The legend of Tartini's "Devil's Trill Sonata"*", Violinist.com, retrieved 21 February 2016, see also in https://en.wikipedia.org/wiki/Violin_Sonata_in_G_minor_(Tartini), 7 July 2020, retrieved 14 July 2020.

[137] Schostakowitsch, quoted in Schmitz/Ure 2019, p. 1044.

[138] Schiller to Goethe, letter from 27 March 1801, cited in Benary 1998, col. 1507.

[139] Spohr 1968, p. 170. Emphasis HWH.

6.6 Preliminary and Final Stage of the Work as a Notationally Completely Realized Idea. Anticipation Form and Score Form

There are two main stages in the transition to the work as fully realized idea in writing: The work as abstract composition and the work as completely articulated and notated composition mediated by the anticipation form. Both result in the score form. This corresponds with the historical state of the art since the French revolution. Since about 1800 the "score" appears as "complete referential text of a work".[140] Between 1789 and 1848 the publication of a score has become an obligatory standard. Somehow disparate with common usage of the term this includes editions of piano works, songs, etc.[141] In the score form a work reaches a *point de perfection*, the fully compositional completion of a work—in the case of notated music (see Chapter 2).

For improvised music this point of completion is much harder to define. It is necessarily transitional, but may manifest itself as a paradoxically 'recurring singularity' in a particularly successful performance. The qualification and then also quantification of fuzziness is complicated here. It becomes considerably more feasible with the technical reproduction of the singular in phonographic form. This allows for a gradual, step-by-step approximation of the intended ideal of a work, through repeated 'takes' of the same improvisation work. The 'takes' are rather not sketches, but more versions of the notated work. These are in principle equivalent in their value, though they might vary in quality.

For music notated as score form there are two stages of completion: as pure, 'naked'structure,[142] as composition without additions *and* secondly as composition with additions. The German term *Tonsatz* ('abstract composition') means just this composition without addition: A polyphonic composition, in which all voices are written out, notated as purely abstract structure. This notation contains, as elaborated, no directions as to its realization as sound. This type of score only exists as thought form, as concept in the mind, but also real in its notation. A typical example for an 'abstract composition' as first stage is J. S. Bach's *The Art of the Fugue*. It represents a completed work, and for the most part was completed in its composition by Bach right up to his death in 1750. His fair copy has the form of a score, but is notated without any hint to execution and instrumentation. For the term *abstrakter Tonsatz* ('abstract composition') there is no evident translation into English. The more general term 'composition' does not address its specificity, and 'arrangement' is not suitable because it is confused with 'arrangement' e.g. for Jazz Big Bands.

---

[140] Röder, in: Jaschinski 2001, p. 185.

[141] In detail ibidem.

[142] Jerold even goes deeper in the layers of 'body' and speaks of "skeletal notation". Jerold 2016, p. 311.

6.6.1 The 'Abstract Composition'

The matter is defined here. For its terminology we shall tentatively adopt one of the following: naked, pure, 'abstract composition', composition in Score Form, but without ornamentation, notated articulation and instrumentation. Either with or without these additions: Both types are the work as score form. The metaphor of body and garment may assist our understanding. Differences between the two exist and thereby fuzziness is evident; as is the necessity for the body to wear clothes—with due consideration of weather and climate phenomena and situational peculiarities and deviations from the rule.

6.6.1.1 Score Form Without Additions

In the Anticipation Form the prospective sound form is consciously and explicitly included in compositional considerations, anticipated and realized as score, thereby enmeshing the ideal and the material. As supplement of the score and the abstract score, in the anticipation form the abstract composition is transformed. At first the anticipation form provides the conceptual compositional anticipation and the intentionally explicit directions regarding the performance and execution in all detail. It provides additions, of which the most important types are: embellishments, more general expression marks, more special directions for a detailed execution of the sounds—*articulation* and *phrasing*—and for instrumentation. For active musicians and those professionally occupied with music the score already provides a preconception and imagination of the sound it might produce.[143] This refines the realization. Composers take this one step further. They can and must have an intimation of the sound *in* the score, based on the yet bare composition and its instrumentation, and thereby prepare the ground for the realization of their imagined sound.

Where nothing else is written, the interpreters can only play and sing according to norms, conventions, rules and traditions of performance. Whether they realize more or less precisely the ideas and intentions of the composer, or whether fuzziness dominates remains to be seen. A high degree of indeterminacy becomes a standard here. A remedy can be found in explicit recording of respective intentions, especially of the imagined sounds. What so far has been a prerogative of interpreters is now constrained in its freedom by directives of the composer. This represents a further

[143]Newer, very precise investigations show that even such in themselves very subjective representations can be objectified. „Notational audiation is the ability to internally 'hear' the music one is reading before physically hearing it performed on an instrument. In earlier studies, the authors claimed that this process engages music imagery contingent on *subvocal silent singing*. This study refines the previously developed embedded melody task and further explores the phonatory nature of notational audiation with throat-audio and larynx-electromyography measurement. […] Notational audiation is a process engaging kinesthetic-like covert excitation of the vocal folds linked to phonatory resources. […] Covert rehearsal with the mind's voice also involves actual motor processing systems and suggests that the mental representation of music notation cues manual motor imagery" (Brodsky/Ginsborg/Henik/Kessler/Rubinstein 2008, n.p. Emphasis HWH).

sharpening of the realization of the work idea through a comprehensive as possible preemptive detailing of performance and execution.

The liberty of the interpreters diminishes with the emergence of the professional concert. A considerable remainder of this liberty since Viennese classicism is the cadence in solo concerts. It is at first improvised, in particular if composer and interpreter are one and the same. In his piano concerts, that he played himself, Mozart never notated the piano voice before its first public performance in fear of plagiarism. But even those are composed in the course of the nineteenth century, hardly by the original composers but mostly by talented interpreters. These imitative re-compositions of an original improvisation are then re-played in a more or less canonical fashion.[144]

Compared to the thought form the anticipation form allows following differentiation: The thought form aims, at this point without much precision, for the whole, while the anticipation form already focusses on details, on the singular and special. In contrast to the thought form, the anticipation form exists not only in the mind—as actually is the case for thought form and perception form—, but exteriorizes and manifests itself objectively in notation and in words. It is fixed in writing, complemented with directions of the conductor by mimic and gesture or with implicit and explicit agreements between musicians in chamber music.

Without this anticipation of the work as sound shape it can hardly be consummately composed or productively realized. Yet again, as with sketching, the process consists of multiple stages and layers. The determining line of distinction between sketch and thought form is the view to dissemination, intended from the beginning and finally realized. Phantasy, imagination, anticipatory imagination is always present at the various stages of corrections, both in this form and later, serving as orientation and target conception for the composition and then later the interpretation. This also applies, while the possibilities of notation for performance and execution might still have been limited. Even for improvised music the anticipation form is essential. Though in those cases it might not be as detailed and comprise only smaller units of time. It is just as essential for newer pop music, which so to say skips the notation and proceeds immediately to the sound or the phonographic form (see Sects. 3.6 and 6.8) In the anticipation form a preview is imagined of the entirety of the work *as well as* its individual parts. In this way the anticipation form takes up the thought form and continues it, only now in a way that the work exists as a whole; in notation in the shape of—at least principally—*fully realized composition.* This may even in relatively abstract execution count as completed and has counted as such for centuries, i.e. as composition with exact determination of tone durations and pitches and nothing else. This 'abstract composition' (see below) is also an abstraction as term and a new, here developed term. Until around 1590 (Gabrieli) and 1608 (Monteverdi), this was a common fact. This historical stage is behind the terminological distinction of (abstract) *composition and addition.*

The sound form might have been taken for granted, but it may also have been considered a secondary concern that its determination might be fuzzy or even vague.

---

[144]In detail, among others see Heister 1983, passim.

But this changed over history. Since the watershed of European music in the year 1590 with the transition to the so-called 'baroque music' or 'music in the age of the *continuo basso*', the determination of performance and execution becomes included *in the* notation form with more and more differentiation and increased intensity. In the nineteenth century and even today the instrumentation constitutes a quite autonomous stage of the production of a score. Always the writing out, the complete filling of all individual voices is bothersome, mostly mechanical work. But it is indispensable for sharpening. The particell is more convenient, but more fuzzy. The arrangement of the parts and the particular instrumentation as distribution of the (abstract) score to various musical instruments and also vocal parts is less and less evident with the growing emancipation of composition from its functional context. Its' emancipation correlates to the growing autonomy and unfolding of the musical work as autonomous work of art. In fact, since this transition phase we find increasing amounts of instructions, both regarding performance and execution as well as instrumentation. Consequently and in the same vein there are increases in quantity and in qualitative differentiation of instructions for the performance and the determinants of recital, articulation and phrasing.

In this regard the anticipation form may be regarded as a historically new stage as compared to the previously available notation and score form. Looking back at the chain of existence forms so far, and historically situated at this new stage it is its' co-specific manifestation. Anticipation form and score form may therefore be conceived of as two stages in the work process. They are intricately interlaced, both temporally in the work process but also factually in all intermediate results as well as the final result. The usual foreshadowing and regress are added to this. The separation of these two existence forms may therefore appear somewhat artificial. But it reflects something real. The abstract composition[145] is not merely an abstraction, but really exists.

6.6.1.2 'Additions'

Additions have more and more become an essential part of the work in its score form since the epochal transition in the 1590s. The differentiation into two types of scores has resulted from that. These additions transform the pure, abstract arrangement into a complete one. Only then does it become an actually completed work, a score that transcends the production of notes in the context of its composition. It anticipates

[145] U. Siegele describes, how "especially in the old times" an "interplay of the *preconcrete* formulation of the score and the more concrete formulation for the specific case" occurred. But evidence dates back into the early nineteenth century. And so [...] still in 1870, when during rehearsals of the *Valkyrie* the harp player of the Munich Opera remarked that it was impossible to play the harp part of the *Feuerzauber* (fire magic), Wagner could reply: ,I am not a harp player. But you can see what I want. It is your job, to arrange your part in such a way, that it sounds just as I imagine it'" (Quoted in Siegele 1968 and Siegele himself, col. 21. Emphasis HWH). Wagner had provided instrumentation in the score and made it fully concrete. But the last word was left to the sound realization – and thereby new fuzziness occurs (see Chapter 3).

the production of concretely adequate sounds and incorporates these into the score form.

The following additions are classified systematically, not historically.

1. Embellishments. These actually are still part of the pure arrangement, but already begin to loosen its abstract and rigid character.
2. Instructions for expression marks. Initially simple with piano and forte, determine the dimension of volume after the determination of tone duration and pitch—but also expressively and often more generally (see Chapter 3. A. Schnabel). Instructions for recital mostly refer to shorter, sometimes a bit longer passages, not to individual tones.
3. For indications of a detailed execution of individual tones there is no collective term, as far as I know. With average accuracy one can for one differentiate *articulation* of a single tone, be it playing style in analogy to the articulation of sounds in spoken language, and the *phrasing* of tones in the context of longer passages or the melodies of each respective instrumental voice or the vocals, in analogy to the 'supra-segmental' of spoken language.
4. This prepares the ground for the transition to the fourth dimension of tone, to tone color. It's capturing is the goal of instrumentation. First beginnings of explicit instrumentation exist early, and singing voices have obviously been determined beforehand and noted since the beginning of polyphony.

6.6.1.3 Sharpening of Details and the Whole. Possibilities and Boundaries of Notation

Some of this relationship between anticipation form and score form is reflected in the notation, structurally and historically. The history of notation is also the history of composition. This can be seen in European music history. Especially the art music was embedded in the dominant literate culture, but it had a large portion of oral transmission and non-written components of its production. Under the premise that it can be written down notation is not peripheral, but essential. What cannot be noted down is hard to compose. Interesting is the circular causality, which leads to a spiral movement of progression. The extended and more precise possibilities of notation lead to the same effects as in composing. The not yet noted or not notable rest is realized in accordance with the standards of performance practices by the musicians, within a framework of rules and the traditions, conventions, capabilities and skills of the interpreters.

A primary concern of the development of notation was to more and more limit the zone of uncertainty and fuzziness, and conversely extend the area where the work idea, mediated by the intention of the composer, can be determined more precisely and completely. Altogether this is a long historical process of sharpening. Thus potentially all dimensions of sound realization are to be determined in composition. At the end point of this development even the final sound product, the sound form, will be determined entirely by the composing subject. This is the case for electro-acoustical realizations of serial music (see Sect. 6.8). This represents a tipping point in the development so far: Notation, in particular the score form, becomes secondary or even dispensable. In principle this is at least a technical, but even an artistic progress.

The extended mandate of the composition as core of the production and thereby extended authority of the composers through more and more driven detail and differentiation of notation appears at first glance a disempowerment of the interpreters. However, these possess own *telos*. On the one hand the pure virtuosity, the freely unfolding consummate singing and playing, on the other hand the virtuosity bound to the score. The connection to the work from the interpreter's view is the shared third, the completed 'score'. The intentions of the work idea and thereby of the compositional production and in inversion the *telos* of its completed sound representation converge here. The interpretation bound to the work then constitutes its own existence form as sound form (see Sects. 6.8 and 7.5).

The development of notation is precisely the reflection of the development of composition: it depicts it, *and* it impacts it actively. At intersection points, such as the introduction of the four staves system about 1025, of modal notation about 1150–1250 and especially the consequent mensural notation, fully developed around 1320, the new and improved style of notation became the starting point of compositional innovation. And even the gliding transition in the late sixteenth century to the new age, in its core still valid notation based on the binary split and multiplication of tone values (instead of the ternary division of mensural notation) made implicit and activated new thinking approaches and opened up new compositional possibilities. At this stage we see the beginning of gradually progressing sharpening by small refinements, which cannot be described in detail here. In passing we might note, that the said history of notation as scientific *representation* of its history seems to invite an application of fuzzy logic methodology and thought.

6.6.2 *Stage I. The Naked, but Integral Body. From Thread to Fabric—Evolving the Polyphonic Composition*

The developments of European melodic polyphony in art music in the longer run, historically speaking, leads from 1 to 1 + n, from the single melodic line to a polyphonic structured fabric with several melodic lines, accompanied, motivated and supported by the development of notation. To this structure then the instrumentation as approximative definition of the dimension of the tone color[146] is applied in a separate step, embellished if required with ornaments, and thus provided with specific instructions for performance and execution. Here the developments fan out. In the older conceptions and possibilities of notation as pure abstract the non-standardized conditions of performance and execution could not be anticipated and considered. Up to the change of tides around 1590 the pure, abstract composition structure was already the clean copy or score. More was not—yet—available. *After* this historical transition the abstract composition must be made more and more concrete, in order to be considered a completed, consummate, perfect score. The two Gabrieli, and Viadana etc. around the 1590s introduced firstly more and more exact instructions and abbreviations for the execution. Secondly more and more precise, *standarized*, that means predictable compositions of ensembles or orchestras, allowed for another type of composition in the sphere of music culture: as example may serve the 24 *violons du Roi* of Louis XIV, a pure string ensemble, or the unbroken and broken consort, i.e. ensembles with either the same or different musical instruments, in the times of Purcell, and suchlike. Here the relationship of system and environment, the context of notation, composition and music culture plays a major role. This shall not be investigated further here.

The anticipation form conveys the concretization of the abstract composition and *is* it simultaneously, and this in a historical-logical progression of differentiation and refinement. Only through and after this differentiation the really fully developed work, represented in the completed score form is achieved, which includes and records all prescriptions for performance and execution. This must then, close to or at the peak of its development, progressed through the stage of anticipation form. The separation of these two existence forms is grounded, as mentioned, historically and in the matter. The imagination of core and cover, understood as the metaphor of body and garments, persists also under the conditions of an integral score form. The anticipation form pre-empts this completed score form: Initially it can be thought at best, but not yet made, as long as internally the possibilities of musical notation, as well as externally the accordingly reliable and precise instruments of execution are not available. Only when conditions on both sides have developed, can the anticipation form be thought, but at this point still without having to make it, and sometimes it never is made. We shall look at some deficient, incomplete types. Once it is realized in notation, it ultimately disappears in the clean copy. It is preserved in the various types of additions to the 'abstract composition'. But an independent, *self-sufficiently material* form is never reached. Some of the difference to its existence as 'abstract composition' remains even after the prevalence of the complete, integral score form

[146]The term 'tone color' implies a peculiar synaesthetic association. This shall only be pointed out here. But the use of color metaphors continues further. Chromatics—literally 'coloring techniques'—are the extension of the original diatonics, the European seven tone system to the twelve tone system. In the isorhythmic composition since the beginning of the fourteenth century „*color*" means the abstract movement of pitch, the diastemics, compositionally autonomous and detached from the organisation of continuous movement. The latter is called *talea*—as 'cut' or section also a spatial-optical metaphor. „*Coloratura*" then is the embellishing 'coloring' of a melody, through figuration dissolving it into smaller tone values, etc.

as general, relatively obligatory norm in the course of the nineteenth century. The composition, at least the notation of embellishments, appears mostly as separate work step following the production of the composition structure. This applies to the chronological order and also to the order of ranking of the compositional value. The particell attests to the practical relevance of this way of thinking and working (see Sect. 6.3). This applies similar or equally to verbal indications of recital style, signs of articulation full of imagery, as well as the instrumentation. It is, since not referring exclusively to the already existing dimensions, but to the new dimension of color, a more clearly demarcated stage in the production process of the work. Main reason for this is the initially mentioned precedence of the dimensions tone pitch and duration in the European art music tradition. (On other traditions and ways of thinking see Sect. 7.4.3.)

6.6.2.1 Basso Continuo and *Real Book*: Notated Chords with Ciphers and Improvised Voicing

The basso continuo, viewed under the aspect of sharpening, constitutes an ambiguous transition form on the way to the completed, *entirely executed* score form. It was the prevalent form between around 1610 and 1740. Recorded are only the bass line, the basso continuo, and the upper voice, the 'melody'. In the developed form the bass line is notated as figured bass or thoroughbass: Numerals and symbols, mostly accidentals, provide the harmonic structure and chord progression. For compositions for ensemble or orchestra featuring more than one 'melody' the voices are written out, but not in the general bass line, usually an organ or a harpsichord. The score form is incomplete. For even embellishments are notated in some cases, but never the middle voices. With instruments capable of playing chords as organ, harpsichord, organ, and lute the voicing of the accompaniment has to be improvised. The very exact execution of the composition does not correspond wholly with the intention of the composer; rules limit the free play and arbitrariness, but characteristic fuzziness is consciously tolerated.

In one variation of the basso continuo, the *basso seguente*, precision and representation of the composition structure are sacrificed for pragmatic reasons. The basso seguente as a notational practice

> was first described by Adriano Banchieri (*Ecclesiastiche sinfonie*, 1607) and then regularly used by Italian composers and organists. Like the basso continuo, the basso seguente *consists of a single staff* and was used (chiefly by organists) for accompaniment of an ensemble, but it is conceptually different from the continuo. In fact, the basso seguente includes not only the bass line but whatever line happens to be the lowest-sounding part in the score. So, in a fugue with ascending exposition, the basso seguente may very well begin with a bass clef and then follow the succession of entries, *changing clefs* according to the voices.[147]

The notation of the basso continuo coincides here with that of the *partimento* (On partimento see Sect. 6.4) (Fig. 6.21)

---

[147] Sanguinetti 2012, p. 11, Emphasis HWH.

Fig. 6.21 'Partimento' (Organo-part) and score with figured bass—a mass of Joseph Haydn Sanguinetti 2012, p. 13, example 2.1

> The *organ part* […] (an excerpt from a mass of Joseph Haydn) looks exactly like the exposition of a partimento fugue; in fact, the main difference between a partimento and a basso seguente is that the latter is a synthesis of an *existing composition*, while the former is *the plan for a composition not yet in existence.*[148]

In contrast to the bass and the upper voice, the 'melody', the middle voices are a secondary 'accompaniment'. The composition of a basso continuo with its emphasis of the outer framing and contouring parts has influenced music long after the 'age of the basso continuo'. It might correspond with a particularity of perception, which also prefers edges and contours in the acoustic perception. Lastly this has an evolutionary background. When, as often done in polyphonic music in the late middle ages or the Renaissance, the tenor, the 'halting' and carrying main vocal part, lies in the middle of the arrangement, it today requires some cognitive adjustment and a conscious reorientation of perception (Figs. 6.22 and 6.23).

The following example shows a late development of the basso continuo, as viewed by a theoretician. The example shows three ways, in which the basso continuo can begin: a, b, c. only the plain accompaniment 'a' corresponds with the level of the—almost achieved—abstract composition. The two versions b and c of the 'mannerly basso continuo' are already furnished with 'mannerisms', which dress the naked composition, so to say make it sophisticated, civilized, polite. Lastly they are embellishments. The basso continuo is at this stage suspended in the composition, but only implemented provisionally in textbook fashion. The bare contour without middle voices can be found in 'a'. Bass line and upper voice. The latter has no primacy here it is derived from the harmonies implicit in the basso continuo (Fig. 6.24).

The actual embellishments, so to say the sequins on the robe of the composition or the necklace and arm bangles, are not fully notated. The basso continuo is not yet a complete, finalized score in its most highly developed state, not only in terms of composition but also of notation: One component, the exact voicing is missing in the 'accompaniment'. It is left for the interpreters to realize it as sound. The basso continuo remains incomplete just as on the other side the particell. The gap of

---

[148] Ibid.

Fig. 6.22 J. S. Bach: Trio Sonata of the *Musical Offering* without voicing
Quoted after https://www.lehrklaenge.de/PHP/Harmonielehre2/Generalbass.php

Fig. 6.23 J. S. Bach: Trio Sonata of the *Musical Offering* with written voicing
Quoted after https://www.lehrklaenge.de/PHP/Harmonielehre2/Generalbass.php

fuzziness is not large, but noticeable. This boundary can only be overcome by a leap to another form of thought and way of composition and the resultant notation. The notation form of standards in Jazz behaves similarly. Only here the chord progression, the changes, and not the bass, are decisive. For that reason the basic tones of the chords are indicated—similar to the ‚basse fundamentale' of Rameau—and furnished with

Fig. 6.24 Heinichen 1728
a plain, **b** and **c** „mannerly“
Quoted from Ferand 1956, no. 29, p. 116

Fig. 6.25 Jazz Standard from the *The Real Book*, 6th Edition 3 May 2013, Thelonious Monk: ballad '*Round (about) Midnight*, 1944
Melody plus chord-ciphers, m. 1–4

additional indicators of alterations b and the notes that complement and extend the chord, such as C#^{7} = C#-E#-G#-B (Fig. 6.25).

It is obvious that with jazz interpreters the voicings, the entire concrete elaboration in it details deviates more significantly both from the original composition as sheet music, the standard, as well as from other elaborations. While in basso continuo music, in the context of primarily and principally notated music, the improvisation

Fig. 6.26 The beginning of an improvisation by Th. Monk himself, m. 1–5
Jazz piano collection—thelonious Monk. https://theloniousspheremonk.weebly.com/jazz-piano-collection-thelonious-monk.html

is reserved for all that the composition lets free, openly, so to speak, uncertainly and fills in gaps. Conversely in jazz the improvisation attains primacy. What is written down in notation is the starting point for a hardly normed, productive, imaginative elaboration and development (Figs. 6.26, 6.27, and 6.28).

The differences of how the original work is realized here are so big, that evidently the original composition is far removed as a starting point. Each individual performance has the tendency to constitute an actual work of art. Something unfolds, which was inherent at best in the original work. On the other hand the changes are retained and in principle remain unchanged, thereby preserving essential matter from its origins, as much as it is enriched by 'dissonances' in the harmonies and the voicing. The identity lies, one might paradoxically conclude, in the diversity. It would be a great challenge for fuzzy logic to quantitatively assess and determine the identities within the differences.

6.6.2.2 Transitional Form 'Tablature'

A principally different style of notation is the tablature. It is a notation purely for instrumental music and denotes fingering:

> tabulature, or tab for short) is a form of musical notation indicating instrument fingering rather than musical pitches. Tablature is common for fretted stringed instruments such as

Fig. 6.27 The same jazz standard, played by Bud Powell, 1950, m. 1–6
Kevin Sun, transcription. https://www.kevinsun.com/2014/12/bud-powell-on-round-midnight.html

> the lute, vihuela, or guitar, as well as many free reed aerophones such as the harmonica. Tablature was common during the Renaissance and Baroque eras, and is commonly used today in notating many forms of music. [...] To distinguish standard musical notation from tablature, the former is usually called "staff notation" or just "notation".[149]

Tablatures skip the function of writing as anticipatory representation of sound, and directly transform script into an instruction for the production of sound—but restricted to a limited range of musical instrument types. It originated from the re-notation of vocal music for musical instruments on separate sheets. Playing after a tablature was the opposite of playing the singing voices directly. Secondly, the tablature differs from the notation for choir or part book: The voices of the composition were arranged vertically. This resembled the score form. The third characteristic was the representation of the composition with ciphers, letters, and notes.[150] On this basis differentiations developed.

> The more the [...] instrumental music [...] moved to the center of musical practice, the more they succeeded in combining the notation of fingering with the notation *en musique*, so that a composition could be understood by mere reading.
>
> Even here there are more differences: The piece can be represented *in corpo,* i.e. only with fingering and sound, or *in partito*, i.e. with exact consideration of voicing, in the score.[151]

In its most highly developed state the tablature reaches the level of an abstract composition. It even surpasses it in view of its reification, since the instrumentation is already

---

[149] See https://en.wikipedia.org/w/index.php?title=Tablature&oldid=937498296, 25 January 2020, retrieved 30 March 2020.

[150] Abridged by Jaschinski 2001, p. 186.

[151] Ibid.

Fig. 6.28 The same jazz standard, played by Oscar Peterson, m. 1–8
Quoted from Hal Leonhard: The Very Best of Oscar Peterson (Artist Transcriptions, Piano) https://www.google.com/search?client=firefox-b-d&q=peterson%2C+transcriptions, retrieved 22 August 2020

implicit in the assigning to a musical instrument. On the other hand the details of voicing often remain fuzzy. Especially tablatures still recorded by hand show the transition from sketch to score form, even when they are intended as final version (Figs. 6.29, 6.30, and 6.31).

Because of the handwriting and the abbreviations this tablature is much closer to the sketch form than to the elaborated score form.

As in all simplified notation, as in the *Fasola* notation for the song books of the Southern Baptists in the nineteenth century in the U.S.A., the tablature is, in that it is limited to only special selected areas or as here musical instruments, not a universal style of notation.

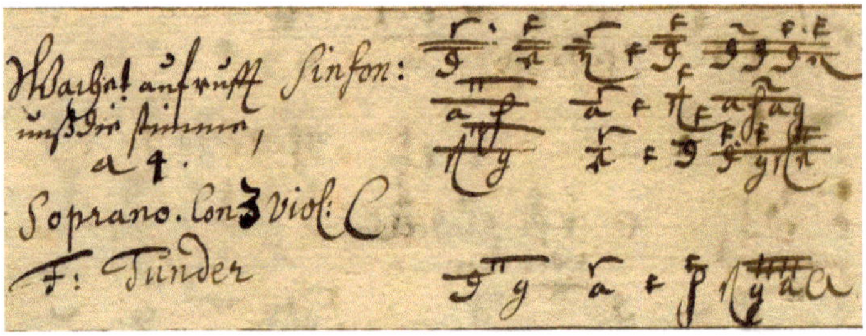

Fig. 6.29 Franz Tunder: Spiritual concert "Wachet auf ruft uns die Stimme" [Wake up, the voice calls to us], Tablature, Autograph
Quoted from Stawiarski, 3 June 2007
https://upload.wikimedia.org/wikipedia/de/a/aa/Tunder_wachet.jpg

Fig. 6.30 Franz Tunder;"Wachet auf ruft uns die Stimme", Code
Quoted from Stawiarski, 3 June 2007
https://upload.wikimedia.org/wikipedia/commons/7/73/Tabulaturbeispiel.png
Left: Duration, octave, name of the note; right: part (voice) 1, 2, 3, 4

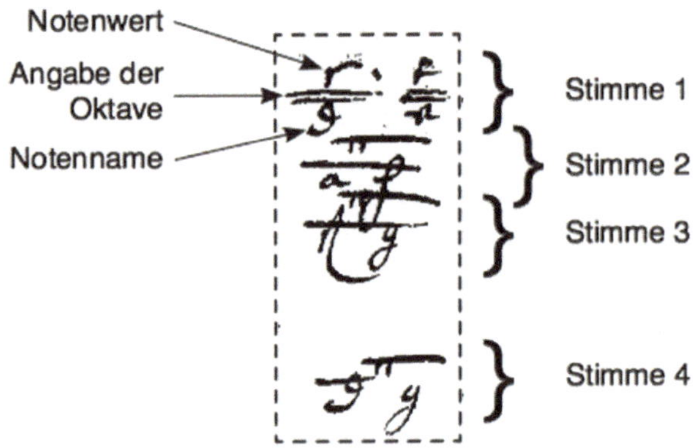

6.6.2.3 The *Obligate Accompaniment* of the Viennese Classic as Fully Notated Voicing

Only with the explicit writing out of the middle voices do we get a *fully elaborated* score, and thereby we get a picture of the score form as we know it. And even there the embellishments are mostly not noted down. They are co-composed in the obligate composition but are placed at the lower end of the table like distant relatives at festivities. This development of composition and notation extended the area of what was to be determined in the composition by the composer, and became a characteristic attribute of the classical style. In the obligate accompaniment the middle voices were composed and noted down. Being so determined by the composing subject and no longer by the interpreting subject,

> the middle voices achieved more freedom and flexibility through the emancipation from the *basso continuo* in the middle of the 18th century – not only the outer [framing] voices, but primarily the so far *improvised, not composed (obligatory) middle voices*; precursors of this development are found in the free polyphonic suite compositions of H. I. F. Biber und J. H. Schmelzer. For G. Adler, who coined this term, the 'obligate accompaniment' is the driver of

Fig. 6.31 Franz Tunder;"Wachet auf ruft uns die Stimme", m. 1f., transcription
Quoted from Stawiarski 3 June 2007
https://upload.wikimedia.org/wikipedia/de/a/aa/Tunder_wachet.jpg

> the development of an autonomously evolved (piano) accompaniment, which is a significant stylistic feature of Viennese classicism.[152]

Most apparent are the transitions in another genre, which from the outset consisted, in line with basso continuo, only of an upper melody voice plus instrumental accompaniment:

> Especially in the (German) piano song of the 18th century the development of the *basso continuo* to the 'obligate accompaniment' can be traced clearly.[153]

6.6.2.4 Embellishments. Transition to the Clothed Body

Embellishments or ornaments are largely independent of instruments and of details of the execution. So they appear logically and systematically *before* the instrumentation

[152] Hilscher 2001, retrieved 6 July 2019, Emphasis HWH.
[153] Ibid.

and before indications of articulation and phrasing, which are also primarily related to the timbre resp. sound color. In principle, ornaments belong to the pitch and tone duration dimension. They are on the one hand a blurring of the "pure" melodic lines and already more than the 'abstract tone set'. On the other hand they are its sharpening, inasmuch as they anticipate the actual tonal realization in the notation. Thirdly, however, they are still only additions to the 'abstract tone set' and not yet the fully developed, complete form of the score, since the execution of the score, with respect to the dimensions of timbre and volume, is Thirdly, however, they are still only additions to the 'abstract tone set' and not yet the fully developed, complete form of the score, since the execution is not defined with respect to the dimension of timbre and volume.

In our context it is important *that* they are noted down at all, typically however mostly only in abbreviations. This can be attributed to the mentioned historical tendency to subjugate the work of the interpreters to the determination through designations of the composers. This results in a structural and epoch-spanning contradiction between composers and interpreters. The composers want to perpetuate their freedom of complete and unrestricted self-determination all the way up to and into the realization of the work in the sound form. The interpreters want to preserve their little freedom of self-determination, which is anyway restricted to the sound form, at least in the productive realization of the work. This includes both the execution of embellishments prescribed in the score form but also execution of those not noted down ad libitum.

The abundance of embellishments makes an overview difficult. Attempts have been made, primarily in German-speaking music theory of the middle of the eighteenth century, foremost by J. Quantz and C.Ph.E. Bach to systematically organize the abundance of types and cases of embellishments and to standardize as much as possible their concrete execution. This has its limits. Indeterminacies and fuzziness are part and parcel of musical embellishments. Especially the representation in abbreviations, which suggests a formulaic nature, leaves gaps, gives ambiguous directions, provoke variations in the execution. Embellishments stand at the transition of composition to realization in sound. They maneuver in this border territory. Theory and notation practice in the middle of the eighteenth century codified the difference of "essential" and "arbitrary" mannerisms. Only the 'essential' ones are noted down at all—or inversely: What is noted down is essential. The others are left to the freedom, the will and the willfulness of the interpreters, to their capabilities and their intentions, and in the best case also to their good taste. This limited space is preserved for the interpreters, until it is then, just as improvisation quite generally, more and more pushed back since the nineteenth century.[154] Embellishments in the close sense can be conceived of as 'micro-figurations'. ‚Figuration' means the dissolution of melodies into smaller tone values, which are used to play around with the melody, thereby embellishing it. The original, fundamental melody surely gets blurred, but the intention towards a completed work, as mentioned, is refined. Figuration is a universal procedure, common both in vocals and even more so in

[154] Cf. e.g. Heister 1983, Heister 1998, Heister 2004.

instrumentals. It is the imaginative, free play with the possibilities of the material as well as the capability and skills of the executing musicians. Part of the nature of the embellishments is their formulaic character, just like in the fine arts and the applied arts, or like in literature the topoi and idioms. Primarily these are material and part of the form. They border on the similarly formulaic 'rhetorical figures', which are primarily part of the content. They constitute a stock of material, to be accessed and used by the executing musicians or the composers. There are tables of embellishments, organized like a dictionary or lexicon for easy reference (Fig. 6.32).

A worthwhile task for fuzzy logic could be to provide a typology, classification, and quantification of identities, differences and similarities of the repertoire of embellishments. A particular challenge is the lack of an actual orientation point that could serve as point of comparison for fuzziness. The note that is embellished might be close to it. But the difference of 'main' and 'side' notes may switch within an embellishment. As orientation point it would therefore be imaginary-real. It becomes real when the embellishments are part of an abstract composition. There the original melody can serve as orientation point, once it is made visible. Embellishments in the realization as sound open a broader field. After all, they are mostly recorded only in abbreviations, or left to the interpreter right from the beginning. The abundance of material again begs for fuzzy logic to sort it by music types—notated versus non-notated music—after epochs, phases, styles, national styles. The art will lie in not randomly distributing embellishments on the playing or the composition like the icing of a cake, but to adapt and adjust them to the specific work. If and in so far this succeeds, the embellishments will, against all purist objections, belong to the essence of the work. Only from a purist perspective it would count as an objection that they moreover and anyway belong to the practice of virtuosity as one of its components.

A rather early example shows that embellishments can be used immediately semantically. Here they even directly transition into the area of 'rhetorical figures'. The „swimming in blood" of the Crucifixus is conveyed by the sounds in a nearly embodied way (Fig. 6.33).

Embellishments can also go in quite another direction. In the following example the embellishments were added to a completed work *subsequently* and on top of that by a stranger's hand. Here tendencies of the ornament gaining independence become apparent. The example shows for one, that the permanent and unmotivated crying or laughing of the text is a semantic motivation for the overflowing figurative singing, which meanders around a simple melodic core. It is appropriate for the described characteristic ambivalent bonding pattern, oscillating between adulation and rejections. Then there are also the now even more excessive embellishments in the *da capo* section, i.e. the reprise of the beginning, motivated by the psychology of reception: The audience must not be bored with simple repetitions, but entertained with variations (Fig. 6.34).

An embellishment can under special circumstances become directly thematic. Beethoven includes in the large fugue of the finale of his piano sonata No. 29 in B-flat Major, Op. 106 („Hammerklaviersonate" a long trill as distinctive, signal-like theme head of the fugue subject. In the introduction the trill is heard three times already, starting with the tone A, as then in the exposition of the theme. Beethoven

Fig. 6.32 List of Embellishments, 1689. Jean-Henry d'Anglebert (1635–1691): Pieces de clavessin (Pièces de clavecin... avec la manière de les jouer. Diverses chaconnes, ouvertures et autre airs de M. de Lully, mis sur cet instrument. Quelques fugues pour l'orgue et les principes de l'accompagnement. Livre premier) (1689).
Estienne Roger, Amsterdam (n.d., c.1700). IMSLP304260-PMLP43812-anglebert_pieces_de_clavecin.pdf. (The extent is extremely modest compared to what can be found today on wikipedia alone, not to mention musical specialist lexica)

Fig. 6.33 Thomas Brewer: Sacred Song for solo voice with *unfigured* bass, c. 1650. Upper staff embellished, under staff plain
Excerpt T. 5–8, quoted from Ferand 1956, No. 25, p. 110

also picks up on the continuation of the theme of the fugue, at first in form of a simple scale. As anticipated counterpoint to the theme of the left hand, Beethoven inserts doubly sequenced embellishment-like figurations of f, d, and b-flat. It is to be considered, that the trill vitalizes a long tone and/or influences tone color (Figs. 6.35 and 6.36).[155]

In a later example from the area of *New Music* dressing and body have literally grown together.

Here the embellishments as 'micro-figurations' are essentially echo effects, reverberations, variational repetitions. The intended meaning is only conveyed by them as figurations, not by an ostensible melodic core.

6.6.3 Stage II: The Clothed Body. Compositional Anticipation and Intentional Determination of Performance and Execution

With the *obligate accompaniment* the highest stage of European notation is reached. Additions and extensions are naturally still possible and necessary, and still not everything has been comprehensively noted down; many things simply cannot possibly be noted down. The score form relates what was determined by the composing subject, in a general form, which applies to all particular performance conditions including

[155]Cf. Gutknecht 1994.

Fig. 6.34 Giovanni Battista Bononcini: Chamber Duet for Two Altos, 1691. Embellishments by Carlo Antonio Benati, c. 1710?
Quoted after Ferand 1956, No. 30, p. 118
"Constantly I cry (laugh) and cannot say / when I will no longer cry (laugh)"
a melody plain, **b** with some, **c** with many ormanents resp. figurations

Fig. 6.35 L. van Beethoven: Piano Sonata No. 29 in B-flat Major, op. 106 „Hammerklaviersonate" Introduction to and beginning of the fugue

Fig. 6.36 Mauricio Kagel: *Schattenklänge* (Shadow Sounds), excerpt, ms. 1–5 https://www.stretta-music.de/kagel-schattenklaenge-nr-142124.html

the particularities of the interpreting subjects as well as for each individual realization. This particular and individual component is not predictable in detail and cannot be determined in anticipation. Therefore necessarily fuzziness remains. Haydn notes the problem accurately during a transition phase:

> The music is composed in view of the performance, mostly with a performance under specific conditions in mind, in a specific situation. Lack of knowledge about these circumstances makes the composer uncomfortable. So Haydn writes in 1768: »I appeal especially to the Master Musicis, in order to promote further my and your own honor, to apply themselves with greatest effort: Should I by any chance have failed to divine their taste preferences in my work, they may not hold it against me, for I know neither the person nor the place; denial of this has, to be truthful, soured my work.« Only in the course of the 18th century the composer could properly assess a general situation, only in the 20th could he disregard the recital altogether (as for example E. Krenek concedes for *Lamentatio Jeremiae Prophetae*).[156]

The term 'performance' comprises the entirety of the work realization. The term 'execution' means mostly the details of the realization of a work. Viewed from the restricted perspective of the 'abstract composition' the matter is clear: The additions to the composition are *not essential* but *accidental*. From the view of the composer the matter is not so clear anymore, but becomes fuzzy. The additions, the 'dressing', are generally speaking also essential. And from the view of the interpreters a tendency of an inversion becomes apparent. In particular the additions, which as a rule are not fully articulated in the composition, give room for the unfolding of productive capabilities and even virtuosity.

The intention of the composer is the work. This applies, to once again speak against all prejudice, mutatis mutandis also to non-written music. Be it maqām or a jazz standard: Merely reciting these would be meagre. It is necessary and sufficient

[156] See in details Siegele 1968 und quoted in Siegele, col. 21.

if it serves as beginning and for setting the mood. Good musicians do not play the pure, bare melody, or the chords, but vary them by 'embellishing' them, giving them shape, playing with and around them, 'diminishing' etc. Only in this way does the work become complete and consummate. If conceptualizing composer and performer are the same person, the variation is a component of their intention, and therefore essential and not accidental.

The work is intended to become 'consummate' resp. 'complete'. In this its work character becomes concrete. This is an essential intention of the producers. 'Consummation' as perfection is absolute. What is achieved always remains relative; in an asymptotical approximation the work becomes the mentioned „entirety complete in itself" (Moritz). The refraction and anticipation of the absolute perfection which can never be achieved is revealed, still in contours and abstractly, in the thought form. A counterpart within the performance is perfection; its minimal form being 'free of errors'. The score form has a hinge function from the viewpoint of antecedent existence forms, the conceptual form, the sketch form, and the thought form. The connection of anticipation form and score form also displays the striving for perfection in two stages. At the first stage it remains the said bare, 'abstract composition'.[157] It contains everything that can be said musically—at a certain historical stage. More is not possible, historically speaking. It is thereby maybe not fully completed, but at least and in any case it is complete.

What is possible at any given time changes over history and develops further. The possibilities and impossibilities of notation—and, of course, composition—have a historical dimension. As always, seemingly unalterable limitations continuously change to transgressable and transgressed boundaries.[158] In musical notation there has been, especially in European music, tremendous progress since the humble beginnings in the ninth century (see Chapter 3). On reaching the second stage the bare 'abstract composition' is not sufficient anymore. The intention for really completed work, as far as possible perfect, remains unchanged. What has changed is our understanding of it. For comprehensiveness and therewith for the intended perfection this must include the notation of all and everything that wants to be said and can be said by the composer.

---

[157] Hegel's 'owl of Minerva' flies at dusk, and so the 'abstract composition' is treated, shortly before the fully articulated composition including instrumentation becomes the rule. On the other hand the described separation of the two stages of the score form remains. Johann Philipp Kirnberger's book-title of his retrospective *Die Kunst des reinen Satzes* (Berlin and Königsberg: G. J. Decker und G. L. Hartung, 1774) could be translated as '*The Art of 'abstract composition*'. Kirnberger studied composition and piano with Johann Sebastian Bach and considered Bach's compositions an ideal. „The rather conservative work was not undisputed among contemporaries, even though it contains theoretical aspects of movement and melody that were still valid for W.A. Mozart, who knew this work." https://de.wikipedia.org/wiki/Die_Kunst_des_reinen_Satzes_in_der_Musik, 3 September 2019, retrieved 3 August 2020, transl. H.-W.H.

[158] For Georg Wilhelm Friedrich Hegel, roughly referenced, borders are negated limits. Cf. St. Herzberg 2019.; in detail see particularily G.W.F.Hegel 2010: C. the limiting of quantity, pp. 166f., and chapter 2 Quantum, p. 168ff.

The anticipation form pre-conceptualizes and limits the expansion of what can be said and done, at first in thought and then materially, finally realizing it in notation. This includes what once could only be conceived of but not realized, that is the 'dressing of the bare body'. The whole then is the complete, sophisticated human. If comprehensiveness and thereby completion includes 'dressing', then this is not only accidental, but instead essential, in spite of being an addition to the 'abstract composition'. In particular embellishment and instrumentation may appear as blurring. And, indeed the contours of the abstract composition are blurred by the 'dressing', by decoration and coloring; especially in excess, which is hard to define specifically, the structures of melody and composition become fuzzier. However, the work idea incorporates said comprehensiveness through additions and their execution in the sound realization. And so the additions serve as sharpening of this idea.

6.6.3.1 Cutting Patterns for the Dressing. Instructions for the Execution of the Music by Notes, Signs, and Words

The compositional determinations of execution through the interpreters bring to an end the process of anticipation *of the sound as core of the realization* and its externalization in the score form, a process which featured continuous *concretization* as well as *concretizations* in the plural. The development goes so far, that every tone gets exactly determined with nuances of sound color and expression (see Sect. 6.7.3).

As much as notation becomes more precise—and it does in the historical process: Fuzziness remains between idea and realization. The *instructions for execution*, which could be considered a superordinate concept, are the interface between composition and realization.[159] They serve as mediation of the transmission. U. Siegele's understanding of the *Vortrag*, only approximately and subsidiary translatable as 'musical interpretation' with a focus on 'prescriptions' or 'instructions for the execution', is very broad. It corresponds with the overarching, general concept of execution and 'realization as sound' used here.

> This area comprises what is not noted, what is not precise, what can be noted suggestively at best; in particular this includes slowing down and accelerating, agogics, dynamics, accentuation, articulation and coloring. Phrasing [Voicing] has, equivalent to tempo, a peculiar position; both are immanent structures that in a recital are to be made perceivable with specific means. […] In this way the field is expanded to include the entire ornamentation, the manners (especially diastemic accents) and rhythmic variations […], yes, and even reaching into the traditional field of improvisation, one could include the virtuoso diminution, alterations of reprises, cadences of soloists. Obviously this transgresses in several ways [in more than one way] the limits of what can be notated; and indeed 'virtuosos' elaborated many things or even had them elaborated. The composition accommodates the situation of each individual performance. Often this is the task of the conductor. The space, its acoustics and the ramifications of the performance are to be considered, as are type and numbers of the force of instrumentation; in a limited number of cases the pitch has to be determined, the

[159] From the abundant literature on interpretation, performance, execution/recital at least three titles shall be mentioned here Danuser 1992, Mazzola 2011, Danuser 2014.

> setting of accidentals is to be organized, the organ as well as the cembalo require registers, the harmonic-rhythmic foundation of the general bass has to be executed, and the text dubbed over precisely. This leads, almost unnoticeably, to parody, arrangement and treatment become means for preparing the composition for its particular recital. Besides, the particular recital is the only place, where concept and theory of musical performance can prove itself and must do so each time.[160]

In the following we shall discuss primarily only the notation for the execution, not the execution itself and the material conditions of the performance. Prescriptions for the execution in this sense refer to the dimensions and components of music, which are not expressly recorded in the abstract composition with its primacy of tone pitches and durations, that means mainly volume and tone color, in a further sense this also pertains to expressive values and to the tempo. On the side of notation they make concrete the *abstract composition.* On the side of realization as sound they make more precise and specify the intentions of the composers and determine more exactly the transformation of the work's idea into its sensual-material reality. They mark the particular and the individual.

There are several axis of the exact internal differentiation of these 'additions'. The classifications are multi-dimensional and not always clearly distinguished from each other. The additions are made taking into account the already determined sound situation of the performance context or with regard to the determining notation, which aims at a result that helps the work to achieve a perfect sound shape that is as adequate as possible to the composed idea. And they themselves as signs move in the triangular constellation of *word, note in a close sense, graphic signs* with respectively varying accentuations. The refined instructions for the performance of the sound realization essentially comprise the following types:

1. Mere abbreviations
2. Instructions for the articulation and voicing, mostly *additional signs* applied directly to the notes
3. Instructions for the execution, mostly in words
4. Determination of the tempo

6.6.3.2 Space and Time Saving Abbreviations

At level zero we have mere *abbreviations*, not actual determinations of the execution, such as the repeat sign. They save notation space and time and are condensed representations still belonging to the abstract composition. Signs are often abbreviations—i.e. *dal segno,* Italian for *from the sign,* usually *D.S.* The symbolic sign significantly reduces the encroaching of the space of the notes and of the deliberations on practical aspects of the realization. It requires some learning and training. Mostly these signs designate the formal process. In fully verbalized form these instructions would be inconvenient and rather longer. The result is comprehensible in regard to the succession of notes. In the following example the S-shaped sign in the second measure means 'dal segno' (see below) (Fig. 6.37).

[160] See Siegele 1968, col. 29.

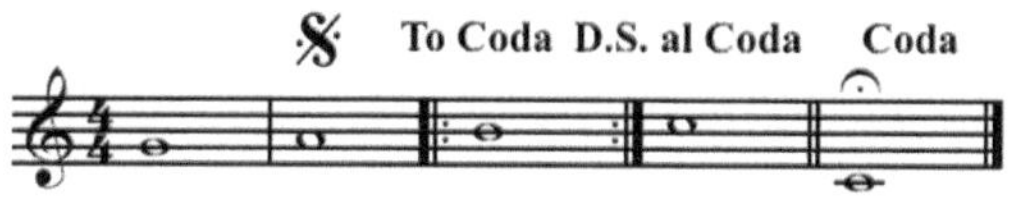

Fig. 6.37 Abbreviations. *Dal segno*
Hyacinth, 17 February 2012, https://upload.wikimedia.org/wikipedia/commons/d/db/Dal_Segno_al_coda_example_short.png, retrieved 13 August 2020

> Use of Dal Segno (D.S.) eliminates the need to write out extra measures [...]. The notes are to be played in this order: G A B B C' A B B C' C.[161]

6.6.3.3 Articulation of Tones and Phrasing of Themes and Tunes

Prescriptions for articulation and phrasing are mostly *additional signs* directly applied to the notes, as the staccato point directly above or below the note head or the ligature combining several notes to a cohesive unit etc. *Articulation*, which includes various instrumental and vocal techniques that specify the respective tone color, primarily relates to single notes, such as the *tenuto* (sustained as opposed to the briefly played st*accato*). Often expressed in words are the instructions for the playing technique, such as the *col legno*, with the wood of the bow of string instruments. In contrast *voicing* applies to larger sections of the 'sound speech', which may be marked by ligatures.[162] In analogy to spoken language these would be 'suprasegmentals', components that go beyond single phonemes. Both types determine how the music will sound in each instance. They are a continuation of instrumentation with further means, a kind of 'micro-instrumentation'. And they are a partial moment of subjectivation, a characteristic of free, albeit work-bound, realization. The dimension of the iconic, the image-like, then gets extended to graphic notation as further differentiation. At the same time this is the transition from notes to image signs. And it prepares the next turnover. Script in itself does not imply sharpening. If you shift the accent within the triangle of *word, note in a close sense, and graphic sign* towards the latter, the specificity becomes fuzzy again (see Sect. 6.7 Graphic Notation and 7.3 Musical Graphic).

6.6.3.4 Expression Marks, Technical and Semantical

Expression marks in a stricter sense usually consist of words. Formulated verbally instead of notational, they constitute a special form of 'text'. They oscillate on an axis between the poles of focusing on technical or on semantic instructions. This

---

[161] Hyacinth, 17 February 2012, https://upload.wikimedia.org/wikipedia/commons/d/db/Dal_Segno_al_coda_example_short.png, retrieved 13 Aug. 2020.
[162] Cf. e.g. Gould 2013.

completes the musical text with a verbal text, by words such as*singing, tender*,[163] phrases such as *somewhat urgent (very little)*,[164] "following the song" (G. Mahler) or, rarely, whole sentences. An important, separate area are expression marks for expressive values—, characters, moods or affects—, such as), *singing, animated* or *very quiet and steady, hardly audible* (A. Schnabel).[165] In R. Schumann's Novelletten, op. 21, 1838, a set of eight pieces for solo piano, expression marks serve as titles and for the segmentation of the pieces, e.g.:

No. 1 in F major Marked and strong

No. 2 in D major Extremely fast and with spirit – In tempo – Intermezzo: Slightly slower, quite delicate – First tempo)

No. 3 in D major Light and with humor – Intermezzo: Fast and wild – First tempo – First tempo – In tempo)

No. 4 in D major Moderate dance/Ball-like. Very lively – More urgent – First tempo – Even faster – First tempo

No. 5 in D major Noisy and festive – Slightly slower – Slightly slower – Very lively – First tempo)

No. 8 in F♯ minor Very lively – Even livelier – As before – Bright and funny – Continued: Simple and melodious – Tempo as in the previous piece – Continued and Conclusion: Lively, not too fast – Gradually lively – Intimate – Tempo I

The significant expression marks used also and especially in instrumental music contribute to the specification and clarification of semantic fields[166] (see Sect. 3.1 the Mahler examples). In addition there are primarily technic-related abbreviations like *rit.* for *ritardando* or also *ritenuto* or signs like *f* for *forte*, in the end also an abbreviation. A boundary case, which belongs to it however, are work and movement titles. Probably the first work in Europe with expression marks combines instructions for the *performance*, an instrumentation specifically determined by the composer, with instructions for a lively *execution* in the interpretation. In this case it is a matter of the elementary binary and overall distinction between loud and soft, a mimesis of the echo effect: A work of Giovanni Gabrieli (1554/1557–1612/1613) for double choir instrumental ensemble from the *Symphoniae sacrae* (1597) is explicitly and programmatically called *Sonata pian' e forte*.

> Venetian polychoral style which arose from architectural peculiarities with regards to St Mark's Basilica. Sonata (at this time) means a piece for instruments. It was probably written to be played as part of a Catholic service at St Mark's, Venice. It was written for 8 instruments divided into 2 groups of 4 and placed in opposing galleries in the cathedral.[167]

The work represents a double progress in compositional organization and notation on the way to a fully notated work in score form: Firstly, the instrumentation is indicated for all parts, which was not usual at the time, since an instrumental-specific

[163] Artur Schnabel: Sonate für Klavier, 1923; II. movement, 3. accolade.

[164] Schnabel 1923, II. movement, 4. accolade.

[165] Schnabel 1923, II. movement, Ms. p. 5, 1. accolade.

[166] Cf. among others Heister 1997.

[167] Cf. https://en.wikipedia.org/wiki/Sonata_pian%27_e_forte, 26 March 17, retrieved 1 July 2020.

Fig. 6.38 Giovanni Gabrieli, 1554/1557–1612/1613: *Sonata pian' e forte,* from *Symphoniae sacrae* (1597), excerpt
Quoted after © 2003–2019 by J. Gedan, https://www.pian-e-forte.de/noten/notbsp/107503.htm

compositional practice was only in the emerging phase. Secondly, the work is the first known work of music published in print that is provided with dynamic instructions. Apparentely it uses a terrace dynamic, which is said to be typical for early music. However, if one considers, for example, measures 62ff, one can assume that not only the two volume levels piano and forte are to be applied here, but that a *decrescendo* is to be made from the initial *forte* to the *piano* in measure 64, because otherwise the passage can hardly be played convincingly.[168] The expression marks for the volume are also used to differentiate between the various types of structures in the 'abstract composition' (Fig. 6.38):

> In the Sonata *Pian e Forte*, the indication *Pia*n appears only when a choir plays alone, the indication *Forte* appears only in double-choir passages, with one exception. This clarifies the difference between one and several choirs.[169]

The progress of differentiation in notation and composition is considerable. More than two centuries later, for example, in Webern's *Variations for piano* op. 27, 1935/1936, in the second movement, each two-tone group receives a new expression mark for the volume; additional articulation marks, such as the *staccato* in measure 1 here, are also often applied (Fig. 6.39).

Roughly speaking, a distinction can be made between technical, pragmatic expression marks such as instructions for the volume, or the pedalization relevant to the volume and timbre of the piano, and primarily semantic ones. Terms for articulation like "legato" or "*martellato*" are transitional types. Of course, the pedalization of the piano has also expressive-semantic values. Artur Schnabel, for example, succeeded in pushing even this technical dimension of the execution in the direction of a verbal text. Thus he writes in the fifth movement of his Sonata for Piano, m. 7: *non legato (not*

---

[168]Ibid.

[169]See https://www.rosenstein-gymnasium.de/e107/Musik/leistungskurs/werke.html, retrieved 1 July 2020.

Fig. 6.39 Anton Webern: *Variations for piano* op. 27, 1935/1936, II. movement, m. 1 Universal Edition, Vienna (1937). Plate U.E. 10881. IMSLP28271-PMLP04634-Webern_-_Variations,_Op._27

noisy) and *in both hands the thumbs sounding*. No expression can also be an intention. For the central number III of his five-movement work *1922 Suite für Klavier* Paul Hindemith demands as general expression mark *Very Quiet Half. With little expression.* At least a bit of expression. Schnabel continues there further too. In the fourth movement of his *Sonata*, right at the beginning, he demands the paradox *without expression*, which corresponds approximately to Paul Watzlawick's ironic paradoxical request "Be spontaneous," but immediately adds *very quiet*. In general, there is also here the double movement in the sense of Hanns Eisler: expression becomes material—material becomes expression. Beethoven used German terms instead of the usual Italian ones. In the already mentioned 2nd String Quartet in F sharp minor op. 10, 1st movement, Schönberg also uses the German and then, as a (back) translation, the Italian tempo indication *moderate, moderato*; both have an affective tinge. This result, as already intended by Beethoven, a slightly stronger expressive aspect compared to the Italian terminus technicus, in which, in turn, semantically expressive elements are sublated that became material—"Allegro", for example, originally means not fast, but joyous.

Another dimension to which expression marks can refer is the optical-gestural aspect of the multimodal music process. The optical, especially the gestural, provides an additional layer to the textual-verbal. In contrast to conducting, the gestures are bound when playing the piano, only relatively free, but the facial mimic is free and more expressive in its own right. And posture and body movement also contribute to the semantic, without the playing having to become 'theatrical'. This goes back a long way, historically and factually. In the tradition of rhetoric, the late baroque poet and Enlightenment philosopher J. Ch. Gottsched, for example, in 1728 referred to the "good recitation" "both to language and to the movements of the body".[170] This also serves to sharpen the sound realization, both in regard to technique and semantics. At the end of the first movement of the Sonata for Piano by A. Schnabel the gestural is implied, at the end of the fifth movement Schnabel says it explicitly.

I: Listen to the rest well! End only afterwards!
V: Keep your hands long over the keys!

[170]Quoted after Siegele 1968, col. 16.

Gregor Piatigorsky, one of Schnabel's partners, needed more than just music to understand the Sonata for Violoncello solo 1931, namely additional words and gestures. He still did not fully understand the work, however.

> This bizarre music fascinated me only with Schnabel's 'obligatory' speech accompaniment. He spoke magnificently about his musical ideas, but without his extraordinary eloquence, his gestures and his performances on the piano, the work was much less convincing.[171]

The *obligatory verbal accompaniment* in oral, spoken form belongs to a preparatory stage of the sound realization, the rehearsal. In written form, like the obbligato accompaniment, it belongs to the work as a score written down in complete form. P. Hindemith mixes gestural, motoric and expressive elements in the above-mentioned work, composed in 1922, entitled *1922 Suite for Piano* op. 26, a series of stylized trendy dances. Thus he writes for the concluding *Ragtime* (No. V), which is rather a kind of instruction manual or instruction for use than an actual instruction for performance and execution:

> Mode d'emploi – Direction for use!!
>
> "Don't pay any attention to what you have learned in the piano lesson. Don't think long about whether you have to strike D sharp with the fourth or sixth [!] finger. Play this piece very wildly, but always very strictly in rhythm, like a machine. Consider the piano here as an interesting kind of drum set and act accordingly. "Shake well before use!".

This is post-dadaistic. Even more far-reaching tendencies to render performance titles independent can be found, pre-dadaistic and with a sometimes somewhat aimless irony, especially with Erik Satie. So for example

> *Comme un rossignol qui aurait mal aux dents* = Like a nightingale with toothache. Think about yourself[172] etc.

The path to the cover as content, to various instructions to the recipients including the expression marks as word husks without sound and sense fruits is paved from here and is trodden after 1945. But already Mozart had made epochal suggestions with regard to movement titles. In the title of his flute quartet in A major, K. 298, IIIrd movement, he combines genre names with expression marks as follows:

> 'Rond*ieaoux*. Allegretto grazioso, mà non troppo presto, però non troppo adagio. così – così – con molto garbo ed espressione'.
>
> = Rondeau/Rondo [with a permutation of the Habsburgic motto acronym AEIOU] Allegretto, graceful, but not too fast, yet not too slow. So lala – with much decency and expression.[173]

---

[171] So Gregor Piatigorsky, *Mein Cello und ich und unsere Begegnungen*, München 1975, p. 173. Emphasis H.-W.H.

[172] https://www.capriccio-kulturforum.de/scherz-satire-ironie-und-tiefere-bedeutung/5306-bitte-recht-cholerisch-kuriose-vortragsbezeichnungen-in-der-musik/, August 2013–December 2014, retrieved 11 September 2015.

[173] Wolfgang Amadé Mozart, New Mozart Edition, series VIII, work group 20, Sect. 2: Quartets with one wind instrument [NMA VIII/20/Abt.2], p. 57 – Emphasis H.-W.H. This complete vowel series, which is superfluous for the genre designation, is also probably a parody of the French vocal accumulations at the end of words, which are then pronounced with a single sound. (Thus) B. Rusam, mail of 20 March 2016.

Fig. 6.40 Wilhelm Busch: Der Virtuos (The virtuoso), 1857, Finale furioso https://www.projekt-gutenberg.org/wbusch/virtuos/virtuos.html, retrieved 24 September 2015

The expression marks refer to the *how*, but at the same time to a *what*. Obviously a little story is told—in connection with music, though without much sympathy for it. In Wilhelm Busch's series of pictures *The Virtuoso. A New Year's Concert* from 1865 Busch narrates the dramatic action after a short verbal introducing the situation only with pictures as well as (sub)titles naming the stages, which are analogies to expression marks.

> Silentium Introduzione Scherzo Adagio Adagio con sentimento Piano Smorzando Maestoso Capriccioso Passagio chromatico Fuga del diavolo Fortissimo vivacissimo Finale furioso Bravo, bravissimo![174]

With the virtuoso, the devil is not absent, as we had already noticed in the case of Tartini, and one of the performance titles invented by Busch as "movement title" is therefore the fugue of the devil. In the penultimate picture, Busch, himself here virtuoso and pre-futurist, draws a

> simultaneous representation of several phases of the pianist's movements, and the notes dissolve into musical notes dancing above the grand piano.[175]

The blurring in the representation is sharpening the mimesis of the represented (Fig. 6.40).

6.6.3.5 Tempo and Character

An even more overspanning, 'suprasegmental' dimension is tempo. It determines the character of sections or entire movements in both semantics and technical effects. It moves between slow and fast within an only rudimentarily 'structured continuum'.

[174]Figures see https://www.projekt-gutenberg.org/wbusch/virtuos/virtuos.html; cf. https://de.wikipedia.org/wiki/Der_Virtuos, 12 September 2015, both retrieved 24 September 2015.

[175]Wikipedia, Ibid.

Relative and absolute tempi have both historically and culturally differentiated implications (see Sect. 7.5.2). The tempo also co-determines the general character of a movement or part of a movement: „Allegro“ for example literally means joyfully, but since the beginning of the nineteenth century it mostly means a relatively fast tempo. Explicit indications with prescriptions of tempo can be found at the latest since 1600 in Spanish tablatures.[176] This again expands the conceptual possibilities of composing—the choice of tempo is no longer left to the intuition or caprice of the executing musicians, instead it is determined beforehand in the composition. Changes of tempo, beyond the mere fading out and slowing down in the finale, are yet another stage; in classical jazz up to free jazz for example they were rather uncommon, since it was primarily dance music—only doubling as escalation or final effect was frequent (see Sects. 7.5.3 and 7.5.4 *Rubato*) Very complex multi-faceted thoughts about the changes in musical understanding of time and its measuring, with far-reaching societal considerations, were developed by U. Siegele.

> The tempo is […] the measure of movement in a recital of a musical piece; it determines the length duration of the unit of a specific musical measurement of time, namely the beat accorded to a specific note, in relation to the general measuring of time. In accordance with the changing understanding of time the occidental polyphony formulated this relationship twofold. The younger style determines the duration of the unit, the beat, adds the beat times to bars, the bars to theoretically limitless musical pieces. Its symbol is the pendulum clock, its instrument the metronome. The older style determines the duration of time of the musical piece itself or a generally limited time period (and thereby follows the more primitive meaning of the word, which is documented in the opposition of ‘time and eternity’ in synchronicity with ‘time and goal’); it splits this limited time into the number of bars it contains, according to the measure of the bar regarding scale and proportion. Its’ symbol is the hourglass. The younger style is capable of changing the tempo in the imagination continuously […]. The older style in contrast changes the tempo in proportioned degrees. It is bound to the church calendar and its annual cycles, the repetition, the type and model. The younger style however relates to secular progress, the unmistakably new, the individual and the original genius. Both types correspond with the respective contemporary systems of tuning, the older using unequal temperament the younger using equal temperament. The transition from the older to the younger style, equally a transition from measuring units of bars to measuring the units of note values, started with mensural theory. It announces itself in the 16th, and develops up until the early 19th century, and is bound to the development of the accentuated bar.—Style of composition and notation, notation and time measure are closely related.[177]

In the context of the industrial revolution and the late enlightenment the compositional subjectivization was augmented. It correlates to the differentiation in the notation. Both naturally influence the realization as sound.

> The slow tempi are stretched out in a recitative style, the faster ones accelerated playfully. *Mesure* and *Mouvement*, bar and movement of bar are now as separate as body and soul, the mathematical division is joined by sentiment. The possibilities of time measures expand and multiply. Still H. Ch. Koch (propped by J. J. Rousseau and deviating from J. J. Quantzen’s four-level scale) in 1802 was able to arrange the time measures into five main types, which

---

176 Bengen, Frobenius, Wolf, and Behne 1998, col. 447.

177 Siegele 1968, col. 26; see also Bengen, Frobenius, Wolf, and Behne 1998.—On this much discussed topic of time in music cf. e.g. Eggebrech 1997, pp. 172–180.

> will allow the adequate movement of each musical work without any significant deviation: »largo« – »adagio« or »lento« – »andante« – »allegro«, at times »vivace« – »presto«. But end of 1826 Beethoven writes to his publisher Schott: »We can almost not have *tempi ordinari* anymore, in which one has to follow the ideas of the free genius.«[178]

The segmentation of tempi from a continuum can only designate relative speeds and durations, for the particular as well as the whole. The tone durations are quantified relatively exact: A half note sounds twice as long as a quarter note. The actual duration however is relative, not absolute. It depends on the respective tempo. The tempi themselves are not objectively quantified or exactly. In productive and receptive realization continua dominate, transitions, in rudiments with a structured continuum. This changes drastically with the introduction of the metronome, patented by the Viennese J. N. Mälzel:

> [The metronome] is a device that produces an audible click or other sound at a regular interval that can be set by the user, typically in beats per minute (BPM). Musicians use the device to practice playing to a regular pulse. Metronomes typically include synchronized visual motion (e.g., swinging pendulum or blinking lights). [...] To get the correct pulse with this kind of visual device, the musician watches the pendulum as if watching a conductor's baton.[179]

And now there is a unified, objective measure. The relation to the pendulum and the clock has mechanised and externalized the quantising of the time continuum. The differentiation goes one level deeper: even the minute is divided. Mälzel's metronome has earlier forms and precursors. As with instrumentation and indications of recital style for volume we are again at the epochal transition of around 1590 and 1607. The background are the laws of the pendulum discovered by Galilei around 1600.

> The French universal genius Marin Mersenne (1588-1648) discovered, that the swinging of a pendulum can not only be used to measure the pulse of the sick, but also the movement of heavenly bodies, the right division into degrees of a sun dial (!) and the determining of the sound of speed, as well as the tone frequency and the tempo of musical pieces.[180]

Beethoven was very delighted with this progress. Yet he only furnished a small part of his works with metronome numbers. And these are again often ambiguous or misleading and disputed. Again this results in some margin for deviations and 'tolerances' as a further field of study for fuzzy logic.

Béla Bartók attempted to empirically and very exactly determine the tempi with the help of metronome numbers. For piano works he even determined the total duration down to the second—he was himself a prestigeous pianist, and was able to determine the times quasi-experimentally ex post. This is precise and perfect. But it restricts the renewed realization of sound and the striving beyond technical perfection. The quantification of musical tempi and duration values is difficult in spite of the reference to the objective time. This can even be seen in the numbers on a standard metronome:

---

[178]Quoted in Siegele 1968, col. 26.

[179]https://en.wikipedia.org/wiki/Metronome, 17 August 2020, retrieved 22 August 2020. Cf. e.g. Helmut Kowar 2001, retrieved 25 August 2020; very detailed in Alexander Evan Bonus 2010.

[180]Helmut Breidenstein 2007.

40 42 44 46 48 50 52 54 56 58 60 63 66 69 72 76 80 84 88 92 96 100 104 108 112 116 120 126 132 138 144 152 160 168 176 184 192 200 208

The steps are significantly different from each other: for very fast tempi the differences become smaller, and the fuzziness grows. Stockhausen attempted to master this problem through logarithmic scaling. Obviously this requires the division into twelve duration values in a 'continuous octave', arbitrary from the view of tempo or rather durations, and only an analogy to the tuning system of equal temperament, which divides the octave into 12 *equal* intervals.

Stockhausen developed [1963] a serial organizational principle at the center of which stood the concept of a twelve-step duration series possessing the same structural properties as the basic twelve-tone pitch series. This became the basis for the entire process of serial organization of *Gruppen* (Misch 1999, 53–54). This duration series, however, is expressed not as single units (which would correspond to single vibrations of a pitch) but rather as metronomic tempos in sufficiently long stretches of time to enable conductors and musicians to change tempo with precision. However, because the resulting "fundamental durations" are not small enough for use in the musical detail, subdivisions corresponding to the transposition of the overtones of a pitch's harmonic spectrum are used [...].

The twelve logarithmic metronomic tempos used in *Gruppen*, covering a tempo "octave" (doubling in speed) from Viertel = 60 to 120 are [...]:

60

63.5

67

71

75.5

80

85

90

95

101

107

113.5

120

The composer recalled that when Igor Stravinsky saw the score for the first time, he wrote that such fractional metronomic values as 63.5 and 113.5 were "a sign of German thoroughness" (Stockhausen 2009).[181]

In spite of all progress of differentiation of what can be noted there remains an irreducible, significant rest of fuzziness.

»Expression may be said to represent *that part of music which cannot be indicated by notes* or, in its highest manifestation, by any symbol or sign whatsoever. It includes all the nuances of tempo, dynamics, phrasing, accent, touch, bowing, etc., by which the mere combination and succession of pitch-time-values is transformed into a living organism.« As W. Apel (Harvard Dict. of Music, 1944) describes the area of the expressive recital.[182]

[181] https://en.wikipedia.org/wiki/Gruppen, 18 August 2020, retrieved 19 August 2020.

[182] Quotation in Siegele 1968, col. 29.

These gaps—and they are many—create space for freedom and play in the interpretation and the realization as sound. Idea and realized idea will never be identical.

6.6.3.6 Deliberate Determination of Tone Color. Instrumentation

An essential concretization of the 'abstract composition' and compositional determination of the respective particular tone color of a work occurs through explicit, notated instrumentation, in tune with the intention of the composer. With this the voices of the 'abstract composition' are distributed and concretized respectively in each particular musical instrument (or vocals). The particular sound is then again defined in more detail especially through the mentioned prescriptions of articulation. An elementary, and as such unremarkable due to it being a naturally effective approach to instrumentation, is the designation of each particular part, the 'voice'. In European melodic polyphony this begins very early, since the voices are often notated together on one page and had to be distinguishable. It is not by chance called 'voice', Latin vox, since this is composed music and art music, which was dominantly vocal up to the of the early renaissance. The notation of the simultaneity of variety in the particular voices as 'parts', characteristic for the score, is a historical accomplishment. It is first employed in instrumental music. The arrangement of voices according to pitch range is familiar to us today for the voice types soprano, alto, tenor, and bass. In orchestral scores this returns *within* the individual instrumental groups, as for example with the violin (I and II), viola, violoncello, bass etc. Ex posteriori this seems evident, but it is not. It requires thinking together via analogies acoustic-temporal and optical-spatial aspects. The similarity of different voices is represented vertically in layers: a temporal identity of differences. The temporal progression is represented horizontally. The time arrow from left to right follows the conventions of newer European scripts. Instead of the *superimposition* of the score the part books presented them in *juxtaposition*, with each voice having its own booklet or book.[183] An overview of the whole is therefore at least difficult. But here some gaps are filled by worded instructions. Interestingly the superimposition of the type 'score' notation already existed earlier, at the very beginning of melodic polyphony in the above mentioned musical theory treatise *Musica enchiriadis* from around 850 or later[184] (see Sect. 6.10). But it took centuries for this notation to prevail in the practices of composition and performance. The modern score arrives late in its developed form. Pioneers are the instruments capable of polyphony, keyboard instruments for one, stringed instruments on the other hand, especially those of the type guitar or lute, somewhat less the string instruments. The differentiation into polyphonic instrumental accompaniment to the often soloist melody accelerates things (see Sect. 6.6.2.2 tablature). The determination of instruments in the score as

[183]See e.g. Jaschinski 2001.

[184]The interplay of superimposition and juxtaposition, sequentiality and simultaneity offers analogies to the three-field system. See Sect. 6.10.

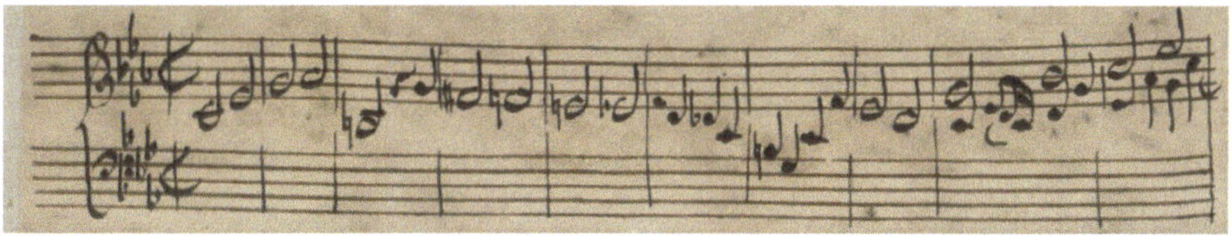

Fig. 6.41 J. S. Bach: *Musikalisches Opfer (Musical Offering)*
Beginning, theme ms. 1–10, Autograph
Berlin State Library, Germany. Quoted after: https://de.wikipedia.org/wiki/Musikalisches_Opfer#/media/Datei:Ricercar_a_6_BWV_1079.jpg

concrete desires and prescriptions instead of adaptation to given circumstances are instrumentation in a strict sense. It appears as model breakthrough only in the time between G. Gabrieli's mentioned *Sonata Piano e Forte* for a double choir ensemble in 1597 and Monteverdi's opera *Orfeo* in 1607; before then, as well as later, the instructions are expressed in words or rather in writing. In the work of Monteverdi the instrumentation gains semantic value behind the material tone color.

> In his published score Monteverdi lists around 41 instruments to be deployed, with distinct groups of instruments used to depict particular scenes and characters. Thus strings, harpsichords and recorders represent the pastoral fields of Thrace with their nymphs and shepherds, while heavy brass illustrates the underworld and its denizens.[185]

As his late *Kunst der Fuge (Art of the Fugue)* is the above mentioned cyclic work *Musikalisches Opfer (Musical Offering)* of J. S. Bachs (1747) an an exemplary 'abstract composition' (Fig. 6.41).

Anton Webern's instrumentation of the *Ricercare* from the *Musikalisches Opfer (Musical Offering)* of J. S. Bachs (1747) aims to use the differentiation of timbres to clearly distinguish the abstract composition' by means of sound sensuality (Fig. 6.42).

Webern's instrumentation transforms the abstract tonal identity of voices resp. parts into concrete differences. This expresses the varying structural and semantic significance, quasi the 'place value' of the individual parts in their progression and at every point of the progression of the work—as in main and second parts—a changing relationship—, and in the spirit of the "durchbrochene Arbeit" of the Viennese classicism, the wandering of the theme through *various* voices or instruments of the composition.

An instrumentation of an 'abstract composition' is at the same time also a treatment and adaption and transformation of the canon of the same work, the *musical offering* (Fig. 6.43).

> The Modern Jazz Quartet used one of the canons (originally "for two violins in unison") as an introduction to their performance of the standard song "Softly, as in a Morning Sunrise". (They feature this on their album Concorde (1955).) The Royal Theme is played on the double bass, with Milt Jackson (vibraphone) and John Lewis (piano) weaving the two imitative contrapuntal voices above.[186]

---

[185] https://en.wikipedia.org/wiki/L%27Orfeo, 21 August 2020, retrieved 22 August 2020.

[186] https://en.wikipedia.org/wiki/The_Musical_Offering, 17 April 20, retrieved 8 August 2020.

Fig. 6.42 Anton Webern: Instrumentation of the *Ricercare* from J. S. Bach's *Musical Offering*, m. 1–5. Excerpt from the score.
Universal Edition, Vienna (1935). Plate U.E. 10277. IMSLP288160-PMLP04550-BachWebernFuga1079

Blurring and sharpening as refinement are intertwined in the instrumentation. The realization as sound of the instrumentation includes both the progressive enlivenment of the 'abstract composition'—through nuances, shading, dark–light contrasts and transitions of chiaroscuro, generally an "art of transition", as formulated by Wagner in regard to instrumentation. As already noted, the aim for aesthetical perfection had primacy over the only abstract technical perfection. So it says in a letter from Leopold Mozart, Milan, 15 December 1770:

Fig. 6.43 MJQ: Canon from *Musical Offering,* m. 1–3
https://upload.wikimedia.org/wikipedia/commons/4/4c/Canon_from_the_Musical_Offering.png
Double bass and keyboard instrument are part of J. S. Bach's instrumental horizon. The vibraphone however transcends it

> Before the first rehearsal with a small orchestra had been completed, there was no scarcity of people, who with satirical tongue already defamed the music as something young, miserable… they claimed, that it would be impossible, that such a young boy, on top of that a German, could write an Italian opera, 'and that he, despite being acknowledged as a great virtuoso, could impossibly understand sufficiently and appreciate the *chiaro ed oscuro* necessary for the theatre.[187]

6.6.4 Changing Cubicle. Sharpening with Turnaround—Transcription as Transformation of the Sound Form into a Notation Form

With the anticipation form and the score form and the completely elaborated composition with instrumentation and prescriptions for the execution, we have the most advanced notational preparation for the sound form. Once this is realized, the written forms disappear at least for this temporally expanded time stretch of realization. These reintroduce the sound form again insofar that the sound form itself can be translated into scripture, i.e. transcribed, even and especially in music types, in which the script does not even exist *before* the sound, meaning traditional music, folklore, jazz, some popular music. This incursion or return of the sound form in the written form marks a sharpening with another starting point. This 'transcription' of already as sound realized music as re-translation or reverse engineering from sound to notation. This does not imply the emerging of new shapes, but rather a permanent holding on to the procedural improvisation. The transcription may imply certain retention of notation as dominant form of recording, the exteriorized retention of music. The procedural, the improvisation is captured and recorded—and of course through this

[187]Quoted in: Studierende des Musikwissenschaftlichen Institutes der Universität zu Köln: Proseminar Opern Mozarts, WS 2001/02: Quellen und Informationen zu 'Termini' in den Opern Mozarts. https://www.rodoni.ch/OPERNHAUS/idomeneo/glossarioted.html#Anchor-Aria-49575, retrieved 13 August 2020.

optical-spatial, thing-like representation it becomes more accessible for teaching and for analysis. And thereby it again becomes easier and quicker to recall past, work-like shapes and use them for further conceptions of new improvisations (see Sect. 3.6).

This reveals almost paradox dialectics of retention through technical re/production. The technical recording and storage is a historical leap. Transcription became easier with the arrival of technical reproduction of music, which allows repeated hearing of recorded improvised resp. not notated music, or allows for special help such as slowed down replay. The digitalization and the subsequent almost automatic transcription present another leap. The mental effort that is necessary even with technically reproduced music is no longer required to transfer the sound into writing, to translate the processual into the thing-like (see Sect. 3.2.3). Hearing works is considerably faster than writing. Simultaneous recording of music in writing, comparable to stenographic recording of speech, almost inevitably leads to fuzzy results. Only the extraordinarily gifted, with a tendency for eidetic memory, can note down at a one-time hearing or from memory a performed piece of music—even an improvisation. Mozart in 1770 demonstrated this probably with the polyphonic *Miserere* of Gregorio Allegri, from the 1630s. Written for two choirs, of five and four voices respectively, singing alternately and joining to sing only at the ending in 9-part, the work which was performed exclusively in the Sistine Chapel, and the score was kept secret. What appears in Mozart's presumed transcription as a playful luxury and the undermining of a kind of censorship and the preservation of exclusivity became a bitter necessity in situations in which notes were not available for making music, but music-making was necessary for survival. In the extreme, this applies to Nazi concentration camps. There, for lack of scores, musicians had to reconstruct entire scores from memory repeatedly. And some of them were able to do so—a special kind of sharpening, which was certainly not always without blurring.

6.6.5 Changing Clothes. The 'Postproduction' of 'Versions'

Production after completion of production results in seemingly or real sharpening between improvement and mutilation of intention. The finalized score in handwritten manuscript form, the 'autograph', can already be a clean resp. fair copy, but often it is not. It is achieved by the clean *copy*—clean and precise, but still containing mistakes—of the copyist.[188] Only now can it serve as printing template or template for engraving. Just like the template itself, it goes through several stages of corrections. These are concerned with always increasing precision, and therefore refinement and sharpening. Corrections are generally an important component of work creation and its refinement. Often composers continue to work on their 'text' in the score form especially *after* the work has been completed. As for example was noted about Carl Maria von Weber:

[188] At a later time, which also has long passed, the transcription of the manuscript in handwriting used a 'machine', specifically the typewriter.

> This involved, emanating from the inner 'completion' [having-finished] of the work, rapidness and security of the production, not a rigid finalization of it with the noting down of the last note. On the contrary, he would in many cases sacrifice the labors of many days, later, especially after the first performances, or in the rehearsals, for a newly attained better conviction. In this manner only the third elaboration [treatment] of the entire last Allegro in the second finale of »Euryanthe« […] satisfied him. […] He did not shy the labour of rewriting the respective part of the score cleanly, without any damage to its fragility.[189]

The subsequent treatment serves in its intention to further approximate the anticipated perfect work. The result oscillates between improvement and mutilation of the intention. This results in a partly apparent, partly actual refinement. The *revision* of an actually accomplished work, completed even in the most advanced state of the score form, has its analogy in the *postproduction* of movies or music recordings. Especially self-editing can best be understood as anticipating and subsequent adaptation to the conditions of realization, both productive and receptive. We could here conceive of the paradoxical term of '*subsequent anticipation form*'. It is usually based on the feedback loop that occurs with and after the realization as sound. This happened extensively in the works of Bruckner and Mahler. Mahler does it by himself. Bruckner is mostly pressured from outside, and he is 'helped'. Usually Bruckner is not so much concerned with the adaptation of the work according to the ideal of his work idea, its' as consummate as possible shape for the realization as sound already in the score form, but rather an adaptation to actual or even just assumed perceptual challenges for the audience—not to forget the perceptual challenges for the musicians[190] (see Sect. 3.5.5).

Here there are no absolute standards, only relative ones. The work idea is one measure. But it can only be reproduced incompletely and remains mostly hypothetical as a reconstruction of work creation and development. Such a circle makes judgements more difficult: For the reconstruction of the work idea several versions are available, which are in principle equivalent. But it is exactly these that are to be measured and weighed. The first version is often the best starting point for comparisons, to see if following versions are improvements or not. The latter is often the case with Bruckner. The potentially differing valuations of the versions could be brought into focus multi-dimensionally through fuzzy logic. Augmentations or diminutions of the instrumentation are a rather minor intervention. Somehow undecided is the quality of versions in the following case. In 1928, Berg arranged the second, third and fourth movements of the six movements of the *Lyric Suite* (1926/1928)—*Andante amoroso, Allegro misterioso, Adagio appassionato*—for string orchestra, with violin I, violin II, viola, violoncello, and contrabass. The homogeneity of the strings' instrumentation is preserved. It is merely augmented and extended in the depth.

> There are both gains and losses in the string orchestra version. That version, performed in such a large hall and with so many players, loses the intimacy of the quartet version, without

[189] Weber 1866, pp. 459f.

[190] For a multi-faceted comment see Tadday 2011; Emans 2016, Teich Geertinger 2016.

> the corresponding gain of the additional colors that winds and brass would provide, yet the great complement of strings adds its own tonal depth.[191]

Adorno differentiated and specified the meaning of this re-instrumentation beyond mere sound:

> If the lyrical nature of the Suite is best fostered in the quartet, its dramatic nature is best fostered in the string tutti; only here are its contours dissolved as completely and enigmatically as the accompanying concept of the sound demands; and only here does the paroxysm attain its full catastrophic force (Adorno 1991, 105).[192]

The original version is insofar superior, because it is simply more complete.

The imperfect perfection of the work is a central intersection of the art process and yet not a final telos and an end point. In procedural arts such as music or theatre it principally has to progress to sensual realization and then to perception. This always demands from the composers to continue their work, new interventions, retouching, changes improvements. Even in the fine arts the artists often cannot attain closure. Turner continued painting in the museum. The detachment from the work, leaving it alone and abandoning it, often seems hard for its creators.

In improvised music without notation refinement and crystallization of poignant shapes occur until the completion of the most highly developed score form in accordance with the "entirety complete in itself" "fully completed unity" in another way, but again in a graded, in its tendency endless process: through trying out and then practicing what was newly discovered. In jazz for example this can often be traced clearly based on the phonographic form, when different recordings—'takes'—of the same standard or even of only a single chorus are available, of which the best or apparently best is selected for release.

In line with the formal content the three basic procedures, as in notated music, that come into effect are repetition, variation and contrast. Analogous on a sort of meta-level are the various takes of the recording of classical, notated and performed works. The basic procedures refer to aspects of the sound realization and can be applied deliberately. The verbal instructions of the sound editor correspond in principle with those of a conductor, only without the gestural mimetic components- as for example 'the oboe a bit louder please', 'the violins with less vibrato', 'all of it more lively' etc. In both cases the comparison of the unreleased versions including the released version, usually done without the aid of fuzzy logic, could gain precision *with* its support.

The modal logic with its realized and unrealized potentials once again comes into play. At first it specifically affects the first four forms of existence. Only the binding, authoritative record as a fair copy that corresponds to the author's intention bans all other possibilities into the Hades of unrealized potentialities; at least preliminarily.

---

[191] Lisa Hirsch: Alban Berg *Three Pieces* from the *Lyric Suite* Lyrische Suite for string orchestra. Quest for Fire at the S.F. Symphony, 26 November 2010, San Francisco Symphony, https://www.sfcv.org/reviews/san-francisco-symphony/quest-for-fire-at-the-sf-symphony, retrieved 2 February 2020.

[192] Adorno 1991, p. 105, Quoted in: https://en.wikipedia.org/wiki/Lyric_Suite_(Berg), 5 July 2020, retrieved 5 August 2020.

For once discarded material can be one hand reactivated, for newer versions of the same work as rewriting, revision etc. or for new, other works. Secondly the consequent transition from the score form to the sound form opens up a new catena of dialectics of the possible and the real. The written work is the realization, from the view of the conception and the idea. Viewed from the sonic interpretation as realization in another medium, the acoustical, it is—only—a possibility, namely for *various* interpretations and thereby new realizations, this is repeated at the stage of receptive appropriation, starting with the reception form, in new shapes and elaborations. Always an important factor are the dialectics of object and subject, of the material-ideally given and its appropriation.

6.7 The Work as a Notationally Completely Realized Idea. Score Form

There is no end result for the notation in the score form. Since the 1950s it is characteristic for graphic notation, that in the note-*script* additional graphic components appear.[193] Also in traditional notation the combination of word, note, and additional signs was constitutive (see Sect. 6.6). Here now the graphic incarnated in the note, itself partially an iconic graphic sign, is enhanced and gains partial autonomy.

The 'graphic notation' resulted thanks to a forking of 'western' music history in the 1950s leading to new trends such as the "sound composition". For one 'graphic notation' served in the attempt to fixate new complex structures and new sounds in writing. It continues the line of thinking which aims to make accessible systematically what has been excluded from musical language so far, together with the attempt to note down everything even more precisely, especially what is vague, in contours, the emphasis on sound before structure. Following serialist thinking, almost insatiably always new 'parameters' were searched for, such as the inclusion of space as a compositional element. These parameters would be systematically analyzed and subjugated to a design according to the composer's will. The search is boundless, just as the matter itself, the musical material and language. But universal feasibility has its limitations. They lie in the feasibility itself. The serialism aims for accuracy and most comprehensive elimination of all fuzziness, both through determination in the composition as well as through absolute predestination of the realization. The 'electronic music' realizes this intention optimally by switching off the disturbingly fuzzy factor of live musicians (see Sect. 6.8).

---

[193] An extraordinarily broad range of contributions on a wide range of mostly new notation topics can be found in the *Proceedings of the International Conference on Technologies for Music Notation and Representation TENOR* since 2015: https://www.tenor-conference.org/proceedings.html#2015, https://www.tenor-conference.org/proceedings.html#2016, https://www.tenor-conference.org/proceedings.html#2017, https://www.tenor-conference.org/proceedings.html#2018, https://www.tenor-conference.org/proceedings.html#2019, https://www.tenor-conference.org/proceedings.html#2020.

Counter-movements, represented by the likes of 'sound composition' "Stochastic Music" "Aleatoric Music" wanted to free and empower the interpreters for a stronger 'creative', productive instead of reproductive agency. This led to the counter-attack, which now would denoted in the 'Graphic Notation' not only new sounds, but also the fuzzy matter. These paragons and inspirations are only contingently sharpening in the classical and direct senses. As sharpening they seem paradox, as new work ideas are specified through and in the fuzziness—once more an 'exact imperfection' (Nono) (see Sect. 7.4.7).

The designations of pitch and duration have been sufficiently developed and refined in European notation. Now especially the new sounds were to be represented as precisely as possible. The new, so far unknown, unimagined and for the interpreters unfamiliar sounds together with the methods of sound production, vocal and instrumental, necessitated new signs. These had not existed in the notation so far. For that

> a wide range of new symbols [were devised] to denote unusual playing techniques on instruments, such as the use of 'breath' sounds on wind instruments, the bowing of parts of string instruments other than the strings themselves, a wide range of techniques on percussion, extended vocal techniques, and so on.[194]

The combination of notes, pictorial signs and words warrants a certain precision. To that end, for example, even components as the movements of the musicians, which had previously been taken for granted, were explicitly ascertained.[195] Insofar this falls into the category of sharpening.

(S)ome additional parameters of sound, such as the time parameter of the musical structure, the positioning of sound in space, that graphic scores, independent of their function as notation, also have a meaning in the visual realm. […] Because of these additional parameters, this graphical notation system was more flexible and polymorphic than the traditional one.[196]

The independent value of the pictorial is ambivalent. It variates the shape of the notation, but is also vulnerable to—renewed—fuzziness.

> (T)he terms "graphical notation" and "musical graphic" […] were coined in an attempt to define this new script in music.*Notations are, generally, systems of signs/symbols, but the musical graphic is a painting, a drawing.*[197] The important difference between them is the fact that a graphic, though musical, *conveys a meaning not because of its use of signs and*

---

[194] Pace 2007, p. 2.

[195] See e.g. Rosenbaum 2014. A comparison to dance notation, which refers to the three-dimensional, could be productive. To my knowledge it has not been done in detail yet. Apart from Egyptian cheironomies, in Europe systematic dance scripts were only developed since the Renaissance. Today mainly two systems seem to compete for hegemony, the Laban notation („Kinetography Laban", 1928) and the Benesh Movement Notation („Choreology", 1956). See e.g. Ann Hutchinson Guest 2005. Always it is unmistakably 'graphic notation'. Cf. e.g. Schwarz 1993; more generally also Dahms et al. 2001.

[196] https://anestislogothetis.musicportal.gr/the_graphic_notation/?lang=e, retrieved 10 July 2020.

[197] See Rosenbaum 2014 and Gieseler 1978, quoted in https://anestislogothetis.musicportal.gr/the_graphic_notation/?lang=e, retrieved 10 July 2020.

Fig. 6.44 Robert Schumann: *String Quartet* no. 1, Op. 41 no. 1 Quoted after Byrd 2018, n.p.

symbols, but because of its special aesthetic quality,[198] while notation is a code [...]. A graphical score is basically judged by the musicians, who comprehend, read and perform the music that lies *beneath* the script. Logothetis himself referred to his system as a "graphical notation system", since his main focus was to broaden the musical script/code and not to provide a score with illustrative elements.

Even the fact that also here, just as in 'sight music', the pictorial aspects often become autonomous, can be justified as sharpening of the whole through optical aspects: The (note) pattern of the work becomes more differentiated and richer, and thereby indirectly also the work in its sound form and perception form. A third function however leads to a transition to blurring: The pictorial is meant to, in contrast to the score form, leave more room for the imagination of the executing musicians.

Fantastic notes without graphic additions have existed already in traditional notation, and they were not as rare as one would expect. To give just one example from an excellent, very differentiated list (Fig. 6.44).[199]

Several quadruple-dotted notes appear in the third movement of Schumann's *String Quartet no. 1*, Op. 41 no. 1. Double-dotted notes are pretty common, but triple-dotted notes are not. Still, even quadruple-dotted notes have been used many times [...].[200]

This type of notation only appears to be absurd, but it is actually rather practical.

Of course the quadruple dotting could have been avoided by using two or more tied notes, but the 32nd note following complements the quadruple-dotted note, filling the 4/4 measure; so it's easy to see the rhythm without counting the dots, and writing it this way is almost certainly the most readable way to describe the intended rhythm.[201]

The example therefore shows a sharpening, not fuzziness or blurring. We will encounter both in the next chapter, especially in complicated superimpositions of tempo in various layers of the composition. Byrd sees the border to concept art already transgressed in some cases.

An important borderline case of music whose interest might be primarily conceptual is the "new complexity" music. To cite just one example, Brian Ferneyhough: String Quartet no. 3 (1987; Peters ed.), II, includes a number of tuplets nested 4 levels deep, and several meters with non-power-of-2 denominators, e.g., 2/10 and 1/12. My personal inclination is strongly towards considering such work as of rather conceptual than musical interest.[202]

---

[198] Anestis Logothetis 1974, quoted in https://anestislogothetis.musicportal.gr/the_graphic_notation/?lang=e, retrieved 10 July 2020.

[199] See Byrd 2018, n.p., retrieved 7 June 2020.

[200] „Extremes of Conventional Music Notation lists six works by major composers that employ them, and no doubt there are others" (ibid.).

[201] Ibid.

[202] See ibid.

A radical consequence of the intricate intertwining of written notation and sound realization, of sharpening and blurring is reached with ‚sound painting‘ (see Sect. 7.6). Again a countermove to this is the ‚verbal notation‘.[203]

6.7.1 New Sounds, New Signs. M. Kagel's **Pas de cinq**

The space of the performance is transformed by M. Kagel into an instrument through the installation of various sound sources in the floor—descendants of the futuristic *intonarumori*. These musical instruments are handled by perambulating musicians (Fig. 6.45).

> *Pas de Cinq* (1967) converts the whole performance space into a single instrument – a collection of 'lanes' of a minimum length of 5m arranged to form a regular pentagon […]. The lanes are built up using 'scaffolding, small platforms and slopes or ramps of differing angles of inclination'.[204] The lanes are then covered with 'the most varied kinds of materials, f or example sheets of metal, plastic and wood, and runners of jute, cloth and linoleum'. […] No specific details are given beyond this […].[205]

The model for the ironic title is the 'Pas de deux' of the classical ballet. One obvious model for the music is the tap dance. Here the walks and steps provide the sounds, as the 'graphic notation' prescribes here (Fig. 6.46).

> Through this space ten pathways are clearly marked; in journeying along these pathways the five performers tap out prescribed rhythms using 'a walking stick, or perhaps an umbrella. […] the sequence of twenty-five panels […] makes up the score proper, indicating the journeys to be taken by the five performers and the rhythmic figures which accompany them. Both acoustic frame and performance activity are clearly defined, though the two remain separate.[206]

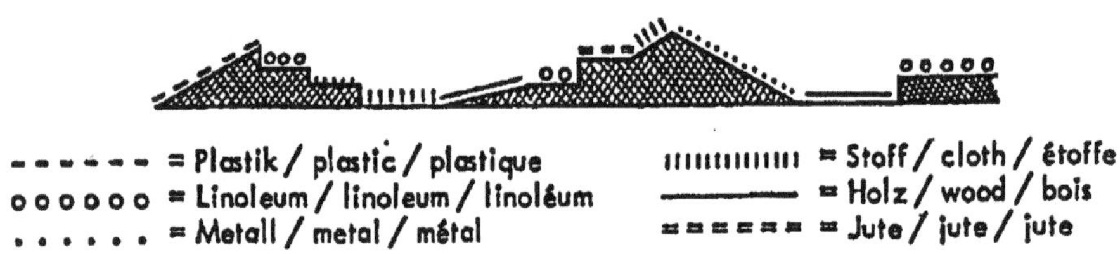

Fig. 6.45 Mauricio Kagel: *Pas de cinq* Score, p. 1
Quoted after McInerney 2007, p. 45

[203]Cf. e.g. Laura Rose Schwartz 2019.

[204]From the instructions which accompany the score, p. 1.

[205]McInerney 2007, pp. 43f.

[206]Ibid.

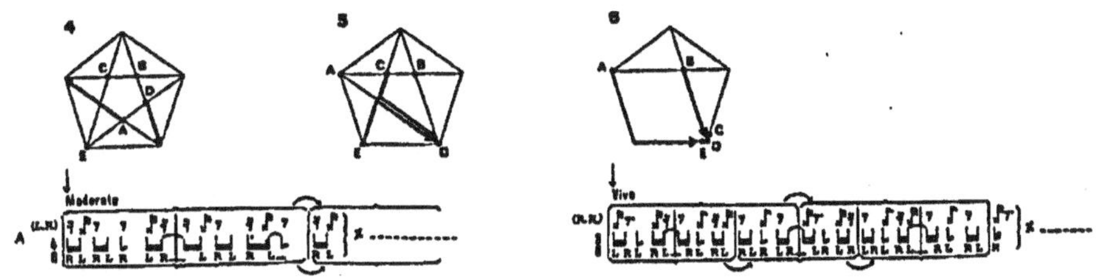

Fig. 6.46 Mauricio Kagel: *Pas de cinq*. Score page, Excerpt
The movements in a pentagon
Quoted after McInerney 2007, p. 44

6.7.2 Sharpening in a Round-About Way. ‚Graphical Notation'

The countless signs of the new notation refer to various dimensions—tone pitch, duration, volume and tone color; there they designate desired outcomes. How these are to be reached is indicated by signage for vocal techniques or playing techniques for each respective instrument. One basic problem of 'graphic notation' is the subjectivity of the signs, the lack of standardization. Because of this differing signs may signify the same (Fig. 6.47).

Characteristically, now there arises the demand for explicit variations of playing techniques such as *vibrato*, which were before left to the interpreters. Pictorial, iconic or verbal signs complement each other (Fig. 6.48).

Even if originally meant as 'precisication', in the main tendency much of the detail is deliberately left to the interpreters: blurring instead of sharpening—aiming "at random" (Fig. 6.49).

The *Dal Segno* is adopted in various manners (Fig. 6.50).

The pleasure of the ever-same, the near-identity of many repetitions then tips over into a repetition forever and ever. The lemniscate, the infinity sign shows that in favor

An indication for playing the highest possible pitch on a string. The note will be indeterminate in nature.[1]

A slight variation on the symbol above which also means to play the highest possible pitch.[2]

Fig. 6.47 Signs for extreme pitches
Tischhauser 2002, p. 10

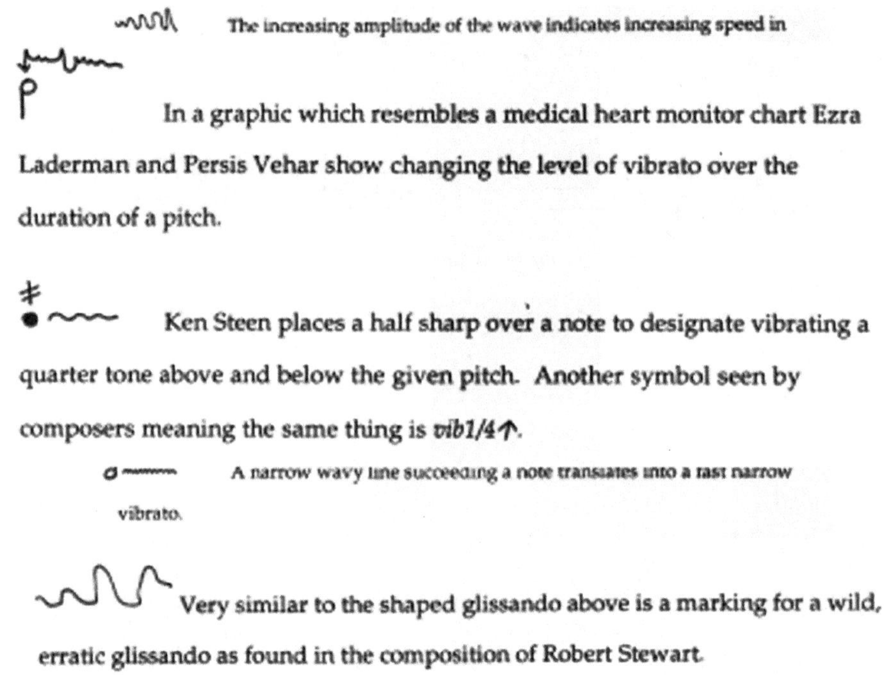
The increasing amplitude of the wave indicates increasing speed in

In a graphic which resembles a medical heart monitor chart Ezra Laderman and Persis Vehar show changing the level of vibrato over the duration of a pitch.

Ken Steen places a half sharp over a note to designate vibrating a quarter tone above and below the given pitch. Another symbol seen by composers meaning the same thing is *vib1/4*↑.

A narrow wavy line succeeding a note translates into a fast narrow vibrato.

Very similar to the shaped glissando above is a marking for a wild, erratic glissando as found in the composition of Robert Stewart.

Fig. 6.48 Signs for various vibratos
Tischhauser 2002, pp. 14f

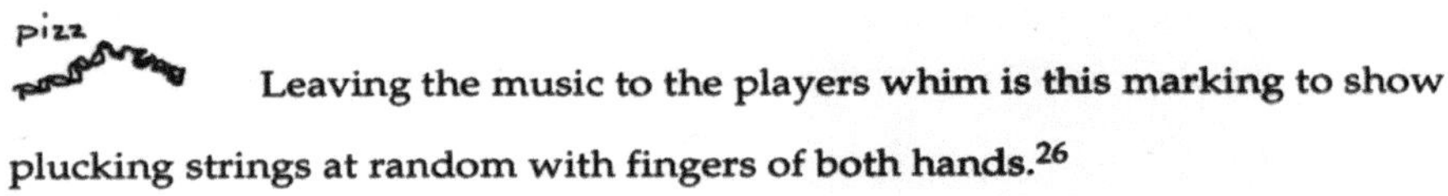
Leaving the music to the players whim is this marking to show plucking strings at random with fingers of both hands.[26]

Fig. 6.49 Pizzicato
Tischhauser 2002, p. 38

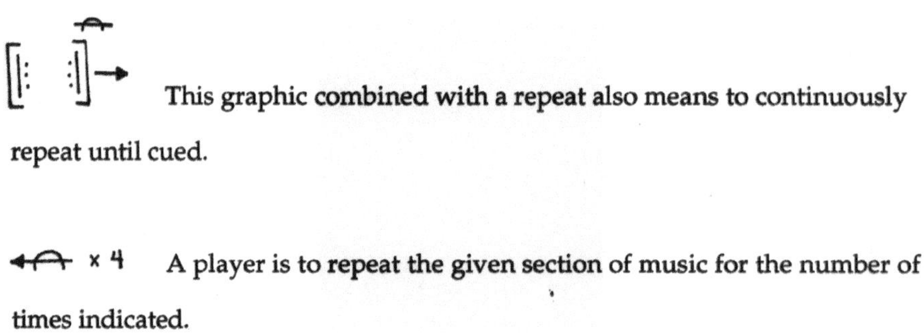
This graphic combined with a repeat also means to continuously repeat until cued.

A player is to repeat the given section of music for the number of times indicated.

Fig. 6.50 *Dal Segno*, reanimated, not at least for the favored repetitions
Tischhauser 2002, p. 44

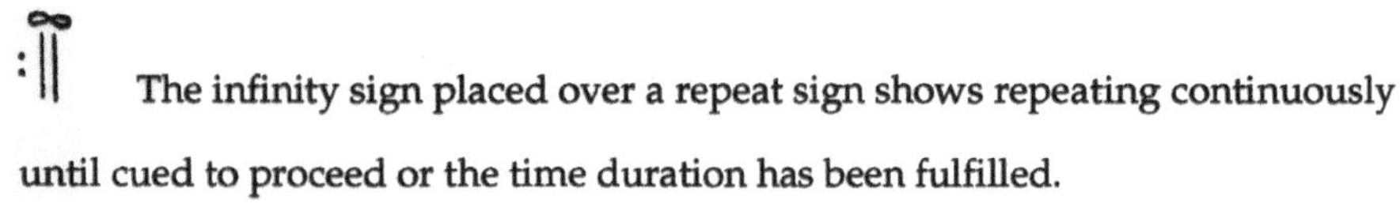
The infinity sign placed over a repeat sign shows repeating continuously until cued to proceed or the time duration has been fulfilled.

Fig. 6.51 Infinity, indefinite
Tischhauser 2002, p. 43

of an anti-Heraclitian river without banks the intention of a 'work of art complete in itself' recedes (Fig. 6.51).

6.7.3 New Freedom from Order to Chaos. Sylvano Bussotti's Siciliano

‚Graphic notation' is principally intended as specification especially of the performance parameters regarding the dimension of tone color, since this appears only secondarily in traditional notation, together with all the new, so far not or hardly recorded sounds. The graphic elements here constitute a sort of notational abbreviation in a second instance. Yet another aim is the general pictorial sharpening of the anticipatory imagination of the sound form of the work. Anestis Logothetis, 1921–1994, a main proponent of graphic notation, names further reasons and goals:

> (T)he compas of modes of musical expression is significantly expanding. However, the notation with which one wants to describe the sonic events is not adequate for this purpose. Several problems arise out of this: not only noise-like sounds can only be represented with great effort or in some cases not at all, but the desire for a *flowing music whose genesis can be experienced again and again* is impossible to realize in this way.[207]

This permanent flowing is the introductorily mentioned river of Heraklit, only without banks, differences without real identity. Any attempt to somehow conserve this identity becomes ambiguous. The singularity of each performance and the conservation of the work at the same time is nothing new.

> In addition, the composer tried in some way to protect on one hand the uniqueness of each performance and on the other hand the *preservation of a piece's principles*, giving the musicians designated flexibility (while performing).[208]

Formulated precisely the work exists rather virtually in a pictorial notation form in this concept. It becomes a hybrid: The stadium of the score form is not reached by it and that is not its intention, but since it is elaborated in detail, it is more a sketch form than a conceptual form. Most likely it resembles the stage of the '*abstract composition*', only not as composition of notes, but as compositing, composition of

[207]Logothetis 1970, quoted in https://anestislogothetis.musicportal.gr/the_graphic_notation/?lang=e, retrieved 10 July 2020.

[208]https://anestislogothetis.musicportal.gr/the_graphic_notation/?lang=e, retrieved 10 July 2020.

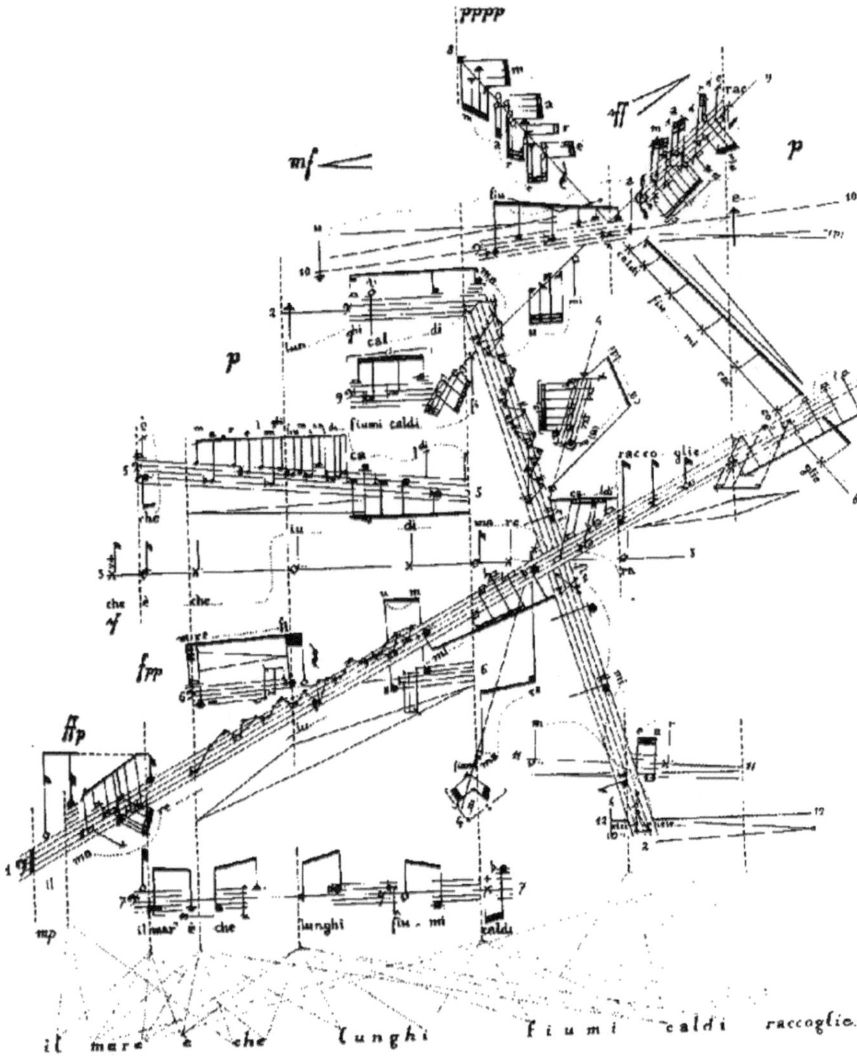

Fig. 6.52 Sylvano Bussotti: *Siciliano* (1962). Excerpt from the score.
Quoted after: Tischhauser 2002, p. 43

graphic elements. That a performance is meant to be unpredictable[209] is ambivalent. In this quite different realization of the anticipation form the performance and the execution are deliberately taken into account, but in a way that fuzziness is already codetermined beforehand (Fig. 6.52):

> What fundamentally differentiates graphic notation from traditional notation is the aforementioned polymorphism, which clearly enables all performers to retain their subjective reaction

[209] Ibid.

> times. The composer takes into consideration the divergences of the different performers in composing and expects a certain degree of *surprise* through the new formalization of musical form in every performance.[210]

The notation as such does at first not seem particularly focused. In a second instance however, the co-intended expressive aspects do represent sharpening by refinement of the work as a whole. Equally the „optical appeal of the image is significant for the expressive force of the work“.[211] Looking even more closely, a decidedly and explicitly political-social content reveals itself. The *Siciliano* is part of a larger whole: *MEMORIA 1962* su testi di A. Braibanti, P.E. Carapezza, P.P. Pasolini con voci e orchestre (to texts of… for voices and orchestra). Movements Part I:

a) Memoria Marcello Elisei

b) per un manifesto antifascista: "Geographie Francaise" Sätze Teil II:

c) Siciliano

d) alla bandiera rossa

e) "la partition ne peut se faire que dans la violence" ("The partition= score can only be done in violence")

In the executive apparatus of part I, among other things a*lettore*, a reader, stands out. The percussion consists only of unusual musical instruments: 4 Flauti a stantuffo (slide whistle, 4 Fischietti da polizia (police whistles), 4 Fischietti da arbitro (referee whistles), 4 Fischietti sirena (siren noise maker),[212] 16 Fischietti diversi richiami per uccelli (various bird whistles). For part II Bussotti requires a baritone pantomime; the orchestra's percussion is divided threefold, here as usual instead of the whistles, but with a marked affinity to military music and music of the banda, the Italian communal brass band. The *Siciliano* itself however is purely vocal, for twelve male voices. The variety of instrumentation may be an indicator of Bussotti's „sublime disregard for practicability“.[213]

> The excerpt consists of the second and last page of the composition. Setting note systems diagonally suggest acceleration of tempo, the luring of the systems into *unrestrained passion*. Where there are notes only on a single note line, it may be intonated freely.[214]

Page two, which can be classed as section four, is based on a single line of text by Paolo Emilio Carapezza: *il mare è che luighi fiumi caldi raccoglia*, which can be rendered as 'the sea is (or might be) the collection of long hot rivers.'[215]

Sylvano Bussotti's Introduction to *Siciliano* emphasizes in multiple variations the "freedom" of the interpreters—a freedom which admittedly hardly appears substantial, but rather as being random and arbitrary.

---

[210] Logothetis 1973, quoted in https://anestislogothetis.musicportal.gr/the_graphic_notation/?lang=e, l.c. A more precise and more detailed interpretation is offered by Attinello 1992.

[211] Karkoschka 1966, p. 96.

[212] Figure: https://www.aliexpress.com/i/32706687201.html, retrieved 20 August 2020.

[213] Attinello 1992, p. 89.

[214] Karkoschka 1966, p. 96.

[215] Attinello 1992, p. 101.

The notation of *Siciliano* permits an *interpretation completely free* from the symbols and their proportions. The vocalists should not interpret the height (pitch) here determined, as written on the staff, according to a temperate reading, but should *favor the spontaneous, individual impulse* in their own register, welcoming sounds and intervals proposed by the score as guidelines or as suggestions. The *tempo is totally free*, conditioned only by the proportions written in each measure which are to be interpreted visually (i.e., spatial notation), and in certain cases by the fermatas. Except for the two concluding caesuras, indicated by II and the final bar, all the measures are indicated with dashed barlines to suggest *freedom and flexibility* even in the coincidence of the twelve soloists.

The oblique interlacing of some voices also suggests the evident mode of acceleration and allows for ambivalence in the structure of intervals and in their rhythmization. The interpreter will *choose* from the two versions the one *he prefers* from his point of view, i.e., between ambivalence about the intervals, or about the rhythms.[216]

The same with fuzzy freedom for"unrestrained emotion", "free" intonation,[217] or "timing […] *approximately* organised".[218] The picture is not spontaneously comprehensible and translatable into sound like a score. It must rather be prepared for the performance and execution. Insofar it has the mentioned status of an 'abstract composition'. But it goes beyond that, since the information for the execution is notated, even though in slightly complicated ciphers. Karkoschka gives the impression that a written intermediate form as performance score is not necessary; a mental preparation and memorization is deemed sufficient.

For practical reasons the study of the score will begin with identifying individual voices according to the ordinal number provided at its beginning and end. In order to more and more inculcate the picture, one should investigate each voice regarding its diagonality (only 3, 7 and 12 are horizontal) and its length (the first is the longest, it marks beginning and end; 11 and 12 are the shortest). After that one should bring to mind the text distribution. The dotted lines that lead from the text syllables at the bottom to the synchronized verticals indicate the sections, in which the respective syllables should be applied. For example the second syllable 'ma' is sung only between the 2nd and the 6th section. The „mare" of the 9th voice in the 7th section is only seemingly an exception: no voice delivers all syllables and letters. And since a caesura is indicated after each letter, one should consider extracting the m from the word 'fiumi', the a from 'caldi' and the r and e from 'raccoglie'. The dotted lines in the notation text help to trace the text syllables below or above the intertwined systems.[219]

In dialectics of necessity and freedom the graphic notation allows for blurring in the realization, in about the same way as determined in aleatorics (see Chapter 7), but on the other hand, encaptic, various components are very specific.

The conductor […] will aim at first to strictly adhere to all vertically defined sections and to realize as exactly as possible the proportions. Only subsequently and from this exact basis are the singers to aspire after a maximum of elasticity and in that be inspired by the depicted notes. As so often with Bussotti the exact analysis of this page reveals that – in spite of the strong graphic effect – the particulars and the whole course are determined relatively exactly. Merely on the middle level, within the sections, the voices have some liberties. The

[216]Quoted in Attinello 1992, p. 107. Emphasis HWH.

[217]Karkoschka 1966, p. 94.

[218]Attinello 1992, p. 102.

[219]Karkoschka 1966, p. 94.

> expressive tendency of these are the again influenced *in parts exactly, in parts suggestively*, by the diagonally slanted systems, by the proportions, the dynamics and altogether by the picture.[220]

The intention is that technical and musical precision will take precedence over visual beauty—which is however not excluded but included.

> As is so often the case with Bussotti, a close study of this example reveals, that – in spite of the strong graphic effect – both details and overall form are relatively precisely determined. It is only at a middle level, within the sections, that the voices are free, but the expressive tendency of the voices is to some extent suggestively influenced by the slanting staves, the proportions the dynamics and in fact by the very picture itself. There is no doubt that the visual effect of the picture is of great significance for the works expressive force.

Attinello provides an interpretation of the structural context and Bussotti's intention:

> The grouping of voices is perhaps the most clearly articulated sign of the actual intent of the work; the progression from unity to total disparity of voices is a progression from predictability to chaos, or perhaps: the coagulation of hesitant fragments into an impassioned synesthesia. The definition and signification of chaos [is] a positive, sensual value point.[221]

6.7.4 Graphic Notation on the Way to Fuzziness

The opposite of sharpening is the case of 'Musical Graphic', in a jesting spirit already used in music and the fine arts in rudiments in the Renaissance especially since absolutism, in earnest then in New Music (*Neue Musik*) since about 1950. The aim here is deliberate blurring. The precision of the notes is sacrificed in favor of the fuzziness of the image, which mostly even reaches vagueness. The indetermination is meant to enable the interpreters to develop their 'creativity'—and at the same time urge them, whether they like it or not. Even closer to musical graphic than Bussotti's *Siciliano* is for example A. Logothetis' *Odyssee* (in detail see Sect. 7.2). The anyway ambivalent graphic notation results in blurring instead of sharpening, when it regresses to concept art. Then these vague implications must be concretized and actually be composed. The translation of the only sketch-like conception towards sound realization is then left to the interpreters. It actually resembles a transcription in the sense of a paraphrase—only in reference to a not yet really ready-composed work. One of the numerous transitions of this type, still within the framework of graphic notation, is demonstrated in a piece by J. Cage, *Solo for Piano* (Fig. 6.53).

> For fragment BE Cage indicates that the numbers refer to events to be expressed and the notes refer to fingers. The hands and forearms are to be used in playing. Here *I used* a range of other sonorities to be obtained from the piano as well as playing notes and clusters.[222]

---

[220] Karkoschka 1966, p. 96. Emphasis HWH.

[221] Attinello 1992, p. 92. Cf. McInerney 2007, pp. 116–119.

[222] Pace 2007, p. 14. Emphasis HWH.

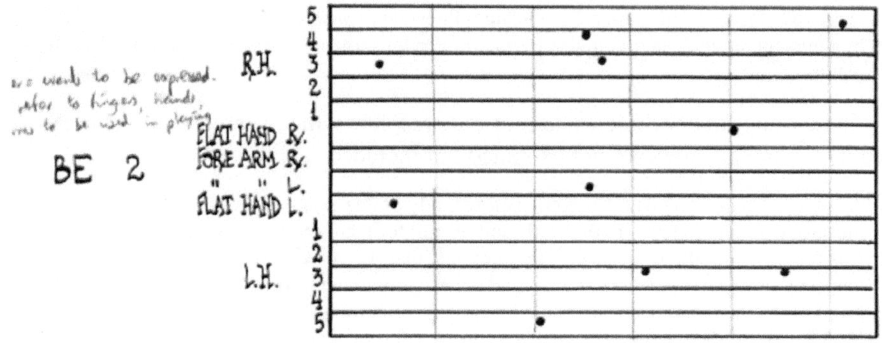

Fig. 16. John Cage – *Solo for Piano*, p. 47.

Fig. 17. Realisation of section 'BE' from Cage, *Solo for Piano*.

Fig. 6.53 J. Cage: *Solo for Piano*, excerpt
Quoted after Pace 2007, p. 14

In this case apparently at first there is a need to transform the mere sketch-like concept into a notational 'realization', before one can even begin to imagine a sound realization at least somehow 'similar' to the pictorial shape.

6.7.5 *Sharpening Notation of Non-notes: Ervin Schulhoff's* **in Futurum** *(1919)*

The note script as prescription emancipates itself, as already elaborated, and to a certain degree gains autonomy as 'sight music' already since the Early Renaissance, as in the mannerism of the so-called „Ars subtilior" with works and even codices in heart form or Baude Cordier's (ca. 1380-ca.1440) chanson about love, *Belle, bonne, sage*, notated in a heart shape, with red notes indicating rhythmic alterations.[223] Here the notation itself, as an optic semiotic system that produces new, primarily optic shapes, attains additional unique semantics (see Sect. 4.4). The prompt 'line circles' (*Linienkreise*) in *Wozzeck* implies that the optical image evoked by the text is optically reflected in the notation, and this more distinctly than in the merely acoustic sound.

A graphic notation of the peculiar type was employed by E. Schulhoff, in the tradition of critical Dadaism (Fig. 6.54).

Fig. 6.54 Erwin Schulhoff (1894–1942): *In futurum. 5 Pittoresken* for piano op. 31 (1919), No. III, excerpt, accolades 1–3
Quoted after https://www.kulturtechno.de/?p=3380, retrieved 24 August 2010

[223] See https://en.wikipedia.org/wiki/Eye_music, 3 July 2020, retrieved 4 July 2020. In detail and with new aspects see Stoll 2020.

The work has more layers than apparent at first. On the face it is simply a funny and witty negation of the usual, of music as *sound* art. But the non-sense carries a counter-sense. The sentence is symmetrically framed by popular dances, Foxtrot, Ragtime, One-Step, and Maxixe. They are associated with the semantic field of modernity, the contemporary, associated with 'Negro music', posited polemically against the cultural reaction. In the same vein the 32-bar song form (here 30-bar)[224] serves as foil, only implied by Schulhoff with significant liberties. The pauses are not simply nothing, but 'specific negation', and altogether the old carnivalesque inversion of the world en miniature:

> Everything is inverted into its opposite: The time measure is 'timeless', instead of notes rest signs and fermatas appear. The upper voice has predesignated the bass clef, the lower the violin clef. Both voices have divergent and absurd time signatures, 3/5 against 7/10. The intervals in the bars result in a continuous, meticulously notated 4/4 rhythm, standing out with its many syncopes, reminiscent of ragtime or cakewalk. Four times the fermatas interrupt the continuous rhythm. In the middle and at the end suns are notated. Between them, *as for a voicing*, commata or caesura there appear exclamation and question marks. The Marshall interval in the second to last system refers both to the general interval with its figural tradition as exaggerated musical sign, as well as to the namegiver '*Marschall* Grosz Metamusician', to whom pure music dadaist 'Colonel Schulhoff' felt connected and congenial. The recital indication *tutto il canzone con espressione e sentimento* ad libitum, *sempre, sin al fine!* is a caricature of the expressive desire of the music of the late nineteenth century and of expressionism: nothing harmonizes, everything is lead ad absurdum [...].[225]

The composition even contains something more. It is evidently given the framing, also a piano piece with two accolades. Schulhoff was an excellent pianist; and it seems as if the right-left commutation is thought from a pianist perspective as imaginary-real crossing over of hands. The rhythm is distinct, but also, aside a percussive non-melody in the shape of repetitions at identical 'non-pitch', there appear in bar 22f and bar 29 even implicated melodic runs, lines and figures. Additionally Schulhoff uses non-notational, iconic-pictorial signs, note-*heads*; they look like an anticipation of the so-called 'smilies'—and insofar like a prophetic »in futurum«. They display in the sense of musical rhetoric affects: firstly (just at m. 12) laughing at the top and crying below, in the final bar (m. 30) however almost a *lieto fine*: both laughing. The faces are equally *note* heads—and they allow, even demand, what so far has possibly been overlooked, a specific tone pitch attribution. The first time it is interestingly identical in both systems, the ominous b—traditionally a key associated with death, as exemplarily demonstrated in the grandiose b in Alban Berg's *Wozzeck*. Rightfully M. Betz emphasizes the 'general interval with its figure-like rhetorical tradition: Its center is among other things the interval as death. The dedication and the as always ironical-satirical self-designation as officer shows how much the military paradigm has influenced the thinking and practice of the opponents of militarisms such as Wilhelminism and Nazism and 'market economy'. The semantic-stigmatic field interval/counter-interval/'field marshal interval' in this perspective has a deeper meaning, that is held back in its latency, next to joke, satire and irony, to implicate

---

[224]But there are 32 1/32 intervals at identical pitch in bar 26.

[225]Betz 1999. Emphasis HWH.

Grabbe's drama.[226] That Schulhoff has interpreted the primarily prosaic-technical designation 'general interval'—all voices shall be silent—political-semantically and exacerbated it into 'field marshal interval', supports this reading: The work without singing and sound is a mute, and thereby more emphatic protest against the million-fold slaughter, the muted deaths in World War I.

6.8 Bypassing Blurs: Direct Composition of the Sound Form Without Notation and Human Interpreters

The already repeatedly discussed problem of fuzziness inevitably emerging in the transition from written realization in notation to the sound realization of the work through the interpretation of the musicians can be elegantly eliminated by removing notation as well as interpretation through live musicians. In analogy to such omission under different medial circumstances the scripturality can be eliminated completely (see Sect. 3.6). Here this possibility is explored from a compositional perspective. With a somewhat one-sided interpretation it says:

> The element of interpretation, for centuries a central characteristic of the art of the musician, becomes a malady and obstacle for the exact music of the modern composer. He is always forced to share the sovereign right over 'his' work with the interpreters. The substitution of the human interpreter through the machine is an act of complete seizing of power by the composer.[227]

The compulsion [urge] to replace human work is old and an essential motif of the development of technology and productivity. Already Aristoteles was talking of automats, which could make slave labor easier or even replace it. In the Hellenistic Age Hieron of Alexandria built music automats. Such automats existed in many advanced civilizations in China, in Byzantium [the Byzantine Empire], in the Islamic areas, but then also in feudal Europe. Even more sophisticated ones originated in the age of absolutism, as in Droz, Vaucason etc.

The strong compulsion to eliminate human labor from music production is of a younger date. It is firstly a delayed reaction to the phonograph of 1877, and begins with the onset of the avantgarde after 1905—which itself ran parallel to the scientific revolution of these years. This compulsion adopts from material production the principle of replacing human labor at first with mechanized, then with automatized objectified work. The limits are defined by profitability. This compulsion culminates in the post-war avant-garde of the 1950s. Here the impulse specifically originates with the serialist compositions. Just like traditional notation was no longer adequate for the compositions impelled to more and more extreme specifications, the traditionally common human labor was considered no longer adequate. For these works achieved such degrees of complexity, that they could impossibly be realized with sufficient

[226]Christian Dietrich Grabbe: *Scherz, Satire, Ironie und tiefere Bedeutung*. Lustspiel, written 1822, alterations until 1827. Premiere in Munich 1907.

[227]Claas Hanson 2015.

accuracy and precision by live musicians. With 'electronic' music, which also has other sources and traditions, an apparatus and an instrument became available for this.

Fundamentally the replacement of human musical labor of realization occurs in many ways, with several procedures, which have a historical signature. The somehow naive assumption, that by delegating the responsibility to machines instead of people the fuzziness is eradicated and perfection achieved, is proved false at every new stage. This concurs with the dialectics and the evolutionary logic of the domination of nature: It becomes progressively more mighty, but with every progress new problems and shortcomings of the domination of nature emerge, often even a sense of impotence, analogous to the here discussed fuzziness.

In music at least two fundamental boundaries seem almost impossible to transgress, in spite of the tremendous technical progress. (1) The technology itself necessarily produces fuzziness: Springs of mechanical musical instruments cause irregular running, the electro-acoustical devices produce static noises etc. (2) The sound remains limited—yet again dialectics of technology. The mechanical flute clocks can only pipe, the electrical piano plays only piano music. Only the electro-acoustical music makes a qualitative leap: It can not only imitate piano and flute sounds, but produce new, yet unheard of sounds in line with the intentions and ideas of the composers. But even at this stage the spectrum of sounds remains limited to what the devices can produce. Apart from that, among other things the sound synthesis fails to replicate the attack and release procedures, which give life to the sounds.

This will only change with the next qualitative leap: Through sampling, sounds from live music realizations can now also be reproduced and integrated into composition. It then sounds almost 'as in reality', though usually these sounds will be layered and reshaped with conventional, electro-acoustically generated sounds. And even then the technical-acoustical boundaries remain: It is 'loudspeaker music'. The limitations of the respective devices are the boundaries of sound, regardless of whether they are mechanical or electro-acoustical or sampled live. Rigidity does not prevent fuzziness. This inescapable technical-instrumental mediation is just as much an inescapable aesthetic fuzziness. Then again music becomes accessible to those without any skills and capabilities of making music, a step towards emancipation and democratization, which has made an enormous leap with the phonograph and its successors and then with the radio. The price paid was surely high: Especially the own singing as well as the instrumental playing of music is in decline. These procedures and media are listed rather cursorily in the following. Each has its own limitations and blurs, apart from the outlined general boundaries.

6.8.1 Automatophones. Barrel Organ and Player Piano, 'Abstract Composition' and Secondary Subjectivity of the 'Pianolist'

The mechanical music instruments or automatophones are the first step on the way to the automation of music realization. They date back to Hellenistic Anitiquity. Though they are *only* music instruments, they no longer require a human player—at least not a musician. The stimulation works mechanically through a spiral coil as in mechanical watches. It therefore already constitutes automation; so the less common denomination *automatophone* is more adequate.[228] The barrel organ, also called hand organ, still requires a player: Of all things the most subordinate, mechanical work of propulsion by turning a crank still requires a player and remains human labour, just as the assistance of the bellows-treaders before the electrification of the organ.[229] The human labour is here cheaper than a costly [elaborate] mechanical design, and anyway available, since the barrel organs need to be transported to respective performance locations on the streets and in back courts. And thereby we again have an executing subject that creates fuzziness. Tempo and proportion of rotations follow the work, but are also subject to human measure and fallibility. This is illustrated by an anecdote from a hiking trip, which is attributed either to Wagner or Verdi. So Verdi hears a barrel organ that totally ruins the tempo of „La donna è mobile". Angered he takes the crank and demonstrates the correct tempo. The next day a sign is attached to the barrel organ reading: „Student of Giuseppe Verdi".

Principally this rarely results in any compositional innovation. Already existing works are permuted into studs on a cylinder instead of being written down and then executed. But the studding also follows a score. Storage and steering medium is the studded cylinder, as used by chimes in towers (specifically the carillons had to be played as a rule). Later also perforated disks and suchlike were used. The sound realization itself is more or less reduced. This also applies to newly composed works—among others C.Ph.E.Bach and Mozart wrote works for a clock flute. And famously in 1813 Beethoven composed for the orchestrion of metronome-inventor Johann Nepomuk Mälzel, the *Panharmonicon* (since 1804), which was also programmed to play tunes by Mozart, Haydn, and others, Opus 91 *Auf Wellingtons Sieg oder Die Schlacht bei Vittoria (To Wellington's Victory or The Battle at Vittoria);* due to technical problems it became an orchestral work in the end. But altogether these few works are a first, still humble beginning of compositions without mediation through the score form or musicians being realized straight away in the sound form.

An orchestrion is not quite an orchestra, a flute is more flexible and richer in color and modes of expression than a flute clock. Although they offer a rich spectrum of sounds, even new ones, tending almost to the orchestral, by combining several

[228] A still valid survey: Bowers 1968 and Bowers, 1972.

[229] For non-mechanized threshing machines or the whims such labour was performed by horses, donkeys, and sometimes even dogs.

musical instruments, this also applies to the pre-electric mechanical musical instruments controlled by pin rollers, pin plates, perforated plates or punched strips, such as the barrel organ, pianola, or all types of orchestrion, all apparatus, which usually contain their own instruments to produce sounds. Be it noises, sound mixtures or tones—none of these instruments exceeds the bounds of musical instruments.

6.8.2 Electrification. Reproducing Piano as Production Instrument: Conlon Nancarrow

Just as the mechanical music instruments introduced so far, the piano turned automatophone also began as a reproducing, not a producing instrument.

For this automatophone, the piano-player or *player piano*, punched tapes instigate the pneumatic movement of the keys. Though the punch tapes are a kind of writing and even notational script, they are not a score, but already its execution as a sort of 'programme' for the automatic piano.

> Pianola was invented in 1895, although it was descended from a whole line of roll-operated musical instruments that stretched back to the mid-1870s. The Pianola was by no means the only *piano-player* in town, but its ease and subtlety of control placed it apart from its competitors, thus ensuring its immediate commercial success, and the survival of its name as a generic term for foot-operated player-pianos.[230]

At first the *piano players* had blinds.

> Early piano-players were fitted in front of the keyboards of normal pianos, and played by means of a set of felt-covered wooden fingers, so that the combination of the two instruments allowed their proud owners to play music from perforated music rolls.[231] [...]. If now *the »player« pedals*, the paper tape movement continues and causes the touch of each finger.[232]

This represents an anthropomorphous mimesis within the process of automatization. Just as in the music automats of the eighteenth century human players are replicated, to give the 'artificial' music-making a more authentic and 'natural' appearance. The body of the player has here already become an abstraction and is reduced to the essential hands. A further development was desanthropomorphized player pianos, where the mechanism was built into a piano or even a grand piano. In the further developments on the one hand desanthropomorphization was continued, but on the other human labour was indispensable. „Ich bin die fesche Lola", [I am the pretty Lola] sings Marlene Dietrich in *Der Blaue Engel* 1930, which rhymes with „und *spiel'* das Pianola" [and play the Pianola]. The *player* piano was rightfully dubbed so: It required a player. The musicians excluded by the automatization return

---

[230] *"You cut the whole roll with razor blades?" Conlon Nancarrow and his Music—A Personal View by Rex Lawson*, https://www.pianola.org/history/history_nancarrow.cfm, retrieved 10 August 2020.

[231] Ibid.; cf. https://www.pianola.org/history/history.cfm, retrieved 10 August 2020.

[232] Krickeberg 2015.

through the back door. Even more than with the barrel organ player the player here has quite some influence on the sharpening of the realization.

The persisting human labor and subjectiviy splits up. The one, purely technical function is the treading of the bellows for the pneumatics as the bellows-treaders with the organ or the players of the harmonium with pedals or with a crank in the case of the barrel organ. This already influences the sound. It results in subjective approximations to something like expression and placing of accents, but simultaneously inaccuracies occur through these methods of stimulation. The new semi-automatic or supplementary subjectivity is, in various shades, a constant of elimination of human labour from the immediate realization through automatization. As with the barrel organ, some types of player piano feature a crank.

> A cheaper version of the Pianola represents the crank piano: Here the cranking person can increase the tempo, but the music gets louder at the same time, whether the person wants or not.[233]

Fuzziness remains indelible; taking into account that subjects often enough have to avoid technically prompted fuzziness. Yet again these are mostly avoided with the electrification of the devices. The propulsion mechanism of the pneumatics could run electronically. In itself this is not essential for the piano, but it decouples the propulsion device and the influencing of components of the performance, thereby separating technology and expression. The other artistic function of the surviving labor in the succession of musicians is, as usually done by interpreters, to ensure a lively execution and an adequate recital. Here it is—only—an addition to the basic and dominant mechanical reproduction. For the "abstract composition" returns in an altered medial form.

> The tape reflects *only the musical text*; the player generates the nuances of the sound volume primarily with the feet, with his hands he provides expressive acceleration and deceleration and grades the sound of bass and descant against each other.[234]

An executing subject is still needed. In a kind of labor division it takes over only accidental components of the complete score form of the work. The essential components the 'abstract composition' is performed by the automat.

> The »automatic« is used here not to replace the people, but just to make even *technically very difficul*t, actually professional works for piano accessible to the music-making lovers.[235]

The execution becomes secondary and is again given life. This functional aspect gained autonomy in the 'pianolist', nothing but a pianist resurrected in technology. He complements what is not noted in the 'abstract composition'.

> The vast majority of *music rolls* for the normal pianola were not recorded at a keyboard, but simply *perforated* by a technician *after being* marked up *in pencil* with reference to the *original sheet music.* Thus there are no changes of tempo, rubato or phrasing on normal

[233] Cf. https://en.wikipedia.org/wiki/Barrel_piano, 2 July 2020, retrieved 20 August 2020.

[234] Krickeberg 2015.

[235] Krickeberg 2015.

> pianola rolls, and instead the *operator* is provided with a tempo control, which must be used to create musical performances. To take the simplest of examples, the famous opening of Beethoven's Fifth, with its dramatic pauses, would rush inexorably past on a standard music roll, if the pianolist were not ready to delay its course appropriately.[236]

A continuation of the *player piano*, a hybrid of the reproducing phonograph and the producing piano, is the *reproducing piano*, pioneered by Welte-Mignon, first introduced in 1905, which was *also pneumatically* controlled by punched tape. It was thereby permitting the recording, notating, and then playback of a particular interpretation. A digital potentiation was then represented by the Bösendorfer SE290 computer-controlled grand piano in 1986 and the Yamaha Disklavier in 1987. However, the sound 'printing' occurring here is not one of the universal soundscape, but only of the piano sound, the restricted equivalent to the phonography form.

Both instrument types were primarily intended as reproducing instruments of existing works, not as production instruments for *new music specifically composed* for this purpose. But due to the fast developing technology they could be repurposed without any technical alterations. The storage on and stirring through punched paper tape permits both options. Especially this highly developed special music instrument type reveals a general problem. The elimination or restriction of human labor does not eliminate fuzziness, but even produces new fuzziness. Even with the Welte-Mignon grand piano, patented in 1904, the volume regulation is crude and too general: bass and descant have the same volume. The sharpening is often transferred, in accordance with automation, from the interpreters to the technicians. They must provide the definitive sound form of the work, which is adequate in regard to both the intentions and realizations of the composers as well as the interpreters (see Sect. 3.3). In the later sound studio the situation differs in terms of technical apparatus, but aesthetically it is similar. This changes only with the even later apparatus of synthesizer up to the computer, where the composers themselves become technicians—and vice versa (Fig. 6.55).

> Despite the ingenuity of its inventor, this automatic recording system was only a partial solution to the problem of recording a pianist's dynamics. Even if the system had been exactly accurate, which it clearly wasn't, there remained the prodigious problem of converting individual note values into the two coding streams that were all the Ampico could respond to. To a musician contemplating *these unwritten processes*, it is clear that the judgment and experience of the roll editors was considerably more important that any individual dynamic readings.[237]

Remarkable is the re-functioning of a reproductive instrument into a work- and instrument-specific production instrument. Attending to this specificity became the specialty of Conlon Nancarrow (1912–1997).[238]

---

[236] https://www.pianola.org/history/history.cfm, retrieved 10 August 2020. Numerous details about the steering possibilities, the rolls and the playing in https://de.wikipedia.org/wiki/Pianolist, 18 September 2019, retrieved 28 August 2020.

[237] Lawson 2013, p. 19.

[238] For details see Phleps 2000. The following elaborations are based on this contribution and are essentially an abridged summary of a section of the article by Phleps.

Fig. 6.55 Punch-nibbling machine
Quoted after *The Pianola Journal*, Vol. 23, 2013, pp. 11–26; quotation p. 19

> He was not aware that other composers had written for the instrument, and was therefore fascinated when he visited London in the 1980s to find out about Stravinsky, Hindemith, and the hundred or so other composers who had written for music roll during the course of the twentieth century. […] Most composers who have written for player-piano have done so on the basis of an incomplete perception of the instrument's range and history. […] All these views of the player-piano are colored by the particular instruments that the composers encountered.[239]

It was not Nancarrow's intention to eliminate the interpreters, but since his works stood no chance of being performed, the automated replacement of musicians was his alternative.

> Nancarrow's politics were of the left, and he fought on the Republican side in the Spanish Civil War, as part of the […] Lincoln Brigade. Returning home in 1939, he moved to New York […], but was refused a passport on account of his political beliefs, and so decided to settle in Mexico, along with Canada one of the two countries to which he could travel without papers. There he remained for the rest of his life, essentially composing music for *player-pianos*,

In contrast to the mechanical music instruments in general, here the *composer* actually skips one existence form, the score form, and realizes his ideas directly in the sound form or rather for the sound form. However, this still happens with the intermediate stage of a now no longer written notation for human instrumental interpretation, but a written instruction for automatic reproduction, which analogous to playing then

[239]S. https://www.pianola.org/history/history_nancarrow.cfm.

really steers the actual realization as sound form. This sound form is restricted to the apparatus with all its inaccuracies; this notwithstanding the intentions are realized without additional fuzziness.

> [I]n 1947, he obtained a secondhand Ampico upright, and also a manual [STATT HAND] perforating machine, [...] (probably a Leabarjan) [...]. The perforating machine was substantially modified after a while, so that more subtle rhythms and accelerations could be achieved, and a second Ampico turned up in Mexico City during the 1950s.[240]

Nancarrow used the reproducing piano [, a reproductive piano,] as automatophone, as *producing* instrument. He was impelled to take into account the technical and thereby phonic boundaries and moved within those limits.

> The Ampico was the first of the main *reproducing piano* systems to come out of America. Invented and improved by Charles Fuller Stoddard over a period of years from 1908 to about 1920, it controlled its dynamics in some quite complicated ways. The Ampico system that Nancarrow used divides the bass and treble sections between E and F above middle C. Each half of the pneumatic mechanism is controlled separately by marginal perforations on its particular side of the music roll. There are fast and slow crescendo and decrescendo perforations, and a series of three other accent levels that can be added together and superimposed upon the general levels created by the crescendo mechanisms. The closer the crescendos reach towards *fortissimo*, the more compressed becomes the dynamic space into which the accented levels must fit. It is a complicated but clever system, and rather too fussy for most of Nancarrow's compositions. As a result, he usually (though not always) restricts his dynamic coding to the use of the accented levels, and ignores the overall crescendos and diminuendos. A notable exception comes in Study no 6.[241]

In an entire series of *Studies*, which brings new honor to the genre of the etude, Nancarrow replaces the heightened technical player virtuosity by a virtuoso compositional handling of the technical apparatus, the player piano. Thereby Nancarrow achieves a 'superhuman' measure both of tempo and of complexity in the layering of canonical counterpoint lines. In this way he claimed to transgress 'time' as the 'last boundary' of music—which, as always with 'frontiers', lastly doesn't exist even in the societal reality. At the same time it relieves humans from a difficult and disproportionately strenuous manual labor and in this regard has utopian potential. As in electro-acoustical music the emancipation from bodily limitations does not by far remove all human limitations.

Nancarrow's work reveals a peculiar dialectic of sharpening and precision. He operates with irrational numbers for proportions of form, with complex, multi-layered superimpositions in the composition or with such rapid tempos that they appear blurred again in perception at the latest, despite all the precision of the machine-made execution, and much more. The "precisiation" (Zadeh) is thus at the same time blurring, but the blurring is at the same time sharpening, since it belongs to the completed total form of the artwork and helps to express its idea.

Pivotal for Nancarrow were most notably the thoughts of Henry Cowell about rhythm in *New Musical Resources*, a book written around 1919 and first published in

---

[240]"You cut the whole roll with razor blades?" Conlon Nancarrow and his Music—A Personal View by Rex Lawson, https://www.pianola.org/history/history_nancarrow.cfm, retrieved 10 August 2020.
[241]Ibid.

1930, which largely went unnoticed. Cowell's basic idea of transferring the proportions of sonic oscillations, creating a relation between overtones and rhythms—a relation evidently existing only through numbers, not in a physical acoustical sense—became the driving impulse, apparent already in *Study No. 1* (released 1951), of Nancarrow's entire work. Serialist conceptions as those of Stockhausen are based on a comparable thought model. In *New Musical Resources* one already finds a reference to the player piano. A systematic work step preceded Nancarrow's motivic-thematic constructions: First he would design the temporal proportions of a new work, which he transferred first with stencils to the cylinder and to music paper, later with a hand driven punching machine For the tempo relations of his canons Nancarrow often chose proportions that by far exceeded the technical possibilities of playing as well as the boundaries of perception.

In *Study No. 33* the two voices stand in the irrational tempo relation $2: \sqrt{2}$, concretely in the relation crotchet $= 140 \times 2 = 280$: crotchet $= 140$ x $\sqrt{2} = 197{,}9899$. A common denominator doesn't exist here, but a virtual point of synchronization after 41 quarters in one and 29 in the other voice, since here both differ by less than 1/384 of a second. For Nancarrow the highest degree of 'temporal dissonance' seemed appropriate for his *Study No. 33* that operates with $\sqrt{2}$.[242]

Technics and material are not arbitrary, but motivated. They have a semantic dimension. (1) Nancarrow's subliminally refers to classical problems of geometry. In the case of Study No. 33, which operates with $\sqrt{2}$, this is the square diagonal. (2) The fact that Nancarrow uses the technique of the canon universally also has very personal reasons. Conlon Nanca*rrow* has the canon twice in his name. The *-row* at the end, as for twelve-tone *row*, was probably just as welcome to him. Permutation and filtering as selection and elimination make this immediately manifest:

C O N l O N	N A N C A r r O w
1 2 3 4 5 6	7 8 9 10 11 12 13 14 15 16
CANON	CANON
1 8 3 5 6	10 11 9 15 7

In contrast, it only becomes clear through complicated metaphorical back and forth transfers between music, text and term in *Study No.21—Canon X* that the *X*, composed of a crossing of two tempo progressions, can also be read as a palindrome of two arrows ▶◀ or ›‹—thus, according to the pun, two nanc-*arrows*. (3) Nancarrow uses gematric procedures to encode his name—numbers for tone letters according to the order in the alphabet, that is:

Conlon Nancarrow $= 3 + 14 + 13 + 11 + 14 + 13 + 13 + 1 + 13 + 3 + 1 + 17 + 17 + 14 + 21 = 68 + 100 = 168$.

168 also contains the sum of the letters of his name: $6 + 9$ ($=1 + 8$).[243]

Die *Study* No. 40 (1975), called *Transcendental Canon*, for two player pianos, is divided into two parts. 40a is a canon in the ratio of the real numbers e: π for only one player piano (Figs. 6.56, 6.57, and 6.58).[244]

---

[242]Phleps 2000, pp. 196f. The ratio 1: 2 is not dissonant, 2: 3 is 'slightly dissonant' etc.

[243]For all this in detail see Phleps 2000, passim.

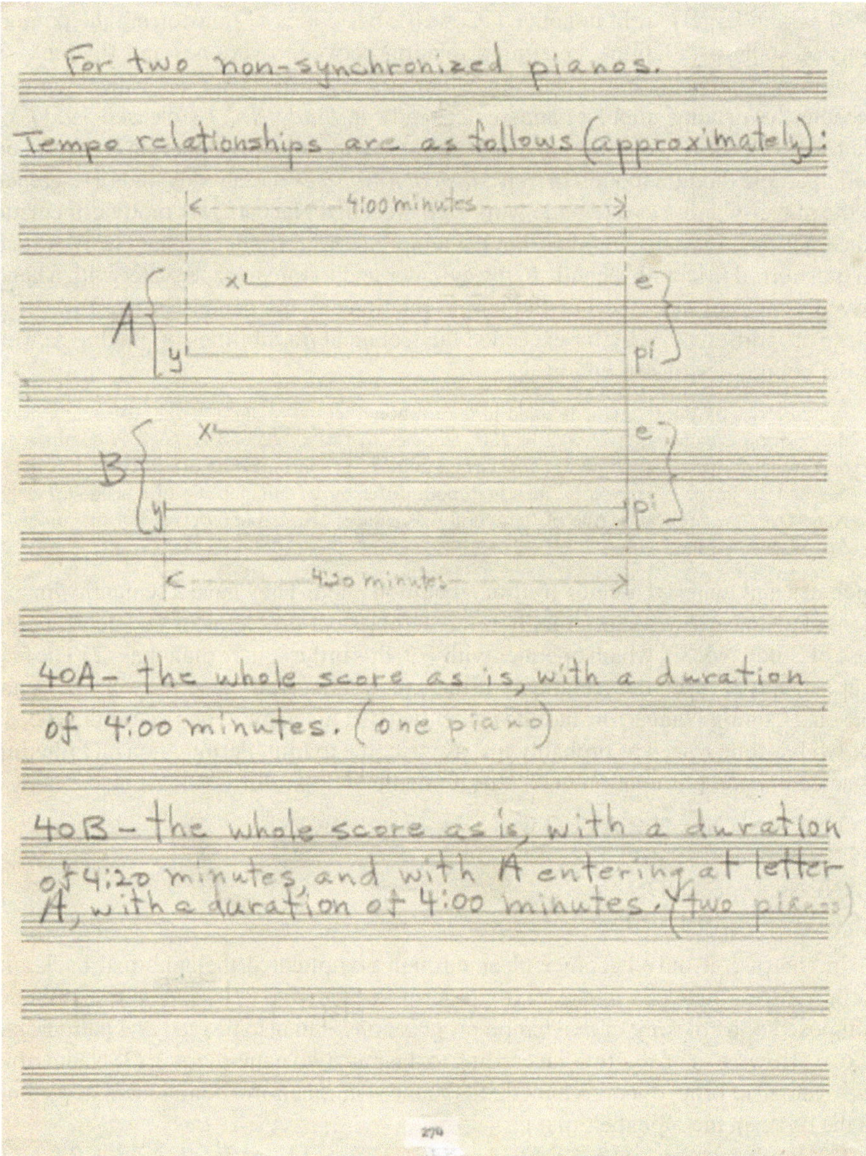

Fig. 6.56 Conlon Nancarrow: *Study* No. 40 a-c
Relations of the parts
https://www.nancarrow.de/Study%2040%20Score/0.jpg (Jürgen Hocker has meritoriously scanned and made available numerous other scores: https://www.nancarrow.de/partituren_autograohen.htm. For further information see https://www.nancarrow.de/)

Fig. 6.57 *Study* No. 40
Score, Autograph, p. 1
https://www.nancarrow.de/Study%2040%20Score/1.jpg

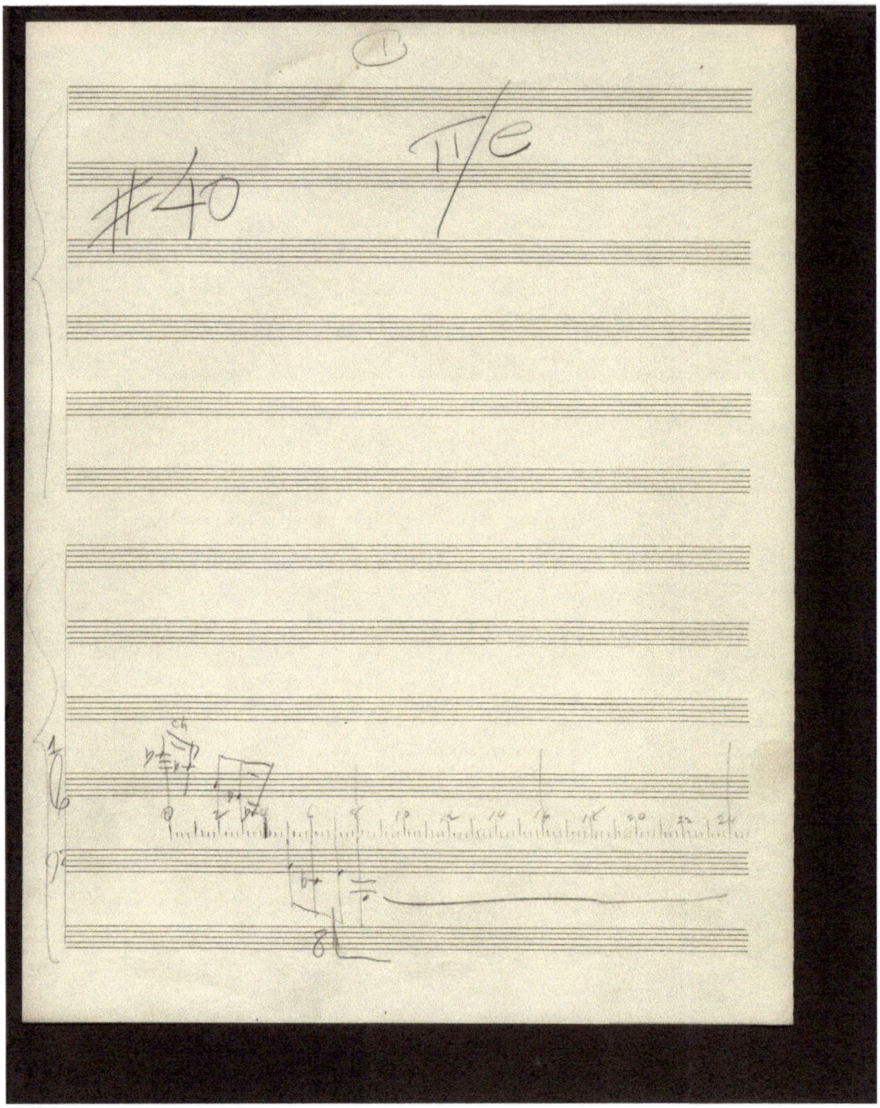

Fig. 6.58 *Study* No. 40
"Punching score", Autograph, p. 1 (Detailed explanation of the punching method, with illustrations, and of the construction of *Study* No. 21: https://www.conlon-nancarrow.de/heisig.htm; see also Phleps 2000)
https://www.nancarrow.de/Study%2040%20Punching%20Score/1.jpg
3rd stave from below: The vertical strokes of the marking as on a ruler or folding rule for measures 1–24 are the notation of a baseline for the punching

> The study b, however, is a canon for two synchronized player pianos. [...] (T)he two pianos play the same score, but the second piano starts approximately 20 seconds away from the first, but a bit faster, in a way that both pianos end up simultaneously.[245]

The super-complex proportions are not really "supranatural", yet superate all natural and technical borders of hands, mind, and instrument. In Study 41 Nancarrow goes further. He then multiplies even the complex proportions of the single parts by combining and superimposing them, in a certain way encaptic or fractal. The *Study* No. 41, as No. 40 for two player pianos, is divided into three parts: 41a, 41b and 41c. The first canon is for the first piano, the second canon is for the second piano, and the third is for both pianos, in the proportions marked in the first page of the score.

On the cover page Nancarrow notes the extremely complex relations in detail (Fig. 6.59).

„Nobody is perfect" (Billie Wilder: *Some like it Hot*), and especially not an apparatus. Nancarrow prepped both of his pianos to improve on rhythmical precision.

> Necessitating absolute rhythmic clarity as his work does, he customised the pianos in a number of ways, two of which are key: the first was to 'tack' the hammers, a practice that had been used in US bar music to help the sound cut through loud speech, and also by Glenn Gould to approximate the timbre of the harpsichord while maintaining the dynamic control of the piano. This was achieved on one piano by metal tacks, similar to drawing pins, supported by leather strips, and on the other by covering the whole hammer with a metal strip. The second customisation was the addition of dampers to the full range of notes (the top 22 on these pianos usually being undamped). The result of these customisations was to produce a thin, sharp note with a crystal clear attack and an absolute minimum amount of sympathetic resonance from unstruck strings.[246]

And he prepped the space for his recordings:

> Nancarrow's studio ... was acoustically designed to have a very modest amount of reverb, most of which was negated by furniture, bookshelves and general clutter.[247]

Even an automatophone has to be 'tuned' to sound rightly.

6.8.3 Film as Sound Carrier—The Optical Sound Track

The sound realization through musicians becomes obsolete also in sound film and cinema, as with the electrical piano, since 1922 when it begins to use the electrified light-and-sound procedures (especially 'Tri-Ergon'). The soundtrack was applied directly to the film next to the pictures. Here the electrification is a technical-medial precondition.

---

[244]Cf. Phleps 2000.

[245]https://en.wikipedia.org/wiki/Studies_for_Player_Piano_(Nancarrow), 4 September 2020, retrieved 10 September 2020.

[246]Cf. Murcott 2013, pp. 3–10; quotation pp. 3f.

[247]Ibid.

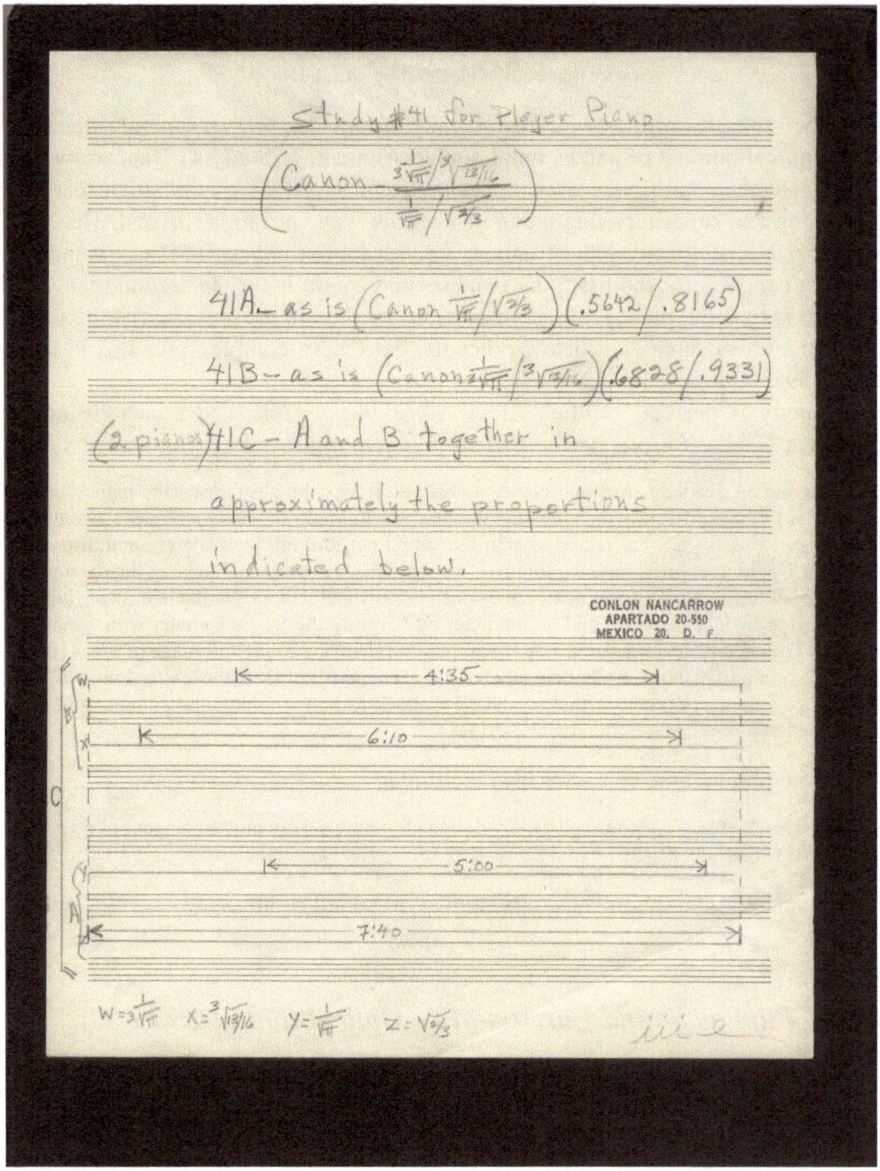

Fig. 6.59 *Study* 41
Cover, Autograph
https://www.nancarrow.de/Study%2041A/0.jpg

> During playback the soundtrack was illuminated from behind and transferred to an electrical photo cell. The varying brightness caused variated electric currents, which through amplifiers and loudspeakers were transformed into audible sounds. The coupling of sound and image guaranteed the compliance with synchronization. Optical sound is still used today for the playback of the sound.[248]

On this soundtrack composers, but also laymen, in particular from the fine arts, could practice composing and record their own music directly as graphics.[249] For the realization of their works they then only needed the apparatus. In the metamorphosis of the notation form—we cannot properly speak of a score form in the actual sense as with the punched tapes of the reproducing piano—one area of fuzziness is eliminated. But again the apparatus limits the realization of ideas. The optical sound, used for sketching, does not produce a very good sound in the translation of graphics to sound, at least in my hearing experience. The recording procedures where the optical film strip serves as sound carrier, is irrelevant for our line of questioning. Relevant is the usage of the optical soundtrack for graphics, which are then translated *directly* into sounds and tones. The traditional notation is so replaced by graphics and the realization through live musicians skipped. The graphics or the film roll exists as optical object independently from its realization as sound. It could be considered graphic notation or even 'musical graphic' that is one without fuzziness in the realization, again aside from technical fuzziness.

The simultaneous presentation of sound form and picture form of a work in sound film is nothing else than its audio-vision form. Strictly speaking the term 'picture form' only applies, if the musical work was created *before* its illustration. If created after the pictures as their accompaniment, we can speak of the *musical form* of pictures. The subsequently composed music for a silent movie, or even the cinema music, is a supplement. In a very liberal sense it is the musical form of a picture work, because the music, whether amplifying or contrapuntal, refers to the optical work (see Sect. 8.4). A technique already used in experimental films of the 1920s was to scratch signs into the soundtrack for the execution as sound.

> *Neighbours* (French title: *Voisins*) is a 1952 anti-war film by Scottish Canadian filmmaker Norman McLaren. Produced at the National Film Board of Canada in Montreal, the film uses pixilation, an animation technique using live actors as stop motion objects.
>
> McLaren *created the* soundtrack *of the film by scratching the edge of the film*, creating various blobs, lines, and triangles which the projector read as sound.[250]

The images are more remarkable than the sounds that sound like the sinusoidal blubbering of early electronic music. In another film by McLaren the procedure

[248]In detail e.g. Stern 2020; cf. e.g. https://www.wasistwas.de/archiv-technik-details/das-lichttonverfahren-veraendert-das-kino.html.

[249]With digitalization this has become impossible. „Today digital optical soundtrack are used, such as Dolby Digital or SDDS, a system by Sony. In contrast to analogous optical sound techniques the sound is not analogously copied onto the film, but digital information, which is captured by a photo cell and then transformed into sound signals in a decoder. This technique allows for higher dynamics, more channels (which allows for a better spatial representation of the sound) and an again improved noise reduction" (Stern 2020, ibid.).

[250]https://en.wikipedia.org/wiki/Neighbours_(1952_film), 14 November 19, retrieved 6 June 2020.

is significantly more complicated. He applies an entire encaptic cascade of special optical objects on photography to the optical soundtrack.

> *Synchromy* (French: *Synchromie*) is a 1971 National Film Board of Canada visual music film by Norman McLaren utilizing graphical sound. To produce the film's musical soundtrack, McLaren photographed rectangular cards with lines on them. He arranged these shapes in sequences on the analog optical sound track to produce notes and chords. He then reproduced the sequence of shapes, colorized, in the image portion of the film, so that audiences see the shapes that they are also hearing, as sound. McLaren had experimented with this technique for creating notes through patterns of stripes on the soundtrack area of the film in the 1950s, working with Evelyn Lambart. Their technique was based on earlier work in graphical sound by German pioneer Rudolf Pfenninger and Russian Nikolai Voinov. The creation of *Synchromy* was documented by Gavin Millar in 1970 in a film called *The Eye Hears, The Ear Sees*. In McLaren's production notes, he stated that "Apart from planning and executing the music, the only creative aspect of the film was the "choreographing" of the striations in the columns and deciding on the sequence and combinations of colors."[251]

The synchronicity is musically rather monochrome. We hear only monophonic, spotty tones of the type synthetic 'electronic music'. The reasons for this reduction probably lie in the available sound generators.

> This animated short by Norman McLaren features synchronization of image and sound in the truest sense of the word. To make this film, McLaren employed novel optical techniques to compose the piano rhythms of the sound track, which he then moved, in multicolor, onto the picture area of the screen so that, in effect, *you see what you hear.*[252]

Expressed more precisely: One hears what the graphic designer externalized using subjective synaesthetic associations and analogies both as image and object. It is a special analogy to graphic notation; but with the important difference: The graphic *is* not notation, but the composition, and can *directly* be translated into sounds. The barren sound is a fundamental boundary. Habituation and frugality makes it less palpable. If one declines to compare it to the richness of instrumental and vocal possibilities of non-automated playing of music, this boundary does not apply. Another way is the expansion of the types of instruments for the sound realization *and* for the picture work to be supplied with sound. With digital technology this kind of realization fundamentally shifts from manual to automatic, and the results on both sides can be considerably more complex.

6.8.4 Composition as Realization. „Musique Concrète" and „Electronic Music"

One impulse for the development of electro-acoustical music, in rudiments in the 1920s and then resurging after 1945, was the compulsion of primarily the composers, to insure the musical work, completed in notation as score form, against changes that

[251] https://en.wikipedia.org/wiki/Synchromy, 30 September 19, retrieved 6 June 2020.

[252] https://www.nfb.ca/film/synchromy/.

cause fuzziness in the sound realization. This already mentioned strong compulsion to exclude musicians from the realization of the sound form here finds its fulfillment. The work is now directly composed and realized as sound. More or less elaborate written concepts exist, but the notation form as independent stage is omitted. Likewise the realization as sound through interpreters is omitted. The composers become conductors, sound designers, instrumentalists and DJs all in one. The recording studio acts as orchestra the first analogous synthesizer as universal instrument, as now the PC with its diverse interfaces. The sound shape of the work has no fuzziness anymore, barring of course the technical blurs. Aesthetically it is so refined, as it was intended, and as was necessary and possible. (On the inner-technical fuzziness see Sect. 3.6 Phonography Form.)

The electro-acoustical music originated from two initially mostly separate approaches that were both opposed and complementary. The „Music concrète“ since 1944/1945 created collages, continuing avant-garde trends from the 1920s as for example W. Ruttmans *Weekend*, featuring real sounds, especially technical ones. The „electronic music“ since about 1949 synthesized and filtered from sine waves 'artificial' new, complex sounds. Both trends would soon merge, significant in this regard is among others Stockhausen's *Gesang der Jünglinge [Singing of the Youths]* in 1955, in which the voice of a singing boy is defamiliarized with „Lobet den Herren“ (*Praise to the Lord*), and amalgamated with own synthetically produced sounds.[253] One consequence of these technical innovations is then a compositionally available continuum between noise and sound, sinusoidal tone and white noise. Thus the *acoustic* reality content potentially increased, but not the *aesthetic-artistic* content; the opposite occurs, in fact.

> Before the advent of electronic music, musical imitations were necessarily constrained by the limitations of musical instruments, whose timbres did not necessarily approximate nonmusical sounds with any precision. Thanks to sampling and synthesis, however, electronic music can produce mimetic representations that are virtually indistinguishable from their real-world counterparts.[254]

This mimesis was largely not realized, though. To the extent that music becomes painting at all in this context, then it is mainly as abstract, non-objective, non-figurative painting—once again as the aforementioned distinction from the technical reproduction of real sounds (see Chapter 4). In how far these new materials and techniques are used to produce truly new *musical* ideas and shapes remains to be seen, but not at this point.

6.8.4.1 Nearly Total Determination. Composers as Interpreters

Only through electro-acoustical music the sound realization becomes fully emancipated from the harassment of having to use live musicians inclusive their inherent subjectivity. Now the composers themselves become the realizers of sound. If now

---

[253] On these well researched trends see e.g. Ungeheuer 2002 and 2005, Supper 1995, Heister 2013.

[254] Demers 2010, p. 172.

any subjectivity is involved, it must be their own. Their work idea, their imagined goal and intention, now directly determine the sound form, mediated by an objective apparatus.

> The electronic music finally executes a long prepared sentence. The increasingly more precise fixation of the score text ultimately aims for the elimination of the recital itself.[255]

But the direct realization of the composition as electro-acoustical music lends 'a new quality' to the live realization.[256] Especially Karlheinz Stockhausen early pleads for a peaceful coexistence.

> Instrumental music could exist next to electronic music. In each area functional work is required, all means shall be used productively: Generators – magnetophone – loudspeakers shall bring forth [produce] what no instrumentalist would ever be able to play [...]; Notation – player – instrument shall bring forth, what no electronical apparatus can ever produce or imitate or repeat. Composing electronic music means: describing the sounding in mechanical and electro-acoustical measures and thinking entirely in machines, apparatus, and circuitry; considering the onetime production and random repeatability [replication/reproducibility] of the composition. Writing instrumental music – now again – means: instigating the actions of the players through optical signs, and directly addressing the living organism of the musicians, their creative and always malleable ability to respond; from recital to recital enabling multiple productions and unrepeatability. Then electronic music and instrumental music would complement each other, the distance between them growing always faster – and only therein would ly the hope, that once in a while a true encounter takes place in a work.[257]

The direct encounter then takes place in 'live electronic music', a formation and development to which Stockhausen has contributed significantly (see Sect. 7.6.2). Overall the tensions between idea and realization, as well as the interplay of fuzziness and sharpening, as relevant for fuzzy logic, are largely suspended. In electro-acoustical music the compositional idea became fully autonomous from the interpreters' realization—surely at the cost of a principally very reduced, in the bad sense of the word 'artificial' sound. What is missing most are the mentioned attack and release procedures of a tone. They are imitated later. The musically optimal course is indicated by the diagonals (Fig. 6.60).

6.8.4.2 „Home Office"-Instrument Synthesizer and Sampling as Expansion of Sound Materials

The sound realization of compositional ideas without the mediation of musicians made another qualitative leap with synthesizers and sampling at the end of the1960s.

The miniaturization of re/production devices—an apparatus instead of a studio—allows individuals to compose all by themselves, even at amateur level. The next qualitative leap comes with digitalization together with the integration of electro-acoustical components in the home computer. Using keyboards, analogous to the keys

[255] Siegele 1968, col. 19.

[256] Ibid.

[257] Quoted in ibid.

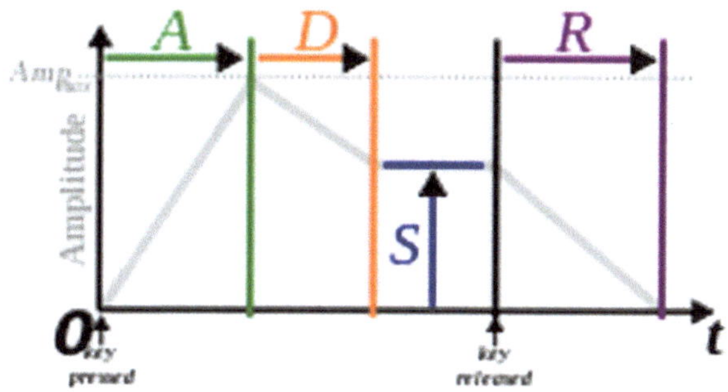

Fig. 6.60 Envelope
Schematic of ADSR
Envelopes control how sounds change over time. They may control parameters such as amplitude, filters or pitch. The most common envelope is the ADSR (attack, decay, sustain, release) envelope:
2. Attack time is the time taken for initial run-up of level from nil to peak, beginning when the note is triggered
1. Decay time is the time taken for the subsequent running down from the attack level to the designated sustaining level
2. Sustain level is the level during the main sequence of the sound's duration, until the key is released
1. Release time is the time taken for the level to decay from the sustain level to zero after the key is released (https://en.wikipedia.org/wiki/Synthesizer, 5 August 2020, retrieved 20 August 2020; see main article Envelope [music])

of keyboard instruments, which complement the typewriter keyboard, once again involves, at a new level, a piece of old-fashioned manual musical labor, similar as with the mouse, though the mouse is not modelled after musical application. In spite of this component of manual work, the blurs do not disappear when using electrified and digital keyboards. Secondly the sound is considerably constrained in comparison with the potential of traditional musical instruments—through the combination of these with electroacoustical sound the potential is then again expanded. Thirdly, the compositional potential is restricted and often resembles the working with formulas and set phrases as in much of traditional and improvised, non-notated music; now elements such as tones and sounds are replaced by pre-formed, pre-fabricated components, so to say building blocks instead of sand, lime, water, etc. (see Sect. 8.2). The analogous synthesizer by Robert Moog was a pioneer apparatus.

> The modularization reached a unified systematization of principles of sound generation and editing and steering from electronic sound studios, all within a compact device of modular design.[258]

What is often forgotten in the enthusiasm about technological progress: The realization of *music* was at first meagre: The synthesizer could only play monophonically:

> Early synthesizers could only play one note at a time, making them suitable for basslines, leads and solos. With the rise of *polyphonic synthesizers* in the 70s and 80s, "the keyboard

[258] Schultz/Seedorf 2016, p. 10.

> in rock once more started to revert to the background, to be used for fills and atmosphere rather than for soloing".[259]

The synthesizer is like a pocket-sized sound studio. As there, the sound comes from the apparatus

> as result of sound editing in three steps: „1. Generators (oscillator, noise generator), 2. Modulators (filter, amplifier, effects), 3. Controller (keyboard, joystick, sequencer and more)" (Enders, 2013, p. 58). Even if the possibilities of sound generation have become more complex and in a sense more abstract in technical terms, the phrases are imbued with an approximation of the music-playing subject and the resounding instrument. This was founded in the mostly automatized functionalities of the new instrument that brought closer together the composer and the intended sound according to the artistic concept, and the conceptualizing subject could often directly handle the devices.[260]

The personal union of composer, realizer and technician became possible—and more and more necessary in view of shrinking public spending and the thereby growing casualization tendencies among composers. The musicians certainly don't benefit from this. They become nearly even more redundant through sampling. Sound sampling, since 1979, is another stage of the sound realization and production, one could call this the "second technical-medial revolution". Crucial is that once again instrumental sounds—in some cases also vocals—expand the restricted sound stock of electroacoustically re/produced music. Presupposed is the digital technology, itself a new stage compared to previous automatization.

> In sampling *natural sounds* are digitalized. These digital waves then form the basis for sound generation. Samplers make possible what was previously reserved for the mellotron that worked with magnetic tape: the *real playback of acoustical instruments* […] Today computers with a sound card can load extensive sampling libraries and use them for computer-based music arrangements.[261]

The 'good old' helps to rejuvenate the new. Sampling of real sounds and concrete instrumental sounds as a new path often leads to astonishing sound results. Not the worst part being that they sometimes sound deceptively similar to an orchestra. Once more the recollection of art-oriented questions about sound results in a productive interplay of development lines of two different systems of apparatus. Especially the digital synthesizers appeared particularly lifeless and overly precise.

> In 1983, Yamaha released the first commercially successful digital synthesizer, the Yamaha DX7 […] (b)ased on frequency modulation synthesis […]- Compared to the "warm" and "*fuzzy*" sounds of analog synthesis, the DX7 was characterized by its "harsh", "glassy" and "chilly" sounds.[262]

The technological progress was here almost an artistic regression. *After* the introduction of sampling the synthesizer industry responded with a procedure that allowed the simulation of actual musical instruments.

---

[259] https://en.wikipedia.org/wiki/Synthesizer, 5 August 2020, retrieved 20 August 2020. Quotation from: Frank (2004).

[260] Schultz 2016, p. 10.

[261] Schultz/Seedorf 2016, p. 10.

[262] https://en.wikipedia.org/wiki/Synthesizer, 5 August 2020, retrieved 20 August 2020.

> Early in the 1990s the first synthesizers with a novel method of synthesizing, physical modelling, reached the market. The sound result of an instrument is not imitated, but a physical model (such as a swinging string) is digitally represented, meaning a model with all its specifications, dimensions, elasticity, tension etc. "exists" in a computer and is stimulated with an equally virtual exciter (like a plectrum) by applying kinetic energy at any given point. In the center lies at first the physical simulation, the sound is only a possibility of mediating it (one could also visually represent the body of the string). Nevertheless PM synthesis has resulted in the most realistic sound results ever, when it comes to *imitating natural instruments*, and this more so the more detail is included in the physical model.
>
> Though the principle has been known for a while, it could only be implemented with the development of the Karplus-Strong-algorithm and its refinement as well as the generalization of the algorithm into digital waveguide synthesizers by Julius O. Smith III et al. For real time processing high-performance digital signal processors (DSP: Digital Signal Processor) were required, which only became available at the end of the 1980s.[263]

The spiral process of progress through a return to the sound of real instruments and the subsequent expansion of the sound stock beyond merely electro-acoustically generated sounds requires innovations in computer science and technology. The return to -relative—nature requires art and technology, in short, culture.

> Digital synthesizers typically contained preset sounds emulating acoustic instruments, with algorithms controlled with menus and buttons.[264]

Through the MIDI standard since 1981 (Musical Instrument Digital Interface, an industrial standard for information exchange between digital controllers and/or instruments) the private home computer became an instrument for composing and a musical instrument.[265] This could be associated with conceptions of fetish. The 'artificial intelligence', embodied in the apparatus, triumphs over the natural human intelligence, the computer as active subject capable of interpersonal relations and with personal traits.

> The informatic artificial intelligence is a perspective for the future of this development and might allow a creative contribution of the computer in the musical, then possibly social cooperation of humans and machine and maybe even the complete usurping of musical production by intelligent machines.[266]

But automatic music production is firstly not a particularly urgent matter, as long as there are enough competent and willing and relatively inexpensive composers (see Chapter 5), and secondly the results so far have mostly not been overly convincing. Seizure of power through the machines might have to be postponed to another day. With digital technology and the computer as foundation for electro-acoustically oriented composition and sound realization there result impressive further differentiations of production possibilities. One may call them 'virtualization', considering

[263] In detail https://de.wikipedia.org/wiki/Synthesizer, 14 August 2020, retrieved 20 August 2020.

[264] https://en.wikipedia.org/wiki/Synthesizer, 5 August 2020, retrieved 20 August 2020.

[265] See Schultz/Seedorf 2016, p. 10. „MIDI was the basis for numerous musical-informatical applications and advancements, as in the area of arrangement, of composition or the automatical musical analysis" (ibid.).

[266] Schultz/Seedorf 2016, p. 11.

that these still have a material basis and the technology has not yet transcended into the immaterial.

> Around 1990, with the phase of virtualization, a development begins that detaches instruments as well as apparatus and processes of music production in the studio more and more from their physical object, from the hardware, and transfers them as software into the virtual environment of the computer. Thereby accessibility of this technology for a mass market is further simplified, since the programs can be executed on nearly any customary computer with the appropriate capabilities such as adequate sound card. Mixer consoles with nearly unlimited amounts of tracks, as well as expensive and heavy effect generators are replicated virtually or newly developed. New possibilities of sound manipulation are created and can be used on the computer or in more recent developments even on the tablet.[267]

More convenient and quick is the sound execution of a work in rather meagre keyboard sounds. At most these reach the level of the sound form of a piano excerpt, which at least uses a proper piano.

> With the simulation of historically and sound-aesthetically relevant instruments, essential aspects of previous developments, as for example modularization, are incorporated into the virtual space. *Instruments or even entire orchestras imitated through sampling techniques achieve a new dimension of authenticity in regard to their sound quality.* Entirely new procedures of sound synthesis, such as wave shaping or granular snythesis, are developed, possible only with digital means. In the virtual reality the physical laws of the real world can be replicated, but also overridden. For example, physical modeling allows a very authentic replication of instruments in their acoustic physic *or can generate new, mechanically impossible* resonating bodies.[268]

The simulation of existing instruments is not so significant or urgent. They are more easily available, now as before not-simulated, even if admittedly not as easily accessible for amateurs as in computerized home music. More exciting is definitely the promised development of new, previously unheard sounds. In strange dialectics the individual isolation of music makers comes with a new, global network, a participation of the re/producing individual in globalized collectives. Whether this is a "new phase" (as Bernd Enders contends) or a spatial and social extension of the use of computers as musical instruments and a special implication of general globalization remains to be seen.

> Along with new distribution methods globalization as impacts composition and production processes, so that artistic expression can also occur in the global communicative community. Samples, MIDI files or scores can be transmitted globally without delay[269] and worked on collectively.[270]

Interestingly such tiered, "interactive", feedback-integrated and encaptic private re/production processes enjoy it when their results culminate in public performances in concert form. A significant difference to the traditional concert lies for one in

[267] Ibid.

[268] Schultz/Seedorf 2016, pp. 9f.

[269] Nearly without delay and mostly safely only in good weather conditions in the server farms and functioning security.

[270] Schultz/Seedorf 2016, p. 11.

the destitution of visual impressions—an orchestra, choir or soloist do better in this regard. This may be abetted by providing additional pictorial accompaniment in many colors and shapes, ideally genetically and structurally derived from the sounds. Furthermore such concerts take place not at a single location, but all over the world. That this happens in "real time" and simultaneously is rather "virtual" than real, and only partially possible: The nature of earth's rotation around the sun and the culture of time zones make an exact practical simultaneity impossible. As with the return to "natural" instrumental sounds—and of vocals, which are of far less concern in the context of electro-acoustical and digitalized music—through sampling the return to nature via technical mediation can be driven considerably further.

> As last development stage Enders names hybridization, in which the music instrument, now as virtual object, and the human as subject are amalgamated through the use of digital electronic technology, in order to restore at least partially the unity, as it was before instrumentalization.[271]

The question whether such fusion phantasies are maybe rather dystopian must not be answered here. A music-instrument-human-body-cyborg is not high on the agenda of the development of music and technology. Vocal music seems not to be envisaged. The differences between compositional subject, work, work idea and intentions, idea and realization become blurred. This kind of identity without differences is like the proverbial night in which all cats are grey.

6.8.4.3 Fusions of Score Form, Picture Form, and Sound Form in Digitalized Audiovision

In view of work realization and the relationship of score form and sound form there is a significant possibility to again combine optics and acoustics directly—not as a stage of the existence forms score form and sound form. Particularly with the conditions of digital technology the directionality of the metamorphoses in the order of existence forms is inverted; instead of progressing from picture to sound, it now moves from sound to picture. And simultaneously, the picture to the sound is generated—or the score form inserted (see Sect. 8.4). With digitalization come new possibilities and procedures, for one to skip the genuine score form, and then to fuse it with the sound form. This occurs always in the mode of the medium, mediated through technical re/production of the acoustical and the optical. Fusion here means, on the one hand, simply to include in the video the score as picture accompanying the sound—an instructive sharpening of perception.

On the other hand, fusion means an—predominantly abstract, not concrete—illustration of the work, now existing only digitally, and the audiovisual playing back of both together (see Sect. 3.6). Here the procedures of experimental films of the 1920s return the simultaneous display of sound form and picture form of a work. A radical digital example is an absurd, formally complex piano work such as Maximiliano

[271] Schultz/Seedorf 2016, p. 12.

Maximilianos (pseudonym) *Faerie's Aire And Death Waltz.*[272] It is, like the works of Nancarrow for the electrical piano, no longer performable by humans, but can only be 'played' by machines. Digital technology allows the automatic conversion of images into sounds in new, more variegated shapes than with the usage of optical sound strips.

> With the introduction of affordable personal computers with processors and applications, which are able to manipulate moving pictures in real time, the so far, in terms of media technology, separate phenomena sound and image could be linked through algorithmic translation of auditive parameters into visual parameters [... and] programs could be developed, that contain generative processes and bring forth independent digital aesthetics. Sound could now be analyzed and dissected into different frequency bands, and thereby deliver input for image-generating systems. In the course of direct programmability by now a variety of complex generating procedures have originated, which continuously expand and enrich the spectrum of expressive possibilities.[273]

One basic characteristic is real time.

> Music visualization or music visualisation, a feature found in electronic music visualizers and media player software, generates *animated imagery based on a piece of music.* The imagery is usually generated and rendered in real time and in a way synchronized with the music as it is played.[274]

> Visualization techniques range from simple ones (e.g. a simulation of an oscilloscope display) to more elaborate ones that often include a number of composite effects. The changes in the music's loudness and frequency spectrum are among the properties used as input to the visualization. Effective music visualization aims to attain a high degree of visual correlation between a musical track's spectral characteristics such as frequency and amplitude and the objects or components of the visual image being rendered and displayed. "Music visualization" can be defined, in contrast to previously existing pre-generated music plus visualization combinations (as for example music videos), by its characteristic of being *real-time generated.*[275]

Stockhausen constructed a continuum of rhythm to pitch—using frequencies with a point of change at 16 Hz—which as a theory is problematic, but as an idea inspiring. Possibly there exists mathematical modelling that analogously creates non-arbitrary, consistent relations between image and sound. Diverse "visualizations" aim in that direction, and the rhythmically jerking sassy light shows demonstrate basic approaches.

> Computer music – music consisting of digital sounds and generated at a computer – has been established in the early 1990s as a common method in live concerts. Starting in 1997 the first visualization software and sufficient processing capacity were available to enable users of the small portable personal computer to edit picture sequences in real time. Following the live improvisation principle of the electronic music, visual artists began to manipulate

---

[272] Video e.g. https://www.youtube.com/watch?v=RbT4lIrdnYk und https://www.youtube.com/watch?v=Da911wDQHpA, Abruf 2 April 2019.

[273] In more detail Tina Frank: *AV-Parameter-Mapping in der Musikvisualisierung*, https://sts.hgb-leipzig.de/kompendium/text/72.html, retrieved 13 August 2020.

[274] https://en.wikipedia.org/wiki/Music_visualization, 14 June 2020, retrieved 17 June 2020.

[275] https://en.wikipedia.org/wiki/Music_visualization, 14 June 2020, retrieved 17 June 2020. Emphasis HWH.

> computer graphics live and later also to generate them. A new kind of real time performance was born. The audio data or also the steering data from the controllers of the musicians was incorporated into the image-generating system and used as triggers for visual impulses.[276]

One—only technical—advancement is variability, which also means arbitrariness,

> the ability of some music visualization systems [...] to create different visualizations for each song or audio every time the program is run, in contrast to other forms of music visualization (such as music videos or a laser lighting display) which always show the same visualization. Music visualization may be achieved in 2D or a 3D coordinate system where up to 6 dimensions can be modified, the 4th, 5th and 6th dimensions being color, intensity and transparency.[277]

Such arbitrariness can also be regarded as aesthetic achievement (see Chapter 7).

> With digital means one can create *arbitrary interrelations* between audio and video, where sounds often steer images.[278]

A starting point for visualization was the declining attractiveness of music without images, at least when music was received not only as accompaniment for other activities—which come with optics—but as autonomous object of interest.

> The first electronic music visualizer was the Atari Video Music introduced by Atari Inc. in 1976, and designed by the initiator of the home version of Pong, Robert Brown. The idea was to create a visual exploration that could be implemented in a Hi-Fi stereo system.[279]

Initially the images to the sounds were not specific to a work, but something pleasant and at the same time schematic.

> The 1984 movie Electric Dreams prominently made use of one, although as a pre-generated effect, rather than calculated in real-time.[280]

This paved the way for more precise and specific correlations.

> Subsequently, PC computer music visualization became widespread in the mid to late 1990s as applications such as Winamp (1997), Audion (1999), and SoundJam (2000) emerged.[281]

Somehow in contrast to the seemingly endless technical possibilities, the reality of perception reached its limits. The often-cited sensory overload does not focus but instead blurs.

> With the increasing popularity of virtual reality, several start-ups have begun working on music visualization although reception has been mixed with one informal poll finding that only 33% of respondents were interested in music visualization for VR.[282]

---

[276]In more detail Tina Frank: *AV-Parameter-Mapping in der Musikvisualisierung*, https://sts.hgb-leipzig.de/kompendium/text/72.html, retrieved 13 August 2020.

[277]https://en.wikipedia.org/wiki/Music_visualization, 14 June 2020, retrieved 17 June 2020.

[278]In more detail Tina Frank: *AV-Parameter-Mapping in der Musikvisualisierung*, https://sts.hgb-leipzig.de/kompendium/text/72.html, retrieved 13 August 2020.

[279]https://en.wikipedia.org/wiki/Music_visualization, 14 June 2020, retrieved 17 June 2020.

[280]Ibid.

[281]Ibid.

[282]Ibid.

As popular as the automatization of the production of the sound form without score form may be: The urge desire for a stable, object-like score is hard to relinquish. Based on the same technology it is reintroduced through the back door. With more advanced digital technology the possibility emerges to automatically generate with the sound its transcription. This is another aspect of ‚real time‘. Now the notation of a ‘score’ does not occur before the sound form, but is actually subsequent to it, in inverted order compared to tradition. Though the composition might have not migrated, the notation has emigrated to the execution and recital apparatus. Margaret Schedel reflects on this and claims to be a

> fact that an instrument becomes a blueprint for the performance or reperformance itself. [Following Tomás 2016] musical work can be seen as «not only defined by the instrument, but more importantly, by the act of playing the instrument. The performer's role [is] to reveal instances of the musical work inherently integrated in the *circuitry*». Tomás is writing about his own electronic instruments, but it is easy to see that the statement « affordances are fully mediated by the embodied relationship between instrument and performer» [Bacon 2015] can apply equally to acoustic instruments.[283]

The musical-instrumental-electroacoustical cyborg goes marching on, even back to past modes of making music. The notational optical reassurance converges with one of the basic functions of ‘visualization’, the

> visualization of specific information in the material, to allow a technical assessment.[284]

Schedel lists an abundance of research on this topic and adjacent fields. The subject matter is probably best labelled as ‘real-time notation’. The titles are already incorporated in the references. Here the relation of the acoustical and the optical is necessary, focused and not arbitrary, in contrast to ‘visualisation’. To this often pertains:

> The correlation of music and video is [...] arbitrary and not always realized in a comprehensive manner.[285]

The beginnings of these technically-automatically mediated visualization date far back. One could mention, based on the electrification of the ‘light organ’ since the 1970s, or László Moholy-Nagy's more complex *Light Prop for an Electric Stage* [...], „conceived in 1922-30, built in 1929-30, and refined throughout the artist's lifetime.“[286] In the course of newer technical developments there was certainly technical progress: The aesthetic-artistic questions of the relation of image and sound seem to have been addressed too little and not thoroughly enough. Both in methods and in the matter further significant sharpening is required. For one the correlations of image and sound tend to be arbitrary. It is not an advantage but a deficit that often different and randomly determined combinations occur. Secondly it is a one-way street, from music, here mostly referred to simply as sound, to the fine arts, which should be more than just ‘visualisation’. A more productive idea could be to look for

---

[283] In detail cf. Schedel 2019, p. 91.

[284] https://de.wikipedia.org/wiki/Audiovisualisierung, 29 May 2019, retrieved 17 June 2020.

[285] Ibid.

[286] Cf. Schedel 2019, p. 93.

a shared third aspect, common to both arts. This shared third aspect would then be represented by algorithms, but not simply resultant, but intentional and deliberately steered.

6.8.4.4 "Inhuman"? Enhancement of the Human Mastery of Nature in Virtuoso Handling of Apparatuses

As becomes apparent, the replacement of human labor in the sound realization of a work through objectified labor can never be total. In spite of all reservations regarding many phenomena that result from this combination, and in spite of the often distinct contrast between complex production procedures and simple products, it nevertheless constitutes progress in the development of material and of technology, as ambivalent as that may be. The instrumental organ projection transcends by the very nature of the matter what is human, what is humanly possible: even here the principle of permanent transgression of borders or rather the continuous transformation of borders into boundaries. Not only in the case of Nancarrow's works that simply cannot be realized by human virtuosos that becomes playable. Generally this applies to the field of electronic music together with all its further developments.

> If this provides the possibility to "make" music, without need of a human player, and if this music can be replayed identically as often as desired, then one might disregard all limitations of the human condition, that is both those of human nature and those of culture, namely of all orders in the fundamentals of music so far. Instead one may introduce conditions, which are not achievable for humans, but very well so for technology. What results is […] un-heard of, but equally in-human, which begs the questions, whether we like it or not, if electronic music is music in the same sense as all previous music. If the answer is in the negative, and in my view it should be, then the term music must be divided or defined afresh. But then it touches on the question of what is human, who has so far carried music as one defining characteristic.[287]

The question of the humanity of music returns again and again. A non-regressive, non-restorative answer surely sounds a bit different. What is primarily forgotten here is that musical domination of nature takes place more indirectly, but on an extended level, still mediated by the human subject and his capabilities and skills. Their unfolding into virtuosity, the perfect and generally consummate execution of the work is now relieved of the bodily, material-technical aspect of sound realization. Even here we can still speak of virtuosity, even if the traditional-artesinal foundation of the immediate sound realization ceases to apply.

This virtuosity refers to the dimensions of work in the studio, with sampler etc. as well as the masterful, ipso facto virtuoso handling of the technical apparatus, which oscillates between new artisanship and engineering, and which have, not only in regard to typewriters or computer keyboards, a material side. The devices could be considered as music instruments on a new scale. Then again composing and realization are interlinked in a new way and in an altered framework. Production

[287] Feil 2005, pp. 754f.

without or production with human interpreters do not have to be mutually exclusive, but might very well be productively synergetic, as in Stockhausen's thoughts about their coexistence. Even in the 'computerized' music circles live performances have not lost their appeal. Similarly the work of the disk jockey could be included in the category of virtuosity—even more so since this involves literally handiwork as well as direct sound realization in dependence on the virtuosity of the DJ. As advanced and elaborate the level of production procedures often are, as shallow is often the standard of what is produced. The works, the work idea on the one hand and the sound results on the other, seem to follow the rules of a restricted code. But this shall not be further discussed here. The perfection of virtuosity on a historical and societal scale, the emancipation from natural limitations, is itself a part of the music-historical process, which partially also resembles a progress.

6.8.5 Without Notes, Text and Graphic. Controlling the Sound Form via Brain Waves

One extreme is reached with the elimination of the score form and the meditation through live musicians in the work of the composer Alvin Lucier. He employs the technical-medicinal advancements of electroencephalography, used since 1875 in animal studies and used with humans in 1924, including biofeedback respectively neurofeedback for immediate sound generation (see Sect. 9.2). The multidimensional sharpening and the growing direct and total determination of the sound form by the composers through the elimination of the score form here reaches a tipping point: The total determination is achieved through a reduction of the ideal and audio-material substance of the work, and a reduction of the compositional subject. Everything is focused, but that everything is almost nothing.

The first work using brain waves to generate sound, *Music for Solo Performer* for amplified brain waves and percussion, was first performed in 1965. Leaving aside the evaluation of the acoustic-sensual product, such a music production constitutes a filtering and crystallization complete with a multiply dialectic, such as that of simplicity and complexity. Peter Niklas Wilson describes the setting of a man in a chair, with his instrument being the brain. His electrically measurable brain activity is registered, amplified and made audible by electrodes and changed into sound by speakers—more specifically into infrasound as alpha waves oscillating at about 10 Hz, which is sound with frequencies below human pitch perception. Yet, as Wilson points out, loudspeakers which are set to this frequency can make diverse percussion instruments oscillate with these impulses through resonance. He names timpani, drums and gongs and describes how they begin to sound as if by magic while the man sits motionlessly on stage. However, Wilson writes, sound can only originate if he is calm because the differential amplifiers which are downstream from the electrodes only capture the alpha waves of his brain activity—and these only occur in a state of total relaxation. If the man on stage is tense, if he wants something, nothing happens. Only if he lets himself go and renounces all he wants does sound occur. In Wilson's words, only absolute passivity can facilitate sound activity, with renunciation bringing gain and lack of will bringing action. He who gives up, Wilson writes, will win, with intention preventing music and silence generating sound. As Wilson concludes, the musical phenomena which this psychoacoustic experimental setup creates may be complex, but it is a complexity which originates from reduction, and "the purer, the simpler the idea, the richer the music."[288] The steering of the realization through ideas could hardly be more direct. And yet there remains the restriction of the ideational in this realization. The arbitrary and the pragmatic, not the conceptually chosen sound generators determine the sounds. Also the ideas themselves are constrained; they are a reduction of cognitive activity to meditative relaxation, i.e. almost-emptiness and almost-nothing.

6.8.6 *Soundless.* Music for Reading

Completely nothing in regard to sound is *music for reading.* As the subtitle reads for Dieter Schnebel's book *Mo-No, Musik zum Lesen* (*Mo-No, Music for Reading,* 1969),

> a book with musical graphics, with the only goal of an imagined sound. Schnebel mixes sound memories in the form of note citations, creative sound conceptions in musical graphics and silently read text fragments.[289]

The egg of Columbus: In this manner one can not only skip the realization through musicians but the entire *sound* realization. *Mo-No* contains a lot of conceptual art and fragments of graphical notation, which here result in fuzziness, indeterminate musical graphics. The book is part of a trend dating back to the avant-garde at the beginning of the twentieth century, which aims to transform both literature as well as music into fine arts, preferably as graphic. This does not include the *pictorial poem*, that typographically arranges lyrics in the shape of images such as a cross, a heart, etc., or the mentioned respective notations of music in mannerist trends already around 1400: This is notation of literature and music in special typo-graphical shape.

> Since 1976 the artist Gerhard Rühm (born 1930), trained as musician, creates *visual music*, which does without an interpreter or rather ascribes this role to the reader. Aside from *readable music*, free sketches on note paper with or without indications of instrumentation, there are superscriptions of notes, in which the notes of a printed musical piece are blacked with pencil, thereby emphasizing the devolution [/course/progression] of musical density visually.[290]

This results in configurations: Text as image, notes as pictures, pictures instead of notes. The problem of the realization of the idea is here also solved 'promptly'. The

[288] Paraphrase of Wilson 2003, p. 15. Emphasis HWH.

[289] Julia H. Schröder: *Grafische Notation und musikalische Grafik*, https://sts.hgb-leipzig.de/kompendium/text/78.html, retrieved 3 March 2020.

[290] Schröder, ibid. See also Schnebel 1993.

graphic *is* the idea a further representation of sound is omitted. Viewing and in given cases reading suffices. This cancels the problem of fuzziness in the realization. A next stage, and surely not the last, is *thinkable music*.[291] This is less than the famous "And the rest is silence", which at least still is something acoustical. This cancels out almost everything, and we are again back to almost-nothing.

6.9 Sharpening of Perception. Anticipative Preparation in the Production and Supporting Activities in the Reception

Not only the productive realization, but also the reception requires sharpening. The production of distinct shapes with anticipation of reception and even rudimentarily of the impact is a general aspect of composing. Here at the threshold of production/reproduction and reception the anticipatory thinking including the respective compositional activity returns at a new stage. The composers imagine an audience, manners of behavior and of hearing, effects, and anticipate them by composing accordingly. Anticipated are not only components of the productive sound realization, but even the perception and possible effects in the receptive realization. The perception is deliberately *steered*—just as certain effects in the recital are consciously planned and notated. There are two concerns: Attention is to be roused *and* the so remembered retained in memory.

6.9.1 „From the Heart—May It Go to the Heart" (Beethoven). Rhetorics and Intensity

This matter of the anticipatory response of recipients is a component of music as rethoric: Music as speech, oration appeal. 'Rhetorical figures' as in word language are anyway a part of musical language. Familiar is the formulation: „Von Herzen – Möge es zu Herzen gehen" [*From the heart – may it go to the heart*]. In 1819 Beethoven inserted it in the autograph of the *Kyrie* movement of his *Missa solemnis* as a kind of motto. The universal and 'all humans' are meant, as in the *Ode to Joy*. The motto is omitted in the transcription, which Beethoven in 1823 handed over to the dedicatee Archduke Rudolph—the work was to be played at his enthronement as Archbishop of Olmütz (Olomouc), but wasn't completed in time. Later Beethoven refined his intention and added, that the work may remain effective for a long time and wrote, that

[291] A new, radical term and book title from Dieter Schnebel: *Denkbare Musik* (1972).

> in the editing of this great mass my main intention was to evoke religious feelings both in the singers and the listeners and make these last.[292]

To this end he employed among others elements of the „stile antico“, the solemn historical style of vocal polyphony and the musical rhetoric, combined with more advanced material.

> According to Cooper Beethoven attempted to make his intention reign through fastidious diligence in regard to each word of the text; after writing down the (Latin, note from website author) text together with its translation (into German, note from website author) and comments, in order to ensure that he was properly understood, he elaborated its full meaning by representing each word as figuratively and vividly as musically possible.[293]

After a concert in the Berliner Singakademie (*Berlin Singing Academy*), on 21 June 1796, he postulated:

> Music should not tease out tears it should spark fire in the spirit of man.[294]

6.9.2 ‚Hookline'. Catchiness, Repetition, Obtrusiveness

The motivations and realizations of the objectives in terms of the anticipated reactions of the recipients and the desired effects are multidimensional and complex. This becomes coarser and more one-dimensional in the hit production. Within larger works, even in 'classics', one can find similar moments, but there they are embedded in a larger context. Verdi's short aria „La donna è mobile“ for the Duke of Mantua in *Rigoletto* or the „Triumphmarsch“ (*Triumphal March*) in *Aida* fall into this category. In both cases Verdi let them appear in context and they are connotated negatively precisely because of their musical forcefulness, which obviously got lost in the reception. In the respective productions in the area of 'Pop' differentiating contexts such as multi-layered meaning are largely absent. In a hook or hook line the appealing function condenses structurally into a shape:

> A hook is a musical idea, often a short riff, passage, or phrase, that is used in popular music to make a song appealing and to "catch the ear of the listener". [...]The term generally applies to popular music, especially rock, R&B, hip hop, dance, and pop [...] and can be something as insubstantial as a 'sound' (such as *da doo run run*) but *ideally* should contain one or more of the following: (a) a driving, danceable rhythm; (b) a melody that stays in people's minds; (c) a lyric that furthers the dramatic action, or defines a person or place.[295]

---

[292]Ludwig van Beethoven: *Briefwechsel*. Gesamtausgabe, Band 5, Brief Nr. 1875, S. 363–364, quoted in https://raptusassociation.org/missagmusikkritik.html, retrieved 12 December 2012.

[293]Paraphrase in https://raptusassociation.org/missagmusikkritik.html, retrieved 12 December 2012.

[294]Quoted in: Franz Welser-Möst/Ronny Dietrich: *Ludwig van Beethoven – Das Prometheus Projekt*, https://www.welsermoest.com/ludwig-van-beethoven-das-prometheus-projekt/, retrieved 13 August 2020.

[295]https://en.wikipedia.org/wiki/Hook_(music), 24 June 2020, retrieved 13 August 2020. Cf. Gary Burns (January 1987). "A Typology of "Hooks" in Popular Records".*Popular Music* 6(1): 1–20.

The hook is best described as a contradictory unity of conspicuousness and catchiness. Conspicuous, that is, sharpened, the figure is especially clearly emphasized against the "ground". At the same time it means higher neg-entropy, more and denser—aesthetic—information. However, the hook should be catchy, i.e. the smooth, familiar, slightly new. This is a general paradox of the hit: it is, in the spirit of Günther Anders, the old-familiar-never-seen-before. This dialectic unity works together to supplement the current attractiveness and give it a long-term recognition value.

> a hook is "a musical or lyrical phrase that stands out and is easily remembered".[296]

Parallels exist in the field of „Classics", especially in works which include what is already popular. In the work of Charles Ives for example, who impressively and with great variety includes memories in his compositions, there keep returning standard songs from popular and church music, which Ives recites as objectified memory. They could be said to work as a kind of multiplication of the memory-inducing Madeleine of Proust and invite to remembrance. Characteristically the melodies are principally defining, namely being what is most easily retained as contours of a more complex shape, the composition as a whole. Repetition as pummeling, as plugging here also has an augmenting effect, at least up until it reaches ad nauseam—which seems to be reached surprisingly late in most cases. The hook has to potentially allow the listeners to sing along, and may therefore not be overly complex. The hit characteristics „triviality and trick" (Carl Dahlhaus), the always-familiar and the somehow-deviant are combined to achieve catchiness:

> In music, incongruity refers to the inclusion of varied or irregular musical and lyrical features, such as mispronounced words or unexpected syllable accentuation. These incongruities are intended to capture the listener's attention and to preserve their level of interest throughout the song, regardless of the simple and otherwise repetitive lyrical content.[297]

The historically outdated poetics of rules experience a resurrection. There are numerous manuals that attempt to teach the writing and composing of hooks. They are mostly directed at amateurs and are recommendable as stimulation to artistic production. Naturally the narrowing down to hit and hook makes such handouts and recipes for idea and success mostly conformist and rather bland.[298] Here a formalized induction through extraction of characteristics from a preferably extensive ensemble

[296]Monaco and Riordan (1980, p. 178). Cited in Burns, Gary (1987). "A Typology of 'Hooks' in Popular Records", *Popular Music*, Vol. 6, No. 1 (January, 1987), pp. 1–20 https://en.wikipedia.org/wiki/Hook_(music), 24 June 2020, retrieved 13 Aug. 2020.

[297]Puiu 2011. Factors of catchiness following Alison Pawley and Daniel Mullensiefen: "Longer and detailed musical phrases, higher number of pitches in the chorus hook, male vocalists, higher male voices with noticeable vocal effort". Puiu, Tibi (3 October 2011). "*What makes a song 'catchy'—Science explains*". ZME Science, quoted in: https://en.wikipedia.org/wiki/Catchiness, ibid.

[298]See e.g. https://www.wikihow.com/Write-a-Catchy-Song#relatedwikihows, June 26, 2020, retrieved 31 Aug 2020. With tiring uniformity, many types are rattled off, e.g. R&B, "Christian" oder How to Write a Rap Punchline; https://www.wikihow.com/Write-a-Hit-Song; Alper Tuzcu: How to Write Memorable and Catchy Song Lyrics, June 12, 2019, Updated May 04, 2020, all retrieved 31 Aug 2020.

of hits might be a promising approach. One study has suggested that the ever-same has significantly more appeal for the memory.

> A 2014 study by the University of Amsterdam and the Museum of Science and Industry in Manchester found "Wannabe" by the Spice Girls to be the catchiest pop song of the last 60 years in the UK.[10] "*'Wannabe' by the Spice Girls Is the Catchiest Song, Study Says". Time. 3 November 2014. Retrieved 1 April 2017.* The study found that having a simple and relentless melody was the key to a song being "catchy". "We found, much to our surprise, that writing a very surprising and unusual hook is not the recipe for long term memorability," musicologist Dr. John Ashley Burgoyne explained. "Actually, the more conventional your melody in terms of the interval patterns that you use; in terms of the rhythms that you use, the easier the song is to remember over the long term.

Especially in the case of *Wannabee* there are most likely other factors involved. In an almost perverted form the desire for the "consummate whole"—a consummation in interplay with perception.

> *Wannabe* felt like it should have been a 3 1/2 to 4-minute pop song, just like every other hit at that time. To create that feeling, the producers cut the song at the knees. The last thing heard in the song is the a cappella line, "If you wannabe my lover," which is also the first lyric of the chorus. The audience has a *natural desire to hear something to its completion.* When they expect a song to go somewhere, they will not feel completely settled until that song resolves itself. *"Wannabe" never resolves*, and therefore creates a situation where the listener cannot get the song out of his head.[299]

This desire is also not satisfied in the frame of the artwork. The deliberately provocative sexual invitation is subconsciously perceived. The dissatisfaction keeps in motion the desire and longing for completion.

> The song title, "*You Can't Take That Away From Me*" immediately raises the question, "What can't you take away from me?" It drives the reader to want to investigate further and find out what the lyric says. It peaks our interest and arouses one's curiosity.[300]

The commercial sharpening tips over into overfocus, but reaches beyond the work, has a touch of advertisement, and is indeed often gladly used in commercials. The sale objective justifies all means.

> According to T.C.W. Blanning: "I would sacrifice everything – rhyme, reason, sense, and sentiment to catchiness. There is… a very *great art in making rubbish acceptable*".[301]

Hook music is intended to anchor in the memory—and naturally is then associated with the advertised product. This means two things: The product music or the product that is advertised with such music as in jingles or as sound environment in commercials.

The peak of sharpening is the closing of the sale.

---

[299] Frank, Jay. *Futurehit. DNA*, quoted in: https://en.wikipedia.org/wiki/Catchiness, 20 June 2020, retrieved 30 June 2020.

[300] E.g. *Vince Corozine*: *How To Write a "Catchy" Song Title,* vincecorozine.com/media/pdf, retrieved 15 April 2020.

[301] Blanning, TCW. *The Triumph of Music: The Rise of Composers, Musicians and Their Art.* p. 164, quoted in https://en.wikipedia.org/wiki/Catchiness, 27 June 2020, retrieved 14 August 2020. Emphasis HWH.

> a hook is repetitive, attention-grabbing, memorable, easy to dance to, and has commercial potential and lyrics. […] The hook is 'what you're selling'.[302]

This refers to both the sales of hits with hooks as autonomous products and to heteronomous functions. For the hooks and their ambience serve as sound environment for advertising, to sell some other goods. Since this area as well as hit and hook are ruled by uncertainty, companies and advertising radio stations are interested in refining these hit effects for advertisement, and to verify this through quantifiable studies. What in the field of art is probing, here becomes testing, to reduce costs and increase efficiency.

> Paul Fisher, vice-president and general manager of CHFI-FM in Toronto. "Our role is to play the songs that people like." Spinning the right songs at the right time determines a station's success, success measured in dollars. Revenue comes from advertising; advertisers want big audiences; audiences go where the good music is. … There are many tools to help grab, build, and gauge audiences but auditorium testing is one of the most effective.[303]

The hooks serve as test material.

> The hooks of a song may be used in market research to assist in gauging the popularity of a song based on the listener's ability to recognize the song's hook. Often radio stations conduct "call out" either on the Internet, via telephone, or a music test (either online or in an in-person setting) to conduct surveys.
>
> Stations may use the services of a professional "hook service" or prepare the materials themselves. In some studies, radio stations play the hook, typically 8–12 bars long, […] for audiences of up to 150 participants. The participants are then asked to rate the song on a scale from "dislike very much" to "like very much". Top 40 stations typically can't wait that long for results and have participants "call out" directly, by listening and rating different hooks. […] Studies such as these inform the radio station how popular current songs are or if the audience is "burned out" of a certain song. Market research based on hooks gives radio stations of all genres awareness of what their audience demographic wants to listen to, and is even used to test the musical boundaries of the audience.[304] Auditorium testing […] usually takes place in a hotel ballroom or large conference room. Between 100 and 150 chosen participants gather for an hour or two to sample, in audio splendour, 600 to 800 songs, carefully marking a scorecard as they go. A $50 thank-you ends the session.[305]

Following the logic of sales the hook attains an autonomous existence form.

> Some groups even release these research hooks on a single's CD release.[306]

The functional transformations of the hook remind almost of the existence forms of a work. It seduces as pars pro toto to listening—and buying—of the entire work. The selling party or its associates investigate the degree of seduction. And the chain continues. The hook itself becomes an object of scientific research and is then

---

[302] Kasha and Hirschhorn (1979), pp. 28–29 (see *Burns 1987)*, quoted in https://en.wikipedia.org/wiki/Hook_(music), 24 June 2020, retrieved 13 August 2020.

[303] Dann-Beardsley, n.d.; see also Burns 1987.

[304] https://en.wikipedia.org/wiki/Hook_(music), l.c.

[305] Dann-Beardsley, ibid.

[306] https://en.wikipedia.org/wiki/Hook_(music), l.c.

methodically investigated by detour, now using online gaming.[307] The catchiness is Janus-faced.

> Catchiness is how easy it is for one to remember a song, tune or phrase. [...] Alternatively, it can be defined by how difficult it is for one to forget it.[308]

This is the link between hook and earworm.

> These (catchy) songs are referred to as earworms due to their parasitic characteristics; their entrance and exit from our mind cannot be controlled and despite our best efforts they may refuse to leave. In that aspect, catchiness, depending on how digestible the music is to the listener, has a level of annoyance unlike anything else. [...] It is concluded that since memory is powerfully affected by repetition that this could be a significant contributing factor to catchiness, though certainly not the only aspect. A song's catchiness may also be due to the auditory "slave system" of our inner ear, much like the visual slave system of our "mind's eye."[309]

6.9.3 ‚Earworm'. Undesirable Remembrance

Hookline und earworm are equivalents in regard to the musical matter, or at least similar, but in function and psychology they are diametrically opposed. The hookline intends to catch the recipient like an angling hook. This is deliberate on the side of the producers. In contrast to this, the earworm comes into being involuntarily, and not only that but even against the intentions of those affected. They usually desire to get rid of the earworm. These contrasts are combined in a hook that becomes an earworm. The externally induced earworms and those that are generated internally in the subject are equally bothersome, especially when the recipients do not want to be hooked on jingles and hits, but wants to be liberated of them.

> an earworm is something more than a catchy tune, more than a song you just want to hear again: it's a song you absolutely can't get rid of.[310]

A basic characteristic is the compulsiveness, the involuntary, the un-intentional occurrence.

> the experience of a short melody getting stuck in the mind and being heard repeatedly outside of conscious control.[311]

---

[307] A European consortium (including Utrecht University and the University of Amsterdam) studies the hook by using online games and the wisdom of the crowd to understand and quantify the effect of catchiness on musical memory. https://en.wikipedia.org/wiki/Hook_(music), 24 June 2020, retrieved 13 August 2020.

[308] https://en.wikipedia.org/wiki/Catchiness, 27 June 2020, retrieved 14 August 2020.

[309] Stafford, Tom, "*Earworms: Why catchy tunes get trapped in our heads*". BBC News, 11 April 2012. Quoted in https://en.wikipedia.org/wiki/Catchiness, 27 June 2020, retrieved 14 August 2020. Emphasis HWH.

[310] Brett Milano https://iconcollective.edu/how-to-write-catchy-song-lyrics/ https://www.udiscovermusic.com/in-depth-features/how-to-write-an-earworm/, 16 April 2020, retrieved 31 August 2020.

[311] Floridou, Williamson, and Müllensiefen 2012, p. 302.

Some musical and textual criteria and preconditions for this are laid out, without an overly distinctive analysis:

> Chart hits are more likely to be earworms than deep cuts. Ditto faster songs over slower ones. If you hear the song on the radio all the time, it's probably earworm-bound. But here comes the insidious part: The actual order of notes does a job on your subconscious, and songs with a "common global melodic contour" – that is, songs whose melodies unfold in a way that matches your subconscious expectations – are the wormiest ones.[312]

Even in the following descriptions the near-identity with the hook is conspicuous:

> [R]epeated sinple melodic and harmonic structures [might] be necessary, but not sufficient conditions to turn a phrase into an earworm.[313] Kellaris (2001) describes the INMI phenomenon as a "cognitive itch" induced by certain musical properties: 1. Repetition within the musical stimulus, which might be a phrase, a motif or a sequence 2. Musical simplicity such as that found in children's songs. 3. Incongruity (an unexpected rhythmic variation).[314]

And here too we find rudimentary approaches to a formalization of the analysis.

> Finkel et al. (2010) established a computational method for examining the common melodic features of earworm tunes as compared to matched control tunes (controlled for artist and popularity). A paper within these conference proceedings (Williamson & Müllensiefen, 2012) details the new method and the future potential to develop an 'earworm formula' based on melodic structure.[315]

The phenomena 'earworm' has been increasingly investigated scientifically since the early 2000s. Liikkanen has given it in 2009 a nicer name:

> Involuntary musical imagery (INMI)[316] refers to a conscious experience of music that is unbidden by the subject. Often this experience consists of the repetition of a short musical fragment of melodic, popular music.[317]

In Freud's (1916/1917) interpretation the 'earworm' serves as expression of unconscious desires.

---

312 Ibid.

313 Quoted in https://de.wikipedia.org/wiki/Hookline 28 August 2013, retrieved 10 September 2013.

314 Ibid.

315 Floridou, Williamson, and Müllensiefen 2012, pp. 302f. Cf. Müllensiefen, Fry, Jones, Jilka, Stewart, and Williamson 2014.

316 Perhaps an intended pun: "INMI" is homophone with IN ME. (Hint from Bernhard Rusam, September 2020.)

317 Liikkanen 2018, p. 7. He gives an exhaustive survey and reviews "nearly fifty studies." "The psychological literature discusses phenomena with features similar to INMI under multiple names such as musical image repetition (Bennett, 2003), earworms, stuck tune syndrome (Farrugia, Jakubowski, Cusack, and Stewart, 2015), music in the brain phenomenon, musical mind-pops (Elua, Laws, and Kvavilashvili, 2012), perpetual music track (Brown, 2006), spontaneous musical imagery (Baruss and Wammes, 2009), intrusive song phenomenon, (Hyman et al., 2013), or involuntary musical recollection (Beaman, Powell, and Rapley, 2015). A full paper has been written on the mere definition of the term (Williams, 2015)" (ibid.).

> The tunes which suddenly come into a man's head can be shown to be conditioned by some train of thought to which they belong, and which for some reason is occupying his mind without his knowing anything about it.[318]

The psychoanalyst Reik differentiates between the earworm as often trivial trigger and its often non-trivial implications. Reik:

> The haunting tune can be trifling and insignificant, but the emotions and problems in its emergence are always meaningful. They reflect the concealed basic demands of the drives and fears of the person and seek to convey his most important interests and drives.[319]

The proximity and the difference to an obsession and to the hallucination are also to be considered.

> The differences between a musical association occurring to a person in the middle of targeted thoughts and of a melody that pursues him can best be compared to that between a fancy or whimsical thought and an obsessive idea. … Their victim does not know and cannot tell us why this particular melody is pursuing him. He very often cannot even identify the tune.[320]

These indications reflect some of the ambivalence between attraction and rejection, first pleasure then frustration, as can be seen with hooks. The objective and subjective conditions for becoming hooked to an earworm are not easy to determine. The web of factors and conditions is manifold.

> Sometimes the acoustical endless loop in the mind occurs after listening once, sometimes it takes days. Equally impossible to calculate is the duration. In a study at the Kasseler Institut für Musik some respondents [/subjects] reported that the earworm disappeared within minutes. Some were haunted by the always returning melody up to three weeks […]. When a listener is particularly familiar with a certain title it increases the chance for it to become an earworm. A study at the Kasseler Institut für Musik showed that 60.5 percent of the pieces that developed into earworms were already familiar to those affected. However, 24.4 percent of the pieces that developed into musical endless loops were new to the respondents. The role of the text should not be underestimated. In the Kasseler experiment three of the five catchiest titles were songs with German texts, whereas instrumental pieces rarely achieved earworm status.[321]

Regarding further factors, there are several attempts to systematize.

> Situation-related factors include a piece being the first music to be heard in the morning («primacy effect», Kellaris 2003) or the most recently heard […]. Other such factors include an individual liking a tune and the anticipation of a pleasant event like upcoming holidays […], exercising, travelling and working alone […], and interacting with people. Finally, Bennett (2002) found that the appearance of earworms was associated with lack of focus

[318]Freud, Sigmund: *General introduction to psychoanalysis*. New York: Garden City, 1952, pp. 97. Quoted in Liikkanen 2018, p. 28 (N.B. The English edition of Freud's ideas was published years after his lectures were first recorded in German during 1916–1917).

[319]Reik 1953, p. 167, quoted in Liikkanen 2018, p. 28; cf. later Kohut, Heinz (1955). Review of the haunting melody. *Psychoanalytic Quarterly*, 24, 134–137.

[320]Reik 1953, p. 166, quoted in Liikkanen 2018, p. 28.

[321]https://www.scinexx.de/wissen-aktuell-11724-2010-05-28.html. Cf. e.g. Hemming 2009, Hemming and Altenmülle 2012, Hemming and Merrill 2015.

and relaxed states. Williamson et al. (2011) summarized and classified the everyday circumstances of INMI through an inductive, generative, grounded theory-based analysis. These authors uncovered four major abstract categories that describe the situations associated with the onset of an earworm. The first category is Music Exposure, which relates to whether the subject was recently and/or repeatedly exposed to music. The second category was termed Memory Triggers. Non-musical memory triggers include a) Association, where stimuli from the environment triggered INMI, which could include a person, a situation, a word or a sound b) Recollection, where the experience of the INMI involved retrieval from autobiographical memory, and c) Anticipation of an upcoming event. The third major category was termed the Affective State and includes the influence of a) Mood, b) Stress, c) Emotion and d) Surprise. The final category was termed Low attention states and includes earworms that were associated with a) Dreams, and b) Mind wandering, where the subject was involved in a monotonous task. Other factors that have been evaluated in connection with earworms include gender and age. With regards to gender, studies have suggested that being female increases the probability of getting earworms […] however other studies dispute this claim […]. Finally, concerning age, correlations have been reported between youth and the frequency of INMI experiences […]. Turning to the relationship between earworms and musical training and musicality, Levitin (2006) suggests that musicians experience earworms more often than non-musicians, and a study of Liikkanen (2012) revealed an association between INMI and musicality. Beaman and William (2010) found a relationship between the frequency of earworms and the subjective importance attached to music, but not with musical training. Personality also plays an important role in the appearance of INMI. A number of studies have looked at the relationship of Obsessive Compulsive Disorder with earworms […]. Neuroticism has been found to increase the frequency of earworms […] as has 'transliminality', the largely involuntary susceptibility to, and awareness of, large volumes of inwardly generated psychological phenomena of an ideational and affective kind […].[322]

Some results show correlation, but these are not yet causalities:

low but significant associations between certain stable individual differences (personality) and everyday musical behaviours, and spontaneous INMI experiences.[323]

Williamson and Müllensiefen refer to the various contexts and raise further questions:

Should we be medicating earworms with OCD drugs?… Should we prescribe Singing to OCD patients? … Musical exposure important (Sacks, 2007) that is recent and repeated (Beaman and Williams, 2010); but so is the activity of non-musical, involuntary memories. State of mental arousal (wakefulness, excitement and stress) and 'mind wandering' – a possible function? (Leverhulme Grant) Individual differences in singing only predict some features of INMI plus ease of singing may predict stickiness: activity of brain areas? Earworms may have structural melodic triggers – why do our minds react to patterns in this way? Wider implications for other spontaneous cognitions (including creativity and rumination), memory processes, music perception and encoding, personality research….[324]

In total the phenomena earworm shows that refinement and sharpening are not per se positive.

---

[322] Floridou, Williamson, and Müllensiefen 2012, pp. 302f.

[323] L.c., p. 308.

[324] Ibid. On clinical-therapeutical matters an abundance of publications exists. See: Julien et al. 2007; Naskar et al. 2017; Orjuela-Rojas and Lizarazo Rodríguez 2018.

6.9.4 *Sharpening Through Motion and Reproductive Participation: Dancing and Singing Along*

Like singing from memory as repetition, singing along also activates and focuses the attention and the memory.[325] Implicit knowledge, skills and capabilities are activated together with significant amounts of the unconscious and the body involved. As the earworm demonstrates, the emotive parts of music reach into the deep, elementary layers of the brain. So it is possible, as Oliver Sacks has reported that people with massive memory loss can reproduce song melodies.[326] In imaginary-real inner, silent singing along the contour and core of the respective work stand out more distinctly than in actual singing along. For a polyphonic composition cannot verily be realized by a solo voice. It is crucial to hear and cognitively comprehend a work and its live interpretation as comprehensively as possible and in its entirety. And moreover it is crucial to perceive especially the new that is the not-yet-heard and remembered, as new and to direct one's attention to this difference.

6.9.4.1 ‚Karaoke'. Commerce and Covering

Singing along and singing from memory, participating and imitating converge in karaoke, which as almost anything is institutionalized and then commercialised. Historically the process is in principle inverted as with the hook and its usage. At first karaoke was heteronomous, advertisement and public relations taverns, restaurants, bars etc.

> Karaoke formalises and makes public something that almost all of us do from time to time, that is to say sing along to a song we know.[327]

Then karaoke became autonomous. The re-singing as singing along becomes the purpose of the exercise. Especially in karaoke competitions the accuracy or replication is quantified. Because an orientation point is available, namely the work to be imitated, the fuzziness is measurable. In the respective shows a temporary cover version is generated, more or less similar to the original, in which only the vocals are replaced.

> In practice the candidates alternately sing one song, each a few lines, and are then evaluated by how /accurately]they hit the tones and kept the rhythm – all this gets visualized […], but only for the viewers in their homes. [Running along is a display which visualizes for the viewers whether the candidates hit the tones or not.][328] So while we can exactly follow

[325] Cf. e.g. Pawley and Müllensiefen 2012.

[326] See particularily Sacks 2007.

[327] Moorey 2012, p. 173.

[328] *Keine gute Idee? Netflix-Nutzer sauer – Show mit Palina Rojinski sorgt für Furore: „Dafür geb ich bald kein Geld mehr aus"*, 24 July 2020, https://www.tz.de/leben/serien/netflix-sing-on-germany-palina-rojinski-nutzer-wuetend-show-prosieben-rtl-tv-fernsehen-joko-klaas-zr-90011696.html, retrieved 21 Aug 2020.

> how Lars catastrophically botches a few wrong notes, he only sees the displayed lyrics for re-singing and has no clue. This should at least be very beneficial for the self-confidence. Afterwards the viewers see who reached what percentage. In exceptional cases this reaches up to 88 percent, mostly it falls in line significantly below that. Whoever performed best automatically proceeds to the next round the overall average determines how much money goes into the jackpot.[329]

Incorporated is the typical neoliberal brutalization of competition.

> After the first three rounds the candidates have to nominate each other for elimination.[330] Eliminations on "Sing On!" are based partly on science, partly in strategy, ... The contestants ... vote for who should be ousted, weighing whether they want to eliminate the better competitors, or want to try to rack up more money for the jackpot at the end.[331]

And this "Theatre of Cruelty" (Artaud) is simply anti-social, but complex in the interweaving of objective technical refinement and subjective economical-social acumen, which blurs the contours of the tactical decisions.

> The mutual elimination may be a classic in TV entertainment, but here it is [...] misguided.[332]

The reviewer considers it a "cruelty", and unnecessary,

> (s)ince it would have been a good opportunity to involve the audience in the studio by giving them a chance to vote. After the third round the computer determines the outcome strictly based on singing performance.[333]

That means the fuzziness is automatically measured.

> There are already many casting shows on German TV, but in most cases some stars or the audience assesses the quality of the performers. Often the audience wonders, why of all things a certain artist progressed to the round. For finally much of that what is heard ends up being a matter of taste. Not in the new Netflix karaoke show "Sing On! Germany". Here the so-called "Vocal analyzer" determines precisely which candidate hit the right notes. And the candidates cannot fool the computer.[334]

The moderator stands out with her fractal narcissism, and advertises with blatant self-praise unperturbed by politeness.

> At the beginning of each episode she delivers her so-called "Palina Performance".

---

[329] Julia Bähr: Karaoke auf Netflix: Die gefährlichste Spielshow der Welt, Frakfurter Allgemeine Zeitung, 11 August 2020, p. 12.

[330] Ibid.

[331] https://variety.com/2020/music/news/netflix-sing-on-tituss-burgess-1234738820/Netflix Karaoke Competition Show 'Sing On!,' With Host Tituss Burgess, to Premiere in September August 18, 2020, retrieved 20 August 2020.

[332] Ibid.

[333] Ibid. The cruelty is superfluous only when seen in the light of the matter, but not in regard to its socialization effects. Democratic participation would comprise more than just mere elimination, which reminds more of Roman gladiatorial games than parlamentarian ceremonies.

[334] *Palina Rojinski moderiert "Sing On! Germany"*, https://www.golde.

"The show is incredibly captivating, one immediately wants to sing along, regardless of whether one can or not, and finally one understands all the lyrics, because they are displayed as subtitles," explains Rojinski.[335]

In the Spanish version it is also „forced fun ".[336]

6.9.4.2 Fuzzy and Exact Synchronisation of Music and Body Movement

The bodily-active appropriation form dancing again changes the perception. For this on the one hand music is indispensable as giver of time and tempo and beat, on the other hand it is devaluated as autonomous work and becomes a supplement. This has a blurring effect in regard to the perception of structures and denotative contents, but focusing in regard to the emotive and comprehensive perception and the effect. External and internal rhythms are synchronized as in listening. Transitions to dancing, that is the co-execution of music, which in parts always is 'motion*, are evident as bodily movements already with simple listening.

> The physical symptoms of listening to a catchy song include "running [it] over in our heads or tapping a foot".[337]

This does not apply to catchy tunes.

> Frith (1996) underlines the intimate connection between musical consumption and production by drawing attention to the innervated muscularity of many of our modes of listening: 'we listen by performing, by reproducing (even if only silently, tentatively) those muscular movements for ourselves, "sympathizing" with a singer by pushing the words up against the top of our mouths when she does' (192). For Frith, there is a literal 'physical sympathy' with the singer, due in large part to the fact that, as organisms with the same glottises and tracheas, we can identify with singers to a greater extent than we can with other musicians, whose dexterity greatly exceeds our own: 'we can sing along, reconstruct in fantasy our own sung versions of songs, in ways we can't even fantasize instrumental technique – however hard we may try with our air guitars – because with singing, we feel like we know what to do' […]. Even as non-musicians, most of us have an extraordinarily precise level of control over the muscles in our voice boxes, the position of our tongue against the palate, and the flow of air from our windpipes, enabling us to produce passable imitations of a range of singers whose records we're familiar with. And for professional musicians, who have developed a high degree of muscle memory in relation to their chosen instrument, listening to recordings of an artist who shares the peculiar sets of involuntary motor responses, as in the case of the sax-player whose lips twitch and quiver when he or she listens intently to another saxophonist playing a solo.[338]

---

[335] https://www.goldenekamera.de/streaming/article230117886/Palina-Rojinski-Sing-On-Germany.html, retrieved 21 August 2020.

[336] Daniel Hart: *Sing On! Spain review – karaoke is best left in the bars Sing for money*, 24 July 2020 https://readysteadycut.com/2020/07/24/review-sing-on-spain-season-1-netflix-series/, retrieved 21 August 2020.

[337] See Bogost, Ian. *How to Do Things with Videogames*, p. 133. Retrieved 23 September 2012, quoted in https://en.wikipedia.org/wiki/Catchiness, 27 June 2020, retrieved 14 August 2020.

[338] Moorey 2012, pp. 175f. See also e.g. Pawley and Müllensiefen 2012.

Movement functions as a common third aspect of music and hearing.

> (P)erception of auditory rhythms is accompanied by neural forms of entrainment, defined here as the process by which oscillations couple with, or alter their period in response to, another (intrinsic) oscillation or a stimulus rhythm.[339]

The complexity of the task heightens the precision.

> [W]e propose that the Synchronize task (which presented one auditory part to track) required a higher level of auditory processing than was required in the Motor task (with no parts to track) and a lower level of auditory processing than was required in the Listen task (with two parts to track).[340]

The sharpening is lastly founded in the fact, that with increased complexity the attention increases. In common dancing to music this tense attention is largely absent,*and* the rhythmical requirements remain moderate. More than in singing along in moving along sometimes not more than the contours of the metre and the rhythm become apparent—in an unintended 'skeletal' form, as composed for example by Strawinsky in the dances of *Histoire du Soldat.* It is remarkable how much fuzziness is allowed for in the practical-bodily musical appropriation in dancing. Contrary to common belief, small children do not have a 'natural' feeling for rhythm, instead it is challenging for them to precisely synchronize their movements with metre and rhythm of the music. Around the age of primary school, that is from 6 to 8 years, this proficiency evolves. But even many adults are not adept at achieving precise synchronization. When dancing partners make fuzzy movements *together* to music, the collaborative harmony is more significant than dancing precision. In group dances it is different. Here a higher degree of snychronization is required, to make the whole performance work. As with the conducting of an orchestra, but with words instead of gestures and mimics, already folklore music, such as the country dance, features an announcer who coordinates the movements of the group; a similar function can be found in leading dancers. The Tiller Girls in the 1920s or the musical ballets for film by Busby Berkeley in Hollywood since the 1930s fascinate through the quasi-mechanical precision of procedures in both sequence and synchronicity. One influence at work here is Fordism and the timing of the assembly belt [/line], exposed in its inhumanity by Charles Chaplin in *Modern Times.* To more precisely investigate such fuzziness,[341] one would have to (1) assess the empirical matter; (2) quantitatively gauge the deviations or rather asynchronicities and dissonances between music and movement; (3) as by now unequivocal in the field of Fuzzy Logic, determine the range of what is not yet regarded as lack of dancing talent. The objective findings

---

[339]Mathias, Zamm, Gianferrara, Ross, and Palmer, p. 1864.

[340]Ibid., p. 1877. In cooperative music-making, which equally requires a high degree of motoric skills, the demands for synchronicity increase. It is achieved through the feedback loop between motor functions and hearing. "Many real-world auditory synchronization tasks—such as group music performance—contain actions that are accompanied by auditory feedback" (ibid., p. 1878).

[341]An insightful multi-dimensional invesitagtion aims for something else than what is here discussed: Howlin, Vicary, and Orgs 2018.

would have to be (4) linked back to the subjective through questioning of the dancers, and (probably more meaningful) various measurements such as muscle tone etc. This would result in (5) for one a gradation of psychophysical „flow" and satisfaction and on the other hand a measure which, to give the scientific enquiry a practical twist, would provide a foundation for determining whether interventions or measures of dance pedagogy and therapy are required. For asynchrony or rather the inability to synchronize indicate (6) beyond the mere appropriation of music through dance a more or less deeper-seated disorder in the relationship to the respective own body both in the partnership and in overall social behavior.

6.9.4.3 Sharpening of Perception Without Involvement of the Motory System

Autonomous activity in principle promotes the precision of perception in the perception process—a principle that tolerates deviations. The sharpening through activity of the recipients works also without motor functions: Preparation for listening, active listening and following, attentiveness are at least equally helpful for an optimal sharpening. And too much motoric function, 'internal' and external, may attract too much attention and be a distraction for listening. The pianist Rudolf Buchbinder gives the problem a specific turn.

> For Rudolf Buchbinder the most important thing is to play in a way that the audience cannot *sing along silently*, but is forced to listen. In this manner it is led to new paths. Buchbinder loves extremes, if they are intended by the composer: A fortissimo in a Beethoven sonata may reach the pain threshold.[342]

Just as perception can be compositorily refined in advance, the memory can be refined subsequently, through the composition of memory processes in renewed sharpening, primarily through citation (see Sect. 8.2). The default mode of perception is obviously, as mentioned, fuzziness, just as in memory the fading away and forgetting. Paradoxically perception is actively promoted when through deliberate blurring fuzziness is composed as specifically artistic effect. And this fuzziness is contradictory especially in perception, which searches for precise shapes, and in a second instance results yet again in sharpening.

6.10 Sharpening Through Blurring. Fuzzy Logic as a Partial Logic of Compositional History

This concluding section of this chapter already forms a transition to Chapter 7, because it shows something of the dialectic of sharpening and blurring with a concrete topic.

---

[342]https://www.clavio.de/das-klavier-im-radio-02-08-10-2011/, retrieved 14 May 14.

The importance of Fuzzy Logic can also be shown in the case example of the development of European melodic polyphony in the context of art music, which can only be sketched here. (Peasant and urban popular music have different conditions and follow their own paths.) For this, in a first step, a norm is violated: Music is no longer just a monophonic tune which serves as a means of transport and minimal adornment of the hallowed ecclesial word. If, in a second step, this tune is now doubled by a parallel tune, it remains static. True polyphony only develops out of movements which are *not exactly parallel*—at first glance appearing less ordered, fuzzier. In this, blurring and renewed sharpening go hand in hand. The logic of difference serves as a stimulus of "creativity" in the interest of a progressive development towards higher complexity.

Melodic polyphony is an achievement particularly of European music.[343] Historically, it developed out of monophony. So far, so trivial. Yet even at this starting point, Fuzzy Logic plays a significant role at several levels and points of development.[344] In the following, readers will need to tolerate a bit of history—but in art, it is difficult to do without it. In simplified terms, we can distinguish approximately twelve historical-systematic stages of development.

6.10.1 The Development of European Melodic Polyphony. Counter Movements Instead of Parallel Movement of the Parts

1. The elementary *fifth* or fourth organum of the ninth century AD developed from the quasi-improvisational practice in the liturgic-sacral music sphere and was the starting point for the development of tuneful polyphony. 'Improvisation', in this case only means that the accompanying tuneful parallel second voice was not noted down. The primary sacral tune with its liturgical texts was subject to strict rules.

The historically earliest example of the notation of European polyphony can be found in the *Musica enchiriadis*, the "music manual", after 850. The dominant voice from the repertoire of the so-called "Gregorian chant" is on top. The rule is simple: the leading voice is doubled in parallel fifths or fourths. The actual driving impulse comes from the differentiation of voices according to direction, the oblique motion and then contrary motion, so far without rhythmic differences (Fig. 6.61).

This means that in contrast to monophony, this constitutes a blurring. Its sense and purpose is once more the 'ornament', the adornment of sacral singing ad majorem Dei gloriam. The dominant voice is the preset, the plainsong, the cantus planus,

---

343 The voices can, just like in sub-Saharan polyphony, be defined in a metric-rhythmical manner.

344 See filtering of a distinct, sharp melodic line from a heterophonic web of lines and timbral blurring of a line as, for instance, in Isang Yun's main-tone technique (see Sect. 7.4.3.2).

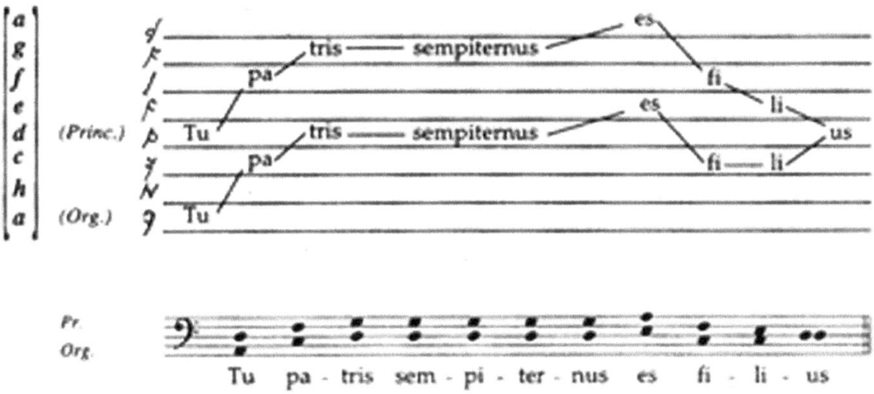

Fig. 6.61 Organum. Parallel motion in fourths—*Musica enchiriadis*, after 850
Anonymous: Musica enchiriadis
Scriptores ecclesiastici de musica sacra potissimum, 3 vols., ed. Martin Gerbert (St. Blaise: Typis San-Blasianis, 1784; reprint ed., Hildesheim: Olms, 1963), 1:152–173. Cf. Electronic version prepared by Stephen E. Hayes E, Peter M. Lefferts, Bradley Jon Tucker C, and Thomas J. Mathiesen A for the *Thesaurus Musicarum Latinarum*, 1993. G SI:163. https://www.chmtl.indiana.edu/tml/9th-11th/MUSENC_08GF.gif; see also https://www.chmtl.indiana.edu/tml/9th-11th/MUSENC_TEXT.html

which progresses equally in monotonous values.[345] The term *cantus firmus* for this originates later in Latin (since about 1235); in Italian, canto fermo (since about 1385). The HMT article is not clear on from when the *cantus firmus*, the fixed, static chant, simultaneously, as the "cantus prius factus", the previously originated chant, denotes especially the choral voice in a polyphonic movement. In any case: The ideas connected to it are, according to Wolf Frobenius, solid, strong, firmly established, never wavering, lasting. He points out its origin from Italian: canto fermo from canto, chant, and fermo, without motion, standing still, established, given, unchanging, firm, lasting.[346] The (Gregorian) chant is, independent of terminological details, the unchanging and dominant point of reference of composing, in music in the same way as the church is in life in connection with feudalism. However, the dual reign of throne and altar, of nobility and church is ideologically shifted, even almost inverted in so far as here, in sacral music, the clergy decidedly holds the scepter in its hand.

The dominant voice is on top. This corresponds to the thinking in social, class dependent ideas of space of "top" and "down". (The ambiguity of Latin altus, however, would need to be considered: 'altus' means both high *and* low.) And the musical 'down' can also mean the basic, the fundamental. And that is what it *is* in figured bass music, which begins with the interfering change in style in Europe from the 1590s, among others in connection with the establishment and development of absolutism.

---

[345] See *Handwörterbuch der musikalischen Terminologie*, article on *cantus firmus*. https://www.sim.spk-berlin.de/static/hmt/HMT_SIM_Cantus%20firmus.pdf, retrieved 19 August 2009.
[346] Ibid.

2. After the octave relationship, the fifth relationship is, even in the physical-acoustic structure of the partial tones respectively "overtones", almost as fundamental as a natural condition. And compared to the mere identity of the octave, the fifth brings a difference into play. The octave remains more abstract at a musical-sensual level. The two voices appear identical, different only because of the pitch level. This does not change fundamentally if the two-part tune is extended to three parts by mirroring along the horizontal axis. The maximum is reached with the four voices compositions of Perotin around 1200 at the Cathedral of Notre Dame in Paris. This all makes the sound somewhat fuller and is in this sense also a trace of something new.
   The parallel organum, however, is already a non-identical doubling: The same at a different pitch is in so far the same thing but already no longer identical.
3. The actual driving impulse, however, is brought on only by the differentiation of voices by their directions of movement, the oblige or the contrary motion, still without differences in rhythm. In this, the composition has to deviate from the schematic parallelism between original and added voice in order not to violate the fundamental rules of the tonal system—also an effect of musical logic.
4. However, nothing musically new originates from these variable parallel formations. This happens only when one consciously defers from the principle of parallels, that is, when in the new, added voice, other intervals are inserted. This constitutes a further blurring of exact parallelism. However, at the same time, it is the beginning of a sharpening in so far as the voices are now differentiated against each other. An additional sharpening as a differentiation of the voices also for perception results from it, also when the voices are composed in a different rhythm. The extension of the number of voices up to the relative optimum of six and the development of the counter movement of voices follows in the next centuries. This counter movement can, still somewhat anachronistically in terminology, already be termed as 'contrapuntal'. Fundamentally, it is governed by the dialectic of sharpening and blurring. This is because at all stages of development, there remains a fundamental relationship of tension between the aims of, on the one hand, achieving a distinct differentiation of the voices against each other and, on the other hand, integrating them in the sense of a 'harmonic' unity.
5. In addition to the internal dynamics and logic of the choral and its performance, particularly at the beginning, a reciprocal action between 'choral singing' as a system and (musical-cultural) environment comes into play. The immediate surrounding of the organum is sacral music. This, however, is in turn embedded into surroundings such as music culture in general etc. Even the mere added parallel voice as such is a tonally mediated sharpening by the difference between instrumental music "from the outside", from a different partial system of music culture, and the dominant vocal sacral art music. Because even for the beginnings of the organum, an influence of the instrumental is assumed as highly likely, as a piece of intersocial appropriation. 'Organum', from Greek 'organon', means instrument, tool. Instrumental music, as a primarily secular thing, was looked down upon in clerical ideology, yet indeed welcome in the feudal, particularly in courtly practice, and equally so in the peasant sphere as dancing music. The

church thus did integrate the deviant in its own way. The sinners, the minstrels, they do have a place, albeit hidden, in the community of believers and find their reflection in song. This vaulting then remains, as it would appear, until the end of this musical period,[347] i.e. up to the 1740s; and well beyond that until today in the respective music itself.

6. At about 1100, the cantus firmus migrates into the *lower voice*; this is also true for the early motet. Here, however, the middle register in the tenor becomes the norm. I need to leave it open here what these changes in register of the musical representation of clerical power meant for society. Georg Knepler, in 1977, cautiously pointed out relationships to the three-field rotation technique with its systematic-permutative change of crops. This temporal and spatial real polyphony thoroughly revolutionized the two-field rotation with a simple change of fields which had been traded down since Roman times and which spread in Europe around 1100. One starting point and nucleus of crystallization for this were the monasteries of the Carolingian period, which also carried importance for music history (Fig. 6.62).[348]
7. This constitutes the starting point of the predominantly hidden "corrosive" artistic work of the composers. In addition to the diastematic differentiation of pitch, we now find the above mentioned *rhythmical* differentiation, especially at Notre Dame since the early thirteenth century. The *motetus* (from the French mot = word), the added voice with often has a different text than the cantus firmus, is not set in an identic, equivalent way, note against note of the cantus firmus, which is now called *tenor*, approximately 'keeping voice'. Rather, tenor and motetus are different from each other both in the duration of tones as well as in pauses. This is the 'note against note', but in the non-identity of the 'punctus contra punctus', that is, the developing counterpoint. Here, again, we have the

[347] This time marks the beginning of the turn towards so-called "homophony". Prominent representatives include the "Mannheim School" (Carl Stamitz et al.). The musical composition is radically simplified and homogenized; the principle of polyphony of more or less equal voices is abandoned, also in the modification of polyphony by means of the adaptation into the harmonically determined thoroughbass. With the transition to the First Viennese School from the 1760s, however, and beginning with Haydn, the musical composition then again becomes more complex and differentiated. Also during this period—being the (high) time of enlightenment and enlightened absolutism—the usually sacral and religious vaulting and imaginary-real harmony is *also* partly replaced and partly added to by means of a mainly secular vaulting in and through music, often directly as religious art, and significantly so through music as well as art in general. This is what the term of bourgeois art-as-religion means. The concert, in addition to the opera, becomes a central societal place for this. Cf. among others Heister 1983: As in the specific case of art-as-religion slowly dissolving since the end of the First World War and respective political-social upheavals, since the structural crisis of the 1970s, the art centered vaulting in general is losing ideological as well as integrative power.

[348] See above score form notation, 6.6. The three-field system is "a regime of crop rotation that was used in China since ancient times" (https://en.wikipedia.org/wiki/Three-field_system, 7 September 200, retrieved 9 September 2020). Yet in Europa, mainly northern the Alpes, it was a revolutionary innovation of high feudalism in agriculture, with its interplay and intertwining of fallow and crop rotation—and there are, surely indirectly, revolutionary consequenes for the composition. Cf. primarily Knepler 1977.

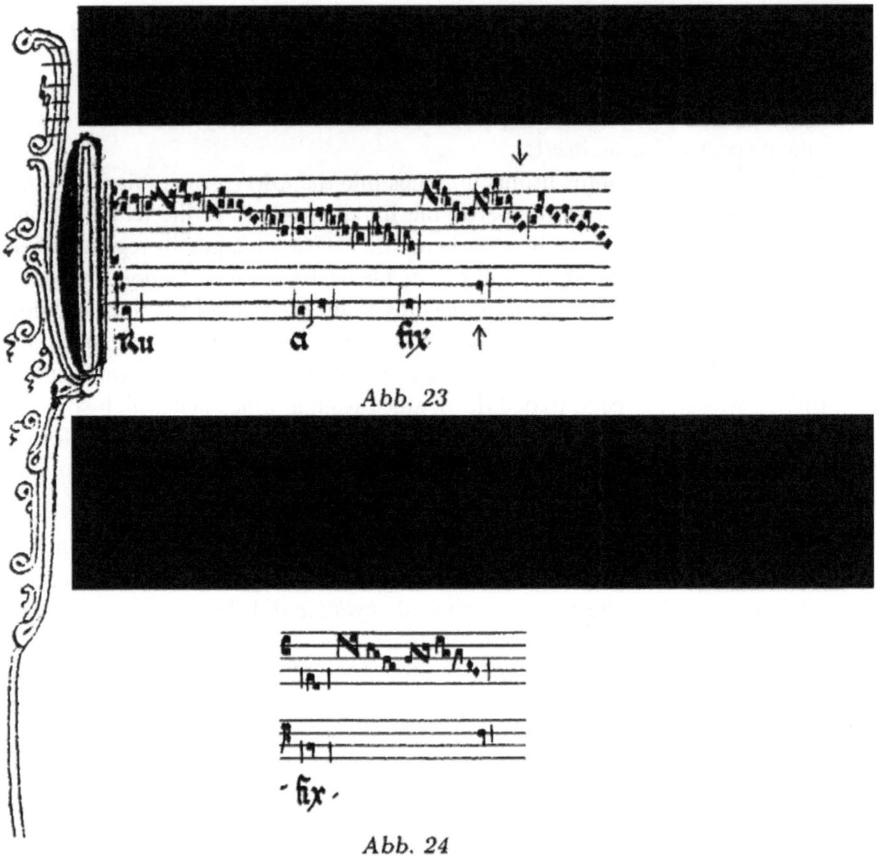

Fig. 6.62 Example from *Magnus Liber Organi* (Great Book of Organum), a compilation at Notre Dame in Paris of the 12th and early thirteenth century, ascribed inter al. to Leonin and Perotin. Movements in different directions and with different rhythms in the two voices. The polyphonic notation is fuzzy. The "zoom" *Abb. 24, below* shows the corrected and precise coincidences of the decisive tones
Quoted from Körndle 1993, p. 80

dialectic of sharpening and blurring. It can be seen as at least an analogy to the role of the cities including craftsmanship, trade and banking capital in the genuinely agrarian feudal society that the cantus firmus now no longer remains untouched but is changed. The process of isorhythm here has a corrosive effect, downright revolutionary. It begins in the French motet of the thirteenth century. A rhythmic pattern *talea* (from Old French *taille*, "section") is repeated throughout a composition in at least one voice. This *talea* is applied to a pattern of pitches. However, the length of the *color* is different from the length of the *talea*. Metric-rhythmical and diastematic 'modules' are designed in such a way that they do not correspond but that the sections always overlap. The *taleae*, particularly in the tenor, are often widely spread in their expanse with long permanent values. Variation occurs automatically, so to speak, because of different overlaps and combinations. The isorhythmical approach thus in a way creates a division of labor in pitch and length development. In this, it does not leave the original choral tune untouched. Furthermore, an additional homogeneity of the composition is created because the cantus firmus tune can appear, in fragmented motives, in all voices, that is, the motivic material is unified. By tendency, this constitutes a blurring. The differences are diminished in the interest of an overarching identity. On the other hand, the motives of the cantus firmus are varied in such a way that the diversity of the individual work is even increased. This constructivist separation and dissociation is both blurring and sharpening, for it negates and overplays as a disjunction the usual parameter conjunction, that is, the coupling of duration and pitch movement, thus blurring relative identities respectively correlations. On the other hand, however, it sharpens the awareness for real musical differences, for duration and pitch movement are experienced as a natural unity in the 'tune' in everyday awareness. Indeed, however, the tune is a unity of these two parameters, that is, of two different things.

8. The next logical-historical further step occurs when the composers choose profane, secular tunes as a basis instead of the sacral cantus firmus, particularly for multi-movement and cyclical masses. Occasionally, this already occurs in the motets of the fourteenth century. The (most likely Burgundy) chanson *L'homme armé* (The armed man) became almost a hit song from the 1430s onwards and was frequently used as a cantus firmus. Musically, this approach often meant a sharpening. For the original, very succinct tune can be discerned and thus sets the cantus firmus voice apart, particularly when it has characteristics as distinct as that of the above mentioned chanson. Sociologically speaking, this procedure can already be counted as a hidden dismissal of clerical power. Both through the text and the context as well as through the musical texture, the inclusion into the feudal system is ensured. Even though the binding power of religion loosens, the socially different remains embedded into the overarching secular-clerical whole.
9. This embeddedness continues to loosen when the composers do entirely without a preexisting cantus firmus and compose the fundamental dominant voice, a type of vox principalis themselves. The mass *mi-mi* by Johannes Ockeghem (written between 1460 and 1480), for instance, plays with this ambiguity of the pitch level mi which, depending on the hexachord, means different notes. Furthermore, to quote Timothy Dickey, "Ockeghem's own chanson […] *Presque trainsi*, a Bergerette in three voices, dates from around 1460, and contains, in its tenor voice (here the lowest) the head motif for the Missa "Mi-Mi," e-A-e–f-e."[349] This mass does not have an actual cantus firmus at all. This as well as the recomposition of the cantus firmus is an expression of increasing individualization and,

[349] In detail at https://www.allmusic.com/composition/missa-mi-mi-missa-quarti-toni-for-4-voices-mc0002378029, retrieved 20 December 2017.

up to a certain degree, also of secularization. In regard of the dialectic between fuzziness/blurring and sharpening, this step is neutral, so to speak. Both tendencies remain in a certain balance at a higher level, for through the densification of motivic work, the uniformity of the composition is heightened while at the same time diversity is intensified. Here, too, Protestantism takes back some of the slowly progressing secularization. In Protestant music, the (Lutheran) choral is used in analogy to the Gregorian choral and as its successor often as cantus firmus, known, for instance, in J. S. Bach.[350]

10. In the developed vocal polyphony, there is a gradual prohibition of consecutive fifths. This begins sporadically as early as 1300 with Johannes de Garlandia, however it is common only from about 1450. Yet even then, this rule permitted many exceptions. The prohibition on parallels has both acoustic-tonal as well as historical-cultural reasons. For on the one hand, European tonal polyphony begins in this manner and a too obvious return to the primitive beginnings is frowned upon. On the other hand, particularly consecutive fifths, as mentioned, suggest themselves because of their partial tone structure and are for this reason not least often used in folklorist-improvisational music practice and are therefore also 'primitive'. The music which, over centuries, had been embedded in a sacral context distinguishes itself from this as a newer, historically higher developed and at the same time as an artistic and rule conformist art. Third, the consecutive fifths in the more complex composition are a blurring of the differentiation of voices and take back the counterpuntal sharpenings.
11. The fundamental fifth relationship of the organum lives on, among others, in the fugue, here already harmonically reshaped and also characterized by the relationship between first and fifth step, by what later will be called tonic and dominant.[351] The second appearance of the theme, the "subject", occurs in a fifth relationship to the first appearance. Even more apparent than in this harmonic dimension is the governing relationship in the terms "dux" and "comes", leader and companion. Strictly speaking, the terminology only refers to the temporal order. However, this is one of the significant characteristics of "nobility": These masters were simply around earlier. From a tonal-harmonic point of view, the first appearance of the "subject" (or soggetto; "theme" is a more recent and for the high time of the fugue an anachronistic term) occurs on the first step, the 'answer' at the fifth, now in a tonal manner—adapted in terms of interval to the relevant key—or in a real manner with unchanged intervals and thus inevitably modulatory. What is remarkable is that the fugue only constitutes a genre as a pair together with the prelude. As an instrumental technique of movement (not a "form"), it has vocal predecessors. And what is even more remarkable is that the

[350]In a fundamentalist rhetoric, Jesus himself can become the cantus firmus by means of metaphorism: „Cantus Firmus | Making Jesus the Center. www.cantus-firmus.com 26 Dec 2016 ... Providing digital books, podcasts, and blog posts that promote *Christ as the central melody* which *gives all other melodies their definition* via ... "; „Dietrich Bonhoeffer also uses the metaphor for love for God in his" Letters and Papers from Prison" (retrieved 17 January 2017).

[351]I have not been able to further trace the origin and exact location of the term dominant = "that which masters".

subject of the fugue, generally speaking, is not so much a motivic-thematically characteristic form which is newly shaped by the composing subject but rather a topos. In so far, the association with the "subject" of the still dominant feudal absolutism seems natural even if not particularly enlightening; things are different with the memory of the cantus firmus, which was also intended supraindividually.

12. In classic Renaissance vocal polyphony, the lines can still be discerned as separate and individual up to a certain degree. Here, a sharpening is again urgently needed, particularly one of listening and perception. From a composition point of view, this can be supported by the evident differentiation of voices. In the sixteenth century, five and six parts become the norm. This decidedly happens at the expense of individuality and succinctness of the individual voice. How far, up to what number of voices the audible differentiation of voices truly goes is contested and has, to my knowledge, not been empirically and experimentally studied. Here, Fuzzy Logic would have an important role to play in determining blurring as well as the marking of boundaries. J. S. Bach's three-part "inventions" already reach a limit. Generally speaking, the problem of perception is probably especially solved by a fast change of focus, by going back and forth between the different components of the musical composition. The 'segmented continuum' of the process of perception is segmented by means of 'chunks' into individual parcels. (Additional) blurring through 'frictional loss' in such 'multitasking', however, is unavoidable.

The limit of perception is somewhat moved by additional differentiation of a non-melodic kind, that is, *in addition* to the fundamental movement of pitch and length. First, a different dimension of the tone, namely volume, may be quite suited to differentiation, however less so for classic polyphony. For here, it sets hierarchical increments because the louder appears to take priority over the quiet—which in fact it does acoustically. Thus, this grading of dynamics, of volumes counteracts the desired relative equal rights and balance of the different voices. Second, the timbral shading of the individual voices can support their differentiated perception as a sharpening. Women's, children's and men's voices are natural givens here, with castrate voices being a questionable cultural extension. The high individual difference of voices, however, belongs to the sphere of interpretation and soundwise realization.

Already at the composition stage, instrumentation can have a differentiating effect. The correspondent to classic polyphonic thinking up to the development of thorough-bass music was the "Spaltklang" in contrast to the "Schmelzklang". The terms of Heinrich Besseler mean two polar types of the tonal realization of the composition, above all through the instrumentation: on the one hand, the voices are clearly differentiated from one another, for example winds against strings, on the other hand they should merge as far as possible in a uniform, homogeneous sound. In instrumental music, this is the later difference between "broken" and "whole" consort in the ensembles of the Elizabethan age which retains its power in Purcell (1659–1695). "Unbroken" means instruments from more than one family, for example a group featuring both string and wind instruments. The counterpart was called "whole consort". A prominent homogenizing counter model are, for instance, the "Quatre vingt violons du Roi", the 24 string instruments from the violin family at the court of the absolutist King Louis XIV. The tendency towards blurring in the area of timbre which becomes discernible here then rules with a change in conception and structure in the "homophony", particularly also in the differentiation of tonal movement with and since the Viennese Classic. Wagner at the latest mixes the acoustic color in such a way that their origin and composition becomes indiscernible. At a higher level of complexity, here the tendency of the structure of singing voice plus instrumental accompaniment returns. The latter is then changed into a harmonic sound web which is highly complex in terms of motives and timbre but hardly polyphonically differentiated any more. Even though it was conceptualized, composed and intended differently, the tendency in perception often goes in this direction.

6.10.2 Poly-Textuality. Sharpening of Sound Perception, Blurring of Sense Perception in Motet and Quodlibet

The problem of perception is exacerbated in polytextual works and even more so when, as sometimes happens, different languages are used. Thus, as early as in the motet of the fourteenth century or in the Renaissance chanson, there are several texts, increased even in several languages. The quodlibet goes even further: It mixes and combines two or more given and usually secular song *tunes*. The sharpening as differentiation has been driven to an extreme here. Yet at the same time it acts as blurring for perception because the complexity is so great that it becomes sensually perceptible only through a reduction.[352]

One extreme example is the *Battalia à 10* from 1673 by Heinrich Ignaz Franz von Biber (1644–1704). This instrumental battaglia (battle) is more than the depiction of the fight itself; Biber also includes the pre- and post-history. The second section, "Die liederliche Gesellschaft von allerley Humor" (The slovenly company with a lot of moods) forms a collage of no less than eight tunes in the form of a quodlibet as a replica of a soldier's song on the eve of the battle. Biber himself said of it that all voices were as dissonant as if several songs were sung at once.[353] The process of montage results in a harshly dissonant, drunk-sounding cacophony—an extreme naturalism risked only again—leaving aside exceptions such as Mozart's *A Musical Joke*—by Mahler or Ives. Biber's mimesis of reality does of course exaggerate and stylize. According to Philip Pickett, the tunes are identifiable as Slovakian, Bohemian, Italian, Austrian and Hungarian folk songs; some can already be found in the dance

[352]Merely as a hint let me name a popular Christmas song. "In dulci jubilo / nun singet und seid froh" etc. two languages follow *one after the other* in such a way that the individual (half) lines connect to each other to make sense and result in a coherent whole text—albeit a bilingual one. This is decidedly simpler than the combination in simultaneity. The problem of the foreign language, of course, remains.

[353]Biber: *Requiem, Battaglia à 10* and other works. New London Consort, David Pickett. Editions de l'oiseau-lyre 436 460-2, 1992, booklet p. 22.

adaptation by Tilman Susato and Michael Praetorius.[354] And one, a "Bergamasca", can be found in the *Quodlibet* of Bach's "Goldberg Variations". Around 1740, J. S. Bach artfully linked the popular song "Kraut und Rüben / haben mich vertrieben ..." with another song as a final culmination of his variations.

A certain amount of multitasking is principally certainly required in vocal music. Because of the self-evident nature and the broad spread of vocal music, this is hardly perceived as a problem—but it is one. It is almost automatic to hear both music and text at the same time, but understanding both is more difficult. With texts in foreign languages, this problem is doubled. Here, empirical studies of perception, with the aid of Fuzzy Logic, could bring insight—for instance on the question if and in how far key words are enough for a rough, "vague" perception of the meaning, in how far and how vocal and musical tune are or can be stored together or separately, completely or in a fragmented manner, in memory. In purely instrumental versions of the quodlibet, the problem shifts. Here, the additional polyphony is eliminated by several different texts. One the one hand, this relieves the pressure on perception and heightens it. On the other hand, however, the differences between the voices are levelled without different texts. In all this, Fuzzy Logic does not become the logic of music history in a straightforward manner. However, it obviously becomes an important mode of thinking in order to describe the incipient stages of change, development and musical progress with all the losses and gains they bring with them in a more precise manner, and to understand them. It even explains it in so far as via the mode of thinking in Fuzzy Logic developments in the process of music history can be embedded into overarching conditions between the reflection of individual societal structures and fundamental, metahistorical anthropological possibilities. Parallels as well as differences to processes of the type of primarily biological evolution can only be mentioned here (see Chapter 9).

References

Adorno, Th.W.: Alban Berg: Master of the Smallest Link. Cambridge University Press, Cambridge, MA (1991)

Anderson, V.: 'Well, it's a vertebrate...': performer choice in Cardew's Treatise. J. Music. Res. **25**(3–4), 291–317 (2006)

Apel, W.: Die Notation der polyphonen Musik. 900–1600. Breitkopf & Härtel, Leipzig (1962) (revised and improved new edition of the English original; The Notation of Polyphonic Music. 900–1600 (1942)

Arnold Schönberg Center, Wien (T 26.01). J. Arnold Schoenberg Inst. **9**(1), pp. 55–77 (1986)

Attinello, P.: Signifying chaos: a semiotic analysis of Sylvano Bussotti's »Siciliano«. In: Repercussions, vol. 1, pp. 84–110 (1992)

Bahle, J.: Der musikalische Schaffensprozess: Psychologie der schöpferischen Erlebnis- und Antriebsform. S. Hirzel, Leipzig (1936)

Benary, P.: Skizze, Entwurf, Fragment. In: MGG2, vol. 8, col. 1506–1519 (1998)

Bengen, I., Frobenius, W., Behne, K.-E.: Tempo. In: MGG2, vol. 9, col. 443–470 (1998)

[354] Ibidem.

Betz, M.: "In futurum" – von Schulhoff zu Cage. In: Archiv für Musikwissenschaft, vol. 56, issue 4, pp. 331–334. Steiner, Stuttgart (1999)

Blackburn, A., Penny, J.: Timbral notation from spectrograms: notating the un-notatable? In: Proc. of the Int. Conf. on New Tools for Music Notation and Representation TENOR (2015)

Blanning, T.C.W.: The Triumph of Music: The Rise of Composers, Musicians and Their Art. Harvard University Press, Cambridge, MA (2008)

Von Blumröder, Ch.: Formel-Komposition—Minimal Music—Neue Einfachheit: Musikalische Konzeptionen der siebziger Jahre. In: Henck, H. (ed.) Neuland Jahrbuch 2 (1981/82), pp. 183–205. Neuland Verlag, Bergisch Gladbach (1982)

Böhme, F.: Tönende Technik. Über das Verhältnis von Komposition und Maschine. In: Phleps, Th., Reich, W. (eds.) Musik-Kontexte. Festschrift für Hanns-Werner Heister (2 vols.), vol. 1, pp. 132–149. Verlagshaus Monsenstein und Vannerdat, Münster (2011)

Bonus, A.E.: The metronomic performance practice: a history of rhythm, metronomes, and the mechanization of musicality. Phil. Diss., Department of Music Case Western Reserve University (2010)

Bowers, D.Q.: Put Another Nickel In. A History of Coin-Operated Pianos and Orchestrions. Vestal Press, New York (1968)

Bowers, D.Q.: Encyclopedia of Automatic Musical Instruments. Vestal Press, New York (1972)

Breidenstein, H.: Mälzels Mord an Mozart. Die untauglichen Versuche, musikalische Zeit zu messen. In: Das Orchester, 2007, issue 11, pp. 8–15. Schott International, Mainz (2007)

Brodsky, W., Ginsborg, J., Henik, A., Kessler, Y., Rubinstein, B.-Sh.: The mental representation of music notation: notational audiation. J. Exp. Psychol. **34**(2), 427–445 (2008). https://doi.org/10.1037/0096-1523.34.2.427

Burns, G.: A typology of 'hooks' in popular records. Popul. Music. **6**(1), 1–20 (1987)

Byrd, D.: Extremes of conventional music notation, revised mid October 2018, https://www.informatics.indiana.edu/donbyrd/CMNExtremes.htm

Coessens, K.: The score beyond music. In: de Assis, P., Brooks, W., Coessens, K. (eds.) Sound and Score: Essays on Sound, Score and Notation. Leuven University Press, Ghent (2014)

Cross, J.: Composing with numbers: sets, rows, and magic squares. In: Fauvel, J., Flood, R., Wilson, R.J. (eds.) Music and Mathematics: From Pythagoras to Fractals. Oxford University Press, Oxford (2003)

Dadelsen, G. von: Verzierungen. In: MGG1, vol. 13, col. 1526–1566 (1966)

Dahlhaus, C.: Arnold Schönberg. Variationen für Orchester, op. 31 (Meisterwerke der Musik. Werkmonographien zur Musikgeschichte, H. 7). Fink-Verlag, München (1968)

Dahms, S. (in cooperation with Jeschke, C., Woitas, M.) (ed.): Tanz (MGG Prisma). Bärenreiter/Metzler, Kassel et al. (2001)

Dann-Beardsley, C.: Eight Seconds to Hook an Audience (Words and Music, vol. 7). ProQuest757538142, retrieved 30 Aug 2020

Danuser, H., Katzenberger, G. (eds.): Vom Einfall zum Kunstwerk. Der Kompositionsprozeß in der Musik des 20. Jahrhunderts. Laaber Verlag, Laaber (1993)

De Benedictis, A.I.: The "dual" form of Igor Stravinsky's Piano-Rag-Music: musical structure as result of a rearrangement process. CIM08 Abstract Proceedings, Thessaloniki (2008)

Decroupet, P.: Totale Organisation und universales Material: Parameter-Denken und die Auflösung traditioneller Kategorien. In: Heister (ed.) Geschichte der Musik im 20. Jahrhundert, Bd. III: 1945–1975, Laaber 2005 (Handbuch der Musik des 20. Jahrhunderts, Bd. 3), pp. 90–96 (2005)

Demers, J.: Listening Through the Noise: The Aesthetics of Experimental Music. Oxford University Press, New York and Oxford (2010)

De Sanctis De Benedictis, F.: Lied by Berio and Lumen by Donatoni: from analysis to interpretation. International Symposium of Musicology "Pitfalls and Risks in Intra-Musical Communication: Terminology, Notation, Performance", held under the auspices of the 10th edition of the Cluj Modern Festival, in April 2013. Musicology Papers, Vol. 28 N. 2, 2013, pp. 35–45. https://www.researchgate.net/publication/284186499 (2013)

Doyle, A.C.: Sherlock Holmes-Geschichten. Der Hund von Baskerville (translated by Trude Fein, selection and epilogue by Andreas Fischer). Deutscher Taschenbuch Verlag/Manesse Verlag, Zürich (1993)
Eggebrecht, H.H.: Zeit. In: Eggebrecht, H.H. Die Musik und das Schöne, pp. 172–180. Piper, München (1997)
Emans, R.: Kompositionsunterricht und ein Wirrwarr von Fassungen. In: „Ei, dem alten Herrn zoll' ich Achtung gern'" (Festschrift für Joachim Veit zum 60. Geburtstag, edited by Richts, K. and Stadler, P. for the Virtuellen Forschungsverbund Edirom), pp. 233–250. Allitera Verlag, München (2016)
Erickson, R.: Sound Structure in Music. University of California Press, Berkeley (1975)
Feil, A.: Metzler Musik Chronik vom frühen Mittelalter bis zur Gegenwart. 2nd enlarged edition. Verlag J. B. Metzler, Stuttgart, Weimar (2005)
Ferand, E.T.: Die Improvisation in Beispielen aus neun Jahrhunderten abendländischer Musik (Das Musikwerk, vol. 12). Gerig, Köln (1961)
Floridou, G., Williamson, V., Müllensiefen, D.: Contracting earworms: the roles of personality and musicality. Poster presented at the 12th International Conference on Music Perception and Cognition (ICMPC), 27th of July. Thessaloniki (2012)
Floros, C.: Gustav Mahler (3 vols.), vol. 3 (Die Symphonien). Breitkopf & Härtel, Wiesbaden (1985)
Floros, C.: Alban Berg. Musik als Autobiographie. Breitkopf & Härtel, Wiesbaden, Leipzig, Paris (1992)
Folkerts, G.: Mikis Theodorakis. Seine musikalische Poetik. von Bockel-Verlag, Neumünster (2015)
Frank, T., Pinch, T.: Analog Days: The Invention and Impact of the Moog Synthesizer. Harvard University Press, Cambridge, MA (2004)
Freud, S.: General Introduction to Psychoanalysis. Garden City Publishing, New York (1952)
Gail, D.: Charles E. Ives' Fourth Symphony. Quellen – Analyse – Deutung (2 vols.). Wolke-Verlag, Hofheim (2009)
Gann, K.: The Music of Conlon Nancarrow. Cambridge University Press, Cambridge (1995)
Gieseler, W.: Zur Semiotik graphischer Notation. In: Melos NZ, vol. 4, pp. 27–33. Schott International, Mainz (1978)
Goldschmidt, H.: Um die Sache der Musik. Reden und Aufsätze. Verlag Philipp Reclam jun., Leipzig (1970)
Gould, E.: Hals über Kopf. Das Handbuch des Notensatzes (german version by Muus, A. and Berger, J.). Faber Music/Edition Peters, Leipzig (2013)
Gratzer, W.: Zur "wunderlichen Mystik" Alban Bergs. Eine Studie. Böhlau, Wien, Köln, Weimar (1993)
Gratzer, W.: Komponistenkommentare. Beiträge zu einer Geschichte der Eigeninterpretation (Wiener musikwissenschaftliche Beiträge, vol. 22). Böhlau, Vienna/Cologne/Weimar (2003)
Gratzer, W., Neumaier, O. (eds.): Arbeit am musikalischen Werk. Zur Dynamik künstlerischen Handelns (klang-reden. Schriften zur musikalischen Rezeptions- und Interpretationsgeschichte, vol. 9). Rombach, Freiburg i. Br. (2013)
Gutknecht, D.: Aufführungspraxis. In: MGG2, vol. 1, col. 954–986 (1994)
Hanson, C.: Die Bedeutung der Technik für die musikalische Avantgarde: Transformation musikalischer Ästhetik und künstlerischen Selbstverständnisses. Kapitel 4, Technik in der Kunstpraxis: 4.1, Musikmaschinen und Instrumentenbau. https://www.diplomica-verlag.de/kunst-kultur_6/index.htm. Diplomica-Verlag, Hamburg (2015)
Hartmann, G.: Die Tonfolge B-A-C-H. Zur Emblematik des Kreuzes im Werk Joh. Seb. Bachs (2 vols.) (Orpheus-Schriftenreihe zu Grundfragen der Musik, ed. by Vogel, M., vol. 80 and 81). Verlag für Systematische Musikwissenschaft, Bonn (1996)
Hegel, G.W.F.: The Science of Logic. Edited and translated by George di Giovanni. Cambridge University Press, Cambridge, UK (1986)
Heister, H.-W.: Das Konzert. Theorie einer Kulturform (2 vols.). Noetzel-Verlag, Wilhelmshaven (1983)

Heister, H.-W.: Virtuosen (1993) Virtuosen, in: MGG, hg. v. L. Finscher, Sachteil, Bd. IX. Kassel, Stuttgart u.a (1998)

Heister, H.-W.: Zur Theorie der Virtuosität. Eine Skizze, in: Musikalische Virtuosität, #hg. v. Heinz von Loesch, Ulrich Mahlert und Peter Rummenhöller (Klang und Begriff Bd. 1), pp. 17–38. Schott International, Mainz (2004)

Heister, H.-W.: Enthüllen und Zudecken. Zu Brahms' Semantisierungsverfahren. In: Heister, H.-W. (ed.) Johannes Brahms oder Die Relativierung der absoluten Musik (Zwischen/Töne, vol. 5), pp. 7–36. Von Bockel-Verlag, Hamburg (1997)

Heister, H.-W.: Sextett für zwei Violinen, zwei Bratschen und zwei Violoncelli Nr. 1 B-Dur Op. 18 / Sextett für zwei Violinen, zwei Bratschen und zwei Violoncelli Nr. 2 G-Dur Op. 36. In: Bockmaier, C., Mauser, S. (eds.) Johannes Brahms. Interpretationen seiner Werke (2 vols.), pp. 119–128 and pp. 242–250. Laaber-Verlag, Laaber (2013)

Heister, H.-W.: „Geprägte Form, die lebend sich entwickelt". ‚Motiv', Teile und Ganzes. In: Bewegtes und Bewegendes. Der Motiv-Begriff in Künsten und Wissenschaften, edited by Heister, H.-W. and Polk, H. (Musik/Gesellschaft/Geschichte, vol. 7), pp. 13–26. Weidler Buchverlag, Berlin (2017)

Heister, H.-W.: Neue Musik nach dem Zweiten Weltkrieg. Ordnung oder Auflösung der Elemente und Engagement- In: Paul, G., Schock, R.: Sound des Jahrhunderts. Geräusche, Töne, Stimmen 1889 bis heute, pp. 313–319. Bundeszentrale für politische Bildung bpb, Bonn (2013)

Heister, H.-W.: 'Real sound,' readymade, handmade. Musical material and the medium between mechanization, automation, and digitalization as an impression and expression of reality. An implicit call for real interdisciplinarity. In: Santi, M., Berner, E. (eds.): Music—Media—History. Re-thinking Musicology in an Age of Digital Media. Transcript, Bielefeld, in print (2021)

Hemming, J.: Zur Phänomenologie des "Ohrwurms". In: Auhagen, W., Bullerjahn, C., Höge, H. (eds.) Musikpsychologie – Musikalisches Gedächtnis und Musikalisches Lernen, vol. 20, pp. 184–207. Hogrefe, Göttingen (2009)

Hemming, J., Altenmüller, E.: When an everyday-phenomenon becomes clinical: the case of long-term 'earworms'. Paper presented at the ICMPC-ESCOM 12 (2012, 23–28 July), Thessaloniki (2012)

Hemming, J., Merrill, J.: On the distinction between involuntary musical imagery, musical hallucinosis, and musical hallucinations. Psychomusicology: Music, Mind, and Brain **25**(4), 35–442 (2015)

Henze, H.W.: Die Englische Katze. Ein Arbeitstagebuch 1978–1982, Fischer, Frankfurt a.M. (1983)

Henze, H.W.: Musik und Politik, Schriften und Gespräche 1955–1984. ed. by J. Brockmeier. Dtv, München (1984)

Henze, H.W.: Reiselieder mit böhmischen Quinten. Autobiographische Mitteilungen 1926–1995. Fischer, Frankfurt (1996)

Henze, H.W.: Phaedra. Ein Tagebuch. In Zusammenarbeit mit Christian Lehnert. Wagenbach, Berlin (2007)

Herwig, H., Kalisch, V., Kortlander, B., Kruse, J.A., Witte, B. (eds.): Übergänge. Zwischen Künsten und Kulturen. Internationaler Kongress zum 150. Todesjahr von Heinrich Heine und Robert Schumann. Verlag J.B.Metzler, Stuttgart and Weimar (2007)

Herzberg, S.: Grenze, I. Philosophisch, Version 22.10.2019, 17:30 Uhr. In: Staatslexikon8 online, URL: https://www.staatslexikon-online.de/Lexikon/Grenze (accessed: 16.08.2020)

Hilscher, E.Th.: Accompagnement, Obligates. In: Oesterreichisches Musiklexikon online, https://www.musiklexikon.ac.at/ml/musik_A/Accompagnement_obligates.xml (25/5/2001)

Hocker, J.: Begegnungen mit Nancarrow (edition neue Zeitschrift für Musik). Schott Musik International, Mainz (2002)

Howlin, C., Vicary, S., Orgs, G.: Audiovisual aesthetics of sound and movement in contemporary dance. Empir. Stud. Arts, 1–24 (2018). https://doi.org/10.1177/0276237418818633

Hutchinson Guest, A.: Labanotation. The System of Analyzing and Recording Movement, fourth edition, illustrations by Doug Anderson. Routledge, New York and Abington (2005)

Jaschinski, A. (ed.): Notation (MGG Prisma). Bärenreiter, Kassel (2001)

Julien, D., O'Connor, K.P., Aardema, F.: Intrusive thoughts, obsessions, and appraisals in obsessive-compulsive disorder: a critical review. Review Clin. Psychol. **27**(3), 366–383 (2007). https://doi.org/10.1016/j.cpr.2006.12.004. Epub 2007 Jan 22

Kalbeck, M.: Johannes Brahms, vol. 4, 2. edition. Deutsche Brahms-Gesellschaft, Berlin. https://www.zeno.org/nid/20007753837 (1915)

Karkoschka, E:, Das Schriftbild der Neuen Musik. Bestandsaufnahme neuer Notationssymbole. Anleitung zu deren Deutung, Realisation und Kritik. Moeck, Celle (1966)

Karkoschka, E.: Notation in New Music: A Critical Guide to Interpretation and Realization. Praeger, New York (1972)

Kim-Boyle, D.: The ontology of live notations through assemblage theory. In: Hoadley, R., Nash, Ch., Fober, D. (eds.) Proceedings of the International Conference on Technologies for Music Notation and Representation (2016). https://doi.org/10.5281/zenodo.923969

Kirnberger, J.P.: Die Kunst des reinen Satzes in der Musik. G. J. Decker and G. L. Hartung, Berlin and Königsberg (1774)

Knepler, G.: Geschichte als Weg zum Musikverständnis. Zu Theorie, Methode und Geschichte der Musikgeschichtsschreibung. Reclam Universalbibliothek, Leipzig (1977) (2nd enlarged edition (1982)

Kohut, H.: Review of the haunting melody. Psychoanal. Q. **24**, 134–137 (1955)

Konrad, U.: Mozarts Schaffensweise. Studien zu den Werkautographen, Skizzen und Entwürfen (Abhandlungen der Akademie der Wissenschaften in Göttingen, Philologisch-Historische Klasse, third series vol. 201). Vandenhoeck & Ruprecht, Göttingen (1992)

Kowar, H.: Metronom. In: Oesterreichisches Musiklexikon online, https://www.musiklexikon.ac.at/ml/musik_M/Metronom.xml (06/05/2001)

Krickeberg, D.: Automatic Musical Instruments in Bar and Salon (Guide Sheet no. 11, Musikinstrumenten-Museum Staatliches Institut für Musikforschung Berlin SIM PK Preußischer Kulturbesitz Berlin, 2nd revised edition. Berlin (2015)

Kross, S.: Johannes Brahms. Versuch einer kritischen Dokumentar-Biographie, vol. 1. Bouvier, Bonn (1997)

Lawson, Rex: On the Right Track: Dynamic Recording for the Reproducing Piano (Part Four). Pianola J. **23**, 11–26. The Pianola Institute, West Wickham, Kent (2013)

Liikkanen, L. A.: Music in everymind: commonality of involuntary musical imagery. Paper presented at the 10th International Conference on Music Perception and Cognition, Sapporo (2008)

Liikkanen, L. A.: How the mind is easily hooked on musical imagery. Paper presented at the 7th Triennial Conference of the European Society for the Cognitive Sciences of Music ESCOM 2009, Jyväskylä (2009)

Liikkanen, L. A.: New directions for understanding involuntary musical imagery. Paper presented at the ICMPC-ESCOM 12, Thessaloniki (2012)

Liikkanen, L.A.: Inducing involuntary musical imagery: an experimental study. Musicae Scientiae **16**(2), 217–234 (2012)

Liikkanen, L.A.: Involuntary musical imagery—everyday but ephemeral (University of Helsinki. Department of Digital Humanities, Studies in Cognitive Science 10). Digital edition https://creativecommons.org/licenses/by-nc-nd/4.0/ (2018)

Logothetis, A.: Kurze Musikalische Spurenkunde. Eine Darstellung des Klanges. In: Melos 1970, No. 2, pp. 39–43. Schott Verlag, Mainz (1970)

Logothetis, A.: Über die Darstellung des Klanges im Schriftbild. In: Impulse für Spielmusikgruppen, pp. 3–9. Universal Edition, Wien (1973)

Logothetis, A.: Zeichen als Aggregatzustand der Musik. Jugend & Volk, Wien (1974)

Mathias, B., Zamm, A., Gianferrara, P.G., Ross, B., Palmer, C.: Rhythm Complexity modulates behavioral and neural dynamics during auditory-motor synchronization. J. Cogn. Neurosci. **32**(10), 1864–1880 (2020). https://doi.org/10.1162/jocn_a_01601

Mazzola, G.: The topos geometry of musical logic. In: Assayag, G., Feichtinger, H., Rodrigues, J.F. (eds.) Mathematics and Music. A Diderot Mathematical Forum, pp. 199–214. Springer, Berlin (2002)

Mazzola, G.: Musical Performance. A Comprehensive Approach: Theory, Analytical Tools, and Case Studies (Computational Music Science). Springer Nature, Switzerland (2011)

Mazzola, G. (with the collaboration of Muzzulini, D., with a contribution of Hofmann, G.R.): Geometrie der Töne: Elemente der mathematischen Musiktheorie. Birkhäuser, Basel, Boston and Berlin (1990)

Mazzola, G., Park, J., Thalmann, F.: Strategies and Tools in Composition and Improvisation (Computational Music Science). Springer, Berlin and Heidelberg (2011)

McInerney, M.J.: Performance and the page: An artist's investigation of the dialogue between the musical event and the written score. A thesis submitted to the University of Plymouth in partial fulfilment for the degree of Doctor of Philosophy, vol. 1: Written Inquiry. Dartington College of Arts. Creative and Performing Arts (2007)

Metzger, H.-K., Riehn, R. (eds.): Bruckners Neunte im Fegefeuer der Rezeption (Musik-Konzepte, vols. 120–122). text+kritik, München (2011)

Meyer, L.B.: Emotion and Meaning in Music. University of Chicago Press, Chicago (1956)

MGG1 (Die Musik in Geschichte und Gegenwart) (The Music in History and Present Day), ed. F. Blume, 14 vols. +2 supplement-vols. + 1 vol. register. Bärenreiter-Verlag, Kassel etc. (1949–1979)

MGG2 (Die Musik in Geschichte und Gegenwart) 2nd, new revised edition, ed. L. Finscher. Subject part, 9 vols. Bärenreiter/Metzler, Kassel and Stuttgart (1994–1998)

Miller, D.: Are scores maps? A cartographic response to goodman. Computer Science. https://www.semanticscholar.org/paper/Are-scores-maps%3A-a-cartographic-response-to-goodman-Miller/b2a0e29fd3856282706ca3639a85d00649ed9682 (2017)

Monson, K.: La Relazione con Hanna Fuchs-Robettin e la «Lyrische Suite». In: Karen Monson, Berg, pp. 237–263. Rusconi, Milano (1982)

Moorey, G.: Everyday reveries: recorded music, memory & emotion. A thesis submitted in partial fulfilment of the requirements of the University of the West of England, Bristol for the degree of Doctor of Philosophy, Faculty of Arts, Creative Industries and Education). University of the West of England, Bristol (2012)

Müllensiefen, D., Fry, J., Jones, R., Jilka, S., Stewart, L., Williamson, V.J.: Individual differences predict patterns in spontaneous involuntary musical imagery. Music Perception **31**(4), 323–338 (2014)

Murcott, D.: Mechanical troubles: performing Nancarrow's player piano studies today. Pianola J. **23**, 3–10 (2013)

Naskar, S., Victor, R., Nath, K., Choudhury, H. A.: "Radio inside my head": a curious case of early onset "stuck song syndrome" or obsessive-compulsive disorder with predominant musical obsession. Indian J. Psychol. Med. **39**(3), 373–375 (2017). https://doi.org/10.4103/0253-7176.207333

Nono-Schoenberg, N. (ed.): Arnold Schönberg. 1874–1951. Lebensgeschichte in Begegnungen. Verlag Ritter, Klagenfurt (1992)

Nyre, L.: Sound Media. From Live Journalism to Music Recording. Routledge, Oxon and New York (2008)

Oesch, H. (with collaboration of Haas, M. and Haller, P.): Außereuropäische Musik, vol. 2 (Neues Handbuch der Musikwissenschaft, vol. 9). Laaber-Verlag, Laaber (1987)

Orjuela-Rojas, J.M., Lizeth, I., Rodríguez, L.: The stuck song syndrome: a case of musical obsessions. Am. J. Case. Rep. **19**, 1329–1333 (2018). https://doi.org/10.12659/AJCR.912402

Pace, I.: Graphic notation. Paper presented at the Lecture, 13-06-2007, Hochschule für Musik, Freiburg i. Br., 35 pp. https://openaccess.city.ac.uk/6476/ (2007)

Pawley, A., Müllensiefen, D.: The science of singing along: a quantitative field study on sing-along behaviour in the north of England. Music Perception **30**(2), 129–146 (2012)

Perle, G.: Twelve-Tone Tonality. University of California Press, Berkeley (1977)

Perle, G.: The Operas of Alban Berg: Vol. 2 Lulu. University of California Press, Berkeley (1989)
Phleps, T.: »Complex, but simple«. Conlon Nancarrows tempo-dissonierende Boogie Woogies und Canons für Player Piano. In: Komposition als Kommunikation. Zur Musik des 20. Jahrhunderts (Hamburger Jahrbuch für Musikwissenschaft, vol. 17), edited by Floros, C., Geiger, F. and Schäfer, Th., pp. 175–205. Lang, Frankfurt a. M. (2000)
Piatigorsky, G., Mein Cello und ich und unsere Begegnungen. dtv, München (1975)
du Plessis, H.J.P.: Parola Scenica: Towards Realism in Italian Opera (Phil. Diss., Faculty of Arts, University of the Witwatersrand, Johannesburg) (english translation). https://core.ac.uk/download/pdf/39665109.pdf (2007)
Rabinovich, G., Slonimski, J.: Towards a galant pedagogy: partimenti and schemata as tools in the pedagogy of eighteenth-century style improvisation. In: MTO, a Journal of the Society or Music Theory **21**(3). https://mtosmt.org/issues/mto.15.21.3/mto.15.21.3.rabinovitch.php (September 2015)
Red'ko, V.G. et al.: Project "Animat Brain": Designing the Animat Control System on the Basis of the Functional Systems Theory. In: Butz M.V., Sigaud O., Pezzulo G., Baldassarre G. (eds.) Anticipatory Behavior in Adaptive Learning Systems. ABiALS 2006. Lecture Notes in Computer Science, vol/ 4520. Springer, Berlin and Heidelberg (2007). https://doi.org/10.1007/978-3-540-74262-3_6
Redlich, H.F.: Alban Berg. Versuch einer Würdigung. Universal-Edition, Wien, Zürich, and London (1957)
Reik, T.: The Haunting Melody: Psychoanalytic Experiences in Life and Music. Farrar, Straus and Young, New York (1953)
Rense, R.: Music, deciphered: a demonic prelude by an ailing chopin, NYT. https://www.nytimes.com/2002/05/12/arts/music-deciphered-a-demonic-prelude-by-an-ailing-chopin.html (May 12, 2002)
Rosenbaum, I.: Scored movement: incorporation of prescribed movement of musicians in musical scores. Doctor of musical arts in composition, University of California Santa Cruz. https://escholarship.org/uc/item/0735k2n4 (2014)
Sacks, O.: Musicophilia: Tales of Music and the Brain. Knopf, New York (2007)
Sanguinetti, G.: The Art of Partimento. History, Theory, and Practice. Oxford University Press, Oxford and New York (2012)
Schedel, M., with Bressan, F.: Notation for an electric stage: twenty years of writing about notation and a thought experiment. With additional commentary about preservation. In: Musica/Tecnologia, Vol. 13, pp. 85–101. Firenze University Press and Fondazione Ezio Franceschini, Florence (2019)
Schmitz, R., Ure, B.: Wie Mozart in die Mozartkugel kam. Kurioses und Überraschendes aus der Welt der klassischen Musik. Pantheon-Verlag, Munich (2019)
Schnebel, D.: MO-NO. Musik zum Lesen. DuMont-Verlag, Köln (1969)
Schnebel, D.: Denkbare Musik. Schriften 1952–1972 (edited by Zeller, H.R.). DuMont Schauberg, Köln (1972)
Schnebel, D.: Anschläge – Ausschläge. Texte zur neuen Musik. Carl Hanser, München (1993)
Schönberg, A.: War-clouds diary. J. Arnold Schoenberg Inst. **9**(1), 55–77 (June 1986)
Schultz, C.M., Seedorf, M.: loop – ein Ensemble virtueller Musikinstrumente zur Vermittlung elektronischer Musik und ihrer technischen Grundlagen (Masterthesis Technische Universität Berlin, Fakultät I: Geisteswissenschaften. Institut für Sprache und Kommunikation, Fachgebiet Audiokommunikation) (2016)
Schulze, H.: Das aleatorische Spiel: Erkundung und Anwendung der nichtintentionalen Werkgenese im 20. Jahrhundert. Hochschule der Künste Berlin (Phil. Diss. 1999). Fink, Munich and Berlin (2000)
Schwartz, L.R.: Repair and self-formation through verbal notation: analyzing self in works by Jennifer Walshe and Pauline Oliveros and an anatomical study on escape [Original Composition]. Submitted to the Graduate Faculty of the Kenneth P. Dietrich School of Arts and Sciences in partial

fulfillment of the requirements for the degree of Doctor of Philosophy University of Pittsburgh (2019)
Schwarz, W.M.: Grafische Notation menschlicher rhythmischer Bewegungen in einem interaktiven multimedialen System. In: Reichel, H. (ed.) Informatik – Wirtschaft – Gesellschaft. Informatik aktuell. Springer, Berlin and Heidelberg (1993) https://doi.org/10.1007/978-3-642-78486-6_77
Shafer, S.: Performance practice of real-time notation. Proceedings of the 2016 International Conference on Technologies for Music Notation & Representation (2016)
Siegele, U.: Vortrag. In: MGG2, vol. 14, col. 16–31 (1968)
Slonimsky, N.: Thesaurus of scales and melodic patterns, Archived 2017-01-09 at the Wayback Machine (1975)
Sonnenfeld, A., Falkenberg Hansen, K.: S-notation: A complete musical notation system for scratching and sample music derived from 'Theory of Motions' (TENOR, in Proceedings of the International Conference on Technologies for Music Notation and Representation) (2016)
Spohr, L.: Lebenserinnerungen [written between 1847 and 1858, first edition Kassel 1860/1861], edited by Folker Göthel. Schneider, Tutzing (1968)
Stern, D.: Filmmusik als Musik. Musikbehandlung im frühen Tonfilm als Modell für Grundsatzfragen der Filmmusik, Manuskript, Einleitung zu Buch Musik und Film: Aneignung der Wirklichkeit. Weidler Buchverlag, Berlin (manuscript, May 2020) (will be published 2021)
Stoll, R.W.: Belle bonne … Amour curtois als Herrscherlob? https://www.sinustext.com/single-post/belle-bonne (2020)
Straebel, V.: Aspects of conceptual composing. In: The Open Space Magazine (10), 69–78 (Fall 2008)
Stürzbecher, U.: Werkstattgespräche mit Komponisten. Gesprächsteilnehmer u.a.: Boris Blacher, Pierre Boulez, Günter Bialas, K. Höller, Hans-Werner Henze, Karlheinz Stockhausen, D. Erdmann. dtv, München (1973)
Supper, M.: Elektroakustische Musik – Live-Elektronik (Bibliography with E. Ungeheuer). In: MGG2, vol. 2, col. 1717–1765 (1995)
Teich Geertinger, A.: Werkbegriffe in Gesamtausgabe und Werkverzeichnis, oder: Wann ist ein Werk keins? Definitorische Probleme dargestellt am Beispiel Carl Nielsens. In: „Ei, dem alten Herrn zoll' ich Achtung gern'" (Festschrift für Joachim Veit zum 60. Geburtstag, edited by Richts, K. and Stadler, P. for the Virtuellen Forschungsverbund Edirom), pp. 265–282. Allitera Verlag, München (2016)
Tischhauser, K.J.: A survey of the use of extended techniques and their notations in twentieth century string quartets written since 1933 by American composers with a selected annotated bibliography and discography (A Treatise submitted to the School of Music in partial fulfillment of the requirements for the degree of Doctor of Music) (2002)
Tomás, E.: Musical instruments as scores: a hybrid approach. In: Proceedings of the International Conference on Technologies for Music Notation and Representation TENOR. https://www.tenor-conference.org/proceedings.html#2016 (2016)
Ungeheuer, E. (Hg.): Elektroakustische Musik (Handbuch der Musik im 20. Jahrhundert, Bd. 5), Laaber-Verlag, Laaber (2002)
Ungeheuer, E.: Elektroakustische Musik 1945–75: Prisma musikalischer Originalität. In: Heister (ed.) Geschichte der Musik im 20. Jahrhundert, Bd. III: 1945–1975, Laaber 2005 (Handbuch der Musik des 20. Jahrhunderts, Bd. 3), pp. 96–106 (2005)
Veitl, A.: Musique, causalité et écriture: Mathews, Risset, Cadoz et les recherches en synthèsenumérique des sons. In: Musique, Instruments, Machines, Paris, OMF-Paris IV Sorbonne, pp. 235–251 (2006)
Wagner, R.: My Life. Dodd, Mead and Company, New York 1911. PART III. 1850–1861 (In German written 1865/1880) (1911)
von Weber, M.M.: Carl Maria von Weber. Ein Lebensbild, vol. 2. Ernst Keil, Leipzig https://www.zeno.org/nid/20007797583 (1866)
Wehnert, M.: Thema und Motiv. In: MGG1, vol. 13, col 282–311 (1966)

Westby, S.K., Glynn, R.: Fabricating performance: reciprocal constructs of dance notation. Nexus Netw. J., 75–94 (January 20, 2018)

Wiggins, G. et al.: A framework for the evaluation of music representation systems. Comput. Music J., 31–42 (March 17, 1993)

Williamson, V., Müllensiefen, D.: Earworms from three angles. Presentation given at the 12th International Conference on Music Perception and Cognition (ICMPC), 26th of July, Thessaloniki. https://www.doc.gold.ac.uk/~mas03dm/papers/VickyDaniel_earworms_ICMPC2012.pdf (2012)

Wilson, P.N.: Reduktion. Zur Aktualität einer musikalischen Strategie (edition neue zeitschrift für musik, edited by Stoll, R.W.). Schott International, Mainz (2003)

Wolf, R.: Die Musikmaschinen von Kaufmann, Mälzel und Robertson. Eine Quellenedition. Preprint 5. Deutsche Museum, München (2012)

Chapter 7
The Principle of Blurring. Conscious, Artistically Produced Blurrings

Abstract Filtering' is the differentiation of the essential and the secondary – depending on context and goal. With music it is the basal stage of sharpening. Thus natural material is processed, transformed into tones, sounds, noises, and systematized. This then serves as material for the second stage, the 'crystallization' (see chap. 6). These filtering processes repeat themselves, now already in the field of music, when chords are filtered out of heterophonic lines, and when in electroacoustic musical tones are worked out of noises or sine tones. The automated filtering by soft computing and FL-oriented methods is analysis of similarities. Our 'leitmotif' BACH serves as a test for internet research, which shows the productivity of Fuzzy Logic. Filtering as highlighting essential information via the Phonographic Form supports the re-sharpening of the memory. The compositional use of the partial-tone-spectrum as in Wagner's Prelude to Rheingold is already a transition to crystallization.

Keywords 'Abstract composition' ('abstrakter Tonsatz') · Aleatorics · (Intentional) blurring · Coloratura · Embellishment as blurring · Electroacoustics · Forms of existence · Fuzziness · Glissando · Graphic notation · Noise · Musical graphics · Polymetry · Polyrhythm · (Tempo) rubato vs. giusto · 'Soundpainting' · Stochastics · "Tone color melody" ("Klangfarbenmelodie") · Tempo · "Visible music"

After similarity and the two types of sharpening, 'filtering and 'crystallization', blurring is the third of the principles of the music process relevant in the context of Fuzzy Logic and of art processes in general. Fuzziness and sharpness are opposites, as are sharpening and blurring as processes. However, because blurring brings intentional artificial fuzziness in parts and different dimensions of the work, on a meta level it is a sharpening of the "fully completed unity" (K.Ph. Moritz, see Sect. 2.2.4). The work becomes richer, more differentiated, more complete in its overall form as a unity of contradictions. However: Blurring is on the one hand a transition stage to a sharpening on a higher level. But on the other hand it is a result of its own right and as such first and immediately self-sufficient: A glissando or portamento is a blurring of a melodic line and a continuum instead of discrete, segmented intervals; a tempo rubato is a blurring as a fluctuation of tempo and deviation from the underlying beat.

H. Heister, *Music and Fuzzy Logic*, Studies in Fuzziness and Soft Computing 406,
https://doi.org/10.1007/978-3-662-62907-9_7

7.1 "Dying" and Concealment

A countermove to the gradual development of musical forms as a sharpening, which becomes perceptible as such in the work as a process, is their likewise noticeable blurring (see Sect. 6.3). Time and again, at least since the First Viennese School, there have been fields of musical dissolution. They are analogous to fading in material reality and in memory (see Chapter 4). These composed processes of degradation, in the dielectric of sharpening and blurring, are not simply decay or even deterioration, but rather components of an overarching formation, which includes the shapeless and the unshaped. In *morendo*, "dying", this type of blurring becomes a texture type of its own, applied on a small scale to individual voices or on a large scale as the conclusion of a movement. Mahler often fully composed it in his compositions, as did his successor Alban Berg, for example.

7.1.1 The "Morendo" Ending in the Prelude of Alban Berg's Three Pieces for Orchestra

Berg's *Three Pieces for Orchestra* op. 6 from 1913/1915 include echoes of the catastrophe of the "Great War", the World War, of which no one at the time could have known that it would be just the first of, to date, two world wars. Like so many of his contemporaries, Berg had initially welcomed it nationalistically as purification, exhilaration, etc., but he quickly recognized this as a grave error. In the case of the orchestral pieces, there may be additional biographical reasons for this (Fig. 7.1).

Fig. 7.1 Alban Berg: Three Pieces for Orchestra op. 6, Revision 1929
Ist. Movement: Prelude, Conclusion. Score excerpt percussion only.
Universal Edition, Wien (1929). Plate U.E. 7396/12194. Philharmonia Series, Plate W.Ph.V. 432.
IMSLP21305-PMLP49158-Berg-OrchesterstuckeOp06fs

The author of the foreword to the miniature score (identified only by the initials F.S.) writes that the *Three Pieces* are "what the first piece, with a touch of shy understatement, pretends to be, namely a single *Praeludium*, a prelude to a tragedy whose end cannot be foretold."[1]
The *Praeludium* is designed as an arch form or formal palindrome, which Berg often used, for micro- or macro-structural shapes. Here the arch is

> opening and closing with unpitched percussion[2]—accelerating at the start, slowing down at the close—and reaching, at the center, a massive climax of activity.[3]

> *Praeludium* begins, as have dozens of pieces since, with soft, unpitched percussion noises (tam-tam, cymbals, bass drum, snare drum), which, via pitched percussion (timpani), introduce "normal" orchestral instruments in whose voices melodic ideas begin to emerge. At the end, there is a corresponding reduction to the single piano tam-tam stroke with which the piece began.[4]

In terms of musical content, it is about the construction and deconstruction of a cosmos out of a chaos that is little more than nothing. A subtlety typical of Berg: on the one hand, the symmetrical rise and fall is found again, fractally, in the performance designation for the volume in all percussion instruments, but moreover mostly supplemented by explicit pp - ppp or p - pp, precisely a "Morendo" becoming weaker, i.e. an asymmetry. On the other hand, Berg precisely shapes the vanishing, the blurring. In the two timpani pair there is a triple dismantling figure. First the sequence of notes f es f es f es f in timpani 1, then timpani 2 shortens an octave lower, varying to the six notes B G B G B G. This is accompanied by timpani 1 with a swirl es f, i.e. with a shortened inversion of the melodic movement whole tone upward instead of downward. This blurs the closely spaced pitches even more. Against this, Berg, almost paradoxically, twice places the expression mark "(clearly)." The penultimate word has timpani 2 with the two-tone motive, now again in the initial direction of motion as a whole tone downward, B-sharp G-sharp. The last word has, now only without a fixed pitch, the great tam-tam with a single note. "The rest is silence."

[1]Michael Steinberg: ***BERG:** Three Pieces for Orchestra, Opus 6*, http://www.sfsymphony.org/music/ProgramNotes.aspx?id=34798, retrieved 17 June 2020. In detail cf. e.g. Redlich 1957, pp. 87–101. George Perle suggests that the third movement, *March*, "captures the impending sense of doom". (Phillip Huscher: *Alban Berg—Three Pieces for Orchestra, Op. 6*, program notes. http://www.sfsymphony.org/music/ProgramNotes.aspx?id=34798, retrieved 15 June 2020.) See also Floros 1992.

[2]For the more general significance of noise in music since the avant-garde after 1905, see below.

[3]Huscher, ibid.

7.1.2 *The Tarnhelm Motif in Wagner's* Ring

A step beyond blurring is concealment. Here, what was previously or usually sharp becomes invisible—in art, normally for a period of time and not forever.[4] Concealment provokes curiosity for the unveiling, a manifestation of sharpening. The "wrapping" of public spaces by the prominent artist Christo represents an intermediate stage, since the contours of the concealed object remain visible. In music, a wall of noise, for instance, functions as concealment, from which the music emerges only gradually. For some rock and pop music, an "ear-splitting volume" is characteristic as something of an "intro" and thus a kind of "wrapping". Musical forms are inaudible there. In a sense, even—specifically composed—silence could be interpreted as a kind of concealment from which the music works its way out as a sound. Intentional blurring for the purpose of concealing and masking happens not only in art but also in life.

> The Dutch censorship reminds us of *impressionist* painting. The authorities artistically dotted over entire stretches of land and residential areas. Defence Ministry bases are located in the […] town of Ede.[5]

So art does return to life. In so doing, the concealment approximates a blurring. Pseudo-artistic 'overpainting' is itself just more camouflage and deception. Though somewhat different, this is similar to mimicry, which is not uncommon in the animal world—the visual adaptation of the animal's appearance to the background, where the "figure" and the "ground" optically blend into each other. And rendering something invisible is one step further, such as with the mythical cloak of invisibility. At best, you just know that something is invisible, but you no longer see it.

Wagner performs an acoustic implementation of this in *Rheingold*. Alberich fortifies his rule over the Nibelungs using a magic helmet that has been created for him. When he puts on the helmet for the first time, the magic spell reads "Night and darkness! Nowhere seen!" (Scene 3). The otherwise basally regulating "invisible hand" of the market is supplemented here by palpable violence, since Alberich invisibly monitors his "employees" and beats them when their performance is poor (Fig. 7.2).

This motif returns multiple times in *Götterdämmerung*. At first, it is less associated with rule and industry and more with the deception and betrayal by the rulers. Thanks to a potion of forgetfulness provided by Hagen, Siegfried forgets his love for Brünnhilde. Hagen's half-sister Kriemhild is promised to Siegfried as his wife if he conquers Brünnhilde for Hagen's half-brother Gunther. Disguised as Gunther, Siegfried then employs the Tarnhelm to violently overpower Brünnhilde on his behalf

---

[4] In detail see e.g. *Hinter dem Vorhang. Verhüllung und Enthüllung seit der Renaissance*, Beat Wismer, Claudia Blümle (eds.). Catalog book, Museum Kunstpalast Düsseldorf 2016–2017. During a painting contest in ancient times, Parrhasius painted a linen curtain to look so deceptively real that his rival Zeuxis thought it was real and only discovered the illusion when he attempted to pull it aside.

[5] http://www.computerbild.de/fotos/50-geheime-Orte-bei-Google-Earth-9215055.html#3, retrieved 17 June, 2018. Emphasis HWH.

Fig. 7.2 Richard Wagner: *Rheingold*, "Tarnhelm" motif
Online-Datenbank der Ring-Motive, Nr. 15. Copyright Auricula (2013). http://www.auricula.de/Datenbank_Ring-Motive.html

Fig. 7.3 Richard Wagner: Götterdämmerung Motif connections
Karl Klindworth (Arrangement): piano reduction. B. Schott's Söhne, Mainz (1900). Plate 26809. IMSLP16827-Wagner_-_Götterdämmerung_(vocal_score)

(Act I, Scene 3). The "Tarnhelm motif" and the "magic deceit motif" are similar to each other: through (Fig. 7.3)

> alternating chords and the rhythm, [...] as well as the harmonically highly ambitious mirror figure [...]. (T)he "Tarnhelm motif" is based on the major third relationship of G sharp minor – E minor, the "motif of the magic deceit" to C minor – E minor (with substituted third E sharp) [...] (B)oth motifs [... are] also audibly linked syntactically in the Gibichung scene [...], the new motif functions as a postscript to the old one (Hagen: "that he ever saw a woman before you"). [...] (Director's note: "Hagen, without leaving his seat, holds Gunter, when he approaches once more, by means of a mysterious wave of his hand").[6]

The reason this chord progression represents camouflage, disguise, concealment can be explained to some extent. (1) The relationship of thirds runs counter to the usual functional harmonic relationships of fifths and thus blurs these links. (2) The fifth relationship of E minor / B minor is indeed established in the second phrase (here on

[6]Christian Thorau: *Semantisierte Sinnlichkeit. Studien zu Rezeption und Zeichenstruktur der Leitmotivtechnik Richard Wagners* (Beihefte zum Archiv für Musikwissenschaft (AFMW-B), Vol. 50, Stuttgart 2003. Franz Steiner-Verlag, p. 134, p. 135. Cf. e.g. https://richard-wagner-werkstatt.com/leitmotive/tarnhelm-motiv/.

the word “ge-naht”). E minor thus occurs as a subdominant of the tonic B minor. This is deceptive, though. For this is not a true harmonic resolution, since the dominant (which would be F sharp major) is missing. (3) G sharp minor remains exterritorial. (4) The creeping, oscillating semitone movement in both the upper and lower voices indicates a possibility of transformation forward and backward: D sharp / E / D sharp and G sharp / G / G sharp. This fits well with the function of an invisibility helmet. From the perspective of Fuzzy Logic, it becomes clear that the harmonic-syntactic blurring is also a semantic sharpening at the same time.

7.1.3 The Hidden Theme in Elgar's **Enigma Variations**

Edward Elgar goes a step further. In his *Enigma Variations* or *Variations on an Original Theme ('Enigma')* op. 36, premiered in 1899, the theme of the 14 total variations is itself a *Tarnhelm.* In the programme for the first performance in 1899 Elgar wrote:

> “The Enigma I will not explain – its ‘dark saying’ must be left unguessed, and I warn you that the connection between the Variations and the Theme is often of the slightest texture. Further, through and over the whole set another and larger theme ‘goes’, but is not played… so, the principal Theme never appears.”[7]

Indeed:

> the principal theme of the piece is never actually heard, and the ‘Enigma’ theme we *do* hear is just a counterpoint on the principal theme.[8]

The Original Theme is thus “original” only insofar as Elgar himself is the author and fashioned it as a kind of acoustic plaster cast that forms the mould to conceal the hidden theme that was actually intended, which serves as a soggetto non cavato. In the

> *Musical Times* in 1900, Elgar clarified the nature of the ‘Enigma’: “Mr Elgar tells us that the heading *Enigma* is justified by the fact that it is possible to add another phrase, which is quite familiar, above the original theme that he has written.”[9]

Elgar further clarifies this in his 1905 biography: “The theme is a counterpoint on some well-known melody which is never heard.”[9] *More or less plausible candidates for this musical “Deus absconditus” are: Auld Lang Syne, God Save the Queen, Martin Luther's hymn Eine Feste Burg, Franz Liszt's symphonic poem Les Préludes, Pergolesi's Stabat Mater (transposed up a tone).*[10] *In a close comparison, it is fairly*

---

[7] *Young composer ‘solves’ Elgar's Enigma—and it's pretty convincing*, 1 May 2019, https://www.classicfm.com/composers/elgar/news/young-composer-solves-enigma/, retrieved 20 June 2020.

[8] Ibid.

[9] Ibid.

[10] See https://www.theguardian.com/music/shortcuts/2017/may/03/has-a-cleveland-policeman-cracked-the-secret-of-elgars-enigma-variations; cf. *Young composer ‘solves’ Elgar's Enigma—and*

certain that Fuzzy Logic could contribute to determining more precisely the degrees of similarity of the "original" as a reflection and blurred prototype.

7.2 Point of Departure. Idea and Conceptual Form. Blurred and Blurrings in the Service of Crystallization—Noises as New Material

Sounds and even noises as disordered mixtures of frequencies are acoustically and musically fuzzier and less specific than precise tones. From a historical-genetic perspective, they are filtered out of the sounds of the environment at a very early stage (see Chapter 5). In this respect, it is a reversion to the elementary, namely to historical 'states' before the filtering and the formation of specifically musical material, especially systems of sound, rhythm and timbre, when sounds have been used more consciously and independently as material in new music since the beginning of the 20th century (see Sect. 7.3). Structurally and aesthetically, this is a blurring. At the same time, though, it represents further development of the elements of music.[11] This is because, in a sense, sounds are now officially independent building materials, equivalent to tones in this regard. Once again, when the artwork is viewed as a whole, this blurring is thus also a sharpening on a new level. For it is possible to realize reality, especially the new reality of the urban soundscape with a high proportion of noise and overall higher loudness levels, more precisely and completely in this way. The mimesis is sharpened (see Chapter 4 *Similarity*, above). Nevertheless, much ado may have been made about noise in music after 1905 since the futuristic proclamation. On the one hand, the artistic merits and progress of the increased use of noise are not always convincing.

> The concrete implementation corresponded [...] to the theoretical demands [of the Futurists] only to a limited extent, even if it was perceived by the general public as a scandalous provocation. During the performances, various noises that were produced by the "noisemakers" most often accompanied relatively ordinary pieces of music.[12]

Expanding and differentiating what can be used as musical material is obviously an achievement. This differentiates the mimesis of the natural and social soundscapes. On the other hand, however, the use of noise as musical material is historically new, especially in terms of quantity and evaluation. For the noisy, the percussive, is a genetic and structural universal of music. Percussion is almost always present in dance music, even in "classical" music, and in world music in any case. Its absence

---

it's pretty convincing, 1 May 2019, https://www.classicfm.com/composers/elgar/news/young-composer-solves-enigma/; Daniel Estrin: Deceptive Cadence. *Elgar's 'Enigma' Still Keeps Music Detectives Busy.* March 15, 2018, https://www.npr.org/sections/deceptivecadence/2018/03/15/593771944/elgars-enigma-still-keeps-music-detectives-busy?t=1592740532540, all retrieved 20 June 2020.

[11] For audio-visual aspects, cf. e.g. Föllmer/Gerlach 2004; Daniels 2009.

[12] Sterneck 1998; cf. Kahn 1999.

is more of an exception, most common in (Gregorian) chorale and Franco-Flemish vocal polyphony and in liturgical music, such as in the Byzantine tradition.

7.2.1 ‚Cluster'. Clumped Chords—Pitches in Transition to Noise

Standing at a juncture between noise and sound is the "cluster". This is a classic example of blurring for the purpose of creating new material. It is not an example of filtering, of the historical creation of pitches from sounds (see Chapter 5, above), but rather an artificial, secondary and late phenomenon, a certain negation of the chord. In relation to the chord, the cluster is the vertical arrangement of pitches, more or less 'naturally' fuzzified—densely packed, undifferentiated or not yet differentiated. In 1919, the US-American composer Henry Cowell coined the term "tone cluster" for complexes of tones, sounding major and minor seconds simultaneously. Such a tonal complex should be treated like a single tone as a musical unit, to be more closely defined by its ambitus (extension in the vertical tonal space), with its variable fringes (Fig. 7.4).

A relationship between chord, motif and cluster (Fig. 7.5):

By filtering we get intervals, here a minor third, which can be arranged successively as 'melody' or simultaneously as 'harmony'. The chord as filtering plus superimposing is a defined, complex unit of vertically layered discrete pitches. Normally—many exceptions granted—it is organized at the core in the form of layered thirds, as can be seen in the following example, fig. 21.2 (Fig. 7.6).

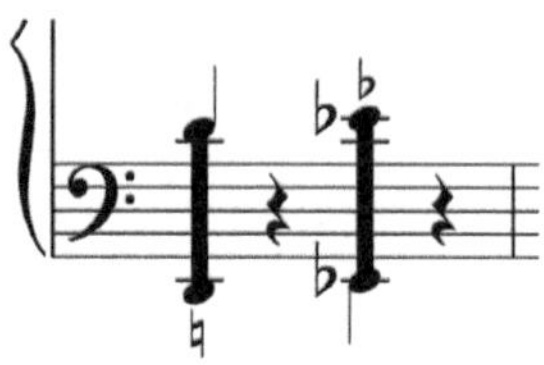

Fig. 7.4 Henry Cowell. Cluster
Jlhughes: Two examples of tone-cluster notation as first used by Henry Cowell, 10 July 2007. https://en.wikipedia.org/wiki/File:Cowell_tone_clusters.png

Fig. 7.5 Left: BACH as Cluster; right: Interval: A C

Fig. 7.6 Chord: B D F A

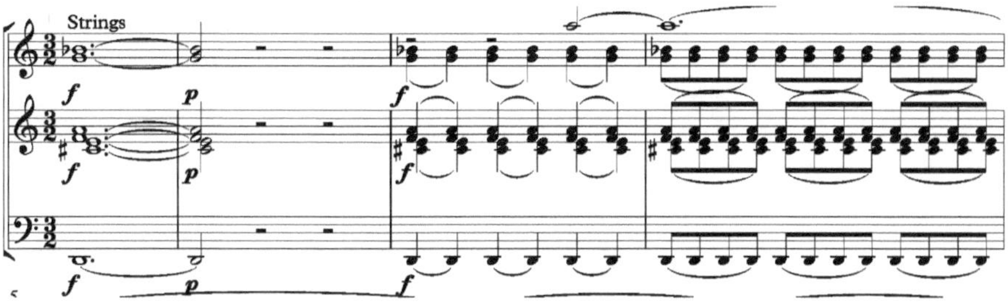

Fig. 7.7 Jean-Féry Rebel: *Les Élémens*, m. 1–3
https://upload.wikimedia.org/wikipedia/commons/5/57/Rebel%2C_les_Elemens%2C

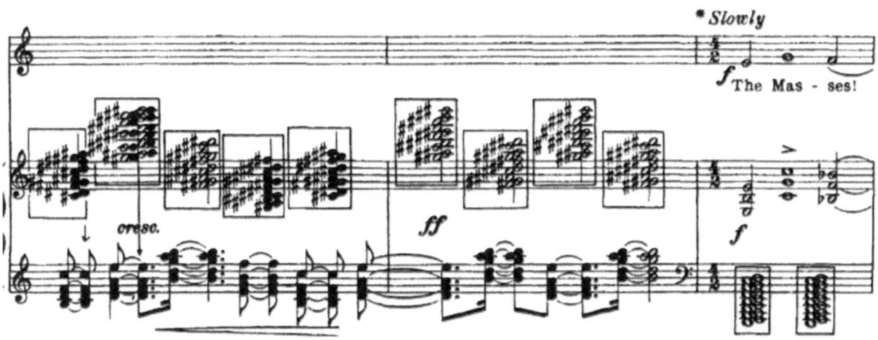

* Preferably for a unison chorus; it is almost impossible for a single voice to hold the part against the score.

Fig. 7.8 Charles Ives: *Majority*, 1921—The first song in Ives's songbook 114 Songs. Redding, Conn.: C. E. Ives, n.d.(1922). IMSLP34908-PMLP66688-Ives-114Songs1922op

Prior to its neutralizing use as mere material, the cluster had been defined since the avant-garde after 1900 as disorder, disorganization, as chaos *after* or *before* a cosmos. The former is the case in what is likely the earliest example of (diatonic) tone clusters in a Western music composition thus far identified, in Heinrich Biber's *Battalia à 10* (1673).[13] In Alban Berg's opera *Lulu*, piano clusters characterize the acrobat, one of Lulu's lovers, with a leitmotif. In this way, he is given negative connotations, with the sense of barbaric, uneducated, pre-musical or anti-musical. This latter instance, the representation of productive chaos, is the case in an orchestral diatonic cluster containing all the notes of the harmonic minor scale, also occurs in the representation of chaos in the opening of Jean-Féry Rebel's ballet *Les Élémens* (1737/1738) (Fig. 7.7).

Within the avant-garde, Charles Ives employed clusters to characterize the fermenting, the productive, the powerful and the promising of the "Masses" (Fig. 7.8).

---

[13] See https://en.wikipedia.org/wiki/Tone_cluster, 15 August 2020, retrieved 18 August 2020.

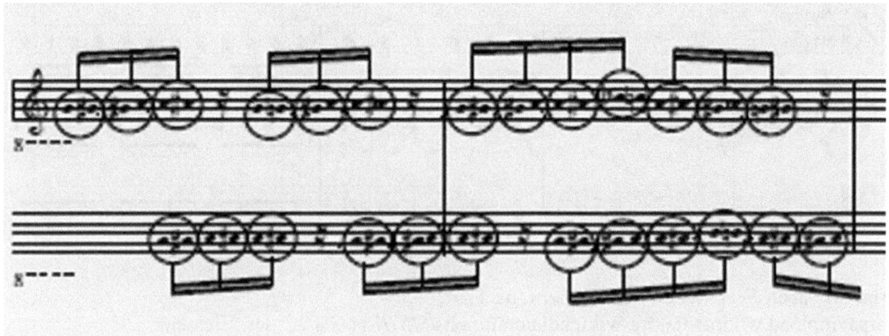

Fig. 7.9 György Ligeti: String Quartet No. 1 (1953/1954), mm. 745–747
Quoted after Julia Kreinin, *"To Arrest the Process": Moving Clusters by György Ligeti and Witold Lutoslawski.* In: *Mitteilungen der Paul Sacher Stiftung*, no. 15 (2002), pp. 36–41; quotation and figure p. 27

A specific derivation of clusters is one that occurs as a blurring of polyphonic lines—"moving clusters". This condenses the polyphonic structure of the texture, but blurs the distinguishability of the individual voices at the same time (Fig. 7.9).

> A "composite chromatic line" can also be termed a "chromatic cluster." When we analyze the overall sound results of the fugato [...], we see a chromatic cluster changing its density (bar 1) or both its density and pitch (the subsequent bars). In the 1960s, Ligeti continued to build his scores like microcanonical structures, where the parts are very close together but are not permitted to coincide.[14]

7.2.2 "Intonarumori". Sirens, Airplane Propellers. Pre-electric Noise Production

What is novel, and initially a sharpening, is the classification of the sounds. The futuristic composer Luigi Russolo began expanding the spectrum of musical instruments in 1913 with self-constructed "intonarumori"—'sound generators', or better 'noise generators'. He wanted to extend the spectrum of sound production beyond the abilities of the orchestra. The parallels to the search for new sound generators in the opposite direction, to more sharpening by differentiating the tones, are evident, especially in the field of microtonality, which was already expanding before the year 1900, with special grand pianos and so on.

Nothing fundamental changes in futuristic noise music. The roar of Futurist propaganda with their 'New à tout prix' drowned out the fact that Russolo's 'intonarumori' were indeed musical instruments that produced sound as material. From the point

[14] Julia Kreinin, "*'To Arrest the Process': Moving Clusters by György Ligeti and Witold Lutoslawski*", in: *Mitteilungen der Paul Sacher Stiftung*, No. 15 (2002), pp. 36–41; quotation and figure p. 27.

of view of poiesis they are traditional, only mechanical, neither electrified nor automated. 'Noise makers' are nothing but a particular form of musical instruments that, like body slaps, drums, sistrums, maracas, rattles etc.; they simply produce percussive sounds but not tones. Russolo first expressed his manifesto *L'arte dei rumori* in the form of a letter to Pratella in 1913 and published it in 1916, along with other texts, in a book with the same title. Russolo classifies the noise in six families or types of noise:

> Roars, Thunderings, Explosions, Hissing roars, Bangs, Booms
>
> Whistling, Hissing, Puffing
>
> Whispers, Murmurs, Mumbling, Muttering, Gurgling
>
> Screeching, Creaking, Rustling, Buzzing, Crackling, Scraping
>
> Noises obtained by beating on metals, woods, skins, stones, pottery, etc.
>
> Voices of animals and people, Shouts, Screams, Shrieks, Wails, Hoots, Howls, Death rattles, Sobs.[15]

Russolo thought that all other noises are only associations and combinations of these fundamental noises. Due to the one-sided concentration on sounds as musical material, the movement and its developments are also known as *bruitismus* (French *bruit* = noise, sound), and in Italian as *rumorismo* (Fig. 7.10).

Another way to produce noise is to enhance the noisiness in existing musical instruments rather than inventing new ones, or else to utilize everyday devices and objects as "intonarumori". This works both on the artisanal-mechanical level and on the industrial-electro-acoustic level. Scratching was developed primarily with the advent of hip hop in the mid-1970s: record players and vinyl LP records were used as musical instruments to produce percussive or rhythmic sounds. Scratching blurs or destroys the integrity of the existing work, but creates a new one in combination with "remix" etc. The closest thing to this is the alienating blurring of the sound through noisiness in the piano by various methods of "preparation," for example with screws between the strings to create distortion effects.[16] Something similar also occurs in stringed and plucked instruments with new playing techniques, such as scratching the strings, tapping the body, etc. Helmut Lachenmann made the composition of ambient noises, which inevitably occur during playing anyway, almost the main focus of his specialty. In rock music, this preparation on an already electrified level—with overdriving, distortion etc.—has become the norm, the "unplugged" the exception. All of this historically and aesthetically presupposes the avant-garde and in particular the futuristic exploration of the intrinsic value of sounds as musical material.

On the one hand, blurring aids in creating new musical material and is a conscious compositional seizure of new technical possibilities. In the pitch and timbre dimensions of music, this blurring means letting go of clear and precise tones. As is so often

[15] *Luigi Russolo (1916): "The Art of Noises, Archived from the original on 2010-11-27", quoted after* 2 February 2020, https://en.wikipedia.org/wiki/The_Art_of_Noises, retrieved 20 June 2020.

[16] Cf. e.g. Paul Dessau's "bedbug piano" with thumbtacks in the hammers for the purpose of sound manipulation, or Cage's piano arrangements; see e.g. Lena Chrisolova: *Präparierte Klänge und Stille im Werk von John Cage*, kunsttexte.de 4/2012–2.

Fig. 7.10 Luigi Russolo: *Risveglio di una Città.* Score, p. 1.
Quoted after https://upload.wikimedia.org/wikipedia/commons/3/3e/Risveglio_di_una_citta.jpg
Quoted after L'Arte# p. 17 21

the case, though, it also involves sharpening on a second level. Reality as a whole becomes easier to represent in this way, not least of all the new, urban soundscape, and it is therefore also easier to process psychologically. The noisy sound of the Futurists stands specifically for the modern, technical, industrial, urban world and soundscapes. Insofar, as mentioned, the blurring on a meta-level is also the sharpening of mimesis of new realities. It is remarkable, however, that Russolo, in his well-known classification of sounds into six types, prominently includes traditional types on the one hand, like the aforementioned wind in his "family of sounds" type 2 as "whistling," "howling;"[17] on the other hand—and this is really innovative—he includes the instrumental production of specific vocal sounds as the sixth type: "Voices of animals and humans: calling, screaming, shouting, howling, laughing, wheezing, sobbing."[18]

What is new about Futurism in relation to music is not the use of sounds as such—that is ancient—but its revaluation (and overvaluation in relation to the tones) on the one hand and its semantic accentuation as an expression of the industrial world on the other hand, with a specific accent on aggressive to military sounds between race car and mitrailleuse. These, too, are not unfamiliar in music to this day, but they have not played such a prominent role. Following the sentence: "Nor must we forget the completely new sounds of modern war,"[19] Russolo's list of the components of urban sounds leads into an enthusiastic essay on this poem, which depicts "in wonderful *parole in libertà*[20] the orchestra of a great battle."[21] These are fine examples of that barbaric aestheticism which already had a glorious history at that time and to which an even greater future opened up in the 20th century, continuing into the 21st. Russolo, with music as his sound weapon, thus also wants to "generously share tonal slaps in the face."[22]

Instead of the Futurists' isolating noises in music, there were after World War I more integrative tendencies, especially in the early USSR. Arseny Avraamov's work *Simfoniya Gudkov* (*Symphony of Sirens*), which premiered in Baku on November 7, 1922, the 5th anniversary of the October Revolution, and then in Moscow in 1923, was a new type of open-air music. In contrast to the Futuristic noise by the musical instruments intonarumori, it takes a completely different approach to the use and design of real sounds with a remarkable interweaving of new and old. The noise was produced not by musical instruments. The foghorns of the Caspian fleet took part, all the factory sirens, two batteries of artillery, several infantry regiments, a machine gun unit, numerous hydroplanes, and finally *choirs*, to which all of the

[17] The well-known score by Russolo, *Il Risveglio della città* (1914), often includes glissandi. The oft-cited page of the score e.g. "Luigi Russolo: 'Intonarumori'." http://www.medienkunstnetz.de/werke/intonarumori/bilder/2/. abb. see also https://en.wikipedia.org/wiki/File:Intonarumori-interno.jpg.

[18] Russolo, *L' arte dei rumori*.

[19] Ibid.

[20] These "parole in libertà," words in freedom, are preferably free of syntax and human sense.

[21] Ibid.

[22] Ibid.

spectators contributed.[23] This special noise music includes human voices. The second performance in Moscow also did not produce

> especially pleasing results: [...] On the one hand, the modulation capability of the 'instruments' that were used was not exactly great, on the other hand the 'compositions' performed were much too complicated. Although the 'conductors,'[24] positioned on high command towers, regulated the use of the various sirens and steam pipes, which were located very far apart, by waving flags, it was not possible to achieve a uniform acoustic impression; the distortions were such that the audience was not even able to recognize the 'Internationale' that was so well-known and familiar to them."[25]

In contrast to the use of sound as musical material, the technically imposed fuzziness is not blurring. It is even detrimental to the sharpenings that should be achieved with the revolutionary musical arrangement. Efforts to "improve the modulation capability and tonal purity of the 'monumental instruments,' as well as the accuracy of joint conducting" continued during the formative phase of the 1920s. Fülop-Müller previously reported on "noise orchestras" with "engines, turbines, sirens, horns, and similar noise instruments," with which "noise orgies" were held in the Moscow Trade Union Palace, for instance, i.e. not as open-air music, followed by "noise symphonies, noise operas, and noise festivals."[26] In any case, the fuzziness of the blurred noise in principle could not and cannot be amended. Futuristic grandiloquence and a cult of technology continue in George Antheil's *Ballet Mécanique* (1924, with Ferdinand Léger and Dudley Murphy).

> Antheil described his "first major work" as "scored for countless numbers of player pianos. *All percussive. Like machines.* All efficiency. No LOVE. Written without sympathy. Written *cold as an army* operates. Revolutionary as nothing has been revolutionary."[27]

The instrumentation extends far beyond Russolo's und Pratella's simple intonarumori:

> Antheil's original conception was scored for 16 specially synchronized player pianos, two grand pianos, electronic bells, xylophones, bass drums, a siren and three airplane propellers.[28]

Interestingly, though, the technology and mechanization, which was supposed to guarantee greater precision, brought about unwanted fuzziness (Fig. 7.11).

> (D)ifficulties with the synchronization resulted in a rewrite for a *single* pianola and *multiple* human pianists.[29]

[23]Fülöp-Müller 1978, p. 163. Further material on this subject can be found inter alia in Smirnov and Sterneck. Emphasis HWH.

[24]A picture appears on page 160 of the conductor directing with signal flags, which appears frequently in literature; this is Avraamov. Fülöp-Müller, p. 160.

[25]Ibid. 163.

[26]Ibid. 162.

[27]Oja, p. 185.

[28]Thomas / Key / Rothe, p. 56.

[29]Ibid. Emphasis HWH. In extenso see Angustin 1960.

Fig. 7.11 Georges Antheil: *Ballet mécanique*, score, ms. 1–3
https://cn.imslp.org/wiki/Ballet_m%C3%A9canique_(Antheil,_George)

So people had to fill the gaps of technical fuzziness once more. More living work instead of less of it: human labour complements and replaces mechanical musical instruments. Their expected precision turned into unintentional blurring, i.e. fuzziness. Compositionally, the noises, so to speak, come to themselves as independent material and thus become futuristic composition with Edgard Varèse's *Ionisation* (1931), probably the first work for percussion orchestra alone. Here, in contrary to Russolo, the composer attaches importance to the fact that this is not mimesis. The technical innovation of electroacoustic music after 1945 does not change this aesthetic position. However, the range of the spectrum of the sounds produced in this way does change. One of its pioneers, Varèse predicted as the future of music,

> when electronic instruments would be liberated from the dictates of conventional music to produce any type of sound."[30] (see Sect. 4.3)

This future became reality not with the first generation of electrified instruments in the 1920s with Thereminvox, Ondes Martenot, and so on, which only or mainly produced tones but with the second generation about and after 1945, with "concrete music" and "electronic music"- (see Sect. 6.9) In pop and rock, noise has been exploding since the 1970s: be it as walls of noise enveloping and preceding the sounds, tones and musical shapes, to "noise music" as a genre, be it as noise *in* the musical shapes by different methods of distorting the "natural", "unplugged" sounds of the instruments. Especially, but not only, in "noise music", the noise of "heavily overdriven amplifiers and synthesizers […] should penetrate, attack and traumatize the human body."[31]

7.2.3 Technical Re/Produced Noise as Musical Material

Since the acoustic world has become technically reproducible, sounds can be used musically and aesthetically as "real tones". As far as mimesis of reality is concerned, there is no longer any need for a detour via onomatopoeia with sharply defined tones. Instead, indexical signs, acoustic indicators of real processes between nature, human sounds and technology, become mimetic and iconic signs in the process of artistic—and at the same time technical—treatment and elaboration. Walter Ruttmann's *Weekend*, 1930, 'quotes' in a grand montage the totality of the soundscape as a soundtrack. Ruttmann mixed real noises with words and music in a sound montage so that all the acoustic appears as an autonomous—imaginary—soundtrack to a non-existing film. *Weekend* is an eleven-minute radio play in the spirit of 'New Objectivity.' Ruttmann, originally a painter, then a filmmaker, used the sound-on-film method here. The radiophonic work, an invisible sound film, consists of mounted

[30]Demers, *Listening Through the Noise*, p. 174.

[31]Weiss 1996, p. 169, quoted in https://de.wikipedia.org/wiki/Bruitismus, 31 December 2018, retrieved 29 June 2020; cf. https://it.wikipedia.org/wiki/Rumorismo, 23 March 2020, retrieved 29 June 2020.

sound film recordings using the Tri-Ergon process, which was the only way to cut image or sound tracks with precision before tape was available. It is a purely acoustic counterpart to Ruttmann's purely visual silent film *Berlin: Symphony of a Metropolis* from 1926.

The advance notice of the radio play *Weekend* in early 1930 describes very precisely the components and progression of the scenes. The work-free weekend, for example, is set against the working week in a semi-documentary form:

> *Ruttmann's photographic radio play /* In the following we present the progression of the radio play outlined by Ruttmann, which acoustically reproduces the events of the weekend from the end of work on Saturday to the resumption of work on Monday.
>
> *WEEKEND*
>
> "1) Jazz of work
>
> Material:
>
> Typewriters Telephone bell Cash register
>
> Various machines Hammering Saws
>
> Files Forging Dictating Commands" etc.[32]

The starting point there are individual film sequences, which, as in the advanced handling of real sounds, become a whole through montage. For example, in a self-commentary from 1927:

> "Every day the shots were developed, and very, very slowly, visible only to me, the first act began to take shape."[33]

Not the ready-made sound objects in themselves but their composition as—mainly horizontal—montage is the essence of the work.

> The post-production is claimed to be the actual and creative process of film.[34]

As the title suggests, music, particularly as rhythm, is the common denominator, integrator and model of the composition:

> He makes the category of the symphonic his own without further ado, all his central formal terms are equally musical terms, and he prefers to describe his montage technique with musical metaphors. Cinematic practice and technical action are presented as being permeated by music; even thinking about film is deeply immersed in music. All of this is essentially part of the romantic, music-focused artistic phase that is typical of Ruttmann.[35]

For the work as a totality, however, the real sounds serve to sharpen the mimesis of reality.

Frisius, in his very detailed analysis extending to the level of sonograms, emphasizes the technical advancement *and* the unusual realism of the work:

---

[32] *Film-Kurier*, Berlin (No. 41), 15 February 1930, cited by: "Walter Ruttmann, 'Weekend'." http://www.medienkunstnetz.de/quellentext/96/.

[33] Cited by Prümm 2011, p. 62.

[34] Ibid.

[35] Ibid. 64.

> "*Weekend* is the first radically experimental composition of montaged sounds. This production anticipated tendencies that would only be taken up again decades later in French 'musique concrète' and in the West German 'New Radio Play'—for example, by Pierre Schaeffer and Ferdinand Kriwet. In Ruttmann's audio film, Schaeffer's early 'musique concrète' and Kriwet's audio texts, approaches to the content-related selection, combination and processing of sounds are evident, as they remained largely singular until the end of the 20th century."[36]

The real sounds of the urban, industrial soundscape appear completely and acoustically vivid in an imageless soundtrack. The artfully assembled, quasi-documentary radio play without a plot is an audio image. It is a glimpse of the later independence of independent musical works in 'musique concrète.' A prototype of this music is Pierre Schaeffer's *Étude aux chemins de fer*, 1948. Technically reproduced, compositionally reprocessed sounds become music. Real tones as readymades are the unstylized, pre-existing material of the 'musique concrète' produced by Schaeffer and others since about 1945.[37] However, they become stylized and assembled into the raw material of the composition. In contrast, they are not used in 'electronic' or 'electro-acoustic' music by Herbert Eimert, Henri Pousseur, Karlheinz Stockhausen, and so on. However, both originally antagonistic currents soon converged. Stockhausen's Gesang der Jünglinge (Song of the Youths) from 1955/1956 represents the synthesis. In Pierre Henry and Pierre Schaeffer, as well, the instrumental is then supplemented by the human voice, first in the jointly produced Symphonie pour un homme seul (Symphony for One Man Alone, 1950). Schaeffer preferred real sounds from the realm of transportation—*Chemin de fer* appears in several of his titles—that are more traditional than the racing cars and airplanes or the military means of destruction in Futurism. In rock and pop music since the 1970s, the amount of noise has increased with a large leap, far beyond the traditional percussion.

7.2.4 "Noise" as Blurry Real Residue and as Ideology

Sounds can also be understood as different types and forms of noise (see Sect. 3.5 as fuzziness in phonography form). It ranges from "white" noise, which includes all audible frequencies as a mixture, to "colorful" noise, i.e. in each case segments from the entire frequency spectrum. In the sound form and phonographic form, noise is a blurring and distortion, which is filtered and selected as musical material is a blurring, but which can lead to a sharpening, as is discussed above.

> [...] noise refers to a statistical model for signals and signal sources [...]. White noise draws its name from white light, [...] although light that appears white generally does not have a flat power spectral density over the visible band. [...] An infinite-bandwidth white noise signal is a purely theoretical construction. The bandwidth of white noise is limited in practice by the mechanism of noise generation, by the transmission medium and by finite observation

[36] See "4. Analyse erfahrungsorientierter Musik". In: Frisius 2005. A broad embedding of the work into various artistic and ideological currents is undertaken by Hagen 2005.

[37] See inter alia Schaeffer 1974.

capabilities. Thus, random signals are considered "white noise" if they are observed to have a flat spectrum over the range of frequencies that are relevant to the context. For an audio signal, the relevant range is the band of audible sound frequencies (between 20 and 20,000 Hz). Such a signal is heard by the human ear as a *hissing sound*, resembling the /h/ sound in a sustained aspiration. On the other hand, the /sh/ sound in "ash" is a colored noise because it has a formant structure.[38]

Even Russolo had not merely placed sounds as material on a "qualitatively equal level with the sounds produced by classical instruments".[39] He had thereby further differentiated urban "everyday sounds and the ubiquitous background noise," both as the subject of musical mimesis and as the product of his intonarumori, which were differentiated according to type. The blurring of tones that had been precisely filtered out and systematized on the material level is, as has been discussed, at once a sharpening of form and content on the mimesis and work levels. These differentiations essentially remain in *musique concrète*, despite the extensive rejection of the mimesis. They have disappeared from postmodern ideology, however. This makes noise and hissing essentially identical and, moreover, asserts that "noise" is a fundamental ontological fact. So not only is the inevitably fuzzy seized upon and intentionally used as musical material. Rather, the inevitable fuzziness is now romanticized and the—intentionally—vague, even imprecise, is transfigured into an independent quality. The screw of blurring is thus driven one turn further and is perhaps overtightened. For, as was mentioned, the enthusiasm for sounds as a component of music, or even as music, had and has a certain justification inasmuch as the urban, city-based soundscape can be reflected more directly in this way. Beyond that, however, there is reason to suspect that it is not the sharpened mimesis, but rather the attractiveness that arises the other way around, from the indeterminacy of sounds as opposed to tones and even more so from the vagueness of the noise—a certain regression. Furthermore, this blurring corresponds to the tendencies to blur boundaries between the real object and the work of art, as is characteristic of many currents of modernism and the avant-garde. Real noise outside of the realm of art can also be rendered musical and emotional. This then even extends as far as the sound of cosmic background radiation. In radio telescopes, the energetic, not strictly acoustic signal is recorded acoustically by substituting for the senses. This in turn entices one to express a half-metaphorical, half-projective interpretation and a formulation such as the following about

the pleasure of hearing the *cosmic noise concert* of the second act of the *Big Bang Overture* in person.[40]

---

[38]"A simple example of white noise is a nonexistent radio station (static)." https://en.wikipedia.org/wiki/White_noise, 30 June 2020, retrieved 2 July 2020. There are different "colors" of noise, e.g. pink, blue, violet, grey, red, green and black. Cf. https://en.wikipedia.org/wiki/Colors_of_noise, 5 June 2020, retrieved 2 July 2020.

[39]See Katja Stopka 2018 in Morat, p. 206f.

[40]https://www.heise.de/tp/features/Vaeter-der-Urknall-Kosmologie-3413652.html, retrieved 1 February 2017. Those who perceive it in this way get two major musical genres at once, namely concert and opera. The overture belongs to the latter, but it does not have "acts" any more than the concert does.

Noise plays a part in the ambivalence of the fuzzy.[41] The burbling of the brook produces a primarily pleasant and beautiful effect, whether it is real or reflected in Beethoven's *Pastoral* or Schubert's song cycle *Die schöne Müllerin;* the murmur of traffic seems less enjoyable—although more pleasant than sudden noises. In the postmodern era, it is precisely the once negative side of noise, especially in the field of technology itself, which is being re-evaluated and fetishized (see Chapter 6. Phonography form). Stopka points out

> that the popularity of noise is paradoxically closely related to digital technology and the entry of the World Wide Web into all areas of life. Paradoxically in that digital technology, which represents technical progress, is seen as a promise to supersede analogue technology, which is hampered by all kinds of malfunctions, and thus to eliminate the vulnerability to failure caused by noise. The notable first successful result, accessible to all, was the noise-free CD, which allows for interference-free sound reproduction. Yet no sooner has the acoustic 'parasite' disappeared than it becomes the mainstay of a new self-conception in contemporary popular and serious music and in literature, wherein the fallback to the pioneering achievements of the art of noise and the ambivalent attribution of meaning to noise around 1900 is obvious. The new imperative is now: Where there is sound, there must also be noise. Once disruptive factors such as vinyl noise or clicks, scratches and cracks are now recognized and useful as artistic elements.[42]

Under the conditions of digitalization, the noise of the good old vinyl record takes on the nostalgic aura of being more authentic compared to the compact disc.

> The mp3 was a political compromise at certain points, and this was also occasionally noticeable in the sound quality. You can demonstrate this in certain pieces of music when they are encoded and decoded and you have trained your hearing a bit. Can you hear the difference? Yes, you can. The mp3 is not perfect.[43] [...] There's a very nice study from the University of Huddersfield in England. There the tested persons were played a recording from a record player. At the same time the same recording was running on a digital server. The testers could switch back and forth, and the participants were asked to say which sounded better: vinyl or digital. 60 percent of the participants said afterwards that vinyl sounded better. What people didn't know, however, was that both signals were identical, with only one code coming from the record to control playback. On the vinyl record there was only a slight crackling sound at the beginning, so that it sounded more realistic. Result: When listening, expectations play a big role. So I reckon that vinyl has to sound better from the turntable – that's why it subjectively sounds better too.[44]

The new, digital "quantization noise" following PCM (pulse code modulation) on the new sound carrier medium and later media is practically imperceptible, and what is more difficult to discern is the actual reduction of frequency range and partial tone spectrum as a result of the digitization. The more extensive reduction of this spectrum in the various "compressions" procedures of the MP3 type etc. have meanwhile also become such common practice and such a standard that the restriction is usually only noticed and critiqued in the optical dimension. In any case, digitization is not a

[41] Cf. Stopka 2018, p. 205.

[42] Stopka 2018, p. 209.

[43] Karlheinz Brandenburg 2020, without pagination.

[44] Ibid.

sharpening in the positive sense in such a way that it would render acoustic-musical elements clearer and more contoured. It is not an actual blurring either, though.

Stopka is making a somewhat short-circuited connection, factually yet justifiably, between the new "noise" used as blurring and the ecstatic "rush":

> New techniques emerge, such as electronic mixing and scratching, from which a completely new body and dance culture develops, as with the rave movement, for instance, where intoxication is closely linked to the acoustic phenomenon of noise.[45]

Postmodern ideology not only considers noise and hissing identical, but moreover claims "noise" as a fundamental ontological fact: Reality is supposed to be "chaos" without a cosmos. In existentialist terms, blurriness, fuzziness, thus becomes a kind of "existential".

> What is real, what is making noise, is chaos […]. The noise of the real is now the noise of chaos: the chaos of the objective, the chaos of nature, the chaos of the world as a whole. And this is with regard to its auditory side: noise. […] Now chaos is the object of art and is the origin of artistic form. […] "Artists today imitate the sound of chaos." (Norbert Bolz 1994, 293) The music of postmodernism as a self-organizing system did not exclude acoustic chaos, but rather assimilated it. Its order grew through the incorporation of chaos. And so the aesthetics of postmodernism can be reduced quite simply to two principles: "*order from noise*, feeding the system with the chaotic; the emergence of noise from collapsing orders." (Bolz 1994)[46]

Mayer rightly objects to such dedifferentiation that noise is not the same as sound because (1) noises are real and differentiated in their artistic appropriation, as Russolo considered and is already discussed here, and (2) because factors shaping the composition with and of sounds are likewise effective.

> As with the laws of unity formation in the visual field, there are also noises in the auditory field, as opposed to "pattern recognition". Admittedly, taking into account their peculiarities compared to "sounds". Noises also have a shape, despite a relatively blurred outer boundary. Compared to other things in the audible world, they stand out from their acoustic environment (ultimately from the "background" of the noise) and are more or less clearly distinguishable (and thus have a shape). Ultimately, each is constituted by a specific context, a characteristic arrangement of properties of form and figure, i.e. they have a shape or a configuration. That the elements do not occur periodically, stochastically, etc., within an acoustic phenomenon of this type does nothing to change the fact that it is accorded "critical points", i.e. particular properties which arise from the specificity of the vibrating substrate and which, even in the case of unclear perception within sequences of events, permit the kind of "summarization" of an acoustic event in the auditory field by which it is identifiable as "this" sound, despite "local changes in detail" or limited perception.[47]

As has already been emphasized, it is clear that the blurring of tones and sounds or their—partial—replacement in the form of noises can represent a sharpening of the reality content of music composed in this way. This runs contrary to the intentions of the composer in the case of Schaeffer. Despite the opposite social perspective,

---

[45] Stopka 2018, p. 208.

[46] Mayer 1996, p. 197f.

[47] Ibid., p. 199f.

the sharpening of the reality content corresponds to the intention both in Karlheinz Stockhausen's *Hymnen*, an electronic and concrete music with soloists from 1967, especially in *Region IV*, and in Luigi Nono's *La fabbrica illuminata* (The Enlightened or Illuminated Factory) for soprano and four-track tape (based *inter alia* on texts from trade union agreements and verses by Cesare Pavese, compiled by Giuliano Scabia) from 1964. There are two fundamental types of real sounds: vocal-physical and technical-industrial.

> a) (V)ocal noise sounds are, one the one hand, stochastic "fields": cheering, laughter, howling of masses of people in Stockhausen, or else expressions of mass outrage, of protest and energetic demands in Nono (where male and female vocal fields are clearly distinguishable). On the other hand, these stand in extreme contrast to the echoes of individual human voices in the gaps constructed by Stockhausen, and the elementary noise of individual inhalations and exhalations or the initially quiet babble of voices from which individual voices stand out, in the huge production halls of the "fabbrica". And there are b) technical noise sounds: On the one hand, these are the downward, steady paths of very intensive – I'll say "engine noises" […], which interrupt the inhalation and exhalation in the final passages of the *Hymns and* create space for changing vocal fields. And then, on the other hand, there are the technical noises from production processes at the Italsider steel and rolling mill, which Nono recorded there.[48]

This differentiation of noises permits a formation of shapes. Through these, simple index signs are transformed into icons and symbol signs; they are mimetic and at the same time artistically independent.

> (A)t times numerous sounds [are] perceptible simultaneously […], that is, very specific sounds in the background can be distinguished from those in the foreground. […] These noise sounds from reality perform […] symptom functions as "found objects". They are "indices" – they refer to entirely real processes of the acoustic expressions of life of large masses of people under particular socio-historical conditions or else to elementary bio-physical necessities of life that occur quite regularly, such as the breathing of a real living person. These mass expressions and actions may seem chaotic. Underlying them, however, are socio-historic and concrete political, economic processes that by no means occur indeterministically. […] The sound of breathing is neither chaos nor noise.[49]

Function as a blurring is not the final word on noises.

7.3 Point of Departure: Score Form. 'Musical Graphics'

Musical Graphics was en vogue for a time within the musical avant-garde of the 1960s. Here, perhaps inspired by Zadeh, the term "vague" would be appropriate. For Musical Graphics itself is generally rather indeterminate. The possibilities for sharpened, precise notation are reduced here. It is therefore similar to 'Graphic Notation', but nevertheless stands in contrast to it (see Chapter 6). It is a

---

[48] Mayer 1996, p. 201.

[49] Ibid. Emphasis HWH.

> form of notation between a sign and a drawing. Musical Graphics is both an instruction to produce sound and an image .... It is based on the fact that autonomously conceived graphic elements can also fulfil a sign function. The main representatives are Roman Haubenstock-Ramati, who also coined the term "Musical Graphics", Earle Brown, who presented the first Musical Graphics in *December 52*, Anestis Logothetis and John Cage, who took inspiration from Wassily Kandinsky or in abstract expressionism (Jackson Pollock and Alexander Calder, among others).[50]

7.3.1 The Pictorial Instead of Notes and Writing as a Blurred or Vague Concept for the Realization

The definitional demarcation from Graphic Notation (see Chapter 6). Emphasizes above all its relatively autonomous character as a work of visual art.[51] A transitional form is "Space Notation". It transforms traditional notation into the pictographic. Here, the graphic-pictorial component is scarcely noticeable. It is

> notation of a piece of music without bars and exact indications of the durations of rests, which are deduced by eye from the arrangement of the notes on the paper. The notes or sounds are notated in a given order, the time at which they sound corresponds to their placement on an imagined horizontal time axis: the greater the space between the notes, the longer the pause between the sounds.[52]

Space Notation employs the mimetic potential of the synesthetically mediated analogies between the acoustic and the optical as correspondences of temporal and spatial distance. Because it intentionally leaves the quantity of these distances undefined, though, it is a blurring compared to traditional notation.[53] By contrast, in one of the earliest Musical graphics, Earle Brown's *December 1952*, there is no chronological sequence. However, it becomes apparent that the similarity correlations between the optical and the acoustic are not arbitrary, but instead are co-determined within the framework of blurring by means of analogies and synesthesia (Fig. 7.12).

On a deeper level, these correlations relate to general patterns of perception,[54] such as up-down / high-low, thick-thin lines / loud-quiet or dense, multi-voice / few-voice or solo voice texture, long-short / duration ratios and others. As a rule, black and white must be sufficient as graphics before and for music, primarily for pragmatic reasons, since multi-color printing was not readily available for reasons of cost. In substance, multi-colored 'musical painting' would be even more appropriate, since "sound colors" are even easier to represent chromatically. Alongside

---

[50] JetztMusik—Glossar—Musikalische Grafik n.d.—The term was stamped by Roman Haubenstock-Ramati before 1959.

[51] Cf. Gieseler 1978. Critique of Graphic Notation: Boehmer 1967, p. 159, footnote 3.

[52] JetztMusik—Glossary Spacenotation. n.d.

[53] See above 6.8. Cf. e.g. Frank 1978, Bavelli 2008.

[54] Cf. to the anthropological background e.g. Wilson 1998, passim.

Fig. 7.12 Earle Brown: December 1952 Quoted after Mathias Wittekopf, http://www.ohrwelt.de (With proposals and plans for a realization. Cf. further e.g. http://www.earle-brown.org and Kokoschka 1966, passim)

pure graphics, there are mixed forms of traditional notation, special characters and graphics, including those by Sylvano Bussotti.[55]

> "As a counterpoint to the most precise regulations possible, Musical Graphics is intended to stimulate without specifying."[56]

Musical Graphics serves to emancipate the interpreters, who are granted significantly greater degrees of freedom and scope for their creations than in traditional notation, as a stimulus for interpretation and tonal realization. It is a concept—almost a kind of written, notated conceptual form—directly rather than mediated by the specifically *musical* notation. Yet in contrast to that and unlike conceptual art, it is an artistically more or less concretized work. However, not only is the execution in detail and on the whole left to the productive, i.e. tonal, realization of the work, but also even elements of the conception as the fulfilment of the idea of the work. First of all, it has the status of a sketch in this respect, even if it appears to be a finished work (at least as a notation): Because nothing is rendered with precise notation. It is thus, second, a special case of a written, notated conceptual form—directly and, above all, figuratively rather than through the mediation of the specifically *musical* notation. In that sense, there are sliding transitions between Graphic Notation and Musical Graphics. Third, both are analogous to the anticipation form, since they visually anticipate and attempt to determine much of the future performance, in particular in terms of the intention, more precisely than traditional notation. Finally, fourth, Musical Graphics is in a certain sense even the pictorial form of a work—not just as a *retrospective* optical formulation of the work but rather in advance, understood precisely as the equivalent of notation. A celebrated pioneering work on

[55]Figures e.g. in Karkoschka 1966, p. 94ff.

[56]Karkoschka 1966, p. 79.

Musical Graphics is Anestis Logothetis's Ballet Music *Odyssey*, 1966, for variable instrumentation. Even as an image, some of the labyrinthine wanderings of Odysseus emerge (Fig. 7.13).

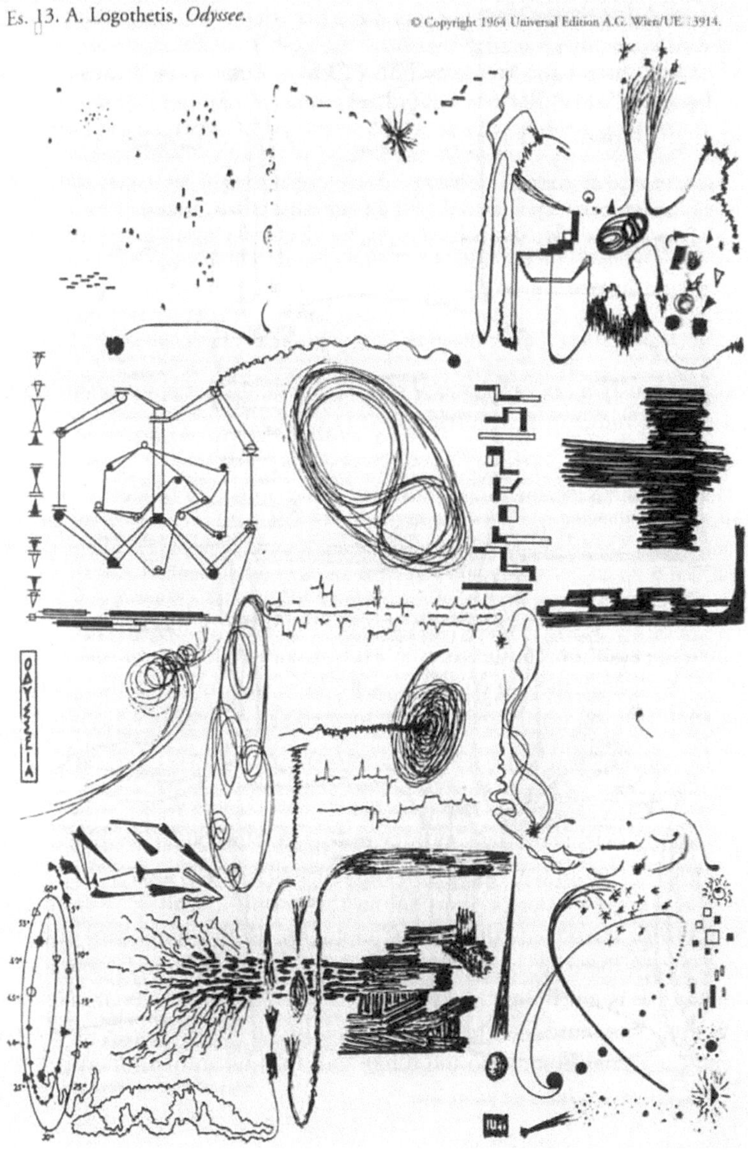

Fig. 7.13 Anestis Logothetis: *Odyssee*
Universal-Edition, Vienna
Quoted after Karkoschka 1966, p. 128

In Musical Graphics, as in Graphic Notation before it, the simultaneity of the optical is exploited:

> In a sonic implementation of Musical Graphics, the page does not have to be read in the traditional reading direction, from left to right. Instead, the eye can scan the page spirally in all directions, continuously or in jumps, can leave out some symbols and repeat others several times. Graphic structures serve as an associative space for sounds.[57]

An Ariadne's thread for this labyrinth of Odysseus's wanderings indicates the course of the work as a whole as a kind of simplified score and thereby translates the visual simultaneity into the succession of a musical process as the "path".[58]

> The lines are printed in red on a transparent sheet of paper so that they remain clearly visible even when the transparency is placed over the drawing. Apart from the "path" itself and information about how it is to be realized by the first group of players, the transparency also includes the instructions for the second and third groups. In this way, these two groups are shown the areas they are supposed to play. While the first group plays the "path", the other two (see Fig. 7.14) play the areas next to the position on the path where the first group is currently located. Each section of the path [...] is marked by the conductor.[59]

As was previously mentioned, the non-arbitrary analogies between the optical and acoustic limit the range of the blurring.

> The possibilities for interpreting each structure in the drawing are different for each instrument. Generally speaking, though, more likely interpretations can be mentioned for some elements: heavy blackening = forte, light blackening = piano. Squares = clusters (white and black keys). Angular lines or surfaces = pitches to be maintained, which are clearly distinguished from one another; round lines and surfaces = glissandi.[60]

The liberation of the interpreters from the intentions of the composers and their realization in the notation implies a certain submission to the intentions of the respective conductors. It is all the more necessary because the improvisational character of execution and performance leads to formulas and simplifications.

> The conductor will be responsible for a concentrated interpretation that never sinks to cheap platitudes.[61]

From the point of view of the score form, Musical Graphics is a blurring. Picture instead of writing, graphical signs—indeed, usually black and white—instead of notation. If it were a matter of sharpening, this would hardly be possible without notation. Bussotti introduces this (see Chapter 6). Rather than providing precise, sharp instructions, there are blurred to vague suggestions for implementation. Much or even most of it is left to the discretion and skill of the interpreters. Nevertheless, Musical Graphics is more than just conceptual art. In its productive realization, the optical really is transformed into the tonal and does not remain merely something

[57] JetztMusik—Glossar Musikalische Grafik.

[58] Karkoschka 1966, p. 128.

[59] Ibid., p. 128f.

[60] Ibid., p. 128f.

[61] Ibid., p. 129.

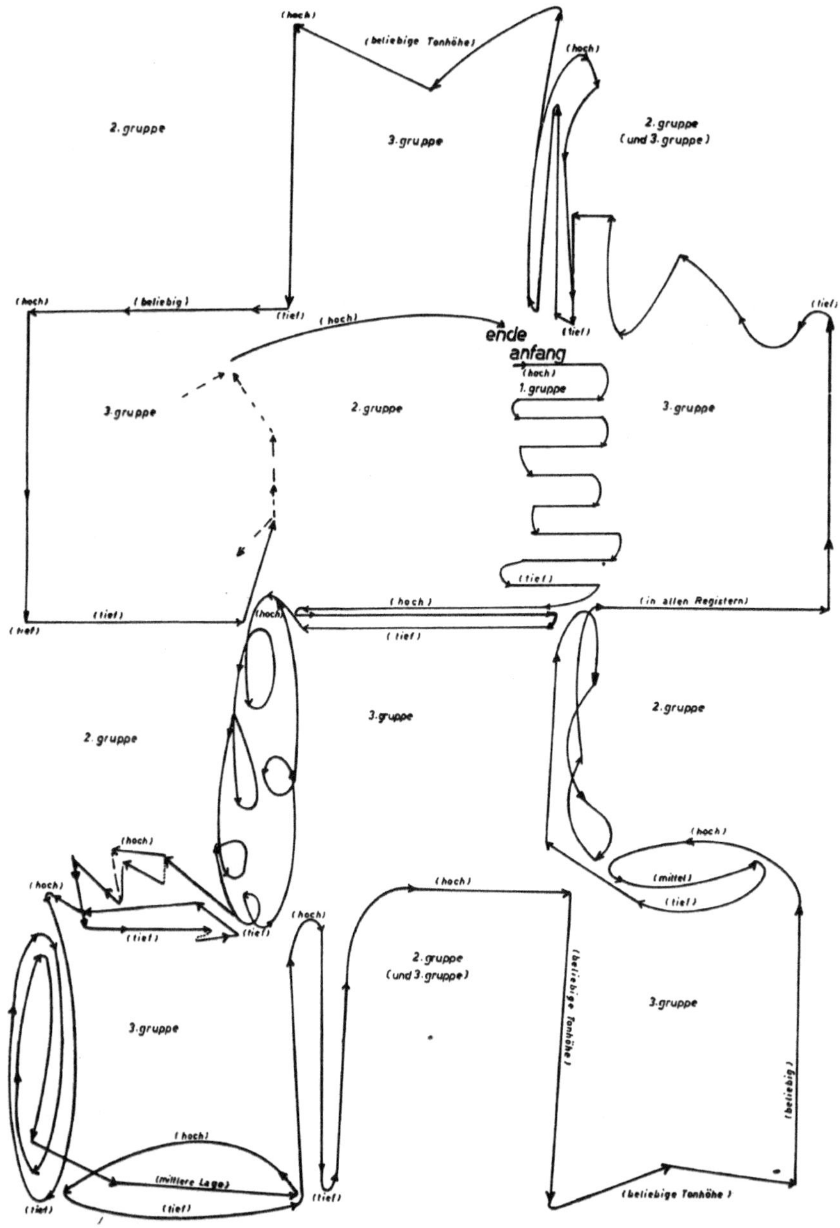

B 42/B 43 LOGOTHETIS: ODYSSEE.
Verlag Universal Edition A.G., Wien

Fig. 7.14 Logothetis: *Odyssee*, Plan
Universal-Edition, Vienna
Cited in Karkoschka 1966, p. 128

said or drawn and thought (see Sect. 3.6 Picture Form). Owing to the indeterminacy of the notation in this form of notation, the range of variants is broad. For it is more image than writing; to a certain extent, it is an anticipated picture form of the work (see Sect. 4.7). The various tonal realizations, which are always interpretations, offer possibilities of comparison for Fuzzy Logic (see Sect. 7.2).

We now find ourselves at the transition from composition to reproduction or the sonic realization of the work. Improvisation takes somewhat different forms, but in principle it behaves the same way.

It does not deny the contradiction between idea and realization, but rather expressly implies the difference between the work and the performance. It does, however, largely dodge the problem of blurring and sharpening. This is because it is mostly a rather "vague" notation of the concept as a composition, and from the outset it implies such a high degree of uncertainty and indetermination in the execution of that which is graphically prescribed that it belongs to the principle of blurring. To a certain extent, the dearth of precision in the composition is offset by the increased liberation of productive capabilities in those doing the reproducing and interpreting. A high degree of improvisation is necessarily involved here, even if reflection, discussion and experimentation with the graphics are possible. The dialectic of blurring and sharpening reaches a new level here. If there are multiple tonal realizations of a work that is notated in this way—as with Earle Brown or Anestis Logothetis—then Fuzzy Logic methods can be employed to quantitatively identify both the differences between these realizations and the differences in the relationship to the "work". The latter is more difficult because, as opposed to conventional musical notation, there are no binding norms for the interpretation and execution of the graphic symbols.

7.3.2 „Soundpainting". Realtime Composing Alias Improvisation

"Soundpainting" was invented in 1974.[62] It is no longer a written notation at all, and needs no words. Gestures take its place. With richer material, this is a return to chironomy, to hand signs, which was originally used as a substitute for notation in the European "Middle Ages" for pitches and durations. Now this chironomy is more general and extended and is used for any kind of performance. The procedure requires lively musicians.

> *Soundpainting*. **Introduction.** Soundpainting is the universal multidisciplinary live composing sign language for musicians, actors, dancers, and visual Artists. Presently (2020) the language comprises more than 1500 gestures that are signed by the Soundpainter (composer) to indicate the type of material desired of the performers. The creation of the composition is realized, by the Soundpainter, through the parameters of each set of signed gestures. **Analysis** The Soundpainter (the composer) standing in front (usually) of the group communicates a series of signs using hand and body gestures indicating *specific and/or*

[62] https://de.wikipedia.org/wiki/Soundpainting, 12 August 2019, retrieved 5 July 2020.

> *aleatoric material* to be performed by the group. The Soundpainter develops the responses of the performers, molding and shaping them into the composition then signs another series of gestures, a phrase, and continues in *this process of composing the piece.* The Soundpainter *composes in real time* utilizing the gestures to create the composition in any way they desire. The Soundpainter *sometimes knows* what he/she will receive from the performers and *sometimes does not know* what he/she will receive – the elements of specificity and chance. The Soundpainter composes with what happens in the moment, whether *expected or not.*[63]

7.3.3 "Visible Music", Air Guitar and Air Drums

Soundpainting is at the same time the transformation of the musical graphic from the thing-like to the process-like, and at the same time a "retour à la nature ", a return to nature, namely to music as a sonic realization.

> At the Darmstadt Summer Courses in 1964, (Earle) Brown put [... *December 1952*] on the music stands of the orchestra and conducted. It has become undeniably clear that his arm and hand movements, and not the score, stimulated the musicians, especially since wave-like progressions and very extended crescendos can only be read from the score with an imagination that can also read Strauss' *Eulenspiegel* theme. What has indeed happened.[64]

The reference to the graphic template, in contrast to sound painting, allows the fuzziness, even vagueness of the realization to be objectified. It is obviously large (see Sect. 7.2).

Existing between imaginary-real music production and mere conceptual art are phenomena like "visible music" or "air guitar", generally the purely visual imitation of making music. In contrast to "visible music," which is not far removed from conceptual art, the playing of air instruments is usually not completely soundless but is instead presented as visual karaoke, as a solo performed to technologically reproduced audible music. In this way, it is easier to exert a certain amount of control over the quality of the air instrument playing, and ultimately to analyze its correctness and adequacy from the perspective of Fuzzy Logic. The term "visible music," coined by Dieter Schnebel in the 1960s, means

> the shift from real-sounding music to the visual. The production of music arises in the imagination or inner ear of the listener or viewer. It is stimulated by body movements, visual materials or even words, on the basis of which the viewer may imagine or associate sounds and musical structures. One could speak of a concept music, a music-in-the-head or a virtual music. So Schnebel's *Mo-No* (1969) is "music for reading", his *ki-no* (1962–67) is "night music for projectors and listeners," *nostalgia* (1960) a "solo for a conductor".[65]

The term Visible Music is additionally used in the context of "Instrumental Theatre". This current of new music draws radial conclusions from the immanent theatricality of the concert. The latent scenic elements and aspects are openly composed and

---

[63] Soundpainting, http://www.soundpainting.com/soundpainting/, retrieved 8 June 2020. Emphasis HWH.

[64] Karkoschka 1966, p. 93.

[65] JetztMusik—Glossary *Visible Music*.

offensively exhibited. Reproduction, "scenic self-portrayal of the music," "mimetic intensification of what is happening on the podium anyway" becomes the central object of the concert experience itself in "Instrumental Theatre".

Production and product, the music-making and the musical result, are radially separated from each other. Thus

> physical actions [...] as elements of the playing are required here [...] in order to produce particular sounds, although they occur separately from the sounds they are needed to produce.[66]

This kind of dismantling only seems to have a defetishizing, illuminating and enlightening effect, since the creating itself may be shown, but its connection to the shoddy result is torn apart—and not only visually. The blurring effect is further intensified when, conversely, audio tapes "reproduce sounds as invisible music without their production being becoming visible".[67] It is a consequence of progressive de-functionalization, which transforms into de-materialization while simultaneously persisting in it, when "the music itself becomes the subject"[68]; ultimately also toward an imagination form without previously objectified forms of existence (see above).

> Audience reactions to an imagined music, descriptions of any actually existing or performed pieces and physical actions that are necessary to produce particular sounds.[69]

Normally, the visual dimension is essential for such events.

> The piece is not intended only to be heard; you have to see it, as well. (Even in traditional music, it was occasionally necessary to bring the impracticable – as well as other things – to light through gestures. For instance, if a cresc. sign is placed above a chord that is to be held in Schubert's last piano sonata, this can probably only be performed theatrically).[70]

For *Sonant for Guitar, Harp, Double Bass and Skin Instruments 1960/....* by M. Kagel, one performance instruction reads:

> Play the whole piece 'as-quietly-as-possible'. The audibility limit of the hall will determine the minimum level of the performance.[71]

This allows for a

> "virtual interpretation" (quoted from Schnebel, 71: "The musicians imitate realization, and do so as close as possible to strings and skins – they're pretending" (Schnebel, 71).[72]

In this kind of air drumming, which we will come to in a moment,

---

[66] Helms 1969, p. 110.
[67] Ibid.
[68] Ibid.
[69] Ibid.
[70] Schnebel 1970, p. 79.
[71] Ibid., p. 71.
[72] Ibid.

> instrumental playing is not only actually playing – playing around: a ball is to be thrown onto the timpani – but also playing in another sense, namely theatre – the instrumentalist plays a role. He has also been doing so before, but only some of it was considered relevant. The realization of music was viewed as abstract from the perspective of the result, as though the performance of a piece of music consisted only in converting the rules of the text into the correct pressing and releasing of keys and pedals (Schnebel, 78).

As with Münchhausen, who lifts himself out of the swamp by the scruff of his neck, it is essentially a matter of the self-rejection of art, of "meta-theatre" and "meta-music," admittedly all within the framework of—ultimately genuinely bourgeois—art. So secondary matters and many of the inevitable fuzziness function as the most important element.

> Waste products of theatre and music [... transform into] works of art. At the same time, insignificant matters are executed compositionally. Steps, gestures, lighting processes are aspects, almost epiphenomena of the theatre, but they are not independent dramatic processes. Because composition turns them into just that, the secondary is pushed into the light ... So preliminary stages of the final products, even production processes of artworks, are themselves already considered artworks. This implies criticism of its concept. A work of art is no longer something that identifies itself by belonging to an approved genre of art, but only by being formed – in this instance: composition – into this type of device. The fact that everything possible can ultimately be composed and thus transformed into art disintegrates the genres of the arts – and themselves.[73]

Composition is ultimately replaced by decomposition. And just as the readymade is designated art by virtue of being exhibited in a museum, everything that is performed in a concert becomes music. Klüppelholz still sees too much organisation and composition in Kagel, and thus ultimately sharpness and precision rather than fuzziness and vagueness:

> Although his models [...] shooed away the earnestness of the concert ritual, such liberation was bought at the price of the compulsion to bring order to what was previously lawless; compositional dominance was not chased away, but merely postponed. When all of the sound characteristics, which had formerly [...] been left to the discretion of the players, had been subjected to rational control by Serial Music, Instrumental Theatre likely turned its back on this, but at the same time it was preparing to take charge of the direction of previously random dimensions such as scene, movement or light.[74]

Another manifestation of visible but inaudible music is more easily accessible to comparative procedures of Fuzzy Logic. In a special type of the picture form of a work, the viewers have to decipher the movements of instrumentalists as representatives of sound. It could occur digitally: One challenge to the "transcription" or the decoding and analysis of moving images is to decrypt what is being played on the "air guitar," "air drums," etc. Around the

> vicarious pleasure in musical performance [...] to mimic playing a rock or heavy-metal guitar solo, [...] here too an industry has emerged in order to facilitate this fantasy. In the first decade of the twenty-first century, air-guitar competitions became an annual fixture in

[73] Ibid., p. 300.

[74] Klüppelholz 1981, p. 261.

> most cities, with Oulu in Finland establishing itself as the global capital of this unlikely subculture: the Air Guitar World Championships have been held there every year since the event's inception in 1996 (Watson 2007). A team of Finnish scientists even went so far as to invent electronic 'data gloves' that link the movements of your fingers to a computer in order to produce a user-generated imitation of a guitar solo (Knight 2005).[75]

Air drumming is the focus in *Adventures of Power,* an American feature film written and directed by Ari Gold, starring Ari Gold.[76] The hero of this melodrama air drums because his father could never afford to buy him drums. Slight irony: He becomes friends with a deaf girl, who reveals that she lost her hearing in her youth because her mother, a former groupie, took her up close to the bands on stage at loud rock shows without hearing protection. With Power, there is no risk of repeating her mother's mistakes. First, because Power isn't likely to become a professional musician, and second, because his music cannot damage her hearing, since she's already lost it. Jazz and rock drummers often have a tendency toward exaggerated, theatrically excessive movements that go beyond the technical. In this way they underscore the sound, just as, conversely, gestures and facial expressions and body movements and postures also underscore and explain what is being said verbally. After winning the air drumming contest, he is asked whether he has been playing the drums properly. He says he has no need for them because he is a drum set himself (Film, 1.23').[77] As far as 'works' are concerned and not simply the simulation of something, the relationship between the intended work and the imaginary-real played work falls within the scope of Fuzzy Logic. The intention is clear, since the individual pieces are recorded on cassette tape, and the air guitarist or air percussionist *only* imitates or simulates the solo instrumental part with hand and body movements without a real instrument.[78] At first glance, this subject seems peripheral, the analysis pointless. It is not. It is actually one of the ways to explore how body movements and—instrumental—music as their sonic manifestation are linked (see Sect. 3.8 Translation of music into sign language, above).

7.3.4 *Text Form Instead of a Notation Form: Concept Art as Mere Verbalization of Ideas Without Realization*

As is explained above, conceptual art attempts to objectify the imagination form of a work that does not exist in a real and sensory way, but only as the verbally formulated concept (see Sects. 3.8 and 6.3). By contrast, and unlike in Musical Graphics, the merely verbal conceptual art is considerably poorer in content and

---

[75]Moorey 2012, p. 173.

[76]Details see https://en.wikipedia.org/wiki/Adventures_of_Power, 19 March 2017, and https://www.rottentomatoes.com/m/adventures_of_power, retrieved 5 October 2017.

[77]See https://www.youtube.com/watch?v=h3TFC4yjquY.

[78]The film features original songs by Ethan Gold and hits by Rush. 40 titles are listed in https://en.wikipedia.org/wiki/Adventures_of_Power, 19 March 2017, retrieved 5 October 2017.

form. It is not a picture form and not a word form, not an ecphrasis or description of the work. It replaces the work. Musical Graphics provides the interpreter with suggestions for the realization of the work. Verbal conceptual art primarily makes prescriptions, preferably to the audience. The realization can actually be dropped right away (see *Soggetto cavato* in the ironic homage *Zukunfts-Brahmanen-Polka* in Sect. 6.1). A logical step toward an "absolute" concrete total artwork is represented by words for perception without any music or sound, musical "concept art," which can also be understood as an absolutisation of commentary and "inscription" in the sense of Walter Benjamin.[79] Dieter Schnebel sets an especially expressive and impressive example in 1969 with his music that is bilingual and that can be started from both sides, i.e. despite the monophonic title, the stereophonic *Musik zum Lesen/Music to Read, MO-NO*, sound in book form. At the same time, music as talking about or describing music and music-like things is a potentiation of the neoclassical concept of "music about music"; it is meta-medial music. Schnebel's gently solipsistic, monochromatic and monotonous neo-Dadaistic opus leaves a lot of emptiness, usually displays few notes, pictures, letters as ciphers and a lot of blank surface or even completely empty pages—a keyboard on which Laurence Sterne and later Mallarmé had already played with virtuosity. With the emptiness and with the lesson, Schnebel is, of course, alluding primarily to Cage, both to his collection of writings *Silence* (which is also expressly mentioned in *Mo-No*) and to *Tacet 4:33*. The double Da, which naturally alludes to one of the essential precursor currents, Dada, can serve as one of the many subtitles. The following section begins on sheet 6 verso, one page each marked with /. The audience itself has to work, even purely mentally reading. There is no talk of sensory-sound realization, no thought of it (Fig. 7.15).

Following two completely blank pages comes:

> / [Blank page] / [Blank page] / Listen! / [blank page] /[80]

Here, too, there are the possibilities for iteration and permutation, i.e. repetition, rearrangement, variation, almost a motivic-thematic work, instead of 15 tones or graphic lines etc. Russolo's sound generators also make an appearance:

> / [blank page] / What this silence sounds like / [blank page] / what lives there, what breathes, crackles, creaks, hums, buzzes, rushes, echoes, reverberates – vibrates, oscillates – sounds /[81]

As far as this hearing is concerned, these are merely instructions. What is somehow heard as an object from any given "soundscape" in the environment of the subjects determines the indeterminacy of chance. Concrete sensory experience, music, musical logic, Fuzzy Logic must once again surrender here.

In contrast to the conceptual form as a stage in the work process, conceptual art is a deliberate act of getting stuck in the approach between escape from the real production of art and anti-art. From the point of view of art, it is a blurring, since nothing

---

[79] In general cf. e.g. Straebel, Volker: *Aspects of Conceptual Composing*, The Open Space Magazine, issue 10, Fall 2008, pp. 69–78.

[80] Schnebel 1969, pp. 6–8 (Book itself without pagination).

[81] Ibid.

Fig. 7.15 Dieter Schnebel 1969: *MO-NO. Musik zum Lesen. DuMont-Verlag, Köln (1969), without pagination* A tone as a note

more emerges from the idea, the thought, the notion than its mere written record, but not an elaboration into a work. (See *Imagination Form* in Sect. 3.8; see *Written Form instead of Notation Form: Conceptual Art as the Mere Verbalization of Ideas Without Realization* in Sect. 7.2.) In "conceptual art", the concept takes on a life of its own without progressing to art as a concrete, shaped object. The form of existence is suspended and isolated from the whole of the work process and is intended to already be the work. The objectification, the condition of the possibility of blurring, is missing. Thus there is not even a 'suspension' in the dialectical sense. Instead, only stages of the music process and the forms of existence that follow the imagination form are deleted or not carried out. In this way, the dialectical contradiction between idea and realization can be denied and repressed, but not eliminated.

7.4 Point of Departure: Anticipation Form. Instrumentation, Ornamentation, Indeterminations

Notation is usually used mainly for sharpening. But it can also, almost paradoxically, be used for the reverse, for blurring.

7.4.1 Notation as Blurring: The Main Theme of the IVth Movement in Tchaikovky's 5th Symphony

There is a curious passage in Tchaikovsky's Sixth Symphony. In the concluding movement IV, *Finale: Adagio lamentoso*, Tchaikovsky notates the main theme in such a blurred way that it is almost unreadable, as (Fig. 7.16)

Fig. 7.16 Tchaikovsky: 6th Symphony "Pathétique", *Adagio lamentoso*, Score, excerpt mm. 1–5 https://en.wikipedia.org/wiki/Symphony_No._6_(Tchaikovsky), 29.4.20. Retrieved 20 June 2020

> a "composite melody"; neither the first nor second violins actually play the theme that is heard.[82]

Only in the sound form does the actual melodic line emerge clearly. It is simply the contour, the sequence of the highest notes in each case, within the two-part movement. They are notated in the score in a kind of zigzag pattern (Fig. 7.17).

Tchaikovsky thus blurs how the melody of the theme is represented within the notation form. At the same time, though, he anticipates a future sharpening, namely that its realization as a sound form will correct this distortion, just as a convex mirror does in anamorphosis, for example.

Such dialectical contradictions can be observed again and again in the following cases.

Fig. 7.17 Tchaikovsky—*Pathétique*—*Adagio lamentoso* Melody of the theme (Stefan Kompacher, 6 September 2007, https://upload.wikimedia.org/wikipedia/de/1/1f/Pathetique5.jpg, retrieved 20 June 2020)

[82] https://en.wikipedia.org/wiki/Symphony_No._6_(Tchaikovsky), 29 April 20, retrieved 20 June 2020.

7.4.2 *Pitch. Embellishment as Blurring*

The following section deals with the execution of music in a compositionally intentional and partially notated way, which enlivens the as yet abstract tone. A wide range of methods can be employed for this purpose. As blurrings, they concretize the tone as a sort of halo, on the one hand by extending its pitch spectrum and on the other hand primarily its timbre, for which a rich palette is available. The dimension of loudness plays a minor role; the dimension of duration almost no role.

7.4.2.1 Ornamentation

Ornamentations or 'embellishments' mainly concern the dimension 'pitch' (see Sect. 6.6.1.4).

> Ornamentations are the *arbitrary* embellishments and modifications to a note or a texture, which are not specified by the composer himself, but rather are undertaken and employed by the performing singer or instrumentalist *in accordance with his own discretion and taste.*[83]

This definition is one-sided: It ignores the fact that ornamentation is absolutely "prescribed" in the score form, just not fully *written out*. Again, we have the 'abstract texture' as a point of reference here. If this is considered essential, then any additions appear to be merely accidental, tending toward superfluous or possibly even detrimental.[84] They are not accidental, though. Ornamentations are partly a component of the completed texture and partly a matter of execution. It is this that ultimately brings the ornamentations to tonal reality and to actual completeness. In principle, they have at least connotative semantics. They reinforce works of expression. They sometimes even exhibit denotative semantics. This is demonstrated by the metamorphosis into the BACH motif in Gustav Mahler's frequent use of the turn motif. With the alteration of tone pitches, the four-tone BACH motif can be produced.

> The auxiliary notes can be chromatically changed by accidentals. An accidental above the turn sign stands therefore for the upper auxiliary note, while below the turn sign, it logically stands for the lower auxiliary note. Both auxiliary notes may be changed at the same time, also mixed with # and *b*. Besides # and *b* a natural sign is also possible.[85]

In the finale of Mahler's 9th Symphony, 4th Movement Adagio, the turn-ornament motif is an integral part of the theme, even in close, semitone proximity to the "original" pitches—i.e.:

Bbb = A - Ab - Cb = B
A Ab B (Bb is missing).

---

[83] Riewe Handwörterbuch 1879, 278, quoted in https://musikwissenschaften.de/lexikon/v/verzierungen/#riemann1882. Emphasis HWH.

[84] Adolf Loos even characterized "ornament as a crime" in reference to architecture in the 1920s.

[85] Riemann 1901, p. 192f.

At the end, the motif returns in multiple variations, transfiguring and transfigured, shortly before the end in connection with a self-quotation from Mahler's *Kindertotenlieder (Songs on the Death of Children)* with the original text "They have only preceded us … on those heights", and very stretched in the durations, e.g. m. 154ff., thus transposing overlapping permutations of BACH.

G Gb / A G# (G) / B Bb A G# (G)

G Gb A G# or B Bb A G#

ABCH or CHBA

Once again we see the configuration 'material becomes expression' (Fig. 7.18).

Even in music that isn't primarily notated, the ornamentations are not just arbitrary, random additions.

Dreydlekh (plural of *dreidel*) or "spins" are musical ornaments of klezmer music, particularly violin, used to produce its characteristic "tear in the voice" sound. The main *dreydlekh* are

(a)

(b)

Fig. 7.18 **a** Mahler: Turn motif
9th Symphony, 4th movement, excerpt from piano score, mm. 1–2.
https://imslp.org/wiki/Symphony_No.9_(Mahler%2C_Gustav)
b Mahler: Turn motif
9th Symphony, 4th movement, excerpt from piano score, mm. 153–157

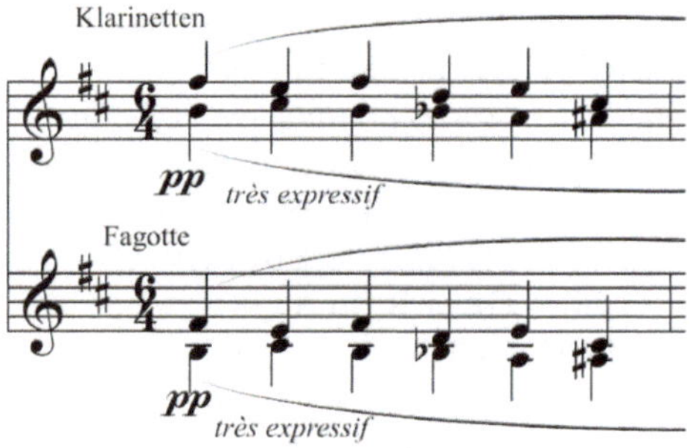

Fig. 7.19 C. Debussy: *Nuages* (Trois Nocturnes, No. I#) https://www.jochenscheytt.de/debussy/debussywerke/nocturnes.html

Fig. 7.20 Claude Debussy: *Jardins sous la pluie* (*Estampes*, No. 3) https://www.jochenscheytt.de/debussy/debussywerke/jardins.html

> *glitshn* (glissandos), *krekhtsn* ("sobs," "wails"), *kneytshn* ("wrinkles"), *tshoks* (a kind of bent notes of cackle-like sound), and flageolets (harmonics).[86]

Apart from that, transitions from the approach to pitch to the approach to timbre can also be seen here (see below 7.4.3). In some cases, the insistence on the abstract texture as the actual texture runs into difficulties. In the following theme from Debussy's *Nuages,* there is no simple, single-line melody that could be embellished. Instead, the doubled melody of Debussy's two-part contrapuntal texture—which deliberately includes three instances of irregular hidden consecutive fifths—actually only forms a melody *together*. And this, in turn, is essentially co-constituted by instrumentation[87] (Fig. 7.19).

The paradigm of the abstract texture can be extrapolated to a paradoxical inversion without much provocation. Debussy quotes a children's song, thereby integrating it into the material. From this perspective, his more complex, rich structuring of it as a new work represents a blurring of the small, clear melody (Fig. 7.20).

---

[86] https://en.wikipedia.org/wiki/Dreydlekh, 26 August 2019, retrieved 30 August 2020.

[87] And likewise the graphic quality: "As a result of the unusual instrumentation, the even and slow quarter note progressions and the hollow fifths, an image is evoked in the mind's eye of pale, gray mountains of clouds sluggishly sliding across the sky." (https://www.jochenscheytt.de/debussy/debussywerke/nocturnes.html).

> *Jardins sous la pluie:* The third and final piece from the *Estampes* is entitled "Gardens in the Rain" and, in the plasticity of its structure, evokes images of pattering rain and children playing together, jumping through puddles and having fun with the water. […] Musically, the rain is represented as continuous motion of sixteenth-note and eighth-note triplets, and the use of melodic excerpts from two French children's songs represents the children frolicking in the rain. […] One of the children's songs is "*Prom'nons nous dans la bois*", also known as "*Do, do, l'enfant do*". Debussy also uses this song in "Jimbo's Lullaby" from Children's Corner.[88]

> Excerpt from the second measure from the half note to the second quarter note in the fourth measure of the children's song. This was placed in the bass clef for better comparability. […] The children's song is in a major key, Debussy's variant in a minor key (G rather than G sharp).[89]

Where the ornamentations are concerned, the composers' point of view stands in contrast to that of the performing musicians. The former place an emphasis on the expression of the idea, the latter emphasize independent or even autonomous tonal interpretation. The moment the tipping point is reached, when ornamentations really begin to proliferate and become arbitrary relative to the structure of the representation and content of that which is represented, depends upon the point of view and must be determined in each individual case. Fuzzy Logic could analyse dimensions and quantities more precisely here. Such a point is probably reached in the following case:

> In Froberger's time [1616-1697], this partita spirit was so abused that not only did at least half a dozen variations have to be performed on special little arias or ariettas, e.g. on a so-called little *Lantürlü* song[90]; but also the allemandes, courants, etc. were so infected with it and did not escape without fractures, crooked leaps and many-tailed notes.[91]

In general, ornamentations have a wide range of types and methods. Most of the time, if not always, they have an expressive function and connotation.

7.4.2.2 Coloratura. Expression and Exhibition of Virtuosity

In coloratura, we are again dealing terminologically with 'color', which, as is mentioned above, is a metaphor with numerous meanings and functions in music. Once again, a transition from pitch to timbre is thereby revealed, and in this case a transition from the sharper to the blurred at the same time.

> Coloratura is an elaborate melody with runs, trills, wide leaps, or similar virtuoso-like material, or a passage of such music. Operatic roles in which such music plays a prominent

---

[88] Performance https://www.jochenscheytt.de/debussy/debussywerke/jardins.html.

[89] Ibid.

[90] See e.g. Rudolf Flotzinger: Johann Heinrich Schmelzer's Sonata "Lanterly", in: Studien zur Musikwissenschaft, 26th Vol. (1964), pp. 67–78, DOI: 10.2307/41465271, https://www.jstor.org/stable/41465271.

[91] Johann Mattheson: Der vollkommene Capellmeister, Hamburg 1730, Part II, Chapter 13, § 131, quoted in von Fischer 1968, col. 1283f.

Fig. 7.21 Thomas, Ambroise 1868. *Hamlet*, piano-vocal score, piano reduction by Vauthrot. Paris: Heugel & Cie, n.d. [1868]
Musical score. The final cadenza from the Valse section of Ophélie's "Mad Scene" in Act 4 (piano-vocal score, p. 292).
1868 *Arranger*: Eugène-François Vauthrot (1825–1871)

> part, and singers of these roles, are also called coloratura. Its instrumental equivalent is ornamentation.[92]

Like ornamentation in general, coloratura lies between the poles of expression and the autonomous pleasure of virtuoso vocal musicianship. It is

> determined by the contrast between an ornamentation that serves only virtuosity and one that also serves drama. […] In Monteverdi's 'Orfeo,' a scene is notated both without ornamentation and with coloration. The development ranges from Stefano Landi's 'Morte d'Orfeo' (1619), which includes a song *with notated coloratura,* to Cavalli, who defined coloratura as a *necessary component* of the aria in Italian opera music. Coloratura thus remained a partially notated, partially improvised integral part of the opera aria for over 150 years. […] Around 1815, Rossini began to notate the coloratura in order to counteract the sometimes excessive practice of the singers.[93]

Coloraturas were still being expressly inserted and notated even in 1868 (Fig. 7.21).

> An example of a coloratura passage from a soprano role. It includes a more difficult variant (top stave) with a leap to a high D (D6). Final cadenza from the Valse in Ophélie's Mad Scene (Act IV) from the opera *Hamlet* (1868) by Ambroise Thomas (piano-vocal score, p. 292).[94]

The instrumental ornamentations are more often attached to the single note. This also happens in vocal coloratura. It creates wider phrases and melodies, though. The

---

[92] https://en.wikipedia.org/wiki/Coloratura, 15 April 2020, retrieved 29 August 2020.

[93] http://www.musikerchat.de/cgi-bin/ultimus/fachbegriffe/musik.pl?id=1451.

[94] https://en.wikipedia.org/wiki/Coloratura, ibid.

ornamentation is built into the figuration. It thus proves itself, even more clearly than instrumental ornamentation, to be a special case of variation and especially of figuration, the dissolution of the theme into smaller note values (see Chapter 8). As a fragmentation or even destruction of the text, coloratura is also a blurring—and thus not a sharpening, not even on a meta level. Vocalese, which uses meaningless syllables as a carrier material, forgoes a coherent text from the outset. In scat singing in jazz, nothing else is possible: coming up with new sounds *and* lyrics at the same time would be too great a task.

Taking it a step further, even the chromatic scale instead of the more elementary diatonic[95] can seem like 'ornamentation' and simultaneously like a blurring of the 'simple' melodic contour, in feedback with the harmonics. Thus in 1901 Hugo Riemann criticized the chromatics in the work of Louis Spohr, who in his opinion provided the decisive impetus for development as far as *Tristan*. Spohr, a composer and violin virtuoso, came from the instrumental sphere. Chromatic semitones as finer transitions are literally at their fingertip on the fingerboard. Although it was intended as a reinforcement of expression, chromaticism, viewed in way, runs the risk of "softening the expression and distorting it into the weepy, plaintive," contrary to the intention.[96] The anti-classical emotive, physical, more naturalistic element appears as a blurring of the artistically clear, especially in its extremely artistic stylization. Using the metaphor "*Tuschmanier*", i.e. frayed instead of clear lines, Riemann builds a bridge to instrumentation, which, in the classical spirit, is generally suspicious of blurring.

> Chromaticism in melodic composition […] is *inwardly related to sound coloration* in that both distract attention from the formative factors of musical composition and bring greater legitimacy to the *elementary active principles*; for chromaticism replaces the striding of the melody with gliding and approaches […] the gradual change in pitch of the constant.[97]

Chromaticism is thus a naturalistic artistic device; it smooths corners and edges while also blurring all sharp outlines of the drawing, rendering them blurry and deliquescent. The fear of the elemental, the sensual, the feminine obviously comes through here.

7.4.2.3 Glissando. Blurred Continuum Instead of Filtered and Articulated Pitch Movement

Like ornamentations, glissando also belongs primarily in the dimension of pitch. It possesses distinct timbral aspects, however. The keyword mentioned above "gliding" in Riemann's 1901 work refers to the proximity of melodic chromaticism to

---

[95]The diatonic scale is older and simpler. Here, this means the use of tones from the scale itself, such as the notes D, E, F sharp, G, A and C sharp in the key of D major, and the notes D, E, F, G, A, B, and C sharp in the "Dorian" mode. The chromatic scale, literally "dyeing (technique)", is the use of tones "foreign to the scale", such as G sharp or A sharp, etc., in D major.

[96]Riemann 1901, p. 192f.

[97]Ibid.

the portamento, i.e. to the micro-glissando. This intentionally transforms discrete, fixed pitches into a continuum. If the blurring-type cluster proceeds from the chord (see Sect. 7.2), i.e. from the vertical, then the glissando starts from the melody, the horizontal. Both tonal space dimensions can be amalgamated by the imaginary-real movement of the texture and mutually transformed into each other. Moreover, this occurs in the new, primarily electroacoustic continuum of noise and tone. In the 1960s, Pierre Boulez pointed out the connection to the mediation of tone and sound, pitch and color, and 1963/1964 the relatedness of vertical tonal complexes, i.e. chords, to the diagonal glissando, which also blurs the distinct and clearly different pitches for the benefit of a 'segmented continuum'—a seemingly realistic paradox—instead of a gliding, fluid continuum as continua normally are imagined. This can also be accomplished pre-electroacoustically, though, using the siren. It is an important musical instrument in Edgard Varèse's (1883–1965) pioneering work *Ionisation* for 41 percussion instruments and two sirens (1931). With him and with the sirens, the glissando takes on a disconcerting quality.

> Their sound thus contained something unheard of, something wild within the musical context, the limits of the tempered tonal system to that which had previously been excluded from it, which might be more reminiscent of their Homeric ancestors than of the Pythagorean soundscapes favoured by von Helmholtz. At least for Varèse himself, the Homeric sirens appear to have been present. In connection with his first siren experiments, which were inspired by Helmholtz, he describes the quality of the glissando: "The beautiful parabolas and hyperbolas of sound the sirens gave me, and the haunting quality of the tones, made me aware for the first time of the wealth of music outside the narrow limits imposed by keyboard instruments" (quoted from Kahn 1999, 86)[98]

This blurring is likewise common in traditional musical instruments. The harp glissando, for instance, which is often quite ample, is a conventional effect, as is the predominantly small-scale glissando on string or plucked instruments in the form of *portamento* or bending, especially on the guitar. There's no lack of imitation in the field of electroacoustics.

> In electronic music, a pitch wheel, pitch bend or bender is a control on a synthesizer to vary the pitch in a continuously variable manner (portamento). The first synthesizer with a pitch wheel was the Minimoog, in 1970. […] Alternatively, pitch bend controllers on synthesizers may be implemented as a joystick, knob, or touch-sensitive ribbon. MIDI represents pitch bend as a 14-bit integer, allowing for 16,384 possible values. General MIDI implementations default to a range of ± 2 semitones.[99]

The clarinet solo at the beginning of George Gershwin's *Rhapsody in Blue* is a wild glissando.

> When Gershwin conceived the ascending clarinet portamento, he knew precisely the effect for which he was reaching: a hyperthyroid, hysterical wail, almost the voice for a hyperthyroid, hysterical era.[100]

---

[98] Welsh 2018, p. 215.

[99] See *"General MIDI System Level 1". The Complete MIDI 1.0 Detailed Specification (document version 96.1, third ed.). The MIDI Manufacturers Association. 2014. In* https://en.wikipedia.org/wiki/Pitch_wheel, 5 May 2020, retrieved 10 May 2020.

[100] Ewen 1991, p. 297.

Fig. 7.22 George Gershwin
Rhapsody in Blue
First edition. New World Music Corp., New York (1924). IMSLP10970-George_Gershwin_-_Rhapsody_In_Blue_(Piano_&_Orchestra)

So this opening fanfare is a blurring, both technically in terms of the material, but artistically it is an intense symbolic expression and thus a sharpening of the idea of the work—it was meant to symbolize the "Jazz Age," and jazz was to become appropriate for concert halls (Fig. 7.22).

> He explained to [the] clarinettist precisely how he wanted that passage to sound. At first Gorman insisted that no clarinettist could produce the effect Gershwin had in mind. The composer, however, was so intransigent that Gorman had to keep on experimenting with various reeds and techniques until, at last, he brought to life Gershwin's music exactly as Gershwin had heard it in his inner ear.[101]

With regard to the instrumental technique of the clarinet glissando:

> Clarinettists combine non-standard fingerings with particular vocal tract configurations to achieve pitch bending […]. Impedance spectra were measured in the mouth of expert clarinettists while they played normally and during pitch bending, using a measurement head incorporated within a functioning clarinet mouthpiece. These were compared with the input impedance spectra of the clarinet for the fingerings used. Partially uncovering a tone hole by sliding a finger raises the frequency of clarinet impedance peaks, thereby allowing smooth increases in sounding pitch over some of the range. To bend notes in the second register and higher, however, clarinettists produce *vocal tract resonances* whose impedance maxima have magnitudes comparable with those of the bore resonance, which then may influence or determine the sounding frequency. It is much easier to bend notes down than up because of the phase relations of the bore and tract resonances, and the compliance of the reed. Expert clarinettists performed the glissando opening of Gershwin's '*Rhapsody in Blue*'. Here, players coordinate the two effects: They slide their fingers gradually over open tone holes, while simultaneously adjusting a strong vocal tract resonance to the desired pitch.[102]

Special effects, a polyphonic glissando, likewise without special musical instruments or electroacoustic means, was achieved by I. Xenakis in *Metastaseis* (1955). For him, this was what it was a matter of,

> pushing melodic-polyphonic continuity to its limits in the form of *superimposed glissandi* in such a way that a mass phenomenon with novel qualities emerged from the meshwork of lines. […] A particular shock effect must have sprung forth from the beginning and the end

[101] Ibid., p. 297f.

[102] See Chen et al. 2009.

> of the work, sections in which a sound surface that could not be heard very clearly grew out of a unison tone through diverging glissandi (with a soloistic division of the strings).[103]

An iconic and mimetic quality is taken on by the glissando in Mathias Spahlinger's *Nude, Descending a Staircase* for Bass Clarinet, Trombone and Orchestra (1997; Donaueschingen Festival 1998), inspired by Duchamp's painting *Nu descendant un escalier no. 2*, 1912. Duchamp sets the picture in motion, quasi-cinematically, but the individual images or frames follow successively, one after the other or side by side, as in the photo series by Muybridge. The movement is deliberately blurred. Spahlinger recreates this as a blurring both in the vertical, through a wider band of sound instead of a monophonic melody, and in the horizontal, by composing an—apparent—continuum rather than a sequence of clearly detached segments. Spahlinger:

> the most diverse methods, as long as they are not analogue and continuous, circular like the wheel, the phonograph, the record, they all have one characteristic in common with film and digital sound recording, which is of interest here: the dismantling of movement into a grid of stationary points in time, which, as narrow as it may be, is never the movement itself, but rather, outwits the senses, simulates them. […] if the duration, the span of time of a movement in space, is to be measured, then the time and point in space must be thought without extension. […] therefore it can be said of an object that it is in motion if it and is not is at a particular point in space at a particular point in time; for this reason, hegel calls motion the "immediate existence of contradiction". […] (specifically musical): pitches in continuous movement, i.e. glissandi, and their representation in steps is the actual theme of the piece. […] the simultaneity […] of the moved and the defined pitch. a lying tone, however brief it is, has an extension in time. confronted with a glissando that slides through it, it is either not yet or no longer identical to the pitch of the glissando.[104]

7.4.3 Tone Color. Instrumentation as Blurring—Color Instead of Drawing, Brush Stroke Instead of Line

The tonal realization of the melodic line changes drastically if it is collective instead of individual chorical instead of soloistic. For untrained singers and players, the line changes almost automatically into a brushstroke with more or less broad edges—an unintentional "aquatint style", as H. Riemann had criticized regarding Spohr's chromaticism. This condition is found in amateur music-making as well as in traditional music and folklore. However, such deviations from the pure tone also exist among professional musicians; they must be all the more careful that the 'sound scape' of deviations from the intended pitch does not become too wide.

[103] See Decroupet 2005b, p. 130.

[104] http://www.swr.de/swr2/service/audio-on-demand/-/id=661264/1cx2cby/index.html.

7.4.3.1 Fuzziness in the Pitch as 'Timbral' and 'Opaque' Polyphony. Heterophony in Traditional Music

The choral execution of melodies, whether vocal or instrumental, objectively and without subjective intervention by the performers, makes the respective individual voices of the composition more voluminous, not only in terms of volume, but also in terms of pitch and tone duration. The primordial collateral octavation of the melody, which is caused by the difference in register between male and female voices, including children's voices, adds further fuzziness. Tiny individual deviations from the imagined, meant and written tone, which are hardly noticeable in detail, broaden the tonal line and transform it into a sound scape—real heterophony instead of monophony.

> heterophony is a type of texture characterized by the simultaneous variation of a single melodic line. Such a texture can be regarded as a kind of complex monophony in which there is only one basic melody, but realized at the same time in multiple voices, each of which plays the melody differently, either in a different rhythm or tempo, or with various embellishments and elaborations.[105]

This objective fuzziness, which is once again inevitable from the point of view of the result, is actually a blurring from the perspective of the compositional intention. For it is aesthetically intended: the sound becomes richer, and in a certain respect even more vivid. This is the configuration 'color instead of drawing'. The coloring is, however, initially largely *monochrome.* Instrument groups are already somewhat more colorful than singing voices, and vocal-instrumental combinations are correspondingly even more colorful. What is *meant*, at least in the beginning, is a melody, the unity of pitch and tone duration sequence, as a line. This is, strictly speaking, blurred. However, the blurring has a certain beauty and liveliness. Bluriness as fuzziness, is here genuine, not a secondary ingredient, but nevertheless at least partially intended and not just a random result. Each individual creates something of his own, which is somewhat different, and yet at the same time fits into the common identity. According to the music-specific harmony as a unity of differences, I and We are united.

Monophony is not an exact melodic line under the elementary conditions given here, as the European musical thinking has it in mind as a guiding idea and corresponding conception, and is additionally reinforced by the notation. In archaic and folkloristic musical practice, it is rather a matter of real heterophony: the simultaneous sound of several slightly different voices. It is, as already mentioned, the initially common thing when several people sing or play together. Thus, on the one hand, it is a prerequisite for the development of a conscious polyphony, but on the other hand it is also an imperfect, fuzzy result of the desired precise monophony. For the voices are different not so much because of an intention, let alone a compositional one, but mainly because of the more or less great differences in the meeting of the melodic *line* meant.

---

[105] https://en.wikipedia.org/wiki/Heterophony, 16 August 2020, retrieved 18 August 2020.

Heterophony in traditional music is a 'timbral' and 'opaque' polyphony. The term 'variant heterophony', which is only reluctantly used today, emphasizes the diversity in unity. The differences, the variants are not actually blurred, but fuzzy. They are *not* composed, but rather realized in a rather coincidentally resulting manner. We could also call it a natural polyphony. The prerequisite is collective, not just individual singing and playing. Here, too, very complex patterns and layering can arise, partly spontaneously resulting, partly planned, as in the following example.

> Thai music is nonharmonic, melodic, or linear, and as is the case with all music of this genre, its fundamental organization is horizontal [...] Thai music in its horizontal complex is made up of a main melody played simultaneously with variants of it which progress in relatively slower and faster rhythmic units. [...] Individual lines of melody and variants sound in unison or octaves only at specific structural points, and the simultaneity of different pitches does not follow the Western system of organized chord progressions. Between the structural points where the pitches coincide (unison or octaves) each individual line follows the style idiomatic for the instrument playing it. The vertical complex at any given intermediary point follows no set progression; the linear adherence to style regulates. Thus several pitches that often create a *highly complex simultaneous structure* may occur at any point between the structural pitches. The music 'breathes' by contracting to one pitch, then expanding to a wide variety of pitches, then contracting again to another structural pitch, and so on throughout. Though these complexes of pitches between structural points may strike the Western listener as arbitrary and inconsequential, the individual lines are highly consequential and logical linearly. The pattern of pitches occurring at these structural points is the basis of the modal aspect of Thai music.[106]

However, even elementary musical practice *filters* out a single voice from heterophony, as an 'idea' and concept within the various deviations. This filtering is also a sharpening. It is the prerequisite for the next stage, namely, *several* distinct, distinguishable and yet related individual voices layered on top of each other.[107]

This is already the starting point for the transition to multi-phony resp. polyphony.

> "The technique of combining simultaneously one main melody and its variants is often incorrectly described as heterophony: polyphonic stratification seems a more precise description, since each of the 'layers' is not just a close approximation of the main melody, but also has distinct characteristics and a style of its own"[108]

In 1963, O. Elschek defined heterophony as one of the three basic types of multi-phony alongside with 'polyphony' and 'homophony'.[109] 'Polyphony'as special manifestation of polyphony means the ,vertical'simultaneity of differentiated, relatively independent melodic or even only rhythmic lines, in contrast to ,homophony'with a mostly harmonically regulated polyphony and with a hierarchical subordination the lines by the guiding line. One step further, relatively conscious differentiation is made within heterophony, and *intervallic distances* between the

---

[106]Morton, David (1976). The Traditional Music of Thailand, University of California Press, quoted in https://en.wikipedia.org/wiki/Heterophony, 16 August 2020, retrieved 18 August 2020.

[107]In detail see Sect. 7.3.1.

[108]Morton, David (1976). *The Traditional Music of Thailand*, p. 21, quoted in https://en.wikipedia.org/wiki/Heterophony, 16 August 2020, retrieved 18 August 2020.

[109]Table in Elscheková et al. 1996, col. 1787.

Fig. 7.23 W.A. Mozart: Piano Concerto c-minor KV 491, first movement, piano and violins m. 211f. (https://upload.wikimedia.org/wikipedia/commons/5/5e/Mozart_K491_first_movment%2C_bars_211-14.png)

voices emerge—a transition to evolved polyphony. The prerequisite for this is that a single voice is filtered out of the heterophonic scape or network, not only ideally and intentionally, but also as a sharpening of the real sound and becomes independent. This was, for example, the European way, exemplary in the (Gregorian) chorale. This historical layer is also preserved as strict monophony. If the monophonic line is supplemented by others, it is a blurring. But from the future point of view, so to speak, of the development of polyphony, this blurring is the starting point of a new, higher order: more complex than the monophonic line, and a more precise structure instead of the relatively disordered heterophonic network of lines. Blurring *becomes* sharpening (see Sect. 7.3.1). After all, heterophony is also found in composed music. Thus a melody can sound both figured *and* unfigured. Thus, the original spontaneous heterophony is raised to a new level of conscious shaping. The aspect of fuzziness is preserved (Fig. 7.23).

Typically, the solo piano has more freedom for deviations from the "naked" melody, i.e. for clothing ornamentation, than the choral strings. When Mozart played himself, he liked to improvise additional figurations.

Especially for the East and Southeast Asian heterophony, I propose the concept of a 'timbral polyphony' as an explanation.[110] It brings a different, but aesthetically just as effective variance into the music as it creates metrical-rhythmic or melodic polyphony. This is because the pentatonic system, which is essentially used in East Asian music, i.e. a system of five tonal steps (instead of the European 7- to 12-step system or the Southeast Asian system of—differently structured—7 or 5 steps), offers relatively little possibility for variation and development in the pitch dimension.[111] For European ears it is difficult to hear the sequence of colors as a consistent, consequent and somehow logical musical process. In East Asian music, especially Chinese music, on the other hand, such a priority of timbre is much closer, especially in traditional music. The 'sound scape composition' of the 1960s thus converges on the sensual surface, although it has other mental prerequisites.

---

[110] I am exceeding my competences, since I am neither a music "ethnologist" nor a specialist for East Asian music. The concept seems plausible from my Yun research (see Heister 1987, Heister 1988a and 1988b). But it would need a close examination.

[111] In Chinese music there are also 12-step scales. However, they hardly serve musical practice but rather universal concepts of order and harmony.

7.4.3.2 Isang Yun's "Main Sound Technique" and Debussy's *Pagodes*

The "main tone technique" (also called "main sound technique") of the Korean composer Isang Yuns (1917–1995) ties in with traditional Korean sound quality, on the one hand almost naturally, but on the other hand quite consciously. Yun was concerned with his own musical-cultural tradition not in the perspective of exoticism but as an equal, mutual intercultural appropriation. In his work, a geographical-cultural reversal of the relation between the foreign and the very own is created, since he created his works as exile works in Europe from 1956 on, first as student, than as exile works. Yun was subjected to multiple persecution, first by the Japanese occupation in his own country and then in exile in the Federal Republic of Germany under the South Korean dictatorship. As one of the early and important representatives of a generation of composers who rose to international prominence after the Second World War, he stands for a caesura in music history insofar as composers from non-European and non-Anglo-American regions are now increasingly and on an international scale, for their part and from their perspective, appropriating achievements of European music and processing them into syntheses that overcome the earlier Euro-American exoticism. Yun teaches European performance apparatuses, techniques, genres with East Asian, especially Korean materials, ways of thinking and attitudes. "The spiritual background and creative starting point […] is the classical teaching of the Tao."[112] The striving for synthesis is often already perceptible in the choice of text and genre, for example in *Der weise Mann*, Cantata for baritone, mixed choir and chamber orchestra (texts by the preacher Solomon and Lao Tse, 1977). The main tone technique includes melodic micro-figurations in the design of the main tone sequence, which are classificatory 'ornaments', as well as tone color variations of the tone. Technically it is first of all a blurring. As with the broad brushstrokes of a bamboo or bristle feather—which can also produce different stroke thicknesses, even extraordinarily fine lines—the melodic line is inked as a network of lines.

> Through a softened concept of twelve-tone composing with elements of thematic work – out of a need for expression – he found his way to "sound composition", the organization of athematic but interiorly individualized soundscapes, among other things through the East Asian tradition of variant heterophony, the polyphonic and simultaneous playing around a model that is identical in its core. The success of his "sound compositions" soon after the appearance of Ligeti and Penderecki in the early 1960s should not belie a fundamental difference. The sensually luminous, physical sound of Yun's works owes itself to the concept of "sound" that he had brought with him: If in the music of Europe it is only the sequence of tones that gains life, whereby the individual tone can be relatively abstract, then the tone itself already lives for us. One can compare our tones with brush strokes as opposed to the line of a drawing pencil. From the beginning to the fading away, every tone is subject to *changes* [emphasis W.-W.S.]; it is equipped with decorations, suggestions, beats, glissandi and dynamic changes, above all the natural vibration of every tone is consciously used as a means of design. The changes of a tone are not so much seen as melody-forming intervals,

---

112 Kunz 1987, p. 95.

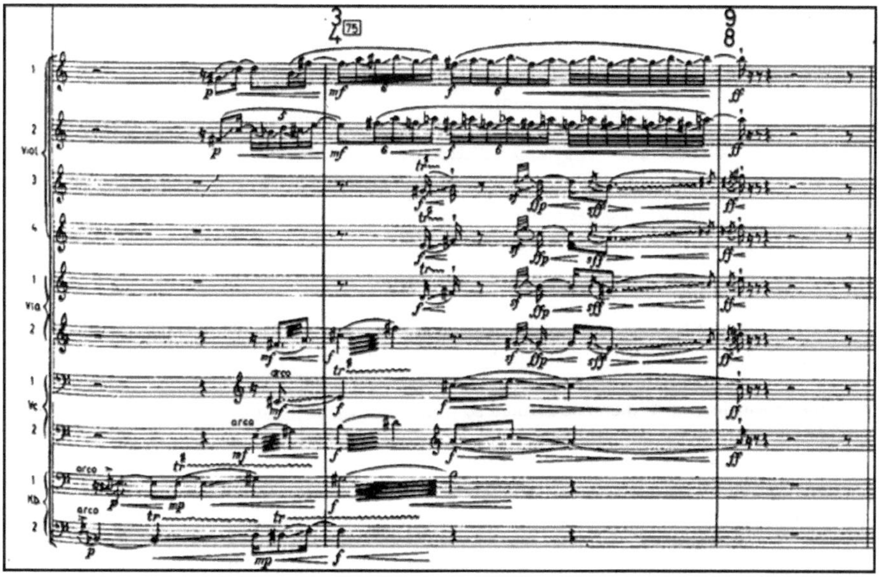

Fig. 7.24 Isang Yun: R*éak* for large orchestra (1966), m. 74–76
Quoted after Yun 2000, sine pag., Fig. 4

but rather in their ornamental function and as parts of the expressive register of one and the same tone" (Yun 1965; in Sparrer 1992).[113]

The traditional European relationship between ornamentation and reference tone, between accidental and substance is suspended, almost reversed into its opposite.

"The changes in pitch of a tone are not so much seen as melodic intervals, but rather in their ornamental function and as parts of the expressive register of one and the same tone" (Yun 1965, 69). The ornament is substance and the "single tone" by itself is already a – ceremonial – progression. It includes the "articulation" of the transient response, a persistent or melismatically animated centre in the middle, and the final phase, which is usually still articulated specifically. The "single tone" […], "main tone" – is then *unfolded chordally to the "main sound"* in the sound compositions.[114]

Every single tone has a "life of its own" (Isang Yun) and as a microcosm is part of a macrocosm (Fig. 7.24).

From the perspective underlying the East Asian concept of tone, Yun developed the term main tone, which he understands not as a single tone but as "a bundle of individual tones". Accordingly, a main tone is a sound with several tones. The bundle of tones, whose individual tones spread horizontally in monophonic pieces, is distributed among the individual instruments or groups of instruments in chamber ensembles or orchestral pieces. The characteristic of the main tones, which has been subtracted from Korean court music, can be characterized by two types: "the fixed centre and the movement around the central tone. Yun attributes these

[113]Sparrer 1992.

[114]Ibid.

Fig. 7.25 Réak for large orchestra (1966), m. 74–76
Only violin 1 and 2

> basic elements of the main tone to the Yin-Yang principle, so that the non-moving element is aligned with the property of Yin and the moving element with the property of Yang. When the main tone [...] is applied to a group of instruments, it is called the main sound. Depending on the musical context, it can be formed by one or more instruments. This results in several layers, sometimes homophonic and rigid, sometimes heterophonic and moving. The main sound technique [...] was already present in a seminal state in Yun's early works, such as *Music for Seven Instruments* (1959), and reached the peak of its development with *Réak.*[115] *Réak* for large orchestra (1966) – the Korean word means "ritual, solemn, festive music" – begins, as is usual in Korean court music, with ritual percussion. ...]. A chromatic total is only reached in the 4th bar. Also the fanning out of the sound to a flat surface can only be perceived from here on.[116]

The detail with only the two violin parts shows intricate finesses of the "main tone technique". The two violin parts are each executed by several players. Both parts play the same and yet not the same, a melody and yet none. Ideally, a tone is meant, most likely the A, which is also semantically occupied by Yun. But the tone is tonally and sensually veiled by ornamentation and figuration, heterophonically blurred by the difference between the two parts, and moreover, in each part again, by portamenti (sliding glissando transitions between the tones instead of clear interval steps), decreasing and increasing volume, fluctuating and iridescent rhythms (Fig. 7.25).

With his "main sound technique"[117] and its focus on intentionally *one* tone—which is in reality a complex of several tones—Yun ingeniously combines, as early as the 1950s, a single tone with an antithesis, the procedures of the dodecaphony of the Western European avant-garde that uses *all* 12 the tones of the chromatic total in a combinatorial manner. In East Asian-Korean thinking, Yun does not understand a melody as a line, but to a certain extent as a brushstroke with vibrating contours. Heterophonic part writing, glissandi, ornaments, dynamic and rhythmic oscillations and fluctuations create this specific effect. What appears to be a network of several voices is meant as a single, semantically significant heterophonic "main tone", which is followed by other main tones as an overarching melodic movement in mostly relatively slow developments.

---

115 Yun 2000, sine pag.

116 Sparrer 1992.

117 Cf. among others Burde 1987, pp. 263–272.

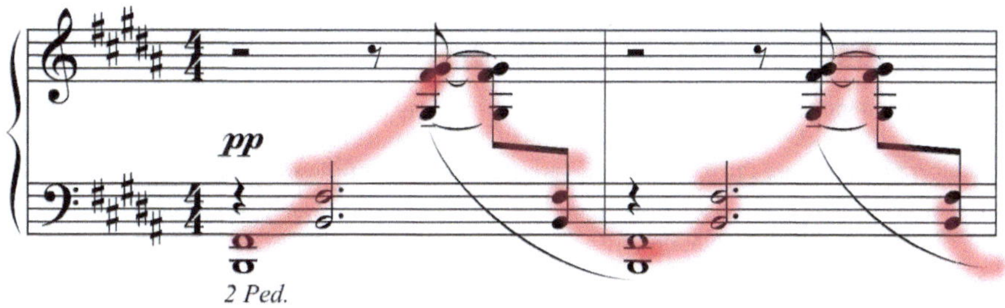

Fig. 7.26 C. Debussy: *Pagodes*
Quoted after Scheytt 2020, fig. 2: Pagodes, m. 1 and 2, Pagode marked

This is fuzziness from the point of view of Europe, but not of East Asia. For Yun does not only compose out the blurring of the intentionally monotone and monophonic melody; it is from the outset and intentionally a regulative and an essential component of the movement.

Again we see the dialectic of fuzziness: from the standpoint of the abstract tone set and even the idea of the main *tone*, Yun blurs the tone, but the many ornaments etc. make it clear, so sharpen it. For the main tone *is* the heterophonic band, the brushstroke with a priori coloring of the tone, not the occidental line with the essentially colorless abstract pitch as the tone and subsequent specific coloring.

A mimesis of the Southeast Asian 'timbral polyphony', especially of the Javanese and Balinese gamelan music, from the European side, is what Claude Debussy, for example, was striving for in his work *Pagodes*, the first of the three pieces of *Estampes* for piano, 1903. Debussy thus also takes visuals as a stimulus. Pagodas are Buddhist temples with a quadrangular or polygonal central building. This may be an analogy to the isotony of the two tunes used in this music. The slendro divides the octave into five equal, isotone steps, pelog in seven in the direction of a whole tone scale. The 3rd or 4th tone thus forms an imaginary-real centre. The isotonicity is denied today, among others by Albrecht Schneider. The question here is secondary in that it is not about ethnomusicology but about the appropriation of foreign music: Debussy conceives of it isotonically, but can represent it on the piano only via the pentatonic. *Pagodes* shows at the same time transitions and mediations between timbre and dense compositional structures—a blurring of the melodic core of heterophony as well as an imaginary 'abstract composition' ('abstrakter Tonsatz') (Fig. 7.26).

> Debussy already gives the most important characteristics in the two opening bars. The notes B flat, F sharp and G sharp are three notes of the B major *pentatonic* (B flat, C sharp, D sharp, F sharp, G sharp) that dominates the piece. The interval of the major second (in the uppermost voice: f#1 and g#1), which is present over long stretches, is also introduced here. The three-layered structure of low, middle and high instruments is also already well recognizable. In the deepest layer, chordal beats on the "1" of the bar provide the metric orientation for the remaining voices. Finally, [...] one can also visually recognize a pagoda per bar in musical notation.[118]

[118] Scheytt 2020.

According to this two-bar exposition, Debussy forms a network of figurations, which motivically refer to the entrance model as variants.

> Debussy [adopts] the multi-layered effect and characteristic movement forms of the gamelan music. [...] Pagodes has a minimum of one and a maximum of four simultaneous layers, but in most bars there are three. The four layers are arranged as follows:
>
> Upper layer: a strongly figured motif in B major pentatonic, which is repeatedly slightly modified; partial melody playing
>
> Middle layer (1): Melody play
>
> Middle layer (2): repeated, ostinato chords or chord breaks
>
> Deepest layer: long fading bass notes, mostly on the "1" of the bar.[119]

7.4.3.3 "Klangfarbenmelodie" (Tone Color Melody). One-Chord- and One-Tone Compositions Between Schönberg and Giacinto Scelsi

Another historical and music-cultural derivation and line of tradition is the explicit composition of tone colors, a trend in avant-garde music around 1960.

In it, instead of the tonal dimensions of the pitch and tone duration, the tone color moves to the heart of the work. With the primacy of color, the composition is designed from the outset to be blurred. The 'abstract composition' is reduced. The sound receives the primacy over the sense. However, ideas of the work are still retained, mainly of impressionistic and pictorial nature. They then recede even further in the next, even broader stream of sound art. The tone color composition of the 1960s was a reaction to serial composition, which was perceived as being too strictly determined, in a sense overly sharp. Electronic music" with its immense expansion of the spectrum of sound possibilities was another point of reference. Ultimately, the composers thus took up the trends of the avant-garde since the beginning of the 20th century, especially the "emancipation of noise" in Futurism (from 1909) and Schönberg's concept of the "Klangfarbenmelodie" (1909/1911). A line of tradition that goes back further is the development and emancipation of the tone color, in its beginnings already during Romanticism since the 1820s with C. M. v. Weber, Berlioz, with even greater importance after 1848 then with Wagner, Mahler, R. Strauss, etc. Approaches to sound surface composition can already be found in the prelude to Wagner's *Rheingold* (1869). There, an E flat major surface starts out from the tone E flat and is enlivened at the beginning by changes in tone color alone, but soon after by motivic figurations (see Sect. 5.3).

Such tendencies towards composing with and of tone colors were intensified, condensed and made independent by A. Schönberg in the third piece of his Five Orchestral Pieces op. 16 from 1909 *Farben, colors*. This far-reaching pioneering work not only means an emancipation of tone color, but even its primacy over the other dimension of the movement. But just as there is no color without any form—even monochrome paintings, such as paradigmatically those by Kasimir Malevich,

[119] In detail Scheytt 2020.

Fig. 7.27 Schönberg, op, 16, 3. Version for 2 pianos by Anton Webern
Only piano 1, m. 1–6.
Hamberger 2012, p. 30 (Cf. Schlothauer 2005/2006, p. 8. Also Schlothauer analyses all dimensions of the piece in extraordinary detail)

receive a formal frame through the frame or border of the painting—so in music there is no color as such. It needs tones or, as the case may be, sounds as a carrier material in order to be realized. And differences in loudness are, intentionally or not, just as implied as different durations, even if they run in irregular, non-metric "rhythms". This is also evident in Schönberg's work. The relatively identical in the constant, kaleidoscopic change of the different tone colors is a chord, C-G♯-B-e-a. It forms the harmonic point of departure and reference. This chord marks the arch form of the piece in that it appears, slightly varied, prominently at the beginning, in the middle at measure 32 and at the end.[120] The chord and its variants (Fig. 7.27)

> appear in alternating coloration and wanders almost imperceptibly through the various groups of instruments in about 60 slow, small changes of tone, only to return to the starting position at the end.[121] Every one of the three inner parts of *Farben* emerges from its own differently orchestrated chordal ribbon, its point of orientation based on the initial chord, itself repositioned through its gradual progression.[122]

> In this example […] we can see the changing voice parts (denoted by circles). This canon continues throughout the movement creating a kaleidoscopic harmonic progression […] In the third section of the piece (mm. 32-44) we find that Schoenberg has creatively inverted the voices of this canon, causing an inversion of the original.[123]

The instrumentation is designed for subtle transitions.

---

[120] See Hamberger 2012, ex. 2.3, p. 26.

[121] https://www.schoenberg.at/index.php/en/joomla-license-3/fuenf-orchesterstuecke-op-16-1909rev-1922.

[122] Therese Muxeneder, 28 April 2020, http://archive.schoenberg.at/compositions/werke_einzelansicht.php?werke_id=228&herkunft=allewerke.

[123] Hamberger 2012, p. 25.

> The four upper voices are alternately realized by 2 instrument blocks, each in half notes. Block a includes (from top to bottom) 2 flutes, clarinet and bassoon 2 (therefore 4 woodwinds), block b English horn, trumpet with mute, bassoon 1 and horn with mute (2 woodwinds and 2 muted brass).[124]

Schönberg's instruction for execution in the score, p. 1 aims at an art of the smallest possible transition:

> The change of chords has to be done so gently that no accentuation of the instruments is noticeable at all, so that it is only noticeable by the different color.

Few chords, many color changes: Schönberg radicalizes Wagner's "art of transition" and creates imperceptible transitions and, as far as possible, color without form (Fig. 7.28).

> In this example of the full chord we can see clearly the composite rhythm between the contrabass and the cello as well as between the two bassoons, the clarinet and the English horn. These interchanges are juxtaposed with the more static flute chords to create an arpeggiated chord with the entrances of each note comprised of a distinct timbral combination. This example also displays how the pattern remains the same, though the pitches gradually change. Similar patterns remain throughout the piece with gradual complexities added. The return of this referential pitch collection can be seen in a few key moments of the piece.[125]

Schönberg repeatedly emphasizes that the piece is unintentional and formless.

> (A)bsolutely not symphonic – precisely the opposite – no architecture, no structure. Merely a bright, uninterrupted interchange of colors, rhythms and moods.[126]

Schoenberg's own program note:

> The music seeks to express all that swells in us subconsciously like a dream; which is a great fluctuant power, and is built upon none of the lines that are familiar to us; which has a rhythm, as blood has a pulsating rhythm, as all life in us has its rhythm; which has a tonality, but only as the sea or the storm has its tonality; which has harmonies, though we cannot grasp or analyze them nor can we trace its themes. All its technical craft is submerged, made one and indivisible with the content of the work.[127]

In reality, the piece is very precisely planned and worked through in its deep structure. Schönberg develops a number of chromatically altered derivatives from the opening chord. The identical thus becomes different in turn. Strictly speaking, it is not just one chord, but several related chords that form the framework. Hamberger shows how Schönberg unfolds a veritable canon with them in the course of the work.

---

[124] Schlothauer 2005/2006, p. 8.

[125] Hamberger 2012, p. 30.

[126] Schönberg to Richard Strauss on July 14, 1909, quoted in Therese Muxeneder, 28 April 2020, http://archive.schoenberg.at/compositions/werke_einzelansicht.php?werke_id=228&herkunft=allewerke

[127] Herbert Glass: Five Pieces for Orchestra, Op. 16, Arnold Schoenberg, About this Piece, without date, https://www.laphil.com/musicdb/pieces/1777/five-pieces-for-orchestra-op-16, retrieved 18 August 2020.

Fig. 7.28 Schönberg: "Farben" op. 16, III, m. 1f.
(New York: Dover Publications, Inc., 1994)
Referential Chord Orchestrated (score not in concert pitch) mm. 1–2
Quoted after Hamberger 2012, ex. 2.2, p. 25

> (The) transposition of the original referential chord in measure 9 of the "Farben" movement. [...] is the outcome of an ingenious compositional technique used by Schoenberg in this movement in which he creates a five-voice canon which Haimo refers to as a "daisy chain" [...] which causes various permutations of the original pitch material throughout the piece. This shifting harmonic canon also contributes to the kaleidoscopic aesthetic of this piece, as the piece is always in a slow yet nearly consistent state of harmonic flux. Haimo describes the canon as consisting of each voice "ascending a semitone... and then" moving "down a whole tone." [...] Each individual musical line changes at a different time creating Haimo's "daisy chain."[128]

The blurring and devaluation of the pitch dimension at the sensual surface is counteracted in the latent pitch order.

> Haimo 2006, 320 also points out an impressive harmonic prescience in the "Farben" movement of *Five Orchestral Pieces*. Though this piece was written well before Schoenberg had developed his twelve-tone compositional technique, the "Farben" movement's pitch content cycles through the entirety of the chromatic scale within the span of the first eleven measures.[129]

The form is the above mentioned, rather traditional ABA form, the so-called "Liedform" or better, in the terminology of Schönberg's disciple Erwin Ratz, the "Adagio-Form"—which, by the way, also forms the framework of the Da-Capo-Aria. In some contradiction to his claims of unconscious and formless creation, Schönberg does compose complex motivic and chordal relationships. Essential is the revaluation of previous hierarchies, a change of perspective and the conception of composition,

> Paradigm shift in the understanding of music: as a fundamental questioning of a centuries-old primacy of pitch over tone color relationships.[130]

Foreground and background change position: what was "figure", the abstract sound set including the motivic-thematic work, recedes and now becomes the "background", and vice versa, the timbres become the figure on which attention is concentrated.[131] Schoenberg explains in a note added to the 1949 revision of the score:

> "The conductor need not try to polish sounds which seem unbalanced, but watch that every instrumentalist plays accurately the prescribed dynamic, according to the nature of his instrument. There are *no motives in this piece which have to be brought to the fore*".[132]

In his "*Harmonielehre*" of 1911, Schönberg conceived the ambiguous term "Klangfarbenmelodie" for this: a sequence like a melody, but not made up of different

---

[128] Hamberger 2012, p. 29.

[129] Ibid., p. 31.

[130] Schmidt 2001.

[131] The reorientation in perception is a problem. In the European perception of East Asian music this has already been pointed out.

[132] David Lambourn, "Henry Wood and Schoenberg", The Musical Times, Vol. 128, No. 1734 (August 1987), pp. 422–27, quoted in https://en.wikipedia.org/wiki/Five_Pieces_for_Orchestra, 13 September 2020, retrieved 15 September 2020.

pitches and tone duration values, but of different tone colors. The perception is at the same time simpler, since colors have a predominantly sensual-emotional effect, and more complex, since, at least from the point of view of European listening habits, a kind of meta-level is necessary to create context.

There are several versions of the work:

Version for large orchestra (1909)
Revised new version for large orchestra (1922)
Version for chamber orchestra (1920)
Version for standard orchestra (1949

The different versions have variable titles:

> Tagebuch 1912: "Akkordfärbungen", Erstausgabe 1912 und Weberns Klavierbearbeitung: „Akkordfärbungen"; Konzert Amsterdam 1914: „Der wechselnde Akkord"; Verlagskatalog 1917: „Farben"; Konzert des Vereins für musikalische Privataufführungen am 13.3.1920: „Der wechselnde Akkord (Der Traunsee am Morgen)"; Kammerorchesterfassung 1925: „Farben (Sommermorgen am See)"; revidierte Partitur 1922: „Farben"; reduzierte Partitur 1950–1952: „Sommermorgen an einem See (Farben)".[133]

Schönberg was reluctant to give the work titles.

> Schönberg had not thought of giving the pieces programmatic titles until the Leipzig publisher C. F. Peters requested them when the work was being printed there in 1912. Schönberg's entry in his diary for January 28, 1912 records that "Altogether, I don't favor the idea – the wonderful thing about music is that one can say everything so that knowing people understand it, and yet one has one's secrets that one confesses to oneself – one does not spread them around – but titles do."[134]

The titles, however, show something of their own interplay between concealing and revealing concrete materials, between blurring and sharpening. "color" is an invariant element in this context.

"Klangfarbenmelodie" is a term newly created by Schönberg in his *Harmonielehre* (1911), initially just for the sequence of different tone colors.

> "In a musical sound (*Klang*) three characteristics are recognized: its pitch, color [timbre], and volume. […] The evaluation of tone color (*Klangfarbe*), the second dimension of tone, is thus in a still much less cultivated, much less organized state […]. Now, if it is possible to create patterns out of tone colors that are differentiated according to pitch, patterns we call 'melodies', progressions, whose coherence (*Zusammenhang*) evokes an effect analogous to thought processes, then it must also be possible to make such progressions out of the tone colors of the other dimension, out of that which we call simply 'tone color', progressions *whose relations with one another work with a kind of logic* entirely equivalent to that logic which satisfies us in the melody of pitches."[135]

---

[133] Haselböck 2016; cf. also Therese Muxeneder, http://archive.schoenberg.at/compositions/werke_einzelansicht.php?werke_id=228&herkunft=allewerke. 28 April 2020. On the synaesthetic aspects cf. e.g. Schönberg/Kandinsky 1981, Nono-Schoenberg 1992, Dittmann 1995, Petzer 2017; even more general Hutter 2006.

[134] Muxeneder, l.c.

[135] Schönberg 1911, p. 503, English translation by Roy E. Carter, quoted in https://en.wikipedia.org/wiki/Klangfarbenmelodie, 28 December 2019, retrieved 27 August 2020. Emphasis HWH.

Strangely enough, Schönberg here forgets the fourth dimension, the tone duration. Independently of this, he leaves open what a 'musical logic' shifted to the dimension of tone color would look like, which would guarantee context, stringency and coherence.[136] At least in the first instance, Schönberg aimed with this piece at a blurring of traditional hierarchies of musical composition, a dethronement of the abstract musical composition to the point of elimination. Against this, A. Webern's orchestration of the six-voice *ricercar* from J. S. Bach's *Musical Offering* explicitly aimed, instead of blurring, at sharpening through tonal color differentiation in order to make motivic micro-relationships more sensual and comprehensible. But Schönberg, too, sought new sharpening in fuzziness, and with the overthrow of previous hierarchies and in self-made chaos, he aimed at a new, higher order using the latent generative depth structure. In this, the tone color takes on a higher, more important position. Schönberg still vaguely envisions a continuum of tone color and pitch, which then returns in a different form in serialism.

> I cannot unreservedly agree with the distinction between color and pitch. I find that a note is perceived by its color, one of whose dimensions is pitch.... If the ear could discriminate between differences of color, it might be feasible to invent melodies built of colors [*Klangfarbmelodien*].[137]

The encapsulated relationship of blurring and sharpening is then repeated in the relationship between serial composition and 'tone color composition' around 1960. In the current of "sound composition"[138] around 1960 with Krzysztof Penderecki's *Anaklasis* (1959/60) and György Ligeti's *Atmosphères* (1961), blurring becomes an aesthetic program. Edgard Varèse (1883–1965), a pioneer, mediates between the avant-gardes of the interwar period and those of the postwar period. "Sound composition" is admittedly not a particularly felicitous term, since music and therefore also composition is usually associated with sound. 'Tone color composition' or 'sound surface composition' is more appropriate. The anti-serial blurring, however, could also be included the other way round in the ancestral gallery of serial sharpening.

> Against the background of newly emerging compositional techniques, the term has also been used since the 1950s as a historical precursor to the "tone color series" of serial music (L. Nono, Karlheinz Stockhausen) or the apologia of electronic music (Herbert Eimert).[139]

The tendency to concentrate on tone color instead of an elaborate, clearly structured composition was already a "paradigm shift" in Schönberg's orchestral piece op. 16, III (Schmidt 2001). It is repeated as a "turn" in turning away from the

---

[136] On problems and the intellectual environment cf. Schmidt 2001: especially Busoni 1907/1916, as well as the theosophy of Emanuel v. Swedenborg, which Schönberg had got to know through Honoré de Balzac's *Séraphita* and which he adapted in an orchestral song of the same name. For discussion cf. inter al. Dahlhaus 1971, Cramer 2002, Hamberger 2012, Zeller 2020.

[137] Herbert Glass: Five Pieces for Orchestra, Op. 16, Arnold Schoenberg, About this Piece, without date, https://www.laphil.com/musicdb/pieces/1777/five-pieces-for-orchestra-op-16, retrieved 18 August 2020.

[138] Cf. e.g. *JetztMusik—Glossar, Klangkomposition*, https://www.swr.de/swr2/programm/article-swr-15682.html, retrieved 21 August 2013.

[139] Schmidt 2001.

complexity of serialism, generally with the current of "sound composition" and especially, in a certain way fractal, in a representative of serialism, Karlheinz Stockhausen. His work *Stimmung für 6 Vokalisten* just from 1968 is a turn to a "meditative style".

> 6 singers intonate a single ninth chord in pure tuning and vibrato-less piano in different articulation forms and overtone colors over the more than one hour duration of the work. Invoked 'magic' names of gods from different cultures and erotic poems by Stockhausen complete the composition.[140]

J. Morthenson also gets along with a single chord, without invoking the gods.

> Jan Morthenson's *Coloratura III* (1963), in which *a single chord* is sustained for the entire duration of the piece, varies only through a *microtonal variation* and through the changing entry of the instruments.[141]

A comparable minimalist regression mixes with historicist exoticism even on the periphery of "occidental" music. In addition, there are usually also ritual-sacral elements, via magic or via moods; for example, Avet Terterjan (1929–1994) in his 8 symphonies since 1969:

> The impression of timelessness results from the minimal change of individual sounds: No. 5 (1976) begins with the tone a flat, which is only varied in tone color over 5 minutes.[142]

This is quite little variation for a fairly long time. In his next symphony, Teterjan enriches the monotony a bit with sampled material.

> No. 6 (1978) is a fantasy about the tone c, which is overlaid by various tape recordings.[143]

As far as the pitch dimension is concerned, the single-tone composition is more consistent. A precursor is Alban Berg. In Wozzeck (1916/1925), in Act the Scene 2 Marie and Wozzeck is an *Invention on a note* (B♮). One note, the death note B♮, forms the core, but Berg then uses the entire chromatic total. The transition to the next scene is a tremendous one-bar crescendo on this one note B♮. In the specific tone color compositions since Scelsi, the one tone is blurred in its sound to a band of sound. Usually, however, this one tone is more than a single one: it encompasses a range of microtonal deviation by at least one quarter tone down and up. In Nono's two-tone composition, it is explicitely two tones, once again a soggetto cavato with the acronym *Carlo eScarpa: Per C.S. (Carlo Scarpa) architetto e suoi infiniti possibilità* (see above in Chapter 1 and 7.1) The connection of tone color composition to older traditions of microtonal composition is conveyed above all by Giacinto Scelsi (1905–1988). His *Quattro pezzi per orchestra (Ciascuno su una nota sola*—Four Pieces for Orchestra (each on a single note)—unfold a colorful spectrum of the respective single note, which, however, is shaded in a quarter-tone, "infrachromatic"[144] manner. Scelsi aims for immersion.

---

[140]Dietel 1994, p. 939. Emphasis HWH.

[141]croupet 2005a, p.

[142]Dietel 1994, p. 944f.

[143]Ibid.

[144]Bandur 2007, p. 1.

> Reiterating a note for a long time, it grows large, so large that you even hear harmony growing inside it .... When you enter into a sound, the sound envelops you and you become part of the sound. Gradually, you are consumed by it and you need no other sound .... All possible sounds are contained in it.[145]

Eastern Meditation serves as legitimation of the reduction of traditional components of the composition and expansion of the tonal component.[146] Scelsi in addition conceived

> artistic creation as a means of communicating a higher, transcendent reality to the listener. In this view, the artist is considered a mere intermediary.[147]

Truly the individual tone as the basis of the composition is used even more consistently by the "spectral music" of the Groupe de L'itinéraire, founded in 1973, with Girard Grisey (1947–1998), Tristan Murail (born 1947) and Michael Lévinas (born 1949), among others. It unfolds the single tone in fluctuating processes of form, which are oriented towards the organic, "biomorphic", into the color spectrum of partial tones.[148]

7.4.4 Musical Chiaroscuro, Sfumato, Clocks and Clouds. from the Transparent Polyphonic Structure to the Opaque "Texture"

The composition of tone colors or sound surfaces was often based on more or less vague ideas of a mimesis of reality, especially of unformed nature; typical are, for example, Ligeti's aforementioned *Atmosphères* with a programmatic background. These are relatively abstract, intangible images of nature. Schönberg was a model for this, with varying precision and concreteness in the changing ecphrastic titles: *Der Traunsee am Morgen – Sommermorgen am See – Sommermorgen an einem See* (The Traunsee in the morning—Summer morning at the lake—Summer morning at a lake).

Such works require a highly developed level of material and technique: The blurring requires prior sharpening. In visual arts representations of clouds in the genre of landscapes are indeed natural, especially in Eastern Asia and Europe (in the latter since the Renaissance). However, musically clouds are only really representable once a high standard of development regarding material, technology and language of music has been reached. Required are loose regulations of musical compositional technique and structure. They are generally to be understood as tendency towards musical prose, and enable a conscious reproduction of fuzziness. This level is given since the time of

[145] Scelsi, interview (1987) with Franck Mallet, quoted in Reish 2006. In detail about Scelsi's innovation ib.; on the ideological background Reish, pp. 151–158.

[146] On Scelsi als "Eastern traveller" Skupin 2012; cf. Ahn 2014.

[147] https://en.wikipedia.org/wiki/Giacinto_Scelsi, 25 August 2020, retrieved 28 August 2020.

[148] Further thoughts on Spectralism and generally e.g. Schäfer and Kursell 2010.

accomplished industrialization in the European centres and the emerging monopolization, as in impressionism and symbolism as well as naturalism on the other hand—otherwiseopposing trends. Exemplary are works such as Charles Baudelaire's prose poem *L'étranger* of 1869 with lines like "J'aime les nuages … les nuages qui passent …là-bas … là-bas … les merveilleux nuages!",[149] or Claude Debussy's movement *Nuages* (Clouds) from the *Trois Nocturnes* (1897/1899)—*Nuages*—*Fêtes*—*Sirènes.* In the last movement with women's choir Debussy employed only wordless vocalizations, like György Ligeti later did in his *Clocks and Clouds.* The representation and imagination of ‚clouds' of all kinds reappeared increasingly in "sound composition" in the time around 1960, only now shifted from the impressionist and programmatic into the structural. Ligeti made these entities more concrete through programmatic titles, as in his orchestra piece *Atmosphères*, prominently employed in Stanley Kubrick's movie *2001: A Space Odyssey* in 1968. In another work, however, Ligeti sought to sublate precise and vague figures, sharpening and blurring into a dialectical unity.

"Fabric" or "texture" is an old metaphor for the multi-voiced structure of compositions. Ligeti re-installs it with the concrete and sensual, by recourse to synaesthetic conceptions and by applying *orchestral sound spectra.* As "texture", i.e. "Gewebe" in a more specific sense, he compiles his dense, interwoven, overlapping, interlooping carpets of tone colors. He uses "micropolyphony" with a sfumato-like blurring of the voices instead of the chiaroscuro with its distinct profile in classical polyphony. However, almost simultaneously he counters this with a counterpoint in the form of poignantly delimited musical gestures and entities, explicitly in 1972/1973 in *Clocks and Clouds.* Without (explicit) knowledge of Fuzzy Logic Ligeti in 1972/1973 composed *Clocks and Clouds* in a rich orchestration and with 12 female voices—the also musically famous 12, as e.g. in equally tempered tuning as in dodecaphony.[150]

> The title of Ligeti's *Clocks and Clouds* refers to an essay by the Anglo-Austrian philosopher Karl Raimund Popper, *On Clocks and Clouds* [1972]. Popper's essay describes two different kinds of processes that occur in nature, one that can be measured exactly ("clocks") and the other, made up of indefinite occurrences that can only be described in a statistical approximation ("clouds")."[151]

Sir Karl R. Popper, foremost known for his fight against objective knowledge and dialectics, for once had a productive idea.[152] Even if he had understood the difference to be a contrary or even contradictory dichotomy—Ligeti in any case configured this polarity dialectically. He designed the polarity as process:

> These transformations are not clearly delineated occurrences of "now clocks" and "now clouds." Instead, through minutely shifting rhythmic patterns, Ligeti presents the listener with a malleable texture whereby the homogeneous character of the musical material allows for

[149] Charles Baudelaire: L'étranger, in *Le Spleen de Paris. Petits poèmes en prose*, I (1869). "I love the clouds … , the clouds that pass by … there …, there…: the wonderful clouds!"

[150] One of the favorite objects of Fuzzy Logic-Studies in music.

[151] Lacoste 2011.

[152] In consideration of the fact that Zadeh had published his concept as early as 1965, a direct connection is at least possible. Cf. e.g. Dubois / Prade 1980, Seising 2005 and 2007.

little distinction between a clearly defined, periodic ticking rhythm and the blurred dissolution into clouds.[153]

Ligeti carries this interference a step further. Even in clearly outlined objects the contours can dissolve— ‚soft composing' instead of hard composing. Ligeti states in a 1978 interview:

I should like to refer to the soft, limp watches of Dali's painting (The Persistence of Memory, 1931), which had associative value in the composition of this piece.[154]

Likewise Ligeti employs microtonality, intervals smaller than half tones, as a partial moment of the dialectics of the clearly outlined and the amorphous. The usage of complementary contrasts in the composition is reflected in the instrumentation.

The composer also creates "clocks" and "clouds" harmonically, moving from harmonies based on standard tuning to those based on non-traditional intervals, thereby creating harmonies that are once clouded, once discrete. Because of their capacity to realize these subtle shifts, five flutes form the backbone of *Clocks and Clouds*' harmonic skeleton. The five clarinets and twelve-voiced women's chorus nearly match the flutes' flexibility in realizing the harmonies as well as combining with them in high-pitched tone color (along with harmonics in the cellos and double basses) in "fluid" textures. In contrast, the other instruments, especially the two harps, four bassoons, two trumpets, and strings produce precise rhythmic patterns.[155]

More common was the less dialectical composition of clouds without clocks.

While Ligeti sought to portray the complex inner life of sound surfaces through a dense interplay of individual voices, from which, in the course of time, so-called micropolyphony developed, Krystof Penderecki (e.g. in *Anaklasis* [1960] or *Threnos* [1961]) relied on static noise bands in which one tone is continuously sustained by one instrument.[156]

More variant-rich in the sonority and in the generative deep structure Iannis Xenakis proceeds in the already mentioned *Metastaseis* (1955). Thus can be found in the "audible surface":

a (rather short) serial-looking *linear polyphony* with strict voice separation according to the position of the string instruments as well as a more extended section in which an underlying layer of tone points (which can also be condensed vertically into clusters) is counterpointed by further layers in which such static sounds are transformed into pizzicato structures or by glissandi into more complex sound surfaces.[157]

---

[153] Quoted in Lacoste 2011, ibid.

[154] Quoted in ibid.

[155] Lacoste 2011, ibid.

[156] Decroupet 2005a, p. 129.

[157] Decroupet 2005b, p. 130.

Fig. 7.29 G.F. Händel: Sonata in F major, 15 Sonate ad Camera, No. 12, second movement, m. 22f.
https://upload.wikimedia.org/wikipedia/commons/5/59/Bariolage_Handel.jpg

7.4.5 *Resulting Patterns as Pre-calculated Blurs—Between Bach/Gounod's* Ave Maria *and Central/East African Xylophone Music*

The following two sections deal with blurred structures, which are not characterized by tone colors, but essentially by pitches and rhythmic-metric processes.

The simplest blurring of a chord is the arpeggio. It is obvious in the harp—hence the name of arpa = Italian harp, as well as in string instruments and strings. The chord is split into a melodic movement, the vertical becomes a diagonal like the glissando, only now segmented, structured and harmonically grounded.

A further development in string instrument playing is the bariolage.

> "the alternation of notes on adjacent strings, one of which is usually open", exploiting "the individual timbre of the various strings." This may involve quick alternation between a static note and changing notes that form a melody either above or below the static note. The static note is usually an open string note, which *creates a highly* resonant *sound*.[158]

The linguistic derivation of the term also includes aspects of the tone color. ‚Bariolage' is an

> "odd mixture of colors", from the verb *barioler*, "to streak with several colors".[159]

In the late 20th century, this aspect is intensified not least in the context of tone color composition. Thus composers explicitly use the technique to enrich the spectrum of tone colors (Fig. 7.29),

> *as an extended technique when used simultaneously in different instruments, or in conjunction with complex rhythmic layering or microtonal tunings. [...]* Elliott Carter adapted the technique to the harp in a solo work actually titled *Bariolage* (1992), which blends the device with trills and a harp technique called *bisbigliando*, "in a profusion of trilling passages and enharmonic unison colorings."[160]

> (T)he second measure is to be played with *bariolage*. The repeated A is played on the open A string, alternating with Fs and Es fingered on the adjacent D string. The notes on the D

---

[158] https://en.wikipedia.org/wiki/Bariolage, 31 August 2018, retrieved 26 August 2020; quotation from Stowell 1990, p.172. „The usual bowing technique required, which also may be used separately from bariolage, is called *ondulé* in French or *ondeggiando* in Italian". (Ibid.) Emphasis HWH.

[159] https://en.wikipedia.org/wiki/Bariolage, ibid.

[160] Whatley 2007, p. 49, n13, quoted in https://en.wikipedia.org/wiki/Bariolage, 31 August 2018, retrieved 26 August 2020.

Fig. 7.30 Springing arpeggios in Mendelssohn's Violin Concerto in E Minor
Quoted after Moat, James; A Study of the Performance Requirements Found in the Sonata for Solo Violin Op 40 by Miklós Rózsa (2017). Student Research, Creative Activity, and Performance—School of Music. 106, http://digitalcommons.unl.edu/musicstudent/106, ex. 3, p. 12

> string (E and F natural) would be fingered as normal (first finger and low second), but the fingerings given above the second measure would be [2040 1040 2040 1040], indicating the switch (*bariolage*) from open A string to the stopped fourth finger on the D string, also playing the note A. The repeated A in m. 2 is played on the open A string, alternating with Fs and Es fingered on the adjacent D string. The notes on the D string would be fingered as normal, but the fingerings given above the second measure would be [2040 1040 2040 1040], indicating the switch from open A string to the stopped 4th finger on the D string, also playing the note A.[161]

This type of figuration simultaneously conceals *and* clarifies the melody—here a simple semitone change between F and E. It is embedded in the figuration and comes all the more forcefully to the fore in the overall result. Arpeggio and bariola as Springing arpeggio or 'spiccato bariola' is connected by a passage in Mendelssohn's Violin Concerto in E Minor (Fig. 7.30).

Both methods can be understood as decorations in a broad sense. From the point of view of abstract composition ('abstrakter Tonsatz') they enliven. However, the ornamentation and figuration technique can be semantified. The violinist and composer Eugène Ysaÿe (1851–1931) in *Six Sonatas for Solo Violin,* Op. 27, composed between 1923 and 1924. *Second Sonata in A Minor, "Obsession,"* quotes of the *Preludio* from Bach's *Partita in E Major*, BWV 1006. This in itself meaningful quotation is enhanced by a second one. Bach is

> sometimes quoted directly, sometimes altered or modified, followed by outbursts (marked *brutalement* in the score), the equivalent of musical cursing, [...] Ysaÿe inserts the *Dies Irae* from the Catholic Mass for the Dead (figure 2). The sonata, completed in July 1923, consists of four movements, *Obsession*; *Prelude, Malinconia, Danse des Ombres; Sarabande*, and *Les Furies*. The Bach quotation appears only in the first movement, but the *Dies Irae* quotation unites the four movements.[162]

The Bach quote on the one hand gets out of hand and determines several components of the work, but on the other hand is configured differently in the new context, in the direction of an additional blurring.

> (Ysaÿe) uses direct quotes, distorted quotes, and borrows musical styles and idioms. Ysaÿe's *Prelude* is modeled on Bach's *Preludio,* using the same tonal areas and harmonic relationships, the same perpetual motion effect, similar textures and violinistic techniques, such as

[161] https://en.wikipedia.org/wiki/Bariolage, 31 August 2018, retrieved 26 August 2020.

[162] Lowell n.d., p. 6.—The *Dies Irae,* connected with BACH, appears in the book repeatedly as a sort of leitmotif. (See above in Sect. 5.2.4, pp. 257ff.)

Fig. 7.31 Eugène Ysaÿe: *Six Sonatas for Solo Violin,* Op. 27 (1923/1924), Second Sonata in A Minor, *"Obsession,* excerpt.
Quoted after Lowell n.d., p. 6

> string crossings, scalar patterns, pedal tones, and *bariolage* bowing. Where Bach implied polyphony within a single melodic line, Ysaÿe hides his melody in a texture of sixteenth notes, c*reating compound lines out of a single monophonic line.*[163]

The manner of quotation corresponds to the programmatic title "Obsession" (Fig. 7.31).

> The *Dies Irae,* combined with the fast tempo and use of accents, lends the piece a demonical madness, a real obsession with Bach and plays into the idea of *the possessed violin virtuoso.*[164] Ysaÿe also creates the impression of being literally bound by Bach, with the real and fictitious quotes, suggesting again that his obsession with Bach is inescapable.[165]

> *Dies Irae* theme played with bariolage bowing technique. Accents are placed on the downbows of each beat to bring out the *Dies Irae* theme.[166]

In an ingenious way, Ysaÿe directly semantizes the technical aspects of playing.

> The body becomes symbolically divided as the right arm moves to Bach and the left had plays the Dies Irae. The Bach quotations, Ysaÿe's brutalement response and *Dies Irae* theme all share similarities in contour and the same initial half step movement. The *Dies Irae* theme appears with the same bariolage bowing technique that Bach used […], further developing the relationship between the Bach and *Dies Irae* quotations.[167]

Another way of blurring chords and at the same time reviving them is the *style brisé*, „the transference to the harpsichord of lute figuration."[168] In French Baroque music the term was style luthé ("lute style"). It means playing a chord as arpeggio, that is one after the other and not simultaneously. It is also practiced with string instruments (Fig. 7.32).

A graph by S. Bruhn tries to show the structuring processes in motives, harmonics and volume of this prelude (Fig. 7.33).

Ch. Gounod (1818–1893) showed plainly the latent upper voice melody resulting from the arpeggios of the style brisé. This happened in several stages and versions. It begins with an improvisation by Gounod, followed by a transcription and an arrangement for melody instrument with accompanying instrument (1853).

---

[163]Lowell n.d., p. 6.

[164]We recall G. Tartini's „Devil's thrill sonata" (See above in Chapter 6).

[165]Lowell n.d., p. 7.

[166]Lowell n.d., p. 9.

[167]Lowell n.d., p. 8.

[168]Armstrong 2011, p. 31.

Fig. 7.32 J. S. Bach: Prelude No. 1 in C major, BWV 846, from Book I of J. S. Bach's The Well-Tempered Clavier (1722), m. 1f

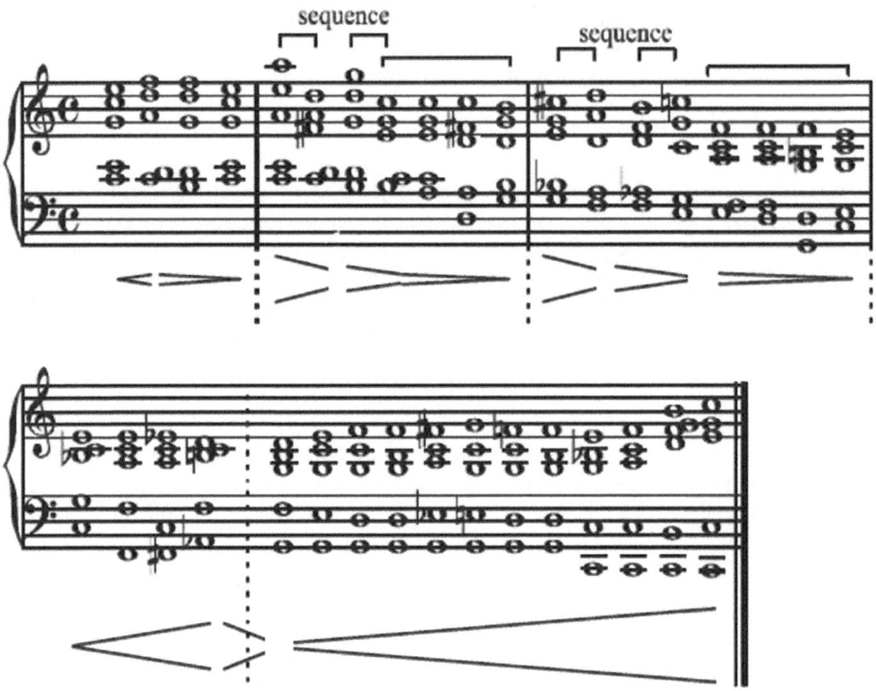

Fig. 7.33 J.S. Bach: WTC I/1 in C major—Prelude. Bruhn (n.d.), p. 52

(H)is future father-in-law Pierre-Joseph-Guillaume Zimmermann transcribed the improvisation and in 1853 made an arrangement for violin (or cello) with piano and harmonium.[169]

The next step was a lyricization of the melody, first with new French, then with the old Latin of the familiar *Ave Maria.*

The same year [1853] it appeared with the words of Alphonse de Lamartine's poem *Le livre de la vie* ("The Book of Life"). In 1859, Jacques-Léopold Heugel published a version with the familiar Latin text. The version of Bach's prelude used by Gounod includes the "Schwencke measure" (m.23), a measure allegedly added by Christian Friedrich Gottlieb

[169]https://en.wikipedia.org/wiki/Charles_Gounod, 1 September 2020, retrieved 20 September 2020.

Fig. 7.34 Gounods instrumental Version for violin and piano over an only very slightly changed version of the Prelude No. 1 in C major, BWV 846, violin and right hand of the piano part, m. 5-7 https://sheetmusic2print.com/products/Gounod-Violin-Ave-Maria. Quoted after: https://www.youtube.com/watch?v=dpF0lr_2Pro, upload 15.02.2018, Screenshot

> Schwencke in an attempt to correct what he or someone else erroneously deemed a "faulty" progression, even though this sort of progression was standard in Bach's music.[170]

Amending a seeming "error" of J.S. Bach was a real error.

> Carl Czerny added an extra measure. He argued that first, it seemed highly improbable, for a composer as conscious of subtle numeric balancing as Bach was, to have written a piece consisting of the uneven number of 35 measures; second, a bass line with a diminished third such as the one in mm. 22-23 seemed melodically incorrect and would therefore require a chromatic link. Czerny therefore inserted a tonic six-four chord to correct Bach's "error," and by doing so destroyed the forceful tension built up here with so much ingenuity.[171]

Apart from that, the number 35 fits much better into J. S. Bach's concept. For the checksum of $3 + 5$ is 8, and that is exactly twice the number of the letters of the name BACH, and he would thus have written his name at an exposed position in the work, namely at the beginning of the WTC I planned as a cycle. The supposed blurring of an "odd" number is in fact sharpening. Gounod used the incorrectly corrected version of the prelude as the basis for his adaptations with the manifest filtering out of a latent melody (Fig. 7.34).

A somewhat less minor change is that Gounod transposes from J. S. Bach's C major to E flat major, most likely to get a better position for the singing voice. Because of the adaptation to the text, J. S. Bach's latent melody is rather fragmented (Fig. 7.35).

The version as *Ave Maria,* transposed to F major, is considerably more elegant. The melody benefits here from the fact that the text is short (Fig. 7.36).

In the 3rd measure, the real, manifest vocal tones at the top of the treble with the syllable break c" g' on "ri-" do not sound at all in the virtual, latent instrumental upper voice contour, but only as the lowest, ostinato note plus fifth, part of the dominant seventh chord C7. The two notes are thus filtered out. In the blurring there is at the same time a sharpening "reduction of complexity": thanks to Gounod a single, simple new melody emerges as a core from the abundance of figurations. Thanks to the dialectic of blurring, this melody can emerge, as in J. S. Bach, usually as a contour

---

[170] https://en.wikipedia.org/wiki/Ave_Maria_(Bach/Gounod), 18 September 2020, retrieved 20 September 2020.

[171] Bruhn n.d. p. 52, footnote 1.

Fig. 7.35 Premier prélude de J.S. Bach (Lamartine: Vers sur un album) (1852), arr. vv, solo vn, orch (1856), originally for insts (1853), with text Ave Maria (1859), m. 4–9
Bibliothèque nationale de France. https://gallica.bnf.fr/ark:/12148/btv1b7200803c.r=Gounod%20Bach?rk=21459;2

Fig. 7.36 Charles Gounod: Ave Maria, m. 1–4.
Ave Maria, mélodie religieuse adaptée au 1er prélude de J.S. Bach (The reduced score for piano and vocal seems to be a smaller version of the arrangement for [1v, pf]/[1v, vn, org, pf]/[1v, vn, org, pf, orch] [1859]; many other arrangements exist). [Free-scores.com]_gounod-charles-ave-maria-16759.pdf. https://www.free-scores.com/download-sheet-music.php?pdf=16759

of the upper voice. "Emergence" here means mainly being perceived.[172] Although the contour tones sound, are compositionally planned and anticipated as a result, they need not be meant as a coherent melody. Musical structures in which such tones are not played, but are heard and are meant as such, go a significant step further. Such resulting patterns can be found as a characteristic especially in the East/Central African Bantu cultures. The technique of *interlocking* three xylophone players in the former court music of Buganda, which was largely destroyed in 1966, results in

[172] In a wider perspective cf. e.g. Quinn 1997.

> the phenomenon of the so-called inherent patterns, […] melodic-rhythmic figures […], which are subjectively generated in auditory perception, and are interest also for the psychology of perception[173]

These figures are by no means accidental, but rather compositionally intended, although, in contrast to patterns of the 'style brisé' type, "none of the playing instrumentalists performs them as such".[174] A numerical notation illustrates the extraordinary complexity of this music.[175] As Kubik sums up:

> The composers of this old music are unknown to us. In the meantime, some rules of composition could probably be crystalized out. […] One also understands that the inherent patterns suggest words and texts; however, the mental processes […] and the experiences of complex numerical relationships behind them remain unexplored. In Kiganda music, little is left to chance. Even that which appears to happen by chance is planned in advance as a possibility.[176]

7.4.6 Musical 'hatching'—Blurring as Structural Condensation. Iteration of Patterns in Minimal Music and Steve Reich's **Drumming**

Blurring often works as an increase in complexity that is difficult to hear and understand, and thus initially appears as chaos instead of a cosmos. But in the final analysis, once again the dialectical result is an indirect sharpening, an enrichment and specification of form and explicit or implicit semantics. This is shown by numerous cases in traditional music as well as in "Classics". But the factual blurring is a necessary transitional stage. For example, there is a corresponding passage in Ch. Ives. In the lower stave is that of Ives, in the upper stave the seemingly exact, simplified and normalized notation stands for the rhythmic-metric shifting and blurring effect that Ives aimed for (Fig. 7.37).

Frequently and generalized to a musical current, such individual cases of blurring through complexity in dialectical tension become simplicity with "minimal music". This is largely limited to simple, diatonic scale excerpts as patterns, which in endless repetition and minimal variation have meditative and trance-like effects. The reduction of the pitch material in the direction of a one-tone tendency is usually articulated in penta- or diatonic form and in this respect also forms a counter position to developed chromaticism, let alone the "twelve-tone". One of the initiators was Terry Riley (born 1935); Philip Glass (born 1937) became particularly popular since the 1970s. In the initial work from 1964, Terry Riley's *In C,* C major is interpreted for up to one and a half hours. High piano sounds give a basic pulse. In addition, 53 melodic rhythmic mini-models of any number of melodic instruments are played as often as desired. Polymetric and microcanonic textures result from the overlapping of the

---

[173] Kubik et al. 1982, p. 14. In detail see ibid., p. 14ff. See also Kubik 1988, 1994, 1998.

[174] Ibid.

[175] Ibid., p. 19ff.

[176] Ibid., p. 21.

Fig. 7.37 Charles Ives: *Three Places in New England,* 2008 edition
3rd movement, Celeste part, mm. 21–26
Quoted after Wild 2018, p. 9.
The rhythms are shifted, 4 against 6, etc. See Wild 2018, p. 9

models. La Monte Young (born 1935) even does without such rhythmic differentiations. *The Tortoise, His Dreams and Journeys* from 1964, as a lifelong ("throughout my lifetime") work in progress, are based on "stationary sounds":

> Notated are systems of frequencies; the basis is a drone, which is produced by a small vibrator ("turtle motor") or by a synthesizer, above which the musicians intonate [...] harmonies in absolutely pure tuning. Microphones and amplifiers are used to additionally clarify overtone effects such as difference and other combination tones.[177]

In order to silence development, change and processuality, La Monte Young here follows, probably only unintentionally, the "occidental" primacy of the pitch dimension. The downright addictive repetition is also a very special, paradoxical modification of the dodecaphonic and serialized thinking in series with its incessant variation of the identical core: here, the thinking in series really only appears as a realization of series, without the intention of the greatest possible density of information, variety, diversity. Thinking and composing in repetitive patterns is a compositional approach that runs counter to the ideal of complexity and innovation of the (European) avant-garde. Admittedly, "material economy" also forms a line of tradition. However, it is not used here to develop diversity and multiplicity out of little, to create sensual richness and abundance in the economical use of resources. Rather, the time and patience of both players and listeners is not spared. The current is less recourse to tradition than a relapse behind modernity and in general the main direction of historical development. (Essential for the tradition of European composing is constant qualitative innovation and development.) Mechanistically generated and unintentionally resulting random results through the sequences of patterns and pieces switch off, objectivistically, in addition compositional and interpretative subjectivity.[178] Folkloristic or exoticist models are added. Three dimensions of exoticism, the "non-European", the historicizing and the intersocial, play a role here. After all, there are parallels

[177]Dietel 1994, p. 930.

[178]The later catchword "New Simplicity" picked up on this tendency towards anti-complexity in the late 1970s. The composers subsumed under it, however, form a subjectivist counter-reaction against anti-individualistic minimalism.

to popular music, especially the later punk and techno types. Influences from non-Western cultures were important, indeed essential for minimalism. Steve Reich, in contrast to the majority of this movement, dispensed with the usual irrational and ritualistic bonusses. He strove for a synthesis, a kind of syncretism that was to be more than exotic.

> The least interesting way to respond to non-Western music would be to imitate the sound of the other music. [...] This is the simplest and most superfluous way, which ends only with musical chinoiseries. No, for a composer it can only be debatable to write music with his own, i.e. Western, sound means in the light of knowledge of non-Western structural principles. ...] Structural ideas of non-Western music must be related to the instruments and the tonal system with which one grew up. The non-Western influence therefore manifests itself in compositional thinking and not in sound.[179]

Through his short, only five-week field studies at the University of Accra in Ghana (summer 1970), Reich felt above all confirmed in his belief that "unplugged", without electroacoustic instruments, a greater sonority of sound could be produced. In *Drumming* (1970/1971) he transferred the concept of electroacoustic phasing to percussion instruments (see Sect. 7.6). There are convergences with black African musical models.

> "Non-Western music in general and African, Indonesian and Indian music in particular will provide Western musicians with new structural models—not just new sound models (which would merely correspond to the old exotic fascination)."[180]

Drumming consists of four parts, each lasting 12–20 min, depending on the frequency of repetition. The parts have different instrumentation and can also be performed individually.

I. four pairs of tuned bongos
II. three marimbas with nine players and two female voices
III. three glockenspiels with four players and with pipes and piccolo flute
IV. all previous instruments together.

The whole piece is based on a single rhythmic pattern, if the individual parts are combined. The 12-order derives diastematically from European, rhythmically-metrically from African traditions[181] (Fig. 7.38).

At first the whole pattern is slowly constituted by Reich gradually replacing silence with rhythmic-metric pulses. Then the pattern is subjected to "phasing", the phase shifting. In interplay of sound concentrations and attenuations, from a simple unison to polyphony and back again, thus polyphonic rhythmic canons, which, however, can no longer be perceived as such by the listener, are created. The ear constructs its own melody out of individual notes as resulting patterns and contour lines, as in African music or the style brisé—which Reich emphasizes instrumentally, however,

---

[179]Reich, Steve: "A Composer Looks East", in the New York Times of September 2, 1973; quoted after Gottwald 1975 p. 3f.

[180]Reich 1993, p. 291.

[181]On this see inter alia Kubik 1994 p. 42f.

Fig. 7.38 The rhythmic core of Steve Reich's pattern of *Drumming*
Quoted after: Heister, H.-W. 2005a (ed.): Geschichte der Musik im 20. Jahrhundert: 1945–1975 (Handbuch der Musik im 20. Jahrhundert Band 3) Laaber-Verlag, Laaber (2005), p. 283

Fig. 7.39 Steve Reich, *Clapping Music*. 6 passes of the pattern, each repeated once.
Steve Reich: Clapping Music [PDF] - the ear reader

in a supporting manner.[182] Trained in African polymetry, Reich dialectically creates rhythmically-metrically complex structures by superimposing very simple initial patterns, as in *Clapping Music* (1972) (Fig. 7.39).

Methods as well as result follow here not least fractal principles, without the corresponding mathematics being applied consciously (see Sect. 7.4.7.1).

> All the complexity unfolds from the simple code in the germ. The linear fractals generated by simple commands represent subtle symmetries in their unfolding. The nonlinear fractals, which are thought to be the germ of mathematical equations, develop a deterministic chaos in which astonishing order is hidden. The musically relevant characteristic of fractal music, self-similarity as the multiplied variation of the themes, has deterministic and random elements, whereby in fractal music somewhat deterministic themes and their variations, which are never repetitive, are structured in a balanced way. Both opposing essences of the music remain in harmony and balance.[183]

[182]Backward connecting lines are shown here to the elaborate polymetric canons of Conlon Nancarrow, which, however, are precisely composed. There are also parallels to Ligeti's "texture" principle, especially in the harpsichord piece Continuum (1968) with its micropolyphonic interweaving of arpeggios, which can be played at a ludicrous tempo. According to Ligeti, the resulting "acoustic illusions" are inspired by Maurits Escher's graphics, but are also reminiscent of Op art, which sets static images in apparent motion.

[183]Yoon 2002, p. 140.

7.4.7 Indeterminacy Instead of "Fully Completed Unity"

"Aleatoric" and "stochastic" compositions often are very close to each other, or aleatorics even appears as a generic term.[184] "Indeterminate music" or "partially indeterminate" music could be a common denominator terminologically. What is actually "indeterminate", both in terms of its procedures and its objects, often remains fuzzy and vague. But it has to be differentiated precisely under the aspect of fuzziness and blurring; with 'stochastics' as a generic term in the following way

> Stochastic processes may be used in music to compose a fixed piece *or* may be produced in performance.[185]

"Stochastic music" a term coined by Iannis Xenakis, in the narrower sense, has nothing to do with sound form or realization. It is a matter of blurring *within* the composing process, generated by procedures based primarily on probability theory, which result in a thoroughly closed and completed work. The aleatoric approach also precedes the performance, but it leaves open places in the work itself, either already or still within the composition process, with regard to the realization of the work. The starting point and background of music history are in both cases two non- or anti-serialist[186] currents since the 1950s: on the one hand, the inclusion of chance in the composition process and the composition itself, and on the other hand, the sound color composition (see Sect. 7.4.3). The motivations are complex. Xenakis turns against the vague and nonreflective fetishisation of "chance" and refers to Parmenides:

> Parmenides implicitly admits that necessity, need, causality, and justice identify with logic; since Being is born from this logic, pure chance is as impossible as not-Being.[187]

Referring again to Xenakis, Yoon also stresses,

> that in the strictly deterministic serial music there are already immanent germs of coincidence, and in coincidence a kind of determination is already hidden. "For if, thanks to complexity, the strict, deterministic causality which the neo-serialists postulated was lost, then it was necessary to replace it by *a more general causality*, by a probabilistic logic which would contain strict serial causality as a particular case. This is the function of stochastic science."[188]

On the one hand, serialistic strict and stringent composition destroys genuine compositional thought and action.

---

[184] Comprehensive and precise to this Frobenius 1995.

[185] https://en.wikipedia.org/wiki/Stochastic, 13 April 2020, retrieved 13 June 2020.

[186] In English, there is no terminological differentiation between the technique of series in the sense of Schönberg's dodecaphony and the 'serialism' since 1949 that is meant here, which attempts to organize not only the pitches but also the tone dimensions of tone duration, tone color, volume (and sometimes even more) in series.

[187] Xenakis 1992, p. 204, quoted in Yoon 2002, p. 53.

[188] Yoon and Xenakis 1992, p. 205, quoted in Yoon 2002, p. 56.

> Through total determination, the composition process gets out of the composer's hands; he has played his role by pre-forming the possibilities of series, the coincidence of automatism prevails.[189]

Secondly, the composers' need for expression, their desire for concrete subjectivity and individuality, therefore, lead to loosening of procedures, even among the serialists themselves, as in the case of P. Boulez' *Le marteau sans maître* (1952/1954).

Thirdly, the experience that the strictly and stringently determined tonal surface structures are not strictly and stringently audible motivates the composers to use right away indeterminate procedures as generative depth structures—especially Xenakis. This could be understood as 'certain negation'. Fourthly, a pendulum in abstract negation swings in the opposite direction of determinism, namely that chance, whatever it may be, serves as a generator—a kind of suicide for fear of death.

7.4.7.1 "Stochastic Music". Calculated Coincidence in the Production Process—I. Xenakis' *Metasteis*

The historical process of composing with stochastic methods has at least three stages.

1. It starts without a computer. Rather handmade, oriented on arbitrariness, lack of intention and subject and without much mathematics, 'coincidence' is at the centre of the US-American version of Cage and Others as determiner of arbitrary individual quasi-compositional decisions, which are, however, externally determined, for example by the Taoist *I Ching*.
2. It is continued with computers, but initially without very developed mathematical methods. They are quickly refined, implemented by means of apparatus and can thus soon be generalized.

> Lejaren Hiller and Leonard Issacson used generative grammars and Markov chains in their 1957 Illiac Suite. Modern electronic music production techniques make these processes relatively simple to implement, and many hardware devices such as synthesizers and drum machines incorporate randomization features. Generative music techniques are therefore readily accessible to composers, performers, and producers.[190]

Lejaren Hiller (with the collaboration of Leonard Isaacson and Robert Baker) attempted to simulate compositional processes on the computer in the pioneering work of computer music, the *Illiac Suite* 1957.

> For this work, the Illiac computer at the University of Illinois generated the pitch, rhythms and other musical parameters. The computer gave the result in an alphanumeric code, which was transcribed *by hand* into string quartet notation.[191]

So there is a free choice, e.g. the instrumentation. In it, the difference between the abstract tone set and the concretely completed score form returns on a new

[189] Yoon 2002, p. 22.

[190] https://en.wikipedia.org/wiki/Stochastic, 13.4.20, retrieved 13 June 2020.

[191] Yoon 2002 p. 74 Emphasis HWH.

level. Determinism and randomness, order and chaos, sharpening and blurring are intertwined.

> According to Hiller and Baker, the development of computer music takes place in two stages. First, they create a state in which arbitrariness prevails, and a higher or lower degree of order is then imposed on this chaos. Although the two stages often merge during the work processes, the two-stage scheme remains basically always valid. Random numbers can be generated mathematically. By relating these random numbers to musical parameters—such as pitch, tone length and volume—you obtain *random music within* the *limits* outlined by the parameters.[192]

The cascade-like procedure leaves considerable room for 'free' compositional decisions. As with Xenakis, the "generate and test" or "random sieve method" is used to

> first generate random numbers, which are then subjected to one or more tests to check their *appropriateness to the musical situation.* [...] If a value breaks one of the rules, it is negated, otherwise it is accepted. The result is recorded and a new value is calculated in the next choice. [...] The programming *composers are free to* describe the characteristics of their desired music. These rules can describe the method of how they choose the pitch and duration of melodies and the rules of harmony, counterpoint, timbre, orchestration or articulation, etc.[193]

In a double movement, determinacy and indeterminacy work together. "Coincidence" appears here less as a function of probabilities than as fate and arbitrariness at the same time—whereby the question of causal factors that are not really connected is excluded from the development of coincidences. Secondly, the scope of subjective, conscious interventions into the automatic of stochastics, which varies according to the chosen initial conditions, is strangely arbitrary, music-oriented, but rather an arbitrarily free choice. Ultimately, the "desired musical context"[194] is again the purpose and goal of the compositional process. There is something helpless or arbitrary about the rebellion of the compositional subject against the external determination objectified in the computer and in the calculation processes. The question is allowed, what's the use here of all the computerized and mathematical effort. It sharpens neither an idea nor its realization, but boils down to a—precisely produced—blurring.

Step 3: In the "Stochastic Music" by Iannis Xenakis, complex statistical procedures are incorporated into the composition. Mathematical precision becomes musical blurring, because his mathematics itself operates with probabilities, i.e. non-deterministically, probabilistically.

As is usually the case when it comes to the connections between mathematics and music, Xenakis' worldview is also centered about an 'universistic', Pythagorean thinking of numbers. Numbers appear idealistically conceived as world-shattering or materialistically representing the world, but both refer to reality. Mathematics

---

192 Yoon 2002 p. 75: Description of the *Illiac Suite* p. 85f.; in each of the 4 movements different mathematical methods are experimented with. Emphasis HWH.

193 Yoon 2002, p. 71.

194 Yoon 2002, p. 110.

thus forms a common third of various sciences, including the arts. For his stochastic composition Xenakis finds various models and analogies in areas of nature.

> It is impossible to describe the trajectory of each gas molecule within a space, but the Maxwell-Bolzmann law describes the distribution of motions within the ensemble of molecules at a given temperature and pressure. In using this as the model for *Pithoprakta*, Xenakis drew an analogy between the movement of a gas molecule through space and the movement of, for example, a violin sound through its pitch range. To construct the seething movement of the piece, he governed the molecules according to a coherent sequence of imaginary temperatures and pressures. The result is a music in which separate voices cannot be discerned, but the shape of the sound mass which they generate is clear."[195]

The list of works, including the mathematical methods Xenakis used in each case, refers to abstract mimetic relationships to areas of reality, precisely under the sign of a rationally reformulated universalist thinking (see Sect. 3.3.4 and Chapter 9).

> Specific examples of mathematics, statistics, and physics applied to music composition are the use of the statistical mechanics of gases in Pithoprakta, statistical distribution of points on a plane in Diamorphoses, minimal constraints in Achorripsis, the normal distribution in ST/10 and Atrées, Markov chains in Analogiques, game theory in Duel and Stratégie, group theory in Nomos Alpha (for Siegfried Palm), *set theory* in Herma and Eonta, and Brownian motion in N'Shima.[196]

Xenakis developed his methods for calculating the not completely determined from around 1953 onwards based on his criticism of serialism with its strict and stringent over-complex structures that were calculated but no longer audible. As a solution to this contradiction, it seemed to him to operate from the outset with—albeit precisely calculated—blurring.

> Linear polyphony destroys itself by its very complexity; what one hears is in reality nothing but a mass of notes in various registers. The enormous complexity prevents the audience form following the intertwining of the lines and has as its macroscopic effect an irrational and fortuitous dispersion of sound over the whole extent of the sonic spectrum. There is consequently a contradiction between the polyphonic linear system and the heard result, which is surface or mass. This contradiction inherent in polyphony will disappear when the independence of sounds is total.
>
> In fact when linear combinations and their polyphonic superspositions no longer operate, what will count will be the statistical mean of isolated states and of transformations ofsonic components at a given moment. The macroscopic effect can then be controlled by the mean of the movements of elements which we select. The result is the introduction of the notion of *probability*, which implies, in this particular case, combinatory calculus. Here, in a few words, is the possible escape route from the linear category in musical thought."[197]

The already mentioned key work *Metastaseis* from 1955 (see Sect. 7.4.4) by Xenakis is regarded as

---

[195] Philippot, M.P.: Xenakis. In: The New Grove Dictionary of Music and Musicians, vol. 20, London 1980, p. 560, quoted in Yoon 2002, p. 93f.

[196] https://en.wikipedia.org/wiki/Stochastic, 13.4.20, retrieved 13 June 2020; cf. https://en.wikipedia.org/wiki/Iannis_Xenakis, 4.6.20, retrieved 13 June 2020, and Yoon 2002, p. 92.

[197] Xenakis 1992, p. 8, quoted in Yoon 2002, p. 93.

> his "first step to [use of] probability theory" in music[...] His theoretical proposal for a solution lies in a different consideration of the relative texture density of a composition: where polyphony as such is no longer perceptible, only the "statistical mean of the interdependent states [...] of the components in their change at a given point in time" is decisive; attention is directed to the "mean value of the movements" that the selected objects can perform, and this mean value requires, in addition to *combinatorics*, the use of *probability theory*.[198]

On a meta-level, the negation of the regular is still mediated by precise, even mathematically calculated determination. So

> the time proportions in which the changes in sound occur are composed precisely, according to the dimensions of the so-called geometric series.[199] In addition to the usual series, such as the Fibonacci series used to calculate the golden section (which are based on integers), Xenakis refers to the "série bleue" from Le Corbusier's writing *Le Modulor* (Paris 1950), which he always had in his sight in his architectural office in the Rue de Sèvres.[200]

In a peculiar dialectic, the technocratic Le Corbusier assumes an anthropomorphic human measure, but then in turn squeezes all kinds of architectural components into the proportions derived from it.

> In addition to the "série bleue" there is a "série rouge": both rows of the modulor assume two average values determined on the human body, namely the "série rouge" assumes an average height of the upright person of 183 cm (resulting in a navel height of 113 cm), and the "série bleue" assumes an average height of the upright person with raised arm of 226 cm (a measure that was also chosen because of the relation to the navel height of the "série rouge").
>
> eben der » série bleue « gibt es eine » série rouge « .[201]

Xenakis follows this line and partly simplifies mathematically precise values, blurring them pragmatically, just as Le Corbusier modified the Fibonacci Sequence. He works with "rounded values

Fibonacci Sequence: 1–1–2–3–5–8–13–21–34–55–89–144–233

Corbusiers "série bleue": 1-2-3-5-8-13-20-33-53-86-140-226. The differences start from 20/21. At the beginning of Metastaseis, the values 33 and 34 appear on different levels as a synthesis, so to speak.[202]

In case of doubt, Xenakis intervened in the mathematically ingenious structure and corrected it in the direction of what was sonically sensible. The calculations and graphic schemes are impressive regardless (Figs. 7.40 and 7.41).

Two decades later, Xenakis achieved a different "reduction of complexity" with the help of precise procedures in the transformation of the graphic-optical into the acoustic (see Sect. 7.3). He was concerned not least of all with a simplification of composition that could also be used in music education.

---

198 Decroupet 2005b, p. 130. Êmphasis HWH.

199 "Thus "series", even if of different origin, also enter into the non-"serialist" non-series music.

200 Decroupet 2005b, p. 131.

201 Ibid.

202 Numbers according to ibid.

Fig. 7.40 Pithoprakta
Xenakis 1992, p. 17
Fig. 1-6. Bars 52–57 of *Pithoprakta*
Score

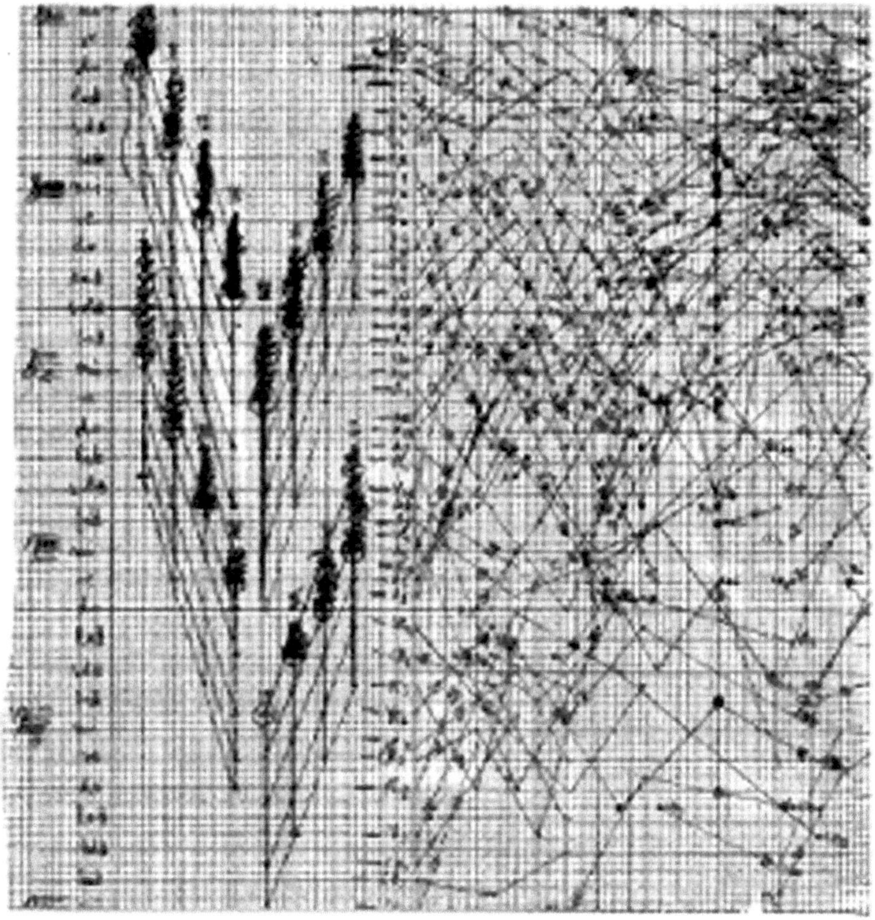

Fig. 7.41 *Pithoprakta*
The graphical scheme shows the complexity and fuzziness of the structure—and this is only the beginning of the graphical representation.
Xenakis 1992, p. 18

> It is therefore necessary to provide tools for the user that are simple and elegant. The difficulty is thus to conceive well-adapted interfaces between thought and music. The *solution was the hand*: the musician gives his commands to the computer through drawings, and not punch cards or programs. The results was UPIC. This computer system allows one to compose music directly by men as of a graphic design, even without knowledge of music or computer science. As you can see, it is a matter of immediate access and of making complete use of computer science. *A child can use it.* He draws houses, flowers, suns, then he coordinated that he hears what he draws."[203] (see Sect. 3.3.4)

[203]Xenakis 1992, p. 25, quoted in Yoon 2002, p. 91.

The mathematics of fractals brings new mathematical aspects. It can be applied to numerous musical phenomena, for example in not under-complex works of minimal music (see Sect. 7.4.6), and it can be used consciously in composition.[204] As far as can be seen, however, they do not bring anything fundamentally new in the subject matter dealt with here under aspects of fuzzy logic and the relationship between sharpening and blurring.

7.4.7.2 "Aleatoric Music". Compositional Integrated Randomness in the Realization Process

As for Xenakis, the change in the scientific world view between quantum mechanics and probability theory was an important stimulus for Stockhausen to make use of aleatory procedures. I use aleatorics (from Latin alea, cube, metonymic for chance) in a limited, more precise sense. In the context of acoustics, Meyer-Eppler defines as aleatoric those processes "whose course is roughly fixed, but depends on chance in detail".[205] What is meant musically is that the composers deliberately leave gaps open for the performers to fill and perform in a self-determined way. They thus refrain from a strict and stringent elaboration of the course of the form in particular. The mathematical and methodical complexity was admittedly applied in a very simplified manner by him, P. Boulez and others. "Aleatoric" meant nothing else than that arbitrary chance phenomena of the interpreters determine the compositionally indeterminate form. This only takes on a concrete form in the one-time realization of a performance. In contrast to the indeterminate and totalizing use of random operations in Cage and his environment

> Pierre Boulez demanded in his lecture Alea – [...] at the Darmstadt Summer Courses in 1957 – that the artist should only accept the "coincidence" he "conducted". Accordingly, the arrangement of certain parts of a piece can be left to the performer's discretion, but the details are completely worked out. Karlheinz Stockhausen first did so in the *Klavierstück XI* (1956), Boulez in his 3rd piano sonata (1956/57).[206]

Stockhausen's instructions for *Klavierstück XI* follow the "text genre *Werkgeneserezept*":

> The musician looks *unintentionally* at the paper sheet on which the 19 groups are recorded and begins with any group he sees first; he plays it at *any* speed, volume, and touch, then looks unintentionally at any of the other groups, and plays them according to the three performance instructions. The term "unintentionally looking from group to group" means that the musician never wants to connect certain groups with each other or skip certain

---

[204] In detail see e.g. Yoon 2002, pp. 158–170.

[205] Meyer-Eppler 1955, p. 22.

[206] *JetztMusik—Glossary Aleatorik.* Once on the website of SWR (Südwestdeutscher Rundfunk, http://www.swr.de/swr2/programm/sendungen/jetztmusik/glossar/-/id=4979696/1m5qq8g/index.html, link now closed.

groups. Each group can be linked with any of the other 18 groups, so that each group can be played with any of the six speeds, basic volumes and forms of touch.[207]

The paradoxical request "Be spontaneous! (Paul Watzlawick) is notoriously difficult to follow. So the liberation of the interpreters, who had to provide for subjectiveblurring left much to be desired. Though it is true that

the overall course of the work depends on chance, but considering the details it is clear right from the start: a multiple choice for interpreters.[208]

It is a somewhat austere manifestation of chance, reduced to spontaneous, arbitrary choices between fixed parts. Going much further in the direction of the productivity of the interpreters is an inversion of the fields of action of chance, the

aleatoric variant, in which the form is fixed, but in certain open spaces of the composition the performer is free to change the details, to improvise associatively.[209]

The relationship between composition and improvisation is then reversed in the *collective improvisation.*

Here the composer takes even more a back seat and the performers are upgraded to co-composers[210]

Similarly, improvisation is generally required in 'Musical Graphics', which Schulze subsumes under 'micro-aleatorics' as the American equivalent of the European 'macro-aleatorics'.[211] It is only there that the perhorresced 'context' tends to be eliminated.[212]

From the point of view of various abolitions—of the subject, the context the semantics—genuinely American minimal or process music is also regarded as an "extreme form" of aleatorics (Schulze 1999, p. 27; see Sect. 7.4.6).

Here, too, the composer no longer exerts any influence on the process after the initial situation has been established. But the result of the process is not uncertain. The intention of the composition is also different: In a gradual process that the recipient physically and psychically has to follow, the artistic means is first gained from an everyday acoustic experience. Fundamental to this process is that no *unexpected change in the course of events* occurs and disturbs the process of *de- and new semantization.* This means nothing less than the opposite of what the composers wanted to achieve by using non-systematic operations.[213]

---

[207] Thomas, Ernst: Was ist Aleatorik? In: Melos 28 (1961), H.7/8, p. 217 quoted in Schulze 1999, p. 28, footnote 24. Emphasis HWH.

[208] Schulze 2000, p. 24. *A musical dice game* of Mozart's time offers something similar with somewhat simpler elements, namely minuet measures that can be combined into a whole by means of permutation.

[209] *JetztMusik—Glossar Aleatorik,* ibid.

[210] Schulze 1999, p. 28, footn, 25; in detail see Müller 1994.

[211] Schulze 1999, p. 28ff.

[212] Cf.: "For however 'open' the forms of the so-called aleatoric music of Boulez, Pousseur, Stockhausen, etc. may have been, the relations between the tones (or notes, respectively) that constitute the musical context are not abolished. Cf. Erdmann, Martin: Untersuchungen zum Gesamtwerk von John Cage. Bonn 1993, p. 7, quoted in Schulze 1999, p. 28, footnote 24.

[213] Ebbeke 1994, col. 444.

The extremes of total determination and total coincidence collapse into each other, as mentioned above, under the banner of the abolition of the subject and a generalizing blurring. In 1958 P. Boulez criticized the US-American in the following way:

> However, the similarities in the genetics of the works are far greater than these differences suggest:
>
> "Chance frees music from accidental subjectivity. Only in this way—according to Cage—can the objectivity of music be regained, which alone guarantees what music has lost since Beethoven, commitment. Cage's idea—this is how extremes meet—coincides very precisely with the ideas of Pierre Boulez, who identified in Joyce's great novels as the most significant moment that the author's person ultimately disappears behind the work. Writing music in this way was also his [Boulez's] goal.[214]

7.4.7.3 Aleatorics and Absurdity

Complexity in the score form can obscure the sparseness of the notation. Exaggerated sharpness thus appears as a blurring

> Often "aleatoric" scores are covered with a very complicated net of arrows, hints, symbols and instructions for execution, which, however, only extend to the area of v i r t u a l relations. Due to their usually complicated graphic form, they then feign real relationships; the score wants to give the impression that the musical form is just as complex as the instructions for making it come about. There is a danger that the composer will not be aware of the separation of action and result notation and will therefore fail to consider the musical results (and their differentiation) as the actual object of compositional work in the face of so many references to action.[215]

Mauricio Kagel's *Transicion II* (1959, Publikation des score 1963) for piano, percussion and 2 tapes follows concepts of aleatoric composition. It

> is a work accentuated in three layers. The first layer is a tape with processed sounds of piano and percussion, the second is given by the direct action of the two instrumentalists during the performance the third arises during the second layer. In this way a continuum is formed, consisting of the tape recordings of the performance in the hall and the first layer played by loudspeakers.[216]

The performers are appointed as interpreters. They are responsible for the production of the tapes. The duration of the work is variable.

> Duration: 8-24'. Instrumentation details: 1 pianist, 1 percussionist (produces all sounds inside the grand piano). It is up to the performer to realize the specified tape recordings and electroacoustic processing before or during the performance. 1–2 magnetophones are required for playback. In addition (depending on composition): electrical filters, reverb unit, ring modulators and tape loops.[217]

---

[214] Gottwald, C.: John Cage und Marcel Duchamp. In: Metzger, Heinz-Klaus (ed.): John Cage I. Musik-Konzepte Sonderband. München 1990, p. 137; as Boulez 1958 quoted in Schulze 1999, p. 29, footnote 30.

[215] Boehmer 1966, p. 103, footnote 12.

[216] Mauricio Kagel, introduction to the work, https://www.universaledition.com/mauricio-kagel-349/werke/transicion-ii-5416.

[217] Ibid.

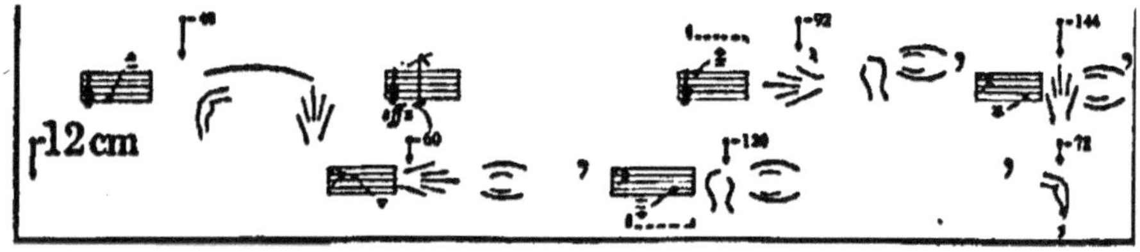

Fig. 7.42 Mauricio Kagel: Transicion II (1959)
Detail page 4a
Quoted after McInerney 2007, Fig. 36, p. 103

The work also has characteristics of 'live electronics'.

Transición II works on three levels. LIVE: The pianist performs on the keyboard while a percussionist performs inside the piano. TAPE #1: Before performing the piece, they prerecord some sections which are to be played back during the performance. TAPE #2: the live performance is also recorded and that soundtrack can be played back. Repetition of any of the musical material is not allowed—instead, as the players continue to perform in the present, the tapes can recall music already played. In addition, the tapes can be manipulated to change timbre and frequencies in order to make recognition of the original sounds impossible.[218]

The tasks for the execution are

extremely difficult, complex and detailed. Like several works at that time, the score presents a meeting of control and flexibility. [...] Parts of the music are extremely tightly controlled while, often at the very same time, other parts present passages of arbitrarily uncontrolled randomness.[219]

The notation is at once almost overly precise and blurred (Fig. 7.42).

In its published version (Universal Edition 1963), the 35-page score is accompanied by 7 pages of mobiles [...] and 16 pages of instructions in German, English and French [...]. (W)hat appears to be a single line of stave for each instrument [...], on close inspection turns out to be a system of five lines referring not to the traditional stave but to a cluster of five adjacent strings (or keys) on the piano (the same cluster of five pitches in the case of both pianist and percussionist).[220]

Notation like notated oscillate between

details of frightening complexity alongside extremely ambitious progressive tempo shifts indicated on a separate stave; other pages display little more than a random scattering of dots with general indications as to pitch and temporal location.[221]

The multiple contradictions between complexity and simplicity, control and arbitrariness are not abolished, but mutually cancel each other out.[222] The contradictions appear fractal to a certain extent, in the parts as in the whole in

[218]http://www.moderecords.com/catalog/127kagel.html.

[219]McInerney 2007, p. 104.

[220]Ibid.

[221]Ibid. Further details p. 104ff.

[222]Cf. McInerney, p. 103.

a perverse kind of logic here – it is as if the formal opportunities provided by the composer serve the same purpose on the macro level as does the meeting of *vagueness* and detail in the individual pages: to retain an absurd level of complexity and fearsome difficulty for the performers while at the same time dissolving any possibility of a formal logic which might merit such *fetishism* being apparent to the audience. [...] Like a *cargo cultist* Kagel undertakes a level of complexity symptomatic of the environment in which he finds himself. *Transicion II*, however, seems to lack the integral logic, the underlying rationale, which justified such work in the hands of his confrères. [...] *Transicion II* is a work which, while making a fetish of control and detail, ends up in a performance whose arbitrariness borders on parody.[223]

7.4.8 Reworking and Arrangement as Blurring

"Versions" and arrangements of a work can be understood as recourse from the sound form to the score form—and encroachment. The various versions of a work are intended to sharpen a relatively identical core after the completion of its production (see Sect. 6.6.4). Often, this may not result in a single, emphasized, ultimately valid version of the type "last hand version", but in different versions of equal value. Either way, however, this kind of self-editing turns into blurring if either foreign arrangers take possession of the original work for their own purposes, or the original composers want or have to adapt their work all too readily to practical performance conditions and constraints, or even want or have to conform to the market. Particularly in the case of arrangements by others, the apparent sharpening often turns into a merely apparently sharpening, and instead of improving the appearance of the work, the result is its mutilation, including the composers' intention and the idea of the work. Many case examples of this are provided by the "Regietheater" (director's theatre). From refreshing and restoring against fading becomes "scratching" and "deconstructing". The blurring and "overpainting" in the sense of the visual artists Arnulf Rainer and Richter are transformed in such a way that the paste-up of the original foreign work is presented as a work of its own.

7.4.8.1 Self-determination and Heteronomy

In 1949 Hans Engel presented examples of important types of editing together with numerous examples, shown here in extracts and with different segmentation and sequence

1. Reworking of own works. Double versions.

[...] Bach's fugues from the Well-tempered Clavier were partly in a more modest form, his piano concertos are adaptations of his own and other works [...]. Beethoven reworked his own works several times, e.g. the Quintet op. 4, after the Octet for Wind Instruments op. 103 in a free manner [...].[224]

---

[223]McInerney 2007, p. 106f. Emphasis HWH.

[224]Engel 1949, col.Engel, Hans; MGG vol. 1, 1949/1951, column.

2. Engel extended the concept of editing in such a way, that already work ideas and not only finished works can become the subject of editing:

> The three Leonore overtures are adaptations of the same plan […].[225]

3. Reduction of the instrumentation I

> Classical symphonies were reworked for chamber music Beethoven himself edited his 2nd symphony (1806) for piano trio like his Septet op. 21.

4. Reduction of the instrumentation II

> Such diminutive instrumentation, arrangements, require masses of […] rework for the "salon orchestra" (in the "Viennese", "Berlin" or "Parisian" instrumentation and larger combinations, cf. light music) […]. They are arranged in such a way *that parts of the instrumentation can be missing* and replaced ("retracted").

5. Expansion of the instrumentation within the range of instrumental music. As early as the 18th century, Mozart added wind instruments to the orchestra in his adaptation of Handel's *Messiah.* Mahler enlarged the apparatus for Beethoven's 5th Symphony. And completely without notation and a new score form of the old work, the number of choral orchestral instruments, especially strings, has been drastically increased since 1848 *within* the sound form of the highly developed concert and opera world. Now this is definitely blurring.

6. Expansion of the instrumentation with text.

> Parodies of instrumental works also belong to the expanding arrangement. Bach has edited instrumental movements […] for choir, and Händel has arranged piano fugues for choir.[226]

7. Orchestration of piano pieces and piano songs

> Popular piano music was orchestrated even Beethoven arranged the funeral march from his Piano Sonata op. 26 for orchestra […]. The development of the orchestra, especially since Wagner, encouraged the instrumentation of piano and vocal music.[227]

H. Wolf orchestrated some of his Mörike songs. The orchestral version is perhaps even more appropriate to the inner greatness of the songs and the colorfulness of the harmonies than the original piano version. The same applies to Hans Werner Henze's new instrumentation of Wagner's five *Wesendonck Lieder* (1976), originally a cycle of piano songs in Wagner's work—Wagner himself orchestrated only dreams of them. Here, the blurring of the semi-'abstract composition' becomes a sharpening of the whole work. It is similar with the aforementioned instrumentation of Mussorgsky's *Pictures at an Exhibition* by Ravel.

8. "Analytical Instrumentation" (Carl Dahlhaus)—the prime example is Anton Webern's orchestration of Johann Sebastian Bach's Ricercar from the *Musical Offering* (1935) (see above)

9. In many recent cases, especially in "postmodernism", the editing of a foreign work passes over to the composition of an own work.

---

[225] Ibid.

[226] Ibid.

[227] Ibid.

> The analysis of *foreign work substance* by means of reworkings can be understood as a process of "forward-looking rear-view mirroring" (Cloot et al. 2010), which gives composers the opportunity to reflect on the historical position of their own work. Paradigmatic for this are works such as Hans Zender's "composed interpretations" of Schubert's "Winterreise" for tenor and ensemble (1993), Schumann Fantasy for large orchestra (1997) and 33 Veränderungen über 33 Veränderungen für Ensemble (2011) [meaning Beethoven's "Diabelli Variations"], in which the original note texts are subjected to constantly changing modes of editing in the sense of a "creative change" (Zender 1993/2004). The fact that in such cases both historical music and the processes of its reception are subjected to critical questioning is also demonstrated by pieces such as Mathias Spahlinger's *adieu m'amour, hommage à Guillaume Dufay* for violin and cello (1982-83).[228]

From the perspective of Fuzzy Logic and from the point of view of the original work, however, it must be asked when the rear-view mirror becomes a distorting mirror, and to what extent the idea and substance of the edited works are suspended in such post-creative changes. In any case, the editing here transforms the work into its quotation form (see Sect. 8.2).

7.4.8.2 „Director's Theatre". Re-/Production as Production

The staging especially of music theatre works (as well as of theatre works in general) can be considered a kind of reworking. With regard to the historical and time-bound nature of the works, it is partly necessarily blurred, since not everything can be reproduced historically "true to the work", partly it is a blurring by deliberately covering up differences, partly it is a sharpening by helping the idea of the work to get a new, contemporary validity. The framework of action and the objective limits of subjective interpretations are determined by the score form of the work. In any case, it is the necessary condition of the staging, even if it does not always determine all details with sufficient precision. The "director's theatre", which has been spreading since the early 1970s, increasingly exceeds its competence as a staging, ultimately encroaching on and attacking the substance of the works, so that their identity tends to be erased. The essential elements of the 'composite work of art' namely the music theatre are in acoustic music vocal and instrumental, word (libretto), and in the visual stage design as a relatively static, thing-like component and scenic action as a process-like component. They all refer to the 'drama', including the not directly sensually tangible action as a focusing center and make it appear sensually as a Gesamtkunstwerk. The scenic action is particularly time-bound and therefore particularly variable, and the stage design including costumes is almost exactly the same and open to updates—within limits that are to be determined in each individual case. The nature of the gestural and mimic movements is based on anthropological universals on the one hand, but on the other hand, the ceremonial movement types of the "Baroque Opera", for example, can only be realized through reconstruction and are therefore necessarily imitated in a blurred manner. Here is the real space for the variations of the staging as performance and execution. The more these variations

[228]Drees 2016, p. 197f.

move away from the original historical location and milieu in the work, the more the action becomes irritated. To transfer Beethoven's *Fidelio*, for example, from its plot and approximate time of origin, the age of the French Revolution, to the Nazi era is not nonsensical, but unnecessary, since the audience is able to make the transfer itself. Inevitably, even necessarily translation is blurring; it is at the same time an editing, often with a tendency to rewriting. The blurring of the wording, however, sharpens the understanding in the reception. Conscious re-instrumentation that changes the tone color may not be necessary, but it is not necessarily harmful. Inevitable are, as I said, changes in the scenic, regarding picture as well as action. With the "director's theatre", however, the staging becomes independent of action, idea and work. Representatives of the sphere of re/production and interpretation of the work set themselves up as producers. In the meantime, these editing tendencies are spreading from the scenic and staged to the musical substance; yet this is ultimately the core of an opera, as indispensable as the scenic is. One gateway was the increasingly popular insertion of pop music into traditional or newer operas.

Meanwhile, under the flag of "intertextuality", the music of a work is partly changed, rearranged, remixed partly adorned with borrowed plumes. The re-costuming of Berg's *Lulu*, for example, into an *American Lulu* (2006/2011) by Olga Neuwirth, including her own music, follows this direction to a pretty large degree.

> What's good for the directors since the dawn of the director's theatre, to move the plot in their productions to other places and in another time, is good for Olga Neuwirth in her fourth opera: inspired by Otto Preminger's film "Carmen Jones", a free adaptation of Bizet's "Carmen", the composer Alban Berg, born in Graz in 1968, moved Berg's opera plot in its paraphrase to the New Orleans of the 1950s and to New York in the 1970s.[229]

Neuwirth's product oscillates between deconstructive processing and own production.

> (H)ere is not the attempt to come as close as possible to Berg's intentions is the focus of interest, but the composition itself is already conceived as an interpretation of the piece [...].[230]

Or rather the other way round: the interpretation appears as a work in its own right. Under the aspect of editing, Berg's *Lulu* is fragmented.

> The prologue of the animal tamer [...] is replaced by a microphone amplified conversation between Lulu and Clarence (formerly Schigolch) in New York; melodramatically accompanied by an imported calliope, a steam-driven keyboard instrument with pipes. The action of the first two acts then follows as a flashback, with speeches by Martin Luther King and poems by the African-American poet June Jordan cut between the scenes. However, the speeches are not experienced in the original, but are reproduced by speakers. The US-American protest movements of the 60s and 70s [...] remain accessories.[231]

---

[229] Peter P. Pachl: Berg-Paraphrasen: Uraufführung von Olga Neuwirths „American Lulu“ an der Komischen Oper Berlin, 1 October 2012, https://www.nmz.de/online/berg-paraphrasen-urauffuehrung-von-olga-neuwirths-american-lulu-an-der-komischen-oper-berlin.

[230] Alban Nikolai Herbst, https://faustkultur.de/865-0-A-N-Herbst-Alban-Bergs-Oper-Lulu-von-Olga-Neuwirth.html; cf. inter al. Olga Neuwirth: Notizen zu American Lulu (2006–2011), http://www.olganeuwirth.com/text28lulud.php.

[231] Pachl 2012, ibid.

Berg's music also becomes an accessory in the end. So

> the Paris scene of the third act is deleted. The new plot of the 25-minute final act [...] only contains marginally Berg quotations. [...] Neuwirth's American Lulu does not die by Jack the Ripper [...]. A final projection shows the New York luxury whore naked, covered with excrement.[232]

Instead of composing an own work, O. Neuwirth participates in the fame of a foreign one. Her adaptation, which is nevertheless a work in itself, is a "destruction" of Alban Berg's opera *Lulu*, which, seen in this light, becomes an "Un-American Activity" in the sense of McCarthyism. "She shouldn't have bothered." From a fuzzy logic point of view, it could be an exciting task to determine, even quantifying, the boundaries and edges up to which the performance form of the work can be considered at least partially identical to the originally intended notation, the sound form including the optical dimension can be considered a realization and reflection of the score form. In the cinematization of operas, the problem is repeated, only at a new level, that of technical re/production.

7.5 Point of Departure: Sound Form. Subjective Vivifications of Tone and Tempo

The tension between perfection and vitality is a fundamental problem of the sound form. Perfection is only really achieved when technical perfection is complemented and thus exceeded by artistic vivification. The vitalization in the realization mainly draws on two components of music: the single tone and more complex, longer structures be they melodic phrases or tempo ratios.

This vivification belongs entirely to the realm of the sound form and to the prerogatives of the performers. However, the composition in score form provides guidelines and suggestions for this. From the point of view of the 'abstract composition', but partially even from that of the score form, there is a blurring. From the overall viewpoint of the idea and the work, however, they are a sharpening. Vivification is an almost congruent term with espressivo, with 'expression'. What 'expression' is at all and especially in music can be argued endlessly—and it has been and is argued endlessly. An ego, a subject expresses its sensations and feelings, more generally or those at the moment of singing or playing, its own or those objectified in the work. For our purposes, we can outline what is meant: Expression is a subjective, particularly individually accentuated vivification. It includes the sensitivities, moods, sensations and feelings of the respective musicians. But its central object is the work. Everything that is objectified and exteriorized in this work, both in terms of the composer's ideas, thoughts and feelings and in terms of the representation of reality, is now exteriorized again and brought from the score form into the living reality of sound. Just as the

[232]Pachl 2012, ibid.; cf. e.g. http://www.ifk.ac.at/index.php/kalender-detail/american-lulu-mediale-transformationsprozesse-im-musiktheater.html.

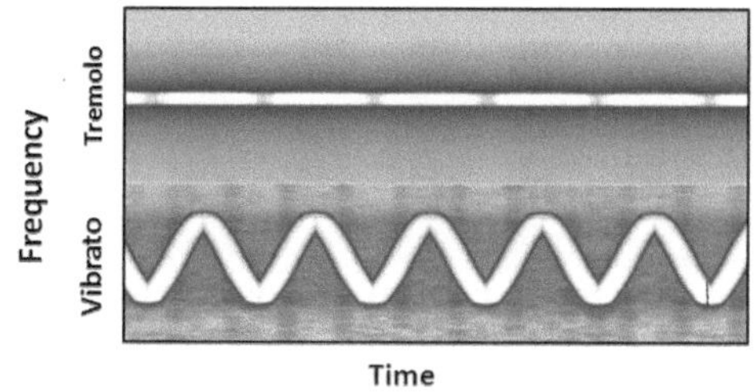

Fig. 7.43 Vibrato and Tremolo
Quoted after Andrew Johnson, 10 August 2009 Aandroyd at English Wikipedia, https://en.wikipedia.org/wiki/File:Vibrato_and_tremolo_graph.PNG (Retrieved 24 August 2020. Cf. https://en.wikipedia.org/wiki/Vibrato, 7 August 2020, retrieved 24 August 2020)

meaning of music requires sound for its full realization, so sound, when understood as vivification and expression includes meaning.

7.5.1 Vibrato. Blurring in the Dimension of Pitches as Vivification

A characteristic means of vivification in performance, whether improvised or after notated music, is the vibrato, a *regularily* pulsating change of pitch (Fig. 7.43).

> The terms vibrato and tremolo are sometimes incorrectly used interchangeably, although (in the classical world) they are properly defined as separate effects with vibrato defined as a periodic variation in the pitch (frequency) of a musical note, and tremolo as a fast repetition of the same note (usually a semiquaver) in order to produce the audible effect of a longer note, especially on instruments which do not have the ability of producing long sustained notes, such as the guitar. In practice, it is difficult for a singer or musical instrument player to achieve a pure vibrato or tremolo (where only the pitch or only the volume is varied), and variations in both pitch and volume will often be achieved at the same time.[233]

Vibrato appears as and is an 'addition' from the point of view of 'abstract composition'. There are various individual variations of execution and instructions for extent and rate calculated in the composition. Numerous variations can be found in US-American string quartet music alone (Figs. 7.44 and 7.45).

On the one hand, vibrato is something traditional, but in traditional music it is often fused with ornaments. So for example in various kinds of Indian Rāga music:

> The term gamaka itself means "ornamented note" in Sanskrit. Gamakas involve the *variation of pitch of a note*, using heavy forceful oscillations between adjacent and distant notes. Each rāga has specific rules on the types of gamakas that might be applied to specific notes, and the types that may not.[234] In Hindustani music *andholan* […] a specific form of ornament (alankar) […] is a gentle oscillation around a note, touching the periphery of an adjacent note

[233]https://en.wikipedia.org/wiki/Vibrato, 7 August 2020, retrieved 24 August 2020.
[234]https://en.wikipedia.org/wiki/Gamak_(music), 7 July 2020, retrieved 30 July 2020.

 The increasing amplitude of the wave indicates increasing speed in vibrato.

 The decreasing amplitude of the wave shows decreasing speed in vibrato. These markings of varying vibrato intensity are used by both James Yannatos and Ramon Zupko.

 The wide wavy line following a pitch indicates a slow wide vibrato.

 The large *V* over a note also designates a slow wide vibrato.

 A narrow wavy line succeeding a note translates into a fast narrow vibrato.

Fig. 7.44 Some types of vibrato in new music
Tischhauser 2002, p. 13

 In a graphic which resembles a medical heart monitor chart Ezra Laderman and Persis Vehar show changing the level of vibrato over the duration of a pitch.

 Ken Steen places a half sharp over a note to designate vibrating a quarter tone above and below the given pitch. Another symbol seen by composers meaning the same thing is ***vib1/4↑***.

Fig. 7.45 Some types of vibrato in new music
Tischhauser 2002, p. 14

> as well as shrutis [the microtones in raga-music] in between. The notes (andolit swars) used for andolan depend on the raga.[235] [...] The Andolan alankar is a gentle swing or oscillation that starts from a fixed note and touches the periphery of an adjacent note. What is special about the Andolan is, that in the course of the oscillation, it touches the microtones or shrutis that exist in between. The note that is being oscillated within an Andolan is known as andolit swar. [...][236]

Related as a blurring that starts at the pitch, the 'dirty intonation' in jazz is yet another, more complex manifestation than the vibrato, but like the vibrato, it is related to the pitch and also expressive. Noise, nasalization and the like are other examples. The core is the blue note, historically-genetically probably less expressive than structurally founded, but secondarily also expressive. In a narrower sense the blue notes are only off-pitchedness, the *desafinado*, i.e. a "detuning". This *out of tune*, however, only applies from the standpoint of purely European or purely African mood systems, not in the hybrid African-American mood.

> blue note is a note that—for expressive purposes—is sung or played at a slightly different pitch from standard. Typically the alteration is between a quartertone and a semitone, but this varies depending on the musical context.[237]

The blue notes are genetically and structurally ambiguous. Their origin is controversial, but not an issue here. From the point of view of both systems of origin, the European and the African, they are in any case a blurring. However, they sharpen the characteristic profile of the blue and the jazz. On the other hand, the vibrato is something secondary and from the late period: it enlivens the already frozen, pure tone. Before that it was an expressive ornamentation, for example as a "tremor" in the clavichord. As 'modern' vibrato, on the other hand, it becomes a continuous vibrato. This weakens its expressive character—once again we have the configuration 'expression becomes material'. This configuration is particularly evident in collective play. The increased 'collective virtuosity' [238] of the orchestras, especially after 1848, led to a de-individualization, normalization and standardization of the vibrato as well. When in special interpretations, the permanent vibrato which historically belongs here is omitted, there are strange alienation effects.

> Gardiners Verdi is an alienated one, a Verdi who has passed through the achievements of historical performance practice, a practice that has had a decisive influence on Gardiner. It is alienated because this performance practice, for example in the vibrato-free playing of the strings, is ahistorical in relation to Verdi: it presents Verdi in a way that was rather unlike the original. For at the time his music was written, it was precisely the *soft-focus technique of the vibrato* that was characteristic of the symphonic music of the 19th century [...]. Applied to Verdi, the historical playing technique becomes on the one hand a manner without philological justification, and on the other hand a way of interpretation that [...] reveals

[235] https://en.wikipedia.org/wiki/Andolan_(music), 16 October 2019, retrieved 30 July 2020.

[236] http://www.itcsra.org/andolan, Introduction to Andolan at the ITC Sangeet Research Academy Archived 21 December 2008, at the Wayback Machine, retrieved 21 August 2020.

[237] https://en.wikipedia.org/wiki/Blue_note, 3 September 2020, retrieved 11 September 2020. Schuller 1968, p. 384: „a microtonal variant"; in detail see pp. 46–52.

[238] Term explicated in Heister 1983.

> enlightening qualities: For all its technical precision, it refrains from any "embellishment" of the music, but instead increases its dissonant tensions to the limit of what is bearable.[239]

Like the dirty intonation, the vibrato also has tone color aspects in addition to the purely pitch-related ones. These components appear radicalized, enriched with other elements of the intonation, when the original expressive character is restored. Methods of avant-garde composition bring out the making as such and the effort of production, especially Helmut Lachenmann[240] and Joachim Hespos. This *practical reflection* of the production in the process of production has an expressive effect in itself: a sharpening of the blurring. It is yet another turn of events when a conductor transfers such avant-garde methods back to the realization of classical works. What is often—and with good reason—concealed and blurred is thus emphasized and sharpened under special circumstances. As in the case of the perfectionist T. Currentzis: Here

> even supposed marginal phenomena of Mozart's composition seem like sensations; one thinks one hears the scratching of the resinous horsehair of the bows on the gut strings. Music like under a burning glass. The conductor comments: "Our radicalness lies in precision, in the strictest discipline. But you can only hear that with the appropriate technique. The magic does not come from the life experience it is in the score".[241]

On the other hand, such perfection does not only include the emphasis on making, but also the—usual and desired—homogenizing and merging of many identical musical instruments into one line, a different manifestation of blurring. Dialectically, however, it brings a sharpening of the second instance—here even in choral music-making (see Sect. 7.4.3.1).

> There are ten first violins. But even at the furious speed of "Fin ch' han dal vino" it still sounds as if they were one.[242]

7.5.2 Modifications of Tempo, Tempi and Beat

Homogenization in the pitch dimension corresponds to homogenization in the duration dimension, especially in the extended "supra-segmental" component of the tempo that goes beyond the single tone. Achieving a common *and*, in principle, uniform tempo that is maintained by all is a long historical process of educating musical collectives that must be constantly repeated and renewed. Through practice

---

[239]Lauterbach, Reinhard: *Wie eine Statue im Fortissimo. Vibratofreier Weltuntergang: Verdis Requiem, dirigiert von John Eliot Gardiner zum Abschluss des Festivals » Wratislavia Cantans « in Wroclaw, Junge Welt,* Berlin, 22.9.18, Feuilleton, p. 11. Emphasis HWH.

[240]Cf. e.g. about *Pression* Sparrer 2005.

[241]Stefan Siegert: *Der Maestro als Edelpunk. Teodor Currentzis ist ein moderner Klassikstar. In der staatlichen Oper der russischen Stadt Perm feilt der perfektionistische Musikdirektor mit seinem erstklassigen Ensemble bis spät in die Nacht an den Aufnahmen, Junge* Welt, Berlin, 31 Jan 2015, Wochenendbeilage S. 4.

[242]Ibid. Refers to Don Giovanni's „Weinarie" in Mozart's opera.

and rehearsal, mere variations in tempo in the temporal sequence, which are insignificant, are gradually eliminated. The same applies to differences in the simultaneity of the ensemble playing.

> Individuals thus achieve a personal performance improvement through the activity in the group. This phenomenon is also inherent in making music in groups: Choir or orchestra experiences confirm this effect. Those who are feeling insecure can orientate themselves on the more confident, while those who are already confident strive to be as clear as possible in their performance because of the group situation. Davidson and Correia (2002) prove that these tendencies towards orientation are not limited to the acoustic, but also include visual impressions. They investigated the movements of musicians while playing and found a regulating function for these rhythmic motion activities: In small ensembles, the performers tend to coordinate their interaction with the help of their performance motions. (see Sect. 6.9.4)[243]

The tempo can be measured quite precisely, also as a sequence of stages, of states, as progress of a relative identity, if necessary, including all persistent uncertainties as—minimized—differences. It can be assumed, that the artistic point de perfection, is not the optimum of metronomic exactness, but an imperceptibly regular rhythm with an imperceptible, microscopic pulsation. The deviations from this center are striking, partly errors, partly blurring and expressive. The technical sharpening is the basis and prerequisite for expressive blurring. The conductor Wilhelm Furtwängler (1886 to 1954) was notorious for his blurred inserts. A concert master once asked modestly: "Please, Doctor, at which point of your lightning do you want us to come in?" Furtwängler's answer is not known, but the reason is: it is precisely the blurring that should ensure a particularly tense attention of the musicians and therefore a higher precision of playing.

7.5.2.1 Tempo, Absolute and Relative. Pulse, Time/Beat, Rhythm—Middle and Deviation

Speed means, as in walking or in cycling, speed. It can be small or large. Unlike here, however, in music it is not km/h, but the number of sound events per unit of time—the more, the higher the tempo, or, reduced to the basic pulse, beat per minute, bpm or M.M. (Mälzel's metronome). Only from a purely technical point of view and with regard to the infantile and juvenile pleasure in the frenzy of speed for its own sake do the Olympic ideals of "Citius, altius, fortius" apply in music, i.e. faster, higher and louder. In musical tempo, this *absolute* tempo of a section or piece can be distinguished from *relative* tempi of individual layers in simultaneity. The latter is the basis for the rubato, and for polymetry and polyrhythm. Extremely fast tempos tend to blur, like hasty speech (see Sect. 7.5.2.4). Yet also the slow motion of extremely stretched slow tempi can cause blurring effects.

Pulse is, factually and terminologically analogous to the continuous beating of the heart, a relatively even continuous sequence. The biological heartbeat itself, however, is two-valued. The systole, the "contraction" of the heart muscle, lasts

---

[243] Lehmann 2005, p. 56.

Fig. 7.46 Schubert: *Gretchen am Spinnrad*
op. 2, D 118, beginning
Eusebius Mandyczewski (ed.): Schubert's Werke, Serie XX: Sämtliche einstimmige Lieder und Gesänge. Band 1. 1811-14, No. 31 (pp.191-96). Breitkopf & Härtel, Leipzig (1894). Plate F.S. 337. IMSLP13971-SchubertD118_Gretchen_Am_Spinnrade

approximately 100 ms with a pulse of 60 bpm, the diastole, the relaxation and blood outflow phase of the 'heart cycle' lasts about 700 ms, i.e. it is considerably longer.[244] The *musical* pulse is filtering, normalizing sharpening the physiological pulse—art as a metamorphosing mimesis of nature. In general, a regular pulse, rather rare in nature, is already the result of a filtering towards art - it also occurs in such working processes for example pounding etc.

> A pulse is one of a series of regularly recurring, precisely equivalent ("undifferentiated") stimuli. Like the tick of a metronome or a watch, pulses mark off equal units in the temporal continuum. [...] A sense of regular pulses, once established, tends to be continued in the mind and musculature of the listener, even though the sound has stopped. [...] The human mind tends to impose some sort of organization upon such equal pulses.[245]

On this basis, which is also the development of a still 'unshaped figure' in the form of regularity, new sharpening and blurring are possible. The two sounds of the heart beating occur frequently in music. In Schubert's *Gretchen am Spinnrad* (from *Faust I*) from 1814, one of his earliest mature piano songs, Schubert overlays these tones in the accompanying system with those of the treading of the spinning wheel (Figs. 7.46 and 7.47).

At one point both are disturbed and come to a halt, an expression of Gretchen's confusion, desire and fear.

The musical pulse can only be latent or manifest as sound. It moves within a general anthropological range of tempo.

> The pulse may be audible or implied. The tempo of the piece is the speed of the pulse. If a pulse becomes too fast it would become a drone; one that is too slow would be perceived as unconnected sounds. When the period of any continuous beat is faster than 8–10 per second or slower than 1 per 1.5–2 seconds, it cannot be perceived as such. "Musical" pulses are

[244]Cf. https://www.kardionet.de/herzzyklus-herzfunktion/.

[245]Cooper, Grosvenor and Meyer, Leonard B. Meyer (1960). The Rhythmic Structure of Music, pp. 3–4. University of Chicago Press, quoted in https://en.wikipedia.org/wiki/Pulse_(music), 30 September 2020, retrieved 30 September 2020.

Fig. 7.47 Schubert: *Gretchen am Spinnrad* op. 2, D 118, stop

> generally specified in the range 40 to 240 beats per minute. The pulse is not necessarily the fastest or the slowest component of the rhythm but the one that is perceived as basic.[246]

Metrum is terminologically ambiguous, but can be combined with pulse, beat and rhythm as follows:

> Meter is the measurement of the number of pulses between more or less regularly recurring accents. Therefore, in order for meter to exist, some of the pulses in a series must be accented—marked for consciousness—relative to others. When pulses are thus counted within a metric context, they are referred to as beats.[247]

Beat means the order of the pulses as beat within meters. These beats, in turn, are ordered as bars, as—various, different—recurring units with a constant accent pattern. *Accent* means primarily stressed versus unstressed. This binary difference is also graded with different strengths and valences in the modern "Akzentstufentakt" (accented measure) (Heinrich Besseler)—a more or a less, which corresponds to the way of thinking of the Fuzzy Logic. The accent, the emphasis is itself multidimensional, is musically marked primarily by volume, but a minimal extension of the duration can also serve this purpose. In the Viennese Waltz, for example, the accentuation is sharpened the beat is blurred as a regular continuous sequence. In its 3/4 time, the first beat is shortened, the second is thus drawn near, and the first is therefore emphasised particularly strongly. The Viennese Waltz is based on the 3/4 time. However, the basic figure consists of six steps and therefore needs two bars (Fig. 7.48).

Rhythm, finally, is, on the one hand, a universality of numerous processes in reality, which take place in and as temporal structure and as movements in a broad sense, from the cosmos to the microcosm, from the movements of the planets and the earthly seasons to biological rhythms and the pulsations of the flagellates. The component of a cyclical repetition, possibly over very long periods of time, resonates with it. Musical rhythm, in contrast, is much more flexible. It remains in the mesocosm according to the human 'metron', i.e. measure. It reproduces such rhythms of reality

[246] https://en.wikipedia.org/wiki/Pulse_(music), 13 May 2020, retrieved 13 August 2020.

[247] Cooper, Grosvenor and Meyer, Leonard B. Meyer (1960). The Rhythmic Structure of Music, pp. 3–4. University of Chicago Press, quoted in https://en.wikipedia.org/wiki/Pulse_(music), 13 May 2020, retrieved 13 August 2020.

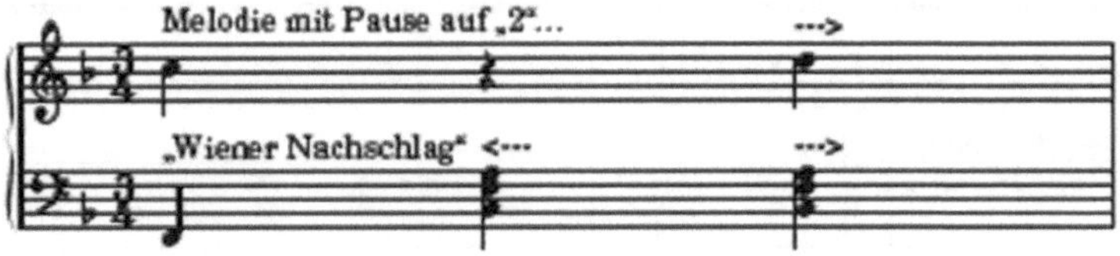

Fig. 7.48 Vienna Waltz
The really audible metrical-rhythmical basic pattern.
Quoted after Dallmann 2015, p. 92

and transforms them. It refers to the dimension of tone duration, is fixed in pulse and metre, or even measure, and moves within this framework as a sequence of different or rarely identical tone durations. Its definition is also controversial. An explicitly normative definition of musical rhythm is offered by W. Seidel 1998:

> Rhythm is the order of movement or times that is immediately and clearly comprehensible to the human mind and whose perception is combined with feelings of well-being.[248]

In our context, it is sufficient to define it as a contradictory unit of standard and deviation, relative identity and difference. It follows the given regular patterns of pulse and meter, and it behaves irregularly and variably to them. In relation to the symmetry of the pulse it is a blurring in relation to the *musical* shape created with it, it is a sharpening of the otherwise shapeless process.

7.5.2.2 Patter Song. Absolute Tempo of Realization up to the Limit of Text Comprehensibility

Such unmeasuredness is the exception. Even within the measured there are sufficient possibilities for blurring and sharpening. The measure of the musical tempo is the well-known bpm, beat per minute. With that the time of the music is related to the general time. What is perceived as fast is subjectively very variable, individually as well as probably historically—acceleration, a phenomenon especially since industrialization, also has a blunting effect here, and in fact, it is actually dulling. A certain middle ground is provided by the resting pulse, which, depending on the individual, lies between 60 and 80 per minute in a blurred zone.[249] Anything above this level has a rather rapid effect, and vice versa. In general, the blurring potential of faster tempos is greater: the more information per time unit, the more difficult it is to capture.

This is exploited by musical genres that are absolutely fast, such as the Patter song—it has a blurry realization tempo, but it is intended to be so in terms of composition. A special case of blurring, already with regard to reception, the perception

[248] Seidel 1998, col. 258.

[249] Siegele sets a somewhat higher basic pulse rate: "J. F. Herbart, who is followed by H. Riemann, bases this center in natural science experiments on the average pulse rate (about 75–80 per minute); the limits within which the rate can vary (about 40 as minimum, 130 as maximum) also depend on this" (Siegele 1968, col. 27).

form, are passages or entire works in which a frenzied, excessive tempo of realization makes an exact receptive realization difficult or even impossible. Already classical examples are the "sound sheets" of saxophonist John Coltrane or his composition Giant Steps with extremely rapid changes in harmony. But even the early Be-Boppers blurred the changes with very fast tempos in order to make plagiarism more difficult for competitors. Especially in music theater, this technique is used to create or intensify comic effects, as in the prototype, the Cavatina of Figaro from Rossini's *Il Barbiere di Siviglia.* In popular music, the type of patter song is similar.

> The patter song is characterised by a moderately fast to very fast tempo with a rapid succession of rhythmic patterns in which each syllable of text corresponds to one note. It is a staple of comic opera, especially Gilbert and Sullivan, but it has also been used in musicals and elsewhere. The lyric of a patter song generally features tongue-twisting rhyming text, with alliterative words and other consonant or vowel sounds that are intended to be entertaining to listen to at rapid speed. The musical accompaniment is lightly orchestrated and fairly simple, to emphasise the text. The song is often intended as a showpiece for a comic character, usually a bass or baritone.[250]

Numbers from Gilbert and Sullivan's comic operas are well known, such as "I am the very model of a modern major-general" from *The Pirates of Penzance.* Especially remarkable is the trio My eyes are fully open from Ruddigore: "this particularly rapid, unintelligible patter / is not generally heard and if it is it doesn't matter."[251] Gilbert & Sullivan thus explicitly ironically and self-ironically reflect on the blurring that is accomplished in reception. Melodramatically executed, i.e. speaking over an instrumental accompaniment, as in rap, is am attraction number in Kurt Weill's musical Lady in the Dark, 1941, Tchaikovsky. It is performed in the circus scene, as a theater within a theater representation of a dream, and is dramaturgically very meagerly motivated with an 'apropos' of the keyword "Russian". A clou is not at least the fact that the irritating blurring of meaning by the rapid tempo only develops its wit when it is articulated sharply and precisely.

> *The singer should be capable of excellent enunciation, to sing the song to maximum effect.*[252]
>
> [Ringmaster] Without the least excuse
>
> Or the slightest provocation
>
> May I fondly introduce
>
> For your mental delectation
>
> The names that always give me a concussion
>
> The names of those composers known as Russian:
>
> There's Malichevsky, Rubinstein, Arensky, and Tschaikowsky
>
> Sapelnikoff, Dimitrieff, Tscherepnin, Kryjanowsky
>
> Godowsky, Arteiboucheff, Moniuszko, Akimenko
>
> Solovieff, Prokofieff, Tiomkin, Korestchenko […]
>
> Stravinsky and Gretchnaninoff

---

[250] https://en.wikipedia.org/wiki/Patter_song, 9 April 2020, retrieved 13 August 2020.

[251] Following https://de.wikipedia.org/wiki/Patter_song, 31 October 2013, retrieval 14 May 2015.

[252] Ibid.

Rumshinsky and Rachmaninoff
I really have to stop, the subject has been dwelt upon enough
[Chorus] He'd better stop because we feel we all have undergone enough.[253]

The blurring itself can then be further intensified. Particularly with Tchaikovsky, during the heyday of this type of musical in the 1940s and 1950s, there was a record hunt among the relevant comedians for the fastest person to speak it.[254] The framework of the competition is that the blurring for perception is driven as far as possible, but balanced by sharpening it to such an extent that the individual names still have to be articulated in a comprehensible way.

7.5.2.3 "Zwiefacher" and Aksak. Bar Changes and Asymmetrically Accentuated Meters

A comparatively mechanical blurring in the dimension of durations are bar changes like in the Bavarian Zwiefachen and in some folklore dances. This change of time does not simply happen, but is intentional.

The Zwiefacher alternates between odd and even time signatures, changing from 3 to 2 beats per bar. The changes may occur regularly – for example, two measures per time signature, may change only once or may change irregularly throughout.[255]

Traditional conventions of notation bring additional blurring (Fig. 7.49).

Zwiefache sheet music written in the traditional way […] is not played the same as modern sheet music. What I mean by this is that traditional Zwiefache sheet music the 3/4 time measures the quarter note is played like a quarternote, and in 1/4, 2/4, or 4/4 time measures, the 1/8 note is played as long as a quarternote.[256]

The blurring is also found in the dance steps.

Fig. 7.49 *Unser alte Kath*, https://www.stammtischmusik.at/noten/boxhamerisch.htm

[253]Quoted in https://lyrics.fandom.com/wiki/Danny_Kaye:Tschaikowsky_(And_Other_Russians), retrieved 13 August 2020.

[254]Cf. the since Schumann not uncommon performance instruction *as rapidly as possible*. „At each performance, Kaye tried to break his previous speed record for reciting this song: consequently, it was intended to be recited *a cappella* (without instrumental accompaniment), as the orchestra could not keep up with him"(ibid.).

[255]Elisa H. Barney Smith, https://web.archive.org/web/20081222013032/http://www.germandance.org/zwiefache.html.

[256]Ibid.

Fig. 7.50 L. Bernstein, music / St. Sondheim, lyrics: West Side Story (1957) Musical example: Wahoofive at English Wikipedia. 11 March 2005. https://en.wikipedia.org/wiki/America_(West_Side_Story_song)#/media/

In terms of dancing, the change between 3/4 and 2/4 time corresponds to an alternation between waltzing and dreher steps. Per bar one waltz step (W, one bar 3/4-beat) or one dreher step (D, 1 bar 2/4-beat) results in half a turn each.[257]

Semantically sophisticated is the regular beat change in the song America in the musical *West Side Story* (Music by Leonard Bernstein, lyrics by Stephen Sondheim, 1957, 1961 movie) (Fig. 7.50).

The 6 quavers are written like a verse, the 3 quarters like a refrain.

Girls: Life can be bright in America

Boys: If you can fight in America

G: Life is all right in America

B: If you're all white in America.[258]

The penetrating repetition of the slowing 3/4 beat is given something scornfully challenging together with the text—on a microscopic scale a critique of the "American Dream". In its energetics and with the alternation of bar and accent pattern, the pattern is reminiscent of a hemiola, but is none in the narrower sense (see below).

The two bar-types alternate and are not superposed, as in a hemiola. [...] The composer's tempo instruction is "Tempo di Huapango".[259]

There are also changes of time in concert music. In the 19th and especially in the 20th century they became more frequent. Stravinsky is prototypical with works such as *Sacré du Printemps* or the *Piano-Rag-Music*. With irregular time changes, irregular 5/8 or 5/4 bars, changing accents and syncopations he artificially imitates ragtime. And he literally translates "ragtime" into musical terms as "torn time" or shredded time (Fig. 7.51).

Boris Blacher's "variable meters", already mentioned above, take this up, but use regular patterns. Olivier Messiaen's "additive rhythms" are even more subtle, in which he also incorporates Indian models. No real blurring, on the other hand, are metrics that appear conglomerate from a Western European point of view, such as the Greek "Rhythmos Kalamatianos", among others as 7/8 time, or other dances like the Zeibekiko with an irregular, asymmetrical accent pattern.

---

[257] Detailed with sheet music examples https://www.stammtischmusik.at/noten/boxhamerisch.htm, retrieved 20 August 2020.

[258] The whole lyrics in http://www.guntheranderson.com/v/data/amerika.htm.

[259] https://en.wikipedia.org/wiki/Hemiola, 15 August 2020, retrieved 29 August 2020.

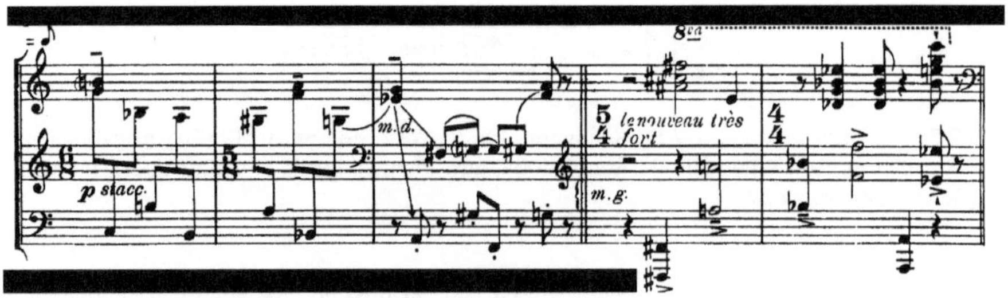

Fig. 7.51 Igor Strawinsky: *Piano-Rag-Music*, mm. 5–9
First edition. J. & W. Chester, London (1920). Plate J.W.C. 2061. IMSLP524047-PMLP180105-PMLUS00694-pianoragmusic

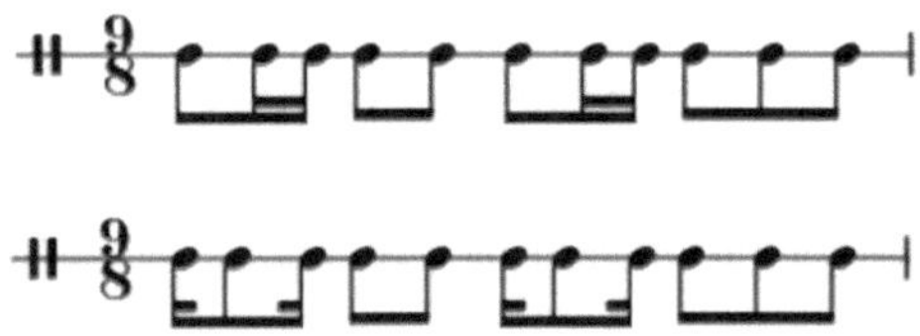

Fig. 7.52 Zeibebiko https://en.wikipedia.org/wiki/Zeibekiko. 9 December 2019. Retrieved: 19 July 2020

Fig. 7.53 Tchaikovsky's symphony „Pathétique“, II. movement
(Stefan Kompacher, September 2007, https://upload.wikimedia.org/wikipedia/de/c/c6/Pathetique4.jpg, retrieved 20, June 2020)

The Zeibekiko usually has a metric pattern of 9/4 or 9/8. But the pattern is not broken down regularly and periodically into 3 × 3 but asymmetrically into 2 + 2 + 2 + 3. Two patterns (Fig. 7.52):

> The Zeibeikiko, as an old dance, is strictly for males. Due to the movements of the dancer, it is sometimes known as the "eagle dance". The dance has no set steps, only certain figures and a circular movement. It takes place in an area little surpassing one square metre and mostly consists of improvised movements.[260]

Equally asymmetrically divided is the Turkish "aksak", also known as the "Bulgarian rhythm" as *alla bulgarese* often used by Béla Bartók, usually as 8/4 from 3 + 3+2 beats. The metric accumulation has a particularly energetic effect, i.e. it is a sharpening on the level of interaction between music and the perceiver. In contrast, the 5/4 bar in Tchaikovsky's 6th Symphony "Pathétique", II movement a disguised, limping, depressive waltz, so to speak. So the 5/4 is a blurred compression of 2 × ¾ (Fig. 7.53).

---

[260]https://en.wikipedia.org/wiki/Zeibekiko, 9 December 2019, retrieved 19 July 2020.

Fig. 7.54 Robert Schumann: 3rd Symphony in E flat major.
Hemiola in the main theme. The theme is virtually in 3/2 time for the first few bars.
Quoted after Dr.C.Gradus, Eigenes Werk, CC BY-SA 4.0, https://commons.wikimedia.org/w/index.php?curid=78384369

Fig. 7.55 Pattern of a Ghanaian gyil (a xylophone)
Quoted after Dr clave, 22 January 2012, https://upload.wikimedia.org/wikipedia/commons/d/d7/Ghanaian_gyil.jpg

The aforementioned blurring in the Viennese Waltz itself contributes to Tchaikovsky's blurring. Changes in time, shifts in accent, crossings of uniformity and regularity in the dimension of duration act above all as an irregularity, as a disturbance and concealment of a uniform metric order. They are therefore blurring, but also sharpening, not least as an activation of attention. In the second instance, these blurring are necessarily not always precisely executed, i.e. they are themselves blurred. A rhythmically particularly attractive variation of the change of beat or accent pattern is the hemiola. Schematically, 6/4 is structured as two times 3/4 or three times 2/4. It can also be interpreted as (Fig. 7.54)

> a special form of a syncopation. In this case, the accentuation scheme of a measure is temporarily broken up in such a way that the semblance of a new, different time signature is created. By linking two three-bars together, the two three-bars are often reinterpreted into one large three-bar.[261]

Here the hemiola has something floating and at the same time compelling, is primarily a ritardando, but with features of an accelerando.

7.5.2.4 Polymetry and Polyrhythm

The hemiola, arranged vertically as two voices, *is* already a simple polymetry (Fig. 7.55).

> A repeating vertical hemiola is known as polyrhythm, or more specifically, cross-rhythm. The most basic rhythmic cell of sub-Saharan Africa is the 3:2 cross-rhythm. […] In the following example, a Ghanaian gyil plays a hemiola as the basis of an ostinato melody. The left hand (lower notes) sounds the two main beats, while the right hand (upper notes) sounds the three cross-beats.[262]

---

[261] https://de.wikipedia.org/wiki/Hemiole, 9. July 2020, retrieved 19 August 2020.

[262] https://en.wikipedia.org/wiki/Hemiola, 15 August 2020, retrieved 29 August 2020.

Polymetry and polyrhythm are in the dimension of the tone duration correspondences to 'polyphony' as 'melodic multiple voices'. In sub-Saharan, black African music, polymetry is widespread as a layering of

> additive rhythms, [...] as in the folksongs of Russia, Bulgaria and the music of the Aka Pygmies of Central Africa [...]. The percussion music of the Aka pygmies, like that of the Brazilian Samba, for example, consists of a web of cyclic rhythmic patterns layered in several individual strands. However, as in the following example, it can be seen that one such strand has a complex internal rhythmic structure [...].[263]

In the music of the Aka pygmies one or the patterns has 24 beats. Yet these 24 beats are not divided evenly, e.g. as 6×4 or 8×3, but in an—in Western categories—"irregular" sequence: $3 + 3 + 3 + 2 + 3 + 3 + 3 + 3 + 2$.[264]

Complex polymetric superpositions, similar to 'phasing' and other operations with the superposition of patterns, produce acoustic emergent, "inherent" or resulting phenomena, which can be attributed something hallucinatory.

> Since there are many beat streams in the implicit metric matrix, the rhythm of a sounded phrase is subject to cognitive re-orientation depending on the flow of beats on which perception is grounded. Within the enigmatic and paradoxical context of the metric matrix, tones are onbeat and offbeat simultaneously.[265]

An alternative perception is the concentration on one layer at a time, like a 'tilting figure'. This layer becomes the ‚figure'the other layer becomes the ground according to the categories of the gestalt psychology.

> Research into the perception of polymetre shows that listeners often either extract a composite pattern that is fitted to a metric framework, or focus on one rhythmic stream while treating others as "noise".[266]

The blurring on the one side is complementary to the sharpening on the other. An example of polymetry in the music of the Viennese classics is at the most audible at the beginning. In Don Giovanni's ball, the bourgeois contredanse resounds in the upper two systems, the courtly-noble minuet in the lower two. (A little later a peasant-plebeian "Teitscher", a forerunner of the Ländler and Waltz, is added as the third.) Mozart thus presents the layering of the three main classes of the feudal order in the musical vertical as a relative harmony of contradictions. The harmony is broken by the dramatic process—an attempted seduction, and ends in chaos. But this is called to order, so to speak, and is embedded in an overarching whole. The metric-rhythmic mixing and blurring is a musical-political sharpening (Fig. 7.56).

Artistically ambitious popular music sometimes goes beyond this in its complexity of polymetry.

---

[263]Gomez et al. 2005, p. 5.

[264]Gómez et al. 2005, p. 5. Die Autoren schlagen präzise mathematische Modelle und Notationen für die „irregulären" patterns vor.

[265]Locke 2011, p. 56.

[266]https://en.wikipedia.org/wiki/Metre_(music)#Polymetre, 18 July 2020, retrieved 29 August 2020.

Fig. 7.56 Mozart: *Don Giovanni* (1788). Dances from Act 1, Scene 5
https://upload.wikimedia.org/wikipedia/commons/1/1b/Mozart_Don_Giovanni_2_dances_together.png

Fig. 7.57 Ferenc Liszt: 3 Études de concert S. 144.
No. 2 „La Leggierezza“, m. 44.
Quoted after IMSLP NLA, series i, vol. 2

> In “Toads of the Short Forest” (from the album Weasels Ripped My Flesh), composer Frank Zappa explains: “At this very moment on stage we have drummer A playing in 7/8, drummer B playing in 3/4, the bass playing in 3/4, the organ playing in 5/8, the tambourine playing in 3/4, […] and the alto sax blowing his nose” (Mothers of Invention 1970).[267]

Even significantly more complex is, for example, the

> song *frame by frame* from the album *Discipline* […] by the band King Crimson […]. At time index 1:07 the two guitars start playing a unison riff in 7/8 time with changing accents. On the fourth repetition Robert Fripp leaves out an eighth note on the riff and plays the original riff in 7/8 time and the riff with one eighth less in 6/8 time. This results in an overlay of the basic riff in 7/8 in the first guitar track with an alternation of 6/8 and 7/8 time in the second guitar track. The drummer plays a 13/8 riff over this, which by shifting the emphasis underlays both the basic 7/8 riff and the 13/8 riff in the second guitar track (6/8 + 7/8 = 13/8). This creates an extremely complex overall rhythmic structure and various harmonic overlays on each pass until the two guitar tracks meet again after a total of 91 quavers (13 repetitions of the 7/8 riff and 7 repetitions of the 13/8 riff), i.e. the “one” (the first beat of a bar) is simultaneous again for both tracks for the first time.[268]

Polyrhythmic figures are considerably more frequent than polymetry. They can arise almost spontaneously, and are in any case a blurring of the usual metric order (Fig. 7.57).

> With polymetre, the bar sizes differ, but the beat remains constant. Since the beat is the same, the various metres eventually agree. (Four bars of 7/4 = seven bars of 4/4.) With polyrhythm, the number of beats varies within a fixed bar length. For example, in a 4:3 polyrhythm, one part plays 4/4 while the other plays 3/4, but the 3/4 beats are stretched so that three beats of 3/4 are played in the same time as four beats of 4/4. More generally, sometimes rhythms are

[267] https://en.wikipedia.org/wiki/Metre_(music)#Polymetre, 18 July 2020, retrieved 29 August 2020.

[268] https://de.wikipedia.org/wiki/Polymetrik, 22 July 2020, retrieved 29 August 2020.

> combined in a way that is neither tactus nor bar preserving—the beat differs and the bar size also differs. See Polytempi.[269]

An increase in polymetry and polyrhythm is finally the 'polytempo', a polyphonic simultaneity of different tempi. It can be realized almost exclusively by machine.

> Multitemporal music is composed using sound streams that have different internal tempi or pulse speed, for example one part at 115 bpm and at 105 bpm at the same time. Multitemporal music was first heard in US-Mexican composer Conlon Nancarrow's work [...]. To overcome the limits posed by a human performer in playing a multitemporal score Nancarrow used two modified player-pianos, punching the rolls by hand. [...] A similar technique, with the tempi similar to each other is rhythm phasing [or phasing (S. Reich, see below)][270]

7.5.3 Rubato. Blurrings in Relative Tempi as Vivification

Tempo rubato, literally "stolen time", is a slight but noticeable deviation from the basic tempo and pulse, either after the beat or before the beat. Bar changes and accent shifts have nothing to do with the tempo or tempo fluctuations of the rubato. Rubato acts as vivification, as expression and as a means of representation of subjectivity. The rubato appears in two basic types. On the one hand, the tempo as a whole fluctuates in relation to the initially established basic tempo of a work or section, which continues to function as a latent pulse; it becomes faster and slower. The accelerando or ritardando as the final signal is the simplest manifestation of this type. They are partly notated, partly legitimized by tradition and custom, like the 'fluctuating tempi' within the musical process. They occur particularly frequently with and since Mahler (see Chapter 3). The deviations from the underlying tempo do not usually occur with caesuras, but rather continuously, and thus necessarily with blurred edges at the transitions. On the other hand, the tempo rubato deviates from the basic tempo only in one part of the composition. Another part maintains the "tempo giusto", the regular pulse. In general, the tempo rubato is a prerogative of the solos over the tutti of the collective—logically, this kind of subjectivity is more easily possible for individuals. This rubato type can also be understood as *'polytempo'*.

> "The Italian singers have a manner of slowing down the singing or deviating from the exact beat movement at will while the orchestra continues to play in the given time," it is said in an old singing manual from France. ...] (T)his style was taken up again by singers and instrumentalists in the 1930s [as] laid-back in [American] popular music.[271]

Yet again, in a particular kind of dialectic, this execution is a blurring from the point of view of the work, at least initially, but a sharpening from the point of view of the interpreters, since they can thus express their subjectivity, and in the best case even help the expression objectified in the work to become tonal. Béla Bartók understood "tempo giusto" and "tempo rubato" as two metric-rhythmic principles

---

[269] https://en.wikipedia.org/wiki/Metre_(music) Polymetre, 18 July 2020, retrieved 29 August 2020.

[270] https://en.wikipedia.org/wiki/Polytempo, 15 August 2020, retrieved 29 August 2020.

[271] Dallmann 2005, Abstract.

and composed the contrast explicitly. In this way, the facts of the tonal realization in performance practice and execution are already taken into account in the composition, anticipated—and regulated, an order of the irregular, the taming of the wild, so to speak.

7.5.3.1 Two Tempi—Tempo Giusto Versus Rubato. Right Versus Left Hand on Piano

Especially with polyphonic musical instruments, and even more so with the piano, which is particularly suitable for this purpose, where the right hand knows what the left hand is doing but does not follow it undifferentiatedly, the polytempo of the two tempi is realized by one and the same person. Especially with the piano—and the piano has been the leading European instrument since the Viennese Classicism—the two basic types mentioned above are reproduced encaptisically, as a historical sequence. Up to and including the Viennese Classicism, rubato meant that the left hand kept the tempo, i.e. the tempo giusto, while the right hand deviated from it, i.e. the tempo rubato. After that, rubato meant a fluctuation of the tempo throughout the entire composition, which was not completely regular in itself. The micro-fuzziness is not imperfect but essential, as an expression of spontaneity and liveliness. In general, it is again a matter of the configuration of relatively abstract technical perfection versus aesthetic perfection. In Romantic piano playing between about 1830 and 1880[272] with an afterglow up into the 1920s, this rubato type was used excessively as a means of expression, often combined with the already mentioned arpeggio, the playing of chords in (rapid) succession of the chord notes instead of simultaneously, as for instance Ignaz Paderewski.

> (T)his form of arpeggiation is indispensable. When subtly applied it creates a body of full and supporting tone and it will also sustain an otherwise empty melodic note in a manner extremely effective and grateful to the ear. Many pianists, in their anxiety to avoid its excessive use, carry to equal extreme the "square stroke," playing unarpeggiated, and exactly together. Certainly nothing can be more *unmusical* or tiresome to the ear.[273]

The decisive feature is the quasi doubled tempo rubato through the non-coordination of the two hands.

> The purpose of noncoordination was to characterize the music, generally by *heightening the expression* and *clarifying the rhythmic structure*. There were – the past tense is appropriate, since the practice has all but vanished – surprisingly few techniques. Melodic notes were delayed or, rarely, anticipated; bass notes anticipated the beat; accompaniment chords were broken or arpeggiated.[274]

---

[272] Around 1885, Adolph Christiani wrote: „[T]he old ad libitum style of interpretation (in rhythmic respect), which only too often resembled the unsteady gait of a drunken man, is giving way to *positiveness and precision*".Quoted in Arnest 2012, p. 9.

[273] Arnest 2012, p. 1.

[274] Ibid.

Again, the blurring from the point of view of the 'abstract composition' and here of a metronomic exactness and technical perfection in the second instance is a sharpening of the expression. Firstly, this blurring can even serve to sharpen the profile of polyphony as „separator in contrapuntal music".

> "an arpeggio may also be employed where the polyphony is to be brought out more distinctly; but only at important points, for instance where one part ends and the other begins at the same time."[275]

The occurrence of this rubato and arpeggio in music is itself specifically distributed

> more in slow pieces than in fast ones (which alone should put to rest any notion that it's a technical problem); more on strong beats than on weak ones; more on dissonances than on consonances; more with wide melodic skips than with narrow ones.[276]

Besides the general sharpening of the expression and the special sharpening in the representation of the contrapuntal lines, there are other reasons for this rubato and the adjacent procedures of vivification. Secondly, due to acoustic conditions, the singing quality becomes clearer, a continuation of the "cantabile" playing, which was already an ideal of the Viennese classics.

> Since the piano's bass notes contain overtones out of which many of the upper notes grow, a pianist can prepare the melodic note by anticipating it in the bass. This softens the melodic attack, giving the note a less percussive, more *singing quality*. [...] Because of this inherent distinction between lower notes and higher ones – there are overtones but not undertones – chords are nearly always rolled from bottom to top [...] This is also why, in a piece such as Chopin's Etude, Op. 25 No.7, romantic pianists played the melodic bass line ahead of the beat even though the melodic right hand, was played behind the beat. This acoustic effect was heightened as piano design changed over the course of the 19th century; the whole instrument became brighter and louder, but especially the bass. A second acoustic effect involved lessening the piano's inherent percussiveness. Swiftly rolling a chord, even a quiet chord, would make the attack less obtrusive by spreading it out over several milliseconds.[277]

Thirdly, under the heading "Beat versus Pulse ", the aim is to accentuate the human body against mechanical, industrial technology. Carl Maria von Weber formulated this explicitly and consciously in 1824:

> "The beat should not be a tyrannical restriction or the driving of a *mill hammer*. On the contrary, it should be to music what the pulse-beat is to the *life of man*."[278]

Once again, the technically perfect dot becomes a spot.

> One effect of any type of noncoordination is to expand the beat, giving it a sense of duration; that is, the romantics favored a pulse-like area over a point-like beat. Weber was especially talking about tempo flexibility, but his words are also appropriate to the nature of the romantic pulse itself, which, like the human pulse, wasn't a dot, but had a beginning, a peak, and an end.[279]

---

[275]Malwine Brée, a pupil of Leschetizky 1902, quoted in ibid., p. 3.

[276]Ibid., p. 2.

[277]Ibid.

[278]Quoted in Arnest 2012, p. 2. Emphasis HWH.

[279]Ibid.

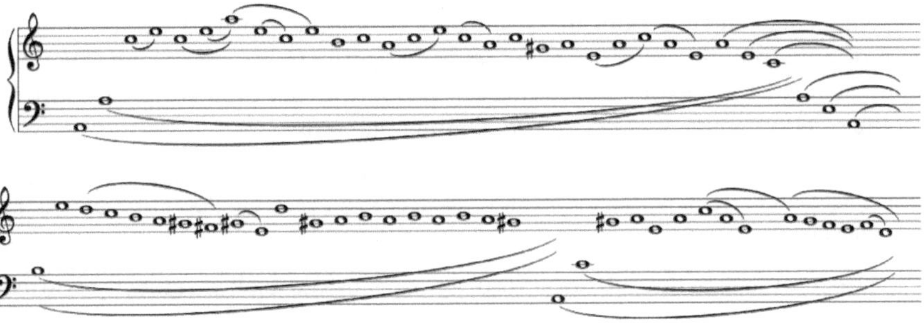

Fig. 7.58 Louis Couperin, Prélude non mesuré à l'imitation de M. Froberger (ca. 1685) Quoted after Odermatt 2018, p. 46, Figure 5.1

Fourthly, this blurring of the tempo coordination as an accentuation sharpens the melodic contour.

> When used melodically, a noncoordination was usually an accent. Earlier pianists were especially fond of marking downbeats by delaying the melody slightly. This clarified the rhythmic structure without resorting to a crude dynamic accent. Especially expressive moments, such as melodic leaps or dissonances, were also often delayed. Even on weak beats, romantic pianists often used a melodic delay as a substitute for – or an intensifier of – a dynamic accent. This gave romantic performances extremely flexible emotional character, with subtle distinctions between different kinds of accents.[280]

Fifth, finally, simulates this arpeggio of the rolled chords and the non-coordination an "orchestral" sound.

> The intention was to sound like a multiplicity of instruments, and we should keep in mind that standards or orchestral precision were much less stringent then than now.[281]

This is a blurring that is *not* at the same time a potential sharpening.

7.5.3.2 *Without Tempo*

An extreme case of rubato, a transition to the zero level of tempo, is singing and playing completely free in tempo, without a continuous pulse, beat and without being divided into regular measures. Recitatives often tend to do this. Music thus comes close to 'prose' instead of metric-rhythmically bound 'poetry', even if the text is metrically ordered, even rhymed. This suspension of a regular tempo is also frequent in introductions, not least in jazz. The actual beginning is marked by a picking up of the tempo and the tempo giusto. In the *prélude non mesuré* before and around 1700 this suspension became an independent genre (Fig. 7.58).

> The enigmatic *prélude non mesuré* was a short-lived genre characterised by rhythmically free *croches blanches* ('flagged white notes', rather than white quavers) and sweeping lines

[280] Ibid.

[281] Ibid., p. 3.

and slurs that were generally notated without specific reference to rhythm or metre. Some of the lines appear to bind the tones into harmonic groups and to articulate cadential and rhythmic units, but the inherent freedom encoded in the notation presents a broad, complex interpretive scope to present-day performers.[282]

Perhaps the blurring, even elimination of a precise indication of beat and tempo is itself another concealment. Possibly it is only a notational blurring, i.e. a pretence of an irregularity and spontaneity that does not exist in the first place.

(M)any notationally unmeasured preludes can be affiliated to notationally measured genres. The *notational dress disguises this musical* fact, but the difference is for the player's eye not the listener's ear.[283]

In this respect, the genre would be an expression of the eye music. One reason is probably the incipient cult of the "genius", which is played off as "nature" against the "culture" in form of "rules" that are objectified in precise notation.

7.5.4 Off-Beat. Generalized Rubato

The tempo rubato in the "classical", pre-romantic combination with simultaneous tempo giusto returns in jazz in a new, generalized form. It is characteristic of the mainly pianistic precursor genres of the jazz, ragtime and boogie-woogie,

that the accompaniment (left hand) remains strictly in time, while the right hand interprets the melody voice with syncopation.[284]

An essential characteristic of jazz, the off-beat or swing, can be understood as a manifestation of the rubato. Swing is difficult, but not impossible to define.[285]

Swing is based on the creation of tension between at least two rhythmic layers, which do not necessarily have to be acoustically perceptible at the same time, but of which, for example, one is only present as a perceived fundamental rhythm (beat). The creation of tension is based on offbeat accentuations, which can arise on the one hand on the level of a macro-offbeat (e.g. "anticipated eighth notes"), and on the other hand on the level of micro-offbeats (impression of minimal tempo changes, of swinging). The creation of tension continues to be based on polyrhythmic superimpositions, as they can occur especially in a "ternary eighth phrasing". Micro-structural nuances in the performance of the aforementioned means of composition are influenced by factors of articulation, accentuation and tempo. They are also influenced by the stylistics of individual genres and the individuality of the interpreters: the swing feeling swing is different from that in bebop; John Coltrane also swings differently from Sonny Rollins.[286]

---

[282]Odermatt 2018, p. 3.

[283]Davitt Moroney. The Performance of Unmeasured Harpsichord Preludes. In: Early Music 4:2 (April 1976), p. 147.

[284]Ekkehard Jost 2003, summarized in https://de.wikipedia.org/wiki/Swing (Rhythmus), 10 June 2020, retrieved 20 August 2020.

[285]Cf. e.g. Heister 1983b.

[286]https://de.wikipedia.org/wiki/Swing_(Rhythmus), 10 June 2020, retrieved 20 August 2020.

The recourse to its origin helps to explain it further. Through the enslavement and deportation of black Africans to the two Americas, the wealth of African music was reduced and only the simplest survived. This includes the fact that the music of the different African tribes intermingled—usually reduced to the simplest, lowest common denominator. This was probably the first acculturation (mixing of different cultures). Thus Kubik (1970, p. 85) believes that the earliest musical acculturation of African-Americans in North America did not originate from the encounter with white music, but was mainly a fusion of different African music styles. This process found its completion in essential elements very early on, although the constant import of new slaves meant that musical material from Africa and the Caribbean was added later. From an acculturation between "white" and "black" music the off-beat and the swing developed. Again, description and explanation are difficult. What is noticeable is that a good part of the melody tones miss the beat by a hair's breadth, the regular counting time or pulse sequence, which is unswervingly kept by the rhythm section. Interpreting this as syncopation is misleading. For the off-beat does not directly question the validity of the underlying metric order: since rhythmic and melodic accent do not coincide, the regular pulse sequence of the beat is not disturbed. However, there is a tension between the rhythmic accent network in the melodies and the underlying metre. And the separation between melody and rhythm group also reflects this tension in the instrumentation type. The melody tones can also hit the beat—so to speak as a cushioning effect. According to duration (1961, p. 133 bibliography # also lat.am. author), there are two forms of off-beat: a retarding one (after the beat) and an anticipating one (before the beat).

The off-beat is not a purely rhythmic phenomenon in the sense that only the duration of the sound is decisive. The rhythmic accents can also be created by intonation, i.e. by hot intonation with its sharp blowing and rapid decrescendo, and by articulation (such as tone strength) and phrasing. The off-beat is the result of the interweaving of African polyrhythm or polymetry (the terms are not quite selective) and European beat rhythm, for which the aforementioned modern "accented timing" with its graded ratio of heavy and light is characteristic. The rhythmic-metric complexity of African music is thereby simplified, not unlike the process of the tonal system. In the process, the strict rules and order of the unalterably kept rhythmic pattern dissolves, not least due to improvisation, which in the course of jazz history has become increasingly important. Only fragments of a different rhythmic order survive. The result is on the one hand a certain disorder and disorganisation, but on the other hand freedom and greater liveliness: only the disintegration of the complicated African rhythm organization opens up the possibility of swing. On the other hand, the accentuated timing, which Schwarzenberg conveys primarily through marching and dance music, is simplified to the relatively uniform impulse sequence of the beat. This beat then forms the lowest common denominator and the strict metric-rhythmic order to which the second rhythmic layer of the off-beat (at first mainly in the melody instruments and later also in the rhythm section itself) can refer in an exciting interplay of bonding and freedom. The impression of the swing is then later intensified in the swing style by combining the various rhythmic layers in the interplay of rhythm and melody group into a unified whole. So here, too, the interaction and competition of solo versus collective plays an essential role. The "rhythm section", in newer jazz especially the bass, provides the tempo giusto. The tempo rubato is mainly taken over by the soloists of the "melody section".

> In order to create the "swing" impression, a rhythm group, for example a pizzicato plucking double bass player, percussive rhythm guitarist or the drummer, must set a beat that is as exact as possible. The other part is then taken over by the solo musician instrumentally or vocally, he "pulls" a slower rhythm through premature syncopations or "brakes" through delayed syncopations, while the rhythm group continues to beat time relentlessly and exactly like a clockwork or metronome. However, both parts play at the same tempo, not faster or slower. In orchestral pieces, when the musicians play the melody voice according to notes and all instrumentalists of a wind section are forced to play simultaneously and exactly, only the rhythm group can create the swing by mostly "braking" (anticipated bass) or more rarely by "pushing". The art of a big band is then that the melody group must keep its own rhythm and not follow the rhythm group' beat pattern.[287]

7.6 Point of Departure: Phonography Form. Electroacoustic Simulation of Liveliness

The perfection of the technical production with the elimination of the blurring of the production by living musicians was and is strived for on the one hand (see Sect. 6.8). On the other hand, this kind of perfection tends to appear unalive, unartistic even inhuman. In return, various technical processes—themselves of course, using electroacoustic technology—strive to restore some of the liveliness in the second instance. This is reminiscent of Heinrich Kleist's metaphor *On the Marionette Theatre*, 1810, in which the marionettes, soulless apparatuses, as quasi-automatons are given something superhuman, almost divine. This was, in the final analysis, already at that time inkling of the function of machinery and industry as a capital as a quasi-automatically self-exploiting value: the "limb man" as a "god" in Kleist's analogy.

7.6.1 "Humanizers". Simulated Vibrato and Other Expression and Animation Effects

The animation within the framework of the phonographic form begins with the sound and smaller units of the work. "Live electronics", related to aleatorics, on the other hand, focuses on larger components and superordinate structures. In principle, the "humanizers" reproduce here all of the vivifying blurring that also occurs in the "unplugged" sound form, perhaps with the somewhat paradoxical difference that in the phonographic form the blurring can be produced much more precisely. Various apparatus effects simulating vividness were already created on early electric musical instruments such as the vibrato, reverberation and delay reverberation etc. on the

[287] https://de.wikipedia.org/wiki/Swing_(Rhythmus), 10 June 2020, retrieved 20 August 2020.

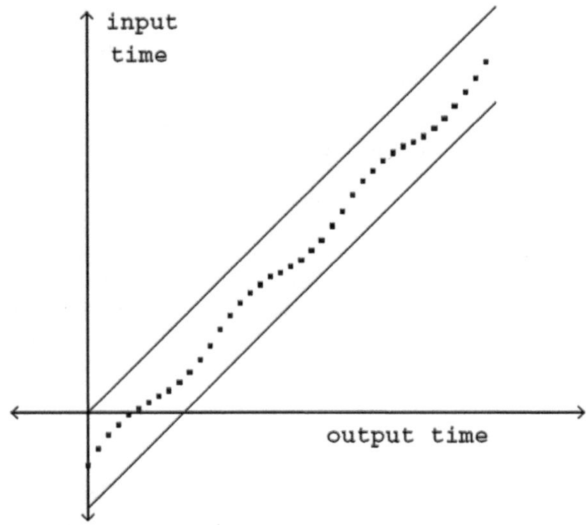

Fig. 7.59 Puckette 2006, p. 204, Figure 7.19: Vibrato using a variable delay line. Since the pitch shift alternates between upward and downward, it is possible to maintain it without drifting outside the strip of admissible delay

"Wurlitzer organ". We can even go back to the pre-electric pipe organ, whose rigid sound was also enlivened with technical tricks. The vibraphone, invented in the time of World War I, originally called "*vibraharp*", takes its name from the vibrato, which is obviously considered to be particularly typical of technically mediated vibrancy. It is specially produced on the electrified vibraphone. With the electro-acoustic technique and completely digital technology, an animation like the vibrato is quite easy to achieve. Special attention is paid to the 'fidelity', a modernized "high fidelity" (Fig. 7.59).

> *Vibrato,* […] that is commonly associated with music performed by skilled singers and instrumentalists in certain musical styles, is an important aspect of *realistic* music synthesis. For synthesized singing particular care must be taken to ensure that the synthesized vibrato behaves *naturally*, especially in the vicinity of portamento pitch transitions. The appropriate fundamental frequency behavior by singers during legato transitions of musical pitch is examined, and an algorithm to produce realistic transitions for vocal vibrato synthesis is proposed.[288]

To a certain extent, it can be done in the twinkling of an eye and with formulas, and there are real arithmetic exercises for it[289] (Figs. 7.60 and 7.61).

A specific, complex process of blurring within the phonographic form was invented by Steve Reich. As early as 1964, he recorded an African-American street preacher, copied his sentence "It's gonna rain" onto tape loops and had two tape

[288]R. Maher: Control of Synthesized Vibrato during Portamento Musical Pitch Transitions, in: Journal of The Audio Engineering Society 2008, https://www.semanticscholar.org/paper/Control-of-Synthesized-Vibrato-during-Portamento-Maher/2a9ad0c2daf726a5476fe47e452ab448801249d2, Abstract, retrieved 29 August 2020. Emphasis HWH.

[289]„5. If you want to make a 6-Hertz vibrato with a sinusoidally varying delay line, and if you want the vibrato to change the frequency by 5%, how big a delay variation would you need? How would this change if the same depth of vibrato was desired at 12 Hertz?" (Puckette 2006, p. 221).

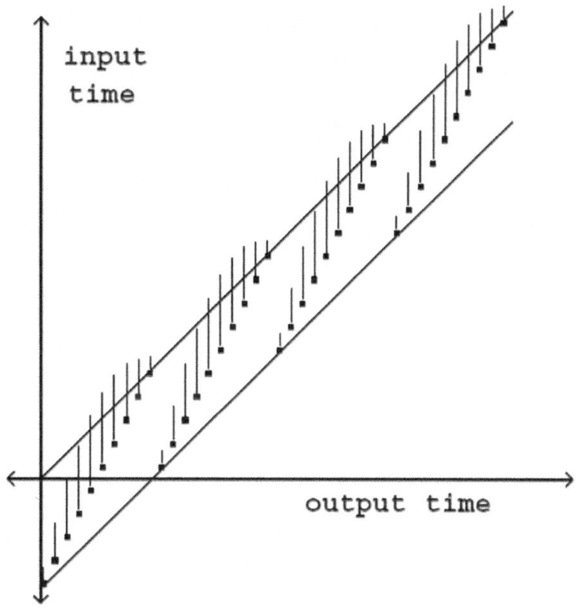

Fig. 7.60 Puckette 2006, p. 205, Figure 7.20: Piecewise linear delay functions to maintain a constant transposition (except at the points of discontinuity). The outputs are enveloped as suggested by the bars above each point, to smooth the output at the points of discontinuity in delay time. Rubato effects and other tempo modifications are also technically no problem

recorders running simultaneously. In the process, they became more and more "out of phase" due to random inaccuracies in the tape speeds or deliberate tempo changes on one machine, so that the aforementioned 'phasing' or phase shifting produced shifting and overlapping effects of increasing complexity, including 'resulting' emergent figures such as new tone colors and harmonies.

> The result of the recordings moving in and out of phase with each other was a transformative process where different timbres, beats, and harmonics were heard; some of which sounded markedly different from the original segment of recorded material. If the sound source had a natural cadence, the phasing created continuously shifting changes to the perceived rhythm as the material drifted in and out of phase. By using additional tracks and loops with identical source material the possibilities for creating a wider range of phasing relationships increases.[290]

After *Come Out* for by now already eight tape recorders (1965), Reich tried to transfer the technique of phase shifting to living musicians as a kind of automatic subject: *Piano Phase* for a piano with two players as well as *Violin Phase* for violin and three tape recorders (1967) and *Phase Pattern* (1970) for four electric organs. The simulated animation in the field of technical re/production is followed by reification in the field of production with musicians.

> Composed phase music features two or more instruments playing a repetitive phrase (part) in a steady but not identical tempo. In the case of gradual phase shifting, initially the tempi of the different instruments will be almost identical, so that both parts are perceived to be sounding in unison and at the same tempo. Over time the phrases gradually shift apart,

---

290 https://en.wikipedia.org/wiki/Phase_music, 29 June 2020, retrieved 20 August 2020.

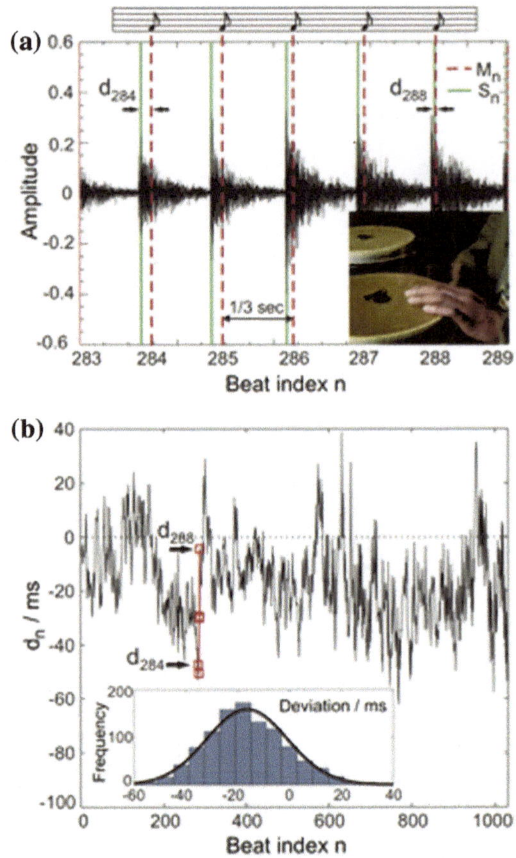

Hennig H, Fleischmann R, Fredebohm A, Hagmayer Y, et al. (2011)
The Nature and Perception of Fluctuations in Human Musical
Rhythms. PLoS ONE 6(10): e26457. doi:10.1371/journal.pone.0026457
http://www.plosone.org/article/info:doi/10.1371/journal.pone.0026457

Fig. 7.61 „Humanizer“. Delay of Beats
Quoted after 10.1371_journal. pone.0026457.001

creating first a slight echo as one instrument plays a little behind the other. This is followed by what sounds like a doubling with each note heard twice. Next, a complex ringing effect arises, after which the phrases eventually return, back through doubling, echo, and unison, to an in-phase position.[291]

[291] Ibid.

Unlike other minimalists, Reich considers the problem of excessive repetition and sometimes even transforms it into its opposite. For in phase shifting, the seemingly identical may be perceived as different even in its hundredth occurrence.[292] However, it is not irrelevant what is repeated. The novelty effect and the information content decreases exponentially with the repetition even of variants of simple few-tone patterns. Insofar as it is an enlivening by blurring, it is a mechanical re-animation. Yet another level of complexity above, the space of the music performance is electro-acoustically reproduced and added to the music recording. Reverberation effects add vividness, as they presuppose a real space for the sound and then suggest it electro-acoustically. Since reverberation and echo

> of a real room are not replaceable, the tape nevertheless was already shortly after its emergence used to simulate the effects of the sounds in a room. Tape Delay, the use of the delay time differences of the tape between recording and playback head for reverb and echo effects[293]

Already since the end of the 1940s, such processes have been used in 'apparatuses' such as the electronic studio and the recording studio, and are today used as an app in the home computer or laptop.

> "Any time that you can do a thing where you have an automatic natural blend, it is absolutely much better. It keeps it out of the pure mechanical world, and now that's what we're getting into. After I discovered my slapback echo I found usually that it gave a type of liveness that was very difficult [to capture]. I never went for, back then people wanted a fairly dead studio, and I wanted to keep it as live as I could and still be able to handle it in there and be able to let each of the members of the band work." Since then, the simulated room has become an indispensable component in the sound concept of the recorded music.[294]

Despite the gaining of independence of the technical simulation, the question of the similarity of simulated rooms with real rooms and their perception is still a reasonable question; it is about the relationship between the iconic image of a room and its direct perception as an indication.

The urgency of the need for liveliness and expression is demonstrated by endless lists of links on the Internet under the keyword "Humanizer" as an app to "Ableton", a bundle of programs for music production including a device.[295] A characteristic example:

---

[292]This was a consideration of Rainer Nonnenmann 2010. See also: Rainer Nonnenmann: Empfangsbereite Antennen. Der Pionier der Minimal Music Steve Reich im Gespräch. In: MusikTexte vol. 121—Mai 2009, pp. 84–87.

[293]Wicke 2011, p. 1075.

[294]Cogan/Clark 2003, p. 90, quoted in Wicke 2011, p. 1075.

[295]Cf. https://en.wikipedia.org/wiki/Ableton, 19 September 2020, retrieved 20 September 2020; see also https://www.ableton.com/. „Ableton Live is a digital audio workstation developed by Ableton [...]. In contrast to many other software sequencers, Ableton Live is designed to be an instrument for live performances as well as a tool for composing, recording, arranging, mixing, and mastering. It is also used by DJs, as it offers a suite of controls for beatmatching, crossfading, and other different effects used by turntablists, and was one of the first music applications to automatically beatmatch songs." https://en.wikipedia.org/wiki/Ableton_Live, 12 September 2020, retrieved 14 September 2020.

> Organized chaos: Create complex patterns with MIDI effects When incorporating sequences, there is no right or wrong answer to the question of whether it is better to program one's own patterns or to let technology do the tricky job. Undoubtedly, *one's own musical abilities* and a fine sense of rhythm are indispensable tools in musical work. However, exciting new possibilities open up when *the computer takes over the helm* and gives the track its shape. [Thus, a] video tutorial shows how to create complex random patterns *without losing control* and thus takes your music to fascinatingly unknown territory. From combining the MIDI effects like Random and Scale for melodic sequences to arpeggiating an arpeggiator. The video conveys a lot of fun and shows simple tricks how you can bring a little bit of *organized chaos into your creative process.*[296]

It would be artistically more "authentic", honest and possibly also more satisfying to leave the technically re-produced music or sounds as sound quality in their own right and to recognize them, and to renounce the deceptive appearance of liveliness. For it cannot become a genuinely human vividness, but remains an instrument-based one. Instead of a realized subjectivity of expression, it remains a *desire* for this. Already with the continuous vibrato, still produced by human hands on musical instruments, expression had become material. The automated "humanizers" produce a double blurring: on the one hand in the technical sense of fuzzy logic, on the other hand ideologically as a deception. If expression is technically simulated or, to put it politely, imitated and imitated, it does not become a real expression of a subject. It is the expression of an avatar. The automat appears as a human being, the human subject appears as a dependent variable of apparatuses and programs: a configuration which, also in this book, appears time and again.

> Vividness and an interpreter-oriented phrasing […] are […] taken over by *humanizer* functions. They lend an *individual interpretation with deliberate inaccuracies* to music produced by digital programs. The individual peculiarities […] are […] included [… -] a masqueraded, seemingly human performance […].[297]

7.6.2 Composition, Indeterminacy and Improvisation: 'Live Electronics'

The technical blurring of the sound form, which we have just discussed, was mainly applied to individual notes and parts of the work. The 'live-electronics' as an attempt to bring the work to life, on the other hand, goes the whole hog. The electro-acoustic composition was originally created to eliminate the ambiguities and blurriness that necessarily result from the realization of the work by musicians. The work was intended to sound purely objective without subjectivity—at best that of the composers was permitted, but which is already very reduced and mediated by the serial procedures and the technical mediation -: instead of liveliness, perfection, even if not perfection in the comprehensive sense. However, this asymmetrical balance of the

[296] https://www.ableton.com/de/blog/organized-chaos-generate-complex-patterns-midi-effects/, retrieved 12 September 2020. Emphasis HWH.

[297] Klotz 2006, p. 362.

dynamic equilibrium of the dialectic of perfection and liveliness was soon shifted back towards liveliness in the new current of "live electronics" on a new medial level, as a subjectivity in performance and execution that was admittedly technically contained. This goes as far as a now technically mediated electrified improvisation.

In live electronics, living musicians interact with impulses from stored electroacoustic sounds, the control of which in turn interacts with what the musicians play or sing. The resulting fields of indeterminacy are blurred from the standpoint of a strict determination of notation and sound form. As blurs, however, they express the intention for the artwork as a whole even more precisely, since the artwork is supposed to encompass both. Composition here tends to be improvisation—which also requires practice and preparation. The " "real-time" fetish reflects a mentality for which everything should always happen quickly and immediately. Conversely, in a kind of spiral movement, there is even a return to notation, even to the score form (see Sect. 6.8.4.3).

> However, the immediacy of the real-time processes is based on numerous time-consuming procedures in which the composer has previously translated his sound ideas into the language and conditions of the respective sound-generating programs. In addition to sound synthesis, new areas of responsibility of the computer have emerged in the field of *score synthesis*, i.e. in the creation and/or elaboration of musical contexts and their summary as a written, graphic or other type of score, which *in turn must then be transformed back into sound.*[298]

Live-electronics works as a vitalization, which—of course—remains secondary and derivative.

> In live electronic concerts, the human element that "live" has always meant for the world of art is intensified in a compensatory way in the face of undisguised technology, which for its part – certainly still in the 1950s – has the nimbus of the inhuman. This symbolic effect of a performance of music with musicians in person, which makes up a large part of the musical aura, plays an important role, especially from the perspective of the viewer. The aesthetics of acousmatic music responds to this in a special way, not least by setting up a *quasi-live performing* loudspeaker orchestra, the *acousmonium.*[299]

This kind of vitalization has features of a 'world upside down'. Things take on an almost spooky life of their own, people are "interactive" and have to adapt to these active things.

> From the perspective of the instrumentalists, the special feature of live electronic concerts lies in the increased challenges regarding their technique they face, since they need to synchronize their playing with the musical performance of sound carriers and/or trigger computer sounds with special triggers attached to the instrument. Also the feedback of being able to listen to oneself, which is so important for instrumental playing, is made more difficult by the frequently used duplication of instrumental sounds through loudspeakers distributed throughout the room. Live-electronic improvisation with instruments requires less of a preliminary conception of the interaction and more of the *player's adaptability to the inherent dynamics of the technical part.* [...] In live electronic concerts, the sounds themselves are often robbed of their originality and also their recognizable identity when

[298] Ungeheuer 2005, p. 99. Emphasis HWH.

[299] Ungeheuer 2005, p. 100. Emphasis HWH.

they are projected into the room and – usually at the threshold of perception – electronically manipulated in real time.[300]

The work exists here mainly as a process and in sound form. It can be this way, but it can also be different—the respective result can of course be fixed afterwards by technical re/production, just like in improvisations. Nevertheless, it is difficult to ask for the realization of an idea, and thus the access of Fuzzy Logic is also difficult.[301]

7.6.3 People as Interface. Towards the Elimination of Professional Composers

By the way live composing since it actually is improvising of course can also be done without live *electronics.* Sound painting" is 'live electronics unplugged', to put it paradoxically (see Sect. 7.3.3).

But this is a byway. The main path goes in a direction that also in "real time" eliminates people and automates music production as much as possible. Human-subjective blurs are replaced by machine blurs. In material production, automation makes sense as a means of relieving people of dangerous, heavy, monotonous work, less so in immaterial production as music production, which is essentially play. A relief from playing is not an urgent social need.[302] In the case of the Brain Wave control as in Lucier, not only the idea but also the compositional subject is extinguished in a more concrete qualitative determination (see Sect. 6.8). Whether this tendency becomes even more pronounced in the following case, or merely takes on other forms, may remain undecided here.

The personal union of composer and performer in electroacoustic music is shifting significantly in live electronics. It is realized in the production of electroacoustic music within the composition process as its conclusion at the transition to the sound realization. Now, however, it is realized within the sphere of realization, within the framework of the sound form. The "interaction" between people and machines becomes more asymmetrical and shifts in favor of the apparatus. In this aleatoric, living performing musicians become handmaids, the composers become

---

[300]Ibid.

[301]Further information e.g. Seifert et al. 2008; https://www.researchgate.net/publication/319857235_Real-Time_Composition_why_it_still_matters_A_look_at_recent_developments_and_potentially_new_and_interesting_applications/link/59be4961458515e9cfd1f67f/download; https://www.researchgate.net/publication/329158193_Composing_and_Improvising_In_Real_Time_13th_International_Symposium_CMMR_2017_Matosinhos_Portugal_September_25-28_2017_Revised_Selected_Papers/link/5d1a5aaa458515c11c093fe4/download.

[302]But it seems to be an urgent need of computer experts. From the abundance of relevant literature only one reference: Hunt, S., Mitchell, T., & Nash, C. (2017, September). Thoughts on interactive generative music composition. Paper presented at 2nd Conference on Computer Simulation of Musical Creativity. https://uwe-repository.worktribe.com/output/881299/thoughts-on-interactive-generative-music-composition (not published). See also the long list in Chapter 5.2.

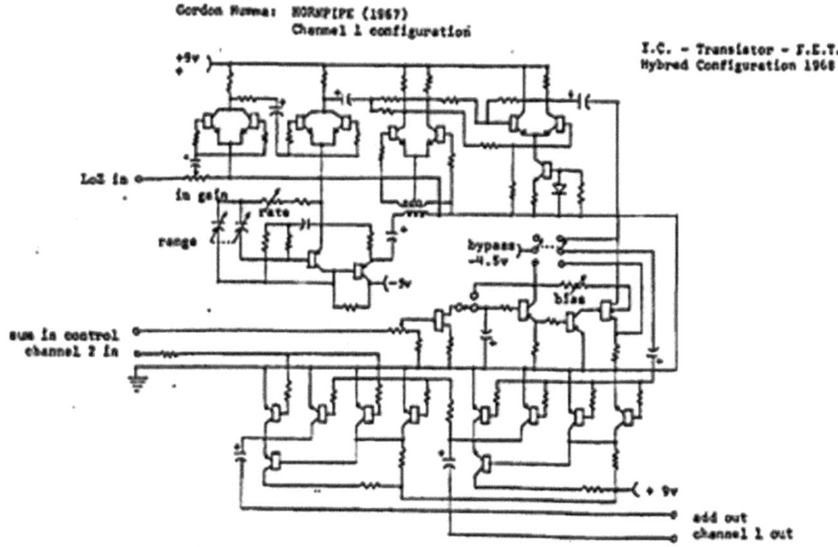

Fig. 5 Gordon Mumma, *Hornpipe (1967)*

Fig. 7.62 Gordon Mumma: *Hornpipe* (1967)
Quoted after: Michael Joseph McInerney 2007, p. 36

"headmaids" (an analogy by Bertolt Brecht) in a complicated, layered triangular relationship,

> based on the intervenient relation between a musician performer, a computer model and pre-interactive material, previously developed by a human composer. The aim is the establishment of a musical composition method, dynamic and in real-time, based on two premises: 1) the human interactive improvisation on notation 2) the computer-generated sound material, resulting from the interaction between computer, composer and interpreter.[303]

Fetish thinking and acting attribute vividness to the apparatus. This paradoxical kind of vitalization reverses relations between persons to relations between things, and vice versa. This leads to alienation. The program of the following work behaves like an inscrutable and hardly controllable black box.

> In *Hornpipe* (1967), the circuit *diagram is the work* – the cybersonic console does not have an independent performer Instead the process of exploration provoked for the performer by the console defines the trajectory of the work in performance. T. he opening sounds played upon the instrument t rigger silent responses in the console which lead to certain circuits becoming unbalanced. ‚While rebalancing, various circuit-combinations occur that produce complex electronic sound-responses'.[304]

Apparatus and program educate the horn player. Once more, they thus illustrate the handover of power to the machinery.

---

[303]Fornari and Maia, n.d.

[304]McInerney and sleeve notes to Mumma's album *Live Electronic Music* (2002) quoted in McInerney 2007, p. 35. Emphasis HWH.

> As these (sound-responses) begin to be heard on the loudspeakers which come out of the console, *a further level of interplay* becomes audible, in which live sounds, electronic sounds and the continuing re-balancing of the internal circuits produce a texture which combines horn sounds, electronic response-sounds and electronic sounds articulated directly by horn sounds. The work becomes an active duet between horn player and cybersonic console, in which *the performer learns* in performance 'which sounds are most unlikely to unbalance and rebalance the cybersonic console.'The work ends when the soloist finds the sustained sound which balances all the response circuits, and brings the console to silence.[305]

Only when the sonic key has been found in a mixture of Indiana Jones, Da Vinci Code and Goethe's sorcerer's apprentice does the automated spook cease to exist. But there are still living musicians in the game. More recent developments since the 1990s also save this "interactive" component.

This includes for example, *Laptronica*,

> a form of *live electronic* music or computer music in which in which *laptops are used as musical instruments*. The term is a portmanteau of "laptop computer" and "electronica". [...] Many sophisticated forms of sound production, manipulation and organization (which had hitherto only been available in studios or academic institutions) became available to use in live performance[306]

As with the synthesizer, the miniaturization of the electronic studio leads to a wide availability. On the meta-level of quasi-*composition* as a kind of "lean production" including the shortening of "supply chains", *live coding* (also known as "on-the-fly programming" or "*just in time* programming") has been operating since the turn of the millennium,

> a programming practice centred upon the use of improvised interactive programming [...], often used to create sound and image based digital media, and is particularly prevalent in computer music, combining algorithmic composition with improvisation.[307]

People are difficult to eliminate from art, and the visibility of people making music is surrogated by other visual stimuli.

> Typically, the process of writing is made visible by *projecting the computer screen in the audience space*, with ways of visualizing the code an area of active research.[308]

In this way the 'Live Electronic' turns inside out the visualization and projects it to the outside, which is inherent in computer composition.

> Componere, in the form of the cut'n paste of the sequencer programs, proves to be no longer a written but an *image*-intensive activity, which tends to fall back on anachronistic analogue visualizations such as envelope curves and frequency progressions, and thus has a high proportion of analogue representation in the service of figurative recognition. They suggest optical figures and sound traces, which are only surface effects of bit structures. Although this is an operation with pure data, they are processed in a way that makes you forget exactly

---

[305] McInerney and sleeve notes to Mumma's album *Live Electronic Music* (2002) quoted in McInerney 2007, p. 35f.

[306] https://en.wikipedia.org/wiki/Live_electronic_music, 29 July 2020, retrieved 24 August 2020.

[307] Ibid.

[308] Ibid. Emphasis HWH.

> that. The user interfaces allow an unconscious manipulation of icons, symbols, envelopes and sequencer functions without having to think of 'music'.[309]

The development splits up. *Live coding* loses its status as an independent concert-like performance and instead serves as accompanying music or even just background music, for living improvised dancers, or

> outside of performance, such as in producing sound for film […] or audio/visual work for interactive art installations. […] Live coding is also an increasingly popular technique in programming-related lectures and conference presentations, and has been described as a "best practice" for computer science lectures.[310]

As "live" instead of liveliness, this "electronic" suffices itself.

7.7 Point of Departure: Perception Form. Mishearing and Misunderstanding Between Fuzziness and Blurring

In perception, as already explained in Sect. 3.3, blurring resp. fuzziness is the default mode and normal case. Blurring as intentional fuzziness on the other hand is a special case. P. Schaeffer intended to go to an extreme in his Musique concrète. It should be as abstract as possible in terms of sigmatics and semantics. The listeners were then to eliminate the unresolved remnants of relation to reality, which could not be erased compositionally (see Sect. 4.3). Here, then, blurring passes over to the erasure of an essential dimension of the work as outlined at the beginning.

7.7.1 Unintentional Parodies: "Lady Mondegreen" = "Laid Him on the Green"

A peculiar type of what S. Freud called "slips" are errors in acoustic reception, i.e. a kind of 'mishearing'. Texts in musical works are misheard, not understood or misunderstood. But the urge for meaning is usually so strong that a blurred but coherent version is aimed for, even if no precise one is possible. The procedures of completion and correction of wrong notes in the oral perception are related to psychodiagnostic procedures such as the Rorschach projection test, in which axially symmetrical inkplot images are to be interpreted as figures, or Schönberg's "Kriegswolkentagebuch" (diary of war clouds), in which he read favourable omens into clouds (see Sect. 6.2). Wordless music can also be used as a projection screen for emotional

---

309 Klotz 2006, p. 362.

310 https://en.wikipedia.org/wiki/Live_electronic_music, ibid., Cf. Collins (2003), Collins (2007). Collins et al. 2018, Emmerson (2007), Gifford et al. (2018), McLean et al. (2010), Manning (2013), Rohrhuber, Julian (2008), Wang and Cook (2006).

impulses and wishful thinking. Hearing something reading something into it and mishearing belong together here.

Heinrich von Kleist achieved a particularly vivid depiction of projective listening by means of subjective trope or text (see Sect. 8.2), with bells as an acoustic trigger in his anecdote *Der Branntweinsäufer und die Berliner Glocken* (1810). The meanings intended in the productive realization of the sound of bells are different from those in the receptive realization. The subject of the anecdote hears the names of types of alcohol in the sound, and he hears the ringing at the same time as an invitation. The naming and appeal function of signs thus coincide. The "brandy drinker" interprets the sacred ringing of bells in a very secular way as something that has a subjective meaning for him he repurposes it so to speak. A quasi-magical effect of the ringing of bells plays a part in this. The irresistible power of music is merely an excuse; the actual power is a drive of another kind. (Of course, it is still a force of nature.) The irresistible force here is not one of the sounds, but the psycho-physical force of alcoholism. Nevertheless, music and text are not random and arbitrary. The release of the instinctive drive through the language of art and the voice of faith is an achievement that mere "sound objects" could not achieve. In terms of psychoanalysis, this is a perfect Freudian slip: the drunkard hears what he wants to hear. The instinct—already misguided—triumphs over precise perception. The projection usually says more about the (mis-)hearing human subject than the sounding musical object. For example, a pause based on syntactic-prosodic reasoning in singing can also be interpreted quite differently.

> In Enrique Iglesias' song 'Hero' I always heard the line 'Would you save my asshole tonight'[311]

The retrospective interpretation of what we can also call a slip lies in an intelligent technical-acoustic interpretation, without getting involved in the underlying psychological issues:

> This confusion is caused by the pause between my – soul of the correct text 'Would you save my soul tonight'.[312]

More in-depth interpretations would have to include the fact that the mishearing along the lines of something anal is no coincidence, but also that the quality of the song might really be a song you can shove up your "a*". The phenomena analyzed in the following are a further, particularly weird, ambivalent interweaving of blurring and sharpening, clarification also in the sense of crystallization and, in the second instance, deliberate blurring. Mishearing—also in the sense of Freud's 'slip'—is the starting point. Of course, the consequences drawn here are in fact only an illusory specification: appropriation as understanding, but through misunderstanding—also a somewhat paradoxical kind of applied dialectics. But it is also an attempt—not

[311] Matthias Guiard, http://einestages.spiegel.de/external/ShowTopicAlbumBackground/a5883/128/10/F.html#featuredEntry, 1.9.2011. Excellent. H.-W. H. Cf. Enrique Iglesias "Hero" on YouTube. Numerous other, often somewhat forcedly intentional, examples in the same context.

[312] Ibidem.

entirely in vain, but also not entirely successful—to construct a formulation and a meaning from the blurring of hearing and understanding, of the auditory and cognitive context of perception—a remarkable interweaving of impression and projection. In English, the term Mondegreen is used for texts that are misheard and misunderstood. The poem line about the slain Earl of Moray "and laid h*im on the green*" was understood by the writer Sylvia Wright when she was a child as "and *Lady Mondegreen"*.[313] Johann Gottfried Herder translated the ballad about the Earl of Moray as *Murray's* murder and Brahms set it to music. An analogous blurring occurs in German.

> There is not only the line: "Ein schöner Ritter war er, / In Wett- und Ringelauf"[314] ('a handsome knight he was/in foot race and ring race"), the last word of which I first read like *Schluckauf*/hiccups in German until I realized what it meant. There is also a grammar that I have not been able to understand, namely "Our Murray's crown was always on top." Our Murray's?[315]

The strange syntax in German can easily be normalized by a permutative syntactic change and erasure of the preceding genitive and is then quite correct: *At all times on top was the crown of our Murray.* (Even more common would be: *The crown of our Murray was always on top.* Of course, this sounds completely unpoetic.) Such quasi-poetic alterations of word orders, as Herder carries them out, are commonplace in music as motivic variants of tone orders. If we take our standard model BACH again, the simplest sequence would be the descending chromatic line C-H B-A. That would also make sense: there are two sigh motifs in a row. Add two more chromatic steps and you get the classic lamento phrase: C B Bb A – Ab G. An identical error, the incorrect segmentation of the phonemes, underlies the prayer, which is communicated and filled with fear through and in religion, when a child changes the wording at the end of the *Hail Mary*: "Bete für uns Sünder, jetzt und in der Stunde unseres Todessamen. (Pray for us sinners, now and at the hour of our *deathseed*)."[316] The wrong word division is here just the manifest linguistic reason. The deeper latent psychological reason is fear. Corresponding German-language examples are difficult to translate, since they are, as in other cases, bound to the respective individual language—but on the other hand, the tonal or graphematical similarities are perhaps especially in the reflection of the foreign language, here

---

[313] *Wortspielereien: Der Todessamen, Die Presse,* Wien, 6 July 2013.

[314] This is once again a false segmentation. The "*Ringel*-auf" is formed according to the pattern of "*Ringel*-Reihen" (Ring a Ring o' Roses)—a circle dance of children. But what is meant here is the "ring race", a type of race; the knightly "Ringe-Stechen" (jousting) should also be considered. Due to its absurd syllable division rules, the New German Orthography leads to unintentionally funny results, such as Ur-Instinct (basic instinct) as Urin stinkt (urine stinks), without regard to morpheme and syllable boundaries.

[315] *Wortspielereien. Ringelauf, ringelab, Die Presse,* Wien, 6 July 2013.

[316] Wortspielereien: Der Todessamen, Die Presse, Vienna, 6 July 2013. The correct term would be: "Tode*s*. / Amen. (Hour of our death/Amen)." It is understandable that a child thinks of death seeds here. Even the Catholic Heidegger (who hided this background in the interest of the greatest possible broad effect) defines human "existence" as "being for death". Little wonder, then, that the germ or seed of death appears to have been sown already with birth.

English, quite vivid and comprehensible. During mishearing and correction of wrong notes in the oral perception, not only words from foreign languages, but also unusual, unfamiliar words, including those of one's own language, including archaisms, are normalized in a more or less folk-etymological way and brought closer to the familiar. In general, the incomprehensibility of the foreign language Latin for lay people is a pan-European general Christian-Catholic problem. The blurring is further aggravated by the blurring in memory of what is merely phonetically, not semantically, learned by heart. A student, probably a folklore researcher, went from village to village in Austria,

> to write down the house prayers. Sometimes the farmers could not reconstruct the exact wording of daily (!) spoken prayers. Apparently, generations in succession did not even know what they were mumbling away to themselves.[317]

Those who do not know Latin acquire the unfamiliar in such a way that they translate or transfer it not only into their own language, but also into their own world, including the individual-private world. This is a specific form of sharpening, actually even of crystallisation in so far as new—even if wrong—meanings are derived from the fuzziness of perception and understanding. This is the case with the omnipresent *Hail Mary.* Gabriele Burger-Scheidlin reports that her father—*the eldest of several brothers* with a mother named Maria—always understood the beginning as follows: "Hail Mary, voll der Knaben (full of boys)".[318] "Voll der Gnaden" (gratia plena, full of grace) is the wording of the German translation: Gnaben (boys) and Gnaden (grace) sound very similar.[319] It would be fuzzy logic applied as a game to find analogous other national language variants here and quantify the similarities. Such appropriation as a misleading sharpening can be combined with a modernization of the traditional.

> Michael Steinhuber reports that in kindergarten someone drew a picture of Ascension Day, with a spaceship. In the cockpit: Pontius *Pilotus.*"[320]

Pilotus and "*Pilatus*" are also almost identical in sound. Moreover, a pilot is undoubtedly more familiar to children than a Pilatus—here it is remarkable that a child in preschool age, by mere hearing, so to speak by hearsay, and precisely in the context of Catholic Latin, gives the pilot the original Latin masculine ending -us, thus reducing the blurring, at least in terms of sound. Even English, which is modern compared to Latin, tempts to mishearing and misinterpretations based on one's own interests.

> Fred Härting reports about the elementary school children who looked forward to the announced *'Landspaghetti'* (country spaghetti) during the school trip, but then only got *'Lunchpakete'* (packed lunches).[321]

---

[317] Ibid.

[318] Ibid.

[319] In addition to this: In Austrian as an eastern central Bavarian dialect, the voiceless, hard consonant is often pronounced voiced, softly.

[320] *Wortspielereien: Der Todessamen, Die Presse,* Wien, 26 July 2013, Spectrum VIII.

[321] Ibid.

Phonetically, this is similar, and even packed lunches could be as tasty as spaghetti. Linguistically, it is a false segmentation, supplemented by imprecise pronunciation: Lunch-Paket and LandS-paget. English and religion are combined, quite appropriately, in the following mishearing, again with the same type of error—wrong segmentation:

> Hartmut Lubich is one of those people who listened to Janis Joplin's famous song "O Lord, won't you buy me a Mercedes Benz" as a child and thought that she demanded a *mercy dispense* from God, which he actually bought for her in the Christian imagination.[322]

The process of wrong segmentation even has its own term in English.

> For us, an Oronym is just a mountain name in topography. In English, it also denotes words that are heard incorrectly, especially through misleading segmentation. For example, I scream instead of ice-cream[323] or, from the collection of the great wordplay expert Gyles Brandreth, who coined the term: Where is the spice centre? – Where is the spy centre?[324]

Wrong segmentation is a frequently occurring type of error in the formal-syntactic analysis of motivic-thematic work. In the case of the wrong text underlay, a small blur makes a big, significant difference. The subjective approximate approach is objectively a serious error. As with music, even minimal variations change the meaning. The anthropomorphic and the own-language-centrist projection also have something to do with the search for pattern recognition and at the same time with the various aspects of pattern formation. Those who only understand roughly or little when listening are almost desperately trying to find a meaningful "suitable pattern". If necessary they invent one that does not exist at all, but only in their mind. In the context of Fuzzy Logic music, it becomes clear once again: the motivations, psychological and social, are not part of the subject and theme of the investigations.

7.7.2 Intentional Parodies, Deliberate Misunderstanding, Rewrites. Walther's and Beckmesser's "Preislied" in Die Meistersinger von Nürnberg

Wagner expands the mishearing to the "mis-singing". Self-deceptive imitation, which wants to deceive and cheat others, is imitated compositionally: the negative figure of Beckmesser publicly disgraces himself by not understanding, incorrectly rehearsing and memorizing a song that he had stolen from his competitor Hans Sachs. In Act II, Wagner describes in detail how from a dream of inspiration Walter's Prize Song is created seemingly ex nihilo. The Meistersinger rules give it a more precise shape.

---

[322] Ibid.

[323] It is just with this (mis-)construction that Kurt Weill composed a ravishing ensemble in his "Volksoper" (folk opera) *Street Scene*, with which the actors react to the New York summer heat. Meanwhile this pun is used for advertising.

[324] *Wortspielereien.* mip, *Die Presse,* Wien, 15.6.13, Spectrum VIII.

SACHS […] I should like,
if I am to teach you the rules,
you to explain them to me anew.
See, here is ink, pen, paper:
I'll write it down for you if you will dictate to me.
WALTHER
How I should begin I scarcely know.
SACHS
Tell me your morning-dream.
WALTHER
Through the good precepts of your rules
I feel as if it were effaced.
SACHS
Then take poetry to your hand now:
many found through it what was lost.
WALTHER
So it might be not dream, but poetry?
SACHS
The two are friends, gladly standing by each other.
WALTHER
How do I begin according to the rule?
SACHS
You make it yourself, and then you follow it.
Think of your beautiful dream of this morning;
of the rest let Hans Sachs take care!
(Walther places himself at the work-bench near Sachs who takes down the song as Walther sings it)
WALTHER
"Shining in the rosy light of morning,
the air heavy
with blossom and scent,
full of every
unthought-of joy,
a garden invited me
to be its guest".
SACHS
That was a "stanza": now see to it
that one just like it follows
WALTHER
Why just like it?
SACHS

So that one can see
that you're choosing an equal as a wife.
Etc.

In Sach's workshop Beckmesser steals the song written down by Walther von Stolzing. Sachs notices the theft and gives it to him as a present. The gift is poisoned. Sachs assumes that Beckmesser will not be able to perform the novel "artwork of the future" properly.

SACHS
(looking on the work-bench)
The poem? I left it here… have you pocketed it?
[…] (A)nd so that people don't think evil of you,
keep the sheet, let it be a present for you.
BECKMESSER
(springing up in joyous surprise)
Good heavens! A poem! A poem by Sachs?
But wait… lest some new harm come to me!
You'll have memorised it pretty well already?
SACHS
On my account have no fear!
BECKMESSER
You'll let me have the sheet?
SACHS
So that you're no thief.
BECKMESSER
And if I were to use it?
SACHS
As you like.
BECKMESSER
But, shall I sing the song?
SACHS
If it's not too difficult.
BECKMESSER
And if I brought it off?
SACHS
I should be very surprised! […]
A song of yours, of that I'm sure,
with that I'll overcome every obstacle: […]
BECKMESSER
(affectionately)
[…] A song of yours, of that I'm sure,
with that I'll overcome every obstacle […]

BECKMESSER
[…] But swear one thing:
that wherever and however you hear the song
you'll never take it into your head
to say that the song was written by you.
SACHS
I swear and vow to you here
never to boast that the song is by me.
[…] But friend, I draw it to your attention
and advise you with all kindness:
study the song properly!
Its performance is not easy,
even if you find the right melody
and get the proper tone!
BECKMESSER
Friend Sachs, you are a good poet;
but where tone and melody are concerned,
admit, no one surpasses me!
So prick up your ears,
and: "Beckmesser,
no one better!"
You can be sure of that
if you let me sing in peace.
But now, to memorise it,
quickly home!
Without losing time […]
But my thoughts
are running away with me:
am I confused
and quite lost?
The syllables, the rhymes,
the words, the lines:
I'm stuck as if by glue,
and yet my heels are itching!
Adieu! I must away!

Beckmesser presents the Prize Song in the finale of the competition at the Nuremberg fairground. As expected by Sachs, it sounds like a parody. Beckmesser's distortions create a kind of pre-Dadaist nonsense lyric. The correction of the distortion, similar to anamorphoses in the mirror, is done retrograde-inverted as equalisation, when Walther performs the correct, namely his original version, on the fairgrounds. A comic-like interweaving: For the audience on the fairground both versions are new, the wrong and the right one. The audience at the opera, on the other hand, already knows the right version and can amuse themselves with the involuntary parody. For Walther's original version, which he subsequently recreates on the fairground, was already heard in the previous act in Hans Sachs' workshop.

Morgenlich leuchtend im rosigen Schein, geschwellt die Luft, von Blüt' und Duft voll aller Wonnen, nie ersonnen, ein Garten lud mich ein. Gast ihm zu sein.	Morgen ich leuchte in rosigem Schein, von Blut und Duft geht schnell die Luft; wohl bald gewonnen, wie zerronnen; geht schnell die Luft.	Shining in the rosy light of morning, the air heavy with blossom and scent, full of every unthought-of-joy, a garden invited me. to be its guest	In the morning I shine in a rosy light, with blood and scent the air moves fast; probably soon won, as if dissolved; in the garden I invited horrid and fine.
Walther	Beckmesser	Walther	Beckmesser

As the text is, so is the music. The differences in both versions are obvious, the parodistic blurring is perfect. Note lines 1 and 2 bring the beginning of the original song by Walther, note lines 3–5 the distortion by Beckmesser.

[redacted]

Fig. 7.63 R. Wagner: "Preislied". *Die Meistersinger von Nürnberg* IIIrd Act
Felix Mottl (ed.). Leipzig: C.F. Peters, Leipzig)(n.d., ca.1910). Plate 982, p. 749 and 774f. IMSLP63567-PMLP16799-Wagner_-_Die_Meistersinger_-_Act_III

References

Ahn, S.-H.: Eastern and western elements in selected works by Giacinto Scelsi and Toru Takemitsu. Thesis Doctor of Musical Arts in Music in the Graduate College of the School of Music, University of Illinois at Urbana-Champaign (2014)

Angustin, J.: George Antheil. Enfant terrible der Musik. Verlag Langen-Müller, Munich (1960)

Armstrong, R.E.: An historical survey of the development of the Baroque Solo Keyboard Suite in France, England, and Germany. Graduate Theses, Dissertations, and Problem Reports. 3023. https://researchrepository.wvu.edu/etd/3023 (2011)

Arnest, M.: Why couldn't they play with their hands together? Noncoordination between and within the hands in 19th century Piano interpretation. https://www.semanticscholar.org/paper/Why-Couldn-%E2%80%99-t-They-Play-With-Their-Hands-Together/dbeb216f1347ea6ac4488 3b00bdc86365082c023?p2df (2012)

Bandur, M.: Scelsi. In: Heister, H.-W., Sparrer, W.-W. (eds.) Komponisten der Gegenwart, vol. 34, pp. 1–44. Edition text+kritik, München (2007)

Boehmer, K.: Zur Theorie der offenen Form in der Neuen Musik. Tonos, Darmstadt (1967)

Bossin, J.: Stravinsky-Antheil-Maschinenmusik. Stravinsky und die neue Ästhetik des Antipathos. In: Neue Gesellschaft für Bildende Kunst Berlin: Absolut modern sein: Zwischen Fahrrad und Fließband – culture technique in Frankreich 1889–1937, pp. 238–242. Elefantenpress, Berlin (West) (1986)

Bruhn, S.: J. S. Bach's well-tempered clavier. In-depth analysis and interpretation. https://edition-gorz.de/bruhn4-engl.html (n.d.)

Burde, W.: »Haupttontechnik« und europäisches Komponieren. In: Heister, H.-W., Sparrer, W.-W. (eds.) Der Komponist Isang Yun, pp. 263–272. Text+kritik München (1987)

Busoni, F.: Entwurf einer neuen Ästhetik der Tonkunst (Sketch of a New Aesthetic of the Art of Music), 1907; 2nd enlarged ed. Insel-Verlag, Leipzig (1916)

Chang, Ph. Ch.-Ch.: Analytical and performative issues in selected unmeasured preludes by Louis Couperin. Vol. 1: text, submitted in partial fulfillment of the requirements for the Degree Doctor of Philosophy. Department of Music Theory Eastman School of Music. University of Rochester Rochester, New York (2011)

Chen, J.-M., Smith, J., Wolfe, J.: Pitch bending and glissandi on the clarinet: roles of the vocal tract and partial tone hole closure, J. Acoust. Soc. Am. https://doi.org/10.1121/1.3177269. Corpus ID: 16352321 (2009)

Cogan, J., Clark, W.: Temples of Sound: Inside the Great Recording Studios. Chronicle Books, San Francisco (2003)

Collins, N.: Generative music and laptop performance. Contemporary Music Rev. **22**(4), 67–79. https://web.archive.org/web/20140514102233 (2003)

Collins, N.: Live electronic music. In: Collins, N., d'Escriván, J. (eds.) The Cambridge Companion to Electronic Music, pp. 38–54. Cambridge Companions to Music. Cambridge University Press, Cambridge (2007)

Collins, N.: Beyond notation: communicating music. Leonardo Music J. **21**, 5–6. MIT Press, Cambridge, MA (2011)

Collins, N., McLean, A., Rohrhuber, J., Ward, A.: Live coding in laptop performance. Organised Sound **8**(3), 321–330. https://web.archive.org/web/20180617115703/ (2003)

Cramer, A.: Schoenberg's Klangfarbenmelodie: a principle of early atonal harmony. Music Theory Spectrum **24**(1), 1–34. https://www.researchgate.net/publication/249980375_Schoen berg 27s_Klangfarbenmelodie_A_Principle_of_Early_Atonal_Harmony (2002)

Dahlhaus, C.: Schönbergs Orchesterstück op. 16, 3 und der Begriff der "Klangfarbenmelodie". In: Dahlhaus, C., Marx, H.J. (eds.) Bericht über den Internationalen Musikwissenschaftlichen Kongress Bonn 1970. Bärenreiter Verlag, Kassel. https://nbn-resolving.org/urn:nbn:de:bsz:14-qucosa2-374743; english as Dahlhaus, C. [1970]: Schoenberg's Orchestral Piece Op. 16, No. 3 and the Concept of Klangfarbenmelodie. In: Schoenberg and the New Music: Essays by Carl Dahlhaus, pp. 141–143. Cambridge, Cambridge University Press (1971)

Dallmann, N.-C.: *Rubato und Laid-back. In:* Sonic – Magazin für Holz- und Blechblasinstrumente, September / Oktober 2015, pp. 106–108 (2005)
Dallmann, N.-C.: Musikterminologie erklärt – Walzer und 3/4-Takt. In: Sonic – Magazin für Holz- und Blechblasinstrumente, Juli / August 2015, pp. 90–92 (2015)
Dalmazzo, D., Ramírez, R.: Bowing gestures classification in violin performance: a machine learning approach. Front. Psychol. **10**, 344. https://doi.org/10.3389/fpsyg.2019.00344 (2019)
Daniels, D.: From visual music to intermedia art, Dieter-Daniels_Von-der-visuellen-Musik-zur-intermedialen-Kunst_From-Visual-Music-to-Intermedia-Art. In: Rainer, C., Rollig, St., Daniels, D., Ammer, M. (eds.) See This Sound. Versprechungen von Bild und Ton / See This Sound. Promises in Sound and Vision Lentos Kunstmuseum Linz, pp. 240–253. Verlag Walther König, Köln (2009)
Danuser, H., Kämper, D., Terse, P. (eds.): Amerikanische Musik seit Charles Ives. Interpretationen, Quellentexte, Komponistenmonographien. Laaber-Verlag, Laaber (1993)
Decroupet Pascal: Maschinenmusik: Fortsetzung futuristischer Ansätze – Neuansätze. In: Heister (ed.) Geschichte der Musik im 20. Jahrhundert, Bd. III: 1945–1975, Laaber 2005 (Handbuch der Musik des 20. Jahrhunderts, Bd. 3), pp. 124–129 (2005a)
Decroupet Pascal: Iannis Xenakis, Metastaseis (1953/1955). In: Heister (ed.) Geschichte der Musik im 20. Jahrhundert, Bd. III: 1945–1975, Laaber 2005 (Handbuch der Musik des 20. Jahrhunderts, Bd. 3), pp. 129–133 (2005b)
Dietel, G.: Musikgeschichte in Daten. Bärenreiter-dtv, Kassel and München (1994)
Dittmann, L.: Schönberg und Kandinsky, In: Litwin, St., Velten, K. (eds.): Stil oder Gedanke? Zur Schönberg- Rezeption in Amerika und Europa, Saarbrücken 1995, S. 216–230 (Schriftenreihe der Hochschule des Saarlandes für Musik und Theater. vol. 3). Pfau Verlag,Saarbrücken (1995)
Dixon, S.: Automatic extraction of tempo and beat from expressive performances. J. of New Music Res. **30**(1), 39–58. Swets & Zeitlinger (2001)
Droseltis, A.: Zufall und Determination in der westeuropäischen Musik um 1960. Dargestellt an Werken von Iannis Xenakis und Karlheinz Stockhausen. Phil. Diss. Fakultät I – Geisteswissenschaften der Technischen Universität Berlin Berlin, D 83 (2011)
Dubois, D., Prade, H.: Fuzzy Sets and Systems: Theory and Applications. New York etc. Academic Press (Mathematics in Science and Engeneering, ed. Richard Bellman, vol. 144) (1980)
Ebbeke, K.: Aleatorik. In: MGG^2, vol. 1, cols. 415–444 (1994)
Elschekovấ, A., Kubik, G., Muyle, R.M., Simon, A., Ziegler, S.; Mehrstimmigkeit. In: MGG^2, vol. 5, cols. 1766–1808 (1996)
Emmerson, S.: Living Electronic Music. Ashgate, Aldershot, Hants (2007)
Ewen, D.: George Gershwin—His Journey to Greatness. Prentice-Hall, Inc., Englewood Cliffs, NJ (1970)
Ewen, D.: George Gershwin. Vom Erfolg zur Größe. Droemersche Verlagsanstalt Th. Knaur Nachf, München (1991)
Finke, G.: Partituren zum Lesen und Schauen. Bildlichkeit als Merkmal graphischer Notation. ZGMTH Zeitschrift der Gesellschaft für Musiktheorie **16**(1), 21–40 (2019)
Föllmer, G., Gerlach, J.: Audiovisionen. Musik als intermediale Kunstform. https://www.medienkunstnetz.de/themen/bild-ton-relationen/audiovisionen/ (2004)
Floros, C.: Alban Berg. Musik als Autobiographie. Breitkopf & Härtel, Wiesbaden, Leipzig, Paris (1992)
Folkerts, G.: Mikis Theodorakis. Seine musikalische Poetik. von Bockel-Verlag, Neumünster (2015)
Fornari, J., Maia, I.L.: Interactive notation: towards a computer model for interactive musical composition. Interdisciplinary Nucleus for Sound Studies (NICS)—Universidade Estadual de Campinas (UNICAMP), Campinas—SP. Brazil. https://www.academia.edu/17380050/Interactive_Notation_Towards_a_Computer_Model_for_Interactive_Musical_Composition (n.d.)
Frank, P.: Visuelle Partituren. In: von Maur, pp. 444–449 (1985)
Frisius, R.: Forum analyse. https://www.frisius.de/rudolf/texte/txhorend.htm (n.d.)
Frobenius, W.: Aleatorisch, Aleatorik. In: Handwörterbuch musikalischer Terminologie-HmT. Ordner 1: A-D. Steiner-Verlag, Stuttgart (1976)

Fülöp-Müller, R.: Geist und Gesicht des Bolschewismus. Darstellung und Kritik des kulturellen Lebens in Sowjetrussland (1926), ed. by E. Siepmann as Fantasie und Alltag in Sowjet-Russland. Bericht eines Augenzeugen (EP 7), Elefanten Press, Berlin und Hamburg (1978)
Garland, T.H., Kahn, Ch. V.: Math and Music Harmonious Connections. Dale Seymour Publications, Palo Alto, CA (1995)
Gess, N., Schreiner, F., Manuela K. Schulz, M.K. (eds.): Hörstürze. Akustik und Gewalt im 20. Jahrhundert. Königshausen & Neumann, Würzburg (2005)
Gieseler, W.: Komposition im 20. Jahrhundert. Moeck, Celle (1975)
Gieseler, W.: Zur Semiotik graphischer Notation. In: Melos NZ, vol. 4, pp. 27–33. Schott International, Mainz (1978)
Gifford, T., Knotts, Sh., McCormack, J., Kalonaris, St., Yee-King, M., d'InvernoM.: Computational systems for music improvisation. https://doi.org/10.1080/14626268.2018.142661 (2018)
Gómez, F., A., Melvin, D., Rappaport, G., Toussaint, T.: Mathematical measures of syncopation. https://www.researchgate.net/publication/228850033_Mathematical_measures_of_syncopation/download (2005)
Gottwald, C.: Signale zwischen Exotik und Industrie. Steve Reich auf der Suche nach einer neuen Identität von Klang und Struktur. Melos / NZfM **1**, 3–6 (1975)
Hagen, W.: Ruttmanns Großstadt-Weekend. H.rstürze. Akustik und Gewalt im 20. Jahrhundert. Tagungs des Graduiertenkollegs ‚Codierung von Gewalt im medialen Wandel" der HUB. https://www.whagen.de/PDFS/11010_HagenWalterRuttmannsGro_2005.pdf (2005)
Hahl-Hoch, J. (ed.): Arnold Schönberg, Wassily Kandinsky: Briefe, Bilder und Dokumente einer außergewöhnlichen Begegnung. dtv, München (1983)
Haimo, E.: Schoenberg's Transformation of Musical Language. Cambridge University Press, Cambridge (2006)
Hamberger, Ch.Ll.: The Evolution of Schoenberg's Klangfarbenmelodie: the importance of timbre in modern music, master of arts. Master Thesis Pennsylvania State University. https://etda.libraries.psu.edu/catalog/16221 (2012)
Haselböck, L.: Zur "Klangfarbenlogik" in Schönbergs Orchesterstück op. 16/3. In: "Kürzen". Gedenkschrift für Manfred Angerer. https://www.academia.edu/29991332/Zur_Klangfarbenlogik_in_SchC3%B6nbergs_Orchesterst%C3%Bcck_op_16_3 (2016)
Heister, H.-W.: Das Konzert. Theorie einer Kulturform, 2 vols. Noetzel-Verlag, Wilhelmshaven (1983)
Heister, H.-W.: Jazz. (Musik aktuell. Analysen, Beispiele, Kommentare, Bd. 5). Bärenreiter-Verlag, Kassel (1983)
Heister, H.-W. (ed.): Geschichte der Musik im 20. Jahrhundert: 1945–1975 (Handbuch der Musik im 20. Jahrhundert Band 3) Laaber-Verlag, Laaber (2005a)
Heister, H.-W.: Synthese Versus Exotismus. Entdeckung der neuen »Dritten Welt« und universalistische Integration des »Exotischen«. In: Heister (ed.), pp. 263–274 (2005b)
Heister, H.-W.: Vereinfachungen. Ritualismus und Minimalismus, In: Heister, pp. 278–284 (2005c)
Heister, H.-W.: La Música. Domination, expropriation, exoticism, appropriation and ambivalence of colonialism. In: Reports of the International Colloquium Music/ology and Colonialism. Centro Nacional de Documentación Musical Lauro Ayestarán, Montevideo / República oriental del Uruguay, Ed. Coriún Aharonián, Montevideo, op. 273–308 (2011)
Heister, H.-W., Sparrer, W.-W. (eds.): Der Komponist Isang Yun, München text+kritik (1987)
Hiller, L. A: Informationstheorie und Computermusik. Zwei Vorträge, Darmstadt 1963. Darmstädter Beiträge zur Neuen Musik, vol. 8. Schott Verlag, Mainz (1964)
Hiller, L.A., Isaacson, L.M.: Experimental Music: Composition with an Electronic Computer. McGraw-Hill Book Company, New York (1959)
Hutter, U.: Klangfarben – Farbklänge. Wechselwirkung zwischen visueller und auditiver Wahrnehmung, zum Zweck der Erlangung des akademischen Grades eines Dipl. Ing. (FH) Telekommunikation und Medien Wien (2006)
Kahn, D.: Noise, Water, Meat: A History of Sound in the Arts. MIT Press, Cambridge (1999)

Karkoschka, E.: Das Schriftbild der Neuen Musik. Bestandsaufnahme neuer Notationssymbole. Anleitung zu deren Deutung, Realisation und Kritik. Moeck, Celle (1966)

Karkoschka, E.: Notation in New Music: A Critical Guide to Interpretation and Realization. Praeger, New York (1972)

Klotz, S.: Kombinatorik und die Verbindungskünste der Zeichen in der Musik zwischen 1630 und 1780. Akademie Verlag, Berlin (2006)

Klüppelholz, W.: Mauricio Kagel 1970–1980. DuMont-Verlag, Cologne (1981)

Kreinin, J.: "To Arrest the Process": Moving Clusters by György Ligeti and Witold Lutoslawski. Mitteilungen der Paul Sacher Stiftung **15**, 36–41 (2002)

Kubik, G.: Zum Verstehen afrikanischer Musik. Reclam, Leipzig (1988)

Kubik, G.: Afrikanische Musik. In: MGG^2, vol. 1, col. 50–194 (1994)

Kubik, G.: Kalimba, Nsansi, Mbira – Lamellophone in Afrika (Veröffentlichungen des Museums für Völkerkunde Berlin, Abteilung Musikethnologie X, Neue Folge 68). Berlin (1998)

Kubik, G., et al.: Ostafrika (Musikgeschichte in Bildern, vol. I, issue 10). Deutscher Verlag für Musik, Leipzig (1982)

Kunz, H.: Die Opern. In: Heister, H.-W., Sparrer, W.-W. (eds.) Der Komponist Isang Yun, pp. 95–137. text+kritik, München (1987)

Lacoste, St.: Los Angeles Philharmonic Association. https://www.laphil.com/philpedia/piec-detail.cfm?id=2742 (n.d.)

Locke, D.L.: The metric matrix: simultaneous multidimensionality in African music. Anal. Approaches to World Music **1**(1), 48–72 (2011)

Lowell, N.: Eugène Ysaÿe and musical dialogue. dornsife.usc.edu (n.d.)

Lubben, J.: Metric vibration: acoustic perspectives on metric dissonance, In: Utz, Christian, Gnadenstätter, Clemens (Hrsg.) Musiktheorie als interdisziplinäres Fach (Music Theory and Interdisciplinarity), 8. Kongress der Gesellschaft für Musiktheorie Graz 2008 (musik.theorien der gegenwart 4). Pfau, Saarbrücken (2010)

Manning, P.: Electronic Electronic and Computer Music, 4th edn. Oxford University Press, Oxford and New York (2013)

McInerney, M.J.: Performance and the page: an artist's investigation of the dialogue between the musical event and the written score. A thesis submitted to the University of Plymouth in partial fulfilment for the degree of Doctor of philosophy, Vol. 1: Written Inquiry. Dartington College of Arts. Creative and Performing Arts (2007)

McLean, A., Griffiths, D., Collins, N., Wiggins, G.: Visualization of live code (https://yaxu.org/visualisation-of-live-code/). In: Electronic Visualization and the Arts, London 2010. Alan Seal, A., Jonathan, P., Bowen, J.P., Kia Ng, K. (eds.) Springer Series on Cultural Computing. https://yaxu.org/writing/visualisation-of-live-code.pdf (2013)

Meyer-Eppler, W.: Statistische und psychologische Klangprobleme. In: die reihe 1: elektronische Musik. Informationen über serielle Musik, pp. 22–28. Universal-Edition, Wien (1955)

Meyer-Eppler, W.: Grundlagen und Anwendungen der Informationstheorie. Springer-Verlag, Berlin, Göttingen. Heidelberg (1959)

MGG^1 (Die Musik in Geschichte und Gegenwart) (The music in history and present day), Blume, F. (ed.) 14 vols. +2 supplement-vols. + 1 vol. register. Bärenreiter-Verlag, Kassel etc. (1949–1979)

MGG^2 (Die Musik in Geschichte und Gegenwart) 2nd new revised edn, Finscher, L. (ed.). Subject part, 9 vols. Bärenreiter/Metzler, Kassel and Stuttgart (1994–1998)

Morat, D., Ziemer, H. (with the collaboration of R. Rutz) (eds.): Handbuch sound. Geschichte – Begriffe – Ansätze. J.B. Metzler Verlag, Stuttgart (2018)

Müller, H.-C.: Zur Theorie und Praxis indeterminierter Musik – Aufführungspraxis zwischen Experiment und Improvisation, Bärenreiter Kassel (1994)

Neue Gesellschaft für Bildende Kunst Berlin (ed.): Absolut modern sein: Zwischen Fahrrad und Fließband – culture technique in Frankreich 1889–1937. Elefantenpress, Berlin (1986)

Nono-Schoenberg, N. (ed.): Arnold Schönberg. 1874–1951. Lebensgeschichte in Begegnungen. Verlag Ritter, Klagenfurt (1992)

Odermatt, A.J.: Historical performance practice of the prélude non mesuré and its relationship to recording and performance. A thesis submitted for the partial fulfilment of the requirements of the degree of Master of Philosophy, School of Music, The Australian National University (2018)
Oja, C.: George Antheil's Ballet Mécanique and transatlantic modernism. In Ludington, T. (ed.) A modern mosaic: art and modernism in the United States. University of North Carolina Press, Chapel Hill (2000)
Thomas, M.F., Key, S., Rothe, L.: American Mavericks. San Francisco Symphony, San Francisco, CA (2001)
Petzer, A.: Farbe Klang Körper. Zur Bewegungsästhetik in Wassily Kandinskys Bühnenkomposition „Der gelbe Klang". Diplomarbeit im Studiengang Pädagogische Ausbildung der Fakultät Musik der Universität der Künste Berlin (2017)
Polth, M.: Zum Verhältnis von Harmonik und Instrumentation ›vor Wagner‹. ZGMTH **11**(1), 37–61 (2014)
Popper, K.R.: Of clouds and clocks. In: Objective Knowledge: An Evolutionary Approach, pp. 2–66. Clarendon, Oxford (1972)
Prümm, K.: Die Montage als alles durchdringendes Prinzip der Stadtsinfonie. Anmerkungen zu Walter Ruttmanns Berlin. Die Sinfonie der Grossstadt und die Folgen. In: Montage AV 20/1/2011, pp. 61–71 (2011)
Puckette, M.: The theory and technique of electronic music. DRAFT, September 24, 2006, sine loco. msp.ucsd.edu › techniques. https://www.researchgate.net/publication/242020079_The_Theory_and_Technique_of_Electronic_Music (2006)
Reich, S.: Einige optimistische Voraussagen über die Zukunft der Musik (1970) In: Danuser et al., p. 291. (1993)
Reish, G.N.: Una nota sola: Giacinto Scelsi and the genesis of music on a single note. J. of Musicological Res. **25**, 149–189. https://doi.org/10.1080/01411890600613827 (2006)
Reith, D.: Elektronische Musik und Algorithmische Komposition. In: Enders, B. (ed.) Neue Musiktechnologie. Schott International, Mainz (1993)
Reuter, C., Enders, B., Jacobi, R.: Lexikon Musikautomaten, CD-Rom. Schott International, Mainz (2000)
Riemann, H.: Geschichte der Musik seit Beethoven (1800–1900). Peters, Leipzig (1901)
Rohrhuber, J.: Artificial, Natural, Historical in Transdisciplinary Digital Art: Sound, Vision and the New Screen. Springer, Berlin and Heidelberg. https://akustik.hfbk.net/publications/rohrhuber_artificial_natural_historical.pdf (2008)
Schäfer, A., Kursell, J.: Kräftespiel. Zur Dissymmetrie von Schall und Wahrnehmung. In: Zeitschrift für Medienwissenschaft. Issue 2: Materialität/Immaterialität **2**(1), 24–40. https://doi.org/10.25969/mediarep/1365 (2010)
Schaeffer, P.: Musique concrète. Von den Pariser Anfängen um 1948 bis zur elektronischen Musik von heute. E. Klett, Stuttgart (1974)
Scheytt, J.: Estampes, No. 1 Pagodes, https://www.jochenscheytt.de/debussy/debussywerke/pagodes.html (n.d.)
Schillinger, J.: Schillinger System of Musical Composition. C. Fischer Inc., New York (1946)
Schillinger, J.: The Mathematical Basis of the Arts. Philosophical Library, New York (1948)
Schlothauer, B.: Analyse von Arnold Schönbergs Orchesterstück „Farben", Opus 16 / 3. Hausarbeit an der Technischen Universität Berlin im Wintersemester 2005/06. https://docplayer.org/237855-Analyse-von-arnold-schoenbergs-orchesterstueck-farben-opus-16-3.html (2005)
Schmidt, M.: Klangfarbenmelodie. In: Oesterreichisches Musiklexikon online. https://www.musiklexikon.ac.at/ml/musik_K/Klangfarbenmelodie.xml, 06/05/2001 (2001)
Schnebel, D.: MO-NO. Musik zum Lesen. DuMont-Verlag, Köln (1969)
Schnebel, D.: Mauricio Kagel. Musik Theater Film. DuMont-Verlag, Cologne (1970)
Schönberg, A.: Harmonielehre. Universal-Edition, Wien (1911)
Schuller, G.: Early Jazz: Its Roots and Musical Development. Oxford University Press, New York (1968)

Schulze, H.: Das aleatorische Spiel: Erkundung und Anwendung der nichtintentionalen Werkgenese im 20. Jahrhundert. Hochschule der Künste Berlin (Phil. Diss. 1999). Fink, Munich and Berlin (2000)
Seidel, W.: Rhythmus, Metrum, Takt. In: MGG², vol. 8, col. 257–317 (1998)
Seifert, U., Kim, J.U., Moore, A. (eds.): Paradoxes of Interactivity. Perspectives for Media Theory, Human-Computer Interaction, and Artistic Investigations, transcript Verlag, Bielefeld (2008)
Seising, R.: Die Entstehung der Fuzzy Set Theorie und ihrer ersten Anwendungen – Ihre Entwicklung bis in die 70er Jahre des 20. Jahrhunderts (Boethius: Texte und Abhandlungen zur Geschichte der Mathematik und der Naturwissenschaften Band 54). Franz Steiner, Stuttgart (2005)
Seising, R.: The Fuzzification of Systems: The Genesis of Fuzzy Set Theory and Its Initial Applications—Developments Up to the 1970s. Springer-Verlag, Berlin etc. (2007)
Skupin, R.: Giacinto Scelsi—homo viator and his musical itinerary. The Academy of Music, January 2012, https://www.researchgate.net/publication/316172352 (2012)
Sparrer, W.-W.: Isang Yun. In: Heister, H.-W., Sparrer, W.-W. (eds): Komponisten der Gegenwart. Loose-Leaf lexikon, vol. 1, 40 pp. Edition Text + Kritik, München (1992)
Sparrer, W.-W.: Helmut Lachenmann: Pression für Violoncello solo (1969/1970), In: Heister 2005 (ed.): Geschichte der Musik im 20. Jahrhundert: 1945–1975 (Handbuch der Musik im 20. Jahrhundert Band 3), pp. 340–344. Laaber-Verlag, Laaber (2005)
Sterneck, W.: Das künstliche Geräusch – Futuristische und konkrete musik, Revised passage from the book Der Kampf um die Träume – Musik und Gesellschaft. https://www.sterneck.net/musik/futurismus/index.php (1998)
Stopka, K.: Rauschen. In: Morat, Z., Rutz, R., pp. 205–209 (2018)
Stowell, R.: Violin Technique and Performance Practice in the Late Eighteenth and Early Nineteenth Centuries. Cambridge University Press, Cambridge (1990)
Stroh, W.M.: Midi-Experimente und Algorithmisches Komponieren – eine Anleitung zum kreativen Programmieren und Komponieren am Computer, mit Diskette(n). midi-pädagogische schriftenreihe, vol. 3. musiklabor Berlin, Berlin (1990)
Supper, M.: Elektroakustische Musik und Computer Musik, Darmstadt (1997)
Tischhauser, K.J.: A survey of the use of extended techniques and their notations in twentieth century string quartets written since 1933 by American Composers with a selected annotated bibliography and discography (A Treatise submitted to the School of Music in partial fulfillment of the requirements for the degree of Doctor of Music) (2002)
Ungeheuer, E.: Elektroakustische Musik 1945–75: Prisma musikalischer Originalität. In: Heister (ed.) Geschichte der Musik im 20. Jahrhundert, Bd. III: 1945–1975, Laaber 2005 (Handbuch der Musik des 20. Jahrhunderts, Bd. 3), pp. 96–106 (2005)
von Fischer, K.: Variation. In: MGG¹, vol. 13, cols. 1274–1309 (1966)
Wang, G., Cook, P.: On-the-fly programming: using code as an expressive musical instrument. In: Proceedings of the 2004 International Conference on New Interfaces for Musical Expression (NIME). NIME, New York. https://soundlab.cs.princeton.edu/publications/on-the-fly_nime2004.pdf (2006)
Weiss, A.S.: Experimental Sound & Radio. MIT Press, Cambridge, MA (1996)
Welsh, C.: Sirene. In: Morat / Ziemer, pp. 211–216 (2018)
Whatley, K.: Rough Romance: Sequenza II for Harp as Study and Statement. In: Halfyard, J.K. (ed.) I: Berio's Sequenzas: Essays on Performance, Composition and Analysis, pp. 39–52. Ashgate Publishing Limited, Aldershot and Burlington, VT (2007)
Wild, Ch.: Charles Ives' three places in New England: an interpretation and a conductor's guide. A Doctoral Thesis, Presented to the Faculty of the Bienen School of Music of Northwestern University in Candidacy for the Degree of Doctor of Musical Arts (2018)
Wilson, E.O.: Die Einheit des Wissens. Siedler Verlag, Berlin (1998)
Xenakis, I.: Formalized Music. Thought and Mathematics in Composition, Rev. edn, Additional material compiled and edited by Sharon Kanach. Harmonologia Series No. 6. Pendragon Press, Stuyvesant, NY (1992)

Yoon, H.J.: Stochastische und fraktale Modelle in der Algorithmischen Komposition. epOs Electronic Publishing Osnabrück Music, Osnabrück (2002)
Yun, Sh.-H.: Zwischen zwei Musikwelten. Studien zum musikalischen Denken Isang Yuns, Dissertation zur Erlangung des akademischen Grades Dr. phil., Seoul Oktober 2000. Ms. (2000)
Yun, Sh.-H.: Die Hauptklangkomposition Isang Yuns unter dem Aspekt der bildenden Kunst. In: Korean Musicological Society (ed.) Umakhak 11, pp. 129–161. Seoul (2004)
Zeller, M.: Planal analysis and the emancipation of Timbre: Klangfarbenmelodie and Timbral function in Mahler, Schoenberg, and Webern. Dissertation submitted in partial fulfillment of the requirements for the degree of Doctor of Philosophy in the Department of Music in the Graduate School of Duke University. https://dukespace.lib.duke.edu/dspace/handle/10161/21042 (2020)

Chapter 8
The Principle of Variation: Similarities, Differences, Development

Abstract Variation is a universal principle not only in music. In the light of FL, it is a intertwining of blurring and sharpening. It is here a gradual change of a given thing (see Chapter 3). In our context it is mainly about the intentional and designed variation. I refer to small-scale motifs, themes or wide-ranging melodies or even polyphonic structures, finally whole work. This given stands as an alternation to its variation in relation to the other as well as its other. Variation is a result of the principles repetition, contrast, alteration, with a wide span between permutation of in themselves unchanged components and developing new ones. Identity and difference, as outlined in Chapter 1, pass through various stages and variations in the thirteen 'Forms of existence' (see Chapter 2) of a work. Fuzziness is a necessary implication here, similarity is both measure and result of the different dialectical movements and counter-movements of fuzzines, sharpening and blurring.

Keywords 'Change ringing' · Centonisation' in Gregorian Chant · Collage · Contrafacture (contrafactum) and parody · Covering and remake · Dance / ballet · Ekphrasis · Film · Media transformation · Music analysis · Palimpsest · Paraphrase · Permutation · Quodlibet, potpourri, medley · Quotation · Readymade · Remix and mashup · Sampling and 'electrified collage' · Sign language · 'Soundalike' · Twelve tone technique

Variation is a universal principle not only in music. In the light of Fuzzy Logic, it is a form of against each other, with each other, and into one another of blurring and sharpening. In this respect, variation is a gradual change of a previously given thing within a fuzzy set (see Sect. 3.1). A distinction must be made between involuntary variation—it always occurs, especially in the case of blurring, but also in sharpening—and voluntary, intentional variation. The latter is decisive here.

Variation as change, alteration, modification necessarily always starts out from something given that is changed, whether it is a small-scale motif, theme or wide-ranging melody, or even a polyphonic composition, finally even a whole work. This given thing, if it does not remain unchanged and stays just as it is, stands as a variety to its variation in relation to the other as *its* other to put it in Hegelian terms.

H. Heister, *Music and Fuzzy Logic*, Studies in Fuzziness and Soft Computing 406,
https://doi.org/10.1007/978-3-662-62907-9_8

8.1 Width and Diversity of the Principle of Variation

In this chapter, some of the relationships developed so far will be put into concrete terms. The starting point was the relationship between identity and difference outlined in Chapter 1, which also forms an axis for the staircase of the thirteen forms of existence. Fuzziness is here a necessary implication. Similarity is both measure and result of the different dialectical movements and counter-movements of sharpening and blurring. Fuzziness itself implies in many respects already dissolution on a meta-level, on which it serves to sharpen. Variation, finally, is a synthesis of sharpening and blurring. It makes concrete, concentrates and unfolds some of what has been analyzed in the preceding movements of contradiction.[1] In the final analysis, variation in art means working with similarities: Variation is both the *same as* and *different from* the varied. And the variations, in turn, are similar to each other. Thus, in the 'variation', the interweaving of blurring and sharpening is finally bundled and raised to a new, higher level.

In the following, some important types of variation will be characterized in more detail and presented from the perspective of fuzzy logic. It is likely that the one or other type is still missing, which actually should have been included. Completeness cannot be achieved. Generally speaking, the principle of variation is universal. In the literal sense of "variation" = "change" it is ontologically subject to everything that exists, as already mentioned at the beginning of the book in Chapter 1 (1.2): Being is becoming. Stasis, identity without difference, non-change is the exception, the emergence from nothing, "creatio ex nihilo", is, like the nothing, with some probability not at all the case: the "Big Bang" is not the first sound, but before that there existed something like, for example, possibly vacuum fluctuations. In physics we find inter alia *magnetic variation*, the difference between magnetic north and true north, and in astronomy „any perturbation of the mean motion or orbit of a planet or satellite, particularly of the moon".[2] In biology genetic variation occurs in two types as somatic and germinal variation; particularly via mutation variation is an essential for genetic diversity, i.e. mainly differences within a species, plus intra specific variation among the different genera. In game variation means „modifications made to a game by a community of players (as opposed to a central authority)"[3] In language resp. linguistics variation is a universal characteristic of languages, including the languages of art: „there is more than one way of saying the same thing", e.g. *elegant variation*, „the gratuitous use of a synonym in prose". The same is not absolutely the same thing and identic by saying it differently: the dialectics of content and form, even more relevant in arts. „Speakers may vary pronunciation (accent), word choice (lexicon), or morphology and syntax […] But while the diversity of variation is great, there seem to be boundaries on variation – speakers do not generally make drastic alterations in sentence word order or use novel sounds that are completely foreign

[1] More detailed and multi-faceted e.g. Pareyon 2011.

[2] https://en.wikipedia.org/wiki/Variation, 19.7.2019, retrieved 13 July 2020.

[3] Ibid.

to the language being spoken."[4] This applies not only to individual variations, but also to collective ones, and this again synchronically and diachronically. "Variatio delectat"—like repetition, variation is one of the fundamental means of rhetoric and thus also of direct relevance in music. The scope and extent of variation is much greater in artistic languages than in communicative language, especially in music. This is why material such as technology, lexis and syntax has a much wider range of blurring and sharpening, of variation possibilities in general. In mathematics inter alia there are

> variations without repetition, a term in combinatorics commonly used by non-English authors for k-permutations of n, and variations with repetition, a term in combinatorics commonly used by non-English authors for n-tuples.[5]

Variation is, as in reality, a general principle in music and art. It permeates many things, the respective production of music and works down to the level of procedures, including, in a broader sense, the historical development.

> Varying is a basic principle of musical and artistic interpretation. At every stage of its development, from the primitive to the advanced civilizations, in the sacred as well as the profane realm, music makes use of the principle of variation if it is passed on orally or in writing. [...] The history of variation begins with the history of music.[6]

8.1.1 Permutation (I)—‚Change Ringing'

A special type of variation is the permutation just mentioned. It is a type of mathematical variation as part of the fields combinatorics and group theory[7] that is particularly important for music. In art it seems a bit mechanical, but is used a lot, not least in dodecaphony (see Sects. 4.6, 5.2, and 6.2). Often in motivic-thematic work there is an incomplete permutation that does not exhaust all mathematical possibilities.

> *Permutations* differ from *combinations,* which are selections of some members of a set regardless of order. For example, written as tuples, there are six permutations of the set {1,2,3}, namely: (1,2,3), (1,3,2), (2,1,3), (2,3,1), (3,1,2), and (3,2,1). These are all the possible orderings of this three-element set. Anagrams of words whose letters are different are also permutations: the letters are already ordered in the original word, and the anagram is a reordering of the letters.[8]

In the case of permutation without repetition, the number of possible permutations of objects is designated factorially by the faculty n! n. In our prime example BACH,

---

[4]Cf. ibid.

[5]Ibid.

[6]Von Fischer 1966, col. 1274. To the broad concept cf. inter alia: Vonderwahl-Nyffenegger 2006.

[7]„Permutations are used in almost every branch of mathematics, and in many other fields of science. In computer science, they are used for analyzing sorting algorithms; in quantum physics, for describing states of particles; and in biology, for describing RNA sequences" (https://en.wikipedia.org/wiki/Permutation, 7.11.2020, retrieved 8 November 2020).

[8]Ibid.

there are thus a total of 24 possible arrangements. Excessive use is made of different types of permutation in traditional British change ringing,[9] a prime example of the relationship between music and mathematics. It has developed since about the tenth century.

> The art of change ringing is peculiar to the English, and, like most English peculiarities, unintelligible to the rest of the world. To the musical Belgian, for example, it appears that the proper thing to do with a carefully tuned ring of bells is to play a tune upon it. By the English campanologist, the playing of tunes is considered to be a childish game, only fit for foreigners; the proper use of the bells is to work out mathematical permutations and combinations. When he speaks of the music of his bells, he does not mean musicians' music—still less what the ordinary man calls music. To the ordinary man, in fact, the pealing of bells is a monotonous jangle and a nuisance, tolerable only when mitigated by remote distance and sentimental association. The change-ringer does, indeed, distinguish musical differences between one method of producing his permutations and another; he avers, for instance, that where the hinder bells run 7, 5, 6, or 5, 6, 7, or 5, 7, 6, the music is always prettier, and can detect and approve, where they occur, the consecutive fifths of Tittums and the cascading thirds of the Queen's change. But what he really means is that by the English method of ringing with rope and wheel, each several bell gives forth her fullest and her noblest note. His passion—and it is a passion—finds its satisfaction in mathematical completeness and mechanical perfection, and as his bell weaves her way rhythmically up from lead to hinder place and down again, he is filled with the solemn intoxication that comes of intricate ritual faultlessly performed.[10]

For example, a sequence with only 4 bells:

Plain Bob

4213 4123 1432 1423 4132 4312 3421 3241 2314 2134 1243 1234[11]

Change ringing achieves completeness and a laus perpetua, the unceasing praise of God, in a modern semi-mechanical way. The former sacral purpose has long since become a game. In the British system of change ringing and especially in method ringing it is a matter of systematically exploiting the combinatorial possibilities of the bells available in each case, if possible to the limit (Fig. 8.1).

> All changes on eight bells, with the musical rows Whittingtons, Queens and Tittums. This is not a call change 'peal', but an example of calling changes for a short period for musical effect. Method ringers can trace their bells' courses visually in a diagram called a *blueline.*)Instead, a system of *change ringing* evolved which centers on mathematical permutations. The ringers begin with *rounds*, which is simply ringing down the scale in order. (On six bells this would be *123456*.) The ringing then proceeds in a series of *rows* or *changes*, each of which is some permutation of rounds (for example *214356*).
>
> In *call change ringing*, one of the ringers (known as the *conductor*) calls out to tell the other ringers how to vary their order from row to row. Some ringers practice call changes exclusively; but for others, the essence of change ringing is *method ringing*.

---

[9]A reflex e.g. with Grieg's "Klokkeklang" ("Bell Ringing"), Op. 54 No. 6 (1891), see Tymoczko, p. 342f.

[10]Quoted in: Benson 2006, p. 315. In Sayers' story the change ringing is, not so solemn, used to kill somebody.

[11]Extensively cf. e.g. http://en.wikipedia.org/wiki/Change_ringing, 17.7.2020, retrieved 20 July 2020; see also mong others Wilson 1965.

Fig. 8.1 Change ringing. Permutation Patterns http://en.wikipedia.org/wiki/Change_ringing

Method ringing In *method* or *scientific ringing* each ringer has memorized a pattern describing his or her bell's course from row to row; taken together, these patterns (along with only occasional *calls* made by a conductor) form an algorithm which cycles through the various available permutations. Serious ringing always starts and ends with rounds; and it must always be *true* — each row must be unique, never repeated. A performance of a few hundred rows or so is called a *touch*; approximately five thousand rows make a *peal* (which takes about three hours to ring). A performance of all the possible permutations possible on a set of bells is called an *extent*; with *n* bells there are *n*! possible permutations. Since 7! = 5040, an extent on seven bells is a peal; 8!=40,320 and an extent on eight bells has only been accomplished once, taking nearly nineteen hours.[12]

[12]Extensively cf. e.g. http://en.wikipedia.org/wiki/Change_ringing, 17 July 2020; among others Wilson 1965.

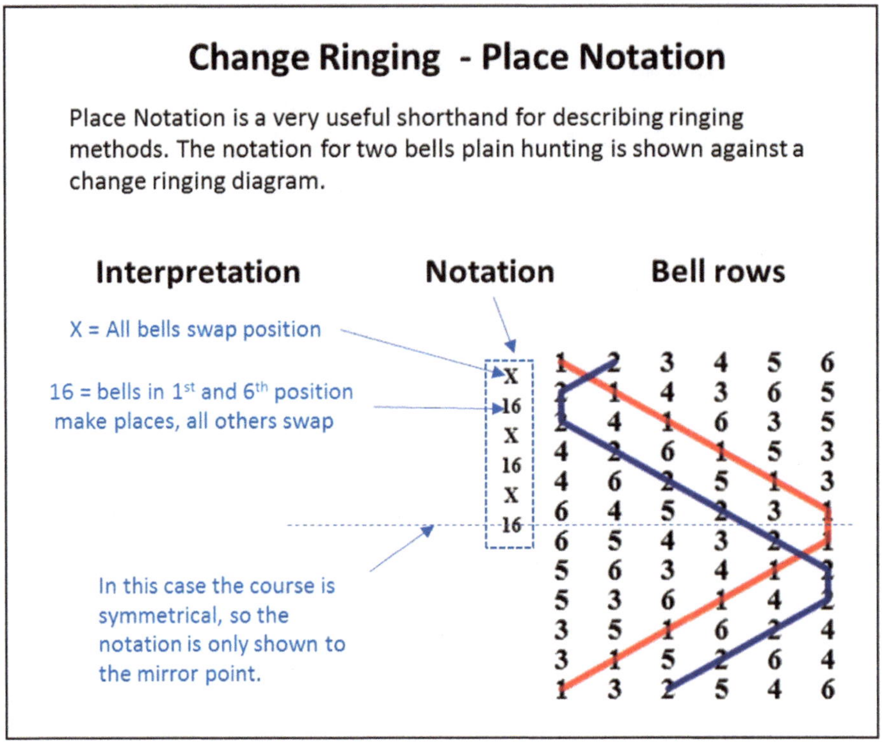

Fig. 8.2 Quoted after https://en.wikipedia.org/wiki/File:Place_notation.png, retrieved 17 July 2020

In accordance with the principles of the numbers faculty, we quickly get into ludicrous temporal perfection. The full Stedman triples[13] series contains 7! variants and takes 3 to 4 h. The procedure is more oriented towards mathematical completeness than musical perfection (Fig. 8.2).

> For some people, the ultimate goal of this system ["method ringing"] is to ring *all* the permutations, to ring a tower's bells in every possible order without repeating — what is called an „extent"(or sometimes, formerly, a „full peal"). The feasibility of this depends on how many bells are involved: if a tower has *n* bells, they will have *n*! [...] possible permutations, a number that becomes quite large as *n* grows. For example, while six bells have 720 permutations, 8 bells have 40,320; furthermore, 10! = 3,628,800, and 12! = 479,001,600. Estimating two seconds for each change (a reasonable pace), we find that while an extent on 6 bells can be accomplished in half an hour, a full peal on 8 bells should take nearly twenty-two and a half hours. (When in 1963 ringers in Loughborough became the first and only in history to achieve this feat on tower bells, it actually took them just under 18 hours. [...]) An extent on 12 bells would take over thirty years![14]

[13] About Stedman, S. https://en.wikipedia.org/wiki/Permutation, 23.6.2020, retrieved 14 July 2020.

[14] http://en.wikipedia.org/wiki/Change_ringing, 26.12.2007, retrieved 2 January 2008.

8.1.2 Permutation (II)—Twelve Tone Technique

Combinatorics or permutation plays a major role in musical counterpoint. We have already outlined some of them.[15] The tone sequences AACH as well as BBCH flatten the original form BACH and can be considered as their defective permutations, i.e. as variations which, however, obscure the essential aspects of the topos' structure. A special case of the permutation is the contrapuntal procedure of the cancer, i.e. the exactly opposite order, called palindrome in literal language, such as *race-car*—ABC CBA. This creates temporal mirror symmetry. It is more difficult to perceive than the virtually spatial inversion. For time runs linearly, as an irreversible process, at least in the area of the universe normally accessible to us in a practical and sensual way. With relativity theory and quantum mechanics, this is somewhat different. Though these two theories and the facts to which they refer influence our view of the world, and thus have an effect on the conception of music, they elude sensory perception. Permutation plays a central role in dodecaphony or Twelve-Tone-Technique. The tone row or series is an ordering of the 12 tones of the equal-tempered chromatic scale (i.e. the 12 pitch classes). Each tone can appear only once. Numerous variation techniques were developed by Schönberg and his successors. They often fall back on traditional counterpoint techniques. The four shapes or the row, the "Grundgestalt", the inversion, the retrograde and the retrograde inversion at can be used on each transpositional level in the complex, so that the series may be used in any of 48 different forms.

There are almost infinite possibilities to combine the 12 series tones. Numerous variation procedures were developed by Schönberg and his successors. They often resort to traditional counterpoint techniques. Thus the row appears in 4 forms: Basic form, inversion, cancer, cancer inversion. Each of these forms can be transposed 11 times, thus appearing on all 12 levels. The series "With this we already have 48 (4×12) pitch arrangements for composing. But that's not all. Schönberg often breaks down the 12-tone series into 2 hexachords (6-tone groups). The pitches of these groups are complementary, i.e. they complement each other in different ways according to the order of intervals and rows, so that the 12-tone row is actually reduced to a 6-tone row. The source material is thus further economized the richness of relationships within the order of pitches is further increased. Anton Webern even goes further and creates rows of 3- or 4-tone groups, which relate to each other in a similar way as cancer or inversion. The twelve-tone row is a freely selectable shape generation regarding the sequence, making full use of all 12 pitches within the octave in the equally tempered tone system. G, U, K, KU = 4 shapes of the series. In a series the 4 shapes 4×12 produce 48 variations. These are repeated again at 11 different transposition levels (13.=1.), thus starting from A, Bb, B, C etc. There are also rows with less or more than 12 tones; this increases the variety of

[15] See above the B-A-C-H in Sects. 4.6.5, further 5.2.4, and 6.2.1; see below in Chapter 9 the *‚musical turn'*—cf. the permuted pairs of bases Adenin (A), Guanin (G), Cytosin (C) Thymin (T) in the DNA; in the RNA variated, quasi transposed—instead of A-G-C-T A-G-C-U (Uracil).

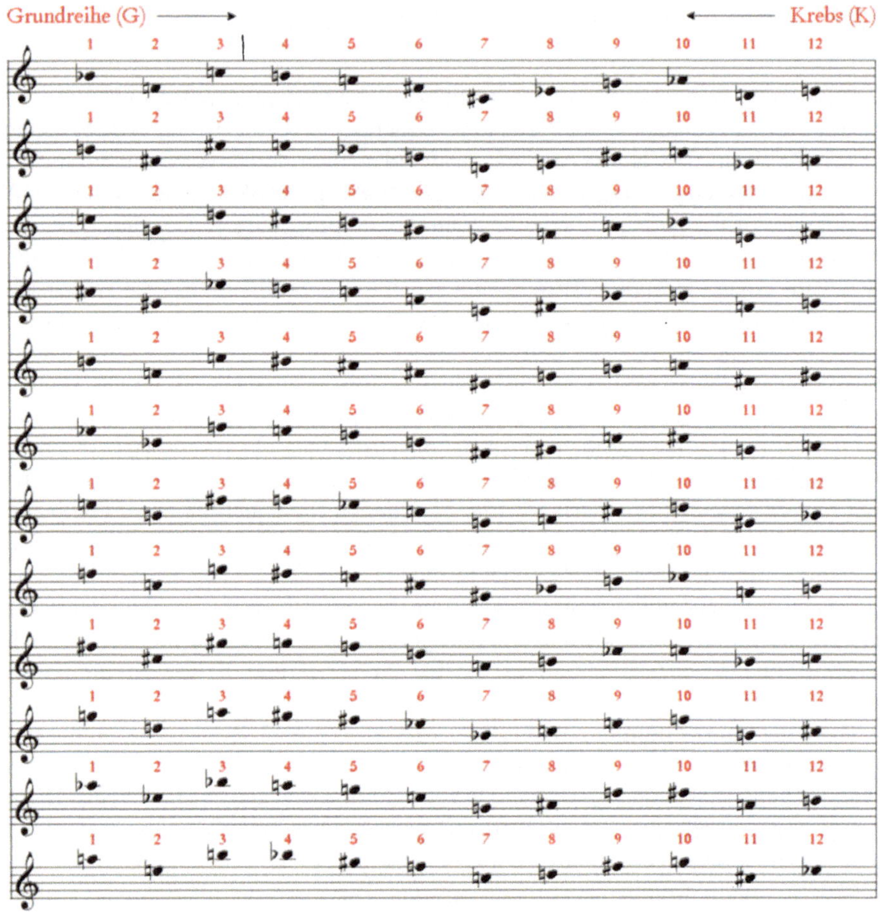

Fig. 8.3 Arnold Schönberg: Piano piece op. 33a: Table of rows 1 (basic rows + cancer forms) Rarus, 28. Mai 2010
https://upload.wikimedia.org/wikipedia/de/6/6d/Sch%C3%B6nberg_op33a_Reihentabelle_1.png

combinations; the same when using other tuning systems, for example quarter tones or microtonality (Fig. 8.3).

On all 12 transposition levels there are $48 \times 12 = 576$ possibilities. $12! =$ 479.001.600: Already in 1923 the composer Joseph Matthias Hauer was enthusiastic about the sheer size of this number as the number of "Melos possibilities".[16] But when it comes to real composing, things become more difficult. Mechanical-mathematical combinations do not yet produce a composition. However, even here it is possible, with certain modifications, to obtain meaningful motifs and shapes (see above in Chapter 6 Alban Berg's *Lulu* Tropes). Once again permutation proves to be

[16] Cf. e.g. Wells 1990, p. 267.

a universal design principle. It combines variation with repetition, deliberate order with random-automatic order, etc.

Actually there is no fuzziness or indeterminacy here. The permutation is mathematically exact, almost mechanical. Nevertheless, even here there is a moment of fuzziness—fuzziness in an admittedly rather broad sense. For the determination and definition of *which* permutation sequence is chosen is arbitrary, subjectively conditioned. The theoretically almost paradoxical conclusion: a rather mechanical filtering serves here the—from the point of view of art and the musical process—productively higher-quality, higher-level crystallization of motivic-thematic shapes. Corresponding procedures are easily computerizable. Whether and to what extent this is necessary and useful will not be further discussed here.

8.1.3 Difference and Identity: Repetition, Contrast, Alteration

The dialectic of identity and difference, which was exposed in Chapter 1, returns here in concrete terms. What was reflected in Chapter 5 as a question of the constancy of shape in connection with motifs and themes is now dealt with on a more general level. Permutation is a reduced type of variation. Its basis here is a finite number of elements. They themselves are not changed, only their temporal sequence. Nevertheless, even this procedure itself, like a kaleidoscope, produces a variety and multitude of new forms. Their similarity to each other is given by the reference to the common basis.

Variation, however, in its fullest sense also includes change of the elements used. Here the question of similarity becomes more complex, and the relationship between core and shell becomes relevant again, and thus also that of constancy and variance. If the definiteness of 'variation' as modification, transformation, metamorphosis, change—of a given!—is the centre, then the two opposing processes of repetition and contrast will form the poles. Neither repetition nor contrast as such is variation in the true sense. They become variation through relationships within the 'variation triangle', repetition, contrast, gradual change. The differentiation points in this direction, there would be

> basically [...] two possibilities of variation: variation by changing the given itself and variation by adding to the original given. Hereby both possibilities can be combined with each other.[17]

This "addition" is the other, the difference to the identical, which is called 'contrast' here. Variation is only in borderline cases incessant change. The dialectic of given and change summarizes the extremely dialectical definition as a compositional „technique where material is *repeated in an altered form.*"[18] The opposite of the contrasting variation is the repetition, the not at all or only very slightly modified repetition. Repetition is on the one hand the antithesis and opposite of variation as change and contrast, but on the other hand it is itself a part of variation.

[17] von Fischer 1966, col. 1276.

[18] https://en.wikipedia.org/wiki/Variation_(music), 7.11.2020, retrieved 7 November 2020.

> However, in order for one part of a series of variations to act as a variation of the other, certain elements must always remain constant.[19]

This also applies to the variation of motifs within a work or part of a work. Repetition is not simply identity and non-change, but often intensification. The sequence in the course of time alone already hierarchizes, simply a first, second etc. appearance. In music, repetition is commonplace. In the literal language it is common in poetry, especially in the form of the choruses. It is a specific rhetorical figure that serves as an affirmation or a component of modern lyric poetry and prose, such as Gertrude Stein's eulogy "Rose is a rose […]"—this is not a simple identity, but rather a metaphor: the first rose is Stein's companion. Otherwise, repetition, especially in the form of the *figura etymologica*, i.e. the use of the same stem in different types of words and inflectional forms or as chains of homonyms, is above all an element of word games, "When flies fly behind flies …" or as pseudo-book titles like "The patient patient".[20] This is neither fuzziness nor mere repetition, but precise differentiation and variation by means of seeming identities. From the outside, from a bird's eye view and in terms of sound, however, it all boils down to something similar, namely repetition. Variations, on the other hand, are more artful. Another facet of the dialectic of constancy and variance is

> the fundamental interaction of preserving and changing musical material has been evident since ancient times in the two principles of framework and coloration [modification of melodies, often improvisational].[21]

Examples of variation run through this book. A wealth of examples is provided, especially since Sect. 4.6 *contains similarities, affinities, structural and procedural relationships*. Fuzzy Logic and Music Analysis. There is therefore almost no need for any additional special evidence or musical examples. I will also present some more systematic and historical facets of the principle of variation. The variation procedures in interaction with the musical facts to which they are applied are manifold and potentially infinite. However, historically and systematically, some main types of variation emerge: variation of rhythm and meter, pitch sequences, melody, key, mode, harmony, counterpoint, compositional techniques such as the relationship between main voice and accompaniment, density of composition e.g. by a changing number of voices, dynamic, loudness, timbre, orchestration or any combination of these etc. The variation e.g. of the otherwise non-primary timbre is constitutive among others in Ravel's *Bolero* or significant in his instrumentation of Musorgsky's *Pictures of an Exhibition*.

At the beginning of his 5th symphony, Beethoven varies the density of the movements and, on a smaller scale, the minore-maggiore ratio, the change of the mode from minor to major or the key. It is indefinite or fuzzy between minor and major at the first occurrence and in the first form of the cross-like four-tone motif G-Eb/F-D, played in unisono, which could be as well Eb major as c-minor (mm. 1–5). The key

[19] von Fischer 1966, col. 1276.

[20] mip, *Die Presse*, Vienna, 2.2.2013.

[21] von Fischer 1966, col. 1277.

Fig. 8.4 Beethoven: 5th Sympony, Ist movement, mm. 10ff
Johann Nepomuk Hummel: Klavierauszug. Les fils de B. Schott, Mainz (n.d. [1827]). Plate 2625. IMSLP93499-PMLP01586-beethoven-hummel_symf-05

is decided and specified as C minor in the recapitulation of the G-Eb, with C in the bass. C minor, now as tonic, is confirmed by the G major as the dominant G-B-d. Meanwhile in a multi-layer progress of variation the two-tone-motif was already melodically varied as Ab G (m. 7f.) and as G D (m. 10f.). And the volume, still one for all parts, is changed blockwise from *ff* mm. 1–5 to *p* mm. 6–11 (Fig. 8.4).

As is often the case, not only one dimension of the compositional structure as pitch, duration, polyphonic densitiy (here from 1 to 3 parts resp. voices) is varied, but several—here the volume variation (ff–p) complements the melodic and harmonic. The timbre, on the other hand, remains the same and changes only in blocks.

8.1.4 Variance Without Constancy

Music shapes the flow of time as a discontinuous experience that tends to be repeated in memory in an imaginary-real way (see above *recollection form*). This will require some interplay of identities and differences. Hans Zender, for example, goes to the extreme of changes without repetition. He even goes so far as to allow the respective uniqueness to spread and have an effect, so to speak, fractally from the overall work to its individual elements: Process takes precedence over object.

> One can make use of the biologically given preconditions [...] by developing strategies, for example, which have the goal that no two figures establish themselves as identical in the forming process. Schönberg's rule that none of the twelve tones may be repeated until all the others have appeared is only the simplest and most basic of such strategies. They all aim to use not memory, but forgetting – the reverse side of memory – to create form.[22]

In this way Zender destroys traditional ways of adopting composition, time and music, and the way they take shape in the course of the process and in retrospect. In this respect, he radically opposes traditions, especially motivic-thematic composition, but also innovations such as, above all, ribbon loop and prayer wheel music and the restoration of the ritual. Both of these form a counter-tendency to the arrow of time in music: the principle of (cyclical) return. However, there is a considerable difference between historical and musical process thinking. In principle, cyclical

[22]Zender 1991, p. 20f.

conceptions of history start from the return of an equal, up to the dystopia of the Nietzschean "eternal return". In music, however, the mere unchanged return is not the rule, but the exception. In the developed sonata form, even the recapitulation is not a mere repetition, but rather brings about a sonata-specific balance of harmonic tensions and, moreover, is, conveyed motivically-thematically, suspension in the Hegelian sense. On the other hand, diversity ultimately becomes a—relative—unity. Even thematic-motivic work, prototype of progressive change, refers in its heterogeneity to an identical core of 'processing identity'. Zender's *Canto V (Continuum and Fragments) for* voices with percussion ad libitum provides an original solution to such dialectics (1972/1974)—it is no coincidence that he once again refers to Heraclitus.

> The basic idea behind the work on Canto V was to write music with the highest possible degree of heterogeneity. All possible contrasts – loud/soft, slow/fast, sung/spoken, continuous/discontinuous, etc. – were to collide as hard as possible without mediation, and the form was to be formed by *randomly combining t*hese contrasts.[23]

Hence, this is 'aleatory' fuzziness (see Sect. 7.4.7).

> This resulted in a series of musically very contrasting fragments for singers: more precisely, an (*unspecified) number of singers* were given identical material. The sequence of the fragments is determined for each singer in a different way, either *by random operations or by a version worked out by the conductor*; the difference in the sequence gives rise to infinite possibilities of vertical relationship in the interplay.[24]

But Zender counters the drift into the indeterminate and fuzzy with structuring contrasts of the second power, which are intended to form a counterweight to the un-heraclitic river without banks.

> However, at the moment when this very contrasting musical event was born, it became clear that the extremely heterogeneous form virtually demanded its opposite, a form that was as uniform as possible – and that only a second, now extremely simple part would be able to touch Heraclitus' basic idea in some way. Thus a "continuum" was created, consisting of a series of very similar particles that could be *repeated at will* by the individual singer, whose simple diatonic and metrics are in stark contrast to the complicated structures of the other part.[25]

8.1.5 Musical Variation Cycles

The principle of variation is on the one hand intensified in procedures such as the modification of figures as in rāga or maqām or in the motivic-thematic work, on the other hand in the genre of variation or 'theme with variations'. 'Variation' as an independent genre with a cyclically sequential form as a sequence of individual segments or movements is a special expression of the principle. It begins in Europe around the sixteenth century.

---

[23]Zender 2004, p. 313.

[24]Ibid.

[25]See Zender 2004, 313. Emphasis HWH.

> A series of variations is created when a musically more or less self-contained section is repeated several times in succession in varied form. There are two types of variation series, which can be distinguished according to their origin: primary or constitutional and secondary or accidental variation series. The first type, which understands variation itself in the narrower sense as a form sequence, arises from the need for change in repetition (double, song or dance variation, etc.). The second type results from the combination of different parts of the form by variably relating the individual parts.[26]

Essentially, the range of variation procedures outlined above is used. A central principle is the *figuration*, especially the fragmentation into smaller rhythmic values and the embellishment, rewriting, stretching of a melody (see above in Chapter 7 as example Fig. 3.1 Coleman Hawkins, *Body and Soul*). The relationship between constancy and variance appears to be doubled: on the one hand, the theme is the constant, the variations are the variable; on the other hand, there are different relationships within the variations, which result in almost permutation-like different types.

1. Cantus firmus variation. Constant: c.f. (chorale, song), which need not be bound to a particular voice and can itself be easily varied.—Variable or new: all other elements such as leading or accompanying voices, harmony, rhythm, formal extension [...].
2. Ostinato variation. Constant: framework tones (dance tenors and basses of the 16th/17th century) or basso ostinato (e.g. ciacona bass), usually also form, often harmony.—Variable or new: all other pieces, whereby the colonel's melody can be new from variation to variation, or can itself be in variable relationship to the theme melody. The framework notes and ostinati can also be subject to more or less pronounced figurations.[27]

The following two types of variations are decisive in Jazz from Swing to Cool, and still constitute a living line of tradition today. When improvising on the changes, the harmony framework, of a song, as a "standard", or a specially created composition, the types that were separated in the development of the European variation are combined with each other.

3. Harmony constant variation. Constant: Harmonics, usually also form.—Variable or new: melody, voice leading, rhythm, tempo, dynamics. In contrast to Type 2, there is no melodic bass or tenor framework. However, the bass often appears as a constant element in the sense of a basso continuo representing the harmony.[28]
4. Melody variation with constant harmony. Constant: Melodic main tones, harmony (at least caesura harmonies), usually also form (form proportions).—Variable: Melodic transposition (coloration, decolorization), rhythm, tempo, dynamics, instrumentation.[29]

In the following type, the constant is already fragmented, its identity fuzzed.

[26]von Fischer 1966, col. 1276.

[27]Ibid.

[28]von Fischer 1966, col. 1276.

[29]Ibid., col. 1276f.

5. Fantasy Variation. Constant: individual parts of the theme (motifs, melody fragments, which are sometimes used as leitmotifs).—Variable: all other elements of the theme, especially the form. Transition from the variation series to fantasy and, technically, not genetically speaking, also to the secondary accidental variation types.[30]

According to a historical tendency, the theme is reduced, so much so that it is not only fuzzed but even erased. In variation in the framework of the series or twelve-tone technique, the variation theme usually recedes and disappears from the sensual realization (This is different in Schönberg's *Variations for Orchestra*; see Sect. 6.3.2.2). The theme functions as a depth structure, the figures of the underlying row, which is hardly perceptible as such.

6. Serial variation. The only constant is the reference to one or more basic forms (tonic, rhythmic, dynamic and other series).[31]

On the other hand, there is the historical tendency to develop and sharpen the theme as a sensually concrete form. A structurally conceived relationship can be made dynamic and put into a temporal sequence. In 1739, Johann Mattheson described the melody as the "body", the harmony as the "clothing" of a multipart composition.[32] From about the 1760s with the Viennese classics, the

> Theme melody of a series of variations [...] should no longer be understood as a sequence of framework tones (as was the case in the majority of German song variations of the 17th and early 18th centuries), but as a natural, songlike whole *that can be heard in the variations at all times.*[33]

The later constant is therefore previously also something that only develops and configures itself. Being was becoming, and will again become becoming in the course of a process such as e.g. the sequence of the variations of the 'forms of existence' shown here. Moreover, the variations already tend to form 'citation forms' of the theme, which may not be considered a work, but nevertheless a self-contained structure (see Sect. 8.2). In a kind of encaptic, nested dialectic, variation contains elements of non-variation, as repetition of elements *within* varying. The differences thus contain identities. Thus, especially in improvised music, i.e. also in jazz, there are recurring phrases and standardized to stereotypical formulas such as the "licks", comparable in some respects to the in principle largely standardized ornaments. In other types of music, too, there is often a reduction, and in some historical phases even predominantly

> the figurations [...] are based on certain often stereotypical formulas, so-called "Klangfüße", as F.W. Marpurg called them in 1765 in his instructions for playing the piano: Sub-notes, tone repetitions, scales, arpeggios, rolls, left and right hand reference, dotting, syncopated shifts, etc.[34]

---

[30] Ibid., col. 1277.

[31] Ibid.

[32] Quoted from von Fischer 1966, col. 1276.

[33] von Fischer 1966, col. 1291. Emphasis HWH.

[34] von Fischer 1966, col. 1291.

The screw of change rotates incessantly: licks are in turn expandable and variable as stock patterns or vocabulary-like structures.

8.1.6 Universal Design Principles of Varying Music Making?

With the genre 'Variation', Fuzzy Logic acquires a corpus of objects of investigation that is already structured in many ways, for example according to types, composers, styles, epochs, and types of music such as "traditional music", "classics", jazz, and popular music. In Javanese music, for example, there are melodic modes that do not crystallize into a figure of the type theme. They form a core, which is played and developed improvisatory in variations.

> *Pathet*, also patet is an organizing concept in central Javanese gamelan music in Indonesia. It is a system of tonal hierarchies in which some notes are emphasized more than others. The word means "to damp, or to restrain from" in Javanese. *Pathet* is "a limitation on the player's choice of variation, so that while in one *pathet* a certain note may be prominent, in another it must be avoided, or used only for special effect. Awareness of such limitations, and exploration of variation within them reflects a basic philosophical aim of gamelan music, and indeed all art in central Java, namely, the restraint and refinement of one's own behaviour." (J. Lindsay 1992). [...] In essence, a pathet indicates which notes are stressed in the melody, especially at the end of phrases (seleh), as well as determines which elaborations (cengkok and sekaran) are appropriate. In many cases, however, pieces are seen as in a mixture of pathets, and the reality is often more complicated than the generalizations indicated here, and depend on the particular composition and style. [...] In an evening of wayang, the pathet of the accompaniment in progresses in throughout the performance. In slendro, the sequence is from nem to sanga to manyura; in pelog, from lima to nem to barang. Very early wayang was in slendro only, but it is now common to combine slendro and pelog in wayang, dance, and klenengan (pure music) performances. The sections of the wayang are also associated with stages of life and a range of emotions; this influences the feeling, playing, and interpretation of pieces in each pathet. [...] In each pathet there are several pathetan, melodies in free meter which are often said to express the essence of the pathet. In wayang, these are based on the melodic structure of solo songs performed by the dhalang, such as suluk or ada-ada. When gamelan is played independently of wayang, pathetan are played instrumentally by the soft elaborating instruments—rebab, gender, gambang, and suling—at the beginning or conclusion of pieces, following the contour of the associated vocal melody.[35]

Thus, it is possible to specifically specify and quantify questions of fuzziness, sharpening and blurring, of relationships between constants and variables on the one hand and the question of possible universals of variation—including attempts to answer them—on the other. The latter in particular would bring musical variation together again with the principle of variation mentioned at the beginning, which applies to different areas of reality.

Here, comparisons of the given and the changed are not only possible in notated music, on the one hand within the form of notation, on the other hand in the relationship of the sound forms to the notation form of a work. Technical reproduction also makes it possible to compare not only different sound forms of a work in terms of

---

[35] https://en.wikipedia.org/wiki/Pathet, 28.7.2020, retrieved 28 August 2020.

its score form, but also different phonographic forms of improvised music or music that existed from the outset only in phonographic form (see 3.6 above). This would concern e.g. (1) different interpretations of a musician, (2) different versions of the same title by different interpreters (cf. Section 8.2.2.8 *Covering*).

Finally, the question of a kind of 'relative universality' of music making could be posed—with regard to improvising both types of music, i.e. jazz and pop/rock, instrumental and vocal on the one hand, with technical means on the other: Are there more general patterns and models of modification and alteration of the given? And are there even—as in the case of pre-technical music-making in the historical process, which would have to be included as an additional field of comparison—similar patterns that remain relatively constant, invariant across different stylistic transformations? This is the question of the relationship between core and contour, or even the mere, particularly variant shell, now structurally and historically placed on a higher level. If the material of the respective investigation is prepared accordingly, Fuzzy Logic can again contribute to the precision, differentiation and quantification of the answers.

8.2 Variation Through Reconfiguration: Quotation Form

The principle of variation goes beyond the genesis and structure of figures within the work. Therefore I use a somewhat extended, generalized concept of variation to embed different forms of existence in the course of the work in the musical process.

8.2.1 Metamorphoses

The dialectics of variation and constancy affect the existence forms themselves, and their whole process constitutes a character both static and ever-changing, within a principally never-ending process of alteration and permutation: Identity through difference and as difference. The same, but as an ending process, with variation *inside* the work. Thus four forms of existence are bundled together in this chapter (8.2–8.5).

8.2.1.1 Forms of Existence as Variations

Like the variation itself, they play a particularly intensive role in both sharpening and blurring. The quotation form is a resurrection of a work, reflected in another work, in whole or in part. In verbal and picture form (8.3 and 8.4), the work is reflected, again as a whole or in parts, transformed in another medium or in another art. For this I use the term 'media transformation'. Finally, the form of imagination (8.5) is the fuzziest or vague form of existence. It could therefore admittedly also be classified in Chapter 3: Fuzziness. However, it finds a more appropriate place here, since it develops on the basis of the verbal and/or picture form.

8.2.1.2 Resurrection in Altered Form: Quotation and Readymade

Even the theme of variation cycles can be captured as a future quotation or ex post simply as a quotation. It is in this certainty the given, the old, the constant and the fixed, the work. By being cited, it becomes the variable, the new and liquid. Through the aforementioned dialectic between these poles, it is also varied at the same time. From the point of view that it is given, the quotation resembles the *readymade* (see Sect. 3.3). In contrast to this however it is no technically reproduced sound or an industrially manufactured product nor is it anonymous. It has name and address, as a work or as part of a work. That the producers and listeners know the origin is part of the matter. Certainly there are also quotations that are recognizable as different and foreign and yet remain unknown in terms of their origin. This is the case, for example with the cantus firmus, which was probably the most frequently cited for masses in the Renaissance, the "fixed" pre-existent "chant" used as the basis of the movement. It is the melody of the anonymous chanson *L'homme armé*. But it was precisely through the quotation, however, that the nameless one made a name for itself and became increasingly well-known and easier to identify, also because of its special qualities of form. Characteristically, cantus firmus masses, which after the Council of Trent (1545–1563) were forbidden in the Counter-Reformation because of the use of secular melodies, but which were still composed, bear the disguising designation *Missa sine nomine*, "Mass without a name". With the—complete—quotation the work remains on the one hand identical, seen in isolation. On the other hand, however, even here it becomes different through its embedding in another work and the suspension of its independent existence. J. Peter Burkholder outlines a systematic catalogue of fourteen types of quotation and of musical borrowing.

> (1) modeling a work on an existing piece, (2) variations on an existing piece, (3) paraphrasing an existing piece, (4) arranging an existing piece for new ensemble, (5) setting an existing piece with a new accompaniment, (6) using an existing piece as a *cantus firmus*, (7) placing existing tunes in succession as a medley, (8) placing existing tunes in a contrapuntal situation or quodlibet, (9) allusion to a style of music as opposed to an individual work, (10) using fragments of an existing piece as developmental material before giving the full quotation (also [...] cumulative setting), (11) using quotations in a programmatic fashion, (12) collage, or using fragments organized around a central theme, (13), patchwork, or the joining of multiple fragments through paraphrase and elision, and (14) extended paraphrase, in which a theme is used in or close to its entirety as the main thematic material for a piece or large section of a piece.[36]

[36]Burkholder 1994, p. 854; quoted in Flinn 2011, p. 16—For further categories cf. e.g. Braun 1974. „The techniques of quodlibet, quotation, and parody are discussed for a selection of fifteen works written between 1908 and 1970. The reasons for borrowing are considered, along with the categories of renewal, homage, humor, and satire" (sine p.).

With a slightly different configuration, we encounter these types in our survey. In the quotation, the cited is reproduced 'literally', 'to the letter', in contrast to the allusion or paraphrase, which in principle allows more or less large deviations from identity.

> Quotation is a near-exact recreation (within 90%) of pitch (both melodic and harmonic) and rhythmic material from a pre-existing piece of music, and paraphrase as a less-exact recreation.[37]

The quotation form can be based on the notation form or the sound form as the cited. In the field of music it is therefore not a question of writtenness and intertextuality in the narrower sense.[38]

As with breaking the balance in general, deviation from the balance promotes development and progress. But paradoxically, recognition—especially of something repeated and not varied—can also surprise, or on the other hand, the recognition of something very strongly changed. Conspicuous figures play a role here, reflecting the problem without having to resort to wordly known things—for example quotations. To cite means to make the strange your own. The quotation as a fragment of a foreign work bursts the flow of the citing work. It enriches its meaning and sharpens the perception. Clarity is created particularly effectively by figures that stand out from the stream of music through conciseness. In a way, the quotation acts like a tower of strength. Although the quotation is usually a fragment, it must itself be sharply contoured. Zofia Lissa emphasizes that

> a recognizable, concise form must have a characteristic sound structure and coherence and its function of pars pro toto, as well as its action in the new context and also on the antefactum from which it originates.[39]

The quotation sets in motion a—asymmetrical—movement of perception and thought in both directions: from the cited to the citing, but also vice versa, although the retroactive effect is weaker. At the same time

> the listener must be able […] to recognize the quotation as such and interpret the reference to the source of the quotation in the new musical work.[40]

This creates a complex three-way relationship between the source work, target work and recipient.

All of the following variants of the differences between source and target work, cited and quotation are potential objects for fuzzy logic analyses. The quotation transforms form and content, both in itself and in the context of the target work, sometimes in a sequence of several works. A striking example is the sequence from the *Contredanse* finale in Beethoven's ballet *The Creatures of Prometheus*, via the "Eroica" Variations for piano op. 35 to the finale of the *Eroica* Symphony—and then the derisive parody in the collage based on the bass theme at Necrotzar's entry in

---

[37] Flinn 2011, p. 5.

[38] From the unmanageable abundance of literature on the citation only a few references: Brendel et al. (2007, Brendel et al. (2011), Gruber (1998), Braem (1992), and Klotz (1992).

[39] Lissa 1970, quoted in Ullrich 2015, p. 47.

[40] Lissa 1970, quoted ibid.

Ligeti's "anti-anti-opera" *Le Grand Macabre*. The latter is also a real second-hand version of Adrian Leverkühn's fictional attempt in *Faustus* to "take back" the hopes of progress in Beethoven's 9th Symphony. Already almost in the spirit of Fuzzy Logic, without knowing it

> Karbusicky created a schema with three axes, consisting of A) the scope of the quotation, B) the degree of semantization as a scale from the direct reproduction of a musical model to forms of semantic metamorphosis (e.g. parody, contrafacture) as well as C) the clarity of the quotation. The scale ranges from the obvious dissimilation to different types of integration into the new context. Possible musical manifestations here are style quotations or even hidden quotations, for example to avoid political censorship.[41]

Numerous partly concealed, partly open quotations can be found in Karl Amadeus Hartmann, Nazi opponent and musical representative of the inner exile in the NS. His first composed reaction to the beginning of the Second World War is the *Concerto funebre* for violin and string orchestra (1939), originally called *Musik der Trauer*. As a "counter-action" to the "fear of what is to come", the music of mourning is nevertheless filled with hope. A later self-commentary on this:

> My Concerto funebre was composed in the autumn of 1939, a time which hints at the basic character and occasion of my piece. [...] The hopelessness for the spiritual at that time was to be contrasted with an expression of confidence in the two chorales at the beginning and end.[42]

In the *Concerto funebre*, Hartmann musically shapes the unity of politically and racially defined victims of Nazism. As *introduction* (*Largo*) or first movement he quotes the *Hussite chorale*, in the II *Adagio* from Hindemith's *Funeral Music for Viola and Orchestra* from 1936 on the occasion of the death of the British ruler George V and in the III *Allegro di molto* he alludes to the cause of the war, the invasion of Poland, with extensive polacca rhythms. Finally, in the IV movement, with the camouflage title *Choral* (*Slow March*), he paraphrases the revolutionary funeral march *Unsterbliche Opfer* (immortal Victims) of the Russian Revolution of 1905 in an interplay of choral tutti and solo replicas. As its ending, he quotes elements of the melody of the Jewish *Eliahu* song, which he used several times, here very abbreviated and reduced to the outline with the characteristic upward quarto leap. Fuzziness as camouflage: The fact that Hartmann incorporates such melismatics into his version of the proletarian funeral march may be noted as a subtlety: as mentioned above, it musically unites the main victims of Nazism determined as "race" and those determined as class. However, later fears of political clarity may also play into it: in the first version, visible through the overpasting, the melody is quoted in an unfigured way (Figs. 8.5 and 8.6).

The camouflage of the workers' song quotation as "chorale" continued to be effective even after Nazism during the Cold War. Rudolf Stephan, who did not know the song, but perhaps had some inkling:

[41]Karbusicky 1992, quoted in Ullrich 2015, p. 48.

[42]Karl Amadeus Hartmann 1965, p. 53.

Fig. 8.5 Hartmann: *Concerto funebre,* IV. movement, Violin solo and violin I orchestra, m. 1-4. Studien-Partitur Edition Schott 5002. B. Schott's Söhne, Mainz. (n.d., ca. 1959)

Fig. 8.6 *Unsterbliche Opfer,* first line
Quoted after Lammel, Inge: *Das Arbeiterlied.* Reclam Verlag, Leipzig 1975. p. 22

> The finale [...], described as a *chorale*, is not an arrangement of a church song, but, like the introduction, the reproduction of an old tune that was only imagined, probably a conduct.[43]

In reality, despite distortion or alienation through melodic changes, metric stretching, insertion of solo lines and dissonant harmonization, this "imagined old tune" is nothing more than the melody of the proletarian revolutionary funeral march *Unsterbliche Opfer, ihr sanket dahin.* The whole movement is his symphonic paraphrase: translation from the plebeian language of the street into that of the hall, but also "concert" as "combat music". The song that was spread during the Russian Revolution of 1905, itself a troping of a military march was also used for funeral ceremonies of the October Revolution. Hartmann had got to know it through Hermann Scherchen, who was an important initiator and friend and who also provided a translation. In 1939, Hartmann, exposed to the Nazis' grip, could hardly dare to quote this song of the workers' movement in an undisguised way. By lifting it up into the symphonic texture, a dialectic of 'Aesopian speech', of the "hidden writing" (Dolf Sternberger)[44] is created: the chorale's cloak of invisibility, which is put over the proletarian song—also musically, especially in the middle section—like the cover with a literary or scientific title for political writings, makes the camouflaged less easily visible.

[43]R. Stephan, *Karl Amadeus Hartmann: Concerto funebre für Solo-Violine und Streichorchester*, in: *Programmheft Musica-Viva-Konzert München* 11. December 1964.
[44]Cf. Ehrke-Rotermund/Rotermund 1999.

8.2.2 *Large Quotation—The Work as a Completely Resurrected Whole*

Usually not a work is cited, but only a part of a work. That a work is cited as a whole is a special case, but it is not so rare. Some types of Bartók's arrangements of peasant songs, for example, preserve the original melody intact and—in principle—unchanged. Otherwise, quotations or 'contrafactures' or 'parodies' of a whole work or at least of whole movements are rarer, but in the 'Parody Mass' and in other genres they were a characteristic procedure in European music between the Renaissance and the late Baroque. In most cases one leaves it at a characteristic excerpt. But already through this pars pro toto the old work receives a "second existence", almost a kind of "second life": Identity in the difference, preservation of the given in and through the change.

8.2.2.1 Imitation: Quotation as Working with Templates

The generalizing term for citing a complete work is *Large quotation,*

> in copyright law, the quotation of a work protected by copyright (e.g. texts, pictures, music, videos) in its entirety[45]

It could also be called '*complete quotation*' or '*full quotation*', in the sense of 'completeness'—the purely quantitative size of the quotation is secondary. Large quotations' are usually of smaller works, e.g. poems in collections or songs in larger works. The new version of the German Copyright Act from 1.1.2008 specifies this to a certain degree and legitimises it further:

> individual works are included after publication in an independent scientific work to explain the content. It is permitted to copy, distribute and publicly reproduce a published work for the purpose of quotation, provided that the extent of use is justified by the specific purpose. This is particularly permissible if 1. individual works are included after publication in an independent scientific work to explain the content[46]

If, as in the grand quotation, the original work becomes a complete quotation, a paradoxical borderline case of the quotation is reached. For, on the one hand, the ideal case of a new form of existence is reached in this way, which is identical with the one it reflects. But on the other hand, quoting actually means that an excerpt from Work 1 is incorporated into Work 2 and Work 1 survives in this new form of existence, even if reduced. The installation in a new whole, however, also means enrichment, for both sides. And yet the identity is imperfect: In the context of the new work, the quoted is changed, more or less intervening.

---

[45]https://de.wikipedia.org/wiki/Gro%C3%9Fzitat, 19.9.2019, retrieved 20 July 2020. Very detailed and informative on the whole complex of quoting, editing, covering, remixing from a copyright perspective: Schulz/Cichon 2004.

[46]https://dejure.org/aenderungen/synopse-UrhG-2008.htmla51, retrieved 18 July 2020.

The form of quotation, in particular and individually, once again intertwines repetition and change, identity and difference, constancy and variance, as outlined at the beginning. There are two historically and aesthetically far apart fields in which the whole quotation is condensed: counterfacing and parody on the one hand, covering and remake on the other. Both are basically non-deceptive imitation and remake, in contrast to *soundalike* and plagiarism (see below).

8.2.2.2 Tailoring: Quotation as Form of Editing

In a somewhat broader perspective, the grand quote is a form of arrangement. Like the quote itself, however, it can also refer to parts of a work only, such as a melody. Here, however, we are dealing with entire works; this can be a song or a movement of a cyclical, multi-movement work. The grand quote, with a largely unchanged repetition of the original, forms one pole of the arrangement, with a 'structured continuum' of transitions to the other pole: there the arrangers variably rearrange the work in such a way that it turns from a work in quoted form into a new, independent work, admittedly still with dependency relations to the original work. Arrangement, quotation, variation, and even composition merge again and again, at both historical and structural junctions. The prerequisite and basis for this is to work with something given, with a clearly defined form. This remains the invariant core of the changes, namely as a quotation.

> All contrapuntal work is more or less arrangement. Polyphony, a prerequisite for counterpoint, is itself the result of arrangement, in that an *existing theme or melody* is embellished by a second, later by several voices, which are first accompanied in parallel in the octave, then in the fifth or fourth, in the second or third. [...] The most important polyphonic genre of the Middle Ages, the motet, developed when the performing singers or players added new voices, even with their own texts, old themes or parts of themes.[47]

The interlocking and interrelatedness of quoting, varying and arranging leads to transitional and intermediate forms and fundamental fuzziness both in the subject matter and in the terms. The term 'arrangement' functions

> as an umbrella term for a whole series of terminologically fuzzy processes, including, among others, the adaptation of a composition for smaller or larger instrumentations (arrangement / reduction or / instrumentation / orchestration), the compositional reorganisation, which aims to create new contexts *using original material* (paraphrase, recomposition, remix, palimpsest), or retouching as a corrective intervention in the original form. (Schröder/Bösche 1994)[48]

I will try to reflect and specify this as much as possible.

> Arrangements can be understood as independent musical creations, which are produced using *pre-existent music* by *changing the original* instrumental / vocal means and / or compositional

[47]Engel 1949/1951, col, 1458. Emphasis HWH.

[48]Ibid.

> structures. [...] they [...] can therefore also be understood as media transfers, i.e. as the transfer of musical information from the original into another medium.[49]

The field of the quotation does not include arrangements for performance purposes (see Sect. 6.7),

> the modification of own works or the completion or reconstruction of compositions handed down in fragments.[50]

8.2.2.3 Re-Dressing: Contrafacture (Contrafactum), Parody

One type of quotation and, at the same time, arrangement is the *rewriting* of existing works, from songs to multi-movement masses. The cited music is preserved more or less intact its text is destroyed, but is still preserved by adapting the prosody to the cited music, like a palimpsest or shadow. Despite all internal differentiation, the facts are uniform and simple, the terminology is inconsistent and complicated; moreover, historically and systematically variable terms play a role in it. The historically unburdened term ‘rewriting’ brings the matter to the point. Out of historically conveyed politeness I use the term ‘contrafactum’. Contrafacture refers to a whole work, which can also be an independent melody. It is generally a special case of text troping, namely the rewriting of existing music, the substitution of one text for another without substantial change to the music.[51] The complementary contrast to text or textualization is ‘setting to music’, the new composition with existing texts setting to music.[52] The non-complementary, asymmetrical contrast is the ‘de-texting’, the removal of the text from vocal works and its transformation into merely instrumental works (see below). From the standpoint of literature, contrafactum is a new musical form of a text, from the standpoint of music it is a form of quotation from a work. Both metamorphoses can occur several times with the same thing, as is often the case with chorales or hymns—corresponding to the multiple setting of a poem to music. Like textualization in general, its special case of re-textualization is a universalism of music.

> Contrafacture as a basic form of the survival of text-bound music is probably as old as vocal music itself. It must have seemed obvious to provide new texts to sages who had attained distribution and popularity by virtue of their conciseness, in order to exploit their conciseness and popularity for concrete purposes (exploitation of the quality of familiarity) or to practise more complicated forms of composition up to the point of independent re-creation.[53]

The contrafactum is an intentional variation, fuzzing and sharpening at the same time.

[49]Drees 2016, p. 197. Emphasis HWH.

[50]Ibid.

[51]https://en.wikipedia.org/wiki/Contrafactum, 29.10.2019, retrieved 20 July 2020.

[52]„A musical setting is a musical composition that is written on the basis of a literary work. The literary work is said to be set, or adapted, to music“ (https://en.wikipedia.org/wiki/Musical_setting, 7.6.2018, retrieved 20 July 2020).

[53]Finscher 1962, col. 816.

While translations meant for singing do not usually constitute intentional "substitution", examples of contrafacta which do constitute wholesale substitution of a different text [...]. An existing tune already possessing secular or sacred words is given a new poem, as often happens in hymns [and chorals]; sometimes more than one new set of words is created over time.[54]

A special case is the contrafacture of an own text by the poet himself.

A lyricist might re-cast his/her own song (or someone else's song) with new lyrics, as in the case of Alan Jay Lerner with the number *She Wasn't You / He Isn't You* from the stage and film versions, respectively, of the musical *On a Clear Day You Can See Forever*.[55]

The addition as a fundamental fuzzing shows in turn different kinds and degrees of sharpening with regard to the appropriateness of the new text to the old music. Thus, for example, the textual transformation of vocal originals in foreign languages ranges from simply literal translation to complete poetic recreation, including re-textualizations that destroy the meaning. Historically, the relationship between the secular and the sacred has played an important role in European music. The usual direction is from the secular to the sacred, i.e. secular model and spiritual text or re-textualization. The opposite is sacrilege and blasphemy, such as the *Dies irae* quotation in the Witches' Sabbath of Berlioz' *Symphonie phantastique.* But especially with the Counter-Reformation the criticism even intensified on the secular-sacral counterfactual. In the age of absolutism then the traces of origin had to be covered up, so that

Contrafactures of songs are only possible where the fact of a worldly model remains unrecognized. The few exceptions confirm this aesthetic attitude, which is hostile to the contrafactum: when, for example, M. L'Affillard [...] has his *Principes trèsfaciles pour bien apprendre la Musique* (1694), accompanied by lascivious songs, followed by a contrafaced parallel edition "aux dames religieuses"[56]

During the Reformation after 1520, numerous sacred songs were reformed and in turn rewritten or counterfactured. Luther formulated this as undressing and redressing:

That is why we *unclothed* such idolatrous, deadly and great texts and stripped them of their beautiful musica and put them on the living, holy word of God ....[57]

Re-textualization, contrafactura, can range from small works such as songs to large-scale ones. Thus

not only secular and sacred melodies served as the framework for the movement, but even whole movements are taken as the model for the arrangement, in the so-called *parody mass*, in which a motet, chanson or madrigal in sections, especially at the beginning of the parts is retained as the model.[58]

---

[54] https://en.wikipedia.org/wiki/Contrafactum, ibid.

[55] Ibid.

[56] In detail von Dadelsen 1962, col. 828.

[57] Quoted in: Finscher 1962, col. 826.

[58] Engel 1949/1951, col. 1459.

The arrangement can proceed from the addition of only one new text to the addition of new music components. In any case, the type of the "parody mass" devours the entire secular composition of the quoted and transformed work lock, stock and barrel, so to speak. For the latter is here taken over as a whole from the secular to the sacred sphere; only the liturgical text is different, including adaptations of the declamation: a transubstantiation of the secular into the sacred. In summary, this is about the procedures and types of quotation, as an arrangement, whether monophonic or polyphonic. What is decisive is that the citation itself has the character of a work, i.e. from the song to the motet or even the movement of a mass.

> Contrafacture and parody are thus sub-areas of the textual-musical or musical arrangement. Like any arrangement, they refer to a model and are thus separated from any free new creation according to musical types or according to formulas and forms typical of the time and genre. They can only be proven if the model and the arrangement have been handed down or can be made accessible.[59]

The obligatory questions about fuzziness, sharpening and blurring can then be based on this. *Contrafacture* was also called parody or paraphrase in the feudal age from the Renaissance to the eighteenth century. In more recent usage, the term 'parody' is reserved for the comic, critical, satirical variation of a work, which then no longer needs to be a full quotation. *Contrafacture* and parody were thus

> not the comic alteration of spiritual statements while preserving the original form of the statement, but the serious, transformative and further-reaching connection to a coined statement, i.e. the transformative imitation of a musical work of art in another context of musical work, usually also a genre context, in which the musical substance is preserved, but the musical form is partly newly created.[60]

A quotation can of course also be meant and made parodistically. In contrast to fuzzing or "painting over", caricatures or parodies exaggerate essential characteristics by means of an exaggerated filtering, so to speak. Such deformations are deliberate, intentional transformations, abstractly technically and from a cognitivistic point of view more imprecise images than 'photography' or phonography, but more precisely in terms of more complex richer and semantic and intensified meaning. Such distortion breaks out of the cycle of forms of existence as a specific variation by setting the deviation of linearly successive levels of similarity as a principle. And as a typus it forms a transition to the *quotation form*; just like parodistic to satanistic perversions such as the entanglement of *Dies-irae*-quotation and the mentioned "Witches' Sabbath" in the finale of Berlioz' *Fantastic Symphony* or the *Brother Jacob* Canon in a sluggish minor in Mahler's 1st Symphony. The back-and-forth of contrafactures as the creation or application of new texts, together with the creation of new melodies, often produces a sequence of variations that are continually creating more and more fuzziness and are not even blurrings; for example, excessive in the post-Reformation period.

---

[59]Finscher 1962, col. 825.

[60]Finscher 1962, col 815.

> Already in the thirties of the sixteenth century, the Reformation core tunes are also counterfaced, and further development is largely exhausted in confusing and difficult to grasp melodic exchanges and endless new texts of popular and proven melodies. Both procedures are foreign to the text and often enough mock Luther's demand. How much trust was placed in the power of the melody alone is shown by notes such as "Aus tiefer Not" (From deep distress) to the texts "Ich will dem Herren sagen Dank" (I will say thank you to the Lord), "Herr wer wird wohnen in Deiner Hütt" (Lord who will dwell in your hut), "Mein Seel lobe den Herren rein" (My soul praise the Lord pure) and "Singet dem Herrn ein neues Lied" (Sing a new song to the Lord); occasionally a note is repeated quite mechanically in the hymnbooks for several successive texts of the same metre and stanzaic form.[61]

The melody thus survives as an invariant core. However, in the relationship of the new texts to the melody, historical and systematic fuzziness becomes increasingly apparent. In this cycle of variation it is therefore often like with the "whisper down the lane" (see 3. fuzziness above).

As in a completely different musical sphere in *remixing*, a disjunction of text and composition or melody also appears in the historical sense in the case of contrafacture or 'parody', and thus at the same time their recombination, which also proceeds separately, and finally, in a 'structured continuum', transitions and stages of an ever increasing dissociation and fragmentation of the original work as a whole.

The contrafactum procedure shows two different basic types. In one case, the quoted original work is preserved mainly as a cantus firmus, in the other case the work's compositional structure is preserved as in a whole quote

> Frye's mass "Summe trinitati" (about an anonymous motet, perhaps by Frye himself [...] shows almost strictly isorhythmic cantus firmus treatment, but quotes the beginning of the motet in all voices as the beginning of all movements. [...] Barbingant's mass "Terriblement" (about a chanson of the same text) uses the chanson-tenor as mass tenor and the chanson-superius as mass-superius. Both pieces, especially the superius, are treated by augmentation, diminution, coloration and insertions, but *their relationship to each other remains basically unchanged*, so that the arrangement of the work preserves the traditional superius-tenor framework duet (with free countertenor).

Two basic possibilities of parody are thus tested. A *system* of the early parody technique the steps of which roughly correspond to the steps of the historical development can be developed from them:

a. Two or more pieces of a model are taken over unchanged and combined with a newly composed voice or several voices (as a quasi-parody without text and genre changes in the chanson composition common from about 1470 to about 1520).
b. The adopted original voices are paraphrased but kept intact in their vertical relationships as the framework of the composition (Barbingant).
c. Instead of complete works or framework movements, individual motifs are extracted from the original, divided up and recombined both vertically and horizontally, but the quotes themselves remain untouched. Most parody masses of the Josquin period use this technique in various forms; it culminates in complicated combination games [...].

[61] Ibid., col. 821.

d. Instead of the fragments of individual works, a coherent section of all the works of the original is quoted as a polyphonic sentence (usually as the beginning of a sentence) and then abandoned or, as in c., disassembled and processed.[62]

Not least with J. S. Bach, the contrafacture and parody of his own works reveals a growing awareness of the independence and intrinsic value of the work of art: paradoxically, almost as inversion, the original work gains a higher status precisely through the varying arrangements.

> Since many secular occasional works of the Leipzig period were spiritually parodied soon after their première, it can be assumed that they were designed for parody from the outset. However, especially Bach's later parodies, in the *Christmas Oratorio*, the Missae breves and the Mass h, cannot be explained as mere labour economics. Obviously, his aim was to further improve sublime movements from earlier times, to place them in a more valid context and thereby preserve them. This heralds a new, forward-looking artistic attitude that is aware of the *uniqueness* of ingenious inspiration.[63]

8.2.2.4 Stripping: Instrumental Tablature and Paraphrase

Another type of quotation and at the same time of arrangement is the transformation of vocal works into instrumental ones. This type of whole-quotation arrangement is a 'de-textualization' of vocal works, especially the adaptation of polyphonic vocal music for musical instruments capable of polyphonic performance, i.e. keyboard and plucked instruments. It is thus a reversal of the transformative movement of contrafacture, the textualization of instrumental music or the re-textualization of vocal music. For this process as for 're-textualization', there is not only no general, uniform term, but no term at all. The term 'de-troping' could serve as an antipole to troping or textualization. The general term *arrangement* applies here, but does not meet the specific:

> traditionally, any adaptation of a composition to fit a medium other than that for which it was originally written, while at the same time retaining the general character of the original.[64]

The subject matter is often referred to as a transcription[65] or 'paraphrase', i.e. rewriting. In Europe vocal music was a priority for over a century, as the church had + considerable reservations + about the secularity of instrumental music. Therefore, as transcription, arrangement, paraphrase

> a large part of [European artificial] instrumental music in general is developed. The music for keyboard instruments and lute in the 14th, 15th and 16th centuries is half of it vocal music, [...] i.e. instrumental arrangement and with effective instrumental ornamentation.[66]

---

[62]Finscher 1962, col. 822ff.

[63]von Dadelsen 1962, column 832ff.

[64]See https://www.britannica.com/art/arrangement, retrieved 20 July 2020. In Jazz and Pop, the term is used totally differently, namely synonymously with "score".

[65]Transcription of course also means the translation of non-notated music into musical notation. See above 6. In this respect paraphrase is probably preferable.

[66]Engel 1949/1951, col. 1462.

These decorations are already a fuzzing of the original composition (see Sect. 7.4). During the transformation from vocals to instrumental, a tablature is used instead of the usual notation—they were 'intabulated'.

> (V)ocal polyphony of the late Middle Ages and the Renaissance, including motets, chansons, and parts of the mass, was *intabulated* (transcribed so as to suggest finger positions rather than pitches) for the use of keyboard and lute players, permitting them to perform singly music written for several singers.[67]

The pure sound-body stripped of its text seems to be more receptive to further and more far-reaching changes in the quoted music through this fundamentally reducing variation than in the case of contrafacture.

> Parallel to the decorative transmission is the fantasy adaptation of songs, which also begins in the 16th century and reaches its climax in the opera fantasies of the 18th and 19th centuries.[68]

Probably best described with a term of the nineteenth century as 'paraphrase', this reworking of a given tends to such radical changes of the original work that it is indeed an 'arrangement', which often transforms into a re-composition, of own as well as other works, but no longer is a quotation. This genre exceeds the limits of the quotation. The "concert paraphrase" or "concert fantasy" as another type of arrangement, often composed for solo piano by Liszt, Schumann, Brahms, Busoni, etc., like the grand quotation, allows an existing work to be revived. The same applies to new orchestrations by Brahms or Schönberg, for example. However, all this is no longer a grand quotation, but an independent piece in which variation, difference, prevails over the aspect of identity. The traces are not fuzzed; on the contrary, as the paraphrase draws on the popularity and prominence of the original. But the original itself is fuzzed. Its fuzziness through the disappearance of the text does not in itself mean a sharpening of the newly composed music. Even scholarly 'transcriptions' as transfers from older notations are suspected of being fuzzed rather than sharpened:

> Performance editions of problematically notated early scores often carry all the earmarks of highly subjective arrangements.[69]

8.2.2.5 Patchwork: Quodlibet, Potpourri, Medley

Transitional forms from quoting the whole to its fragmentation during quotation constitute the genres quodlibet, potpourri and medley. In these, only several different—smaller—works, mostly songs and usually only the melodies, are strung together to form a new composition. In the case of the quodlibet they are also combined simultaneously in polyphony vocally or instrumentally and as vocal music even in multi-text works (For examples see Sect. 6.10.2). In the quodlibet,

---

[67] https://www.britannica.com/art/arrangement, retrieved 20 July 2020.

[68] Engel 1949/1951, col. 1462.

[69] https://www.britannica.com/art/arrangement, retrieved 20 July 2020.

(Latin: "what you will") [...] several well-known melodies are combined, either simultaneously or, less frequently, sequentially, for humorous effect. *Quodlibet* can also refer to an amalgamation of different song texts in a vocal composition. While simultaneous combinations of two or more melodies go back to the 19th century (motets using, for example, a chant melody and a secular tune), quodlibets were especially popular in the 15th and 16th centuries. [...] An English example is the *Cries of London* by Orlando Gibbons. Perhaps the best-known quodlibet is the finale of J.S. Bach's *Goldberg Variations* for harpsichord (published 1741). Terms related to quodlibet technique include *fricassée* (French: "hash"), *ensalada* (Spanish: "salad"), *centone* (Italian: "patchwork"), and, in later centuries, *medley* and *potpourri*.[70]

8.2.2.6 Disguise: Soundalike—Imitation, Surrogate

A special case of quotation is the use of well-known works, mostly in excerpts, for advertising purposes in jingles. Here too, the familiarity and recognition effect of the quotation is used, but for non-artistic purposes.

> Jingles are a form of sound branding. A jingle contains one or more hooks and meaning that explicitly promote the product or service being advertised, usually through the use of one or more advertising slogans. Ad buyers use jingles in radio and television commercials; they can also be used in non-advertising contexts to establish or maintain a brand image. Many jingles are also created using snippets of popular songs, in which lyrics are modified to appropriately advertise the product or service.[71]

For this purpose, the *soundalike* is particularly popular with buyers of advertising. Different from the previously introduced large quotation types, the *soundalike* is different. It is

> a recomposed melody or a song that sounds *similar* to the original. A soundalike is mostly used by the advertising industry for cost reasons to avoid a much more expensive purchase of rights for the original title.[72]

The *soundalike* sounds like a quotation, but it is not a real quotation, but a surrogate. It is only similar, but without admitting it: different with the appearance of identity. The adaptation to the advertising purpose is done by means of contrafactures or re-texturalization. It does not necessarily have to be a grand quotation parts of a work do too. Fuzziness is intended here for a perception as *misunderstanding*: The surrogate is intended to come as close as possible to the original.

---

[70] https://www.britannica.com/art/quodlibet, retrieved 20 July 2020.

[71] https://en.wikipedia.org/wiki/Jingle, 22.5.2020, retrieved 22 July 2020.

[72] http://filmlexikon.uni-kiel.de/index.php?action=lexikon&tag=det&http://filmlexikon.uni-kiel.de/index.php?action=lexikon&tag=det&id=6035id=6035, 22.8.2011, retrieved 1 December 2011. Emphasis HWH. This fits in with some tendencies to use Fuzzy Logic to make music production cheaper through automation.—In contrast to this, a charming parallel of soundalike that is not meant seriously is „Anguish Languish, an ersatz language constructed from English language words, [...] created by Howard L. Chace [...] 1956). It is not really a language but rather a homophonic transformation created as a work of humor. [...] His story "Ladle Rat Rotten Hut" is "Little Red Riding Hood" re-written with *similar-sounding* words substituting for the original folk tale. A professor of French, Chace wrote "Ladle Rat Rotten Hut" in 1940 [...]" https://en.wikipedia.org/wiki/Anguish_Languish, 10.11.2013, retrieved 11 November 2013.

> We constantly hear music in advertisements that seems familiar to us. Popular songs by artists are imitated, these usually come away empty-handed. A lousy trick, but completely legal.[73]

Within the framework of demand and supply, there are specializations right away on the surrogate of the quotation

> lots of soundalikes on request or even ready-made in the sound archive. [... So] some professional musicians even specialize exclusively in soundalikes – including their own legal advice.[74]

The soundalike is therefore a deceptive plagiarism de facto, but not one de jure. Fuzzy Logic could be used here too to measure the blurriness. But it is hardly worth it.

8.2.2.7 Cut Adjustment: 'Adaptation'

Adaptation as an "adaptive" arrangement is a very general matter; it has some overlaps with *arrangement* as a term. In particular, whole pieces of music are almost always arranged and given a quotation form. In the *soundalike* process, the tonal difference is made to disappear as far as possible under copyright and legal regulations. But as with *adaptation*, it is intended to remain in the consciousness of the recipient and is an essential component in the quoted work. Both are about feeding on the popularity and prestige of the quoted work, in the case of the soundalike directly and un-artistically for sales purposes—advertising for the sale of any other goods—, in the case of the adaptation above all of "classics" indirectly, first artistically and then also commercially for the sale of one's own artistic product.

Adaptation is a historical phenomenon as an adaptation,

> in which the tension between historical distance and current appropriation becomes an aesthetic attraction of its own.[75]

Following Duke Ellington's "Suites" based on works of the "classics" such as *The Nutcracker Suite*, 1960, and phenomena such as Jacques Loussier's *Play Bach*, 1959 from 1965 on

> so-called classical adaptations temporarily also became a separate genre of popular music, where they were quite popular especially in rock music within the framework of classic rock until the 1970s[76]

[73]Dominic Holzer: *Soundalike. Songs in Werbung*, https://www.br.de/puls/musik/aktuell/soundalikes-in-der-werbung-100.html, 29.4.2015, retrieved 29 July 2020.

[74]Ibid. Cf. as reference to the extensive discussion of the subject matter e.g. P&P Marketing-Assistenz: *Was ist eigentlich ein Soundalike?* 28.2.2019, retrieved 29 July 2020. The embedding in the even wider field of *advertising* is only referred to generally.

[75]*Adaption,* Wicke/Ziegenrücker 2007, pp. 14–17; quotation p. 14.

[76]Ibid., p. 15.

Fig. 8.7 Jingle
Quoted after Robert Weedon: Audio Identities: Radio station themes, 2009
http://www.imagedissectors.com/article/46, retrieved 28 July 2020

Adaptations act not only as advertising for a product but also as advertising for a special medium. After 1945, jingles were often used for channel identification

> The media's obsession with brand recognition and image have notably increased in the last few years. The BBC in particular uses musical idents on television to convey a distinctive theme which somehow identifies the channel [...], However, it is noteworthy that while musical themes for the primarily visual medium of television are all over the place, radio theme tunes and jingles have gradually been going out of fashion.[77]

It is not uncommon for the jingle to be an adaptation of "Classics" works in order to profit from their still lasting prestige, in the following case however at the same time as a remix.

> Classic FM, the only commercial contender on our list was the last nationally-available radio station to have had its own theme [...]. Classic FM launched in 1992 with a mock-Baroque orchestral piece by David Arnold, which sounded like a cross between a Bach *Brandenburg Concerto* and Handel's *Water Music.* What was particularly exciting about this was that the station used its theme tune as the basis of its jingles, which I almost hesitate to call jingles – more 'variations on a theme', usually based on this motif deriving from the final bars.[78]

So the channel had a whole series of variations developed (Fig. 8.7).

> Other parts of the piece were utilised for other jingles, but ultimately all evidently derived from the same source, creating an integrated feel to the station.[79]

8.2.2.8 Cut-Out: Self-Quotation, Remake, Covering

The self-quotation is a type of quotation the remake in turn is a type of self-quotation as a grand quotation. The self-quotation is something different from the aforementioned self-arrangement of a work in such a way that composers take up and rework their *own* works (see Sect. 6.7.3). "Versions". Composition/Production after completion of the production—apparent and real sharpening between improvement and mutilation of the intention. This arrangement objectifies the processual nature of the individual work and thus becomes object-like. The self-quotation strengthens the relationships of the individual works to others and the overall context of an oeuvre.

---

[77] See e.g. Wicke/Ziegenrücker 2007, p. 15. In details Weedon 2009.

[78] Weedon 2009, ibid.

[79] Ibid.

›The [fictional] composer Alwa [in Berg's opera *Lulu*] finds it 'impossible' to write an opera about the irresistible no man's child, the protagonist of the play. When the real composer Alban Berg implants the glittering opening chords of his *Wozzeck* at this point, he does not just use this self-quotation as an ironic and ambiguous commentary. He thus also aligns himself with a number of great opera composers from Mozart to Richard Strauss, who offered initiates similar cross-references between their works. Of course, Berg does not leave it at that, merely to reuse the element from the older score in the newer one and to incorporate it without variation and without comment. After the dynamically somewhat more marked, but otherwise literal adoption of the motif, he subjects it to a transformation by stretching the note values, repeating the chordal constellation – without the glissando – and finally dissolving the gesture into a *downward tumbling movement.*[80]

From the point of view of fuzzy logic, questions of blurring or similarities between "original" and further processing are also relevant here. Under the conditions of technical re/production, the remake generally is a 'remaking' of a successful work, especially in film, and here especially self-quotation in the context of sound and phonographic form. In contrast to *covering*, the same interpreters are responsible for making and re-making. Remake

in the music industry is the [...] new recording of a previously published musical work by the same performer. The reasons for this may be due to improvements in recording technology (mono recordings are replaced by new recordings in stereo, analogue recordings are replaced by digital recordings), but also to an arrangement adapted to the changed musical taste or a change of the record company.[81]

There are various forms and motivations for this "reinterpretation, which is committed to the original as far as possible".[82]

The *re-issue* (or *re-release*) is the temporally later publication of the same music title by the same interpreter – and this in unchanged form. [...] When *remixing*, previously recorded material is also used. For example, when stereo technology was introduced, many titles recorded in stereo were still mixed in mono because stereo recordings sounded different from mono recordings when they were played back in mono. If the conserved master tapes are later remixed in stereo and released, it is not a remake.[83]

Going beyond the simple remake, is a re-creation as an aesthetically slightly eerie resurrection of the original. For the Charlie Parker "Biopic" *Bird*, 1988, the music producer took

the original Parker recordings [...], extracted Parker's saxophone voice from the sound environment in a complicated process of filtering and sound processing, and had renowned jazz musicians re-record the backing tracks.[84]

The self-quotation suddenly turns into a quotation, the remake musically into a cover version.

---

[80]Ender 2011, p. 35.

[81]https://de.wikipedia.org/wiki/Remake_(Musik), 17.8.2017, retrieved 4 July 2020.

[82]*Remake,* Wicke/Ziegenrücker 2007, p. 589.

[83]Ibid.

[84]Ibid.

A special case and a variant of the grand quotation especially in the more recent popular music is the already mentioned *covering*. As subject matter and term it is used mainly in the field of popular music.[85] The cover version quotes the work as a whole, but it is (re-)produced by other authors, unlike the *remake*. Thus, in addition to mere replication and reproduction, a new work of one's own is added, thus creating differences in identity. Since the interpreters are different and the sound is therefore also different, this is not plagiarism with the intention of deception. The cover version essentially presupposes the phonographic form and starts from it for the arrangement (see Sect. 3.5 Phonographic form). The new version, which quotes the original work and at the same time varies it, however, often falls back on the notation form and still contains many elements of the 'handmade'. This will change step by step with further developments like *remix* or *mashup*. The result is an aesthetic-historical double movement: The handmade recedes in favor of the readymade, and the whole quote, which in principle and at least in outline preserves the integral form of the quoted work, fades and finally dissolves. On the basis of the sampling procedures and intensified with digitalization, they fragment, reshape and alter the original more and more until the whole quotation turns into a partial quotation and finally into a collage (see below).

> As early as the 1960s, a form of cover version was created in the form of the so-called *tribute albums*, which, as a tribute to the original, place it in a different, often more up-to-date stylistic context, but in contrast to an adaption, perform an adjustment of the musical material. The originals are therefore not merely presented in a different stylistic guise, but rather as they might have sounded if they had been created in this context from the outset. This practice aims at the timelessness of the templates. In the mid-1980s, with the development of *house* and *techno*, a [...] wave of cover versions [...] began. The originals were often played with using digital mixing techniques to demonstrate the variability and inexhaustibility of the musical material of the originals.[86]

In rap, the quotation often serves as a crystallization core or launch pad. The rap or hip hop version, for example, takes *the entire original work* as a quasi-instrumental accompaniment to the new 'melodramatic' text-melody structure oscillating between speaking and singing. The original is here, turned differently incorporated as a spolia, as a prey-like quotation—once again the 'fly in amber'. Although in popular music mainly the phonographic form became the primary form of existence, there is still a pre-technical core, in vocal music a kind of song, generally conceptions and notes, verbal or written, in the manner of a sketch form (see Sects. **3.5** and **8.1**). A reference point is the 'song', the unity of melody and harmony, as a kind of generative depth structure, like the jazz standard. Like the latter, they are often still published in notation form, such as the Beatle Song Book, etc.—beginners can thus first rehearse the work instrumentally and vocally. All of this plays a subordinate, vanishing role (see Sect. **3.5** *phonographic form*). But it can become a reference point for comparative analysis in the light of fuzzy logic with regard to similarity etc. alongside and in addition to the covered phonographic form of a work. When comparing the various

[85] Cf. e.g. Pendzich 2004. .

[86] *Cover-Version*, Wicke/Ziegenrücker 2007, p. 174f.

covered phonographic forms of the same work, the earliest historical sound work is, in principle, the reference object.

A special kind of covering is based mainly on the phonographic form as a 'replay' of what has already been played: de facto, arrangements that have been around in jazz for a long time, for example by Duke Ellington, or important solos are played as literally and "true to the original" as possible, "live" in contrast to the Parker film example mentioned above. The arrangements as musical notation should serve additionally for precision. Historicism, Alexandrian late-period tendencies thus also penetrate into the technically modern. The sound form itself appears as a work. A deliberate, almost paradoxical extreme is Mark Coleman's *Playback* with the replay not only of solos but of whole pieces: a reflected identity instead of the more or less blurred similarity to the original and a special kind of resurrection and "Second Life".

8.2.2.9 Combinations: Remix (I) and Mashup

The remix begins with the arrangement or 'adaptation' of a grand quotation, continued with the decomposition and recombination of two grand quotations.

> A remix is a piece of media which has been *altered or contorted from its original state* by adding, removing, and changing pieces of the item. A song, piece of artwork, books, video, poem, or photograph can all be remixes. The only characteristic of a remix is that it appropriates and changes other materials to create something new. Most commonly, remixes are a subset of audio mixing in music and song recordings.[87]

As a general method, remixing can also be applied in other arts and also draws near the collage.

> A remix may also refer to a non-linear re-interpretation of a given work or media other than audio such as a hybridizing process combining *fragments of various works*. The process of combining and re-contextualizing will often produce unique results independent of the intentions and vision of the original designer/artist. Thus the concept of a remix can be applied to visual or video arts, and even things farther afield.[88]

As far as quotation and integrity of the quoted is concerned, it starts innocuously.
Remix

> at the end of the 70's with the maxi single sound carriers and in the course of the 80's with the CD single sound carriers the need arose to fill the additional playing time with substance. This led to the trend to publish additional versions of the existing sound tracks on the single sound carriers in the form of remixes in addition to the A-side production. Typically these were versions with special rhythmic accents ("dancefloor version", "disco version"), versions suitable for radio with emphasis on the melodic voices and reduction of the bass line, drum line or e-guitar track ("radio edit", "radio mix") [i.e. at the same time 'adaptations'] or versions with extended playing time by repetition or variations of

[87] https://en.wikipedia.org/wiki/Remix, 21.10.2020, retrieved 21 October 2020. Cf. e.g. Wicke/Ziegenrücker 2007, *Remix*, p. 590f.

[88] https://en.wikipedia.org/wiki/Remix, ibid.

> the existing motives ("extended version"). Originally, remix was therefore defined as the making of changes to a pre-existing music production in which the melody or at least parts of it remain recognisable as a characteristic element of the original version in the remix. This is still the more frequent case of a remix, since the producer usually starts from a vocal line, dismantles the original version (original) from this starting point and constructs it anew.[89] Early pop remixes were fairly simple; in the 1980s, "extended mixes" of songs were released to clubs and commercial outlets on vinyl 12-inch singles. These typically had a duration of six to seven minutes, and often consisted of the original song with 8 or 16 bars of instruments inserted, often after the second chorus; some were as simplistic as two copies of the song stitched end to end.[90]

Here the original work is even still negated, preserved and cancelled as a whole quotation. Meanwhile, the procedures for which 'remix' stands as a core are almost omnipresent as

> aesthetic strategies of repetition, remix, mashups or loops can be found in all musical areas from serious experimentation to commercial productions and the popular avant-garde.[91]

Genetically speaking, they have a special place of origin. According to Grossman

> they are the result of a musical strategy of self-assertion: they come from – formerly known as the Third World – colonized states, from the ghettos of the poor and powerless, who, without regard to author, work and originator, use the phonographic means of production and their archives in order to instrumentalize them for a new music of (re)production – in the literal musical sense.[92]

Seen in this way, the appropriation by the poor is almost an "expropriation of the expropriators", admittedly only in the limited area of popular music production. One approach of the remix is the simple de-textualization of the original. But it does not stop there.

> Modern remixing had its roots in the dance hall culture of late-1960s/early-1970s Jamaica. The fluid evolution of music that encompassed ska, rocksteady, reggae and dub was embraced by local music [...] popularized stripped-down instrumental mixes (which they called "*versions*") of reggae tunes. At *first, they simply dropped the vocal tracks*, but soon more sophisticated effects were created, dropping separate instrumental tracks in and out of the mix, isolating and repeating hooks, and adding various effects like echo, reverberation and delay.
>
> [...] From the mid-1970s, DJs in early discothèques were performing similar tricks with disco songs (using loops and tape edits) to get dancers on the floor and keep them there.

The opposite path is also taken: The original vowel remains, the instrumental is changed.

> After the rise of dance music in the late 1980s, a new form of remix was popularised, where the vocals would be kept and the instruments would be replaced, often with matching backing in the house music idiom.[93]

---

[89] See Schulz/Cichon 2004, p. 3.

[90] https://en.wikipedia.org/wiki/Remix, ibid.

[91] Großmann 2015 quotation p. 207. See also: Großmann 2011.

[92] Ibid.

[93] https://en.wikipedia.org/wiki/Remix, 21.10.2020, retrieved 21 October 2020.

Interestingly, in the struggle in the market for the "survival of the fittest", some artists sacrifice the homogeneous integrity of their work so that it at least lives on as a varied and arranged work, but still somehow recognizable as a whole.

> Underground mixing is more difficult than the typical official remix because clean copies of separated tracks such as vocals or individual instruments are usually not available to the public. Some artists (such as Björk, Nine Inch Nails, and Public Enemy) embraced this trend and outspokenly sanctioned fan remixing of their work; there was once a web site which hosted hundreds of unofficial remixes of Björk's songs, all made using only various officially sanctioned mixes. Other artists, such as Erasure, have included remix software in their officially released singles, enabling almost infinite permutations of remixes by users. The band has also presided over remix competitions for their releases, selecting their favourite fan-created remix to appear on later official releases.
>
> Remixing has become prevalent in heavily synthesized electronic and experimental music circles.[94]

In this way, "wandering melodies" (Wilhelm Tappert) as they are often found in pre-industrial folk and popular music are created at a high-tech level, comparable to the 'standards' of jazz, but now within the phonographic form of the works.

> Remixes have become the norm in contemporary dance music, giving one song the ability to appeal *across many different musical genres or dance venues*. Such remixes often include "featured" artists, adding new vocalists or musicians to the original mix.[95]

Often remixes are contrafactures, not only re-textualizations, but expanded by supplements and adding new verses.

The remix is also widely used in hip hop and rap music. An R&B remix usually has the same music as the original song but has *added or altered verses* that are rapped or sung by the featured artists. It usually contains some if not all of the original verses of the song however they may be arranged in a different order than they originally were.

> [...] In the early 1990s, Mariah Carey became one of the first mainstream artists who *re-recorded* vocals for a dancefloor version, and by 1993 most of her major dance and urban-targeted versions had been *re-sung*, e.g. "Dreamlover". Some artists would contribute *new or additional vocals for the different versions of their songs*. These versions were not technically remixes, as entirely new productions of the material were undertaken (the songs were "re-cut", usually from the ground up).[96]

The remix remains at this historical and systematic level in a peculiar ambivalence between quoting and destroying the original work, between negating, preserving and suspending. The aesthetic issue, the question of the work's character, is economically relevant; both are reflected legally as copyright problems. At the same time, the question of identity and difference, of similarities becomes once again relevant.

> Because remixes may borrow heavily from an existing piece of music (possibly more than one), the issue of intellectual property becomes a concern. The most important question is

---

[94] Ibid.

[95] Ibid.

[96] Ibid.

> whether a remixer is free to redistribute his or her work, or whether the remix falls under the category of a derivative work according to, for example, United States copyright law. [...] There are two obvious extremes with regard to *derivative works*. If the song is substantively dissimilar in form (for example, it might only borrow a motif which is modified, and be completely different in all other respects), then it may not necessarily be a derivative work (depending on how heavily modified the melody and chord progressions were). On the other hand, if the remixer only changes a few things (for example, the instrument and tempo), then it is clearly a derivative work and subject to the copyrights of the original work's copyright holder.[97]

Despite the term that accents the "mashing", the *mashup* also preserves, at least in its original form, essential components of the original work. It is a new work,

> a creative work, usually in a form of a song, *created by blending two or more pre-recorded songs*, usually by superimposing the vocal track of one song seamlessly over the instrumental track of another[98] increasing the tempo and pitch while adding or reducing gaps to make it flow. [...] To the extent that such works are "transformative" of original content, in the United States they may find protection from copyright claims under the "fair use" doctrine of copyright law.[99]

Within the relative identity *and* in contrast to it, differences are emphasized in the context of this quotation. As a variation, the mashup conveys both poles.[100]

> The attraction of the mashup is that mostly titles are mixed to a new one, which belong to different styles, genres, milieus, ensemble forms, images etc., but despite this distance, they work surprisingly effectively together musically. In the mashup, you work with large-scale samples that adopt whole form parts such as verses and choruses from their templates and combine them with each other. Mashups are thus studies of genre boundaries and what makes them up or not. They thrive on a balance between the discovery of musical congruence between the templates through their combination and the simultaneous contextual distance between these templates, on the basis of which one does not expect such effective functioning to be possible.[101]

Terminologically and factually, the mashup is not always very precisely distinguishable from the remix. It appears as a reducer and special type of general remixing. In its beginnings in the mid-1990s, the interweaving of isolated soundtracks of two to n works creates a new one,

> which is basically a cover version of at least two songs.[102]

So in a new work we have the quotation form of two existing works. With the *megamix* there can even be more, but sure enough it can be connected with fragmentation tendencies at the same time.[103]

---

[97] Ibid.

[98] Laut https://de.wikipedia.org/wiki/Mashup_(Musik), 22.5.2020, retrieved 31 May 2020 called "A vs. B": a simple permutation.

[99] https://en.wikipedia.org/wiki/Mashup_(music), 22.5.2020, retrieved 4 July 2020.

[100] See below the turnaround to the other pole, the dissolution of the quotation.

[101] https://de.wikipedia.org/wiki/Mashup_(Musik), 22.5.2020, retrieved 31 May 2020.

[102] Wicke/Ziegenrücker 2007, *Mashup,* p. 418.

[103] Cf. *Megamix*, Wicke/Ziegenrücker 2007, p. 426f.

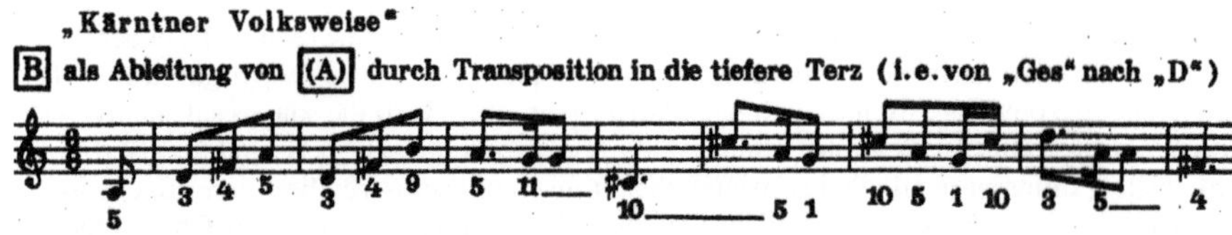

Fig. 8.8 „Kärntner Volksweise"in Berg's Violin Concerto in its embedding into the twelfe-tone-row Quoted after Redlich 1957, p. 272

8.2.3 Embedding: Smaller Work as Complete Quotation Within a Larger Work

One of the famous full quotations is that of a Bach chorale in Alban Berg's violin concerto. For this commissioned work, Berg interrupted the composition of his opera *Lulu* from February to August 1935 because he urgently needed money due to the restrictions imposed by the Nazism - his works were not yet performed in NS-Germany. He did not begin work on the concerto immediately. Spurred into writing he was by the death of 18-year-old Manon Gropius, the daughter of Alma Mahler and Walter Gropius. Berg dedicated the concerto "To the memory of an angel". The quotation of the chorale refers to this dimension of the concerto.

8.2.3.1 Foreign Matter: Bach Chorale in Berg's Violin Concerto

The concerto consists of two movements, each divided into two sections:

I. a Andante (Prelude)
I. b Allegretto (Scherzo)
II. a Allegro (Cadenza)
II. b Adagio (Chorale, Variations, Reminiscence I.b, Coda)

The first movement treats and means life, the second death.

Another *completely* quoted tonal work in the concerto is a Carinthian folk song with components of yodel, referring to a girl. It is embedded in the second section of the first movement, and returns briefly before the coda in the second movement.[104]

This chorale was added to the composition relatively late. Berg was obviously looking for something sacred; on 8 June 1935 he wrote to Willi Reich (Figs. 8.8 and 8.9)

> "Please send me (on loan) the Matthew Passion (score or piano score) and, if you have it, a collection of chorales (I need a chorale melody for my work: discretion!).
>
> Reich sent him, among other things, a collection of 60 Bach chorales […]. Berg chose the final chorale "Es ist genug" (It is enough) from cantata no. 60 ("O Ewigkeit, du Donnerwort") (O eternity, you word of thunder), and said: "Isn't that strange: the first four notes of the chorale melody (ascending whole tones), correspond exactly to the last four notes of the

[104]Cf. Redlich 1957, p. 281.

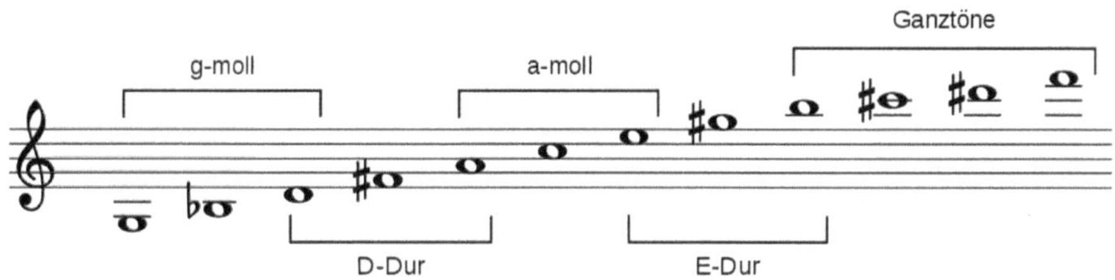

Fig. 8.9 A. Berg: Violin Concerto, 12-Tone-Row
Quoted after https://de.wikipedia.org/wiki/Violinkonzert_(Berg)

Fig. 8.10 J. S. Bach: Cantata no. 60 ("O Ewigkeit, du Donnerwort")—Chorale complete in Berg's Violin Concerto
Quoted after Redlich 1957, p. 273

> twelve-tone row, with which I am building the whole concerto? (reported by Willi Reich) (Fig. 8.10).[105]

This pre-stabilized harmony between the intrinsically atonal twelve tone row and the tonal quotation is twice-weird. On the one hand, the four-tone whole-tone row is already tonal extraterritorial in J. S. Bach's work and, with the ambitus of a tritone as the traditional "diabolus in musica", as the devil in music, it at least suggests something negative. The melody of Bach's chorale is itself a grand quotation. He

[105]https://www.capriccio-kulturforum.de/index.php?thread/5549-berg-alban-violinkonzert-dem-andenken-eines-engels/, 19.11.2013, retrieved 14 July 2020.

composed a four-part setting of the melody by Johann Rudolph Ahle to the hymn by Franz Joachim Burmeister. The row itself on the other side has

> a strong tonal undercurrent: the first three notes of the row make up a G minor triad; notes three to five are a D major triad; notes five to seven are an A minor triad; notes seven to nine are an E major triad; and the last four notes (B, C♯, E♭, F) and the first (G) together make up part of a whole tone scale. The roots of the four triads correspond to the open strings of the violin, which is highlighted in the opening passage of the piece. The resulting triads are thus fifth-related and form a cadence, which we hear directly before the row is played by the violin for the first time.[106]

Berg's ingenious interweaving of atonal and tonal elements even lets the BACH cipher shine through like on a palimpsest. The four chords g minor, d major, a minor, e major

> imply the note sequence B (B♭) – A – C – H (B♮) which forms the BACH motif.[107]

g – g (B♭)– d'
d'– f – A'
a'– C"– e"
e"– g"– H (B♮)"

Berg leaves Bach's composition untouched. However, he instruments it according to the state of his material, i.e. not historically, but still as a vignette with a clearly defined sound. The four-tone head of the melody leads over to the second and last part, *Adagio*: transposed as a pre-echo no less than three times in the viola. It then begins immediately with the explicit quotation of the chorale, first monophonic in the solo violin—marked as *CH* in the score, as "chorale main voice" and even textualized.

Then follows in the third clarinets the third chorale line. The solo violin is answered by the woodwind choir, following the original Bach movement. Berg continues the movement, alternating between solo violin and clarinets in the manner of a traditional 'Alternativo'. He does not spurn a kind of soundalike, since the (Fig. 8.11)

> strange timbre [of the woodwind collective] – clarinet, saxophone, bass clarinet – simulates a delicate organ stop.[108]

The chorale itself then becomes the theme of two variations.[109] Quotation and quoted, chorale by Bach and by Berg comment on each other—with the solo violin as a kind of "lyrical self" of the work (Fig. 8.12).

> *Poco più mosso, ma religioso* (a little more driven, but [!] religious) is the general headline as expression mark. More will follow. The *doloroso* (mournful) on the farewell keyword "nun gute Nacht" (now good night) "is remarkable. (m. 143f.) A few bars later, from m. 149

[106]https://en.wikipedia.org/wiki/Violin_Concerto_(Berg), 22.4.2020, retrieved 14 July 2020. See Redlich 1957, 272ff.

[107]https://en.wikipedia.org/wiki/Violin_Concerto_(Berg), 22.4.2020, retrieved 14 July 2020. See Redlich 1957, 272ff.

[108]Redlich 1957, p. 281. Emphasis HWH.

[109]Variation cf. Redlich 1957, p. 281.

Fig. 8.11 A. Berg: Violin Concerto, *Adagio* (m. 136) mm. 134–141

Fig. 8.12 A. Berg: Violin Concerto mm. 142–145 and mm. 149–152

onwards, Berg composes discreetly but clearly, which is the essential reason for this, i.e. his mourning. On the keyword mm. 150 "Jammer", lament, begins a descending chromatic sequence, the traditional topos of the "lament phrase", actually with a quart ambitus, here as Bb to E with a tritone ambitus. It is then turned to a final F. On the one hand, this preserves the quart ambitus – and two decisive tones are accentuated: Bb and F. referring to *B*erg and *F*uchs-Robbetin. At the same time, slightly masked by the transposition Bb-E, the tritone H-F (in English B-F) resounds, i.e. *H*anna *F*uchs-Robettin. Berg thus quotes not only the BACH-chorale, but at the same time, and *blurred* by the chorale, quotes as a kind of incantation his secret love and beloved *H*anna *F*uchs-Robettin, as extensively and – to her – overtly a decade before in his *Lyrical Suite.* (See above 6.1.1.6 HF and AB. A. Berg's *Lyric Suite*) In addition to the tone letters, he gives another hint, the expression mark *molto espressivo e amoroso* – a little strange if it would refer to a sacred chorale of death and surrender. The "Requiem for an Angel" is below the manifest programmatic latently a requiem for him and their love. At the end, m. 153f., the 1st clarinet affirms "Es ist genug", again with the ambitus F-Bb, then repeated and extended as an "echo", *molto espressivo* and then *dolce espressivo* (!) of the solo violin ["Es ist, es ist genug."].

At the end the 1st clarinet affirms "Es ist genug", repeated and extended as an "echo" of the solo violin ["Es ist, es ist genug."]

Then the variation begins. All in all, the quotation here is closely interwoven both with the row as a basic pre-thematic figure and also woven into the music, in variations as well as in quasi-official part *Variations.* Although clearly detached and marked as a quotation, it becomes an essential part of the material and meaning substance of the work (Figs. 8.13 and 8.14).

Fig. 8.13 A. Berg: Violin Concerto mm. 153–156

Fig. 8.14 A. Berg: Violin Concerto m. 161

8.2.3.2 New Embedding: Mahler/Uri Caine

The US-American pianist's[110] concept album *Urlicht/Primal Light* translates Mahler's symphonic and orchestral song into the sphere of jazz. It contains a complete quotation of the *Adagietto* from Mahler's 5th Symphony.

> It is the only piece on the CD in which Uri Caine spells out the complete original from bar to bar.[111]

Caine's "spelling out" of Mahler is admittedly a peculiar and strange "reading". It produces, right through the quotation, something new, which, as mentioned as a quality of the quotation triangle, also lets the quoted appear in a somewhat new light (see Berio, *Sinfonia* below).

> He does not simply "jazz up" Mahler, which would mock the greatness of his works. Instead, he worms himself inside the songs and harmonies and uses them as a starting point to create a related, but new, synthesis of his music. Jewish folk melodies, cantorial renditions, free jazz, and classical violin are all merged in a whole that transcends the parts [...].[112]

Like in a caricature, but without parodic or critical intention, Caine even sharpens characteristics of Mahler's music.

> At first, only the piano plays, extremely slowly, exaggerating Mahler's stretching of time once again. Only with the consequent [...] do the winds and strings join in. They too exaggerate, for they sharpen Mahler's technique of counter-cantilena with the means of jazz, when, for example, the violinist's counter-cantilena – Caine's melody line is in the clarinet – steps out of beat and harmony. [...] At the end of the theme [...] Caine counteracts the "somewhat pressing" of the original with a ritardando, the "flowing" section with the stopping of time: what seems like a small breakthrough in Mahler's work, or at least a breaking open of the theme, is translated by Caine in a collective improvisation. When returning from the middle section to the main section [...] Caine uses such a collective improvisation analogous to Mahler's "fields of dissolution" at a formal interface. Caine also borrows the compositional technique of "broken melody" from Mahler and illustrates it by "exaggeration". If the theme of the middle section in Mahler is first divided between first violins and violas, the melody in Caine's work is played by: violin (also the viola section), then clarinet, trumpet, clarinet, trumpet, later soprano saxophone and violin duet, even the piano contributes three notes to the melody.

[110]Uri Caine Gustav Mahler/Uri Caine: *Urlicht/Primal Light*. Winter & Winter. München, 1997. 910004-2.

[111]Becher 1999; cf. more generally Schäfer 1999.

[112]Loewy 2014, quoted in https://en.wikipedia.org/wiki/Urlicht_/_Primal_Light, 13.6.2020, retrieved 23 July 2020.

In three respects, Caine thus quotes Mahler on a meta-level at the same time. He was inspired by the material, technique and structure of Mahler's work:

> Material, because, as in the first movement of the fifth, it sometimes only needs a clever intonation and an offbeat to strike a jazzy note. Technique, because the layering of independent lines, i.e. Mahler's polyphony, is a characteristic of the entire history of jazz. Structure, because the path from Mahler's fields of dissolution to collective improvisations is a particularly short one.[113]

8.2.3.3 Mask: Style Quotation as Work

A borderline case of the quotation form should be mentioned briefly. Stravinsky's *Piano Rag* 1919 does not quote a specific work, but rather a genre. He himself emphasized that this is a "portrait".

> Stravinsky himself declared that his pieces were not true ragtime, saying in 1963 that 'the *Histoire* ragtime is a concert portrait or snapshot of the genre in the sense that Chopin's *Valses* are not dance waltzes but portraits of waltzes.'[114]

Stravinsky was right in his assessment that the small, concise work was not a "true ragtime". But the fuzziness is a blurring. He fuses fragments of this piano style with his own style of this phase and makes both styles resemble each other.

> Stravinsky's knowledge of stylistic jazz properties was at first limited to scores … However, he had managed to hear live jazz bands by the time he finished *Piano-Rag Music*. … Piano-Rag Music neatly reconciles the ostinati, shifting accents, and bitonal passages of Stravinsky's Russian manner with the repetitive patterns, syncopations and colorful harmonies of the jazz idiom. Indeed, the piece is remarkable for its bitonal passages (with, for example, the left hand in C and the right in F sharp major), chord clusters, a three-note basso ostinato in the middle sections, and a percussive piano style. Moreover, Stravinsky achieves an improvisatory feeling by dispensing with meters entirely during half of the piece, and employing only irregular ones in the remainder.[115]

Strictly speaking improvisation is a characteristic of jazz, but not of ragtime. Stravinsky's dismantling of ragtime, which he disassembles and recombines some elements into a montage-like image, is not a naturalistic portrait but a cubist one. The stylistic quotation, in turn, is in a certain proximity to the type of "in the style of", which often serves as an homage, sometimes also as its opposite, a parody; this was already the case with François Couperin's Le Parnasse ou L'Apothéose de Corelli.Grande Sonade, en Trio from 1724 and Concert instrumental sous le titre d'Apothéose composé à la mémoire immortelle de l'incomparable Monsieur de Lully from 1725. Johannes Kreidler had his own style imitated by a Chinese composer and formalized by an Indian programmer as a commissioned work for ensemble, sampler and moderator (2009) for a small fee.

---

[113] Becher, ibidem.

[114] https://interlude.hk/stravinsky-meets-modern/, retrieved 1 Nov. 2019.

[115] https://www.allmusic.com/composition/mc0002362194, retrieved 1 Nov. 2019.

Here, too, the question arises under the aspect of fuzzy logic as to the extent to which such style variations and quotations were used to reproduce the given and the original in a sharp, fuzzy, blurred or vague way.[116]

8.2.4 Palimpsest—Parts of the Work as a Changed, but Maintained Combination of Sound and Idea

As a rule, quotations are a pars pro toto, a piece of recognizable old in the new, in which the past and reality shine through like a palimpsest.[117] They are parts in the new whole, and not the old as a whole, as in a grand quotation.

8.2.4.1 Brahms' Sonata for Violoncello and Piano no. 2, op. 99

Brahms has quoted a student song by Brahms at least twice, once in full and once in fragments. Initially largely in its entirety in the *Academic Festival Overture* in C minor op. 80 composed for the University of Breslau. The original is itself a contrafacture (see above 8.1.2.3). So we again have multi-layering as a palimpsest phenomenon.

> The template *Wir hatten gebauet* is better known as *Ich hab mich ergeben* [...] and was taken from the *Commers Buch für den deutschen Studenten* [of 1861], which Brahms owned himself. August von Binzer actually composed the song taken up by Brahms on the occasion of the dissolution of the Jena Brotherhood (1819). The today common text *Ich hab' mich ergeben* (I have surrendered) is by Georg Friedrich Maßmann [1820].[118]

Brahms' final theme in op. 99 is the student song *Ich hab' mich ergeben*, a full quotation from the *Academic Festival Overture* op. 80. Here Brahms only takes the two opening bars,

> to continue with a self-invented twist. [...] A musical idyll is presented, which soon proves to be dangerously fragile.[119]

Brahms conjures up the "the good old student days" that have become proverbial through another student song, the lost, indeed lost youth. Brahms composes his search for the lost era and its rediscovery as a glorification, albeit a sceptical and fragile one. The expressive content is conveyed and intensified through compositional technique. This is how Kontarsky interprets

---

[116]Hiekel/Utz 2016, p. 137.

[117]On the matter and concept of palimpsest see e.g. Genette (1993), Heister (2011), Heister (2012), Rotermund (2011), and Thym (2012).

[118]Kontarsky 2013, p. 719. The following comments are based on this contribution.

[119]Ibid.

> the simultaneous increase in harmonic stability of the *song quotation*, contrasting with the occasional loss of its primary thematic function. [...] Doesn't the one here determine the other? As in a memory of something beautiful, but long gone, from one's own youth? The actual thematic event *fades away* (one remembers more and more vaguely),[120] but the circumstances surrounding it transfigure – until a smoothed out and purified of all real irregularities basic feeling of something beautiful remains. ... Johannes Brahms links what is actually heard with a component that points to the past. [...] The composer is by no means concerned with achieving an actual state (which, in the awareness that this state is not attainable, would also be difficult to concretize), but about the invocation of a paradise, which in reality never existed in this way. Similar to Heinrich Heine he uses a "song from ancient times" which, as a musical cipher, describes the painful disparity between everyday reality and one's own wishful thinking.[121]

Another quotation comes into play here. Brahms fades over the given song quote with his own, the cipher F-A-E from his youth.

> Is it bold to ask what for Brahms the lost treasure was? The rondo theme before us deliberately reduces the quoted song to a circling around the tones a and f (T. 1 and T. 2), in order to then show the alternation of e and present it as central double tone (T. 3 / T. 4, T. 5 / T. 6). Does the chorus have a parallel to the well-known tone cipher f-a-e from the composer's youth? The sequence of tones represented an abbreviation of the life motto "Free but lonely!" of the Hof- und Staatskonzertmeister Joseph Joachim, with whom Brahms was indeed a close friend for years. The two met in 1853 in a student environment (Hannover and Göttingen), in which the "Wir hatten gebauet" (We had built) composed by August von Binzer was certainly also performed. Is this where the circle is complete? Does the composer use the motto f-a-e as a background for the adaptation of the student song template?[122]

Brahms sharpens the specific character of the imaginary-real of memory as a fuzzing via instrumental effects:

> Perhaps even the coda can be understood in this way: Once again the theme appears in the cello part, clear and distinct, but "pizzicato". What we hear is no longer the theme, but its imagination that has become unreal. Brahms' invoked own youth, which here has clotted into a longing for timeless happiness, all this has transfigured into utopia – and the knowledge of it into the quiet wisdom of old age that happiness in itself contains a utopian component.[123]

However, this does not mean that the sequence of citation forms and contra-invoices is complete. The original also served as a social contrafactum, with the analogously fitting title *Wir hatten gebauet ein stattliches Haus* (We had built a stately house), newly functionalized: It

> was sung when the parliament of West Germany adopted the West German constitution of 1949.[124]

The melody was even internationally transformed in various contrafactures:

---

[120]See Sect. 3.4 *Recollection Form*.

[121]Kontarsky 2013, p. 719f.

[122]Ibid., p. 720.

[123]Ibid.

[124]https://en.wikipedia.org/wiki/Wir_hatten_gebauet_ein_stattliches_Haus, 2.6.2020, retrieved 28 July 2020. Detailed information about the history of the song Widmaier 2011.

> To the same tune, there are Christmas lyrics by Abel Burckhardt written in 1845.[1] The Swedish adaptation from 1851[1] by Betty Ehrenborg-Posse is a well-known Christmas song.[125]

Censoring interventions based on the background of the problem of sacral versus secular, including additional recourse, were not absent here either.

> It was originally used by the Swedish free church movement. The lyrics includes childhood references that were removed when the song was included in the hymnal Den svenska psalmboken 1986, and replaced with an adapted version of the lyrics that was originally published in 1906 in the song book *Sjung, svenska folk!*.[126]

The original text, on the other hand, was in turn socially contrafaced again and functionalized nationally:

> "Ich hab' mich ergeben" (English: "I have surrendered myself", lit. ""I surrender""), originally titled "Gelübde" ("Vow") [...] was one of the unofficial national anthems of West Germany from 1949 to 1952, when the "Deutschlandlied" was officially reinstated.[127]

The emphasized German-National is also here internationally usable nationally:

> The national anthem of the Federated States of Micronesia, "Patriots of Micronesia", uses the same tune, [...] as does the Estonian song "Mu Isamaa armas" ("My beloved native land" by Martin Körber) which used to be Estonia's official flag song until 2009 when it was replaced by Gustav Ernesaks's "Mu Isamaa on minu arm" ("My homeland is my love").[128]

Via the common denominator nationalism/patriotism, a conscious or unconscious memory and allusion may even have inspired the original text:

> The second stanza includes the words "Land of the Free", similar to the well-known words of "The Star-Spangled Banner", written eight years earlier.[129]

8.2.4.2 Disguise: Plagiarism and Counterfeiting

Plagiarism is a quotation that is not identified as a quotation.

> Music plagiarism is the use or close imitation of another author's music while representing it as one's own original work. Plagiarism in music now occurs in two contexts—with a *musical idea* (that is, a melody or motif) or *sampling* (taking a portion of one sound recording and reusing it in a different song).[130]

---

125 https://en.wikipedia.org/wiki/Wir_hatten_gebauet_ein_stattliches_Haus, ibid.

126 https://en.wikipedia.org/wiki/Ich_hab%27_mich_ergeben, 29.3.2020, retrieved 28 July 2020.

127 Ibid. .

128 Ibid.

129 Ibid.

130 https://en.wikipedia.org/wiki/Music_plagiarism, 23.7.2020, retrieved 27 July 2020. Detailed and differentiated according to types of music and numerous musical and legal examples ibidem.

It is characteristic of genuine plagiarism that (1) the authors are kept secret—fuzziness as erasure—and that (2) if it is a plagiarism of a whole work, the imitation is staged as the original (see above in Chapter 5 Filtering).[131] The development of copyright law has expanded and clarified the concept of plagiarism. The law has difficulty keeping up with technical and economic changes.

> The baroque age continued to practice the treatment of foreign or own material. Heinrich Schütz, for example [...] has, following the practice of the time, incorporated foreign works into his own collections several times.[132]

This is already going in the direction of plagiarism. As a precautionary measure, Schütz claimed that he could also do otherwise.

> In the Symphoniae sacrae III of 1650 Schütz arranged a work by Al. Grandi [...], in the 2nd part of the Symphoniae two madrigals by Monteverdi for the concert [...]. He added: "However, no one should be equally suspicious of my other works as I am not eager to adorn my works with foreign plumes." Until 1750 it was common practice to adapt and take over other people's works. Bach and Handel have to a large extent arranged their own works and those of others. Bach borrowed themes from German, Italian and French composers, *in rare cases with specification of the authors* [...].[133]

When someone reproached Handel with his intellectual thefts, he replied coolly according to a widespread anecdote: "But these bastards don't know what to do with a good melody."

> However, such adopted themes were [...] not individual inventions, but often traditional types. One only spoke of plagiarism when a composer had the work of another composer performed entirely under his own name.[134]

Used slightly ironically, the aforementioned permutation possibilities of "only" 12 notes show how unlikely merely coincidental matches between two works are. The plagiarism of a work overlaps with a type of forgery, the fake, the forgery of a work of art. The purpose of the fraud is similar both times: mainly fame and/or money. Both times, an imitation pretends to be the original. However, unlike plagiarism, a fake does not necessarily have an original. The counterfeiters often do not copy a single work, but only a particular style of the respective artist. An attempt is made to make similar things appear identical. The forgery practically exploits modal logic: A possible work is transformed into a real work—even if this transformation is only illusory. The issue is relevant not only in the field of the visual arts, where a lot of money is involved—as is often the case with music plagiarism—but also in the field of music, where, for example in the forgery of autographs, it is primarily a matter of scientific truth, both aesthetic and historical. The field of Fuzzy Logic here is the comparison with *similar* works of the artists. In this comparison, where the question of identity no longer applies, it is all the more important to consider fuzziness, similarities, possibilities and probabilities.

[131] Niedermayer 2007; Schmidt 1992.

[132] Engel 1949/1951, col. 1459f.

[133] Ibid., col. 1460.

[134] Ibid., col. 1461.

8.2.5 Collage—Fragments of Works as Semantic and Syntactic Building Blocks

In the types of quotation sketched so far, the basic principle has always been that the quotation should remain concise in its form and that the context should also be so sharply defined that its origin is comprehensible. The principle tolerates exceptions. In the following cases, this integrity is increasingly blurred.

8.2.5.1 Rag Rug: Reactivated Memory and Living on of the Artwork in Fragments

The collage is another special form of quotation.[135] However, it primarily negates the quoted and only secondarily eliminates it at a higher level. The transition from quotation to collage becomes a transition to an almost no longer quotation. The essence of the collage is the combination of different, heterogeneous, given materials. The demarcation from the montage is not always very clear-cut. The focus of the montage lies in the process of assembling and 'composing' of the materials, even of homogeneous or similar ones, in some analogy to industrial processes and techniques as a historical specificity.[136] Collaging as working with prefabricated formulas and "gluing them together" to form a new whole is a universal way of making music—and of other arts. In traditional music as well as in art music, the heterogeneity of the material used is usually not emphasized,[137] apart from exceptions such as the quodlibet mentioned above. An investigation using fuzzy logic methods would be all the more productive. This would allow the small, often imperceptible differences in particular to be recorded more precisely through mass comparisons. On the other hand, explicit quoting generally emphasizes the difference between quoted and quoting work. In modern collage, too, heterogeneity is deliberately emphasized and exhibited as a specific quality.

> Earlier traditional forms and procedures such as the quodlibet, medley, potpourri, and centonization differ from collage in that the various elements in them are made to fit smoothly together, whereas in a collage clashes of key, timbre, texture, meter, tempo, or other discrepancies are important in helping to preserve the individuality of the constituent elements and to convey the impression of a heterogeneous assemblage.[...] What made their technique true collage, however, was the juxtaposition of quotations and unrelated melodies, either by layering them or by moving between them in quick succession, as in a film montage sequence.[138]

This 'avant-garde' collage essentially began with Dadaism during and shortly after the First World War. In it, quotations are made, but at the same time the quotation is levelled and its concise form fuzzed. On the other hand, however, sometimes

---

[135] Cf. e.g. Burkholder 2001.

[136] Cf. e.g. Wescher 1987.

[137] One example of many: Cuneo 2017.

[138] https://en.wikipedia.org/wiki/Sound_collage, 3.10.2020, retrieved 3 October 2020.

something of the original work survives, especially if the fragments are somewhat more extensive.

With the following types of quotation, we move away from the quotation of *one* work in another work, or *as* a different one in the case of a grand quotation, which is, for example, as a cover version, a quotation form of the respective work. Now the work in quotation form as a possible stage in the sequence of its forms of existence is transformed into a work *made of* quotations. The new work combines the quotation form of *many* works. So we are leaving the guiding point of pursuing the work as a 'processing identity'. Instead, we analyse the facts from the point of view of the new, quoting work and the quoting—and the variation. Initially, the varied identity of the work is still preserved. But in the 'articulate continuum' of quotation types between grand quotation and collage, an alteration finally takes place: Fuzzing to deliberate unrecognizability instead of recognizability. When everything becomes a quotation, namely when it is broken down into the smallest snippets, nothing tends to be truly quotable. The collage in the sense of the avant-gardes since the beginning of the twentieth century carries out this tendency, as a 'sampling collage', as we might call it, in "Classics" as in "Pop".[139]

The collage as an *artistic* method of assembling existing, preferably already aesthetic-artistic material, is however considerably older. On the one hand it is, like the arrangement, almost one of the universals of music production (see below centonization). On the other hand, it is, even in its modern form, older than the technical-industrial media. It is a reflection of the social assertion of big industry and thus of 'mechanization' and historicism—both located in the late nineteenth century. Heterogeneity is already accentuated in Mahler since about 1885 and Charles Ives since about 1905. And from there, in retrospect, lines going back into the past stand out more clearly, which also rely on open heterogeneity; the multi-textual and even multilingual motets of the Middle Ages, using musical quotations, are an example of this. Nevertheless, these works were also made into a "self-contained whole" in the second instance. The principle of collage cannot therefore be identified with a specific type, the collage based on the technical re/production of sounds.

8.2.5.2 'Centonization' in Gregorian Chant

A particular example of the generality of the collage is the centonisation in the early unisonous Christian (Gregorian) chant: the composition of new melodies from already existing components, from shorter melodic sections and formulas. These are literally "set pieces" (cento = literally spot, patch), patchwork elements, prefabricated and pre-existent formulas. They are built into the originally "pneumatic",

[139]Cf. e.g. Cada 2011. Hubert Steins: *Prinzip Stillstand Loops und Schleifen in der elektronischen Musik* von s Sendung: Montag, 4. July 2016, https://www.swr.de/swr2/programm/sendungen/essay/swr2-essay-prinzip-stillstand/-/id=659852/did=17506476/nid=659852/y2j3d3/index.html, retrieved 31 July 2020.

ecstatic-flowing melos, transforming it, shaping it and developing it further.[140] The primordial tradition is thereby opened up to additional, subjective-compositional access. As far as "protomelodies" of ancient origin are concerned, the historical difference would thus be directly given. If in any case, not only in the "Gregorian", but even more so in the ancient Roman and Beneventian chant or in the Byzantine liturgical chants, melodies are "usually composed of a number of stereotypical phrases, cadenzas, formulas and figures",[141] then the awareness of historical differences—still—becomes at most tentatively perceptible and intentionally relevant. The immanent historicity is here rather obscured and covered up: it is partly reflected in a planned and theoretical way, partly simply fuzzed by practice—expression becomes material. Thus the difference between earlier and later, old and new becomes blurred, even on the background of a religious, aesthetically relevant ideology from the all-time valid, revealed absolute truth. Although this truth does not openly tolerate its own historicity and historical change, it adapts itself covertly and undiscussed to changing historical conditions.

The primordial tradition is thereby opened up to additional conscious, subjective compositional access. The concept of 'centonization' can be generalized beyond its original field of origin and application: namely, as a musical process of creating something new by combining the existing—mainly formulaic. Analogies can also be found in traditional music such as rāgas or maqāmen or ancient Greek nomoi. And the use of formulas in the text as well as in general in the musical performance of epics, from Bhagadavita or Homer, reaches up to the present day.[142] In the cento of the chorale, the quotation does not always have to go back to a fixed and notated melody. Formulas can also be quoted in their turn. The notation of the chorale has been rather fuzzy for centuries (see Chapter 3). The written tradition must therefore be supplemented by the oral tradition at least into the tenth century.

> The constraints by which the text and its inherent structures are determined form a web of restrictive principles, a "formulaic system" [Treitler 1975]. This web allows the singer [...] to *compose* at high speed, which is based on the principle of economy. For a concrete idea in a given metric environment there is only one formula available.[143]

The realization of the idea is thus sharp, but also limited in its possibilities. The "formulaic system" creates as a "generative system"

> such a high constancy of performance that the result can be compared to memorizing the songs.[144]

The—paradoxically anonymous—quoting in the Cento is similar in procedure to procedures based on fragmented sampling, however different everything else is from

---

[140] Critical about the term in relation to the chorale among others: Hiley (1993), Treitler (1974), Treitler (1975), and Treitler (1992).

[141] See Schier 1995.

[142] Cf. e.g. Parry 1971. As an example of the role of epic formulas from another culture cf. e.g. Sellmer 2017.

[143] Schier 1995, ibid., col. 547.

[144] Ibid., col. 548.

the material to the context. It is a fuzzing of the past and—at the same time—a sharpening of the present new. However, within the respective piece of music this is a transition to an explicit historical differentiation of the former. But this is not a real quotation process, since no works are quoted, but at most a borderline case: that what is quoted, is what already exists and thus the past, here a sacred tradition.

The presence of the former is ultimately the presence of history. The quotation is the guarantor and also the sign of it.

8.2.5.3 Sampling and 'Electrified Collage'

The procedural basis of the musical collage is sampling. Whether analogue or digital—the decisive factor is the aforementioned preformed, prefabricated components that justify its classification in the field of quotation (see Sect. 6.9).

Here the aspect of quotation in a broad sense, of working with the given, is accentuated.

Sampling, as collecting, is a universal in music production.

The craft of making music, whether improvising or composing, oral or written production, has always included the 'collecting' of *shaped* material, of formulas, phrases, set pieces, catch phrases, patterns on the one hand, and on the other hand of beautiful passages, special, peculiar figures from one's own production or that of others. The creation of such stockpiles, the formation of a corresponding repertoire, is equipment on the side of the material, which is then drawn on with the respective state of individual and collective musical technique. Even great composers, whether J. S. Bach or Beethoven, have created such small collections. A relatively late example are the "Kinotheken" of the film score composer Giuseppe Becce (1877–1973).

> In the years between 1919 and 1929 he published his *Kinobibliothek* (later shortened to: *Kinothek*) – a multi-volume collective work which he understood as a *construction kit* for the composition of silent film music. The Kinothek contained *set pieces* for all kinds of typical silent film scenes. Depending on the film scene, the pianist could now use the programmatic titles of the Kinothek (such as: Popolo giocondo / Merry People, Fuga comica / Comic Fugue, Marcia solenne / Solemn Parade, etc.) and adapt the music to the film by directly playing the Becce pieces or by varying them. Becce even made a record of his cinema pieces as a help for film people who could not read music. The Kinothek became a great success – but it was also massively criticized because it would destroy the meaningful connections between the scenes through the modular fragmentation of the sequence of scenes and lead to a kind of industrial stereotyping and standardization of film music. The original intention of the Kinothek (like similar projects), however, was exactly the opposite – it was about liberating the musical accompaniment from the arbitrariness of the "man at the piano" its upgrading to art; it is no coincidence that the Kinotheken were created parallel to the social and artistic rise of cinema or film. In 1927, in cooperation with Hans Erdmann, the *Allgemeine Handbuch der Filmmusik* (General Handbook of Film Music) was compiled, a collection of about 3,000 pieces of music suggested to accompany film situations described in key words.[145]

Thus from the collage of selected fragments, sampled here, the whole of the accompanying music to the respective film is created: not a quotation form of a musical

[145]Kieler Beiträge zur Filmmusikforschung 1, 2008, pp. 178–183.

work, which does not exist here as a whole anyway, but the musical form of this film. Collecting in the form of *sound files, snippets* etc., of sound snippets of all kinds and storing them on sound carriers, synthesizers, computers etc. is only a new technical form. But that ONLY is an understatement: it is a huge qualitative leap in the development of music production (see Sect. 6.9). The building blocks in Ruttmann in the 1920s or in the musique concrète in the 1940s were not yet digitized, but they were sampled, even if it wasn't called that at the time. The collage continues to develop from an originally handcrafted, pre-industrial basis. An important further stage is the collage on the technical basis of electrification.

> [...] The first documented instance of sound *collage created by electronic means* is the piece "Wochenende" (in English, "Weekend"), a collage of words, music and sounds created by film-maker and media artist Walter Ruttmann in 1928. (see Sect. 4.3)

The third, current stage develops on the basis of automation and digitalization and as their media-artistic reflection. To put its specificity terminologically, it could be called a 'sampling collage'. Sampling is a superordinate procedure which is necessary for all other procedures and the resulting new genres mainly of pop music,

> a sample is generally understood as a digitally captured sound file. Often, samples are taken—captured—from prerecorded music and used in the composition and/or performance of a new form of music; in other words, individual drum sounds, loops, vocals, and so on.[146]

Sampling is ultimately an implication and consequence of the technical re/production of the acoustic world in general, immanent in this reproduction, and does not only appear digital, but already in the 1920s on the basis of radio and film technology.

Drummer Chris Cutler, for example, emphasizes this with sovereign cool nonchalance towards copyright law

> (I)t is only what is not recorded that belongs to its participants while what is recorded is placed inevitably in the public domain.[147]

Sampling today forms the basis and starting point for practically all procedures, techniques and genres, especially in popular music outside traditional genres such as hits etc.—although it is now penetrating there as well, and this furthermore without too much respect for the authors' rights. Sampling, for its part, is a layer of electroacoustic music production both sound synthesis and sound collage of the musique concrète type together with its radiophonic predecessors, as in Ruttmann's work in the 1920s. The collage-based trends after World War II were a counter-movement to the rigorous 'centralization' and homogenization of composition in serialism. With regard to the question of the similarity in the relationship between music and reality (see Chapter 4), both trends, in each case one-sided, reflect something of real developments. In the collage it is the experience of a fragmented world and the endangerment or even dissociation of the 'indivisible' individual, the ego and the subject. In serialism, it is the increasing organisation of technology, economy and

---

[146] Emsley 2011, p. 2.

[147] Quoted ibid.

society, together with the idea of a rational planning of society as a whole. For the first time, samples became more widely usable in the synthesizer, which could not only reproduce, but also produce. The possibilities were expanded dramatically with digitalization and computerization, including new features such as Ableton etc. As in popular music, the fragmentation of the given has been pushed as far as possible, for example since the 1980s.

> *Micromontage* is the use of montage on the time scale of microsounds. […] The technique may include the extraction and arrangement of sound particles from a sample or the creation and exact placement of each particle to create complex sound patterns or singular particles (transients). It may be accomplished through graphic editing, a script, or automated through a computer program. […]
>
> *Granular synthesis* incorporates many of the techniques of micromontage, though granular synthesis is inevitably automated while micromontage may be realized directly, point by point. It […] may be compared to the pointillistic paintings of Georges Seurat.[148]

Sampling is thus also the essential *technical* basis of collage since the 1970s.

> Recipe 1: Take an old church hymn, the national anthems of two countries from Central Europe and South America, a four-bar piece from an Italian madrigal, a very distinctive passage from the work of a friend's composer or something from your own piece – now mix the ingredients with homemade ingredients.
>
> Recipe 2: Take the four or six final bars of all piano sonatas by a famous 19th century composer. Use them to make a new piece, there are no homemade ingredients.
>
> In Recipe 1, different existing materials are integrated into a texture of their own, "glued" onto a sound layer of their own. In the history of music since 1960, this type is most frequently encountered; the best known are Bernd Alois Zimmermann's opera Die Soldaten (1963/64) and other of his works as well as Luciano Berio's Sinfonia (1968/69).
>
> Mauricio Kagel, *Ludwig van. Hommage von Beethoven* (1970); John Cage, *Europera I-V* (1987–1991).[149]

8.2.5.4 Transition from Large Quotation to Collage: Mahler/Luciano Berio

A complex interweaving of whole quotation, covering and collage is the III movement of Luciano Berio's Sinfonia, 1968/1969 for orchestra and eight amplified voices. The basis here is therefore still the notation form, not the phonographic form.

The five movements:

I. [untitled]
II. *O King*
III. *In ruhig fliessender Bewegung*
IV. [untitled][150]

---

[148] https://en.wikipedia.org/wiki/Sound_collage, 3.10.2020, retrieved 3 October 2020.

[149] See *JetztMusik—Glossar: Collage.* JetztMusik—Glossar, SWR2 (Südwestrundfunk, Baden-Württemberg, Bundesrepublik Deutschland), https://www.swr.de/swr2/, link is disabled.

[150] But it begins again with a Mahler quotation—the chorus taken from the end of the "Resurrection" symphony, the choir „Auferstehn", 5th movement.

V. [untitled]

Berio quotes for his movement III. the movement III., the Scherzo *In ruhig fließender Bewegung (With quietly flowing movement)* from Mahler's Symphony No. 2. In Berio's *Sinfonia* the orchestra plays an arrangement or ‚adaption'of Mahler's symphony,

> a slightly cut-up, re-shuffled and occasionally re-orchestrated version.[151]

Mahler's movement for itself is already a paraphrased self quotation: It is based on Mahler's own setting of *Des Antonius von Padua Fischpredigt* (*St. Anthony of Padua's Sermon to the Fish,* July/August 1893), No. 6 in the cyclic work *Lieder aus "Des Knaben Wunderhorn*", 12 Songs from "Des Knaben Wunderhorn" for voice and orchestra.[152] Mahler's movement functions in Berio's movement as 'framework' or is

> used vessel-like [– a wealth of other quotations (from Bach to Berio himself) are mounted in the empty spaces.[153]

The 'whole quotation' includes the form quotation, and even the quasi-quotation of the position of the movement in the cyclical context of the work:

> The *Sinfonia* models the form of the Mahler movement then in two ways:
>
> (1) The structure of the movement is virtually identical. The length, climaxes, and even shifts in tonal center/organization of the aggregate happen at cognate points in each piece; and
> (2) The Sinfonia movement occupies the same space in the Sinfonia as the original movement does in the Mahler symphony – the third, or "scherzo" spot.[154]

Berio preserves Mahler's movement proportions. His additions, however, introduce a number semantic aspect:

> The addition stresses the proportional importance of the fateful number eleven standing for imperfection, which in turn is related to the meaning of Mahler's scherzo.[155]

The necessary imperfection—the 11 instead of the perfect 12—reminds us of basic questions of fuzzy logic. In addition to a wealth of musical pieces, Berio also quotes a number of literature pieces, Joyce'*Ulysses,* and mainly from Beckett's novel *The Unnamable,* political slogans, uses expression marks, and phonetic material.

---

151 https://en.wikipedia.org/w/index.php?title=Sinfonia_(Berio)&oldid=958839017, 25.5.2020, retrieved 14 July 2020. Some of the many essays to this work: Krieger/Stroh 1971; Kühn 1972; Budde, 1972; Emons 1998. Ausdrücklich hingewiesen sei auf die besonders aspektenreiche und methodisch neuartige Analyse von Finn 2011.

152 Detailed to the relationship of the "Fischpredigt" to the Mahler scherzo Tibbe 1971.

153 Dietel.

154 Flinn 2011, p. 113.

155 Altmann 1977.

> Of the more than 1400 words of spoken and sung text in the Berio movement, more than two- thirds are from Beckett. The rest derive from various sources including Joyce; the spoken phrases of Harvard under-graduates; dialogues between Berio, friends, and family; and slogans written on the walls of the Sorbonne during the May 1968 riots, to which Berio was witness. While the concept of *Sinfonia* is not strictly political, the inclusion of this last material informs the movement with added ramifications, the death- resurrection motive expanding into a metaphor for the socio-political state of the sixties.[156]

As with Mahler's music as structural basis, the Beckett quotations

> appear in the movement in the relative position they occupy in the book, the earlier passages earlier in the movement, the later toward the end, and so on.[157]

Some things are recognizable, but many things remain unrecognizable and can only be perceived with the help of explicit explanations (Fig. 8.15).

Interestingly, even within the very heterogeneous quotations, Berio quotes and models according to the principle of similarity.

> The quoted fragments are often chosen because they bear *a rhythmic or melodic likeness* to Mahler's scherzo. For example, Berio uses a violin line from the second movement of Alban Berg's violin concerto with chromatically descending sixteenth notes two measures before a similarly descending line appears in Mahler's scherzo. This is then accompanied by another violin descent, taken from Johannes Brahms' violin taken from Johannes Brahms' violin concerto […]. The text from Beckett at this point begins, "So after a period of immaculate silence there seemed", but, instead of continuing the quotation ("a feeble cry was heard by me"), Berio substitutes the words "to be a violin concerto being played in the other room in three quarters" and then, after the Berg quotation, alto 2 insists on "two violin concertos", at the point where Berg is interrupted by Brahms.[158]

Berio's quotation themselves are loaded with specific semantics—not to explain here in details.

> A partial list of musical quotations used in the third movement of *Sinfonia* in order of their appearance:
>
> Schoenberg's *Five Pieces for Orchestr*a, fourth movement, "Peripetie"
>
> A brief quotation from the beginning of Mahler's Symphony No. 4 […]
>
> Claude Debussy's *La mer*, second movement, "Jeux de vagues"
>
> Mahler's Symphony No. 2 (Resurrection), third movement (the only quotation that is ongoing) entering in bar 7, from where it continues to the end of the movement, though not always audibly
>
> Paul Hindemith's Kammermusik No. 4
>
> Ravel's *Daphnis et Chloé*, flute solo from the Pantomime
>
> Debussy's "Jeux de vagues" returns
>
> Berlioz's idée fixe from the *Symphonie Fantastique* (played by the flutes and oboes)
>
> Ravel's *La Valse* (orchestra plays octave motif with piccolo playing a chromatic scale)
>
> Stravinsky's *Le Sacre du Printemps* (the "Dance of the Earth" sequence at the end of the first tableau)

---

[156] Hicks 1982, p. 211; cf. Altmann 1977.

[157] Hicks 1982, p. 213.

[158] Ibid.

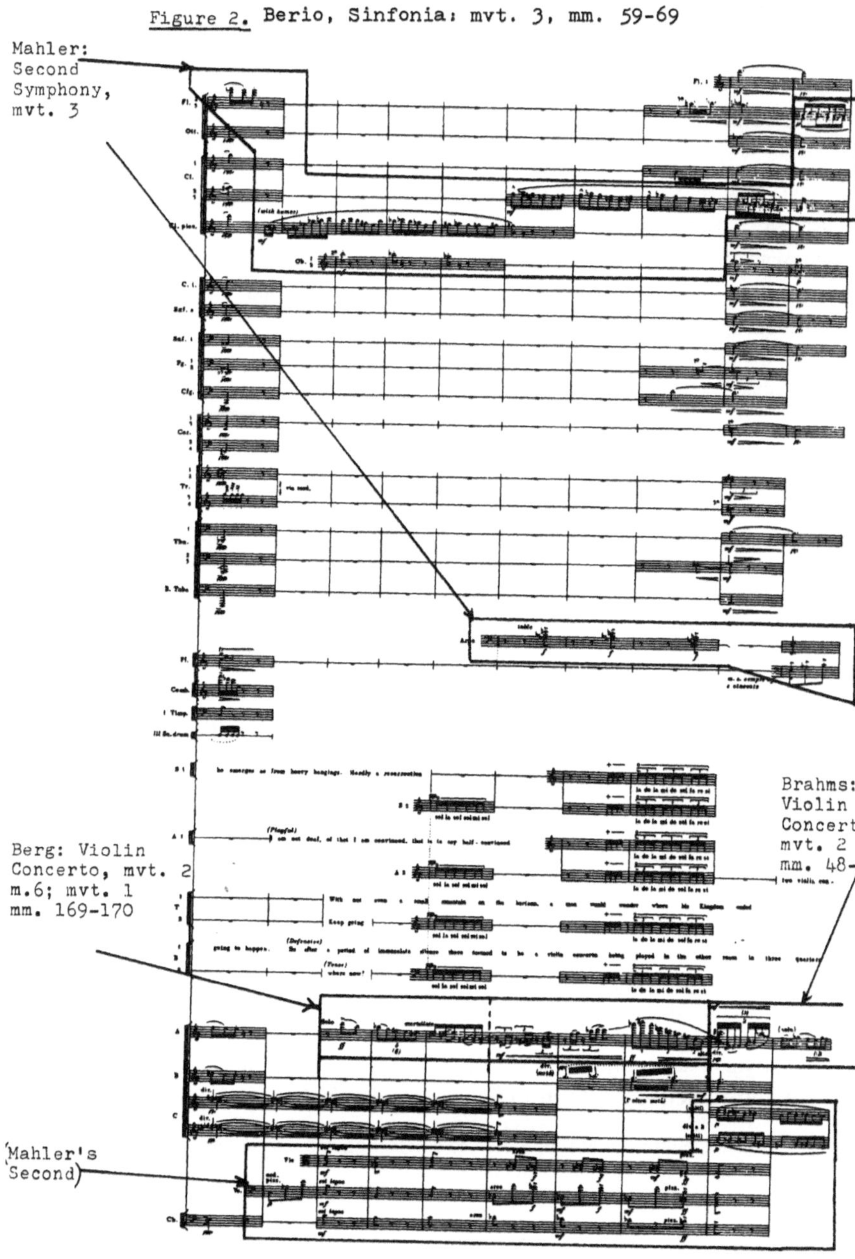

Fig. 8.15 Luciano Berio: *Sinfonia*
Score excerpt with comments, quoted after Hicks 1982, Fig. 2, p. 201

Stravinsky's *Agon* (upper oboe part from the "Double pas de quatre" as well the castanet part from the "Bransle Gai")

Richard Strauss's *Der Rosenkavalier* (one of the waltzes composed for the opera)

Another extract from Ravel's *La Valse*

A chorale by Johann Sebastian Bach

The end of the second movement of Bach's First *Brandenburg Concerto*

Alban Berg's *Wozzeck* (the drowning scene late in the third act)

Beethoven's *Pastoral Symphony*, second movement (melody stated with the clarinets)

Resumption of Hindemith's Kammermusik No. 4 in the solo violin

Another quotation from Beethoven's Pastoral Symphony

Brief recapitulation of the opening of the movement: Schoenberg's "Peripetie", Debussy's *La Mer* (this time from the third movement "Dialogue du vent et de la mer")

Boulez's *Pli Selon Pli*, very first chord of the entire piece from the first movement ("Don")

Anton Webern's Cantata No. 2, Op. 31, fifth movement (opening)

Karlheinz Stockhausen's *Gruppen* for three orchestras[159]

At one point Berio quotes longer from a work.

> At measure 169, Berio changes the texture and meaning of his piece completely by dropping another continent onto this new world and bringing in most of an entire section of Igor Stravinsky's *Le Sacre du Printemps*. Specifically, Berio uses a large excerpt from the final section in Part I, "The Dancing Out of the Earth."[160]

Here, as in some other cases, Flinn tries to capture identities and differences quantitatively in the sense of fuzzy logic (Fig. 8.16).

Berio quoting Stravinsky, shown graphically. Note the subsuming of Mahler material by Stravinsky material.

This quotation, like an aspect of the *Sinfonia* in general, brings out the ironic and parodic component by thwarting expectations.

> by bringing in a completely unrelated piece that seems upon first hearing to be a natural outgrowth of the rhythmic motive that grew from the acceleration of the Mahlerian source material (see ex. 5-29) but soon manifests itself as something different than expected, the composer is subverting our expectations.[161]

Hicks's analysis has a somewhat startling but debatable punch line. Assuming that the bulk of the quoted texts are from Samuel Beckett's *The Unnamable*, he concludes

> we will see that the movement is best viewed as a setting and interpretation of that text: it is a book turned into music.[162]

---

[159]https://en.wikipedia.org/w/index.php?title=Sinfonia_(Berio)&oldid=958839017, 25.5.2020, retrieved 14 July 2020, mainly based on and mixed up with Hicks 1982, pp. 212–221, here slightly abbreviated, without marking each cut.

[160]Flinn 2011, p. 123.

[161]Flinn 2011, p. 125.

[162]See Hicks 1982, p. 207.

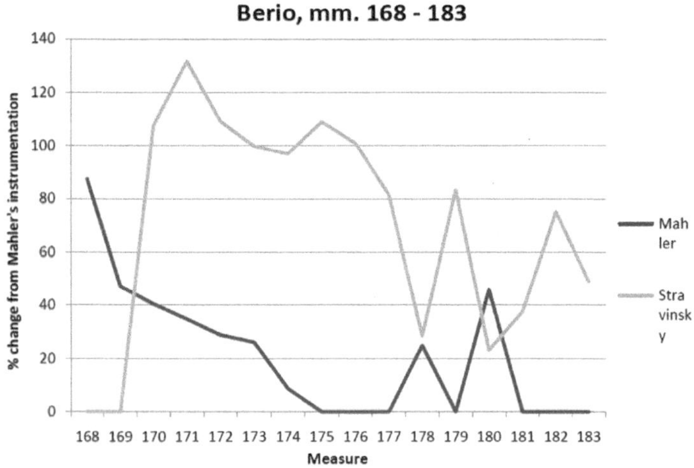

Fig. 8.16 Berio: Sinfonia
Quoted after Flinn 2011, p. 123. Ex. 5–14

The 3rd movement of Berio's *Sinfonia* would, seen in this light, not only be a quotation form of the 3rd movement of Mahler's Symphony No. 2, but also a musical form of Beckett's work *The Unnamable.*

8.2.5.5 Caleidoscope: Remix (II), "Conceptual Writing"

Berio's work represents a relatively early stage of the progressive decomposition of quotation material to the point of completely fuzzed unrecognizability. Instead of sharpening, fuzzing is the predominant method. One extreme of collage was developed by Johannes Kreidler. He turns the screw of fragmentation one turn further, and now combines readymade and handmade, technically reproduced music (snippets) and live music in his *Dekonfabulation* (2007/2008) for accordion, percussion, female speaker and tape.

> For some years now, my compositional work has concentrated on the idea of creating "music with music"; what is meant is the dissection of existing, *more or less recognizable sound files*, which are combined with what is played live, or the mutual overwriting of the two. Depending on duration, speed, density and context, one perceives "pure" sounds with a certain parametric order or "meaningful" sounds with experiential values. In *Dekonfabulation* [...] many excerpts from different music and texts are heard, accompanied by the instrumentalists playing and speaking live. I intentionally used mainly bad pop music as material that is worth being "expropriated" and whose simple contexts are – fragmented – suitable for perceptual play. According to the Duden dictionary, "confabulation" means: a report based on deception of memory about supposedly experienced events.[163]

[163] http://www.kreidler-net.de/werke/dekonfabulation.htm, 13.2.2013, retrieved 30 March 2013.

Pop music went just as far in quotation destruction with its further developments of remix and mashup now as a fragmented collage, as did comparable "Classics" phenomena such as the aforementioned *microcollage*.

> Art of Noise took the remix styles to an extreme—creating music entirely of samples. [...] In the 1990s, with the rise of powerful home computers with audio capabilities came the mash-up, an unsolicited, unofficial (and often legally dubious) remix created by "underground remixers" who edit two or more recordings (often of wildly different songs) together. Girl Talk is perhaps the most famous of this movement, creating albums using sounds entirely from other music and cutting it into his own.[164]

In the further development of the remix, the given music turns from a quotation into a non-quotation—a change to the opposite by the accumulation of quotations.

> Especially in recent times [...] remixes are increasingly appearing on the market in which the original work on which they are based is no longer musically recognizable at all. This phenomenon can be explained by the fact that the appeal of a music title no longer depends only on the reputation of the performing artist or the degree of recognition of the recorded works, but is also significantly influenced by the choice of a renowned producer (in the technical-creative sense). Thus, some disc jockeys ("DJ's") of well-known discotheques and clubs have emancipated themselves from pure record players to famous sound directors ("producers") and have given the music productions they created an unmistakable character through an individual, recognizable "sound". The recording industry takes advantage of this bonus and commissions these producers to produce remixes of pre-existing productions. In the broader sense, these remixes sometimes leave only the title line of the original, while in favour of an own sound, possibly new rhythms and new arrangements, the sound tracks of the original production are used only rudimentarily or even not at all. In contrast to *remixes in the narrower sense*, where at least the above-mentioned condition of remixing existing soundtracks is complied with, the *remix in the broader sense* distances itself from this premise and leaves existing sound recording material largely unnoticed.[165]

This is a manifestation of the general configuration 'expression becomes material'; previously semantically filled music becomes syntactically functioning building material. A main reason for this is to avoid copyright and problems with plagiarism allegations. An aesthetic reason for this is that the 'micro-collage' with the reduction of given forms to mere material can release productive impulses and lead to the production of new, individual forms. This is always controversial and often leads to lengthy processes. Copyright law reflects this fact:

> *New text, new melody.* If a remix no longer contains any protected elements of the original version, but merely sound or rhythmic parallels, uncharacteristic set pieces (e.g. unrecognizable samples) or if it only bears the same title as the original, one can assume that it was created using the original work as a source of inspiration [...]. This is [...] only possible, however, if no recognizable, protectable melody parts of the original are contained in it [...]. If such a free use exists, the remix is subject to a copyright completely independent of the original version [...]. The author of the original work has [...] no longer any right to this new work. Therefore, his consent is not required for the exploitation of the remix [...]. In these cases, the copyright holder is the producer alone.[166]

---

[164] https://en.wikipedia.org/wiki/Remix, 21.10.2020, retrieved 21 October 2020. Emphasis HWH. .

[165] Schulz/Cichon 2004, p. 11.

[166] Schulz/Cichon 2004, p. 12; the references to §§ of the UrhG are omitted.

The previous work becomes processable mere material for new works.

> In all these musics there is little sense of 'the album' or 'the single' as a finished product; instead all music seems malleably subject to further sampling, remixing, or other manipulation to realize yet more, different, music. Both 'authenticity' and 'authorship' are at issue here in a discourse that sees such boundaries as at best contingent[167]

In the "Classics" segment Johannes Kreidler reacted ironically and critically to the plagiarism persecution with a work that is a piece of realized conceptual art, an asymptotic approach to the limit of absolute composition, completely only made up of quotations, he achieved

> with his 33-second *product placement* (2008), which consists of 70,200 third-party quotations: The piece also includes requesting and filling out the corresponding registration forms, which were brought to the Berlin headquarters of GEMA [Gesellschaft für musikalische Aufführungsrechte] on 12th September 2008 in an "action with trucks, columns of paper, carriers and press".[168]

This critical-satirical fuzzing is action and conceptual art, destruction and construction at the same time. The quotations are made completely unrecognizable aesthetically, but remain technically and legally preserved. Finally, a parallel to these variation and quotation procedures is "*conceptual writing*" in literature (see above *Appropriation Art*, Plunderphonics 3.3 Real Tones and Existing Pieces of Music as Readymade Material of a Work).

> We use the term Conceptual Writing in the broadest sense, so that it intersects other terms such as: allegory, appropriation, piracy, flarf, identity theft, sampling, constraint and others. Conceptual Writing, in fact, might best be defined not by the strategies used but by the expectations of the readership or thinkership.[169]

So it is not conceptual art, but it borders on it. It is subject-focused, similarly reduced and an art without much of its own productive output. Further capers towards l'art pour l'art and solipsistic self-referentiality brings the „allegorical writing".

> In allegorical writing [...] the use of source materials in reframing or mixing. The primary focus moves from production to post-production. This may involve a shift from the material of production to the mode of production, or the production of a mode.[170]

Much is fuzzy here, including the difference between sign and signified, art and reality.

> there is no distinction between manipulation and production, object and sign, contemplation and consumption.[171]

---

[167]Cook/Pople 2004, p. 487. .

[168]Kunkel 2011, p. 481.

[169]Place/Fitterman 2009, 2010, p. 12.

[170]Ibid., p. 18.

[171]Ibid.

8.3 Variations Through Media Transformations (I): Word Form

As in the sound form, so too in the word form and in the picture form, there is a double dialectic: on the one hand, in the subject, the sharpening of memory and even crystallization as the formation of new forms, and on the other hand, as in every realization, different kinds of fuzziness in the object.

The ominous encoding melody in *The Lady Vanishes*, like with *Augustin* in Schönberg's Second String Quartet, was about sharpening something once quite sharp that over time had been overlaid by other experiences and thus became fuzzy, as far as possible by reconstructing the original state.

8.3.1 Recrystallization and Reconfiguration in a Detour Through Fuzziness and Transformation into a Different Art

In the type of the word form as well as the picture form and then the imagination form, on the other hand, the fuzziness is there from the beginning, tending even to be a mere vagueness. In a laborious process a certain sharpening and crystallization must be achieved here. Basically, the musical work can be at least partially and in some components verbally expressed and formulated, if necessary with the help of artificial language such as musical notation. For the language of words became a metalanguage for all other languages, both artificial and artistic, after it had prevailed over the language of gestures as the primary medium of communication with signs, most probably already in the early days of human history and with the emergence of human language. With reference to the writer Butor, J. F. Klarmann shows that artistic experience is also mediated by word language:

> Butor makes [...] clear that our experience with painting contains a considerable amount of verbal content. The picture viewer never sees pictures alone, seeing is never pure vision. We heard about the pictures, we read art criticism, our gaze is "surrounded by a *halo* of commentary".[172]

When asked whether language and images are "incommensurable", Butor says

> that words and texts are always already present, where images are perceived and received, that a complex structure of texts surrounds the individual image and influences the way we see [...]. According to Butor, the experience and interpretation of works of visual art is linguistically mediated because Western culture is a culture of writing.[173]

Butor presents the facts correctly, but his reasoning is too narrowly defined and inadequate. Not only for the visual arts, but for all the arts, the cooperation of words is

[172]Klarmann (undated), p. 125. Emphasis HWH; see above the *Halo-Effect*. .

[173]Ibid., p. 30f.

necessary. For language—by which I mean the language of words—is, as mentioned, a general metalanguage for all other languages. It can also be used to talk about works in the respective artistic language. This does not mean, however, that everything and above all the same is said and can be said in the language of words as in the artistic language. The difference between the work in its two most highly developed forms of existence, the score form and the sound form, and its word form is not only fuzziness, but a real gap. Logically, many things cannot be translated from music into the language of words, especially the dimension where a sense of tone color and tonal resource plays a role—otherwise there would be no need for the specific musical language. The word form is thus lacking a decisive element. From the music as a mediated unity of sense and sound it can at best catch the sense, but not the sound, not even with the help of onomatopoeic means. It is a crutch, crash barrier and railing on the way to the work as a goal, not the work itself and not even in a sensually adequate reflection. But it is an indispensable aid.

8.3.2 *Word Form—Afterwards and Afore: Music Criticism, Music Analysis, Introduction to Works*

The word form can be presented orally or in writing. On the one hand, it reflects what is remembered, namely the work in sound or phonographic form. Under this aspect we could also call it 'report form'. Concert, opera, recording and, more rarely, radio criticism is its medium. On the other hand, the word form can refer directly to the notation form and describe what is written in the score. Music analysis in the narrower sense is ultimately only a special, intensified form of the word form. In both cases, it follows the perception form in a temporal and systematic way. Often it is embedded in a circular process: The appropriation of the verbalized work can in turn prepare the perceptual appropriation of its sound form—concretely in such a way that the reading of an introductory text in concert or opera guides precedes the concert or opera visit. The description in advance, a priori so to speak, has a reception-guiding effect on the listeners and their perception, and ex post after listening to the work it has a reception-guiding effect on memory. Fuzziness can cancel itself out in this circular process, so that, seen as a whole, it can lead to a sharpening.

Since the emergence of the press in the course of the Enlightenment and the rise of the bourgeoisie, a considerable portion of word forms in music criticism has been clustered in magazines and newspapers. This refers mainly to the sound form, less often to the score form. It therefore refers to the work *and its* interpretation and describes both—usually the latter more so, since with the development of a canon, in classics as well as pop, the relatively new has priority over the relatively old and familiar. Music criticism has a claim to objectivity even in polemics, but necessarily always operates with conscious and intentional or unintentional subjective (un-)sharpness. It is objectively a kind of 'variation' of the work or a cycle of variations *on* the work, regardless of subjective intentions regarding 'faithfulness to the original'. This *subsequent* verbalization complements the *preceding* verbalization such as introductions to works of all kinds, in program notes, program booklets, independent "guides" for concert hall and opera house. Finally, a preliminary verbalization is also the (non-pictorial) program of a work (see Sect. 4.5 'Program Music': 'Music Form' of texts and 'Music Form' of images).

8.3.3 Ekphrasis: Word Form Versus Program Music as 'Music Form' of Texts and 'Music Form' of Pictures

In both cases, a description of the work that is as precise as possible is indispensable for a concrete, as meaningful as possible impression or idea of the work. The word form of a work is a conscious, often artistic and as vivid a description as possible. In recent tendencies in art history, a traditional term, the 'ekphrasis', has been used again for this purpose. Since Greek antiquity, the 'ekphrasis' has been an established literary form between fiction and science, which tries to put real works of art, preferably those of the fine arts, into words as periphrasis, paraphrase, description. The phenomenon of ekphrasis in Western, European culture is usually traced back to Homer's description of the shield of Achilles in the *Iliad* (book VIII, v. 18.483-18-608).[174] Also developed as a rhetorical exercise, the ekphrasis is subject to a historical change. On the one hand, it is broadly defined and related to real facts such as works of art, on the other hand it is narrowed down to descriptions of works of fine art. Perhaps the most reasonable extension of the term is a medium extension of the term, as a description of works of all arts, both in prose—scientific or literary—and in poetry. On the one hand, synaesthetic effects result.

> The degree of vividness distinguishes the ekphrasis from the factual report. It is a literary visualization strategy: the ekphrase tries to "turn the listener into a spectator" (according to Nikolaus von Myra) and to suggest a quasi synaesthetic, holistic experience.[175]

"Ut pictura poesis erit" (Horace) as painting will be poetry; the described object is imaginatively "put before the eyes" (so Quintilian in his rhetoric doctrine). The concept of ecphrasis can be differentiated within the field of music: musical ekphrasis versus "program music" (see Sect. 4.4).

> The two genres belong to the same species: both involve purely instrumental music that has its raison d'être in its response to a definite literary or pictorial source; both have variously

[174]This poetic ekphrasis was a starting point for the later *blazon* in the heraldic of the feudal era, since Thirteenth/Fourteenth century. "a blazon is a formal description of a coat of arms, flag or similar emblem, from which the reader can reconstruct the appropriate image. The verb *to blazon* means to create such a description. The visual depiction of a coat of arms or flag has traditionally had considerable latitude in design, but a verbal blazon specifies the essentially distinctive elements. A coat of arms or flag is therefore primarily defined not by a picture but rather by the wording of its blazon (though in modern usage flags are often additionally and more precisely defined using geometrical specifications)" (https://en.wikipedia.org/wiki/Blazon, 30.4.2020, retrieved 5 February 2020).

[175]Cf. https://de.wikipedia.org/wiki/Ekphrasis, 18.1.2020, retrieved 20 February 2020.

> been described as "illustrative" or "representative" music [...] "Program music" narrates or paints, suggests or represents scenes or stories (and, by extension, events or characters) that enter the music from the composer's mind. Musical ekphrasis, by contrast, narrates or paints stories or scenes created by an artist *other* than the composer of the music, and in another artistic medium.[176]

The synaesthetic is complemented by the synergetic. Different arts interact, a back and forth almost like a ping-pong match, or a cascade of media changes and transformations.

> In this way, a painting may represent a sculpture, and vice versa; a poem portray a picture; a sculpture depict a heroine of a novel; in fact, given the right circumstances, any art may describe any other art, especially if a *rhetorical element*, standing for the sentiments of the artist when they created their work, is present.[177]

Of course, such translations are not without frictional losses. However, it is precisely the fuzziness during reflection that can lead to what we have already noted several times: an animation.

> A descriptive work of prose or poetry, a film, or even a photograph may thus highlight through its *rhetorical* vividness what is happening, or what is shown in, say, any of the visual arts, and in doing so, may *enhance the original art* and so *take on a life of its own* through its brilliant description.[178]

The word form of musical works as their translation into words is thus the opposite or counterpart to the translation of words or images into musical works. This is a matter of the program music mentioned, of music as description. The basic problem here is conversely the same as that of how and to what extent the corresponding word and picture forms are realized musically and are adequate to the ideas they stimulate or generate (See Chapter 4).

8.3.4 *Program as Ekphrasis: Charles Ives' Remarks on His Orchestral Pieces* Three Places in New England

Sometimes these opposites touch each other; namely when, in rather rare cases, the program for the work already describes it in such detail that it evokes an idea even before it is perceived, certainly a fuzzy one, but more than a vague one. This condition is fulfilled, for example, by the description that Charles Ives provided for his work. Ives composed *Three Places in New England* mainly between 1911 and 1914, but with sketches dating as far back as 1903 and last revisions made in 1929 (I *The "St. Gaudens" in Boston Common (Col. Shaw and his Colored Regiment)*, II *Putnam's Camp, Redding, Connecticut*, III *The Housatonic at Stockbridge*).

---

176 Bruhn 1999, p. 553f.

177 Cf. https://en.wikipedia.org/wiki/Ekphrasis, 26.2.2020, retrieved 28 February 2020.

178 Ibid.

The second piece, *Putnam's Camp, Redding, Connecticut* was created from two short theater orchestra pieces composed by Ives in 1903. These pieces, *"Country Band" March* and *Overture & March: "1776"*, were completed in 1904. Lyman Brewster, Ives' uncle, had asked him to compose the pieces for his play *Major John Andre* which was never performed due to Brewster's untimely death. In the early fall of 1912, Ives began tinkering with these compositions again.[179]

Ives wrote a program into the score, describing first the real historical place:

Near Redding Center, Conn., is a small park preserved as a Revolutionary Memorial; for here General Israel Putnam's soldiers had their winter quarters in 1778-1779. Low rows of stone camp fireplaces still remain to stir a child's imagination. The hardships which the soldiers endured, and the agitation, of a few hot-heads, to break camp and march to the Hartford Assembly for relief, is part of Redding history.[180]

Then the special story of the movement:

Once upon a '4 July,' some time ago, so the story goes, a child went here on a picnic, held under the auspices of the First Church and the village cornet band. Wandering away from the rest of the children past the camp ground into the woods, he hopes to catch a glimpse of some of the old soldiers. As he rests on the hillside of laurels and hickories the tunes of the band and the songs of the children grow fainter and fainter; – when-"mirabile dictu"– over the trees on the crest of the hill he sees a tall woman standing. She reminds him of a picture he has of the Goddess Liberty – but the face is sorrowful – she is pleading with the soldiers not to forget their "cause" and the great sacrifices they have made for it. But they march out of camp with fife and drum to a popular tune of the day. Suddenly, a new national note is heard. Putnam is coming over the hills from the center, – the soldiers turn back and cheer. – The little boy awakes, he hears the children's songs and runs down past the monument to "listen to the band" and join in the games and dances.[181]

Burkholder and Wild emphasize the multi-perspective nature of Ives' composing, which incorporates the viewpoint of the recipient.

Ives frequently removes his personal voice from the music, choosing instead to use a collage technique to "convey the sense of viewing both real and imagined events through the eyes and ears of a participant." (Burkholder 1995, p. 389)[182]

The numerous quotations of patriotic tunes are intended as a reminder and visualization of actions and actors and historical sites.

---

[179]https://en.wikipedia.org/wiki/Three_Places_in_New_England, 28.2.2020, retrieved 8 May 2020.

[180]Quoted in: Mortensen 2002. "Putnam's Camp, near Redding, Connecticut, was established as a historic landmark by the Connecticut legislature in 1887 and named in honor of the American Revolutionary War General Israel Putnam, who set up a camp in the area during the winter of 1778–79. The site has been preserved as a historic treasure because of Putnam's important role in the Revolutionary War, especially the Battle of Bunker Hill. Fourth of July celebrations are often held at the site due to its historic significance" (https://en.wikipedia.org/wiki/Three_Places_in_New_England, 28.2.2020, retrieved 8 May 2020).

[181]Wikipedia 2020, ibidem.

[182]Burkholder, quoted in Wild 2018, p. 19, and Wild.

> Eventual quotations of *Yankee Doodle*, *The British Grenadiers*, and, in the final measures, *The Star-Spangled Banner*, are indeed fun to play and hear. In addition to implementing all these borrowed tunes, the composer is also borrowing from himself, as much of the music resembles his 1903 composition *Country Band March*. Interestingly enough, amidst all the borrowed tunes, it is Ives' original march melody that seems catchiest of all. And by using so many song quotations in the first two movements of *Three Places in New England*, Ives' music reflects an operatic approach to populating a scene with characters – in his imagined historical settings, *his characters express themselves through song.*[183]

As another "American in Paris" with the first performance in Paris, 1931,

> Ives' music was not just interesting because it was composed by an American, it also fascinated them because the music really described America. Although the listeners didn't understand all the cultural references, Ives was calling attention to American ideals, issues, experiences and perspectives. The combination of such songs conjured up images of the fight for freedom in America. International recognition solidified the image of Ives as an American composer, especially strengthened by his use of borrowing from typically American-sounding pieces.[184]

The quotations are a pars pro toto, with ragtimes, plantation songs, songs of the American Civil War etc. signifying landscapes and land, history and patriotism, repression and struggle for freedom, and at the same time on a meta-level memories. Usually Ives does not quote the works, i.e. the songs, completely, but rather in fragments and in collages (see Sect. 8.2 *Quotation Form*). A little list:

I

Old Black Joe

Massa's in de Cold Ground

Marching Through Georgia

The Battle Cry of Freedom

Reveille

Deep River

II

The British Grenadiers

Marching Through Georgia

The Girl I Left Behind

Arkansas Traveler

Massa's in de Cold Ground

The Battle Cry of Freedom

Yankee Doodle

Columbia, Gem of the Ocean

Hail, Columbia

---

[183] Wild 2018, p. 16. Emphasis HWH.

[184] https://en.wikipedia.org/wiki/Three_Places_in_New_England, 28.2.2020, retrieved 8 May 2020.

Tramp, Tramp, Tramp[185]

In this way, Ives realizes what an at most accurate programmatic musical ekphrasis should achieve as a mimesis of reality: Perfected in the work as and in music, in word form prepared in his own preliminary remarks.

> Following a work of music entails [...] the same basic activity as following a story: the interpretation of a succession of events as a meaningful configuration. Musical plot characters (or "agents") are defined by the composer and identified by the listener through attributes that "may be located in various musical elements—for example, in instrumentation, tempo, texture, interval vocabulary, metric design, rhythmic motive or style, or harmonic support," and "these actors may then be understood as actors in the unfolding of a plausible series of actions and events" (Newcomb 1998). The difference between such musical agents and agents in representational painting, literature, or film is that these genres are forced to attach attributes to specific human figures and to flesh out these figures with many ancillary details. Music has no need to do so; just as it can present thematic generality (in what Newcomb calls "archetypal plots") without having to attach it to any specific situations or settings, so it can present shifting constellations of attributes to construct plausible agencies—among them human—left undefined. "The actions and events can [...] remain purely musical ones and still imply referential narrative patterns, stories, or plots" (Newcomb).[186]

8.3.5 Facilitating the Imagination Through Verbal References to Pictures and Known Works

It is all the easier to convey reasonably sharp ideas about the work if the music's reference to—known—images *and* the allusions to well-known music form starting points for the imagination.

> A Respighi triptych *Trittico Botticelliano, Il tramonto* and *Vetrate di Chiesa.* As the title suggests, the first work (written a year before the final instalment of the 'Rome' trio [*I Pini di Roma,* 1924]) is a three-part extravaganza in itself, drawing inspiration from three of Botticelli's most iconic paintings: *'La primavera'* (which also spawned Debussy's *Le printemps), 'L'adorazione dei Magi', and 'La nascita di Venere'.* The *score was new to all of us* in the editorial team, but such is the infectious energy of Respighi's opening depiction of spring (all glittering triangle, high woodwind and chirruping violins paying cheeky homage to Vivaldi) that snatches of the jaunty faux-medieval dance-tune that kicks it all off have been being whistled around the office all week. The orchestration is equally vivid in the chiaroscuro sound-world with which Respighi conjures up the mysteries of Epiphany, with the Liège woodwind on evocative form, whilst the sonorities which envelop Venus rising from the waves surely owe more than a little to Debussy's *La mer*. The engineering throughout all of this is quite superb, creating the impression of shepherds and troubadours wandering in and out of focus against a background of distant birdsong in the first 'painting' and replicating the ebb and flow of the surf in the last.[187]

---

[185] Ibid. Here, too, it would be worthwhile using Fuzzy Logic to specify the degrees of variation in their interweaving of fuzzing fragmentation of the original works and their recombination into a new whole in an original work.

[186] Bruhn 1999, p. 564.

[187] Katherine Cooper: Recording of the Week, Respighi from John Neschling and the Orchestre Philharmonique de Liège, 23.2.2018, retrieved 16 March 2019.

In an entanglement of work description and meta-level of a predictive description of perception, it says in addition:

> An educated audience listening to a performance of Respighi's *Trittico botticelliano*, either in the concert hall *without access to a program* that would give the movements' headings, or on the radio *without the benefit of some introductory remarks*, will smile with recognition at the opening sounds of the initial movement. The first sixteen measures constitute a (modified but immediately identifiable) quotation from one of the most memorable phrases in the first of Vivaldi's *Four Seasons* violin concertos, "Spring." [...] Respighi thus presents us with a multiple ekphrasis. By means of something like the musical equivalent to indirect speech, the composer draws our attention to the season of blossoming flowers (and love) as the suggested object of representation. Moreover, the extraordinary amount in this movement of what musicians refer to as "timbral coloring," achieved above all with the aid of tuned percussion instruments like glockenspiel, celesta, xylophone, etc., recommends the movement as one in which music "paints." Finally, the movement comprises two additional musical "quotations" (as Sternberg employs the term): an allusion - through characteristic rhythmic and melodic figures -to the (musical) genre of a "Renaissance dance," and a troubadour hymn to spring, the latter a signifier linking the time of blossoming even more unambiguously with idealized love. For the (admittedly rare) listener capable of identifying even these more esoteric mimetic inserts, the imagined listening experience, unencumbered by verbal information, might yield sufficient hints to identify the source text. As a painting in Renaissance style about spring and idealized love, Botticelli's "Spring" may well be the first that comes to mind. [...] (T)his compound of second-order mimeses thus serves as an intra-linguistic (i.e., musical within the language of music) marker alerting the audience to the (extra-musical) object of re-presentation, a specific work of visual art.[188]

As already explained with regard to perception, there are obviously many (literacy) prerequisites for the understanding and the realization. In case these are implied, on the basis of these two descriptions, the word form provides here a fuzzy, but nevertheless empirically and sensually enriched idea of the musical work meant. Yet of course a core aspect of music, the sonority is omitted here.[189]

8.3.6 *Fictional Musical Works in Word Form: Descriptions of Fictional Works with Reference to Real Ones in Thomas Mann's Novel* **Doktor Faustus** *and Elsewhere*

Parallels to the problem of verbalization of music can be found, for example, in psychoanalysis or literary studies. In principle, the forms of existence in the music process also exist elsewhere in the art process of the various arts. In dreams, nothing less than an entire cascade of specific forms of existence develops. Psychoanalysis assumes "at least three dream versions": dreamed, remembered, reported dream. A fourth form of existence brings further changes, namely when "a dream that has

[188] Bruhn 1999. Emphasis HWH. See also Bruhn 2000.

[189] Poetry can be „described as ‚dry music', with a definite tempo (meter) and phrasing. All it lacks is melody." Niki Kantzios 2018. If that is all....

already been reported is told again at a later time".[190] A sharpening is paradoxically done in a detour through fuzziness and unintentional and intentional subjective fuzzing. The word form generalizes, abstracts intensifies. As a rule, a complete description of a work is not even aimed at. Verbalization is content with outlines, contours, concise and characteristic details. The rest remains fuzzy—not only out of inability, but because music itself says something that cannot be fully translated. It is already a lot when essential features and the most important stages of the work process are described. In the best case, the verbal form evokes an idea of the work described. It will be clearer, more concrete, more detailed, and generally sharper if the reader is already familiar with the work. But even from an unknown work, style, idiom, form and progressions can be reconstructed mentally on the basis of the words, at least in outlines and broad strokes. Even more so in the case of perception, however, prerequisites play an important role here, not least literacy prerequisites, including knowledge of other works of the composer, knowledge of the context with regard to style, musical language, idiom, and additionally, on the other side of the language of words, denotations and connotations of the words used for the description. The result is usually, as mentioned above, vague to at best fuzzy, and moreover difficult to objectify for cross-checks on degrees of accuracy.

Like real musical works, fictional works can also become the object of literary reflection. In this respect they belong to the category of the 'word form'. However, these fictional works are usually collages or montages of already existing real works. In this respect, they are related to the quotation form within the word form (see Sect. 8.2; see above program music and 'Ekphrasis'—3.5 and 4. Similarity). In the modern novel, the work-describing ekphrasis is continued in particular for musical works, especially in artist novels. The word form of the—fictional—works acquires an independent aesthetic value. Examples include Romain Rolland's three-volume work about a fictional composer *Jean-Christophe* (1904/1912) (*Kinder- und Jugendjahre, In Paris, Am Ziel,* Marcel Proust's *À la recherche du temps perdu* (1913/1927), Hermann Hesse's *Das Glasperlenspiel* (The Glass Bead Game, 1931/1943), and Thomas Mann's equally fictional *Doctor Faustus. The life of the German composer Adrian Leverkühn, told by a friend (1943/1947).* In this novel Thomas Mann frequently describes works by Leverkühn.[191] He collages, assembles and synthesizes for it mostly different real works of different composers.

> *Meerleuchten* is described as resembling Debussy and Ravel—presumably, *La Mer* in particular. I imagine it sounding like Zemlinsky's *Seejungfrau*, Schreker's Chamber Symphony, or Schoenberg's *Pelleas und Melisande*, all of which show Debussy's influence. The young Leverkühn was also somewhat under the spell of Richard Strauss; in 1905, shortly before *Meerleuchten* was composed, he was present for the première of *Salome* in Dresden, and subsequently attended the work Austrian première's in 1906 — at least, claimed to do so, that trip having constituted a pretext to visit the syphilitic-Satanic prostitute Esmeralda.

[190] So Hau 2009, p. 375 und 276. See below 3.7 the absurd project of filming dreams by the actress H. Schygulla.

[191] From the hardly manageable abundance of literature on Thomas Mann and also specifically on his quotation techniques, we should at least mention: Bergsten 1963; Schlee 1981; Puschmann 1983; Wehrmann 1988; Osterkamp 2001.

> The settings of Verlaine bring to mind Debussy and perhaps also Stravinsky. The *Brentano Lieder* suggest Mahler, although the description of Leverkühn's "strict" method points toward Schoenberg. *Gesta Romanorum*, for marionette theatre, calls to mind Stravinsky's *Histoire du soldat* and *Petrushka*. The *Apocalipsis* makes one think variously of Schoenberg's *Moses und Aron* (the Dance Around the Golden Calf), Honegger's *Jeanne d'Arc* (the episode for shouting chorus), Stravinsky's *Oedipus Rex* [...]. The later chamber music is inspired by Schoenberg. The *Ensemblemusik* resembles the instrumentation of *Pierrot lunaire* and the Serenade; the Quartet suggests Schoenberg's Third and Fourth Quartets; the String Trio is modeled on Schoenberg's work in the same form, which he discussed with Mann in October 1946, while the novel was in progress. The *Lamentation of Dr. Faustus*, described as a revocation of Beethoven's Ninth Symphony, seems to draw variously on late Beethoven, Mahler, Schoenberg, and Berg. The progression from the sardonic "choral scherzo" of Faust's descent into hell to a desolate Adagio lamentoso echoes the progression from *Rondo Burlesque* to *Adagio* in Mahler's Ninth. The "lamentoso" marking points toward Tchaikovsky's *Pathétique*. The description of the ending follows Adorno's account of Berg's *Lyric Suite*: "One instrument after another falls silent. The viola alone remains, but it is not even allowed to expire, to die. It must play for ever; except that we can no longer hear it."[192]

Mann often basically used very vivid, musically and technically underpinned descriptions of the fictitious works that the music sociologist and philosopher Theodor W. Adorno (1903–1969) had worked out for him, supplemented by analyses of actual works.[193] Adorno uses real works as models, such as the violin concerto by his composition teacher Alban Berg for Leverkuh's Violin Concerto. Here, comparisons and analyses of fuzziness are possible to some extent, also with other descriptions of the works to which Adorno refers. However, he mixes this ekphrasis, which refers to a specific work, with descriptions of other works, so that by means of montage an imaginary-real musical work of art is created as a literary artefact. The musical work is as a fictional work the synthesizing quotation form and at the same time the 'fiction form' of a work or, as in most cases, of several works. The inversion in the work *Apocalipsis cum figuris* of Mann's novel hero Leverkühn[194] after Dürer's woodcut cycle does not fall under the concept of the picture form. It is program music, here specifically the musical form of a picture. A borderline case of the fiction form and a challenge for fuzzy logic is in the ping-pong of intertextualities a renewed inversion including media transformation, the *real* musical realization of *fictional* literary works. Konrad Boehmer, for example, composed *Apocalipsis cum figuris* for

[192] https://www.therestisnoise.com/2016/03/an-adrian-leverk%C3%Bchn-companion.html, 6.3.2016, retrieved 20 July 2020.

[193] Cf. e.g. Heister 2013.

[194] Cf. e.g. Ernst Osterkamp: *APOCALIPSIS CUM FIGURIS. KOMPOSITION ALS ERZÄHLUNG*, Vortrag am 20.11.1997 in der Seoul National University, https://www.google.com/url?sa=t&rct=j&q=&esrc=s&source=web&cd=&ved=2ahUKEwjglLPCh_bsAhV07eAKHeiED40QFjAAegQIARAC&url=http%3A%2F%2Fs-space.snu.ac.kr%2Fbitstream%2F10371%2F85338%2F1%2F2.%2520%25E3%2580%258C%2520%25ED%2598%2595%25EC%2583%2581%25EC%259C%25BC%25EB%25A1%259C%2520%25EB%25B3%25B8%2520%25EB%25AC%25B5%25EC%258B%259C%25EB%25A1%259D%2520%25E3%2580%258D%2520-%2520%25EC%259D%258C%25EC%2595%2585%25EC%259E%2591%25ED%2592%2588%25EC%259D%2584%2520%25EC%2586%258C%25EC%2584%25A4%25EB%25A1%259C%2520%25ED%2592%2580%25EC%2596%25B4%25EC%2593%25B0%25EA%25B8%25B0.pdf&usg=AOvVaw1v14UPp3ExPt8lXloIvAPn (sic!); see also e.g. Zywietz 2009.

4 percussionists, 2 pianos, 3 pop vocalists and 4 loudspeakers (1984) based on the fictional work of Adrian Leverkühn in Thomas Mann's Faustus. Boehmer follows Faustus/Levekühn/Adorno/Mann and claims a total *conversion into the opposite.* In the introduction to the score he writes:

> "Apocalipsis cum figuris" [...] turns all established esthetic norms on their head topsyturvy. Everything which was considered' beautiful' in traditional music stands in Leverkühn's work for the 'ugly', the 'banal', whereas all outbreaks of modern music – screaming, roaring, glissandi, sharp dissonants – stand for the 'beautiful'.[195]

Boehmer decided to put into music Thomas Mann's description of that non-existing work. Scoring, texts and sound material are a postmodern collage mix.

> Three pop singers, 2 pianos and 4 percussionist act against a wall of organized noise (a 4 track tape). The entire material on the tape consists of vocal sounds and quotations from texts by the 'apocaliptic' apostle John, Karl Marx, Marquis de Sade, Pierre Louys, Giacomo Leopardi and others. The three singers yell and scream as if they were the devils (which may not be missing in a good apocalypse!) After having produced isolated sounds, they gradually introduce more continuous textual passages, before finally singing a tasteless, Elton-John-like hymn in C major. These texts are based upon a verbal ebullition by the Russian composer Alexander Scriabin, who by time and while believed that he was God.[196]

Among others we find also Bengt Hambraeus, *Apocalipsis cum figuris secundum* Dürer (1987), or Hans Werner Henze based on Leverkühn's Violin Concerto with three movements referring to characters from the novel Violin Concerto No. 3: 1st Esmeralda, 2nd Echo, 3rd Rudi S. (1996/2002). A circle of existential forms and media transformations is thus closed: here again we are dealing with the *musical form of literature* or literary works, which are word forms of—only verbally existing—musical works.

8.4 Variation Through Media Transformation (II): Picture Form

The picture form and the score form are complementary to each other: this is a form of existence of the work *before* its productive realization as a sound form the picture form is one of the forms of existence following the sound form. Both are the visual appearance of the work intended as significant sound, a regulation there, here with the picture form a kind of postscript. Analogous to the order of precedence in the relationship between opera libretto and music, the prima la musica applies here, first the music, then the picture. The picture form manifests itself in two phenotypes: as still and moving pictures, i.e. on the one hand genres of the fine arts such as painting and graphic art, and rarely sculpture, on the other hand dance and film, both "motion pictures". In the following section, we will only deal with the forms of existence of

195 www.kboehmer.nl, retrieved 21 July 2020.

196 Ibid.

specific, individual works that may occur in the process of realizing a work. In order to illustrate the contrast, I will briefly sketch other types of music-picture reference. The question of fuzziness/sharpness in the relationship between picture form and musical work is difficult to answer. However, as soon as there do exist *several* picture form versions of *a* work, comparisons are also possible from the point of view of the Fuzzy Logic. In terms of its significance for the musical process as well as its socio-cultural significance, the picture form differs considerably from the word form.[197] Firstly, it hardly enters into the—usually incomplete—circular process of the forms of existence.[198] Introductions to works or performance reviews have a fairly regular effect on the receptive and productive realization, often even guiding the reception of the work, as explained before. This applies, to a rather lessened extent, even to the less publicly received sketch form, which has at least a stimulating effect. The picture form, however, remains outside. This has, secondly, art-ontological reasons: Works of visual art are material, not processual like music, and thus as the ekphrasis of a work necessarily less precise, rather vague than fuzzy—this will be explained in the following.

8.4.1 Demarcations: Pictures with Music, Pictures Based on and About Music, Picture Form of Works, Music Form of Pictures

The typology of the picture form is analogous to that of the word form. Thus the three main types can also be found in connection with the word form: description of musical facts of all kinds, adoption of principles of form and structure especially in lyric poetry (for example Kurt Schwitters' *Sonate in Urlauten*) as well as the reversal as a musical form of literary works (see Sect. 3.6).[199]

(1) There are countless pictures that have music as their subject and theme. A whole branch of research, the musical iconography, deals with this. It is mainly concerned with the subjects depicted, the contents of the pictures such as musical

[197]The desire for the removal of the limitations, for example in the total work of art, is just as urgent and as often expressed as its realization is problematic. So for example Schiller in continuation of Horace's "Ut pictura poesis erit", poetry should be like painting: "Music in its highest perfection must become form [...]; the fine arts in their highest perfection must become music [...]; poetry in its most perfect manifestation must, like the art of music, move us deeply, but at the same time, like sculpture, surround us with calm clarity. It is precisely this that shows the perfect style in all art, that it knows how to remove the specific limitations of the art without removing its specific advantages, and by a wise use of its originality gives it a more original character". (Friedrich Schiller: On the aesthetic education of man in a series of letters. Part IV, Letter XXII (1794). With a slightly different translation see e.g. Paul Halsali, transl.: https://sourcebooks.fordham.edu/mod/schiller-education.asp.

[198]I would like to thank Inge Rubbert for the inspiration to this thought (18.5.2020).

[199]Besides many publications on the subject of picture and music, Von Maur 1985 is still a standard work; see also e.g. Behne 1994; Böhme/Mehner 2000.

instruments, ensembles, opera and concert halls, audience, etc.[200] (2) Very numerous are also pictures that treat music as a stimulus and model for design.[201] In contrast to the first type, music is therefore not primarily content-wise the subject of representation in the wide sense, but is of interest primarily in terms of form and generality: with regard to formal aspects in the narrower sense, such as structures, blueprints and rhythms on the one hand and/or on the other hand with regard to affective analogies, impressions, moods and effects, which are conveyed particularly through colours.[202] A particularly popular model is the "Fugue", and thus of course J. S. Bach.[203] The polyphony as simultaneity of different things and the relative invariance of the fugue theme and the similarity of its motivic splits among themselves and with the theme (or themes or "subjects", soggetti) facilitates the transformation of contemporary art into spatial and planar art. However, this type is not concerned with the picture form of *individual* works.[204] In general, especially the abstract visual arts of the twentieth century refer to music, for example through the mediation of 'rhythm' or 'composition'.[205]

Paul Klee devoted himself very intensively to this reference of paintings to musical structures—Klee himself was musical on a professional level.[206] In Paul Klee's work, the mediation sometimes takes place through the inclusion of notes. These, in turn, represent a temporal progression in a spatial sequence, and numerous Bauhaus artists also worked in this direction. Wassily Kandinsky, who came into contact with Schönberg through the *Blauer Reiter* (artists' group and magazine) worked even more strongly with tendencies toward the total work of art.[207] The sculptor Anthony Caro attempted to come particularly close to the processual nature of music, emphasizing the multi-faceted nature of his sculptures and comparing the temporal development implied therein with musical composition:

> I have been trying [...] to eliminate references and make truly abstract sculpture, composing the parts of the pieces like notes in music. Just as a succession of these make up a melody or Sonata, so I take anonymous units and try to make them cohere in an open way into a sculptural whole. Like music, I would like my sculpture to be the expression of feeling in terms of the material, and like music, I don't want the entirety of the expression to be given all at once.[208]

(3) A special case, but an independent type, is the picture form of *fictional* works, which are treated by the title as if they were real musical works. This special type forms a transition from references to music in general and to individual works. As

[200] See e.g. Hammerstein 1984.

[201] With inspiring explanations e.g. Keiner 2014.

[202] This also includes the broad field of synaesthesia. Cf. e.g. Hutter 2006.

[203] See e.g. Würtenberger 1979; Bach 1985; Salmen 1986.

[204] Numerous examples can be found especially in Von Maur 1985.

[205] Sündermann/Ernst 1981.

[206] From the large number of publications on Klee, the following should be mentioned: Vitali et al. 1986; Klee-Zentrum (ed.) 2004; Kieser 2004.

[207] Cf. e.g. Kandinsky 1998.

[208] So Caro in Rubin 1975, p. 99, cited after Dittmann 1983, footnote 24, p. 278.

with the word form, there are also fictional musical works: the visualization of music or works that do not exist as real musical works. In the visual arts, a *new* music-related work of art is created without reference to an old, existing *musical* work whose picture form it would be. This includes, for example, the series *Broadway Boogie* (1942/1943) or *Victory Boogie* (1944, unfinished) by Piet Mondrian. For their part, the paintings are programmatic works with concrete subject references, whose titles sound as if they were actually composed boogie-woogies. Here the visual artist creates a new, fictional musical work of art in picture form, with artistic means, with rhythm, structures and forms and colors. The fiction induced—via the verbal title—*is* the work. Because there is no reference from work 2 to work 1 in the sense of mimesis or reflection, fuzzy logic has few possibilities here, for example to start with questions about similarities etc.

(4) Very rarely, however, is the picture form directly related to individual works. It is much more common in the verbalization of music, especially in non-fictional genres such as music criticism etc. (see Sect. 3.6).

(5) Considerably more often than the picture form of musical works is the reverse, namely the musical form of pictures.[209] Especially typical and well-known is Modest Mussorgsky's piano cycle Pictures of an Exhibition, later colourfully instrumented for orchestra by Maurice Ravel (see above program music 3.6; see Sect. 4.3).[210]

8.4.2 *Individual Works in Picture Form*

The core of the picture form is, as already mentioned, works of fine art that relate to a particular musical work of art. The point of reference can be directly the form of notation, the sound form—this is most common—or the phonographic form. The depiction of concrete musical works and thus of processual sequences with often greater temporal extension is, as mentioned, decidedly more difficult in the visual arts than ekphrasis with words. If a work is to be reconstructed in a reasonably sharp way on the basis of the picture form—and then again in terms of its verbal language—this can actually only be done through verbal mediation. For on the one hand—as already mentioned—the language of the visual arts, unlike the language of words, is *not* a general metalanguage. On the other hand, visual art is material. This forms a limit for the translation or reflection of musical works that are processual. Moreover, different arts have different abilities to represent stories, actions processes.

> Narrative structures are inherent in all the arts, but their potential and thus the possibilities of their reference in the visual arts, music and literature are categorized differently. Werner

---

[209]For example, currently as a combination of the musical form and picture form of a text in audio-visual form: "Among the interesting media changes in recent times are [...] new musical interpretations of picture book texts [such as] the picture book *Lola rast* [...]. An episode of this current 'Struwwelpetriade' was integrated into new musical contexts as rap in the animated short film *Die schöne Anna Lena* [...] in 2013. Oberhaus/Oetken 2017 quote p. 10.

[210]See e.g. Fink 1988.

> Wolf classifies the three arts on a scale between maximum and minimum work narrative, which is particularly interesting for the approach to synaesthetic interfaces. […] The fictional narrative is considered genuinely narrative, whereas the picture, especially the monophase single picture, does not act as a story-telling device (in the sense of linking events in a temporal sequence), but "at best can hint at stories by means of a plot phase" [Wolf 2002]. Wolf classifies the potential of music as "quasi-narrative", since music, especially instrumental music, cannot tell stories on its own terms in such a way that "they can be recounted or re-narrated in another medium and thereby become *recognizable*" [Wolf 2002].[211]

The argument of recognizability is convincing, even if in Franz Liszt's Faust Symphony, for example, the three characters Faust, Mephisto and Gretchen are characterized musically and thematically so vividly that someone with a certain amount of musical experience recognizes them immediately. Otherwise, a meta-linguistic verbal inscription is at least helpful if not necessary. The "narrative" abilities of music are considerable,[212] and certainly greater than those of the fine arts.

8.4.2.1 Structure and Sculpture: Heinrich Neugeboren's Representation of Bars 52–55 of J. S. Bach's Eb Minor Fugue, BWV 853 (*the Well-Tempered Clavier*, Part I)

Sculptures with reference to a single, specific musical work of art are obviously very rare. After a long search I have found only one.[213] Heinrich Neugeboren (Henri Nouveau), 1901–1959, also drew a graphic representation of the entire fugue in C major, WK I, No. 1 (1928/1944).[214] His sculpture of the three-part E flat minor/Dis minor fugue from WK I, on the other hand, which is particularly demanding in terms of compositional technique, represents only a part, but one that is significant for Neugeboren, in which J. S. Bach's fugue type is condensed. Both times, Neugeboren hopes that the transformation into the visual "will increase the possibility of understanding".[215] The concept of his *Sculptural Representation (Stereometric view)*—Design 1928, Reconstruction 1966/1967—Neugeboren regards as three-dimensional (Fig. 8.17).

> addition to the planimetric representation, whereby the apparent rise and fall of the voices was extended to an actual one.[216]

The code conversion is complicated

1. Horizontal: the constructive course

---

[211] Oberhaus/Oetken 2017. Quote p. 11. Emphasis HWH.

[212] Detailed arguments and evidence e.g. Bruhn 1999, p. 574. See above the commentaries to Ives, *Three Places in New England.*

[213] Jack Ox painted three triangular pictures of a work by J. S. Bach in such a way that the triangles can be combined to form a three-dimensional tetrahedron.

[214] Von Maur 1985, p. 36.

[215] Neugeboren 1929, quoted in Von Maur 1985, p. 36.

[216] Ibid.

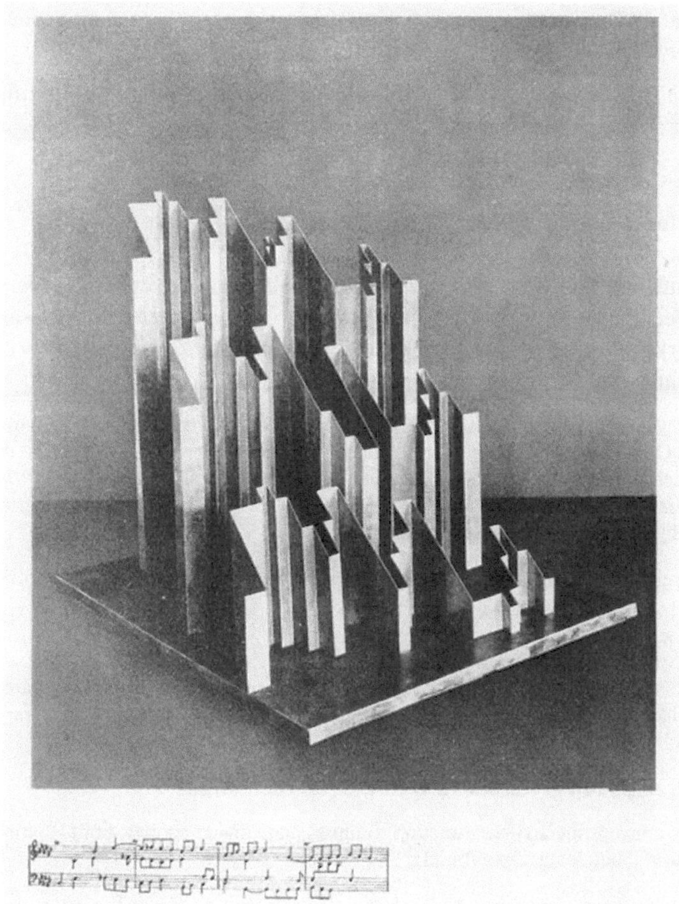

Fig. 8.17 Heinrich Neugeboren: fugue in C major, WTC I, no. 1
Quoted after von Maur 1985, p. 36, ill. 16

> 2. Vertical: the distance of each tone from a fundamental that is the same for all three voices [...]
> 3. from front to back, both at the base and increasing at an angle of 45 degrees: the distance between the voices.[217]

In the overall sense of the sculpture, Neugeboren thinks of a monument, not without German national overtones. Realized as sculpture

> these four bars, these signs of Bach's greatness of combination, which as music are unperformed, dead and performed transient, would be transformed into architecture (= frozen

[217]Ibid.

> music) and made tangible, an imperishable document of intangible and incomprehensible greatness of mind.[218]

Neugeboren's three-dimensional music translation is presentable as a sculpture, but less evident as the picture form of a work.

8.4.2.2 Max Beckmann's *Begin the Beguine*

Max Beckmann's painting *Begin the Beguine* (1946) is a veritable reversal of the methods used for by Neugeboren. The title is also concrete and refers to an existing musical work that is to be made famous. However, the painting is primarily concerned with the content of the music and less with its formal aspects (Fig. 8.18).

> After being persecuted by the Nazis for his "degenerate" artistic innovations, Beckmann fled his native Germany for the Netherlands. This canvas was painted there just before the artist finally received the traveling papers necessary for emigration to the United States "*Begin the Beguine*" is the title of a Cole Porter song that had become a big band standard after being recorded by jazz clarinetist Artie Shaw in 1938.

The picture shows at least three perspectives. On the one hand Beckmann is concerned with depicting the situation, with the "Sitz im Leben" of such songs, and specifically already as dance music.

> Characters typical of Beckmann's pictorial repertoire are present: the blond cigarette girl, the shackled woman in a negligee, and staring birds. the figures appear to be *attempting to dance but are unable.*[219]

Secondly, Beckmann gives his view in the representation.

> The scene conveys the artist's attraction to music and cabaret society, but also hints at an ambivalence about American culture.[220]

Beckmann's painting is thus a borderline case in that it is not his sole and perhaps not even his primary goal to translate the envisaged musical work into music. But, thirdly, he actually paints aspects of Cole Porter's work, both structures and moods:

> The quick, rhythmic disharmony of the song is *echoed* in Beckmann's strong colors and composition.[221]

8.4.2.3 Jack Ox: Alpin Landscape, St. Florian's Architecture and Measures 89–109 from Bruckner's, 8th Symphony's Score

Transitional forms of structural reproductions of music in general to concrete works are, for example, pictures based on works by J. S. Bach, which sometimes come

[218] Ibid.

[219] Ibidem. Emphasis HWH.

[220] Ibid.

[221] Ibid.

Fig. 8.18 Max Beckmann (1884–1950)
Begin the Beguine, 1946
Quoted after *I got rhythm. Kunst und Jazz von 1920 bis heute.* Redaktion JazzZeitung, 4. November 2015. https://i1.wp.com/www.jazzzeitung.de/cms/wp-content/uploads/2015/11/BECKMANN_Begin-the-Beguine_a06df4.jpg

close to graphic transcriptions of polyphonic lines (see below 'audio score').[222] In recent art, Jack Ox, among others, developed an innovative interpretation of musical works such as Anton Bruckner's 8th Symphony (1983/1991). Just like the abstract pictures without any special reference to musical works, Ox's illustrations are very

[222]Numerous illustrations in von Maur 1985.

complex and ornamentally condensed. She uses a very complex, multi-nested cascade of transformation and translation processes from the acoustic to the optical. First of all, she defines her conception of the picture form as "translation".

> Since 1976 my paintings have been formulated through a set of visual equivalents employed effectively to "translate" a composition written in the aural "language" of music into a visual "language". The musical vocabulary used by the composer is put into a new vocabulary, one that uses visual images instead of sounds. However, the original structure of the piece is retained along with the meaning.[223]

This translation transmits the sound and *meaning* of the work, structures and meanings. Here both score form and sound form are the starting point.

> Not only do I, as a painter, function in a way similar to an orchestra conductor [...]. My activity could be described as a re-orchestration, followed by a visual performance. The actual act of painting produces a work which expresses feeling– mine and the composer's –but is structured by a mathematical system. [...] I do not listen to the music and "paint my feelings". I work from the score, making a traditional musicological analysis. I give systematic consideration to the composition's thematic, melodic, harmonic, rhythmic, and dynamic elements. During this period of immersion and analysis I listen carefully to various performances of the music, noting the interpretations of different conductors.[224]

With a surprising twist, Ox calls her paintings an optical "re-orchestration" of the acoustic work.

> I do not paint music; I paint a composition which was written in the language of music. [...] My paintings are basically re-orchestrated versions of the original composition– re-orchestrated, that is, for the eyes. The first drawing make is called the "orchestration sketch". Rather than appoint different instruments to different lines, I assign them to different images and locations within the picture. This drawing, and all the others produced before the painting is painted, are my working scores. The finished painting is *a visual performance*. In effect, I play the part of conductor; I thereby make decisions based on my interpretation of the score.[225]

Ox, like Klee, also includes notes in her work interpretation. The picture is actually a sequence of pictures. It is based less on classical paintings than on comic strips with a series of individual frames, comparable to film—and a score, but without its vertical complexity.

The paintings are significantly wider than tall, like a broad strip. And the long rectangle is subdivided by many narrow vertical units. That gives the impression of a musical "score", being to "read" them from left to right following an imaginary time-line. Ox, however, interlaces the sequence of pictures and thus approximates them again to a uniform, elongated painting. Especially the dynamics and the richness of colors of Bruckner's music are reflected in a series of paintings. In doing so, she uses a whole series of paintings to depict the chronological sequence of the work, 13 in all. In outline, the stages of "re-orchestration" up to "visual performance":

---

[223] https://www.academia.edu/5406848/The_Systematic_Translation_of_Anton_Bruckners_Eighth_Symphony_into_a_Series_of_Thirteen_Paintings, retrieved 2 May 2020.

[224] Ibid.

[225] Ibid.

> The first drawing make is called the "orchestration sketch". Rather than appoint different instruments to different lines, I assign them to different images and locations within the picture. This drawing and all the others produced before the painting is painted, are my working scores. The finished painting is a visual performance. In effect interpretation of the score.[226]

Ox's coloring is semantic and motivated by the harmony of the work. In her "translation" she differentiates herself from traditional synaesthetic assignments of colors and tones or keys (as in A. Scriabin and others).

> Harmonic movement and qualities are portrayed through a mathematically determined glazing system. I have superimposed a circle of fifth intervals over a twelve-step color wheel. As Bruckner modulated through keys on the circle orchestral chord manifests itself with an appropriate percentage of complementary color layered in, making the original color greyer. Therefore, if a chord is very consonant [...] then it will be a pure hue from the color wheel. If, however, it is a dissonant as it can be within the same sensibility, then it will be 50% the color from the wheel and 50% its complement. As the relative dissonance decreased, so does the percentage of complementary color (40%, 30%, 20%, 10%). The value (relative lightness or darkness) of the glaze is also controlled mathematically in order to portray rising and falling pitch and dynamic levels.[227]

Primarily, Ox particularly translates the themes of the Bruckner symphony visually. The 13 paintings of the series include (Fig. 8.19)

> all of the Symphony's themes plus the last thirteen measures (in which many themes from all four movements are layered). [...] The earlier paintings are less complicated than the later ones. The entire series of paintings represents a journey of unfolding ideas as the various themes develop out of one another and reappear in later movements.[228]

The illustration hints at the complex projection and layering technique. It contains (1) score (2) contours, courses of melody (3) architecture, which is itself already stylized-abstract (4) stripes of landscape.[229]

> These thematic ideas are combined with a similar development of architectural motives taken from St. Florian's Monastery and romantic landscapes based on Alpine views. [...] These two basic visual elements [...] are the two basic stylistic elements in Bruckner's music. The architectural images from St. Florian's represent Bruckner's frequent use of Baroque counterpoint. Of course, views of St. Florian's are also appropriated here because of the ‚composer's close relationship with the monastery in life and in death. The Alpine landscapes embody the chromaticism in Bruckner's harmony and also the romanticism of his overall musical style.
>
> I paint these images onto fiberglass strips which are cut into widths determined by time values (durations) of notes. These strips are arranged on the wall, shifted up and down according to the rising and falling of melodic lines. When the strips are returned to their original rectangular formation, the painted images, now fractured, visually portray the various melodies. Often there will be a collaging together of two or more images.[230]

---

[226] http://www.jackox.net/pages/bruckner/BrucknerText.html2, retrieved 2 May 2020.

[227] Ibid.

[228] Ibid.

[229] On Ox in general cf. e.g. Fuhrmann 1994; von Maur 1985, p. 281 and *Visuelle Partituren*, p. 444; https://en.wikipedia.org/wiki/Jack_Ox, 24.4.2020, retrieved 30 April 2020.

[230] http://www.jackox.net/pages/bruckner/BrucknerText.html2, retrieved 2 May 2020.

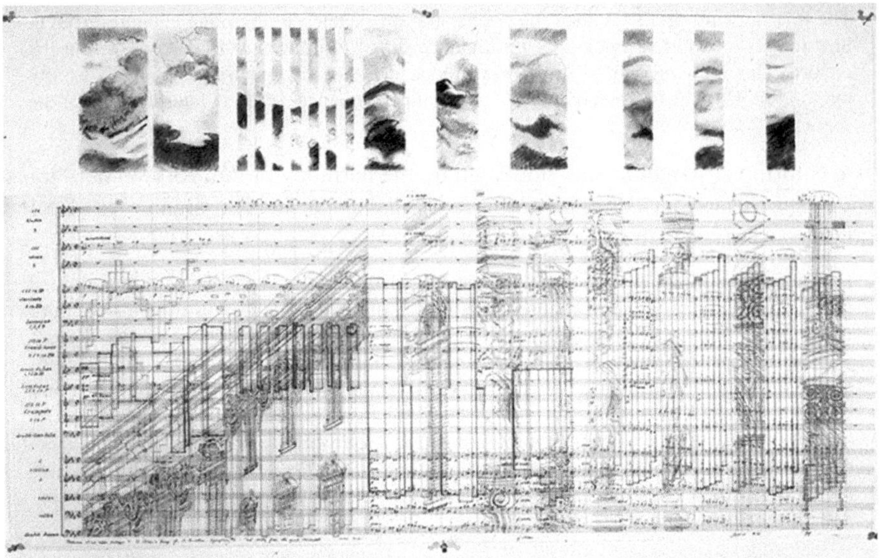

Fig. 8.19 Jack Ox: Bruckner's 8th Symphony, m. 89–109
Charcoal pencil on rag paper, electrostatic transfer on mylar and lacquer pens on clear mylar
Reduction of the delicate coloration with lilac and light orange to black and white only
https://intermediaprojects.org/pages/Bruckner.html

8.4.2.4 Jorinde Voigt's Graphic Cycle with the 32 Beethoven Piano Sonatas

Jorinde Voigt, born in 1977[231] has a professional musical education and has made music a subject and theme on several occasions. In general, she proceeds in an extremely reflective manner.

> For each theme Voigt establishes [...] a new, special algorithm and defines a structure according to which the work is created, but within which there is still room for creative, spontaneous decisions: "For example, when two points have a reference, but you don't want to resort to a convention, which would be the case, for example, if you simply draw a straight line. Instead, you say that every possible way of connecting is right, and that's part of the algorithm: to draw the line in some spontaneous way. The corresponding information is fully disclosed in order to ensure that the viewer can understand the creative process and the logic of the works.[232]

Music-theatrical picture forms: Epic-theatrical picture form in moritat, reified static stage design and processual scenic action.

> "It is very important to me that the process remains transparent. For example, by labeling everything what it's all about, and by making it very clear that it's a thought model and

[231] For the reference to J. Voigt in early May 2020 I thank Inge Rubbert.

[232] Andreas Pein: *Treffen mit Jorinde Voigt*, https://wissensarten.net/klangwelten/treffen-mit-jorinde-voigt/, retrieved 8 May 2020.

> not just an artistic drawing. I don't really have anything to do with drawings it's ultimately nothing more than an exalted kind of writing."[233]

So far, the peak of her graphic adaptation of musical works has been an entire cycle of works in the form of a picture cycle: All 32 piano sonatas by Ludwig van Beethoven in 32 drawings. Unlike J. Ox, she aims for a complete picture form of each work. She achieves this with radical condensations. Musical processuality, movement, development are transformed in several layers and condensed in one picture each. Because of the multitude and minute size of the characters, they can only be deciphered very vaguely as illustrations in the book. They need the real space of an exhibition or a museum. The format should be just sufficient for a detailed study of the pictures and a musical performance: Ink, pencil on paper, each 86.5 × 140 cm. J. Voigt was particularly interested in emotional values. Therefore, 'inscriptions', i.e. sentence titles, tempo and character information, performance names for the transformation came to the fore.

> The 32 part work titled „Ludwig van Beethoven, Sonata 1-32"was born from the thought of how it would be possible to make this music the theme without it becoming an illustration or interpretation. The biggest challenge form me was to develop a *notation* that would extract the emotional spectrum that is engrained in Beethoven's score. „Extract Intonation + Dynamic": For this I wrote down the intonation and dynamic notations from the original score for each sonata, the movements and bars, and translated them from Italian into english in order to translate the theme into an international language. [...] These extracted progressions are the basic structure of my work. [...] The composer Beethoven portrays an extreme emotional spectrum. Beethovens notations concerning the expression and attitude in his score are such a trail for me, a historic one, that can be reproduced in the present tense at any moment.[234]

In the picture production, reading the score and writing, listening and drawing are intertwined. Voigt thus refers to notation and sound form of the musical work.

> The individual „trail" of the intonations and dynamic notations is written along a freely drawn line with corresponding bars. This line is drawn intuitively while *listening* to the corresponding movement of the sonata and tries to summarize the characteristics of the melody and the progressions of the motifs in this line while also indicating the relationships of the different movements to each other.
>
> For each sonata Beethoven wrote between 2 and 4 movements. Consequently there are 2 to 4 of these „trails" on my drawings.[235]

Supplemented by further graphic and at the same time semantic components this results in a picture form which, taking into account the gap between the acoustic and the optical, is an independent reflection of the works. The dynamics and energy of Beethoven's composition become obvious (Fig. 8.20).

> Internal centres + external centres

---

[233] Ibid.—Further information: Kaiser 2013.

[234] Jorinde Voigt: *CONCEPT Ludwig van Beethoven Sonate 1 to 32,* [June 2012]. http://jorindevoigt.com/blog/concept-ludwig-van-beethoven-sonate-1-to-32-konzept-ludwig-van-beethoven-sonate-1-bis-32/, retrieved 8 May 2020.

[235] Ibid.

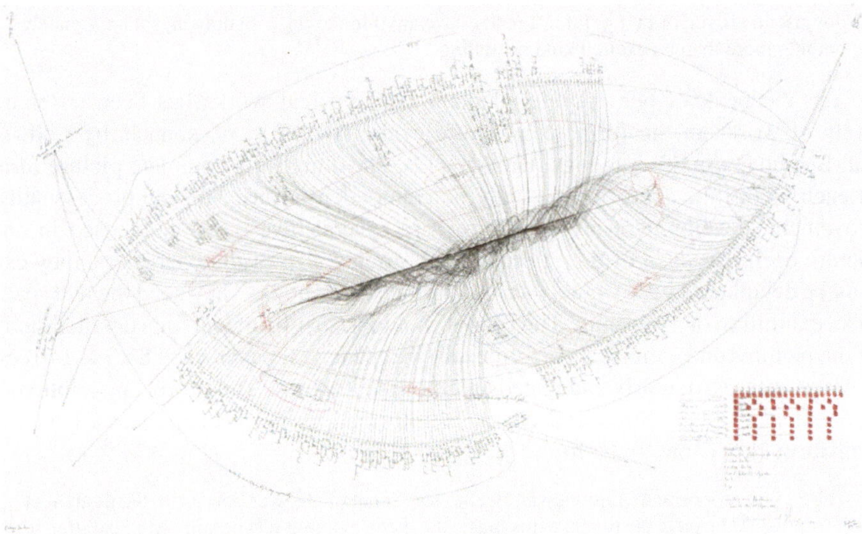

Fig. 8.20 Jorinde Voigt: Ludwig van Beethoven—Sonate Nr. 31 (op. 110)
Moderato Cantabile molto espressivo 3/4—116 Takte Allegro molto 2/4—158 Takte Adagio ma non troppo 4/4—213 Takte Adagio ma non troppo 12/16–20 Takte Fuga Allegro ma non troppo 6/8–87 Takte L'istesso tempo di Arioso 12/16–23 Takte L'istesso tempo della Fuga poi a poi di nuovo vivente 6/8–77 Takte Himmelsrichtung N-S; Rotationsrichtung/Rotationsgeschwindigkeit; 2 Interne Zentren; 4 Externe Zentren; Extract Intonation + Dynamic; Beat; Loop Jorinde Voigt Toronto 2012 86,5 × 140 cm Tinte, Bleistift auf Papier Unikat Signiert
http://jorindevoigt.com/blog/category/artwork/beethoven-sonate-1-32/

The space that is defined by this configuration is defined through 2 to 8 „internal centres" and 4

„external centres".

„Internal Centre" stands fort he „inner compass", „external centre" stands for any possible external

point of reference such as geographical or social and so on.

The „internal centres" are connected through an axis.

Every single mark on the „trail" is connected through a spontaneously drawn line with this axis.

Rotation direction/rotation velocity

The axis is describes as being in rotation so that it forms a so called vortex in which the entirety of all

intonations and dynamic changes are mirrored as an extract.

North South axis

The starting and finishing points of the „trails" as well as their internal centers are marked with a North

South direction, that point to the external centers. The individual strands of the lines which transfer the

individual elements of the trails into the extract are notated subject to their rotation.

The different dynamics of the bars of the individual movements are noted in the bottom part of the

drawing as synchronized trails. Musically speaking a meta-rhythm is thus constructed from this

structure that is unique to each sonata.

Loops are inserted per bar into these beat-trails in order to touch on the aspect of *infinite time.*

8.4.2.5 Music-Theatrical Picture Forms: Epic-Theatrical Picture Form in the Street Ballad, Reified Static Stage Design and Processual Scenic Action

A further expression of pictures not based on music but on music that borders on the picture form are the panels to the street ballad. They depict the action in outlines and, in a sense, strophically in individual stations. They are thus above all pictures of the key words and the material-content side of the "Gesamtkunstwerk" street ballad, not pictures of the music as a whole. Nevertheless, they belong in the context of the picture form.

Similar, but differently accentuated, are stage sets for works of music theatre. They are one of the optical components of how the work is staged and thus acoustically and optically realized. The stage design sketches, if necessary also quite naturalistically, the ambience of the dramatic-musical action, the historical and social space, its character and moods. Clearly a special case of the picture form of a work is the other optical component of the 'staging' with moving pictures, the scenic action of the performers. It borders again on dance, which also represents music (including text), in a more stylized way. There can even be a cycle of musical forms of pictures and picture forms of musical works. Mussorgsky composed his aforementioned piano cycle *Pictures at an Exhibition* based on pictures by his friend Hartmann, while Kandinsky in turn designed stage sets for a theatrical performance of Mussorgsky's work, i.e. its picture form.

8.4.2.6 Audio Scores: Rainer Wehinger—Ornette Colemans *Free Jazz*

Audio scores are a special type of visual form related to the notation form. They are subsequently produced scores or transcriptions without notes, which graphically visualize music and thus promote a sharpened, analytical perception of the works. Musical Graphics' made before, i.e. before the productive realization as sound form, are score equivalents or replacements with aspects of the conceptional form. Preforms of the audio score are compressed representations of musical analyses.

The following graphics, however, do not provide much more than outlines of form and tension (Fig. 8.21).

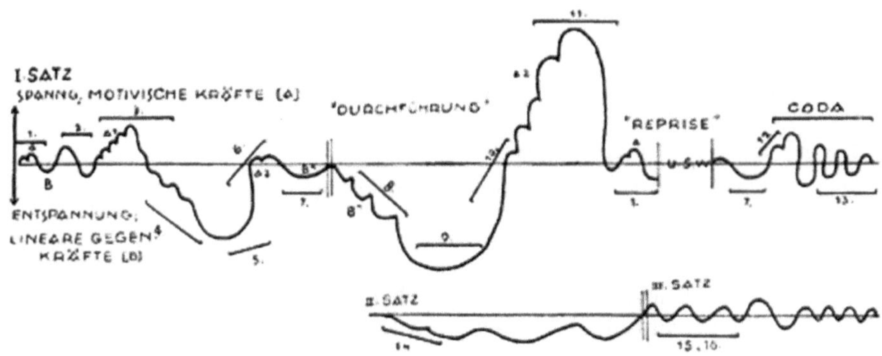

Fig. 8.21 Joseph Haydn: Piano Sonata in E flat major Hob. XVI:49
Quoted in Meischein 2019, p. 17 (Meischein 2019, p. 17—a quotation from Hans Mersman: *Versuch einer Phänomenologie der Musik.* In: Zeitschrift für Musikwissenschaft 5, 226–269. Further and continuative aspects in Soorani 2018)

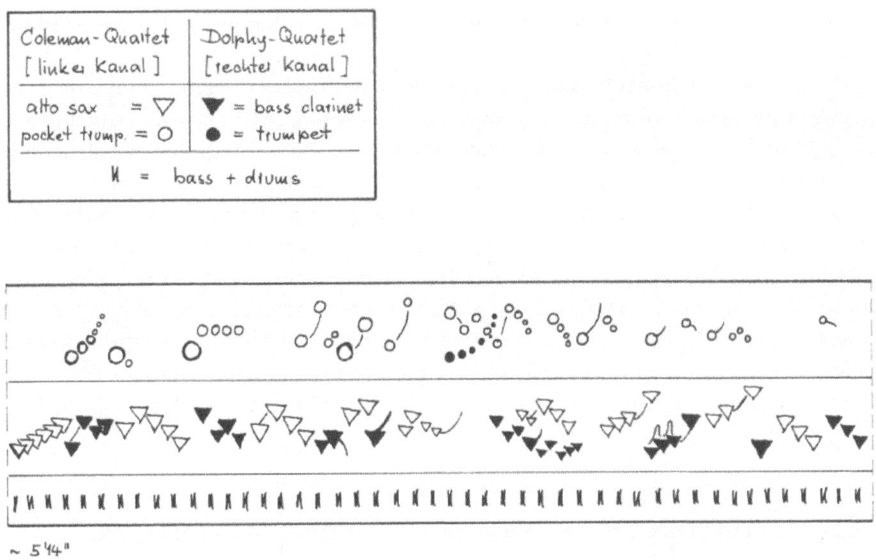

Fig. 8.22 Rainer Wehinger: Audio score, a segment of Ornette Coleman's *Free Jazz*
Quoted after Heister 1983, p. 106

Rainer Wehinger wrote a complete audio score for György Ligeti's electroacoustic work Artikulation (1958) in 1970, and an excerpt from Ornette Coleman's epochal work Free Jazz in 1983[236] (Fig. 8.22).

[236]For further examples see http://sts.hgb-leipzig.de/kompendium/text/78/6.htmltextbegin, retrieved 8 May 2020.

8.4.3 Title Pages of Sheet Music Editions, Sound Carrier Covers and Advertisements as Picture Form of Works

The picture form of a work also includes, in connection with the score form, the title pages of sheet music editions and, in connection with the sound form, illustrated advertisements, advertising in various media and, above all, record or MC, CD and DVD covers (The record offers the best visual possibilities because of its large format). Here, issues of fuzzy logic concerning the adequacy, similarity and appropriateness of image and sound are relevant—and have hardly been asked systematically so far.[237] Many copies of this type of picture form on the packaging of sound carriers refer directly to the stored work. This already begins with the Edison phonograph with its rollers. With the phonograph it was usually one piece, with the double-sided record with A and B sides it is already two pieces of music. Either way, the pictures, simply being pictures, can only represent a particularly significant aspect. In this, the cover converges with the other traditional line, the title page of sheet music editions of a work. This too can represent an aspect of the work notated here.

Parallel to the cylinders of Edison phonographs sheet music was distributed with

> beautiful cover pages, highlighting scenes from the songs, as *Cheyenne (Shy Ann)*.[238] Song, words by Harry Williams, music by Egbert Van Alstyne (Detroit: Jerome H. Remick, 1906).[239] Just this "Shy Ann" is shown riding on a horse and slightly smiling, but pursued by men who do not seem to have the best intentions; these may mean "Cheyenne", however, they are at the most only *"fuzzily"* identifiable as those wearing cowboy clothing and stetsons.[240]

The song had already become a hit as an interlude in a musical in 1905.

> Harry Williams and Egbert Van Alstyne's *Cheyenne* started the cowboy music craze in 1906. Written by the *Navajo* composers, *Cheyenne* also first gained fame on stage in the two-act music comedy *The Earl and the Girl*, which ran in England between December 1903 and September 1904 before adding new songs, including *Cheyenne*, in 1905.[241]

Both the play and song, on the other hand, are based on a novel. This is about a rescue story between cow-boy and cow-girl (Fig. 8.23).

> Although sometimes mistaken as part of the Indian song craze, *Cheyenne* does not include any reference to the Cheyenne tribe but is instead a love story between a cowboy and a cowgirl known as Shy Ann. In fact, the story line clearly recounts the meeting between the Virginian and Molly Stark Wood from Owen Wister's novel after her stagecoach got stuck midstream and the cowboy carried her out on his horse.[242]

---

[237] One of the exceptions is Lock/Murray 2009. Here it must be enough to point out this gap Lacuna in research.

[238] The girl's name sounds like a slightly degrading folk-etymological translation of the name of the Indian tribe.

[239] Ill. Amundson 2017, p. 78.

[240] Amundson 2017.

[241] Ibid.

[242] Ibid.

Fig. 8.23 *Cheyenne*
Words by Harry Williams, music by Egbert Van Alstyne
(Detroit: Jerome H. Remick, 1906). Sheet music
Quoted after Amundson 2017, p. 78

The publication in notation form in 1905 emphasizes the Western aspect of the song in the picture, along with a chase. This in turn, as far as can be seen, has nothing to do with the lyrics of the song. In other words: Every rotation of the kaleidoscope of the various forms of existence and especially of the picture results in new fuzziness. Often, as in this case, a sharpening does not even seem intended.

> The cover depicts neither the Wister story nor the song's lyrics. It has a bright red border around a western scene with the word "Cheyenne" in white letters crossing through the border and illustration at the top. [...] On the top right of the cover illustration are the words "Shy Ann" in parentheses and "song" below them. The illustration itself has a western setting with dirt, a little grass, and rocks in the foreground and distant mountains in the background.

> In the center a dark-haired cowgirl rides a mad, dashing horse whose eyes bulge out. The woman, presumably Shy Ann, is no eastern spinster but sits astride the horse, holding the reins in her left hand and a whip in her right. She wears a large cowboy hat, a red blouse with sleeves rolled up to her elbows, and a kerchief around her neck. A six-gun is holstered around the top of her blue knee-length skirt, and tall leather boots with fringes wrap her legs. Although the girl's face is forward, her eyes look back to a group of three mounted cowboys in hot pursuit, one with a similar whip held high in the air.[243]

What the lyrics tell is the violence that is essential to the Western. The song's lyrics begin with three "yips!" then go on, and complement the novel's rescue story with an abduction and forced marriage story.

> Way out in Old Wyoming long ago,
> Where coyotes lurk while night winds howl and blow,
> A cowboy's lusty voice rang out "Hello"
> And echoed thro' the valley down below
> Then came back a maiden's answer sweet and clear,
> Cowboy tossed his hat up in the air
> Said he "I've come to take you right away from here,
> Cheyenne, they say, is miles away but they've a preacher there."
> Then She just drooped her eye,
> She was so very shy,
> So shy, oh my, and then he made reply.
> Oh! Oh! Oh!
> *Chorus*
> Shy Ann, Shy Ann, hop on my pony,
> There's room here for two dear,
> But after the ceremony,
> We'll both ride back home dear, as one,
> On my pony from old Cheyenne.
> They rode that night and nearly half the day,
> Cheyenne was sixty seven miles away,
> But when at last they galloped up the street,
> The cowboy's pride was really hard to beat,
> On his arm his future bride a carrying,
> But beneath the little churches dome,
> Said she "I feel like turning back, not marrying,
> His face got red, and then he said,
> "You will or you'll walk home,"
> If you ride back today, you'll honor and obey,
> I do, I do, then he was heard to say, Oh! Oh! Oh!
> *Repeat Chorus*[244]

[243] Amundson 2017, p. 78.
[244] Amundson 2017, p. 79.

Among the at least 21 different recordings of the song, two of the famous Bill Murray from 1906, listed as a "cowboy song", stand out because of their musical reference to the Indian motif—they best represent the song.

> Although the sheet music calls for an *allegro moderato* tempo, roughly 116–120 beats per minute, Murray sings it a bit slower, more like 90 beats per minute. Prominent throughout is a tom-tom beating through both the verse and the chorus. This feature, most associated with Native American music, is perhaps why *Cheyenne* has been characterized as an Indian song.[245]

A second version supplements the "soundtrack" with naturalistic, sound-painting real tones, which are imitated instrumentally (see Sect. 4.2).

> Billy Murray made his more famous recording with Victor in July 1906. It also features a prominent tom-tom, a band accompaniment, and western sound effects. The record also has a faster pace at about 120 beats per minute, as the tempo markings indicate. Then, accompanying the opening notes, war whoops and yells are heard before Murray ever sings a note. As he sings the first line a shot rings out, and as he finishes the second line of the verse about howling winds a slide whistle sounds. The whoops and yells, as well as the tom-tom, return between the chorus and the second verse, which features the sound of galloping horses. The cries return after the second chorus as the band plays the melody for the last time.[246]

A sound carrier advertisement picks up the connection between cowboy and Indian girl in the picture.

> The trade journal *Talking Machine World* ran an advertisement for the American Record Company's release, made from the Victor pressing, with the picture of a cowboy and Native American woman (sitting sidesaddle) in a cameo illustration. The adjacent text simply says "Cheyenne" in large letters at the top and lists the record number and that it was sung by Billy Murray. […] A cowboy romance, depicting life on the Western Plains. Introducing the hoof-beats of the pony, Indian yells, and other effects.[247]

The word form as an advertising description or better naming of some characteristic features thus complements the picture form. The latter even extends to the shop window decoration, which is supposed to trigger "western" nostalgia:

> Finally, the July issue [of *Talking Machine World*], in an article on the value of window displays, wrote: "We will take, for example, 'Cheyenne,' the cowboy love song which is so popular just now. A window filled with the regalia of the western plains, saddles, spurs, rifles and revolvers, and if the Indian side is to be represented, Navajo blankets, bows and arrows arranged in an artistic manner will collect a crowd of the curious immediately."[248]

Among the numerous imitations of the song, one that is particularly weird stands out.

---

[245] Ibid., p. 80.

[246] Amundson 2017, p. 80.

[247] Ibid.

[248] Ibid., p. 81.

Like today's reality and talent show television programs, the success of *Cheyenne* spurred the creation of similar songs almost immediately. *Talking Machine World* lists a *Cheyenne March and Two-Step* [...] and [... a] *Cheyenne Medley.* On December 5, 1906, none other than Billy Murray recorded by far the strangest take-off on the hit cowboy song. Inspired by Upton Sinclair's popular muckraking novel *The Jungle* set in Chicago's meatpacking area, Murray sang *Cheyenne: Parody*, which tells the tale of a sick horse bound for the slaughterhouse. Murray, accompanied by orchestra but thankfully without sound effects, sings on Victor records:

Way out in old Wyoming lives a man,
Who owned a sick old pony named Shy Ann
This pony was so old it couldn't stand
And it had every sickness in the land.
But one day the beef trust sent its agent down
He paid thirty cents for old Shy Ann
Said he we'll ship this pony off to Packingtown
She may be sick
But she'll look slick
When packed into a can
We'll take poor old Shy Ann
We'll put her in a can
We'll can Shy Ann, she'll look like potted ham.
Oh, Oh, Oh,
Cheyenne, Cheyenne, you sick old pony
We'll take you,
And bake you
And make you into baloney
And the folks who eat you won't know
You're that pony from old Cheyenne.[249]

8.4.4 Moving Picture Form of Works I: Dance/Ballet

Dance as well as film, are processual by nature, cross the boundaries of the tangible of the optical-spatial visual arts. A ballet music with a choreography, in which the music was composed *on the basis* of an already existing one or *together* with it, does not initially fall under the concept of the 'picture form' of a work. This applies, for example, to the story ballets such as those in the period between Tchaikovsky/Petipa and Stravinsky/Fokin or Nijinsky. If, however, and as soon as the original already existing choreography is replaced by a new, work, this new 'setting to dance'[250]

[249] Amundson 2017, p. 81. "This version surely tested the old saying that imitation is the sincerest form of flattery" (ibid.). But the "imitation" here is a parody with sadistic traits. It adds another variation to the blurring fan.

[250] Term in analogy to the "dubbing" of texts.

or 'balletization' (Eberhard Rebling 1963)[251] becomes a picture form of the work. This often happens because the choreography can be seen as the equivalent of music theatre direction and production. The musical work then receives the same status as any other work that is danced.

The original choreography is incarnated in the musical work as a part of content and form. This also constitutes a component of judging whether the new choreography is appropriate to the work. In the case of several choreographies, the degree of adequacy can also be examined using fuzzy logic methods. Of greater complexity is the question of the appropriateness of the picture form in a *new ballet* in relation to any existing *old* works that are not intended as ballet music. For here the reference of the dance to the musical work is not intended from the outset, but is arbitrarily produced afterwards. Here, too, only the comparison between different settings to dance of the same work makes it possible to examine the fuzziness/sharpness. This type of ballet has been spreading since the 1930s with pioneers such as Frederick Ashton, Leonin Mjassin, Georges Balanchine, and in the 1960s especially by John Cranko and later John Neumeier.[252] From the aspect of the picture form, it is probably more appropriate to refrain from an action. As in the case of the static image, here too analogically exists a ballet-like picture form of works as sound carrier covers. On the Internet you can find lots of them e.g. for Igor Stravinsky's *Sacre du Printemps (The Rites of Spring)* from 1913, a key work also with regard to the later settings to dance.[253] To list them here and even to analyze them under the sign of Fuzzy Logic would take us too far afield. Even a cursory glance reveals that there is a relatively invariant, stable core around which variable versions with different types and degrees of fuzziness accumulate. However, the limit of the static image remains, since in a work of over 30' duration it can only be an extremely shortened pars pro toto.

8.4.5 Moving Picture Form of Works II: Translation of Musical Works in Sign Language

One type of the picture form of works to be mentioned is their translation into sign language, the "prose" equivalent of the stylized "poetry" of dance. There are already numerous examples of this, especially for simpler music. There are now several sign language interpreters who try to translate the acoustic musical language into the optical gesture language for deaf people during songs etc. This seems to be a promising trend. It is also more complex: The composer Helmut Oehring, child of

[251] See Rebling 1963.

[252] For details see among others Dahms et al. 2001, especially pp. 143–150. Numerous examples in Piper-Enzyklopädie Musiktheater.

[253] https://www.google.com/search?q=sacre+du+printemps+strawinsky+animated&client=firefox-b-d&sa=X&tbm=isch&source=iu&ictx=1&fir=zFBGtH6af26KEM%253A%252CsHgZoJ94dzZAlM%252C_&vet=1&usg=AI4_-kQ3X129JrTwy41Z3Cecimn9bFhhHQ&ved=2ahUKEwim3tL07bXpAhXB6eAKHSL3AFEQ9QEwC3oECAcQBw&biw=1267&bih=617imgrc=zFBGtH6af26KEM (sic), retrieved 15 May 2020.

deaf parents, twice as often translates his works and performances as sound plus sign language. The optical-acoustic linking, whether live or technically reproduced, makes it easier for hearing people to perceive the gestures as picture form. Conversely, the visibility of the music-making *also* makes it easier for deaf persons to associate sound and gesture to be imagined. There are also new technical devices, even for professional musicians—

> electronic music visualization in order to enhance the music listening experience for the deaf and hard of hearing. Richard Burn [...] at Birmingham City University is currently researching a device that displays detailed visual feedback from electronic instruments. These visuals will provide information on the specifics of what is being played, such as the pitch and the harmonics of the sound. This allows deaf musicians to better understand what notes they are playing [...] Researchers from the National University of Singapore have also created a device that will enhance musical experiences for the deaf. This technology combines a music display and *haptic chair* that integrates sound qualities from music into vibrations and visual images that correlate to the specific qualities found within the music. The visual display shows various shapes that change size, color, and brightness in correlation with the music. Combining this visual display with a haptic chair that vibrates along with the music gives a more all-around experience of music to those hard of hearing. Music visualization can also be used in education. The Cooper Union in NYC is using music visualization to teach deaf children about sound. They have developed an interactive light studio in the American Sign Language and English Lower School in NYC. This consists of an interactive wall display that shows digital output created by sound and music. Children can *trigger the playing of instruments with their movement*, and they can watch the visual feedback from this music. They are also able to view a "talking flower" wall, in which each flower can transform sound into light based on the specific frequencies of the sounds.[254]

8.4.6 Moving Picture Form of Works III: Film

The processual film as a medium of the picture form of a work develops in principle similar types as those that arise in static images. This is thus (1) music as object and theme, for example the filming of a musical performance in a feature film. (2) Music as a stimulus and model for design, for structures and/or moods often plays a considerable role, for example in assembly processes, since film, like music, takes place in time. The structural accentuation appears relatively independently in experiments, especially in the 1920s, in the "Abstract Film" (Fischinger, Ruttmann etc.). (3) In contrast to the visual arts or literature, it is impossible to create a cinematic picture form of *fictional* works that are treated by the title as if they were real musical works. This is because, in principle, film, as a reproduction of visible reality, is dependent on what occurs in it.

(4) As a picture form of individual works, film is, since it is processual, much more suitable than a static picture, whether painting or drawing.

Film as a medium of technical re/production can also function as a meta-medium, which in turn 'represents' other media. As a sound film, it can then even take the

---

[254]https://en.wikipedia.org/wiki/Music_visualization, 14.6.2020, retrieved 17 June 2020. To the here unmentioned problems of the precise correlation of music and pictures see Sects. 6.8.3 and 6.8.4.3.

sound form itself into its work process and show performances, whether real or fictional, incorporated into the film plot, as a whole, acoustically and visually (see Sect. 3.5 'Audiovisual form' as a special form of the phonographic form). The filming of a work performance, i.e. the audio-visual variant of the sound form, is a special and transitional case of the picture form. With a good choice of camera perspective, zoom, etc. on the leading voices of an orchestra, and even more so with the depiction of facial expressions and gestures of singers, much of the form and content of the work can be conveyed. Here, the analysis of *different* audio-visual forms of a work is immediately a possible subject of Fuzzy Logic.

On the other hand, *no* meaningful and adequate picture form of a work results if the sound form is accompanied by any kind of pictures of anything—popular are landscapes, animals and plants. This takes up with the everyday perception of music: In this perception, if listening is not music-centred and not attentive, pictures are often associated, preferably with personal memories and the like. This illustration follows the subjectivity logic of the director, not a musical logic. Fuzzy Logic could easily come into play here. With well-known musical works there are often several versions of these illustrations. Fuzzy Logic could examine the types and degrees of inadequacy and fuzziness.

8.4.6.1 Debussy's *Afternoon of a Faun*

In addition to the filming of a performance of a work, the animated film is well suited as a picture form of a work. Here, on the one hand, the experimental "abstract film" offers some Materials and methods of realization, while on the other hand, new digital programs often produce remarkable results. A pioneer was the Walt Disney Company in 1940 with the rightly famous episodic film *Fantasia*, an illustration of prominent classical music works.[255] The film is technically advanced, aesthetically, however, mostly retarded, because it did exactly the wrong thing, namely the illustration of works with more or less arbitrary images,[256] with a climax like Mickey Mouse as a sorcerer's apprentice in Paul Dukas' work of the same name after Goethe's ballad.[257] Of course, the remake should not be missing, 2000, still with the same restricted aesthetic. If it really aims at representing the work in picture form, the work-based animated film, like the abstract film, operates with graphic figures, predominantly

[255]Cf. e.g. *FANTASIA! John Aldred Tells the Story of Disney's Masterpiece*, https://web.archive.org/web/20110818032035/, http://www.amps.net/newsletters/issue15/15_fanta.htm, retrieved 1 November 2013; detailed with critical discussion https://en.wikipedia.org/wiki/Fantasia_(1940_film), 25.3.2019, retrieved 27 March 2019; on media association and marketing https://en.wikipedia.org/wiki/Fantasia_(franchise), 31.12.18, retrieved 27 March 2019; in addition https://de.wikipedia.org/wiki/Fantasia, 20.3.2019, retrieved 27 March 2019, More in relevant literature on film history and film aesthetics.

[256]For example in Fantasia 2000 the use of Ottorino Respighi's *I Pini Della* Via *Appi*a as background music to flying Wales. See above chapter 4.

[257]After all, we could say, people listen to musical works that they would otherwise never perceive without a picture.

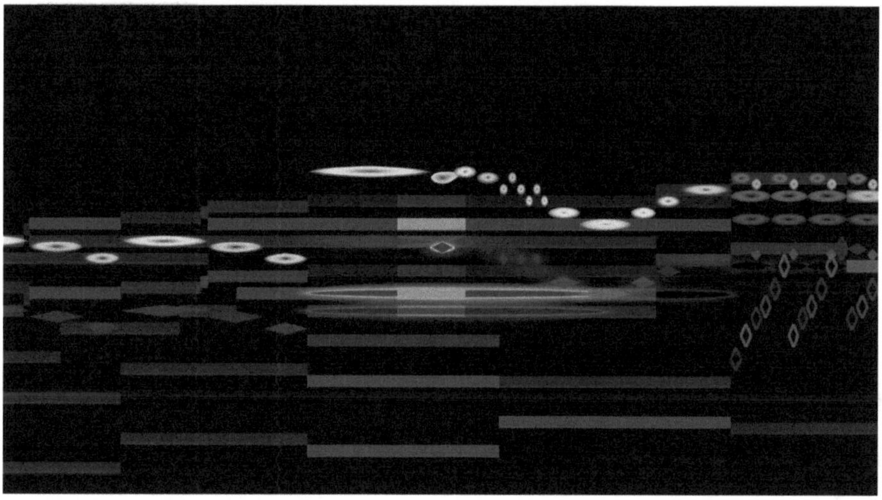

Fig. 8.24 Claude Debussy: *Prelude to 'The Afternoon of a Faun'*
There are three versions by this animator. The simplest is probably the best https://www.youtube.com/watch?v=Oi226-KTt5s, upload 4 August 2016 (Q: Why did you do multiple versions of this piece? A: I did the first version [...] http://www.youtube.com/watch?v=-YazhxBA7oo ... in 2012, just a year or two after I'd started developing renderers [...] http://www.musanim.com/renderers/ [...] http://www.youtube.com/watch?v=IJ4qJ8mW6-o, Between 2012 and 2016, my palette of visual effects expanded [...] I posted the second version as soon as I completed it [...] http://www.youtube.com/watch?v=S0A48ZXk6ww ... but subsequently discovered that the color calibration of my computer was very badly adjusted, and the Voronoi tessellation I used for the background [...] http://www.musanim.com/Voronoi/ ... was much too prominent; so I made a third version [...] https://www.youtube.com/watch?v=Oi226-KTt5s. For similar, but different concepts see e.g. Harmonic Coloring, A method for assigning colors to musical pitches offers an impressive and in principle convincing approach [http://www.musanim.com/HarmonicColoring/])

geometric signs. Basically, it is a new edition of the audio score in digitalized audio-visual form.[258] In contrast to the abstract film, the natural starting point here is the color film. It raises the possibilities to at least two directions. On the one hand, sound colors can be visually transformed, on the other hand, the "chromaticism", the division of the octave into 12 semitones, can be precisely and in detail. The melody then becomes not only a line, but a colored line (Fig. 8.24).

> T)he flute (which begins the piece) is shown in "walking ellipses," and the brass is shown with an "ellipse aura." In the first version of the video [...], the other shapes mean this:
>
> rectangle: strings
>
> rhombus (filled): woodwind
>
> rhombus (hollow): pizzicato strings
>
> In the second and third versions of the video, the other shapes mean this:
>
> octagon: clarinet

[258] A turn further in the kaleidoscope of media transformations, we see the graphic notation of films, their 'paper' or image form. See in detail e.g. Wurm 2010.

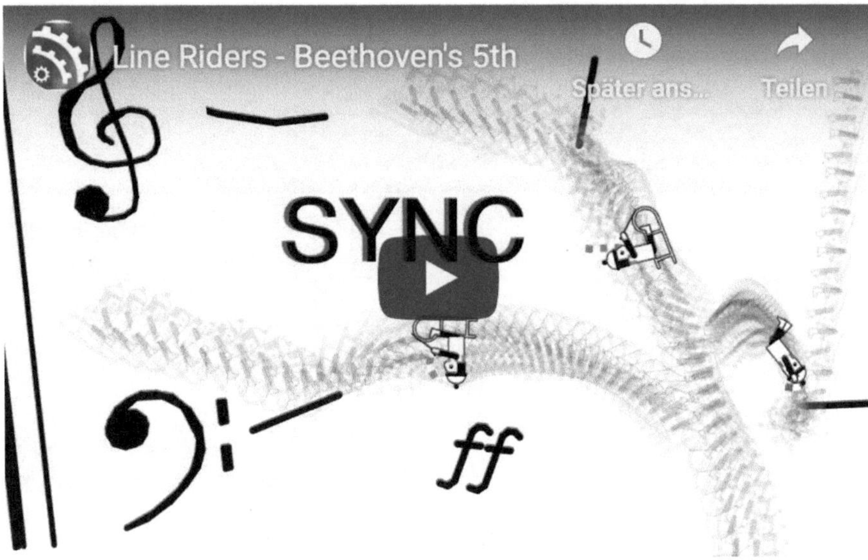

Fig. 8.25 Beethoven: Symphony No. 5, first movement, beginning https://thekidshouldseethis.com/post/line-riders-beethovens-5th-doodlechaos

> rhombus: strings (hollow for pizzicato)
>
> inverted ellipse: oboe, bassoon
>
> circle (hollow): harp
>
> triangle (hollow): crotales
>
> In all versions, the notes flicker when the strings are playing tremolo.
>
> […] The colors are assigned by a system called "harmonic coloring" in which the twelve pitch classes, C, C-sharp, D, etc., arranged on the "circle of fifths," are mapped to twelve hues, arranged on the "color wheel." This allows you to see changes in harmony and tonality.[259]

8.4.6.2 Beethoven's 5th Symphony, First Movement as Animated Film

A structurally quite passable filmic approach to the first movement of Beethoven's 5th Symphony—which is admittedly only offered as a shortened appetizer—is presented by an animation that is based on a game. The pretty playful lightness is of course not quite appropriate to the objective weight of the work. The concrete figures make it seem funny (Fig. 8.25).

The short film contains only a rather shortened version of the movement. It is the music-related application of an internet game,

> originally created in September 2006 by Boštjan Čadež (also known as "fšk"), a Slovenian student.[260]

[259]http://www.musanim.com/HarmonicColoring/.

[260]https://en.wikipedia.org/wiki/Line_Rider, 10.5.2020, retrieved 15 May 2020.

The polylines and figures are suggestively guided by the flow of the music.

> **These bodies sled, skid, and sail across a *choreography* of lines, coming from nowhere. [...] Any time a character leaves the screen they are being directed by unseen lines to the next location.**[261]

Just as Disney transforms the highly differentiated music of the world into his own, reduced world of pictures with a restrictive code of pictorial idiom, so too, of course, do the procedures of the new digital level tend towards a certain uniformity that corresponds only roughly to the diversity of the works. Admittedly, appealing results are not excluded in both cases. Since the picture form here is part of the audio-visual form, it can have an effect on the further reception of a work, in that the sequences of pictures stimulate either rather arbitrary pictorial associations or analytically appropriate ideas about the work.

8.5 Variations Through Receptive Realization: Imagination Form

The imagination form is the form of existence with the widest gap between the objective and the subjective. For its objective basis is already derived from the word form and the picture form and thus already tends to be rather fuzzy. The subjective effort of sharpening—the goal is to create as precise imaginations as possible—therefore quickly encounters its limits (see Sect. 3.8). In contrast to the tangible word form and picture form, the imagination form is purely subjective. It is, like the perception Form and recollection form, a mental, inner image of the work. Viewed in this light, these could also be regarded as manifestations of the picture form. But in this case, in contrast to these two forms of existence, we are dealing with processes independent of the sound form. As with verbal and picture form, fuzziness is the highest goal to be achieved here. Only secondarily can or could the imagination form become more objective, for example by formulating the ideas of a work and communicating them to others. This can be done acoustically or optically—but in this way it regains the character and status of verbal and picture form. The object and starting point of the imagination form are essentially these two forms of existence. A third possibility is that the imagination form is formed on the basis of the work in notational form. However, this requires a considerable amount of professional training.

Thus, in the imagination form quite a bit of what has been developed so far converges categorically. I can only outline what the most important determinations

[261] https://thekidshouldseethis.com/post/line-riders-beethovens-5th-doodlechaos, retrieved 15.05.2020. In spite of digital technology, the implementation requires a high degree of quasi-manual work, something that is often cut back on—at the expense of the integrity of the musical work: „This is not the full length of the piece. I spliced together parts to make it possible for me to finish the project in a practical amount of time. I don't think my sanity could handle 3 more months of this song" (ibid.).

of this form of existence are therefore I will mainly stick to the field of theory. Methodically and empirically this would need to be continued by psychophysical investigations, not least by brain research. Activity patterns etc. are a reflection of just this form. Research here is still far from coming close to a plausible result. But it is on its way. Fuzzy Logic in the embedding developed here could help to develop and specify questions, interpret and sharpen results. There are essentially two different, even contradictory types of the imagination form. On the one hand it refers to concrete, designed works, whether fictional or real. On the other hand, it refers to works that are not concrete, not designed, but merely available as concepts.

8.5.1 Mental Reconstruction of the Realized Work from Fragmentary and Vague Information—Optimum Fuzziness

Even those who know little about music know how to form an idea from pictorial descriptions of a work or from verbal ones. The latter, the verbal or report form, occurs more frequently and is in principle more precise. It can be transmitted orally or in writing. If this communication is already fuzzy, then the fuzziness increases to the point of vagueness, first and foremost because *the imagination form* is not objectified-exteriorized and not objective and it cannot be. The attainable optimum here is virtually fuzziness. This form of existence is the subject and perception aspect of the object-like picture and word form. This also includes that which readers read as constructs from the work descriptions mentioned, for example Thomas Mann, i.e. from the verbal form of fictional works or gather from pictures that represent a musical work (see Sect. 3.6).

From and in the imagination the work results as an imaginary-real object: imaginary, as such not material, real, as represented and objectified in the brain. In the imagination form, the imagination itself is a combination of expectation and paradoxical anticipation ex post.[262]

8.5.2 Mental Construction of a Non-real Work—Imaginations of "Conceptual Art"

In the case of conceptual art, the result is at best a minimal work. The concept *is* already the work. This non-real work means a deliberate getting stuck in the mere approach to art between escaping from the real production of art and anti-art. Here, then, the imagination form gets stuck at the level of conception and anticipation form, that is, *before* the crystallization and sharpening of the work. In contrast to

[262]See extensively e.g. Jeong 2012; cf. Williams 1995.

conceptual art, the conceptual form, the work as idea and concept, is a temporary form of existence of the work, which is further developed and sharpened (see Chapter 6). From the concept art concept, on the other hand, no "in itself complete whole" as a work of art emerges. It remains a conceptual form without the realization of the concept or even the will to do so. The composers do not concretize the idea verbally, in notating as score or in pictures. Since there is no sensual-material realization, it is of secondary importance or even irrelevant and a question of convention and institution to which art form conceptual art is assigned. If conception or the immobilized form of conception as such is declared art, there is no tension between idea and realization on the object level, therefore there is neither blurring nor sharpening—it simply evades this. The idea itself *is* the realization. This is the opposite of Hegel's concept of art as the "sensual appearance of the idea". Following the pattern of Andersen's fairy tale *The Emperor's New Clothes*, "Invisible Art" and "Inaudible Music"[263] are then the goal. If the object and design are omitted, the mere thought alone seems to be a work of art.[264] blurring would be only rudimentary and only experimentally, possibly, verifiable in the relationship of the receiving subjects to the conceptual art object.

> I wanted to get away from the physical aspect of painting. I was more interested in recreating ideas in painting.[265]

But Duchamps still worked with a "visual product". It was the context that created the art and then the—initially concealed—authorship of the artist. The object was the concept, in a certain reversal of the later usual replacement of the object or product by the concept. The product was not one's own, but an externally produced one, in this respect an art object, not an art work.

> Duchamp places the urinal on a pedestal, thereby monumentalizing and museumizing it. He also tilts his Fontaine by 90°. In so doing, he applies a kind of contrapuntal gimmick, which is then only made compositionally productive when serialized.[266] So he is doing not nothing at all. The poiesis is not entirely eliminated, though. He does not leave the real thing as it was but rather firstly eliminates its utilitarian function; secondly, with his signature as "Mutt"[267] he transforms it into a work, albeit a simulated one, and thus defines art by virtue

---

[263] According to Nolte 1991, among others p. 199.

[264] Cf. Nolte 1991, passim.

[265] Duchamp 1946, quoted after Moure 2009, p. 116.

[266] More on rotation and mirroring processes such as this in Springer 2004. For row rotation, see e.g. Eimert 1964.

[267] See Tomkins 1999 (1996). The pseudonym is a type of portmanteau: Duchamp condenses company name 'J.L. Mott Iron,' where he purchased the urinal (214), and amalgamates it with the name from the comic strip Mutt and Jeff; the R. stood for Richard, which in French slang meant 'moneybags.' In the discussion at the time, the object was often referred to euphemistically-obfuscatingly as "bathroom accessory" (217) or even "drinking fountain" (219), possibly precisely because of the title "Fountain." What Tomkins characterized as "bigotry" was also still relevant later. "Psycho (1960) broke numerous taboos. Yet this is not directly about the depiction of violence [...]. Psycho (is) the first American film in which the flush of a toilet could be seen and heard—which at the time was enough to make censors blush" (*7 Filme, die Tabus brachen.* https://wieistderfilm.de/7-filme-die-tabus-brachen/).

of his existence as an artist; and thirdly he transposes this work into the artistic context of an exhibit.[268]

A minimum of poiesis is thus essential for simulating art, although mimesis is missing.

The idea with the urinal was good. But the countless repetitions with variations not of the ideas but only of the objects, also by Duchamps, devaluate them continuously. He himself attached importance to the fact that his "readymade" was not a special "objet trouvé", but an industrial, technically (re-)produced object; he was keen to have this potential "multiple" reproduced as replicas. In contrast to the elegant urinal, his bottle dryer is simply ugly. And Warhol's "Brillo" repetitions then merely reproduce advertising as if it were art. Despite his "ideas", Duchamps himself continued to create self-made visuals, such as installations like the Large Glass, etc.[269]

But the concept "idea" instead of art was the most often imitated, not least by John Cage, who for his part made a career in music culture and had and still has many successors. It saves the artists the trouble of the work involved and the production and is therefore understandably popular with some of them. Incidentally, the concept is also a new edition of the Bohème, for whom "life" as an artist was and still is far more important than their activity as producing artists. Productive artists or artists doing their job creating art.

In order not to let the conceptual form exist as a mere idea only in the composer's head, it must be objectified somewhat. Even conceptual artists must at least write down the concepts that they do not execute. This is usually done verbally or figuratively, rarely with notes, and if so, then only with isolated, unconnected notes; for if performed, these would again result in a work that is not merely conceptual. We could therefore speak of deficient verbal, pictorial and notational forms. The work process is reduced, and the stage of the later form of existence, the imaginary form, is short-circuited with the early form of conception. The dissipation and elimination of the representational object and representational material is not compensated by great ideas. Rather, a vanishing of artistic ideas tends to come along with it. The conceptions that do without sensual realization are usually so poor that their "sensual appearance" would hardly make the "idea" any better. In the following conceptual art piece, as in general, the point is to keep the artists' workload as low as possible. *In* the "work", i.e. the verbal text of the quasi-program, a classical woman-mother role is more than obvious: she serves the husband-son. The pianist himself does, strictly speaking, nothing (Fig. 8.26).

Mom applauds him, pushes the pedal for him, actually does humiliating verbal actions like saying a word 134 times (like multiple writing of a sentence once a popular punishment at school), and as the culmination of conformism she prays for him. This piece of conceptual art contains as a verbal form sentences that imply instructions, rules and commands. As absurd as it would be: In principle, this concept could even be realized as mute, pantomimic music, as music without sound—an old,

[268]For more on the major controversies surrounding the exhibition in New York in 1917: Tomkins 1996, p. 212ff.

[269]See Tomkins 1999 (1996).

Mother may do her makeup.	Mother may tidy her hair.	Mother may doze off.
Mother may applaud lightly.	Mother may yawn audibly.	Mother may scream.
Mother may rap Pianist's knuckles.	Mother may give Pianist a dirty look.	Mother may read Harper's Bazaar.
Mother may tap foot.	Mother may scratch.	Mother may nod to Pianist.
Mother may nod to audience.	Mother may purse lips.	Mother may look disgusted.
Mother may beam.	Mother may hold pedal down for Pianist.	Mother may adjust seat for Pianist.
Mother may hum or sing quietly to herself.	Mother may become desperate.	Mother may make a scene.
Mother may be above it all.	Mother may say "tisk" very rapidly 134 times.	Mother may pray for Pianist.

Seite 3 aus „Composition for Pianist and Mother“
Copyright 1964 Alvin Lucier, Middletown, Connecticut

Fig. 8.26 Alvin Lucier: *Composition for Pianist and Mother*, 1964
Quoted after Ronald J. Kuivila: Erkenntnis, Intuition und Poesie. Bilder und Aktionen in Luciers Musik. In: MusikTexte 156, February 2018, p. 59

once avant-garde concept that has lost some of its freshness and novelty value after a century due to countless repetitions.[270] For example with Peter Ablinger: *KLÄNGE AUF PAPIER* (*Atelierstudien* 1999), "different materials and objects put on paper, different versions and formats".[271]

> 'Klänge auf Papier', when translated into English, is reminiscent of a piece by Alvin Lucier, "Sounds on Paper", [...is] 'music for untrained score readers' [...]. Here it is not notes that are left on paper, but objects. Objects that are supposed to stimulate the reader to read them like a score. *Imagine*, for example, the sounds and noises that these objects could trigger or contain.[272]

Ablinger even introduces the term that is decisive in his "Idea of a 'music as imagination'".[273] Objects, things replace signs—Duchamps revisited (see below in Sect. 8.4.3 or please refer to Sect. 4.3). And it seems essential that music without sound, texts without poetry, in total something by no means artistically designed, leads to a sensually perceived idea.

> 'prose scores' [...] are not instructions with which one produces the "real" thing, a performance for example. The text IS the real thing, the play. And they are also not poetry. There is nothing sophisticated or poetically complex in these texts.[274]

However, in Conceptual Art, the rule that represents the musical notation as an instruction for the realization of the composed and notated pseudo-work is usually shifted outward towards the object-subject relationship, i.e., to the fact that the non-composer gives orders to the recipient. The following concept *Out* by Istvàn Zelenka shows a particularly nice combination of constraint and illusory freedom.

> and on 5 *freely chosen* consecutive days, at 5 different, *independently determined* times of day between early morning and late evening, you will play per se per day one of the 5 sequences of this composition with *random* order of the individual pages.[275]

These commands correspond in outline to Paul Watzlawick's paradox "Be spontaneous!"

This type of art remains somewhat solipsistic. To vary Hans Fallada's novel title *Everyone Dies Alone* about resistance during the Nazi era: "Everyone imagines alone". It's all just "vague" in the sense of Zadeh, and not even fuzzy. Therefore, there is not much for Fuzzy Logic to think and do here, and it may therefore end here.

---

270 Such as Ablinger's *COMPOSITION BEYOND MUSIC/MUSIC BEYOND COMPOSITION*. Variation of the workshop title "Composition beyond Music", which I led at the Darmstadt Summer Courses 2014, https://ablinger.mur.at/docs/BeyondMusic_deutsch.pdf, retrieved 21.4.2020. See also Schnebel 1969.

271 http://ablinger.mur.at/kl_papier.html, retrieved 2 May 2020.

272 Ibidem.

273 Ibidem. Emphasis HWH.

274 See ablinger.mur.at/werke.htmlb12, retrieved 2 May 2020.

275 *Out* for 1 pianist, 2008, Ms. In: Houben, E.-M., Zelenka, I.: *1 Milieu—a book not only for reading*, Zurich: Edition Howeg 2009, pp. 125–130.

8.5.3 Imagination Form as the Translation of Conceptual Art Concepts into Subjective Realizations

The concept as a trigger is sufficient and must suffice for anticipatory perceptions. These are imaginary as object perception, imaginary-real as imagination within the subject. In contrast to the imagination form of concretely designed works, that of un-designed, merely conceived works is quite arbitrary. The relations are difficult to objectify. It would at least be possible for someone to take the trouble to write down what he or she thinks of the "stimuli" of non-art producing artists, and how he or she imagines the result of a concept more specifically. From this, experiments, test arrangements could be derived, and incidentally also for the imagination form of real works (see also Sect. 3.9). With regard to the imagination form of conceptual art, a research question that is methodologically difficult to realize would be relevant. To what extent does the concept that the work represents lead to certain forms of imagination in the recipient? And how, in turn, can the relationship of these forms of imagination to the (verbal or pictorial) concept be methodically grasped and investigated? Recipients can essentially only articulate the conceptual forms of imagination verbally, orally or even in writing by means of questioning. The film actress Hanna Schygulla has tried to replace brain scans and reproduce her dreams on video instead of just telling them. This is somewhat absurd, since filming requires an external object to be recorded. It is reminiscent of the story of Baron von Münchhausen, who, according to his own account, pulled himself out of the swamp by his own tuft of hair, even together with his horse.[276] Apart from this, and more modestly related to the verbal or picture forms of imaginary forms, Fuzzy Logic would, in corresponding investigations, mainly record degrees of approximation, the similarities of different individual versions with each other.

References

Altmann, P.: Sinfonia von Luciano Berio: Eine analytische Studie. Universal Edition, Wien (1977)

Amundson, M.A.: Talking Machine West: A History and Catalogue of Tin Pan Alley's Western Recordings, 1902–1918 (Popular Music Series Vol. 2). University of Oklahoma Press, Norman, OK (2017)

Bach, F.T.: Johann Sebastian Bach in der klassischen Moderne. In: von Maur, K. (ed.) pp. 328–335 (1985)

---

[276] Hanna Schygulla: *Dream protocols.* 1 February–30 March 2014, Academy of Arts, Berlin. Press kit. Schygulla is concerned with individual and collective memory. "On the basis of short memos I relived some dreams with open eyes in front of my running video camera" [... says Schygulla]. In the *Traumprotokolle*, filmed in 1979, she deals with her dreams and finds intense, sometimes surreal-looking images. *Traumtunnel* (2005) confronts the viewer with the themes of loss, death and reconciliation. With *Hanna Hannah* (2006/2007) she enters the open labyrinth of the Holocaust Memorial. More realistic than the fiction that one can reproduce what is thought on film as if it were external reality, are pictorial or artistic designs of dreams. Cf. e.g. Dieckmann 2012.

Becher, C.: Primal Light. Mahler im Spiegel des zeitgenössischen Jazz. In: Heister, H.-W. (ed.)Wo die schönen Trompeten blasen – vokale und instrumentale Semantik im Werk Gustav Mahlers (Musik und. Eine Schriftenreihe der Hochschule für Musik und Theater Hamburg. Neue Folge, vol. 11), pp. 335–344. Weidler Buchverlag, Berlin (2011)

Behne, K.-E.: Gehört-Gedacht-Gesehen. Zehn Aufsätze zum visuellen, kreativen und theoretischen Umgang mit Musik. ConBrio Fachbuch, München (1994)

Benson, D.: Music: A Mathematical Offering. Cambridge University Press. Cambridge. http://www.maths.abdn.ac.uk/~bensondj/html/maths-music.html (2006)

Bergsten, G.: Thomas Manns Doktor Faustus. Untersuchungen zu den Quellen und zur Struktur des Romans (Studia Litteraria Upsaliensia 3). Svenska bokförlaget, Stockholm (1963)

Böhme, T., Mehner, K.: Zeit und Raum in Musik und Bildender Kunst. Böhlau-Verlag, Köln, Weimar, Wien (2000)

Braem, P.B.: Zitat und Zitieren in den Gebärdensprachen der Gehörlosen. Zeitschrift für Semiotik **14**(1–2), 79–109 (1992)

Braun, W.R.: Three Uses of Pre-existent Music in the Twentieth Century. Phil. Diss, University of Missouri-Kansas City (1974)

Brendel, E., Meibauer, J., Steinbach, M.: Aspekte einer Theorie des Zitierens. In: Dies. "DIES." AUF ENGL. (eds.): Zitat und Bedeutung (Linguistische Berichte, Sonderheft 15), pp. 5–25 (2007)

Brendel, E., Meibauer, J., Steinbach, M. (eds.): Understanding Quotation. De Gruyter, Berlin and New York (2011)

Bruhn, S.: A Concert of Paintings: "Musical Ekphrasis" in the Twentieth Century. Paper presented at the Fifth World Conference of the International Association for Word and Image Studies, in Claremont, California, March 1999. A published Version in: Poetics Today **22**, 3 (Fall 2001), pp. 552–605. Porter Institute for Poetics and Semiotics (1999)

Bruhn, S.: Musical Ekphrasis: Composers Responding to Poetry and Painting (Interplay, vol. 2). Pendragon Press, Hillsdale, NY (2000)

Budde, E.: Zum dritten Satz der Sinfonia von Luciano Berio. In: Stephan, R. (ed.) Die Musik der sechziger Jahre, pp. 128–144. B. Schott's Söhne, Mainz (1972)

Burkholder, J.P.: The Uses of Existing Music: Musical Borrowing as a Field. Notes, Second. Ser. **50**(3), 851–870 (1994)

Burkholder, J.P.: Collage (The New Grove Dictionary of Music and Musicians, second edition, edited by S. Sadie and J. Tyrrell). Macmillan Publishers, London (2001)

Butor, M.: Die Wörter in der Malerei. Suhrkamp, Frankfurt (1992)

Cada, N.: Strategien der Repräsentation. Chris Ofili und das Konzept des Samplings (Phil. Diss. Ludwig-Maximilians-Universität München). urn: nbn:de:bvb:19-160500 (2011)

Cook, N., Pople, A. (eds.): The Cambridge History of Twentieth-Century Music. Cambridge University Press, Cambridge (2004)

Cuneo, D.: "This Is Not a Quote": Quotation Emplotment (sic!), Quotational Hoaxes and Other Unusual Cases of Textual Reuse in Sanskrit Poetics-cum-Dramaturgy, Section 3: Adaptive Reuse of Untraced and Virtual Texts. In: Freschi, E., Maas, P.A. (eds.): Adaptive Reuse. Aspects of Creativity in South Asian Cultural History (Abhandlungen für die Kunde des Morgenlandes, vol. 101), pp. 219–239. Harrassowitz Verlag, Wiesbaden (2017)

von Dadelsen, G.: Parodie und Kontrafaktur seit 1600. MGG1 **10**, col, 826–834 (1962)

Dahms, S. (in cooperation with Jeschke, C., Woitas, M.) (ed.): Tanz (MGG Prisma). Bärenreiter/Metzler, Kassel et al. (2001)

Dieckmann, L.: Dream dramaturgy and self-reflection. Bildstrategien romantischer Traumdarstellungen im Spannungsfeld zeitgenössischer Traumtheorie und Ästhetik. http://dx.doi.org/10.16994/bab (2012)

Dietel, G.: Musikgeschichte in Daten. Bärenreiter-dtv, Kassel and München (1994)

Dittmann, L.: Perspektivität und Polyperspektivität der Skulpturen Anthony Caros. In: Jaeger, P., Luthe, R. (eds.) Distanz und Nähe: Reflexionen und Analysen zur Kunst der Gegenwart (on the occasion of Walter Biemel's 65th birthday), pp. 271–290. Königshausen + Neumann, Würzburg (1983)

Drees, S.: Bearbeitung. In: Hiekel / Utz (eds.), pp. 197–212 (2016)
Duchamp, M.: The great trouble with art in this country (1946). In: Moure, G. (ed.) Marcel Duchamp: Works, Writings and Interviews. Ediciones Poligrafia, Barcelona (2009)
Ehrke-Rotermund, H., Rotermund, E.: Zwischenreiche und Gegenwelten. Texte und Vorstudien zur„Verdeckten Schreibweise" im „Dritten Reich". Wilhelm Fink *Verlag*, München (1999)
Eimert, H.: Grundlagen der musikalischen Reihentechnik (Series Information über serielle Musik). Universal-Edition, Wien (1964)
Emons, H.: Berios Sinfonia und Mahlers 2. Sinfonie: Re-Komposition als ästhetische Idee. In: Musikwissenschaft zwischen Kunst, Ästhetik und Experiment (Festschrift Helga de la Motte-Haber zum 60. Geburtstag), edited by R. Kopiez, pp. 151–160. Königshausen & Neumann, Würzburg (1998)
Emsley, J.: The Laptop DJ Handbook: Setups and Techniques of the Modern Performer—Course Technology. Cengage Learning, Boston, MA (2011)
Ender, D.: Übermalung? Montage? Wucherung? Strategien und Wege kompositorischer Selbstbearbeitung in neuer Musik. In: ÖMZ (Österreichische Musikzeitschrift) 1/2011: Transformationen, pp. 29–37. Lafite-Verlag, Wien (2011)
Engel, H.: Bearbeitung. MGG1 **1**, col. 1458–1466 (1949/1951)
Fink, M.: Musik nach Bildern: Programmbezogenes Komponieren im 19. und 20. Jahrhundert (Innsbrucker Beiträge zur Musikwissenschaft, vol. 13). Helbling, Innsbruck (1988)
Finscher, L.: Parodie und Kontrafaktur. Bis 1600. MGG1 **10**, col. 815–826 (1962)
von Fischer, K.: Variation. MGG1 **13**, col. 1274–1309 (1966)
Flinn, J.W.: Reconstructive Postmodernism, Quotation, and Musical Analysis: A Methodology with Reference to the Third Movement of Luciano Berio's Sinfonia. Phil. Diss., Graduate School of the University of Cincinnati in the Division of Composition, Musicology and Theory of the College-Conservatory of Music (2011)
Fuhrmann, A.: Klang und Bild: Die visuell-musikalischen Konzepte der Jack Ox. In: Neue Zeitschrift für Musik, vol. 155, no. 4, Frau & Musik, pp. 46–48. Schott International, Mainz (1994)
Genette, G.: Palimpseste. Die Literatur auf zweiter Stufe. Suhrkamp, Frankfurt a. M. (1993)
Großmann, R.: Musikalische Wiederholung und Wiederaneignung. Collagen, Loops und Samples. In: Bense, A. et al. (eds.) Musik im Spektrum technologischer Entwicklungen und Neuer Medien, pp. 207–218. EPOS, Osnabrück (2015)
Großmann, R.: Reproduktionsmusik und Remix-Culture. In: Saxer, M. (ed.): Mind the Gap. Medienkonstellationen zwischen zeitgenössischer Musik und Klangkunst, pp. 116–127. Pfau-Verlag Saarbrücken (2011)
Gruber, G. 1998: Zitat. MGG2 **9**, col. 2401–2412 (1998)
Hammerstein, R.: Musik und Bildende Kunst (IMAGO MUSICAE. Internationales Jahrbuch für Musikikonographie). Bäremreiter-Verlag, Duke Univ. Press, Durham, NC (1984)
Hartmann, K.A.: Concerto funebre. In: Thomas, E. (ed.) Kleine Schriften, p. 65. Schott-Verlag, Mainz (1965)
Hau, S.: Traum. In: Gehrig, G., Pfarr, U. (eds.) Handbuch psychoanalytischer Begriffe für die Kunstwissenschaft, pp. 373–383. Imago/Psychosozial-Verlag, Gießen (2009)
Heister, H.-W.: Jazz (Musik aktuell. Analysen, Beispiele, Kommentare, vol. 5). Bärenreiter-Verlag, Kassel (1983)
Heister, H.-W.: „Das ist Polyphonie." Zur Natur in und von Mahlers Tonsatz. In: Heister, H.-W. (ed.) Wo die schönen Trompeten blasen – vokale und instrumentale Semantik im Werk Gustav Mahlers (Musik und. Eine Schriftenreihe der Hochschule für Musik und Theater Hamburg. Neue Folge, vol. 11), pp. 205–228. Weidler Buchverlag, Berlin (2011)
Heister, H.-W.: „… ein Traum das Dasein ist" – „vorbei ist die Musike …". Aspekte der Intertextualität und der Palimpsest-Polyphonie im V. Satz von Mahlers Lied von der Erde. In: Berger, C., Schnitzler, G. (eds.) Bahnbrüche: Gustav Mahler (Series Voces), pp. 247–283. Nomos, Freiburg i. Br. (2012)

Heister, H.-W.: Sechseck zwischen Hollywood und New York: Brecht – Weill – Adorno – Thomas Mann – Schönberg – Eisler. In: Marx, L., Koopmann, H. (eds.): Exilsituation und inszeniertes Leben, pp. 57–98. Mentis-Verlag, Münster (2013)
Hicks, M.: Text, Music, and Meaning in the Third Movement of Luciano Berio's Sinfonia. Perspect. New Music. **20**(1/2), 199–224. https://www.jstor.org/stable/942413 (1981/1982)
Hiekel, J.P., Utz, C. (eds.): Lexikon Neue Musik. J.B. Metzler, Stuttgart/Bärenreiter, Kassel (2016)
Hiley, D.: Western Plainchant: A Handbook. Clarendon Press and Oxford University Press, Oxford and New York (1993)
Hutter, U.: Klangfarben – Farbklänge. Wechselwirkung zwischen visueller und auditiver Wahrnehmung. Diploma thesis diploma course (FH) Telecommunications and Media at the University of Applied Sciences St. Pölten (2006)
Jeong, E.-H.: The application of imagery to enhance "flow state" in dancers, Doctor of philosophy School of sport and exercise science. Faculty of arts, education and human development Victoria University, Wellington (2012)
Kaiser, F.W.: Jorinde Voigt and Music—For Example, Beethoven's Sonatas, Jorinde Voigt and Music, Franz W. Kaiser 2013_EN.pdf. http://jorindevoigt.com/blog/wp-content/.p-content/uploads/Portofolio-No-22-II-WV-2010-794-BIS-WV-2012-171.pdf (2013)
Kandinsky, W.: Über das Theater, edited by J. Boissel (assisted by J.-C. Marcadé). DuMont, Köln (1998)
Kantzios, N.: Ekphrasis: Writing about music. https://floridawriters.net/ekphrasis-writing-about-music/ (2018)
Karbusicky, V.: Zitat und Zitieren in der Musik. Zeitschrift für Semiotik **14**(1–2), 61–77 (1992)
Keiner, L.: Audiovisionen.Versuche der Sichtbarmachung von Musik. Kontexte, Theorien und die Suche nach einer Traditionslinie. diploma thesis University of Vienna (2014)
Kieler Gesellschaft für Filmmusikforschung (ed.): Kieler Beiträge zur Filmmusikforschung www.filmmusik.uni-kiel.de (2008ff.)
Kieser, E.: Analysen von Bildern und Plastiken der Moderne. Delvaux, Kandinsky, Klee, Mondrian, Moore, Picasso, edited by D. and K. Stein. Könighausen & Neumann, Würzburg (2004)
Klarmann, J.F.: Der illustrative Diskurs: Michel Butor und die Malerei (Inaugural dissertation for obtaining the doctorate of the Neuphilological Faculty of the Ruprecht-Karls-University of Heidelberg). http://archiv.ub.uni-heidelberg.de/volltextserver/15528/1/Zur%20Ver%C3%B6ffentlichung%20Diss.pdf (2013)
Klee-Zentrum (ed.): Paul Klee Bern. Hatje&Crantz, Ostfildern (2004)
Klotz, V.: Zitat und Montage in neuerer Literatur und Kunst. In: Weisstein, U. (ed.) Literatur und bildende Kunst. Ein Handbuch zur Theorie und Praxis eines komparatistischen Grenzgebietes, pp. 180–195. Verlag Schmidt, Berlin (1992)
Kontarsky, M.: Sonate für Violoncello und Klavier Nr. 2 F-Dur Op. 99. In: Bockmaier, C., Mauser, S. (eds.) Johannes Brahms. Interpretationen seiner Werke (2 vols.), pp. 713–721. Laaber-Verlag, Laaber (2013)
Krieger, G., Stroh, W.M.: Probleme der Collage in der Musik, aufgezeigt am 3. Satz der Sinfonia von Luciano Berio. In: Musik und Bildung, vol. 3, pp. 229–235. Schott-Verlag, Mainz (1971)
Kühn, C.: Das Zitat in der Musik der Gegenwart. Verlag Wagner, Hamburg (1972)
Kunkel, M.: Gibt es neue Typen des Engagements in der neuesten Musik? Positionen und Negationen. In: Phleps, T., Reich, W. (eds.) Musik-Kontexte. Festschrift für Hanns-Werner Heister, pp. 472–487. Monsenstein und Vannerdat, Münster (2011)
Lissa, Z.: Ästhetische Funktionen des musikalischen Zitats. In: Greimas, A.J. et al. (eds.) Sign, Language, Culture, pp. 674–689. Mouton, The Hague, Paris (1970)
Lock, G., Murray, D. (eds.): The Hearing Eye. Jazz & Blues Influences in African American Visual Art. Oxford University Press, Oxford (2009)
Maur, K. von (ed.): Vom Klang der Bilder. Die Musik in der Kunst des 20. Jahrhunderts, pp. 328–335. Prestel, München (1985)

Meischein, B.: Das Hören sehen lassen. Anregungen zu Kulturtheorie und Geschichte musiktheoretischer Graphiken. ZGMTH. Zeitschrift der Gesellschaft für Musiktheorie **16**(1), 7–20 (2019)

Niedermayer, B.: Automatische Erkennung von Cover-Versionen und Plagiaten in moderner Pop- und Rockmusik (Magisterarbeit zur Erlangung des akademischen Grades in der Studienrichtung Informatik). Institut für Computational Perception, Linz (2007)

Nolte, Jost: Kollaps der Moderne. Traktat über die letzten Bilder (11989). Droemersche Verlagsanstalt Th. Knaur Nachf., München (1991)

Oberhaus, L., Oetken, M. (with the collaboration of Susanne Stamm) (eds.): Farbe, Klang, Reim, Rhythmus. Interdisziplinäre Zugänge zur Musik im Bilderbuch. transcript Verlag, Bielefeld (2017)

Osterkamp, E.: „Apocalipsis cum figuris". Komposition als Erzählung. In: Röcke (ed.), pp. 321–343 (2001)

Pareyon, G.: On Musical Self-Similarity. Intersemiosis as Synecdoche and Analogy. Acta Semiotica Fennica XXXIX. Approaches to Musical Semiotics 13. International Semiotics Institute at Imatra. Semiotic Society of Finland, Helsinki (2011)

Parry, A. (ed.): The Making of Homeric Verse: The Collected Papers of Milman Parry. Clarendon Press, Oxford (1971)

Pendzich, M.: Von der Coverversion zum Hit-Recycling. Historische, ökonomische und rechtliche Aspekte eines zentralen Phänomens der Pop- und Rockmusik. Lit Verlag, Münster (2004)

Piper-Enzyklopädie Musiktheater. ed by Carl Dahlhaus in cooperation with the Forschungsinstitut für Musiktheater der Universität Bayreuth (director: Sieghart Döhring). 6 vols. + register-vol. Piper-Verlag, München (1986–1997)

Place, V., Fitterman, R.: Notes on Conceptualisms. Ugly Duckling Presse, no place (2009/2010)

Puschmann, R.: Magisches Quadrat und Melancholie in Thomas Manns Doktor Faustus. Von der musikalischen Struktur zum semantischen Beziehungsnetz. AMPAL-Verlag, Bielefeld (1983)

Rebling, E.: Ballett. Sein Wesen und Werden. VEB Breitkopf & Härtel, Leipzig (1963)

Redlich, H.F.: Alban Berg. Versuch einer Würdigung. Universal-Edition, Wien, Zürich, London (1957)

Röcke, W. (ed.): Thomas Mann: Doktor Faustus 1947–1997 (Publikationen zur Zeitschrift für Germanistik 3). Lang-Verlag, Bern etc (2001)

Rotermund, E.: Choralzitat und Chopin-Allusion. Schumanns Heine-Vertonung op. 24, Nr. 8 – ein faszinierendes Unikum der deutschen Kunstlied-Geschichte. In: Phleps, T., Reich, W. (eds.): Musik-Kontexte. Festschrift für Hanns-Werner Heister, pp. 750–756. Monsenstein und Vannerdat, Münster (2011)

Salmen, W.: Reflexionen über Bach in der bildenden Kunst des 20. Jahrhunderts. In: eds. On behalf of the Neue Bachgesellschaft Hans-Joachim Schulze and Christoph Wolff: Bach-Jahrbuch vol. 72, pp. 91–104. Merseburger, Berlin (1986)

Schäfer, T.: Modellfall Mahler. Kompositorische Rezeption in zeitgenössischer Musik. VERLAG?, München (1999)

Schier, V.: Cento. MGG2 **2**, col. 544–549 (1995)

Schlee, A.: Wandlungen musikalischer Strukturen im Werke Thomas Manns. Vom Leitmotiv zur Zwölftonmusik (Europäische Hochschulschriften I/384). Lang-Verlag, Frankfurt a. M. and Bern (1981)

Schmidt, K.-H.: Zitate in musterverarbeitenden Forschungsprozessen. Zeitschrift für Semiotik **14**(1–2), 111–121 (1992)

Schnebel, D.: MO-NO. Music to Read. DuMont, Cologne (1969)

Schulz, P. F., Cichon, C.: "Remixes" und "Coverversionen" – Urheberrecht und Verwertung. 11/2000; published in 3/2004 unter der Artikelnummer 8976 auf den Seiten von jurawelt.com publiziert. www.jurawelt.com/artikel/8976 (2000)

Sellmer, S.: Methodological and practical remarks on the question of reuse in epic texts. In: Freschi, E., Maas, P.A. (eds.) Adaptive Reuse. Aspects of Creativity in South Asian Cultural History (ABHANDLUNGEN FÜR DIE KUNDE DES MORGENLANDES. Im Auftrag der

Deutschen Morgenländischen Gesellschaft ed. by Florian C. Reiter. vol. 101), pp. 355–372. Harrassowitz Verlag . Wiesbaden. https://www.academia.edu/36625491/Methodological_and_practical_remarks_on_the_question_of_reuse_in_epic_texts (2017)
Soorani, F.: Musicalization in Qualified Media of Poetry and Painting (Inter)mediality in šāhnāmeh and Its Illustrations, Phil. Diss. Eberhard Karls Universität Tübingen (2018)
Springer, P.: Das verkehrte Bild: Inversion als bildnerische Strategie. Aschenbeck und Holstein, Delmenhorst (2004)
Straebel, V.: Aspects of Conceptual Composing. Open Space Mag. **10**, 69–78 (2008)
Sündermann, H., Ernst, B.: Sound—Colour—Sign Is Aimed Specifically at the Gestural as the Pivot of Mediation. Klang-Farbe-Gebärde Musikalische Graphik. Anton Schroll & Co, Vienna and Munich (1981)
Thym, J.: Song as memory, memory as song. Archiv für Musikwissenschaft **69**(3), 263–273 (2012)
Tibbe, M.: Lieder und Liedelemente in instrumentalen Symphoniesätzen Gustav Mahlers. Emil Katzbichler, Munich (1971)
Tomkins, C.: Marcel Duchamp: Eine Biographie. Hanser, Munich and Vienna (1999)
Treitler, L.: Homer and Gregory: The Transmission of Epic Poetry and Plainchant. Musical Quarterly **60**(3), 333–372 (1974)
Treitler, L.: Centonate' Chant: Übles Flickwerk or E pluribus unus? J. Am. Music. Soc. **28**(1), 1–23 (1975)
Treitler, L.: Reading and singing: On the genesis of occidental music-writing. Early Music. Hist. **4**, 135–208 (1984)
Treitler, L.: The „unwritten“ and „written transmission“ of medieval chant and the start-up of musical notation. J. Music. **10**(2), 131–191 (1992)
Tymoczko, D.: A Geometry of Music. Harmony and Counterpoint in the Extended Common Practice (Oxford Studies in Music Theory). Oxford University Press, Oxford (2011)
Ullrich, Anna Valentine: Gebaute Zitate. Formen und Funktionen des Zitierens in Musik, Bild und Architektur. transcript Verlag, Bielefeld (2015)
Vitali, C. et al.: Paul Klee und die Musik. Exhibition catalogue Schirn – Kunsthalle Frankfurt a. M. (1986)
Vonderwahl-Nyffenegger, H.: Variation – Variante – Variable: Prinzipien der Veränderung in musikalischen und literarischen Texten des 19. und 20. Jahrhunderts. Diss. University of Zurich, Faculty of Arts. https://doi.org/10.5167/uzh-163488 (2006)
Wehrmann, H.: Thomas Manns „Doktor Faustus“. Von den fiktiven Werken Adrian Leverkühns zur musikalischen Struktur des Romans (Europäische Hochschulschriften I/979). Verlag Lang, Frankfurt a.M. et al. (1988)
Wells, D.: Das Lexikon in Zahlen. Fischer Taschenuch-Verlag, Frankfurt a. M (1990)
Wescher, H.: Die Geschichte der Collage. Vom Kubismus bis zur Gegenwart (1st ed. 1968). DuMont, Köln (1987)
Wicke, P., Ziegenrücker, K.-E.: Handbuch der populären Musik. Geschichte – Stile – Praxis – Industrie (enlarged edition) Schott Music GmbH, Mainz (2007)
Widmaier, T.: Ich hab mich ergeben. In: Populäre und traditionelle Lieder. Historisch-kritisches Liederlexikon. http://www.liederlexikon.de/(2011)
Wild, Ch.: Charles Ives' Three Places in New England: an Interpretation and a Conductor's Guide, A Doctoral Thesis, Presented to the Faculty of the Bienen School of Music of Northwestern University in Candidacy for the Degree of Doctor of Musical Arts (2018)
Wilson, W.G.: Change Ringing: The Art and Science of Change Ringing on Church and Hand Bells. Faber, London (1965)
Williams, J.: Psychophysiological Correlates of Dynamic Imagery. Thesis submitted for the degree of PhD at the University of Warwick. http://go.warwick.ac.uk/wrap/2312 (1995)
Wurm, B.: Vom Wissen (in) der Film-Notation. Ivan Ladislav Galeta: Auf-Zeichnungen. In: Zeitschrift für Medienwissenschaft. Vol. 3: Aufzeichnen. 2010, no. 2, pp. 42–55. https://mediarep.org/handle/doc/2671 (2010)

Würtenberger, F.: Malerei und Musik. Die Geschichte des Verhaltens zweier Künste zueinander. Verlag Lang, Frankfurt a.M (1979)
Zender, H.: Happy New Ears. Herder-Verlag, Freiburg i.Br. et al (1991)
Zender, H.: Messiaens Aktualität – das Haiku-Denken. Positionen **13**(1992), 46–48 (1992)
Zender, H.: Die Sinne denken. Texte zur Musik 1975–2003, (ed.) J. P. Hiekel. Breitkopf & Härtel, Wiesbaden etc.. 2004
Zywietz, M.: „Das Geheul als Thema – welches Entsetzen!“ – Zum Oratorium „Apocalipsis cum figuris“ in Thomas Manns Roman „Doktor Faustus“. Die Musikforschung **62**(2), 140–149 (2009)

Chapter 9
Perspectives and Prospects

Abstract Summarizing I demonstrate how productive the interplay of FL and music is. In the historical progress of music through fuzziness and sharpening FL is though not the logic of history, but a logic in this history, not at least via pendulum swings between blurring and renewed sharpening. Analogies and parallels between music and life or scientific-technical procedures and the technical and psychosocial use for e.g. biofeedback justify to speak of a ‚Musical Turn' in some sciences. This includes that music is an active component of the world view, e.g. by representing the time arrow. Qua mimesis of chaos and anomy, cosmos and harmony music is a model for dialectics. FL and musicology correspond with their respective 'inner interdisciplinarity'. In the interdisciplinary as sum and synthesis FL and musicology encounter each other, and probably also a 'Fuzzy Logical Turn' should be declared. With all this, music and its analysis contribute to a concrete, widespread ‚Philosophy of Fuzzy Logic'.

Keywords acoustics · „biofeedback" · biology · cosmos and harmony · „experimental music" · Fuzzy Logic as logic in the history of music · ‚Fuzzy Logical Turn?' · interdisciplinarity · mimesis · musical history · ‚musical turn' · music as model for dialectics' · philosophy of Fuzzy Logic' · psychology · science · sound · technology · world view

It had been the idea, the aim and the objective of this book to confront the musicological perspective with a central question of Fuzzy Logic, the search for functions of fuzziness.

Therefore, symbolically the tunnel was to be dug from two sides and it was questionable whether the two diggings would meet in the middle. Now the music-related tunnel section has been extended considerably and has been worked out much more precisely. This is a result of my profession as a scientist with a focus on music; I have naturally considered and described all questions from the perspective of music, and from the analysis of the musical process. But I hope that readers who approach the topic from the fuzzy logic perspective and are interested in questions of music analysis will find in this book many suggestions and a lot of information that will quickly lead them deep into "fuzzy-musical-logical" analyses.

H. Heister, *Music and Fuzzy Logic*, Studies in Fuzziness and Soft Computing 406,
https://doi.org/10.1007/978-3-662-62907-9_9

The dialectic of fuzziness and sharpening, which is at the center of the topic, supplemented by the newly developed concept of 'blurring' as a conscious artistic production of fuzziness, proved to be very productive scientifically. The variety of productive blurs in the process of art and especially in the process of music could be qualitatively more precisely identified and differentiatedly analyzed. My structuring of the musical process into thirteen forms of existence proved to be a helpful instrument in the precise work of tracking down the many conscious and unconscious blurs on the one hand, but also the deliberate blurring and sharpening processes on the other, and I hope that this instrument can continue to serve the exploration of the topic. A multitude of other research questions have arisen that could be clarified more precisely with the help of the quantifying methods of fuzzy logic. In case researchers of Fuzzy Logic should turn their attention to these topics in the next few years, the further analysis of the musical process could also benefit from it.

In conclusion, additional connections between fuzzy logic and further musical and artistic subject areas will be sketched in order to provide it with an even higher significance beyond its present areas of focus—by tendency even as a kind of theory similar to meta theories for the sciences, humanities and social sciences, as mentioned in the preface.

9.1 Productivity I. Historical Development and Progress of Music Through Fuzziness/Blurring and Sharpening—Fuzzy Logic as Logic in the History of Music

The dual movement of blurring and sharpening, and particularly of crystallization, can be found both as a driving force as well as a result in history. The loss of expressiveness of musical passages, be it intervals, chords, motives or others, and sometimes the entire loss of concrete semantics, is a very general phenomenon, almost a universal, and with parallels to phenomena in the history of the language of words. In regard to music, Eisler has found the formula for an important dimension of this dual movement: expression becomes material—material becomes expression.

9.1.1 Pendulum Swings in Musical History Between Blurring and Renewed Sharpening: Expression Becomes Material—Material Becomes Expression

Thus, the diminished seventh chord, for instance A—c—e♭—f♯—a, changed in the harmonic tonality from an expression of pain, fright and the like into a generally mere syntactical changeover point for quick modulations, that is, changes of key, during

Fig. 9.1 Ferenc Liszt: *Bagatelle sans tonalité* 1885, measure 1–5 Neue Liszt-Ausgabe. Serie 1, Band 14. Budapest: Editio Musica, 1984. Plate Z.8858. IMSLP108764-PMLP13500-Liszt_NLA_Serie_I_Band_13_11_Vierter_Mephisto-Walzer_S.216a_scan

the 18th century. On the other hand, it occasionally maintains some of its original force. Thus, in Carl Maria von Weber's *Wolf Glen* scene from *Der Freischütz*, 1821, it appears as a signet of the infernal and its personification. This corresponds with the configuration 'material becomes expression'. The process of desemantication occurs in a gradual manner, in almost unnoticeable transitions, and is thus a typical object for fuzzy logic. The same applies to the tritone, that is, the diminished fifth respectively the augmented fourth such as, for instance, c-f♯ or b-f'. In the system of the 'modi' in the medieval high feudal music, the relatively unsingable tritone was a foreign body in the system because it consisted of three whole tones and not of the diatonic sequence of whole and half tones. It was seen as the "diabolus in musica", as the devil in music. It was used countless times with this meaning. One example among many is Ferenc Liszt's chain of *Mephisto Waltzes*. The tritone interval b-f' is central up to the last variant of these waltzes and has an explicit tonality breaking effect (Fig. 9.1).

At the same time, however, the tritone is used quite matter-of-factly in tunes and, because of its leading tone effect, especially as part of the dominant seventh chord, for instance in G-*B*-d-*f*. The tritone transposition of a series as part of the twelve tone technique, as for instance in Schönberg's suite for piano op. 25, 1921, is also entirely without semantic qualification. The same is true for the production of an entire twelve tone series of two complementary hexachords (six-tone group).[1] The latter are often found to be in a tritone distance to each other.

Complementation is the separation of pitch-class collections into two complementary sets, with one containing the pitch classes which are not part of the other one. More restrictively, complementation is "the process of pairing entities on either side

[1]Complementary to and competing with Schönberg, composer Joseph Matthias Hauer "discovered the 44 tropes, pairs of complementary hexachords, in 1921 allowing him to classify any of the 479,001,600 twelve-tone melodies into one of forty-four types". (https://en.wikipedia.org/wiki/Trope_(music)#20th-century_music, July 1, 2017, retrieved December 20, 2017).

Fig. 9.2 Arnold Schönberg: *Moses und Aron* Quoted after https://en.wikipedia.org/wiki/Combinatoriality, 7 August 2020, retrieved 20 August 2020

of a center of symmetry". (Marianne Kielian-Gilbert)[2] The term "'combinatorial' appears to have been first applied to twelve-tone music by Milton Babbitt" in 1950 (Fig. 9.2).

Combinatorial tone rows by pairing complementary hexachords. The f-b and c'-f♯ is a tritonus.

First line: c♯ d g♯ f♯ g f /b a b♭ c e♭ e.

The hexachords together contain the chromatic sum of all semitones in the octave: c c' d e♭ e f f♯ g g' a b♭ b.

There are further analogies for the polarity of desemantication and semantication, the interplay between sharpening and blurring. For instance, this relatively general process can be found also in geology.[3] Sedimentation and erosion change the outline of landscapes and level their profiles. By contrast, volcanism and tectonic uplift have a constructive effect. And processes of crystallization, used here, as mentioned above, as a metaphorical term, provide an answer to the tendency towards the amorphous, for instance through melting.

In this, however, not all of these processes occur as an endless game of ping pong, as a toing and froing or a pendular movement—which, incidentally, will also eventually come to a standstill if there is no new energy input. Rather, there are irreversible processes, just like in the cosmological context of thermodynamics.

The arrow of time of such processes asserts itself, for instance, in the irreversibility of grammaticalization (Meillet 1912; also grammaticalization or grammaticization), a process of language change by which words representing objects and actions (i.e. nouns and verbs) become grammatical markers (affixes, prepositions, etc.). It thus creates new function words by a process other than deriving them from existing bound, inflectional constructions, instead deriving them from content words. For example, the Old English verb *willan*, 'to want', 'to wish' has become the Modern English auxiliary verb *will*, which expresses intention or simply futurity.[4] Grammaticalization is thus also the use of an independent linguistic expression, generally a word, for grammatical purposes, in other words the transition from a lexeme to a

[2]Kielian-Gilbert, M.: "Relationships of Symmetrical Pitch-Class Sets and Stravinsky's Metaphor of Polarity", *Perspectives of New Music* 21, p. 210, quoted after https://en.wikipedia.org/wiki/Combinatoriality, 7 August 2020, retrieved 20 August 2020.

[3]Cf. here on the aspect of analogies to other areas of reality and science, Heister 2016 and Heister 2016a.

[4]Extensively covered in https://en.wikipedia.org/wiki/Grammaticalization, Dec 17, 2017, retrieved Dec 17, 2017.

grammatical formative.[5]One of the roles of a diminished seventh chord as a means of modulation is analogous to this. This transition has a systematic and a historical dimension, as Lehmann points out: According to him, on the diachronic axis, grammaticalization is a process of language change, while on the synchronous axis, it is the transformation of one construction into another, with both of them ranged in a scale.[6] They are thus also an object of Fuzzy Logic.

Grammaticalization is a kind of "automatization" and therefore persistent and hardly reversible. The counter tendency, a "de-grammaticalization", therefore hardly ever occurs—in contrast to the expression-material-dialectic in music. To quote Lehmann again, "the suffix *-ism* can be liberated from derivatives such as capitalism, communism etc. and be made independent as a noun which is used in the form of *the isms* (emphasis in the original)."[7]

The case of phonetic *erosion*[8] (phonological attrition or phonological reduction) is of a somewhat different nature. The phonetic-phonological loss of morphemes respectively words is at least partly revocable. The Old French word for 'bee', ap (from Latin apis) is reduced in such a manner that, for better comprehension and distinction, recourse was taken by substitution to the diminutive form—abeille, with regular palatalization of the intervocal p between a and another vowel into b. The blurring is thus answered by a different sharpening.

The dialectic of Fuzzy Logic with the mostly multiple encaptic, that is, nested interplay between blurring and sharpening, thus does not only characterize the different dimensions of the musical process and the musical artwork and its forms of existence but also *music-historical* developments. As has already been demonstrated in numerous cases, fuzziness and sharpening develop in part in an objective manner and by themselves, yet they are also in part consciously manipulated and used.

9.1.2 Analogies Between Music and Life with Movements in Space: Purposefully With or Without Parallels—Sharpening and Blurring

Diversions from the seemingly or actually precise, differences instead of identity with the previous, variations instead of repetitions, movements away from worn paths occur as a driving force of developments in general and of progress in particular also beyond music and other arts in different spheres of reality. For instance, analogies – no more, but also no less—show themselves between types of movement in nature, even among lower forms of life, and some types of movement in music.

[5]Cf. in detail https://www.christianlehmann.eu/ling/ling_theo/index.html?https://www.christianlehmann.eu/ling/ling_theo/grammatikalisierung.php, retrieved Dec 17, 2017.

[6]Ibid.

[7]Ibid.

[8]Nota bene, a metaphor from geology; see Heister 2016.

However, the main direction of the respective primary and dominant principles will most likely be an inverted one for each case. In nature, it is primarily about an optimizing sharpening. As optimization and precisely not maximization, also not as precision, this always includes a certain measure—depending on the individual case—of fuzziness. Also in this respect, art is more open and liberal. On the one hand, it reaches a precision which can hardly be achieved otherwise. On the other hand, it permits the intentional blurring in the interest of beauty which does not exist as such in nature or technology (At most in forms such as "planned obsolescence" and, of course, in many ways in society in between ideology and lie).

Neither here nor there, however, does movement without aim or direction, the "vague" in Zadeh's sense, constitute the productive counterpoint to rigidity. The vague and flexible is rather a starting point which initially needs sharpening, particularly in the form of 'filtering'.

Human movement in natural space is often without direction. Without orientation at distinctive points or lines, not least because of different gait lengths between right and left, we often literally walk around in circles—and these circles are in turn also imprecise.

Particularly in the forest, we lack a horizon as well as further-reaching vistas which would offer a kind of sight line.In this regard, nature is ahead of us sometimes, as for example in the case of the simple level of single cell organisms. Flagellates, e.g. mastigophorans via phototaxis, can autonomously direct and correct their movements. Orientation is not provided by pathways or trails; rather, light constitutes, so to speak, the terminus ad quem, the aim; characteristically, there is no binary yes-no-situation but rather increments, that is, zones of *lesser* and *greater* brightness: the specific "something more /something less" of Fuzzy Logic.

A transitional form between free floating movement and movement that is limited to paths, thus in the analogy between a freely floating tune and strict parallels between two or more voices, can also be found among flagellates. Some organisms such as blue or pebble algae can only move in parallel to their axis, like on rails—they may only reach their goal, as far as can be determined, not directly but by detours—but they do reach it. They need to have their "rails" run in the right direction or bump off obstacles. The condition for this, however, is that on these rails, they are at an angle to the source of the stimulus or happen to arrive there because of accidental outer impulses. In this instance, therefore, it can be questioned whether this kind or relatively rigid 'sharpening', which is at the same time dependent on outer circumstances, is truly and literally productive.

9.1.3 Quadrature of the Circle and Rules of the Game—Keeping Them, Consciously Breaking and Changing Them

Arnold and Wilhelmine Keyserling point out that the human cerebellum is able to synchronize rhythms of movement and to integrate them as strategies into our thinking.[9] This is evident, for instance, in behaviors in dancing such as in the twosome or in larger groups, for instance in discotheques. Aside from individual resonance, stochastic-statistical determinants also are at play here, something which is already evident in "collective intelligence" in the animal world. Either way, the cerebrum is capable of rebelling against such adaptations or to look for other ways. In the production of works of art, particularly in composing, there are similarities but also differences to rules of the *game*. In art, rules are specific to the type of art, the genres and forms, and furthermore specific to epochs and regions. One the one hand, however, it is part of the art to *not* always play by the rules. Rather, they are systematically and consciously broken. On the one hand, this *breaking of rules* is necessary to ensure the functioning of a given system, of a style as such, tangible as a kind of adaptation to the environment and the circumstances. Thus, the *breaking of rules* becomes a decisive driver of further development, of progress (Unintentional violations of rules, however, are usually mere mistakes). The breaking of former rules leads to a new system of rules. In this, we may indeed think of these states "far from equilibrium" from which new things, new orders also originate. The *breaking of rules*, however, is on the other hand only productive when it is coupled with a partial keeping of these rules. We can, once again, speak of a dialectic here.

Therefore, on the one hand, it is important to filter, organize, standardize, on the other, to differentiate, blur, specify and utilize fuzziness. As always, this dual movement between the alternatives is not completely symmetrical: The tendency to minimize blurring dominates because an aim is to be reached as effectively as possible.

Art has even more room to maneuver. Here, too, sharpening predominates, particularly as crystallization, yet blurring as a counterpart has a high significance as well. Art initially starts out from its individual historical-societal differentiated system with all its numerous rules. They are partly conscious and also consciously traded down, partly procedural and partly only semi or preconscious, simply taken for granted—this is how one does things. They only become changeable and usable for further development if the actors make themselves aware of the existence of these rules as well as of the rules themselves and when they change them from non-reflected background norms into conscious settings. They are then comparable to the control variables in control loops in cybernetics. These are in principle, that is, in the literal

[9] Arnold and Wilhelmine Keyserling: *Ars Magna, I. Astrologie. Die Planeten im Enneagramm.* http://schuledesrades.org/palme/books/arsmagna/?Q=1/3/29/8, retrieved July 2, 2017. Despite the esoterical, unscientific background, this is a justified reference to socially relevant achievements of the sensory apparatus and the brain in the physiological realm, that is, a level well below conscious action.

sense of the rule, not arbitrary and random but are conscious settings—thermostats are set to temperatures which are pleasant and suitable for humans, or, as the case may be, also for frozen meat.

The work of art as a process pursues an external "aim", namely the effects and functions, predominantly joy and societal eudaimonia. Effectivity does also play a role in this. However, the individual end of the work does not have to be reached as an aim as quickly and effectively as possible. While internal relationships and connections should be as consistent, coherent, 'logical', the journey itself is a reward: The more attractive and, in a deeper sense, the more logical it is, the more enjoyable it is perceived to be. Detours, fuzziness and blurring therefore are part and parcel of this process.

The—human—art lies in reflecting and correcting the obedience to seemingly natural phenomena, literally to set disorganized lines straight, or perceiving the circle as an intentional form and frame of reference, as for example in round dances, or in contrary squaring the circle in quadrille or squaredance.

Analogies between types of movements of single cell organisms and parts in a polyphonic musical composition might seem esoteric and far-fetched, but they could turn out to be quite real and practical. A recent Spanish study discovered that noise caused by air traffic could be dramatically reduced by applying minimally different movements during take-off and landing procedures that employ Fuzzy Logic because the resulting interferences of the sound waves can cancel each other out.[10] In European music history the parallel movement or counter-movement of the voices of a composition, their differentiation—and homogenization—blurring and re-sharpening are an essential motor of development (see above Sect. 7.3).

In all this, Fuzzy Logic does not become the logic of music history in a straightforward manner. However, it obviously becomes an important mode of thinking in order to describe the incipient stages of change, development and musical progress with all the losses and gains they bring with them in a more precise manner, and to understand them. But indeed Fuzzy Logic is a logic *in* the history of music. It even explains it in so far as via the mode of thinking in Fuzzy Logic, developments in the process of music history can be embedded into overarching conditions between the reflection of individual societal structures and fundamental, metahistorical anthropological possibilities. Parallels as well as differences to processes of the type of primarily biological evolution can only be mentioned here.

[10]See *Fuzzy logic and grey science,* 17 March 2009, Universitat Politècnica de Catalunya, https://www.alphagalileo.org/PrintView.aspx?ItemId=56202&CultureCode=en, retrieved August 1, 2009.

9.2 ‚Musical Turn' in Natural Sciences, Liberal Arts, and Social Sciences. Generalizations and ‚Philosophy of Fuzzy Logic'

Based on these as well as similar insights and research results, I hereby declare: After so many "turns"[11] in the last decades, the 'musical turn' is long due—in the social, cultural and art sciences from history to linguistics, but also in the natural sciences.[12] This musical turn relates to different areas and aspects of reality and encompasses at least the following dimensions:

1. The acoustic and the musical as technical and psychosocial instruments
2. Scientific procedures and devices as instruments of music production
3. Unconscious role models. Musical parallels to scientific-technical processes
4. Music as an active component of the world view
5. Music as a historical-social seismograph
6. Music as a model of dialectics

9.2.1 Pragmatics and Practice: Acoustical and Musical as Technical and Psychosocial Instrument

I begin with relatively trivial fields of reality. The omnipresent background music of the imposed "Muzak" type or the self-chosen type of the portable music player or smart phone is only mentioned here. The worrying effects of such music have been studied to some extent. An increased empirical grounding as well as quantification not least with the means of Fuzzy Logic would be a desideratum. This is also true for the more specialized application in medical respectively clinical treatment. Their unscrupulousness in regard to psychological, cultural and other effects is remarkable.[13]

A less worrisome dimension of the musical turn is not to do with artistically formed gestalts but predominantly with sound, the acoustic as a primarily physical phenomenon and simultaneously a basis for music. Thus, sound and particularly ultrasound are increasingly used for all kinds of questions and problems, especially technical ones, for instance material testing, for cleaning objects, for medical screenings of the invisible, for instance in pregnancy, etc. In a passive-receptive fashion on the part of the examiner, this began with the stethoscope from 1816 onwards. The medical auscultation was followed by the technical one. Here, the course seems to have gone from the physiological application to the technical one; elsewhere, it

[11] The latest one might be the "animal turn", cf. Schulze /Conrad 2017, p. 11.

[12] The *acoustic turn*, which followed the *pictoral* or *iconic turn*, has been called e.g. by Petra M. Meyer as early as 2008. Gerhard Paul has presented significant evidence in two extensive 2013 volumes on the history of sound in the 20th century.

[13] In detail e.g. Schneck and Berger 2006.

is often the other way around. In the 1950s, the sound of crank shafts was used as an indicator; it was listened to with the aid of a stethoscope and in this manner, defective points were identified. Electroencephalography began as early as 1875 in animal studies but was only used in humans in 1924.[14] And it took several decades again until the medical application reached musical application. The composer Alvin Lucier used processes and instruments from this area for his musical experiments since the 1960s and generated sounds through the deduction of his own brain waves—with the method and process somewhat more exciting than the aesthetic result. A parallel to this can be found in the Mindball Games in which a ball rather than sounds is controlled. It is described as a game for two in which players control the movement of a ball across a table by means of their brain waves—namely through relaxation and focusing. It constitutes a further development of similar games designed in the 1970s which included biofeedback respectively neurofeedback.[15]

Also from the side of Fuzzy Logic, it would need to be determined which relationships exist between brain waves and sounds—be they similar or different. Such interferences between nature, technics, science and art link up to areas such as bionics but also prosthetics, and finally the vast field of robotics, artificial intelligence and lastly the dubious "transhumanism". Also in the field of Fuzzy Logic inspirations and models are used—modified and transformed, as in "Nature-Inspired Design"

> of hybrid intelligent systems based on *nature-inspired optimization* and their application in areas such as intelligent control and robotics, pattern recognition, medical diagnosis, time series prediction and optimization of complex problems."[16]

Radio astronomy or seismography are already older procedures which ultimately operate in a "multisensorial" fashion. Purely acoustic diagnostic methods, on the other hand, are relatively new. Just like in other diagnostic methods, this medical application immediately constitutes a broad field of application of Fuzzy Logic, for casuistry, case distinction and similar. A pharmaceutical company has an interest in the differentiation of coughing sounds:

> Fraunhofer IDMT develops acoustic cough recognition for Boehringer Ingelheim. [...] The sound of a cough can provide an initial indication of what therapy is required for treatment. Working on behalf of Boehringer Ingelheim, scientists in the Oldenburg Project Group for Hearing, Speech and Audio Technology at the Fraunhofer IDMT have developed an app that analyzes cough noises and provides an assessment of whether the cough is a dry cough or a productive cough. The free app with the name "Silometer Husten-Tester" is available for iPhone and Android.[17]

---

[14]The rather questionable electroshock therapy in psychiatry began during the First World War and then more intensively in the 1930 s. Electroshock therapy, today euphemistically called electroconvulsive therapy (ECT) had precursors in numerous studies in humans since the discovery of electricity, particularly in the 18th century.

[15]See https://en.wikipedia.org/wiki/Mindball, January 10, 2015, retrieved August 6, 2015.

[16]Cf. Kacprzyk /Castillo / Melin 2017, Preface: Nature-Inspired Design of Hybrid Intelligent Systems, http://www.springer.com/978-3-319-47053-5, retrieved Feb. 05, 2017. Emphasis H.-W.H.

[17]Press release December 9, 2013; https://www.idmt.fraunhofer.de/en/Press_and_Media/press_releases/2013/New_app_recognizes_type_of_cough_by_sound.html, retrieved January 7, 2018.

In terms of a scientific environment, this special application is embedded into a vast field of computer based acoustic procedures of investigation—and thus immediately also a field of Fuzzy Logic.

Computer-based recognition of speech and acoustic events is a research field of the Project Group for Hearing, Speech and Audio Technology of the Fraunhofer IDMT. Among other things, the scientists develop systems for automatic recognition of emergency situations in care facilities or public spaces. Other application fields include voice control in man-machine interaction and acoustic monitoring of industrial manufacturing processes.[18]

These cases may suffice here as reference to the musical turn from this side.

9.2.2 Musical "Biofeedback", "Mindball" and "Experimental Music". Music as Science?

The electrification of medical diagnostics by means of electroencephalography began as early as 1875 in animal studies. It was, however, used in humans only in 1924. And it took another four decades until it was applied in music. The already mentioned composer Alvin Lucier used it for his musical "experiments" since the 1960s. The first work using brain waves to generate sound, *Music for Solo Performer* for amplified brain waves and percussion, was first performed in 1965.[19] (See above Sect. 6.8.5) Leaving aside the evaluation of the acoustic-sensual product, such a music production constitutes a filtering and crystallization complete with a multiple dialectic, such as that of simplicity and complexity. The same procedure can be found in the 2003 game "Mindball" with its precursors "Brainball" and, as early as 1974, "Will Ball" which use biofeedback. Concentration and the advertised relaxation here are methods of competition.

> Mindball Game is a game that goes against the conventional competitive concept, instead of activity and adrenalin it is attention focus that marks the truly successful Mindball Game player. The goal is to be more focused and to some extent relaxed than your opponent and thus move a physical ball away from you into your opponent's goal. Mindball Game is contained in a table. On the table top a ball is moving back and forth during the game. The players wear headbands with electrodes which are connected to the table. The electrodes are wired up to a biosensor system. The system that is used to measure the body's biological signals registers the electrical activity in the brain - so called EEG (Electroencephalogram […]).[20]

The simple visible movements of the ball are doubled or completed by making the invisible visible to the audience:

> The players and the audience can follow the players' mental process during a game as the players' brain activity is presented graphically if a monitor is attached to the game table.

---

[18] Ibidem.

[19] See Wilson 2003, p. 15.

[20] http://www.mindball.se/mindball-game, no date, retrieved August 15, 2015.

> Mindball Game is an exciting and social game where the audience can follow the match by watching the ball on the table, the graphics and the players, thus giving biofeedback (neurofeedback) to the players.[21]

This procedure is generalized for other electronic games as a further step in the automatization in the invention of an Electronic system prearranged to replace in the electronic games the existing manual controls with equivalent controls directly commanded by the brain.

> The present invention consists in a multi-functional electronic device, having the purpose, when added to an electronic game, of replacing in its base programme the voluntary manual interference for the playing of the same game, with a voluntary interference of not manual character, but directly drawn from cerebral electrical impulses, thus rending the game directly influenceable by the will of the player.[22]

Ultimately, this process is transferred—entirely in the sense of the musical turn—from the model of music and games to painting, here also as "brainpainting", developed in 2010 as a form of rehabilitation for those affected by motor skills damage.[23] Composers such as Lucier count among what is termed "experimental music". Here, an adaptation, even a sameness with science is proclaimed for art. The justification for this remains somewhat vague. Experiments were "necessary to question one's own preconditions", as US composer Alvin Curran put it.[24] With this stance, he saw himself "in the scientific tradition of the experiment as a gamble, an undertaking which can also go wrong, where expectations can be fulfilled or disappointed."[25] The open-ended musical experimental setup led, particularly, in the 1960s, to an integration of the element of chance in aleatorics or to "open forms".[26] In these regressions towards "conceptual art" in relation to the entire work, parts of the work remained elaborated down to the last detail. The blurring remained partial and did not become total. For the concept of art production, however, it appears more appropriate not to leave the final result of the work up to the discretion and ad hoc decisions of the interpreters. The "open ending" is not a characteristic of art but of games.[27] As these experiments, aside from the dimension of the open ending, usually lack the other dimension, that of confirming or falsifying hypotheses and thus gain new insights, this type of approach of music to science is usually rather illusory. Nevertheless (or even because of this), the concept still enjoys popularity.[28]

---

21 Ibidem.

22 Ibidem.

23 http://worldwide.espacenet.com/publicationDetails/biblio?FT = E&DB = &&CC = EP&NR = 0177075&KC = A2&locale = en_EP, retrieved August 16, 2015.

24 Quoted in *JetztMusik - Glossar Experimentelle Musik,* June 22, 2009, https://www.swr.de/swr2/programm/sendungen/jetztmusik/experimentelle-musik/-/id = 659442/did = 5002204/nid = 659442/14oftfj/index.html, retrieved October 7, 2017.

25 Ibidem.

26 On sharp criticism cf. Boehmer 1967.

27 Cf. e.g. Heister 2016b.

28 In detail e.g. Kreuzer 2012. Cf. also further literature at http://www.oefg.at/legacy/text/arge_wissenschaftkunst/wissenschaft_kunst/wissenschaft_kunst.html, retrieved Dec. 24, 2017.

9.2.3 *Chimera. Musical Parallels to Scientific-Technical Procedures*

In the same way that science can offer procedures, instruments or models to music, music in turn can act as scientific-technical procedure. There is a remarkable parallel, mediated together with the *chimera* myth, in the interlacing and fusion of different form types and the combination of different animal species. There probably have been blends and blurrings into mythical creatures and monsters in the human imagination right from the beginning; they have been documented in objectifications in cave paintings or small sculptures since the late Paleolithic, for instance the Lion-man of the Hohenstein-Stadel in the Swabian Jura around 30,000 B.C. In the myths of the early advanced civilizations of Mesopotamia, Egypt etc., the combination of human body and animal head up to the Sphinx is almost the rule. Figures such as the centaur, the harpy, the siren, the chimera and many more also feature in Greek mythology. The mixing of distant species has been made possible much later than the interbreeding of related species such as mules and hinnies, and only by means of genetic engineering.

With music everything is easier. Without knowledge of a potential reference to biological species, there is a group of cases which is special in regard to the classification developed here. Blurring as encapsis, that is, an interlacing of different musical forms is a general problem of genetics as well as living conditions in general in the context of the manipulation of the genome, including the mixture of biological species which originate from it. The programmatic-musical 'symphonic poem' which was, to a significant degree, developed by Ferenc Liszt already interlaced movement types of classical multi-movement symphonies in a single movement piece. Starting from different premises, Brahms as well as Gustav Mahler transferred this procedure back into multi-movement symphonies. Thus, Brahms drafted the finale of his fourth symphony as an interlacing of more static passacaglia with a recurring base as well as melodic models as a structure of variations and the dynamic sonata form—a kind of squaring the circle.

Retrospectively, in 1937, Arnold Schönberg called his *Kammersymphonie* Opus 9, completed in 1906, a "true turning point" of his composition style, the last work of his first period which only consisted of one consecutive movement. It still bore a certain similarity with the first string quartet op. 7 which also combined the four movement types of the sonata form. This form concept of a single theme, Stefanie Kreuzer writes, which interlaces the moments of sonata movement and sonata cycle, was studied by Schönberg in Beethoven's "Great Fugue", Schubert's "Wanderer fantasy" and Liszt's sonata in B minor. She also points out that multidimensional form thinking in almost seamless movements (exposition, scherzo, development, adagio, recapitulation), an overflow of motivic-thematic material (19 themes) and a complex harmony (major-minor tonal, whole tone as well as quartal harmony) in opus 9 open up the kind of multi-perspectivity which constitute this "turning point" for the composer in his artistic development.[29] This technique means more than a

[29]Second chamber symphony (in E♭ minor) for small orchestra op. 38 [Setting for two pianos op. 38B] (1906–1916/1939). Schönberg, *"Wie man einsam wird"*, 1937, and comment by

mere collage or montage. For instead of a juxtaposition or overlaying, this constitutes an interlacing of form types. Blurring here changes into crystallization, not unusual for the principle of variation in the sense developed here in the context of Fuzzy Logic and music. Here, however, music is not a suitable model for societal states of affair. For what is harmless in art, a multi-pronged, productive and progressive development of form types, can be highly problematic in life.

> Dr. Gómez-Lobo: Just a question of conceptual clarification or for any other reason for my own mind. I think that from what I read about *chimeras* there's a sense in which there are no real chimeras, that there is an incorporation of something coming from a different species into the life of whatever organism that is.
>
> In other words, the geep, strictly speaking, is a goat, is a goat. Now, this may prove to be useful in trying to draw boundary lines, and as Ben just said, for instance, just using pig valves clearly does not threaten the identity of the host. So the real question is here what kinds of mixers are such that the identity of the host is being threatened. But that brings us back, of course to the need to identify in humans what is that core.[30]

The question of classification and contours and boundaries thus gains significant ethnical and anthropological topicality.

9.2.4 The Creation, *Becoming Instead of Being* in *and* as *Music, and the Arrow of Time—A Mimesis of Nature. Music as an Active Component of World View*

In spite of such parallels, it would still be an exaggeration to claim that science was mirroring music in its procedures or model objects such as the genetically engineered creation of chimera mentioned here. Generally speaking, the relationships and processes of mirroring occur in the opposite direction. Composers try, on the one hand, to imitate natural sounds from wind and thunder to bird-"song" in and as music. On the other hand, composing is about a relatively abstract mirroring of natural processes, including mathematical procedures, which is mediated by procedures of production and composition, such as e.g. in "Fractal Music", under the banner of "New Age" ideology or fascinated by procedures and phenomena of the "fractal" such as in György Ligeti, Messias Maiguashca and others. Once again, a comparison between the original and the copy, the model and the mimesis could be productive here in the light of Fuzzy Logic.

The archaic value of music in universism and particularly in occidental Pythagoreanism is in a way a precursor to the modern 'musical turn' in regard to both nature and society. The Pythagorean musica mundana, the "world music" with the tertium of the 'number', is not a true cosmological causa efficiens, that is, an efficient cause in the

---

Therese Muxeneder, http://www.schoenberg.at/index.php/de/joomla-license-sp-1943310035/kammersymphonie-nr-2-op-38-a-38b-1906-1939, May 10, 2017, retrieved August 1, 2017.

[30]http://bioethics.georgetown.edu/pcbe/transcripts/march05/session6.html, retrieved November 8, 2017.

Pythagorean-Platonic and then in a rather weakened manner still in the Aristotelian-scholastic sense.[31] However, this world music is indeed a 'copy', thus however something caused rather and something primarily at work. Yet in this, it is not even a mathematical-musical reflection of the world but that of a certain mythological, pre-scientific *image* of the world.[32] The archaic 'universism' is still commonly known as something highly virulent in the form of religious or "holistic" views of the world. Sometimes, the musical turn gains the character of a revenant. For especially in art, advanced techniques and antiquated content, the use of modern science and unscientific or antiscientific explanations of the world can be immediately connected, as for instance in Michael Ranta's *Yuen Shan* (1995–2007)—*Ein elektroakustisches Ritual in vier Teilen.*

> "Yuen Shan" is an homage to the mountain by the same name in Taiwan where Michael Ranta lived for several years. The work, which is inspired by Taoism, consists of four chapters which, on the one hand, present the four epochs of the *history of the earth* (from the origin of the world up to today) and other the other hand the four cycles of human life *according to the Hindu tradition*: 1. birth, childhood, learning, 2. experience, family life, 3. old age, wisdom, 4. withdrawal, asceticism, death.[33]

The mimesis of outer nature is always mediated through societal conditions, from the state of the art and the qualification of workers to everyday consciousness up to a consciously formed worldview. An example which needs to be at least sketched here is when becoming rather than being *in* and *as* music is presented. It is quite telling that this kind of presentation of natural phenomena and processes only unfolds with the dynamization of the world view and particularly with the dynamization of the economy and society and especially with the historicization of cosmology in and *after* the classical phase of enlightenment after the middle of the 18th century under the banner of "natural history". In music, the depiction of the coming into being of the cosmos is initially still embedded into the religious mythology of "creation". The libretto to Haydn's oratorio *The Creation* of 1796/1798, first performed in 1799, is based on *Genesis*, the Biblical book of Psalms, and John Milton's *Paradise Lost.* (See above 6.3.2) Here, fuzziness is the starting point as that which is unformed, as "chaos" in the sense of the Old Testament which has been upgraded for some time as a

---

[31] Nor is mathematics: Numbers or numerically proportioned tones are not the cause of planetary movements but their reflection. This is still quite unclear in Pythagorean and possibly still in Ptolemaic thinking, maybe even vague in Zadeh's sense. The orientation at the ideal of beauty and perfection for which the circular movement stands here obstructs closer insight into the reality of nature. The equations by Kepler and the ellipses are much close to this, even if Kepler himself still believed in the reality of "spherical harmony".

[32] Cf. among others Heister 2009.

[33] Quoted after *Newsletter der Abteilungen für Musik der Gegenwart und Sound Studies am Musikwissenschaftlichen Institut der Universität zu Köln*, email to distribution list on October 18, 2017. The generic term "ritual", that is, the regression to the sacral and the rigidly repeated, takes back the achievements of art as secular part of the world view into the sacral. This matches the recommended regression in perception to the immersive instead of the concentrated with recommendations such as bringing seating cushions, sleeping mats or similar things to a concert in order to create an appropriate frame for the contemplative air of the work. There obviously is blurring as a conscious regulative in reception, too.

scientific term. In Haydn's work, order still originates from chaos by divine intervention. Haydn's century, following on the discoveries of Newton, held the view that an orderly universe—particularly the mathematically governed motion of the heavenly bodies—attests to divine wisdom. Deism was a quasi-fuzzy compromise between strictly secular and traditional sacred determinism. Haydn, a naturally curious man, may have had an amateur interest in astronomy, as while in England he took the trouble to visit William Herschel, ex-composer and discoverer of Uranus, in his observatory in Slough.[34] The creation of light is depicted first with a soft pizzicato note from the strings, followed by a sudden surprise fortissimo C major chord on the word *Licht* (Light).[35] This moment created a sensation during a public rehearsal, not the premiere itself, as Frederik Samuel Silverstolpe, a Swedish diplomat, wrote:

> At that moment when light broke out for the first time, one would have said that rays darted from the composer's burning eyes. The enchantment of the electrified Viennese was so general that the orchestra could not proceed for some minutes.[36]

"Enchantment" and "electrification" may mark the two poles of archaic magic and modern enlightened technology. The enlightened version of the origin of the universe was developed by Kant as early as 1755 in his *Universal Natural History and Theory of Heaven*. The scattered matter in a kind of original nebula in the universe moves because of gravitational force by attraction and repulsion in the direction of a state of ordered balance. This "nebular hypothesis" was then limited to the origin of the solar system by Laplace in his 1796 *Exposition du système du monde*. Leaving out Theism, Haydn was therefore indeed at the height of contemporary science.

The dynamics of cosmic-natural as well as societal developments was then really only musically developed by Beethoven based on the experiences of the revolution and the Napoleonic Wars after the model of the French revolutionary music, particularly Gossec, namely as a self-organization of matter and not as a divine "no sooner said than done" of the type of "let there be light". His First Symphony in C major of 1800/1801 transfers Haydn's cosmological model of the transition from nothing into being through becoming into independent instrumental music. The meandering introduction goes against the grain of the usual musical logic of beginning with the key-defining tonic. Beethoven, however, goes on a rather long harmonic detour in his very long introduction until from the – well thought out – disorder, the key is established and order crystallizes and he reaches the actual beginning with the basic key and then the exposition of the first theme. According to the rules of building forms and tuneful progression, this constitutes a blurring. From the perspective of the presentation of national and societal processes, it is a clarification. Aside from the indirect mimesis of natural processes and the abstracting mimesis of becoming and development in general, this beginning has a further dimension as it also constitutes a grandiose biographic beginning. With this, Beethoven begins as a composer

[34]See https://en.wikipedia.org/wiki/The_Creation_(Haydn), December 10, 2017, retrieved December 12, 2017.

[35]See ibidem.

[36]Quoted from ibidem. Emphasis H.-W.H. (See above Sect. 6.3.2.1).

of symphonies, that is, the largest secular musical genre aside from the oratory, and transcends his former limitation as a composer of piano and chamber music.

The discovery of the "time arrow" also for nature, of irreversible processes against the world view of Newtonian mechanics, was significantly driven by thermodynamics since the 1820s. In 1824, in Beethoven's lifetime, N. L. S. Carnot assumed that caloric energy was retained; Julius Robert Mayer then discovered the *changeability* of energy forms in 1841, and in 1854, Rudolf Clausius linked Mayer's and Carnot's ideas. The historicization of physical nature which originated from here was supplemented by the leap in the historicization of biological nature, which had already been driven by the enlightenment through Buffon, with Darwin's *Origin of the Species by Means of Natural Selection, or the Preservation of Favoured Races in the Struggle for Life* (1859), which was published in German as early as 1860.

In his *Rhinegold*, Wagner radicalizes Haydn's *Creation* both musically and ideologically. The work presents the prologue of the plot with a focus on Siegfried, the hero, as a *prelude* in the tetralogy of *The Ring of the Nibelung*. It was first performed in 1869; Wagner wrote the text in 1851/1852 and began first sketches on the music in 1852. Wagner's own description of how the idea of the sound originated in a half dreamlike manner was already discussed in the conceptual form. (see above 2. and 6.3) In this context, this idea in Wagner constitutes the first crystallization of figures from a previously nebulous and almost somnambulistic state. The second crystallization is both a musical subject as well as its presentation as a musical process. The elementary beginning of Wagner's *The Rhinegold* is a copy of an origin of the world—not a creation of the world, despite the *Ring* trilogy being rife with gods, goddesses, heroes and heroines. Music shows—and *is*—the becoming of a nature which is initially disturbed and almost destroyed through culture (here criticized as capitalist) but ultimately changes into a higher state. (See 5.3.3) A sound, an abysmally deep E♭, is the starting point as the elementary. Gradually, the E♭ major chord builds up via the also elementary fifth[37] into a composed-out segment fitting into the partial tone spectrum, as part of a pendular movement with a certain upwards tendency. The chord as a timbral composition ante letteram, realized prior to this term being coined, is enough for a time in a changing and altogether growing instrumentation until nature is dissolved into culture via the tones from the dominant and the dominant seventh chord. Music circles in gracefully swimming movements just like the Rhine maidens and describes these two harmonic functions. The—paradoxically put—monumental idyll in a way forces an immersive music experience, the diving into the "full waving of the depth of the water"—thus a scenic instruction in the libretto, and Wagner's waking dream image also describes this submersion. "M'immerso nel immenso" (I submerge myself into the infinite), one could call this auditive impression of *Rhinegold*, borrowing a paraphrase from Giuseppe Ungaretti's "M'illumino /d'immenso" (I enlighten myself /by the immense). This seems quite topical: "Datumsoria: The Return of the Real" & "The Art of Immersion" are announced in the September 2017 newsletter from the ZKM in Karlsruhe as a double exhibition opening and as part of that year's programmatical focal points of the ZKM on the digital revolution (Email

[37] See above Sect. 6.10.1, p. 395, the organum in parallel fifths and fourths.

distribution list dated September 1, 2017). It is reassuring that the real reappears and apparently has learned to reemerge from the "deep water".

Bruckner, a Wagnerian, then uses empty and repeated fifths in several of his symphonies, particularly in the 2nd (1871/1872) and 3rd, which correspond to the naked fifths of the *Rhinegold* prelude, as an indefinite-incipient, fertile chaos with the potentiality of becoming, a necessary further development. Only from this do the themes then unfold. It is a kind of "reluctant realism": The conservative of all things, the socially rather backward, designs the origin and becoming, the transition from the not-yet-being instead of the traditional and usual immediate positing of the beginning with the main theme. The 8th symphony begins with an F which is indefinite in key, "in a tonally ambiguous manner with a theme rhythmically reminiscent of the main theme of the first movement of Beethoven's Symphony No. 9 in D minor"—[...] according to theorist Heinrich Schenker this passage is '*like the beginning of the world*'".[38]

9.2.5 Seismograph. Music as a Historical-Social Document

Furthermore, and as a fifth element, the musical turn means that the reflection of the societal and the sacral in music receives—or should receive—adequate appreciation as an expression of historical circumstances and developments. As a mimesis of material societal states and states of consciousness, music is a special form of the reflection of history. As a historical document, as a source, it mainly reports on the subjective point of view, of mentalities, psychological-social sensitivities, attitudes and responses etc. Music, in this assertiveness, is a "seismograph"—a term by Theodor W. Adorno which in turn is a metaphor from the area of geology.

Music does not simply archive what has been but how it was perceived and felt. Through and from music as well as other arts, we therefore receive additional information, learn different and more things than from economic, legal, diplomatic and other documents. It would therefore be highly interesting, albeit also very challenging and complex, to attempt, using Fuzzy Logic methodology, to correlate scientific presentations e.g. of the First World War, the "Great War", with literary, pictorial and musical depictions. Even the methodological-methodical research design alone should yield new and unexpected information and insights.

The design of this is mediated via diverse and diffuse everyday experiences which, as a rule, are themselves mainly "vague" and "fuzzy", but also via complex and already shaped images of the world. In one way or another, sometimes a sudden sharpening, a downright lightning-fast insight into the state of the world will develop. A US-american concert goer reacted in an emphatic and precise, in her turn seismographic manner to (new) music as a reflection of the historical-societal in spite of some naïveté. After the exiled Ernst Krenek had played his new and rather dissonant

[38] https://en.wikipedia.org/wiki/Symphony_No._8_(Bruckner), February 2, 2017, retrieved March 17, 2017, emphasis H.-W.H.

piano concert as a first performance in the Boston Symphony Hall in the afternoon concert on November 4, 1938, the old lady was reading in her playbill. After a while she looked up and addressed her husband by remarking that things in Europe must be in a terrible state. She had said everything there was to say about the music she had just heard: the fear in this music, the confusion, the shock, the tension, the despair, the emptiness, the philosophy in "there was nothing one could hold on to".[39]

9.2.6 "West-Eastern Divan Orchestra" and "Mozart Effect"—Chaos and Anomy, Cosmos and Harmony. Music as a Model for Dialectics

In spite of the 'musical turn', music is usually not a model for science, as mentioned above. It is, however, as a sixth point, a model of 'harmony', as a unity of different and even contradictory things as dialectic thinking since Heraclitus' times has recognized. As initially discussed, Fuzzy Logic has some affinity to this. For instance, analogies to the matter-energy or the wave-corpuscle contradiction within one and the same 'object' in music are e.g. ambivalences in a harmonic progression or a part of the form. In this, transitions and possible points of change can be more precisely determined by means of Fuzzy Logic. Musical turn here generally means that insight in and through music mutually penetrate and amplify each other. A tonally-real dialectic of the dissolution of contradictions and the unification of differences is one of the main abilities of music. For in music, different voices can sound simultaneously. This seems trivial as it constitutes the norm, particularly within the framework of European and US music—but it is not. For such a polyphony does not work in the language of words, at the most in Dadaist or generally avant-garde poetry between Hugo Ball and Hans G Helms. In everyday communication, verbal polyphony only leads to chaos. When making music, however, polyphony is an exponentiation of a dialectic which is idiosyncratic to music. For in and through music, the I and the We are simultaneously confirmed and activated; when making music together, individuality and communality are unified. "Music says we", said Theodor W. Adorno once; and, as I would add, at the same time music says "I".

Collaboration and interaction in general can even react with the societal world via common practice or at least mediated through intensive receptive appropriation. Music as an art can give a glimpse of how society, the economy, the social could be designed in a joint and conscious manner. The 'Serialism' after World War II till the early 60ies was an important model in this direction: Composers as Olivier Messiaen, Karlheinz Stockhausen, Karel Goeyvaerts, Henri Pousseur and others

[39] As reported by the exiled publisher and concert organizer Heinsheimer 1953, p. 58 and 62f.; quote in p. 63. Heinsheimer quotes here from a line in the finale of the Brecht/Weill opera *Rise and Fall of the City of Mahagonny,* 1930: "Und weil es nichts gibt, woran man sich halten kann"—fuzzy by the way: He uses the preterite "gab" instead of the original present "gibt".

tried, to organize all musical dimensions (called "parameters") as pitch, duration, dynamics, timbre in a very strict, rational, planned way.

This modelling music as the utopian other, however, can only work when art remains the other, the non-commonplace. Art is different, at least in principle. The principle tolerates many exceptions which tear down the border between art and reality, between art and "life", and thus of course create illusions. It makes more sense instead to endure the art specific other state, the limbo between the imaginary and the real. To put it briefly, imaginary-real means: Something is not real in the actual, material and objective sense, but it is nevertheless effectively and subjectively real. If, for instance, citizens at a philharmonic concert do not feel themselves to be singular and by tendency hostile subjects of the market in their appropriation of the music but instead as people who become brothers and sisters, then this is what they are—at least for the duration of the concert. The peacemaking effects of the *West-Eastern Divan Orchestra* led by Daniel Barenboim certainly reach even beyond the performances. In them, the collaboration and coexistence of young Arab Palestinians and Jewish people appears in a concrete manner and as already realized within this framework.

Generally speaking, it would be ideal if everyone who is present in the concert assembly participated actively and joined in the performance in order to realize social harmony. But an attentive, sympathetic reception is enough. Thus, the Frankfurt orchestra musician Carl Gollmick writes in the year of the 1848 revolution that an orchestra was never more independent than in the symphony because it constituted its own state for each member and wherein each member is a free citizen and a small king for himself. He stated that with its thousands of invisible threads, all members were united as a federation in which all earthly matters were shed and where each heart felt the religion of tonal art more intensively than ever before. He suggested that mortal enemies should play in a symphony together as they would have to be brothers, surrounded by secret appeal. Skeptically-realistically, he concluded that this would last as least as long as the symphony.[40] So much also for the so-called "Mozart Effect", an illusionary idea about the quasi-automatic positive effect of music. That music, particularly that by Mozart, automatically furthers the intelligence, performance in school and others evidently is also not true. Games and art in any case have better things to do than to improve performance in school. Overall, music is an exceptional model for the effect of dialectics. This is very firmly part of its nature. It takes very different elements of reality, of time and sound, of the material and the imaginary, of the emotional and the cognitive etc. into its inner anthropological stratification and brings the contradictions contained within into a processual resolution.[41] Music promotes the sense for the contradictions of reality, sometimes consciously, more often probably unconsciously and therefore even more effectively. Even at the level of a conformist everyday consciousness, we sometimes get a glimpse of such states

---

[40] *Allgemeine musikalische Zeitung* 50, 1848, p. 129; quoted after R. Schmitt-Thomas 1969, p. 708f.

[41] As affirmation of me *and* we, we *and* the others, music is ambivalent: it can "incite intergroup hatred" and "valued identities" of groups, but also "reduce intergroup hostility"—see Gollmick 1848 said—"e.g., via cross-cultural musical collaboration and contact". (Harwood 2017, p. 1).

of affairs and insights, such as in a sports report in which a skier was compared with Mozart because of her skills. She said that if she wanted to build up her energy before a race and visualize the course she listened to music such as *Get Lucky* by Daft Punk, or Ellie Goulding, or NoNoNo. These songs, she said, formed the beat of her mental race—and that this also worked with Mozart's music because his compositions were abstract but still (sic!) had a rhythm. Mozart inspired her because *chaos and harmony were in balance* in his works.[42]

All these phenomena once again justify the introduction of the 'musical turn'. Thus, the intuitive and unconscious is also made conscious to some extent. This makes a way of thinking which is inspired by music and aims to achieve a deeper, more intensive appropriation of reality available in cooperation with Fuzzy Logic and in combination with the cognitive as well as practical appropriation. In this way and dimension, music gains a status which is comparable to a kind of 'meta-science'. As mentioned, it can serve as a model for different areas of reality both methodically and in an exemplar fashion, as mimesis and as a model, especially a methodically-procedural one.

Like Fuzzy Logic, 'music' also appears as a form of thinking in connection with the musical turn which also forms a partial meta level parallel to both science in general and individual disciplines. Not least the use of musical terms *outside* of music as metaphors for social states of affair points to the paradigmatic relevance of music. In the following, I introduce several examples for biological-psychological issues, from matters of the body to the affective.[43] As Rudolf Bilz[44] points out, parables from music seem particularly suitable for characterizing vital and emotional connections. He gives the example of using the terms crescendo and accelerando in connection with sexual intercourse to refer to an increase in the intensity of movements as well as an accelerated pace. There are musical metaphors also for sleep, as Bilz describes by saying that if in nature's score there were tempi for falling asleep, they would not, for instance, be *furioso* but rather *senza espressione* or, if the term existed, *monotono*.[45] As the nice invention of the musical term *monotono* shows, even metaphorical music can be transferred back to actual music. It is then however only valid for an imagined, fictional music which we have touched on in connection with the quotation form (See Chapter 8).

---

[42]As reported in *Sport. Jungstar Shiffrin. Überall ist Tiger Mom, Der Spiegel*, Hamburg, 51/2014, p. 106ff.; quote on p. 107. Emphasis H.-W.H.

[43]Thus for instance Breidenbach 2015.

[44]Bilz 1967, p. 9ff.

[45]A term used by Jakob von Uexküll: *Bedeutungslehre,* Leipzig 1940. Uexküll also speaks of impulse tunes which move a subject in a certain way ("Impulsmelodien [...] die ein Subjekt in bestimmter Weise bewegen"; quoted ibidem).

9.3 Productivity II. Interdisciplinary Cooperation Transgressing Disciplinary Boundaries and 'Inner Interdisciplinarity'. Fuzzy Logic, Music and Musicology

Fuzzy Logic, music and the science of music meet under the banner of interdisciplinarity. Interdisciplinarity as a sum and a synthesis of theory and practice respectively application has been in the air (again) for a long time.[46] It appears all the more urgent because of the constantly progressing and unavoidable specialization in the division of labor as a means of compensation. When applied *within* musical culture and the sphere of music, Fuzzy Logic has been successfully at work for a while.

9.3.1 Interdisciplinarity as sum and synthesis. Musical contribution to "The Philosophy of Fuzzy Logic"

Declarations and trite declarations of intent are not enough. However, the realization of interdisciplinary cooperation is hard work. It probably is most successful in groups in which those who struggle with what they do not understand in the respective foreign discipline can receive personal clarification where needed. In itself it remains difficult nevertheless. Especially in the relation of Fuzzy Logic and music, the arrogance of the natural sciences including mathematics has an effect here.[47] Mathematics can, however, be counted as a kind of meta science among the natural sciences. These think of themselves as the center of the world these days. This is not entirely justified: For not everything can be understood adequately by means of natural science and mathematical methods. Let me only mention the pertinent 'reductionism' problem as well as more complex approaches and ways of thinking such as consideration of qualitative differences of different layers of reality, emergence or spontaneous self-organization. On the other hand, this arrogance is not entirely unjustified as the advances in the epochal upheavals became particularly apparent at the latest with Max Planck's discovery of the Planck effect quantum or Planck constant and then the theory of relativity. This evidence is amplified through the technological progress which is based on it in spite of all ambivalence for which the atomic bomb, nuclear reactors, the deep sea dive boat batyscaphe (Auguste Piccard) and "renewable energies" are particularly significant.

Nevertheless, and in spite of the 'musical turn', knowledge or at least the acknowledgement of the musical as well as its theoretical processing in general may not be indispensable for the realization of numerous areas of reality. In order to understand the law of falling bodies one does not need to know about the descending fifths in

[46] Cf. e.g. Synergie-Forum http://www.zflprojekte.de/synergie/doku.php, retrieved March 3, 2017; Petzer 2012, etc.

[47] Cf. on the other hand see Zadeh 1999.

music; and in order to understand a musical piece of art, one does in turn not need to understand physics, in spite of the fundamental acoustical-physical level of the musical process. However, acknowledging music is useful for the knowledge of the world, not least as a sharpening of attention. From a different perspective and in a somewhat more general manner, Werner Heisenberg, for instance, continued to emphasize the "beauty" of the mathematical as well as the physical as a significant part even of specifically *scientific* insight. In so far, the science of music including the perspective on music is significant for Fuzzy Logic, on the one hand because of specific qualities of music, on the other hand as part of an interdisciplinary array from the point of view and as part of the framework of Fuzzy Logic.

Fuzzy Logic is, as mentioned and explicated, *a logic of non-identity*. It tends towards productivity and innovation as a logic of variation and change. Its modes and methods of thinking, based on Lotfi Zadeh's concept, will continue to affect the research and study of music and other arts in all its structures and processes.

Based on studies on the optical, Jori Cat points to graduations *within* Fuzzy Logic such as different kinds of sets, fuzzy, rough and near sets, as well as "values of imprecision"[48] in which, for instance, aspects relating to the theory of probability as well as modal logic meet. Also in connection with music, it is important to focus on the acoustic, to reveal last but not least a little bit of „the philosophical background relevant for issues such as representation, categorization and reasoning".[49] The program and title of online journals with titles such as *Archives for the Philosophy and History of Soft Computing* (since 2013) then also point explicitly in the direction of a "philosophy of Fuzzy Logic".

9.3.2 'Fuzzy Logical Turn'?

The generalization through linking Fuzzy Logic with further musical and artistic universes of discourse is a double-sided process: the integration of foreign disciplines into Fuzzy logic and, conversely, the integration of Fuzzy Logic into different disciplines.

Digit or image, number or artistic imagination, Brownian motion or movement in time and space of a musical work often seems worlds apart. And they are.

But at the same time, they are parts of one world. And the confrontation or combination of these different areas of reality in thinking and the different logics (including "emotional logic") are part of the world view, of the scientific, artistic, also religious and pre-, un- or antiscientific ideas which we conceive. This always occurs in a historically concrete manner, differentiated by social position, degree of education, societal-political habits, national and regional specifics and many more. Instead of a "unified science" based on an ultimately idealistic positivism, the—mediated—unity of sciences is an aim to strive for.

[48] **Cat** 2017.

[49] Ibidem.

It may be a somewhat unrealistic question whether in such a context, Fuzzy Logic, aside from its unquestioned qualities as an independent theory, might be established as a kind of 'meta language' or meta theory for principally all sciences, be they natural sciences, humanities or social sciences. Terminologically it would probably be better to refer to it as a scientific discipline which transcends the individual subjects. For in any case, its approaches are already effective at a meta level compared to the individual disciplines, mainly at the level of methodology but simultaneously also as a method and with its own methodical repertoire. In such a function as an interdisciplinary methodological-methodical approach, it would need to (further) develop both relevant trans- and interdisciplinary as well as individual discipline-specific concepts in and of itself.

In regard to the question about its potential value as a 'meta science' for other disciplines, it would be comparable, maybe at a lower level in terms of the degree of generality and generalizability, with communication theory, semiotics, information theory, cybernetics (which in the meantime has been dethroned in favor of informatics), "general system theory" (based mainly on Ludwig von Bertalanffy), probability theory and finally logics (including multi-valued logic).[50]

9.3.3 'Inner Interdisciplinarity'. Answered, Processed, Open and New Questions

In the context of music and Fuzzy Logic, the clarification of such elementary questions as the moods mentioned in the beginning as well as more complex ones such as problems of storage and searches and similar has historically largely been dealt with (see above in Chapters 1, 3, 5, passim). However, even at this fundamental level, among others an analysis of pitch and rhythm under the banner of Fuzzy Logic at an international, global level has not been undertaken yet. Further current and only partially resolved tasks for Fuzzy Logic in the area of music are the numerous and diverse clarifications of similarities in tunes, motives, themes between different works, such as, for instance, in questions of plagiarism and copyright, and in the individual forms of existence of a single musical respectively artistic process—a never-ending story. Furthermore, Fuzzy Logic also continues to be productive for the—multidimensional—systematization of musical genres, styles etc., so far utilized mainly for problems of storage and finding—as incidentally for everything that can or must be rubricated, classified, typologized or systematized. Methodically, the problem is similar for instance in medical diagnostics or in zoology and botany up to questions of evolution and particularly the human-animal relationship. Here, in spite of the results and achievements up to now, much remains to be studied. The dialectics of the boundary always play a central role: separation and transition, breakage and continuance, line or more or less broad band.

[50]Carnap already spoke of logical tolerance and metamathematics ["logische Toleranz" and "Metamathematik"] (Seising 2005, p. 4).

In the future, it will be important to analyze and productively apply the points of contact and intersection, the similarities, analogies and isomorphisms between the principles (including substantiation and diversification) of Fuzzy Logic on the one hand and the respective disciplines of art sciences in a systematic manner; and to do this from the point of view of methods as well as the respective sections of reality which are researched and depicted through them. The latter include ideal, conceptual realities or assemblages such as art as the imaginary real. The potential non-applicability of Fuzzy Logic or in turn of methods of the sciences of art need particular attention as well.

It is difficult to find truly convincing algorithms with which to measure similarities in an exact and quantitative manner, particularly those referring to the arts/music-reality-relations. Quantitative and qualitative points of view are opposites, of course. However, they are not contradictory but complementary. As a rule or tendency, the qualitative questions and answers should in a second step of research be transformed into quantitative results—not as a scope in itself but as a more precise version. However: in order to quantify something, this something must be recognized, conceptualized and shaped for further research. And this something is something real, it is not immediately and simply given but rather it must be explored. In the case of music in particular, this is my duty and my part of the collaboration with Fuzzy Logic. I provide preparations as, so to speak, the Propylaea to the Parthenon. But you do not reach the Parthenon without going through these Propylaea. For Fuzzy Logic applied to music, it is necessary to understand music, and indeed more than only superficially and in regard to its common sense qualities. To produce and to formulate such algorithms or mathematical formulas in general would transcend my mathematical capabilities—the indispensable *'inner interdisciplinarity'* especially of social and art sciences, as musicology here reaches not objective but subjective-personal limits. This means a complex, encaptic variety of achievements from outside, even well outside of one's own discipline.

I define 'inner interdisciplinarity' as the cooperation of different and even disperse disciplines, in regard to methods as well materials, inside of a discipline, asking for connections between the three main realms of reality, i.e. nature, society, mind. To evolve and to practice this 'inner interdisciplinarity' would, by the way, in return not be inappropriate for other sciences and humanities, too. One way or another: Here, real interdisciplinarity is called for, especially from those coming from mathematics, informatics and computer sciences and technologies in general. In respect to music and from the view of musicology, they often stay restricted to very elementary matters and questions. (From the other side, much of my deliberations may appear naïve or primitive.) In any case, collectively and concertedly we would proceed faster and better. Scientific progress demands such cooperation beyond sheer declarations. Interdisciplinarity as cooperation beyond the borders of the respective individual scientific disciplines normally stands in the foreground. It is supplemented by the above-mentioned *'inner interdisciplinarity'*. What constitutes an "auxiliary science" in this is usually not absolute but relative. Thus, numismatics is an auxiliary science of historiography, for instance with regard to images of rulers. The relational status can, however, just as well be reversed. Then archeology and historical science become auxiliary sciences of numismatics, for instance to date coins. This is to a certain degree normal anyway, perhaps even always the case. But it is not reflected enough. This 'inner interdisciplinarity' is visible in particular for relatively heterogeneous and relatively young disciplines such as musicology. It includes inter alia philosophy and sociology, psychology and economics, numismatics and watermark- or paper-research, history and physics (acoustics) etc.—and also, from now on, Fuzzy Logic. It is always less about the adoption of methods and terms from these scientific approaches in detail.

The detailed appropriation of methods and terminologies from various scientific disciplines is, as said in the Preface, important but probably not the main factor. What is decisive is rather that we let ourselves be inspired by the fundamental thoughts and ways of thinking of the others in a productive manner. This applies both to the sciences, from music and other arts, as well as to Fuzzy Logic. It means (1) thinking in contexts and universal symmetrical or asymmetrical interactions, (2) the remembrance that insight just as artistic or other practice has 'fuzziness' as its boundaries, and (3) the most precise analysis possible of just this fuzziness in its respective conditions. Not least in this way, initially seemingly uncrossable borders progressively approximate permeable barriers—with both barriers and borders meant as mental-subjective *and* material-objective facts.

All these problems must be resolved in an interaction of empirical and theoretical aspects and methods. This is an aspect accentuated by Rudolf Seising, also referring to Zadeh, mentioned in the Preface. That is in general trivial, but not in this special case. For Fuzzy Logic needs a little help—I suppose—for its enhancement as theory and its further expansion to new fields of research and practice, even from sciences at the periphery.

I am certain that some problems and questions will arise during the course of such research which I would not even be able to name right now, both from the side of musicology and art sciences as well as from the side of Fuzzy Logic.[51] In an ideal case, this productive interdisciplinary is reciprocal in such a way that not only the musicologists learn from the sciences, but also vice versa. In view of the relative marginality of music sciences in the canon of disciplines, the latter might seem a pious hope. Marginal or not: There are dialectics of centre and periphery in globalization and on the way to a world society. New developments and impulses do neither necessarily nor exclusively originate from the center: Musicology can help seemingly or truly "harder sciences", last but not least Fuzzy Logic.

However, I am also sure that already known as well as unknown problems in the future will be resolvable at least principally, with necessarily always only asymptotic approximation to the respective "absolute" truth, and helped along by a cooperation

[51] In conclusion I quote from the of a book by an expert on Fuzzy Logic: "Meantime, I published [...] but the work on the present book was progressing very slowly. The readers will see that there are a lot of open problems in the book, which I cannot solve at the moment, but I hope that the readers will be more successful than me in this enterprise. [...] I do not cherish any illusions that the present book is perfect. However, I do believe that the present book holds some interesting ideas that open opportunities for future research. I would encourage researchers to pay more attention in this direction." (Atanassov 2016, Preface, p. VII.).

between the sciences related to music and other arts on one side and on the other Fuzzy Logic.

References

Atanassov, K.T.: Intuitionistic Fuzzy Logics, Studies in Fuzziness and Soft Computing. Springer, Berlin (2016)

Bilz, R.: Die unbewältigte Vergangenheit des Menschengeschlechts. Beiträge zu einer Paläoanthropologie (Theorie vol. 2). Suhrkamp, Frankfurt a. M. (1967)

Boehmer, K.: Zur Theorie der offenen Form in der Neuen Musik. Tonos, Darmstadt (1967)

Breidenbach, B.: 'Stimmung': Conceptualising attunement in music, philosophy and literature, symposion music and literature: Critical polyphonies. Durham Castle, Durham University, 2nd July 2015

Cat, J.: Fuzzy pictures as philosophical problem and scientific practice. A study of visual vagueness (Studies in Fuzziness and Soft Computing, ed. by J. Kacprzyk, vol. 348). Springer Wissenschafts-Verlag, Heidelberg etc. (2017)

Harwood, J.: Music and intergroup relations: Exacerbating conflict and building harmony through music. Rev. Commun. Res. **5** (2017)

Heinsheimer, H.W.: Menagerie in Fis-Dur. Musikalische Impressionen aus Amerika. Pan Verlag, Zürich und Stuttgart (1953)

Heister, H.-W.: Weltenmusik und Menschenmusik. Ein Knotenpunkt in den Traditionslinien der universistischen und mathematisch-quadrivialen Musikauffassung. In: Banse, G., Küttler, G., März, R. (eds.) Die Mathematik im System der Wissenschaften (Abhandlungen der Leibniz-Sozietät der Wissenschaften, vol. 24), pp. 193–216. trafo, Berlin (2009)

Heister, H.-W. (ed.): Schichten, Geschichte, System. Geologische Metaphern und Denkformen in den Kunstwissenschaften (Zwischen/Töne, ed. H.-W. Heister, vol. 6). Weidler Buchverlag, Berlin (2016)

Heister, H.-W.: Einleitung. Aufbau- und Abbau-Prozesse, Wandlungen, Verwerfungen in Natur und Gesellschaft. In: Heister, H.-W. (ed.) Schichten, Geschichte, System. Geologische Metaphern und Denkformen in den Kunstwissenschaften (Zwischen/Töne, vol. 6), pp. 11–42. Weidler Buchverlag, Berlin (2016)

Heister, H.-W.: Spiel und Musik. Eine systematische Skizze. In: Croatian Musicological Society. Int. Rev. Aesthet. Sociol. Music. (IRASM) **47**(2) (December 2016), pp. 171–194 (2016)

Kacprzyk, J., Castillo, O., Melin, P. (eds.): Nature-Inspired Design of Hybrid Intelligent Systems. Springer, Heidelberg etc (2017)

Kreuzer, S. (ed.): Experimente in den Künsten. Transmediale Erkundungen in Literatur, Theater, Film, Musik und bildender Kunst. transcript, Bielefeld (2012)

Meyer, P.M. (ed.): Sound History. Acoustic Turn. Wilhelm Fink-Verlag, München (2008)

Paul, G., Schock, R.: Sound des Jahrhunderts. Geräusche, Töne, Stimmen 1889 bis heute, pp. 313–319.. Bundeszentrale für politische Bildung bpb, Bonn (2013)

Petzer, T., Steiner, St. (eds.): Synergie. Kultur- und Wissensgeschichte einer Denkfigur (Trajekte-Buchreihe). Wilhelm Fink Verlag, München (2016)

Schmitt-Thomas, R.: Die Entwicklung der deutschen Konzertkritik im Spiegel der Leipziger Allgemeinen Musikalischen Zeitung (1798-1848). Phil. Diss. Johann Wolfgang Goethe-Universität zu Frankfurt am Main (1969)

Schneck, D.J., Berger, D.S.: The Music Effect. Music Physiology and Clinical Applications. Jessica Kingsley Publishers, London and Philadelphia (2006)

Schulze, S., Conrad, D. (eds.): Tiere: Respekt - Harmonie - Unterwerfung (Ausstellungskatalog Museum für Kunst und Gewerbe Hamburg). Hirmer-Verlag, München (2017)

Seising R.: Die Fuzzifizierung der Systeme. Die Entstehung der Fuzzy Set Theorie und ihrer ersten Anwendungen - Ihre Entwicklung bis in die 70er Jahre des 20. Jahrhunderts (english 2007). Steiner, Stuttgart (2005)

Wilson, P.N.: Reduktion. Zur Aktualität einer musikalischen Strategie (edition neue zeitschrift für musik, edited by Stoll, R.W.). Schott International, Mainz (2003)

Zadeh, Lotfi A.: From computing with numbers to computing with words—From manipulation of measurements to manipulation of perceptions. IEEE Trans. Circuits Syst. **45**(1), 105–119 (1999)

Bibliography

Adak, S.: Time-dependent spectral analysis of nonstationary time series. J Am Stat Assoc **93**, 1488–1501 (1998)

Adorno, Th.W.: Alban Berg: Master of the Smallest Link. Cambridge University Press, Cambridge, MA (1991)

Adorno, Th.W.: Einleitung in die Musiksoziologie – Zwölf theoretische Vorlesungen. Suhrkamp, Frankfurt a. M. (1962) (extended as: Einleitung in die Musiksoziologie, Rowohlt, Reinbek (1968)

Adorno, Th.W.: Philosophie der neuen Musik [1949]. Europäische Verlagsanstalt, Mannheim (1958)

Agon, C., Andreatta, M., Assayag, G., Amiot, E., Bresson, J., Mandereau, J. (eds.): Mathematics and Computation in Music. Third International Conference, MCM, Paris 2011. Proceedings. Springer, Berlin and Heidelberg (2011)

Ahn, S.-H.: Eastern and western elements in selected works by Giacinto Scelsi and Toru Takemitsu. Thesis Doctor of Musical Arts in Music in the Graduate College of the School of Music, University of Illinois at Urbana-Champaign (2014)

Ahrendt, P.: Music genre classification systems - a computational approach. Ph.D. thesis, Technical University of Denmark, DTU (2006)

Allan, H., Müllensiefen, D., Wiggins, G.A.: Methodical considerations in studies of musical similarities. In: Proceedings of the Eighth International Conference on Music Information Retrieval, pp. 473–478. Vienna (2007)

Allen, P.E., Dannenberg, R.B.: Tracking musical beats in real time. In: International Computer Music Conference, Glasgow, Scotland, pp. 140–143 (1990)

Alonso, M., David, B., Richard, G.: A study of tempo tracking algorithms from polyphonic music signals. In: Proceedings of the 4th COST 276 Workshop, Information and Knowledge Management for Integrated Media Communication, Bordeaux, France, pp. 1–5 (2003)

Amundson, M.A.: Talking Machine West: A History and Catalogue of Tin Pan Alley's Western Recordings, 1902–1918 (Popular Music Series Volume 2). University of Oklahoma Press, Norman, OK (2017)

Anderson, V.: 'Well, it's a vertebrate…': performer choice in cardew's treatise. J. Musicol. Res. **2**(3–4), 291–317 (2006)

Angustin, J.: George Antheil. Enfant terrible der Musik. Verlag Langen-Müller, Munich (1960)

Apel, W.: Die Notation der polyphonen Musik. 900–1600. Breitkopf&Härtel, Leipzig (1962) (revised and improved new edition of the English original; The Notation of Polyphonic Music. 900–1600 (1942)

Aramaki, M., Barthet, M., Kronland-Martinet, R., Ystad, S. (eds.): From Sounds to Music and Emotions. In: 9th International Symposium CMMR 2012, London, UK, June 19–22, 2012, Revised Selected Papers. Information Systems and Applications, incl. Internet/Web, and HCI, vol. 7900 (2012)

Arber, A.: Sehen und Denken in der biologischen Forschung. Rowohlt, Reinbek (1960)

H. Heister, *Music and Fuzzy Logic*, Studies in Fuzziness and Soft Computing 406,
https://doi.org/10.1007/978-3-662-62907-9

Arcos, J. Ll.: Music and similarity based reasoning. In: Seising, R., Sanz, V. (eds.) Computing in Humanities and Social Sciences (Studies in Fuzziness and Soft Computing, Vol. 273), pp. 467–478. Springer, Heidelberg (2012)

Arenas-Garca, J., Larsen, J., Hansen, L.K., Meng, A.: Optimal filtering of dynamics in short-time features for music organization. In: Proceedings of the 7th International Conference on Music Information Retrieval, Victoria, Canada, pp. 290–295 (2006)

Armstrong, R.E.: An Historical Survey of the Development of the Baroque Solo Keyboard Suite in France, England, and Germany. Graduate Theses, Dissertations, and Problem Reports. 3023 (2011). https://researchrepository.wvu.edu/etd/3023

Arnest, M.: Why Couldn't They Play With Their Hands Together? Noncoordination Between and Within the Hands in 19th Century Piano Interpretation (2012). https://www.semanticscholar.org/paper/Why-Couldn-%E2%80%99-t-They-Play-With-Their-Hands-Together/dbeb216f1347ea6ac44883b00bdc86365082c023?p2df

Arnold Schönberg Center, Wien (T 26.01). J. Arnold Schoenberg Institute 9(1), 55–77 (1986)

Asai, Y.: Vom Akkord zur Polyphonie. Zum Kompositionsprozess von Anton Weberns «Das Kreuz, das mußt' er tragen» op. 15/1, in: Mitteilungen der Paul Sacher Stiftung, no. 29, pp. 29–34, April 2016

Asendorf, C.: Die Künste im technischen Zeitalter und das utopische Potential der Kybernetik, In: Hagner, M., Hörl, E. (eds.) Die Transformation des Humanen. Beiträge zur Kulturgeschichte der Kybernetik, pp. 107–124. Suhrkamp, Frankfurt a. M. (2008)

Assayag, G., Dubnov, S.: Using factor oracles for machine improvisation. Soft. Comput. **8**(9), 604–610 (2004)

Assayag, G., Feichtinger, H. G., Rodrigues, J.F. (eds.): Mathematics and Music. A Diderot Mathematical Forum. Springer-Verlag, Berlin, Heidelberg, New York, Barcelona, Hong Kong, London, Milan, Paris, Tokyo (2002)

Assemlal, H.E., Tschumperlé, D., Brun, L.: Efficient computation of PDF-based characteristics from diffusion MR signal. In: Metaxas, D., Axel, L., Fichtinger, G., Székely, G. (eds.) MICCAI 2008, Part II. LNCS, vol. 5242, pp. 70–78. Springer, Heidelberg (2008)

Atanassov, K.T.: Intuitionistic Fuzzy Logics, Studies in Fuzziness and Soft Computing, Springer, Berlin (2016)

Attinello, P.: Signifying chaos: a semiotic analysis of Sylvano Bussotti's »Siciliano«. Repercussions **1**, 84–110 (1992)

Aucouturier J.-J., Pachet F.: Improving timbre similarity: how high is the sky. J. Neg. Results Speech Audio Sci. **1**(1), 1–13 (2004)

Aucouturier, J.J., Pachet, F.: Representing musical genre: a state of the art. J. New Music Res. **3**(1), 83–93 (2003)

Bahle, J.: Der musikalische Schaffensprozess: Psychologie der schöpferischen Erlebnis- und Antriebsform. S. Hirzel, Leipzig (1936)

Bakker, T., ap Sion, P.: The order of things. Analysis and sketch study in two works by Steve Reich. Zeitschrift der Gesellschaft für Musiktheorie **16**(1), 99–122 (2019). https://doi.org/10.31751/1003. Accessed 29 Nov 2019

Baldizzone, R.: Spontaneità organizzata ad 'arte'. Il metodo compositivo di John Lewis. In: Philomusica on-line. Rivista del Dipartimento di Musicologia e Beni Culturali (Università degli Studi di Pavia), vol. 6, N° 1 (2007). https://riviste.paviauniversitypress.it/index.php/phi/article/view/06-01-LAU01/101

Ball, H.: Erklügelte Nervenkultur. pendo Verlag, Zürich (1991)

Bandur, M.: Scelsi. In: Heister, H.-W., Sparrer, W.-W. (eds.) Komponisten der Gegenwart, vol. 34, pp. 1–44. edition text+kritik, München (2007)

Banse, G., Reher, E.-O.: Ambivalenzen von Technologien - Chancen erhöhen, Gefahren mindern, Missbräuche verhindern. In: Sitzungsberichte der Leibniz-Sozietät der Wissenschaften zu Berlin, vol. 112, pp. 19–41. trafo Verlag, Berlin (2011)

Banse, G.: „Nicht so exakt wie möglich, sondern so genau wie nötig!“ - Das Einfachheitsprinzip in den Technikwissenschaften. In: Sommerfeldt, E., Hörz, H., Krause, W. (eds.) Einfachheit als Wirk-, Erkenntnis- und Gestaltungsprinzip (Sitzungsberichte der Leibniz-Sozietät der Wissenschaften, vol. 108), pp. 93–104. trafo Verlag, Berlin (2010)

Barck, K., et al. (eds.): Ästhetische Grundbegriffe. Historisches Wörterbuch in 7 Bänden. Verlag J.B.Metzler, Stuttgart and Weimar (2005/2010)

Barthet, M., Fazekas, G., Sandler, M.: Music emotion recognition: from content- to context-based models. In: Aramaki, M., Barthet, M., Kronland-Martinet, R., Ystad, S. (eds.) From Sounds to Music and Emotions. CMMR 2012. Lecture Notes in Computer Science, vol. 7900. Springer, Berlin, Heidelberg (2013). https://doi.org/10.1007

Bates, E.: Glitches, bugs, and hisses: the degeneration of musical recordings and the contemporary musical work. In: Washburne, C., Derno, M. (eds.) Bad Music. The Music We Love to Hate, pp. 212–225. Routledge, New York and London (2004)

Bellos, D.: Was macht der Fisch in meinem Ohr? Sprache, Übersetzen und die Bedeutung von allem. Eichborn-Verlag, Köln (2013). (English as: Is that a Fisch in Your Ear? Translation and the Meaning of Everything. Penguin Group, 2011)

Benary, P.: Skizze, Entwurf, Fragment. In: MGG^2, vol. **8**, col. 1506–1519 (1998)

Bengen, I., Frobenius, W., Behne, K.-E.: Tempo. In: MGG^2, vol. **9**, col. 443–470 (1998)

Benjamin, W.: Einbahnstraße. Berlin, Ernst Rowohlt Verlag (1928)

Benson, D.: Music: A Mathematical Offering. Cambridge University Press, Cambridge (2006). https://www.maths.abdn.ac.uk/~bensondj/html/maths-music.html 2008. Accessed 17 June 2020

Berenzweig, A., Logan, B., Ellis, D., Whitman, B.: A large-scale evaluation of acoustic and subjective music-similarity measures. Comput. Music J.**28**(2), 63–76 (2004)

Betz, M.: “In futurum” – von Schulhoff zu Cage. In: Archiv für Musikwissenschaft, vol. 56, issue 4, pp. 331–334. Steiner, Stuttgart (1999)

Biles, J.A.: Genjam in perspective: a tentative taxonomy for genetic algorithm music and art systems. Leonardo 36(1), 43-45 (2003)

Bilz, R.: Die unbewältigte Vergangenheit des Menschengeschlechts. Beiträge zu einer Paläoanthropologie (Theorie vol. 2). Suhrkamp, Frankfurt a. M. (1967)

Binet, L., Müllensiefen, D., Edwards, P.: The Power of Music. Admap, October 2013, 10–13 (2013)

Blackburn, A., Penny, J.: Timbral notation from spectrograms: notating the un-notatable? In: Proceedings of the International Conference on New Tools for Music Notation and Representation TENOR (2015)

Blakolmer, F.: Wie Bilder lügen. Die Frühägäer und ihre Götter. In: Forum Archaeologiae 54/III/2010. n.p. (2010)

Blanning, T.C.W.: The Triumph of Music: The Rise of Composers, Musicians and Their Art. Harvard University Press, Cambridge, MA (2008)

Blaukopf, K.: Werktreue und Bearbeitung. Zur Soziologie der Integrität des musikalischen Kunstwerks (Musik und Gesellschaft, vol. 3). Braun-Verlag, Karlsruhe (1968)

Blübaum, D. et al. (eds.) Gestochen Scharf! Die Kunst zu reproduzieren (catalog). Kehrer, Heidelberg (2007)

Boehmer, K.: Zur Theorie der offenen Form in der Neuen Musik. Tonos, Darmstadt (1967)

Böhme, F.: Tönende Technik. Über das Verhältnis von Komposition und Maschine. In: Phleps, Th., Reich, W. (eds.) Musik-Kontexte. Festschrift für Hanns-Werner Heister (2 vols.), vol. 1, pp. 132–149. Verlagshaus Monsenstein und Vannerdat, Münster (2011)

Bonus, A.E.: The Metronomic Performance Practice: A History of Rhythm, Metronomes, and the Mechanization of Musicality. Phil. Diss., Department of Music Case Western Reserve University (2010)

Boone, Chr.: Mashing: toward a typology of recycled music. Music Theory Online **19/3**, 1–14 (2013)

Borup, H.: A History of String Intonation (n.d.). https://www.hasseborup.com/ahistoryofintonationfinal1.pdf

Bossin, J.: Stravinsky-Antheil-Maschinenmusik. Stravinsky und die neue Ästhetik des Antipathos. In: Neue Gesellschaft für Bildende Kunst Berlin: Absolut modern sein: Zwischen Fahrrad und Fließband – culture technique in Frankreich 1889–1937, pp. 238–242. Elefantenpress, Berlin (West) (1986)

Bowers, D.Q.: Put Another Nickel in. A history of Coin-Operated Pianos and Orchestrions. Vestal Press, New York (1968)

Bowers, D.Q.: Encyclopedia of Automatic Musical Instruments. Vestal Press, New York (1972)

Breidenbach, B.: 'Stimmung': Conceptualising Attunement in Music, Philosophy and Literature, Symposion Music and Literature: Critical Polyphonies. Durham Castle, Durham University, 2nd July 2015

Breidenstein, H.: Mälzels Mord an Mozart. Die untauglichen Versuche, musikalische Zeit zu messen. In: Das Orchester, 2007, issue 11, pp. 8–15. Schott International, Mainz (2007)

Brodsky, W., Ginsborg, J., Henik, A., Kessler, Y., Rubinstein, B.-Sh.: The mental representation of music notation: Notational Audiation. J. Exp. Psychol. **34**(2), 427–445 (2008). https://doi.org/10.1037/0096-1523.34.2.427

Bruderer, M.: Automatic recognition of musical instruments. Master thesis, Ecole Polytechnique Fédérale de Lausanne (2003)

Bruhn, S.: J. S. Bach's well-tempered clavier. In-depth Analysis and Interpretation (n.d.). https://edition-gorz.de/bruhn4-engl.html

Bruhn, S.: A Concert of Paintings: "Musical Ekphrasis" in the Twentieth Century. Paper presented at the Fifth World Conference of the International Association for Word and Image Studies (March 1999) (1999). https://www-personal.umich.edu/~siglind/ekphr2.htm. Accessed 4 Feb 2017

Bruhn, S.: Musical Ekphrasis: Composers Responding to Poetry and Painting (Interplay, No. 2). Pendragon, Hillsdale, NY (2000)

Brümmer, L., Kanach, Sh., Weibel, P. (eds.): From Xenakis's UPIC to Graphic Notation Today. ZKM, Karlsruhe (2011)

Burde, W.: »Haupttontechnik« und europäisches Komponieren. In: Heister, H.-W., Sparrer, W.-W. (eds.): Der Komponist Isang Yun, pp. 263–272. text+kritik, München (1987)

Burkholder, J. Peter: All Made of Tunes: Charles Ives and the Uses of Musical Borrowing. Yale University Press, New Haven, CT (1995)

Burney, C.: Carl Burney's der Musik Doctors Tagebuch einer Musikalischen Reise, vol. 2: Durch Flandern, die Niederlande und am Rhein bis Wien. Hamburg (1773)

Burns, G.: A typology of 'hooks' in popular records. Popul. Music. **6**(1), 1–20 (1987)

Busoni, F.: Entwurf einer neuen Ästhetik der Tonkunst (Sketch of a New Aesthetic of the Art of Music), 1907; 2nd, enlarged ed. Insel-Verlag, Leipzig (1916)

Calvin, W.H., Ojemann, G.A.: Einsicht ins Gehirn. Wie Denken und Sprache entstehen (english 1994). dtv, München (2000)

Cason, R.J.S., Müllensiefen, D.: Singing from the same sheet: computational melodic similarity measurement and copyright law. Int. Rev. Law Comput. Technol. **26**(1), 25–36 (2012)

Castillo, O., Kacprzyk, J., Melin, P. (eds.): Nature-inspired design of hybrid intelligent systems (Studies in Computational Intelligence vol. 667). Springer, Heidelberg (2017)

Cat, J.: Fuzzy Pictures as Philosophical Problem and Scientific Practice. A Study of Visual Vagueness (Studies in Fuzziness and Soft Computing, ed. by J. Kacprzyk, vol. 348). Springer Wissenschafts-Verlag, Heidelberg etc. (2017)

Chang, Ph. Ch.-Ch.: Analytical and Performative Issues in Selected Unmeasured Preludes by Louis Couperin, Volume 1: Text, Submitted in Partial Fulfillment of the Requirements for the Degree Doctor of Philosophy. Department of Music Theory Eastman School of Music. University of Rochester Rochester, New York (2011)

Chaudhuri, A., Mandaviya, K., Badelia, P., Ghosh, S.K.: Optical character recognition systems for different languages with soft computing. Studies in Fuzziness and Soft Computing, Series Ed.: J. Kacprzyk, Vol. 352, Springer Wissenschafts-Verlag (2016)

Chemillier, M.: Ethnomusicology, ethnomathematics. The logic underlying orally transmitted artistic practices. In: Assayag, G., Feichtinger, H.G., Rodrigues, J.F. (eds.): Mathematics and

Music. A Diderot Mathematical Forum, pp. 161–184. Springer, Berlin, Heidelberg, New York, Barcelona, Hong Kong, London, Milan, Paris, Tokyo (2002)

Chen, C.-M., Chien, P.-C., Tsai, M.-F., Yang, Y.-H., Lin, Y.-C.: Exploiting latent social listening representations for music recommendations. In: ACM International Conference on Recommender Systems (RecSys), September 2015

Chen, C.-M., Tsai, M.-F., Lin, Y.-C., Yang, Y.-H.: Query-based music recommendations via preference embedding. In: Proceedings of the ACM Recommender Systems (RecSys), pp. 79–82, September 2016

Chen, C.-Y.: Early Chinese Work in Natural Science. A Re-Examination of the Physics of Motion, Acoustics, Astronomy and Scientific Thoughts. Hong Kong University Press, Hong Kong (2000). https://www.questia.com/read/120363874/early-chinese-work-in-natural-science-a-re-examination

Chen, J.-M., Smith, J., Wolfe, J.: Pitch bending and glissandi on the clarinet: roles of the vocal tract and partial tone hole closure. J. Acoust. Soc. America. DOI: https://doi.org/10.1121/1.3177269 Corpus ID: 16352321 (2009)

Chen, Y.-A., Wang, J.-Ch., Yang. Y.-Hs., Che, H.: Linear regression-based adaptation of music emotion recognition models for personalization. In: IEEE International Conference on Acoustics, Speech and Signal Processing (ICASSP), May 2014

Chen, Y.-H., Huang, Y.-S., Hsiao, W.-Y., Yang, Y.-Hs.: Automatic composition of guitar tabs by Transformers and groove modeling. In: Proceedings of the International Society for Music Information Retrieval Conference (ISMIR), October 2020

Chen, Y.-W., Yang, Y.-Hs., Chen, H. H.: Cross-cultural music emotion recognition by adversarial discriminative domain adaptation. In: Proceedings of the IEEE International Conference on Machine Learning and Applications (ICMLA), December 2018

Chew, T.-L. Tracking the Literary Metamorphosis in "Das Lied von der Erde" (2009). https://www.mahlerarchives.net/archives/daslied_lit.pdf. Accessed 14 July 2012

Chin, Y.-H., Wang, J.-Ch., Wang, J.-Ch., Yang, Y.-Hs.: Predicting the probability density function of music emotion using emotion space mapping. IEEE Trans. Affective Comput. **9**(4), 541–549 (2018)

Chou, S.-Y., Yang, Y.-H., Jang, J.-S., Lin, Y.-C.: Addressing cold start for next-song recommendation. In: Proceedings of the ACM Recommender Systems (RecSys), pp. 115–118, September 2016

Coessens, K.: The score beyond music. In: de Assis, P., Brooks, W., Coessens, K. (eds.) Sound and Score: Essays on Sound, Score and Notation. Leuven University Press, Ghent (2014)

Cogan, J., Clark, W.: Temples of sound: inside the great recording studios. Chronicle Books, San Francisco (2003)

Collins, N., d'Escriván, J. (eds.): The Cambridge Companion to Electronic Music. University Press, Cambridge, MA (2007)

Collins, N., McLean, A., Rohrhuber, J., Ward, A.: Live coding in laptop performance. Organised Sound **8**(3), 321–330 (2003). https://web.archive.org/web/20180617115703/

Collins, N.: Live electronic music. In: Collins, N., d'Escriván, J. (eds.) The Cambridge Companion to Electronic Music, pp. 38–54. Cambridge Companions to Music. Cambridge University Press, Cambridge (2007)

Collins, N.: Beyond notation: communicating music. Leonardo Music J. **21**, 5–6. MIT Press, Cambridge, MA (2011)

Collins, N.: Generative music and laptop performance. Contemporary Music Rev. **22**(4), 67–79

Competitive Reinforcement Learning. In: Butz M.V., Sigaud O., Pezzulo G., Baldassarre G. (eds) Anticipatory Behavior in Adaptive Learning Systems. ABiALS 2006. Lecture Notes in Computer Science, vol. 4520. Springer, Berlin, Heidelberg (2006)

Conklin, D., Witten, I.: Multiple viewpoint systems for music prediction. J. New Music Res. **24**, 51–73 (1995)

Conklin, D.: Music generation from statistical models. In: Proceedings of Symposium on AI and Creativity in the Arts and Sciences, pp. 30–35 (2003)

Connor, E.: Separating trash from treasure: how the brain assigns value to objects. Neurosci. News February 9, 2018 (2018)

Cont, A., Shlomo, D., Assayag, G.: Anticipatory model of musical style imitation using collaborative and competitive reinforcement learning. In: Anticipatory Behavior in Adaptive Learning Systems: From Brains to Individual and Social Behavior April 2007, 285–306 (2007). https://doi.org/10.1007/978-3-540-74262-3_16

Corey, J., Benson, D.H.: Audio Production and Critical Listening. Technical Ear Training. Technical Ear Training Software, 2nd edn. Routledge, New York and London (2017)

Correia, J., Ciesielski, V., Liapis, A. (eds.): EvoMUSART: International Conference on Evolutionary and Biologically Inspired Music and Art. Computational Intelligence in Music, Sound, Art and Design. 6th International Conference, EvoMUSART 2017, Amsterdam, The Netherlands, April 19–21, 2017, Proceedings (2017). https://link.springer.com/book/10.1007%2F978-3-319-55750-2

Coutinho, E., Cangelosi, A.: A neural network model for the prediction of musical emotions. In: Nefti-Meziani, S., Grey, J.G. (eds.) Advances in Cognitive Systems, pp. 331–368. IET Publisher, London (2010)

Coutinho, E., Cangelosi, A.: Musical emotions: predicting second-by-second subjective feelings of emotion from low-level psychoacoustic features and physiological measurements. Emotion **11**(4), 921–937 (2011)

Cramer, A.: Schoenberg's Klangfarbenmelodie: a principle of early atonal harmony. Music Theory Spectr. **24**(1), 1–34 (2002). https://www.researchgate.net/publication/249980375_Schoenberg27s_Klangfarbenmelodie_A_Principle_of_Early_Atonal_Harmony

Cross, I.: Music analysis and music perception. Music Anal. **17**(1), 3–20 (1998)

Cross, J.: Composing with numbers: sets, rows, and magic squares. In: Fauvel, J., Flood, R., Wilson, R.J. (eds.) Music and Mathematics: From Pythagoras to Fractals. Oxford University Press, Oxford (2003)

D'Avis, W.: Können Computer denken? Eine bedeutungs- und zeittheoretische Analyse von KI-Maschinen. Campus, Stuttgart (1994)

Dahlhaus, C.: Arnold Schönberg. Variationen für Orchester, op. 31 (Meisterwerke der Musik. Werkmonographien zur Musikgeschichte, H. 7). Fink-Verlag, München (1968)

Dahlhaus, C.: Musiktheorie. In: Ibid. (ed.): Einführung in die systematische Musikwissenschaft, pp. 93–132. Arno Volk Verlag, Köln (1971). Reprint in: Danuser, H. et al. (eds.): Carl Dahlhaus, Gesammelte Schriften in 10 Bänden, vol. 2., pp. 209–232. Laaber Verlag, Laaber (2000–2008)

Dahlhaus, C.: Schönbergs Orchesterstück op. 16, 3 und der Begriff der "Klangfarbenmelodie". In: Bericht über den Internationalen Musikwissenschaftlichen Kongress Bonn 1970, Dahlhaus, C., Marx, H.J. (eds.) Bärenreiter Verlag, Kassel. https://nbn-resolving.org/urn:nbn:de:bsz:14-qucosa2-374743; english as Dahlhaus, C. [1970]: "Schoenberg's Orchestral Piece Op. 16, No. 3 and the Concept of Klangfarbenmelodie" In: Schoenberg and the New Music: Essays by Carl Dahlhaus. Cambridge, Cambridge University Press, pp. 141–143 (1971)

Dahms, S. (in cooperation with Jeschke, C., Woitas, M.) (ed.): Tanz (MGG Prisma). Bärenreiter/Metzler, Kassel et al. (2001)

Dallmann, N.-C.: Musikterminologie erklärt – Walzer und 3/4-Takt. In: Sonic – Magazin für Holz- und Blechblasinstrumente, Juli/August 2015, pp. 90–92 (2015)

Dallmann, N.-C.: Rubato und Laid-back. In: Sonic – Magazin für Holz- und Blechblasinstrumente, September/Oktober 2015, pp. 106–108 (2005)

Dalmazzo, D., Ramírez, R: Bowing gestures classification in violin performance: a machine learning approach. Front. Psychol. **10**, 344 (2019). Doi: https://doi.org/10.3389/fpsyg.2019.00344

Daniels, D.: From visual music to intermedia art, Dieter-Daniels_Von-der-visuellen-Musik-zur-intermedialen-Kunst_From-Visual-Music-to-Intermedia-Art. In: Rainer, C., Rollig, St., Daniels, D., Ammer, M. (eds.) See This Sound. Versprechungen von Bild und Ton/See This Sound. Promises in Sound and Vision Lentos Kunstmuseum Linz, pp. 240–253. Verlag Walther König, Köln (2009)

Daniels, D.: Kunst als Sendung. Von der Telegrafie zum Internet. C. H. Beck, München (2002)

Dann-Beardsley, C.: Eight seconds to hook an audience (Words and Music, vol. 7). ProQuest757538142. Accessed 30 Aug 2020
Danuser, H. (ed.): Musikalische Interpretation (Neues Handbuch der Musikwissenschaft, vol. 11, 12 vols., edited by C. Dahlhaus, continued by H. Danuser). Laaber-Verlag, Laaber (1992)
Danuser, H., Kämper, D., Terse, P. (eds.): Amerikanische Musik seit Charles Ives. Interpretationen, Quellentexte, Komponistenmonographien. Laaber-Verlag, Laaber (1993)
Danuser, H., Katzenberger, G. (eds.): Vom Einfall zum Kunstwerk. Der Kompositionsprozeß in der Musik des 20. Jahrhunderts. Laaber Verlag, Laaber (1993)
Danuser, H.: Gesammelte Vorträge und Aufsätze, ed. Hans-Joachim Hinrichsen, Christian Schaper und Laure Spaltenstein, vol. 1, pp. 471–487. Edition Argus, Schliengen (2014)
De Benedictis, A.I.: The "dual" form of Igor Stravinsky's Piano-Rag-Music: Musical structure as result of a rearrangement process. CIM08 Abstract Proceedings, Thessaloniki (2008)
de Bieberstein Ilgner, P.: Von der fragmentierten Rezeption zum totalen Werk. Intermediale Aspekte im Werk von Gerhard Rühm. Phil. Diss., Universität Bremen (2014)
De Poli, G., Rocchesso. D.: Computational models for musical sound sources. In: Assayag, G., Feichtinger, H. G., Rodrigues, J.F. (eds.) Mathematics and Music. A Diderot Mathematical Forum, pp. 57–285. Springer-Verlag: Berlin, Heidelberg, New York, Barcelona, Hong Kong, London, Milan, Paris, Tokyo (2002)
De Sanctis De Benedictis, F.: Lied by Berio and Lumen by Donatoni: from analysis to interpretation and performance. Paper for the International Musicology Symposium The Pitfalls and Risks of Intra-Musical Communication: Terminology, Notation, Performance, Academia de Muzică "Gheorghe Dima" Cluj – Romania, 10 April 2013. https://drive.google.com/file/d/0B3MEy62wz8QDS05UVjZ5Q3Z4SzA/view. Accessed: 20 Feb 2020
Decroupet Pascal 2005b: Iannis Xenakis, Metastaseis (1953/1955). In: Heister (ed.) Geschichte der Musik im 20. Jahrhundert, Bd. III: 1945–1975, Laaber 2005 (Handbuch der Musik des 20. Jahrhunderts, Bd. 3), pp. 129–133 (2005)
Decroupet, P.: Totale organisation und universales material: Parameter-Denken und die Auflösung traditioneller Kategorien. In: Heister (ed.) Geschichte der Musik im 20. Jahrhundert, Bd. III: 1945–1975, Laaber 2005 (Handbuch der Musik des 20. Jahrhunderts, Bd. 3), pp. 90–96 (2005)
Decroupet. P.: Maschinenmusik: Fortsetzung futuristischer Ansätze – Neuansätze. In: Heister (ed.) Geschichte der Musik im 20. Jahrhundert, Bd. III: 1945–1975, Laaber 2005 (Handbuch der Musik des 20. Jahrhunderts, Bd. 3), pp. 124–129 (2005)
del Corral A., León T., Liern V.: Compatibility of the different tuning systems in an orchestra. In: Chew, E., Childs, A., Chuan, C.H. (eds.) Mathematics and Computation in Music. MCM 2009. Communications in Computer and Information Science, vol. 38. Springer, Berlin, Heidelberg (2009). https://doi.org/10.1007/978-3-642-02394-1_9
Deliège, I., Sloboda, J. (eds.): Perception and Cognition of Music. Psychology Press, Hove (East Sussex) (1997)
Demers, J.: Listening Through the Noise: The Aesthetics of Experimental Music. Oxford University Press, New York and Oxford (2010)
Deutsch, D. (ed.): The Psychology of Music, 2nd edn. Academic Press, Boston (1999)
Deutscher, G.: The Unfolding of Language. The Evolution of Mankinds Greatest Invention (1. ed. 1988). Random House, London (2005)
Díaz-Báñez, J.M., Farigu, G., Gómez, F., Rappaport, D., Toussaint, G.T.: El Compás Flamenco: A Phylogenetic Analysis, In: Proceedings of BRIDGES: Mathematical Connections in Art, Music and Science, Southwestern College, Winfield, Kansas, July 30–August 1, 2004, pp. 61–70. Deutsch als Mathematische Strukturanalyse der Flamenco-Metrik. PDF https://cgm.cs.mcgill.ca/~godfried/publications/winfield.pdf
Dibelius, U.: Moderne Musik nach 1945 (enlarged ed., part 1–3). Piper, München (1998)
Didi-Huberman, G.: Ähnlichkeit und Berührung. Anachronismus und Modernität des Abdrucks. Dumont, Köln, Archäologie (1999)
Dietel, G.: Musikgeschichte in Daten. Bärenreiter-dtv, Kassel and München (1994)

Dittmann, L.: Schönberg und Kandinsky, In: Litwin, St., Velten, K. (eds.) Stil oder Gedanke? Zur Schönberg- Rezeption in Amerika und Europa, Saarbrücken 1995, pp. 216–230 (Schriftenreihe der Hochschule des Saarlandes für Musik und Theater. vol. 3). Pfau Verlag, Saarbrücken (1995)

Dixon, S., Mauch, M., Tidhar, D.: Estimation of harpsichord inharmonicity and temperament from musical recording. J. Acoust. Soc. America **131**(1), 878–887 (2012)

Dixon, S.: Automatic extraction of tempo and beat from expressive performances. J. New Music Res. Swets Zeitlinger **30**(1), 39–58 (2001)

Döhl, F., Wöhrer, R. (eds.): Zitieren, appropriieren, sampeln. Referenzielle Verfahren in den Gegenwartskünsten. transcript, Bielefeld (2014)

Döhl, F.: Mashup in der Musik. Fremdreferenzielles Komponieren, Sound Sampling und Urheberrecht. transcript, Bielefeld (2016)

Dolscheid, S., Shayan, Sh., Majid, A., Casasanto, D.: The thickness of musical pitch. Psychol. Sci.**24**(5), March 2013. https://www.researchgate.net/publication/236090333_The_Thickness_of_Musical_Pitch/link/542956870cf26120b7b5bdfa/download

Domschke, W., Scholl, A.: Heuristische Verfahren (Jenaer Schriften zur Wirtschaftswissenschaft 08/2006)

Dong. H.-W., Yang, Y.-Hs.: Convolutional generative adversarial networks with binary neurons for polyphonic music generation. In: Proceedings of the International Society for Music Information Retrieval Conference (ISMIR), September 2018

Doolittle, E.L., Brumm, H.: O Canto do Uirapuru: Consonant intervals and patterns in the song of the musician wren. J. Interdiscip. Music. Stud. **6**(1), art. 12060103, pp. 55–85. Spring 2012

Doolittle, E.L., Gingrasb, B., Dominik M. Endresc, D.E., Fitch, W.T.: Overtone-based pitch selection in hermit thrush song: Unexpected convergence with scale construction in human music. PNAS November 18, 2014. 111(46), 16616–1662 (2014)

Douwe, D.: Wie wir träumen. Galiani, Berlin (2015)

Downie, J.S.: Evaluating a simple approach to music information retrieval: conceiving melodic n-grams as text. PhD thesis, Faculty of Information and Media Studies, University of Western Ontario, London (Ontario), Canada (1999)

Doyle, A.C.: Sherlock Holmes-Geschichten. Der Hund von Baskerville (translated by Trude Fein, selection and epilogue by Andreas Fischer). Deutscher Taschenbuch Verlag/Manesse Verlag, Zürich (1993)

Draaisma, D.: Das Buch des Vergessens. Warum unsere Träume so schnell verloren gehen und sich unsere Erinnerungen ständig verändern (Netherland. 2010). Galiani, Berlin (2013)

Draaisma, D.: Halbe Wahrheiten: Vom seltsamen Eigenleben unserer Erinnerung. Galiani, Berlin (2016)

Draaisma, D.: Warum das Leben schneller vergeht, wenn man älter wird. Von den Rätseln der Erinnerung (Netherland. 2001). Piper, München (2006)

Droseltis, A.: Zufall und Determination in der westeuropäischen Musik um 1960. Dargestellt an Werken von Iannis Xenakis und Karlheinz Stockhausen. Phil. Diss., Fakultät I – Geisteswissenschaften der Technischen Universität Berlin Berlin, D 83 (2011)

Drösser, C.: Fuzzy Logic. Methodische Einführung in krauses Denken. Rowohlt, Reinbek (1994)

du Plessis, H.J.P.: Parola Scenica: Towards Realism in Italian Opera. Phil. Diss., University of the Witwatersrand, Johannesburg, Faculty of Arts (2007)

Dubnov, S., Assayag, G., Lartillot, O., Bejerano, G.: Using machine-learning methods for musical style modeling. IEEE Comput. Soc. **36**(10), 73–80 (2003)

Dubnov, S., El-Yaniv, R., Assayag, G.: Universal classification applied to musical sequences. In: Proceedings of ICMC, pp. 322–340. Michigan (1998)

Dubnov, S.: Spectral anticipations. Comput. Music J. June 2006, 1–39 (2006). https://doi.org/10.1162/comj.2006.30.2.63

Dubnov, Sh., Bar-Joseph, Z., El-Yaniv, R., Lischinski, D., Werman, M.: Synthesis of Sound Textures by Learning and Resampling of Wavelet Trees (2002)

Dubois, D., Prade, H.: Fuzzy Sets and Systems: Theory and Applications (Mathematics in Science and Engeneering, ed. by Richard Bellman, vol. 144). Academic Press, New York etc. (1980)

Dubois, D.: Uncertainty in Knowledge Representation and Reasoning: A Partial Bibliography (n.d.). https://www.irit.fr/~Didier.Dubois
Dümling, A.: Der Komponist Artur Schnabel. In: Dissonanz/Dissonance 1987, no. 12, pp. 10–20. Zürich (1987)
Dupuis-Désormeaux, N.: Tutti! - Music Composition as Dialogue. Phil. Diss., York University Toronto, Ontario (2017)
Ebbeke, K.: Aleatorik. In: MGG^2, vol. 1, cols. 415–444 (1994)
Edelman, G.M., Tononi, G.: Gehirn und Geist. Wie aus Materie Bewusstsein entsteht. C.H. Beck Verlag, München (2002)
Edelmann, B.: Permutation, Spiegelung. Palindrom. Das Bratschenkonzert von Karl Amadeus Hartmann. In: Komponist zwischen den Fronten und zwischen den Zeiten. Bericht über das musikwissenschaftliche Symposium zum 100. Geburtstag in München, 5.–7. Oktober 2005, ed. Inga M. Groote und Hartmut Schick (= Münchner Veröffentlichungen zur Musikgeschichte Bd. 68, pp. 25–82. Verlag Schneider, Tutzing (2010)
Eggebrecht, H.H.: Zeit. In: Eggebrecht, H.H.; Die Musik und das Schöne, pp. 172–180. Piper, München (1997)
Eiden, W.: Präzise Unschärfe. Informationsmodellierung durch Fuzzy-Mengen. Ibidem-Verlag, Stuttgart (2002)
Eigen, M., Winkler, R.: Das Spiel. Naturgesetze steuern den Zufall. Piper, München und Zürich (1985)
Einstein, A.: Geometrie und Erfahrung. Erweiterte Fassung des Festvortrages gehalten an der Preußischen Akademie der Wissenschaften zu Berlin am 27. Januar 1921, p. 3 f. Julius Springer, Berlin (1921)
Elscheková, A., Kubik, G., Muyle, R. M., Simon, A,, Ziegler, S.; Mehrstimmigkeit. In: MGG^2, vol. 5, cols. 1766–1808 (1996)
Elsner, J.: Maqām - abstrakte Einheit - konkrete Vielfalt. In: Phleps, Th., Reich, W. (eds.) Musik-Kontexte. Festschrift für Hanns-Werner Heister (2 vols.), vol. 1, pp. 198–214. Verlagshaus Monsenstein und Vannerdat, Münster (2011)
Emans, R.: Kompositionsunterricht und ein Wirrwarr von Fassungen. In: „Ei, dem alten Herrn zoll' ich Achtung gern'" (Festschrift für Joachim Veit zum 60. Geburtstag, edited by Richts, K. and Stadler, P. for the Virtuellen Forschungsverbund Edirom), pp. 233–250. Allitera Verlag, München (2016)
Emmerson, S.: Living Electronic Music. Ashgate, Aldershot, Hants (2007)
Enders, B.: Vom Idiophon zum Touchpad. Die musiktechnologische Entwicklung zum virtuellen Musikinstrument. In: Musik/Medien/Kunst, ed. B. Flath, pp. 55–74. Transcript, Bielefeld (2013)
Enders, B., Weyde, T.: Automatische Rhythmuserkennung und -vergleich mit Hilfe von Fuzzy-Logik. In: Systematische Musikwissenschaft IV(1–2), pp. 101–113 (1996)
Engel, H.: Bearbeitung. In: MGG^1, vol. 1, col. 1458–1466 (1949/1951)
Erickson, R.: Sound Structure in Music. University of California Press, Berkeley etc (1975)
Ethier, G.: Techniques of hypermetric manipulation in Canadian blues. Canadian J. Traditional Music, 33–53 (2001)
Fabian, D., Timmers, R., Emery Schubert, E. (eds.): Expressiveness in Music Performance. Empirical Approaches Across Styles and Cultures. Oxford University Press, Oxford (2014)
Fauvel, J., Flood, R., Wilson, R.: Music and Mathematics. From Pythagoras to Fractals. Oxford University Press, Oxford (2003)
Feil, A.: Metzler Musik Chronik vom frühen Mittelalter bis zur Gegenwart, 2nd enlarged edition. Verlag J. B. Metzler, Stuttgart, Weimar (2005)
Ferand, E. T.: Die Improvisation in Beispielen aus neun Jahrhunderten abendländischer Musik (Das Musikwerk, vol. 12), Gerig, Köln (1961)
Finke, G.: Partituren zum Lesen und Schauen. Bildlichkeit als Merkmal graphischer Notation. In: ZGMTH Zeitschrift der Gesellschaft für Musiktheorie, vol. 16. no. 1, pp. 21–40 (2019)

Finkel, S., Müllensiefen, D.: Involuntary Musical Imagery - Investigating Musical Features that Predict 'Earworms'. Talk given at the Third International Conference of Students of Systematic Musicology (SysMus10), University of Cambridge, 14th of September (2010)
Finscher, L.: Parodie und Kontrafaktur. Bis 1600. In: MGG^1, vol. 10, col. 815–826 (1962)
Fischer, K. von: Variation. In: MGG^1, vol. 13, col. 1274–1309 (1966)
Fitch, W.T.: The biology and evolution of rhythm: unravelling a paradox. https://doi.org/10.1093/acprof:oso/9780199553426.003.0009 (2011)
Fladt, H., Heister, H.-W.: Zur Konstituierung des musikalischen Kunstwerks. In: Die Musikforschung vol. 33, pp. 473–486 (1980)
Flath, B.: Sound & Image. Eine experimentelle Untersuchung zum Einfluss von Klangqualitäten auf die Wahrnehmung eines Produktimages im Kontext von Fernsehwerbung. epOs-Music, Osnabrück (2012)
Flexer, A., Pampalk, E., Widmer, G.: Hidden Markov models for spectral similarity of songs. In: Proceedings of the 8th International Conference on Digital Audio Effects, Madrid, Spain (2005)
Flinn, J.W.: Reconstructive Postmodernism, Quotation, and Musical Analysis: A Methodology with Reference to the Third Movement of Luciano Berio's Sinfonia, Phil. submitted to the Graduate School of the University of Cincinnati in the Division of Composition, Musicology and Theory of the College-Conservatory of Music (2011)
Floridou, G., Müllensiefen, D.: Involuntary Musical Imagery While the Mind Wanders: An Experience Sampling Study using Bayesian Networks. Presentation given at the biennial meeting of the Society for Music Perception and Cognition 2013, 9th of August, Ryerson University, Toronto (2013)
Floridou, G., Williamson, V., Müllensiefen, D.: Contracting earworms: the roles of personality and musicality. In: Poster presented at the 12th International Conference on Music Perception and Cognition (ICMPC), 27th of July, Thessaloniki (2012)
Floros, C.: Alban Berg. Musik als Autobiographie. Breitkopf & Härtel, Wiesbaden, Leipzig, Paris (1992)
Floros, C.: Gustav Mahler (3 vols.), vol. 3 (Die Symphonien). Breitkopf&Härtel, Wiesbaden (1985)
Folkerts, G.: Mikis Theodorakis. Seine musikalische Poetik. von Bockel-Verlag, Neumünster (2015)
Föllmer, G., Gerlach, J.: Audiovisionen. Musik als intermediale Kunstform (2004). https://www.medienkunstnetz.de/themen/bild-ton-relationen/audiovisionen/
Foote, J., Uchihashi, S.: The beat spectrum: a new approach to rhythm analysis. In: Proceedings of the IEEE International Conference on Multimedia and Expo, Tokyo, Japan, pp. 224–228 (2001)
Foote, J.: Audio retrieval by rhythmic similarity. In: Proceedings of the 3rd International Conference on Music Information Retrieval (2002)
Fornari, J., Maia, I. L.: Interactive Notation: Towards a Computer Model for Interactive Musical Composition. Interdisciplinary Nucleus for Sound Studies (NICS) – Universidade Estadual de Campinas (UNICAMP), Campinas – SP – Brazil (n.d.). https://www.academia.edu/17380050/Interactive_Notation_Towards_a_Computer_Model_for_Interactive_Musical_Composition
Frank, P.: Visuelle Partituren. In: von Maur 1985, pp. 444–449 (1985)
Frank, T., Pinch, T.: Analog Days: The Invention and Impact of the Moog Synthesizer. Harvard University Press, Cambridge, MA. Revised Edition (2004)
Franke, H.W.: Kunst kontra Technik? Wechselwirkungen zwischen Kunst, Naturwissenschaft und Technik. Fischer Taschenbuchverlag, Frankfurt a. M. (1978)
Franke, H.W.: Computergrafik. Computerkunst. Bruckmann, München (1971)
Franke, H.W.: Welten aus Bits und Bytes. In: VDI Nachrichten Magazin 9/92, 13–17
Franke, H.W., Jäger, G.: Apparative Kunst. Vom Kaleidoskop zum Computer. DuMont, Köln (1973)
Franklin, J.A.: Predicting reinforcement of pitch sequences via lstm and td. In: Proceedings of International Computer Music Conference, Miami, Florida, Volume 2004 (2004). https://hdl.handle.net/2027/spo.bbp2372.2004.015
Freud, S.: General introduction to psychoanalysis. Garden City Publishing, New York (1952)
Fricke, St: Collage. In: MGG^2. vol. 2, col. 937–944 (1995)

Friedman, S.M. (1960): Ein Aspekt der Musikstruktur. Eine Studie über regressive Transformation musikalischer Themen. In: Oberhoff, B. (ed.) Das Unbewußte in der Musik, pp. 189–210. Gießen (2002)

Frieler, K., Fischinger, T., Schlemmer, Lothwesen, K., Jakubowski, K., Müllensiefen, D.: Absolute memory for pitch: a comparative replication of Levitin's 1994 study in six European labs. Musicae Scientiae **17**(3), 334–349 (2013)

Frisius, R.: Forum Analyse (n.d.). https://www.frisius.de/rudolf/texte/txhorend.htm

Frobenius, W.: Aleatorisch, Aleatorik. In: Handwörterbuch musikalischer Terminologie-HmT. Ordner 1: A-D. Steiner-Verlag, Stuttgart (1976)

Frost, S.: Striking the wrong note. J. Art Record Prod. (01) | Industry Perspective (2007). https://www.arpjournal.com/asarpwp/content/issue-1/

Fucks, W.: Mathematische Analyse von Formalstrukturen von Werken der Musik (mit Diskussion). In: Arbeitsgemeinschaft für Forschung des Landes Nordrhein-Westfalen, pp. 39–114. Westdeutscher Verlag, Köln und Opladen (1963)

Fucks, W.: Gibt es mathematische Gesetze in Sprache und Musik? In: Frank, H. (ed.) Kybernetik – Brücke zwischen den Wissenschaften, pp. 171–183. Umschau Verlag, Frankfurt am Main (1964)

Fucks, W.: Nach allen Regeln der Kunst. DVA, Stuttgart (1968)

Fucks, W., Lauter, J.: Exaktwissenschaftliche Musikanalyse. Westdeutscher Verlag, Köln und Opladen (1965)

Fucks, W.: Mathematical analysis of formal structure of music. IEEE Trans. Inform. Theory **8**(5), 225–228 (1962)

Fülöp-Müller, R.: Geist und Gesicht des Bolschewismus. Darstellung und Kritik des kulturellen Lebens in Sowjetrussland (1926), ed. by E. Siepmann as Fantasie und Alltag in Sowjet-Russland. Bericht eines Augenzeugen (EP 7), Elefanten Press, Berlin and Hamburg (1978)

Gail, D.: Charles E. Ives' Fourth Symphony. Quellen – Analyse – Deutung (2 vols.). Wolke-Verlag, Hofheim (2009)

Gann, K.: The Music of Conlon Nancarrow. Cambridge University Press, Cambridge (1995)

Gardiner, J.M., Kaminska, Z., Dixon, M., Java, R.I.: Repetition of previously novel melodies sometimes increases both remember and know responses in recognition memory. Psychon. Bull. Rev. **3**(3), 366–371 (1996)

Garland, T.H., Kahn, Ch.V.: Math and Music Harmonious connections. Dale Seymour Publications, Palo Alto (1995)

Genette, G.: Mimologiken. Reise nach Kratylien (french 1976). Suhrkamp, Frankfurt a. M. (2001)

Genette, G.: Palimpseste. Die Literatur auf zweiter Stufe (french 1982). Suhrkamp, Frankfurt a. M. (1993)

Genz, H.: Wie die Naturgesetze Wirklichkeit schaffen. Über Physik und Realität, Reinbek (2004)

Gess, N., Schreiner, F., Manuela, K., Schulz, M.K. (eds.): Hörstürze. Akustik und Gewalt im 20. Jahrhundert. Königshausen & Neumann, Würzburg (2005)

Gieseler, W.: Komposition im 20 Jahrhundert. Moeck, Celle (1975)

Gieseler, W.: Zur Semiotik graphischer Notation. In: Melos NZ, vol. 4, pp. 27–33. Schott International, Mainz (1978)

Gifford, T., Knotts, Sh., McCormack, J., Kalonaris, St., Yee-King, M., d'InvernoM.: Computational systems for music improvisation (2018). https://doi.org/10.1080/14626268.2018.142661

Goebl, W., Pampalk, E., Widmer, G.: Exploring expressive performance trajectories: Six famous pianists play six Chopin pieces. In: Lipscomp, S.D., et al. (eds.) Proceedings of the 8th International Conference on Music Perception and Cognition, Evanston, IL, 2004 (ICMPC8), pp. 505–509. Causal Productions, Adelaide, Australia. https://www.researchgate.net/publication/239717974_EXPLORING_EXPRESSIVE_PERFORMANCE_TRAJECTORIES_SIX_FAMOUS_PIANISTS_PLAY_SIX_CHOPIN_PIECES/link/53e094110cf2aede4b4cd925/download (2004)

Goldschmidt, H.: Musikverstehen als Postulat. In: Faltin, P., Reinecke, H.-P. (eds.) Musik und Verstehen. Aufsätze zur semiotischen Theorie, Ästhetik und Soziologie der musikalischen Rezeption, pp. 67–86. Arno Volk, Köln (1974)

Goldschmidt, H.: Um die Sache der Musik. Reden und Aufsätze. Verlag Philipp Reclam jun, Leipzig (1970)

Gómez, F., Melvin, A., Rappaport, D., Toussaint, G.T.: Mathematical Measures of Syncopation (2005). https://www.researchgate.net/publication/228850033_Mathematical_measures_of_syncopation/download

Goto, M.: An audio-based real-time beat tracking system for music with or without drum-sounds. J. New Music Res. **3**, 159–171 (2001)

Gottwald, C.: Signale zwischen Exotik und Industrie. Steve Reich auf der Suche nach einer neuen Identität von Klang und Struktur. In: Melos/NZfM 1975, no. 1, pp. 3–6 (1975)

Gould, E.: Hals über Kopf. Das Handbuch des Notensatzes (german version by Muus, A. and Berger, J.). Faber Music/Edition Peters, Leipzig (2013)

Gouyon, F.: A computational approach to rhythm description: Audio features for the computation of rhythm periodicity functions and their use in tempo induction and music content processing. PhD thesis, Universitat Pompeu Fabra, Departament de Tecnologia, Barcelona, Spain (2005)

Gouyon, F., Dixon, S.: A review of automatic rhythm description systems. Comput. Music J. **29**, 34–54 (2005)

Gouyon, F.: A computational approach to rhythm description: Audio features for the computation of rhythm periodicity functions and their use in tempo induction and music content processing. PhD thesis, Universitat Pompeu Fabra, Departament de Tecnologia, Barcelona, Spain (2005)

Gratzer, W. (ed.): Perspektiven einer Geschichte abendländischen Musikhörens (Schriften zur musikalischen Hermeneutik, vol. 7). Laaber Verlag, Laaber (1997)

Gratzer, W., Neumaier, O. (eds.): Arbeit am musikalischen Werk. Zur Dynamik künstlerischen Handelns (klang-reden. Schriften zur musikalischen Rezeptions- und Interpretationsgeschichte, vol. 9). Rombach, Freiburg i. Br. (2013)

Gratzer, W.: Komponistenkommentare. Beiträge zu einer Geschichte der Eigeninterpretation (Wiener musikwissenschaftliche Beiträge, vol. 22). Böhlau, Vienna/Cologne/Weimar (2003)

Gratzer, W.: Zur "wunderlichen Mystik" Alban Bergs. Eine Studie. Böhlau, Wien, Köln, Weimar (1993)

Gratzer, W.: Zur "wunderlichen Mystik" Alban Bergs. Eine Studie. Böhlau, Wien, Köln, Weimar (1993)

Gredig, M.: Tiermusik. Zur Geschichte der skeptischen Zoomusikologie. Königshausen&Neumann, Würzburg (2018)

Grekow, J., Raś, Z. W.: Detecting emotions in classical music from MIDI Files. In: International Symposium on Methodologies for Intelligent Systems ISMIS 2009: Foundations of Intelligent Systems, pp. 261–270. Springer Verlag, Heidelberg (2009). https://link.springer.com/chapter/10.1007/978-3-642-04125-9_29

Grekow, J.: An analysis of the harmonic content - main parameters in the AKWET method. In: Proceedings of II Conference on Technologies of Knowledge Exploration and Representation, TERW (2007)

Grekow, J.: Broadening musical perception by AKWETS technique visualization. In: Proceedings of the 9th International Conference on Music Perception and Cognition, ICMPC9 (2006)

Gruber, G.: Zitat. In: MGG^2 Sachteil, Band 9, 1998, col. 2401–2412.

Gülke, P.: "... immer das Ganze vor Augen". Studien zu Beethoven. J.B. Metzler Verlag, Kassel et al. (2000)

Günther, G.: Beiträge zur Grundlegung einer operationsfähigen Dialektik, 3 vols., Meiner, Hamburg, 1976–1980 (1980)

Günther, G.: Idee und Grundriss einer nicht-Aristotelischen Logik, Meiner, Hamburg 3. Aufl.1991 (1991)

Günther, G.: Identität, Gegenidentität und Negativsprache (Vortrag beim Internationalen Hegel-Kongress, Belgrad 1979. In: Hegeljahrbücher 1979, pp. 22–88

Gutknecht, D.: Aufführungspraxis. In: MGG^2, vol. 1, col. 954–986 (1994)

Hagner, M.: Vom Aufstieg und Fall der Kybernetik als Universalwissenschaft, In: Hagner, M., Hörl, E. (ed.) Die Transformation des Humanen. Beiträge zur Kulturgeschichte der Kybernetik, Frankfurt a. M., pp. 38–71 (2008).
Hahl-Hoch, J. (ed.): Arnold Schönberg, Wassily Kandinsky: Briefe Bilder und Dokumente einer außergewöhnlichen Begegnung. dtv, München (1983)
Haimo, E.: Schoenberg's Transformation of Musical Language. Cambridge University Press, Cambridge (2006)
Halpern, A., Müllensiefen, D.: Effects of timbre and tempo change on memory for music. Q. J. Exp. Psychol. **61**(9), 1371–1384 (2008)
Halpern, A.R., Müllensiefen, D, Wiggins, G.: Modelling Memory Responses in a Memory Recognition Task. In: Paper given at 10th International Conference for Music Perception and Cognition, Hokaido University, Sapporo, 25–30th of August 2008 (2008).
Haluška, J.: Mathematical Theory of Tone Systems. Bratislava, Basel und New York (2004)
Hamberger, Ch.Ll.: The Evolution of Schoenberg's Klangfarbenmelodie: The Importance of Timbre in Modern Music, Master of Arts, Master Thesis Pennsylvania State University (2012). https://etda.libraries.psu.edu/catalog/16221
Hanson, C.: Die Bedeutung der Technik für die musikalische Avantgarde: Transformation musikalischer Ästhetik und künstlerischen Selbstverständnisses. Kapitel 4, Technik in der Kunstpraxis: 4.1, Musikmaschinen und Instrumentenbau. Diplomica-Verlag, Hamburg (2015). https://www.diplomica-verlag.de/kunst-kultur_6/index.htm
Hargittai, I., Hargittai, M.: Symmetrie. Eine neue Art, die Welt zu sehen, Reinbek (1998). (Englisch 1994)
Hartmann, G.: Die Tonfolge B-A-C-H. Zur Emblematik des Kreuzes im Werk Joh. Seb. Bachs (2 vols.) (Orpheus-Schriftenreihe zu Grundfragen der Musik, ed. by Vogel, M., vol. 80 and 81). Verlag für Systematische Musikwissenschaft, Bonn (1996)
Hartmann, G.: Carl Philipp Emanuel Bach bei seinem Namen gerufen, In: Musik und Kirche 59.1989, H. 4, pp. 199–203 (1989)
Harwood, J.: Music and intergroup relations: Exacerbating conflict and building harmony through music. In: Music. Review of Communication Research, 5, 1–34 (2017). https://doi.org/10.12840/issn.2255-4165.2017.05.01.012. Accessed: 30 Jan 2018
Haselböck, L.: Zur "Klangfarbenlogik" in Schönbergs Orchesterstück op. 16/3. In: "Kürzen". Gedenkschrift für Manfred Angerer (2016). https://www.academia.edu/29991332/Zur_Klangfarbenlogik_in_SchC3%B6nbergs_Orchesterst%C3%Bcck_op_16_3
Hau, S.: Traum. In: Gehrig, G., Pfarr, U. (eds.) Handbuch psychoanalytischer Begriffe für die Kunstwissenschaft, pp. 373–383. Imago/Psychosozial-Verlag, Gießen (2009)
Heidenreich, S.: Form und Filter - Algorithmen der Bilderverarbeitung und Stilanalyse, In: zeitenblicke 2 (2003), Nr. 1 [08.05.2003], https://www.zeitenblicke.historicum.net/2003/01/heidenreich/index.html. Accessed 4 Jan 2010
Heinsheimer, H.W.: Menagerie in Fis-Dur. Musikalische Impressionen aus Amerika, Pan Verlag, Zürich und Stuttgart (1953)
Heisenberg, W.: Schritte über Grenzen. Gesammelte Reden und Aufsätze. R. Piper & Co., München (1971)
Heister, H.-W.: Homöostase historisch: Zu Fragen der Regelung des musikalischen Tonsatz-Systems, Beitrag zum Kolloquium 1948–2008: 60 Jahre Kybernetik – eine nach wie vor junge Wissenschaft?! im Rahmen der Jahrestagung der Deutschen Gesellschaft für Kybernetik, Berlin am 22. November 2008 (not published)
Heister, H.-W.: Kybernetik, Fuzzy Logic und die Existenzformen der Musik. Ein vorläufiger Entwurf. In: Innovation aus Tradition – Festschrift Hermann Rauhe zum 80. Geburtstag, pp. 93–115. Schott International, Mainz (2010)
Heister, H.-W.: Musikprozess und Fuzzy Logic, in IRASM [International Review of the Sociology and Aesthetics of Music], vol. 43/2 (Dec 2012), pp. 377–401 (2012)
Heister, H.-W.: Zur Codierung und Decodierung von Bedeutungen im Musikprozess. In: Musik und Psychoanalyse hören voneinander. Was die Psychoanalyse von der Musik zu lernen hat,

Psychosozial 35. Jg., Nr. 130, 2012, Heft IV, S. Gießen. psychosozial-verlag, pp. 33–52. – Wiederabdruck In: Johannes Picht (ed.): Musik und Psychoanalyse hören voneinander. Bd. 1, pp. 37–68. Psychosozial-Verlag, Gießen (2013)

Heister, H.-W.: "Fuzzy Clouds": Sfumato versus chiaroscuro in music. In: Seising, R., Trillas, E., Moraga, C., Termini, S. (eds.) On Fuzziness. A Homage to Lotfi A. Zadeh – Volume 1 and 2, Studies in Fuzziness and Soft Computing, vol. 298, pp. 1249–260. Springer, Berlin and Heidelberg (2013)

Heister, H.-W. (ed.): Geschichte der Musik im 20. Jahrhundert: 1945–1975 (Handbuch der Musik im 20. Jahrhundert Band 3). Laaber-Verlag, Laaber (2005)

Heister, H.-W. (ed.): Schichten, Geschichte, System. Geologische Metaphern und Denkformen in den Kunstwissenschaften (Zwischen/Töne, ed. H.-W. Heister, vol. 6). Weidler Buchverlag, Berlin (2016)

Heister, H.-W. (ed.): Wo die schönen Trompeten blasen – vokale und instrumentale Semantik im Werk Gustav Mahlers (Musik und. Eine Schriftenreihe der Hochschule für Musik und Theater Hamburg. Neue Folge, vol. 11). Weidler Buchverlag, Berlin (2011)

Heister, H.-W., Polk, H. (eds.): Bewegtes und Bewegendes. Der Motiv-Begriff in Künsten und Wissenschaften (Musik/Gesellschaft/Geschichte, vol. 7). Weidler Buchverlag, Berlin (2017)

Heister, H.-W., Sparrer, W.-W. (eds.): Der Komponist Isang Yun, München text+kritik (1987)

Heister, H.-W.: 'Real Sound,' Readymade, Handmade. Musical material and the medium between mechanization, automation, and digitalization as an impression and expression of reality. An implicit call for real interdisciplinarity. In: Santi, M., Berner, E. (eds.): Music – Media – History. Re-Thinking Musicology in an Age of Digital Media. Transcript, Bielefeld, in print (2021)

Heister, H.-W.: „Geprägte Form, die lebend sich entwickelt". ‚Motiv', Teile und Ganzes. In: Bewegtes und Bewegendes. Der Motiv-Begriff in Künsten und Wissenschaften, edited by Heister, H.-W. and Polk, H. (Musik/Gesellschaft/Geschichte, vol. 7), pp. 13–26. Weidler Buchverlag, Berlin (2017)

Heister, H.-W.: Das Konzert. Theorie einer Kulturform (2 vols.). Noetzel-Verlag, Wilhelmshaven (1983)

Heister, H.-W.: Einleitung. Aufbau- und Abbau-Prozesse, Wandlungen, Verwerfungen in Natur und Gesellschaft. In: Heister, H.-W. (ed.) Schichten, Geschichte, System. Geologische Metaphern und Denkformen in den Kunstwissenschaften (Zwischen/Töne, vol. 6), pp. 11–42. Weidler Buchverlag, Berlin (2016)

Heister, H.-W.: Enthüllen und Zudecken. Zu Brahms' Semantisierungsverfahren. In:Heister, H.-W. (ed.) Johannes Brahms oder Die Relativierung der absoluten Musik (Zwischen/Töne, vol. 5), pp. 7–36. Von Bockel-Verlag, Hamburg (1997)

Heister, H.-W.: Geschlechterverhältnisse als Modell. Musik-Anthropologie: Gegenstände, Themen, Forschungsperspektiven. In: Stahnke. M. (ed.) Musik - nicht ohne Worte. Beiträge zu aktuellen Fragen aus Komposition, Musiktheorie und Musikwissenschaft (Musik und. Eine Schriftenreihe der Hochschule für Musik und Theater Hamburg, vol. 2), pp. 13–46. Von Bockel-Verlag, Hamburg (2000)

Heister, H.-W.: Invariance and variance of motives: a model of musical logic and/as fuzzy logic. In: Seising, R., Sanz, V. (eds.) Computing in Humanities and Social Sciences (Studies in Fuzziness and Soft Computing, vol. 273), pp. 423–450. Springer, Heidelberg (2012)

Heister, H.-W.: Jazz. (Musik aktuell. Analysen, Beispiele, Kommentare, Bd. 5). Bärenreiter-Verlag, Kassel (1983)

Heister, H.-W.: Konstanz und Wandel. Zu Hartmanns konzertanten Werken zwischen 1930 und 1955. In: Karl Amadeus Hartmann. Komponist im Widerstreit, ed. U. Dibelius, pp. 147–173. Bärenreiter, Kassel u. a. (2004)

Heister, H.-W.: La Música. Domination, Expropriation, Exoticism, Appropriation and Ambivalence of Colonialism. In: Reports of the International Colloquium Music/ology and Colonialism (Centro Nacional de Documentación Musical Lauro Ayestarán, Montevideo/República oriental del Uruguay, Ed. Coriún Aharonián, Montevideo, pp. 273–308 (2011)

Heister, H.-W.: Mimetische Zeremonie - Gesamtkunstwerk und alle Sinne. Aspekte eines Konzepts. In: Heister (ed.) Mimetische Zeremonien - Musik als Spiel, Ritual, Kunst (Musik und. Eine Schriftenreihe der Hochschule für Musik und Theater Hamburg. Neue Folge, vol. 7), pp. 143–185. Weidler Buchverlag, Berlin (2007)

Heister, H.-W.: Natur, Geist, Gesellschaft. Musik-Anthropologie im Gefüge der Musikwissenschaft. In: Petersen, P., Rösing, H. (eds.) 50 Jahre Musikwissenschaftliches Institut in Hamburg. Bestandsaufnahme - aktuelle Forschung - Ausblick (Hamburger Jahrbuch für Musikwissenschaft, vol. 16), pp. 65–82. Peter Lang GmbH, Internationaler Verlag der Wissenschaften, Frankfurt a. M. (1999)

Heister, H.-W.: Natura, spirito, società. Antropologia musicale e nuova musicologia. In: Musica/Realtà, vol. 22, issue 65, pp. 25–54 (Part I); Musica/Realtà vol. 22, issue 66, pp. 115–150 (Part II). Edizioni Unicopli, Milano (2001)

Heister, H.-W.: Neue Musik nach dem Zweiten Weltkrieg. Ordnung oder Auflösung der Elemente und Engagement. In: Paul, G., Schock, R. (eds.) Sound des Jahrhunderts. Geräusche, Töne, Stimmen 1889 bis heute, pp. 313–319. Bundeszentrale für politische Bildung bpb, Bonn (2013)

Heister, H.-W.: Sextett für zwei Violinen, zwei Bratschen und zwei Violoncelli Nr. 1 B-Dur Op. 18/Sextett für zwei Violinen, zwei Bratschen und zwei Violoncelli Nr. 2 G-Dur Op. 36. In: Bockmaier, C., Mauser, S. (eds.) Johannes Brahms. Interpretationen seiner Werke (2 vols.), pp. 119–128 and pp. 242–250. Laaber-Verlag, Laaber (2013)

Heister, H.-W.: Spiel und Musik. Eine systematische Skizze. In: Croatian Musicological Society. International Review of the Aesthetics and Sociology of Music (IRASM), vol. 47, No. 2 (Dec 2016), pp. 171–194 (2016)

Heister, H.-W.: Synthese versus Exotismus. Entdeckung der neuen »Dritten Welt« und universalistische Integration des »Exotischen«), In: Heister, H.-W. (ed.), pp. 263–274 (2005)

Heister, H.-W.: Vereinfachungen. Ritualismus und Minimalismus. In: Heister, H.-W., pp. 278–284 (2005)

Heister, H.-W.: Virtuosen. In: MGG^2**, vol. 9**, col. 1722–1732 (1998)

Heister, H.-W.: Weltenmusik und Menschenmusik. Ein Knotenpunkt in den Traditionslinien der universistischen und mathematisch-quadrivialen Musikauffassung. In: Banse, G., Küttler, G., März, R. (eds.) Die Mathematik im System der Wissenschaften (Abhandlungen der Leibniz-Sozietät der Wissenschaften, vol. 24), pp. 193–216. trafo, Berlin (2009)

Heister, H.-W.: Zeitliche und szenische Polyphonie, dargestellt an Ausschnitten aus "Götterdämmerung" (III/2, "Siegfrieds Todesgesang" und "Trauermarsch beim Tode Siegfrieds") und "Falstaff" (II/2, 2. Teil). In: Marinelli Roscioni, C. (ed.) Quaderni dell' I.R.TE.M. (Istituto di Ricerca per il Teatro Musicale) 11, Serie 3: Atti, N. 5: Tempo e spazio: Problemi di un rapporto tra opera e televisione, pp. 115–136. Rom (1991)

Heister, H.-W.: Zur Theorie der Virtuosität. Eine Skizze, in: Musikalische Virtuosität, ed. by von Loesch, H., Mahlert, U., Rummenhöller, P. (Klang und Begriff Bd. 1), pp. 17–38. Schott International, Mainz (2004)

Heister, H.: The Sympathetic Imagination in the Novels of J.M. Coetzee. Empathy and Mirror Neurons (Phil. Diss.), Humboldt-Universität zu Berlin, Kultur-, Sozial- und Bildungswissenschaftliche Fakultät (2015). https://edoc.hu-berlin.de/bitstream/handle/18452/17838/heister.pdf

Heister, J., et al.: dlexDB – eine lexikalische Datenbank für die psychologische und linguistische Forschung. Psychologische Rundschau 62(1), 10–20. Hogrefe Verlag, Göttingen (2011)

Helms, H.G.: Voraussetzungen eines neuen Musiktheaters. In: Musik auf der Flucht vor sich selbst. Acht Aufsätze (Reihe Hanser 28), pp. 92–115. Hanser-Verlag, München (1969)

Hemming, J., Altenmüller, E.: When an everyday-phenomenon becomes clinical: The case of long-term 'earworms'. Paper presented at the ICMPC-ESCOM 12 (2012, 23–28 July), Thessaloniki (2012)

Hemming, J., Merrill, J.: On the distinction between involuntary musical imagery, musical hallucinosis, and musical hallucinations. Psychomusicol.: Music Mind Brain **25**(4), 435–442 (2015)

Hemming, J.: Zur Phänomenologie des "Ohrwurms". In: Auhagen, W., Bullerjahn, C., Höge, H. (eds.) Musikpsychologie – Musikalisches Gedächtnis und Musikalisches Lernen, vol. 20, pp. 184–207. Hogrefe, Göttingen (2009)

Henckels, P.: Ich war kein Musterknabe. Eines Lebenskünstlers lachende Weisheit, Lothar Blanvalet, Berlin 3rd ed. (1966)

Henze, H.W.: Die Englische Katze. Ein Arbeitstagebuch 1978–1982, Fischer, Frankfurt a. M. (1983)

Henze, H.W.: Musik und Politik, Schriften und Gespräche 1955–1984, ed. by J. Brockmeier. Dtv, München (1984)

Henze, H.W.: Phaedra. Ein Tagebuch. In Zusammenarbeit mit Christian Lehnert. Wagenbach, Berlin (2007)

Henze, H.W.: Reiselieder mit böhmischen Quinten. Autobiographische Mitteilungen 1926–1995. Fischer, Frankfurt (1996)

Hermann, T., Hunt, A., Neuhoff, J.G. (eds.): The Sonification Handbook. Logos Verlag, Berlin (2011)

Herre, J., Allamanche, E., Ertel, C.: How similar do songs sound? Towards modeling human perception of musical similarity. In: Proceedings of the IEEE Workshop on Applications of Signal Processing to Audio and Acoustics, pp. 83–86 (2003)

Herrera, P., Sandvold, V., Gouyon, F.: Percussion-related semantic descriptors of music audio files. In: Proceedings of the 25th International AES Conference, London, United Kingdom (2004). https://ieeexplore.ieee.org/xpl/conhome/8891805/proceeding

Herrmann, D,: Mathematik im Mittelalter. Die Geschichte der Mathematik des Abendlands mit ihren Quellen in China, Indien und im Islam. Springer Spectrum, Berlin und Heidelberg (2016)

Herwig, H., Kalisch, V., Kortlander, B., Kruse, J.A., Witte, B. (eds.): Übergänge. Zwischen Künsten und Kulturen. Internationaler Kongress zum 150. Todesjahr von Heinrich Heine und Robert Schumann. Verlag J.B.Metzler, Stuttgart and Weimar (2007)

Herzberg, S.: Grenze, I. Philosophisch, Version 22.10.2019, 17:30 Uhr. In: Staatslexikon^8. https://www.staatslexikon-online.de/Lexikon/Grenze. Accessed 16 Aug 2020

Hesse, H.-P.: Gehör, IV. Psychoakustische und psychophysikalische Grundlagen. In: MGG^2, vol. 3, col. 1103–1118 (1995)

Hicks, M.: Text, music, and meaning in the third movement of Luciano Berio's Sinfonia. In: Perspect. New Music **20**(1/2) (Autumn, 1981 - Summer, 1982), 199–224 (1981). https://www.jstor.org/stable/942413. Accessed 03 Sept 2018

Hiekel, J.P., Utz, C. (eds.): Lexikon Neue Musik. J.B. Metzler, Stuttgart und Bärenreiter, Kassel (2016)

Hiller, L.A: Informationstheorie und Computermusik. Zwei Vorträge, Darmstadt 1963. Darmstädter Beiträge zur Neuen Musik vol. 8, Schott Verlag, Mainz (1964)

Hiller, L.A., Isaacson, L.M.: Experimental Music: Composition with an Electronic Computer. McGraw-Hill Book Company, New York (1959)

Hilscher, E.Th.: Accompagnement, Obligates. In: Oesterreichisches Musiklexikon online. https://www.musiklexikon.ac.at/ml/musik_A/Accompagnement_obligates.xml,25/5/2001. Accessed 16 Aug 2020

Hinsche, F.: Ästhetik der musikalischen Interpretation. Phil. Diss., Philosophische Fakultät der Heinrich-Heine-Universität Düsseldorf, D61. Düsseldorf (2017)

Hirt, K.M.: When Machines Play Chopin. Musical Spirit and Automation in Nineteenth-Century German Literature (Interdisciplinary German Cultural Studies, vol. 8). De Gruyter, Berlin and New York (2010)

Hochstein, S.: „Runnin'" und „Empire State of Mind": Zwei Modelle aus den Subgenres Independent- und Mainstream-Hip-Hop, Hausarbeit für die Erste Staatsprüfung im Fach Musik, Hochschule für Musik und Theater Hamburg, Juni 2010, Ms. (2010)

Hocker, J.: Begegnungen mit Nancarrow (edition neue Zeitschrift für Musik). Schott Musik International, Mainz (2002)

Holz, H.H.: Speculum mundi. Schriften zur Theorie der Metapher, spekulativen Dialektik und Sprachphilosophie. From the legacy edited by Jörg Zimmer. Aisthesis-Verlag, Bielefeld (2017)

Honolka, K.: Leoš Janáček. Sein Leben - sein Werk - seine Zeit. Belser Verlag, Stuttgart and Zürich (1982)

Hörz, H., Liebscher, H., Löther, R., Wollgast, S. (eds.): Philosophie und Naturwissenschaften. Wörterbuch zu den philosophischen Fragen der Naturwissenschaften, Berlin 1983. Dietz (1983)

Hottmann, K.: Männlichkeitskonstruktion in der Oper am Beispiel von Hugo von Hofmannsthals und Richard Strauss' Arabella. In: Grotjahn, R. et al. (eds.): Musik und Gender. Grundlagen – Methoden – Perspektiven (Kompendien Musik, vol. 5), pp. 183–194. Laaber Verlag, Laaber (2010)

Howlin, C., Vicary, S., Orgs, G.: Audiovisual aesthetics of sound and movement in contemporary dance. Empirical Stud. Arts, 1–24 (2018). https://doi.org/10.1177/0276237418818633

Hu, X., Yang, Y.-H: The mood of Chinese pop music: representation and recognition. J. Assoc. Inf. Sci. Technol. **68**(8), 1899–1910 (2017)

Hu, X., Yang, Y.-Hs.: Cross-dataset and cross-cultural music mood prediction: a case on Western and Chinese pop songs. IEEE Trans. Affect. Comput. **8**(2), 228–240 (2017)

Huang, Y.-S., Chou, S.-Y., Yang, Y.-H.: Pop music highlighter: marking the emotion keypoints. Trans. Int. Soc. Music Inf. Retrieval **1**(1), 68–78 (Sept 2018)

Huang, Y.-S., Yang, Y.-Hs.: Pop music transformer: beat-based modeling and generation of expressive Pop piano compositions. In: Proceedings of the ACM Multimedia, October 2020

Hung, Hs.-Tz., Wang, Ch.-Y., Yang, Y.-Hs., Wang, Hs.-M.: Improving automatic Jazz melody generation by transfer learning techniques. In: Proceedings of the Asia Pacific Signal and Information Processing Association Annual Summit and Conf. (APSIPA ASC), Nov 2019.

Huron, D.: Sweet Anticipation: Music and the Psychology of Expectation. MIT Press, Cambridge (2006)

Hutchinson Guest, A.: Labanotation. The System of Analyzing and Recording Movement, fourth edition, illustrations by Doug Anderson. Routledge, New York and Abington (2005)

Hüther, G.: Das Geheimnis der ersten neun Monate. Unsere frühesten Prägungen. Beltz Verlag, Weinheim and Basel (2011)

Hüther, G.: Die Macht der inneren Bilder. Wie Visionen das Gehirn, den Menschen und die Welt verändern. Vandenhoeck & Ruprecht, Göttingen (2006)

Hutter, U.: Klangfarben – Farbklänge. Wechselwirkung zwischen visueller und auditiver Wahrnehmung, zum Zweck der Erlangung des akademischen Grades eines Dipl. Ing. (FH) Telekommunikation und Medien Wien (2006)

International Symposium of Musicology "Pitfalls and Risks in Intra-Musical Communication: Terminology, Notation, Performance", held under the auspices of the 10th edition of the Cluj Modern Festival, in April 2013. Musicology Papers, vol. 28, N. 2, 2013, pp. 35–45 (2013). https://www.researchgate.net/publication/284186499

Jakubowski, K., Müllensiefen, D.: The influence of music-elicited emotions and relative pitch on absolute pitch memory for familiar melodies. Q. J. Exp. Psychol. **66**(7), 1259–1267 (2013)

Janich, P.: Handwerker und Mundwerker. In: Wehr, M., Weinmann, M. (eds.) Die Hand. Werkzeug des Geistes, pp. 271–292. Spektrum Akademischer Verlag, Heidelberg (2005)

Jao, K., Lin, Y.-P., Yang, Y.-H., Jung, T.-P.: Using robust principal component analysis to alleviate day-to-day variability in eeg based emotion classification. In: Proceedings of the International Conference on IEEE Engineering in Medicine and Biology (EMBC), Aug 2015

Jaschinski, A. (ed.): Notation (MGG Prisma). Bärenreiter, Kassel etc (2001)

Jeong, E.-H.: The application of imagery to enhance "flow state" in dancers, Doctor of philosophy School of sport and exercise science. Faculty of arts, education and human development Victoria University, Wellington (2012)

Jeßulat, A.: Erinnerte Musik. Der Ring des Nibelungen als musikalisches Gedächtnistheater (Wagner in der Diskussion, vol. 8). Königshausen & Neumann, Würzburg (2013)

JetztMusik – Glossar, SWR2 (Südwestrundfunk, Baden-Württemberg, Bundesrepublik Deutschland). https://www.swr.de/swr2/. Link is disabled.

Juliano, B.A., Bandler, W.: Tracing Chains-of-Thought, Fuzzy Methods in Cognitive Diagnosis, vol. 7. Springer Wissenschafts-Verlag (2017)

Julien, D., O'Connor, K.P., Aardema, F.: Intrusive thoughts, obsessions, and appraisals in obsessive-compulsive disorder: a critical review. Rev. Clin. Psychol. **27**(3), 366–383 (2007). https://doi.org/10.1016/j.cpr.2006.12.004. Epub 2007 Jan 22

Jürgensen, F., Knopke, I.: A comparison of automated methods for the analysis of style in fifteenth-century song intabulations. In: Parncutt, R., Kessler, A., Zimmer, F. (eds.) Proceedings of the Conference on Interdisciplinary Musicology (CIM04), p. 98 f. (2004). https://www-gewi.uni-graz.at/staff/parncutt/cim04/CIM04_paper_pdf/JurgensenKnopke.pdf

Kacprzyk, J., Castillo, O., Melin, P. (eds.): Nature-Inspired Design of Hybrid Intelligent Systems. Springer, Heidelberg et al. (2017).https://www.springer.com/978-3-319-47053-5

Kaehr, R., Ditterich, J.: Einübung in eine andere Lektüre: Diagramm einer Rekonstruktion der Güntherschen Theorie der Negativsprachen, Philosophisches Jahrbuch, 86. Jhg., pp. 385–408 (1979). https://www.vordenker.de/ggphilosophy/kaehr_einuebung.pdf. Accessed 23 Juli 2009

Käfer, B.: Paläolithische Knochenflöten im Ostalpenraum und dem nordöstlich vorgelagerten Lößgebiet (Magisterarbeit). Hochschule für Musik und darstellende Kunst Wien (1998)

Kahn, D.: Noise, Water, Meat: A History of Sound in the Arts. MIT Press, Cambridge (1999)

Bronner, K., Hirt, R., Ringe, C. (eds.): (((ABA))) Audio Branding Academy Yearbook 2012/2013. Nomos Verlagsgesellschaft, Baden-Baden (2013)

Kalbeck, M.: Johannes Brahms, vol. 4, 2. edition. Deutsche Brahms-Gesellschaft, Berlin (1915). https://www.zeno.org/nid/20007753837

Kandel, E.: Das Zeitalter der Erkenntnis. Die Erforschung des Unbewussten in Kunst, Geist und Gehirn von der Wiener Moderne bis heute. Siedler Verlag, München (2012)

Kanitscheider, B.: Kosmologie: Geschichte und Systematik in philosophischer Perspektive. Reclam, Stuttgart (1984). (2., extended ed.. 2002)

Karkoschka, E: Das Schriftbild der Neuen Musik. Bestandsaufnahme neuer Notationssymbole. Anleitung zu deren Deutung, Realisation und Kritik. Moeck, Celle (1966)

Karkoschka, E.: Notation in New Music: A Critical Guide to Interpretation and Realization. Praeger, New York (1972)

Kautny, O.: Talkin' All that Jazz - ein Plädoyer für die Analyse des Sampling im Hiphop. In: von Appen, R., Doehring, A., Helms, D., Phleps, T. (eds.) Online-Publikationen des Arbeitskreis Studium Populärer Musik e.V. (ASPM) vol. 9 (2010). https://www.aspm-samples.de/Samples9/Kautny.pdf. Accessed 20 Oct 2011

Keizer, G.: The Unwanted Sound of Everything we Want. A Book About Noise. Public Affairs, New York (2010)

Kim-Boyle, D.: The ontology of live notations through assemblage theory. In: Hoadley, R., Nash, Ch., Fober, D. (eds.) Proceedings of the International Conference on Technologies for Music Notation and Representation (2016). https://doi.org/10.5281/zenodo.923969

Kim, Y., Schmidt, E., Migneco, R., Morton, B., Richardson, P., Scott, J., Speck, J., Turnbull, D.: Music emotion recognition: a state of the art review. In: Proceedings of ISMIR (2010)

Kirnberger, J.P.: Die Kunst des reinen Satzes in der Musik. G. J. Decker and G. L. Hartung, Berlin and Königsberg (1774)

Klaus, G.: Kybernetik - eine neue Universalphilosophie der Gesellschaft? (Zur Kritik der bürgerlichen Ideologie, vol. 27). Akademie-Verlag, Berlin (1973)

Klaus, G., Liebscher, H. (Eds.): Wörterbuch der Kybernetik, Berlin (1976)

Klaus, G., Buhr, M. (Eds.): Philosophisches Wörterbuch, 7., berichtigte Aufl. Berlin 1970 [ursprünglich 1964], 2 Bde (1970)

Klotz, S.: Kombinatorik und die Verbindungskünste der Zeichen in der Musik zwischen 1630 und 1780. De Gruyter Akademie Forschung, Berlin (2006)

Kluge, R.: Skalen und Stimmungen des bronzezeitlichen Glockenspiels aus dem Grabe des Zeng Hou Yi. Zur Deutung eines tonometrischen Befundes. In: Ereignis und Exegese. Musikalische Interpretation – Interpretation der Musik. Festschrift für Hermann Danuser zum 65. Geburtstag, pp. 231–242. Argus Verlag, Schliengen (2011)

Klüppelholz, W.: Mauricio Kagel 1970–1980, DuMont-Verlag, Cologne (1981)

Klüsener, S.: Qualitative Heuristik. Strukturierendes Entdecken. Ruprecht-Karls-Universität Heidelberg, Geographisches Institut, Heidelberg (1998). https://www.kluesener-net.de/Deutsch/Pdf/Qualitative_Heuristik.pdf. Accessed 7 Aug 2017

Klüsener, S.: Qualitative Heuristik. Strukturierendes Entdecken. Ruprecht-Karls-Universität Heidelberg, Geographisches Institut (1998).https://www.kluesenernet.de/Deutsch/Pdf/Qualitative_Heuristik.pdf. Dead link

Knees, P., Pampalk, E., Widmer, G.: Artist classification with web-based data. In: Proceedings of the 5th International Conference on Music Information Retrieval, pp. 517–524. Barcelona, Spain (2004)

Knepler, G.: Geschichte als Weg zum Musikverständnis. Zu Theorie, Methode und Geschichte der Musikgeschichtsschreibung. Reclam Universalbibliothek, Leipzig (1977) (2nd enlarged edition) (1982)

Koelsch, S. et al.: Bach speaks: a cortical language-network serves the processing of music. Neuroimage **17**, 956–966 (2002)

Koelsch, S. et al.: Music, language and meaning: brain signatures of semantic processing. Nat. Neurosci. **7**, 302–307 (2004)

Kohut, H.: Review of the haunting melody. Psychoanal. Q. **24**, 134–137 (1955)

Konrad, U.: Mozarts Schaffensweise. Studien zu den Werkautographen, Skizzen und Entwürfen (Abhandlungen der Akademie der Wissenschaften in Göttingen, Philologisch-Historische Klasse, third series, vol. 201). Vandenhoeck & Ruprecht, Göttingen (1992)

Korhonen, M.D., Clausi, D.A., Jernigan, M.E.: Modeling emotional content of music using system identification. IEEE Trans. Syst. Man Cybern. Part B Cybern. **36**(3), 588–599 (2005)

Kornauth, G.: Plunderphonics – Plagiarismus in der Musik. https://plagiatsgutachten.de/wp-content/uploads/2011/06/Kornauth_Gaar_Past_Plunderphonics.pdf (2011)

Körndle, F.: Das zweistimmige Notre-Dame-Organum „Crucifixum in carne" und sein Weiterleben in Erfurt (MÜNCHNER VERÖFFENTLICHUNGEN ZUR MUSIKGESCHICHTE, Begründet 1959 von Thrasybulos G. Georgiades, Herausgegeben seit 1977 von Theodor Göllner, Band 49) HANS SCHNEIDER, TUTZING (1993). https://epub.ub.uni-muenchen.de/28365/1/Franz_Koerndle_Das_zweistimmige_Notre-Dame-Organum_%E2%80%9ECrucifixum_in_carne%E2%80%9C_und_sein_Weiterleben_in_Erfurt.pdf. Accessed 13 Feb 2018

Korzybski, A.: Science & Sanity [1933]. European Society for General Semantics. https://esgs.free.fr/uk/art/sands.htm

Kosko, B.: Die Zukunft ist fuzzy. Unscharfe Logik verändert die Welt, München und Zürich (1999)

Kostek, B., Czyzewski, A.: Representing musical instrument sounds for their automatic classification. J. Audio Eng. Soc. **49**, 768–785 (2001)

Kostek, B., Wojcik, J., Holonowicz, P.: Estimation the rhythmic salience of sound with association rules and neural network. In: Intelligent Information Systems, Gdansk, Poland. Advances in Soft Computing (AINSC, volume 31) (2005a)

Kostek, B., Wojcik, J.: Machine learning system for estimation rhythmic salience of sounds. Int. J. Knowl.-based Intell. Eng. Syst. **9**(4), 275–284 (2005b)

Koza, J.R.: Genetic Programming: On the Programming of Computers by Means of Natural Selection. MIT Press, Cambridge (1992)

Krämer, R.: Algorithmic music analysis: a case study of a prelude from David Cope's "From darkness, light". Phil. Dissertation, University of North Texas (2015)

Kranenburg, Pv., Backer, E.: Musical style recognition—a quantitative approach. In: Parncutt, R., Kessler, A., Zimmer, F. (eds.) Proceedings of the Conference on Interdisciplinary Musicology (CIM04) (2004). https://www-gewi.uni-graz.at/staff/parncutt/cim04/CIM04_paper_pdf/Kranenburg_Backer_CIM04_proceedings.pdf

Krause, M.: Original oder Illusion: Ziel der technischen Reproduktion von Musik? In: Kopiez, R., Barthelmes B. et al. (eds.) Musikwissenschaft zwischen Kunst, Ästhetik und Experiment. Festschrift Helga de la Motte-Haber zum 60. Geburtstag, pp. 319–330. Königshausen und Neumann, Würzburg (1998)

Kreinin, J.: "To arrest the process": moving clusters by György Ligeti and Witold Lutoslawski. In: Mitteilungen der Paul Sacher Stiftung, no. 15, pp. 36–41 (2002)

Kreuzer, S. (ed.): Experimente in den Künsten. Transmediale Erkundungen in Literatur, Theater, Film, Musik und bildender Kunst. transcript, Bielefeld (2012)

Krickeberg, D.: Automatic Musical Instruments in Bar and Salon, Guide Sheet No. 11, Musikinstrumenten-Museum Staatliches Institut für Musikforschung Berlin SIM PK Preußischer Kulturbesitz Berlin, 2nd revised edition (2015)

Kross, S.: Johannes Brahms. Versuch einer kritischen Dokumentar-Biographie, vol. 1. Bouvier, Bonn (1997)

Kruse, R., Gebhardt, J., Klawonn, F.: Foundations of Fuzzy Systems. J. Wiley & Sons 1994. Geleitwort: Lotfi A. Zadeh, Berkeley, California, March 1994 (1994)

Kruse, R., Gebhardt, J., Klawonn, F.: Foundations of Neuro-Fuzzy Systems. Wiley, New York (1997)

Kubik, G. 1998: Kalimba, Nsansi, Mbira – Lamellophone in Afrika (Veröffentlichungen des Museums für Völkerkunde Berlin, Abteilung Musikethnologie X, Neue Folge 68). Berlin (1998)

Kubik, G.; Afrikanische Musik. In: MGG^2, 1994, vol. 1, col. 50–194 (1994)

Kubik, G.: Zum Verstehen afrikanischer Musik. Reclam, Leipzig (1988)

Kubik, Gerhard et al: Ostafrika (Musikgeschichte in Bildern, vol. I, issue 10). Deutscher Verlag für Musik, Leipzig (1982)

Kühn, C.: Kompositionsgeschichte in kommentierten Beispielen (Bärenreiter-Studienbücher Musik, 9). Bärenreiter, Kassel et al. (1998)

Kunkel, M.: Gibt es neue Typen des Engagements in der neuesten Musik? Positionen und Negationen. In: Phleps, T., Reich, W. (eds.) Musik-Kontexte. Festschrift für Hanns-Werner Heister, pp. 472–487. Verlagshaus Monsenstein und Vannerdat, Münster (2011)

Kunz, H.: Die Opern In: Heister, H.-W., Sparrer, W.-W. (eds.) Der Komponist Isang Yun, pp. 95–137. text+kritik, München (1987)

Kurth, F., Gehrmann, T., Müller, M.: The cyclic-beat spectrum: Tempo-related audio features for time-scale invariant audio identification. In: Proceedings of the 7th International Conference on Music Information Retrieval, pp. 35–40 (2006)

Lacoste, St.: Los Angeles Philharmonic Association (n.d.). https://www.laphil.com/philpedia/piecdetail.cfm?id=2742

Lawson, R.: On the right track: dynamic recording for the reproducing piano (Part Four). Pianola J. 23, 11–26. The Pianola Institute, West Wickham, Kent (2013)

Lee, C.-T., Yang, Y.-H., Chen, H.-H.: Multipitch estimation of piano music by exemplar-based sparse representation. IEEE Trans. Multimedia **14**(3), 608–618 (2012). (第二屆美律電聲論文獎 – 銅質獎)

Lehmann, Chr.: Klassifikation (n.d.). https://www.christianlehmann.eu/ling/epistemology/concepts/index.html?https://www.christianlehmann.eu/ling/epistemology/concepts/klassifikation.php

Lehnel, M., Engel, P., Rohrmeier, M., Menninghaus, W., Jacobs, A.M., Koelsch, S.: Reading a suspenseful literary text activates brain areas related to social cognition and predictive inference. PLoS ONE **10**(5), e0124550 (2015). https://doi.org/10.1371/journal.pone.0124550

Lehwark, P., Risi, S., Ultsch, A.: Visualization and clustering of tagged music data. In: Proceedings GfKl 2007, Freiburg i. Br. Conference Paper · January 2007 (2007). https://www.researchgate.net/publication/221649327_Visualization_and_Clustering_of_Tagged_Music_Data/link/56f4daab08ae81582bf0a868/download

Leman, M.: Expressing coherence of musical perception in formal logic. In: Assayag, G., Feichtinger, H.G., Rodrigues, J.F. (eds.) Mathematics and Music. A Diderot Mathematical Forum, pp. 185–198. Springer, Berlin, Heidelberg, New York, Barcelona, Hong Kong, London, Milan, Paris, Tokyo (2002)

León, T., Liern Carrión, V.: Music and Similarity Based Reasoning. In: Seising, R., und Sanz, V. (eds.): Computing in Humanities and Social Sciences (Studies in Fuzziness and Soft Computing, vol. 273), pp. 451–468. Springer, Berlin and Heidelberg (2012)

León, T., Liern Carrión, V.: Obtaining the compatibility between musicians using soft computing. In: Hüllermeier, E., Kruse, R., Hoffmann, F. (eds.): International Conference on Information Processing and Management of Uncertainty in Knowledge-Based Systems, IPMU 2010 Part II, CCIS 81, Dortmund, Germany, June 28 -July 2, 2010, pp. 75–84. Springer, Berlin and Heidelberg (2010)

Lerch, A.: On the Requirement of Automatic Tuning Frequency Estimation (2006). urn:nbn:de:kobv:83-opus-20822 https://doi.org/10.14279/depositonce-2037

Lerch, A.: Software-based extraction of objective parameters from music performances. Phil. Diss., TU Berlin (2008)

Lesaffre, M., Tanghe, K., Martens, G., Moelants, D., Leman, M., De Baets, B., De Meyer, H., Martens, J.P.: The MAMI query-by-voice experiment: collecting and annotating vocal queries for music information retrieval. In: Proceedings of the 4th International Conference on Music Information Retrieval, Baltimore, Maryland, and Library of Congress, Washington, DC, pp. 65–71 (2003)

Levens, U.: Computergestütztes Modellieren mit Musiknetzen. In: Diekert, V., Weicker, K., Weicker, N. (eds.): Informatik als Dialog zwischen Theorie und Anwendung. Fschr. für Volker Claus zum 65. Geburtstag, pp. 44–57. Vieweg+Teubner Verlag, Wiesbaden (2009)

Levy, M., Sandler, M.: Lightweight measures for timbral similarity of musical audio. In: Proceedings of the First ACM Workshop on Audio and Music Computing Multimedia (AMCMM), pp. 27–36. ACM, New York (2006)

Li, T., Ogihara, M., Li, Q.: A comparative study on content-based music genre classification. In: Proceedings of the 26th International ACM SIGIR Conference on Research and Development in Information Retrieval, pp. 282–289. ACM, New York (2003)

Li, T., Ogihara, M.: Detecting emotion in music. In: Proceedings of the Fifth International Symposium on Music Information Retrieval, pp. 239–240 (2003)

Lichtenberg, G.C.: Sudelbuch A. [A 1], erstes bis fünftes Heft (1765). Georg Christoph Lichtenberg. Schriften und Briefe, ed. And commented by Wolfgang Promies., vol. 1. Hanser Verlag, München (1968)

Lidy, T., Rauber, A.: Evaluation of feature extractors and psycho-acoustic transformations for music genre classification. In: Proceedings of the 6th International Conference on Music Information Retrieval, pp. 34–41 (2005)

Ligges, U.: Transkription monophoner Gesangszeitreihen (Diss.), Universität Dortmund, Fachbereich Statistik. Dortmund, Germany (2006)

Liikkanen, L.A.: How the mind is easily hooked on musical imagery. Paper Presented at the 7th Triennial Conference of the European Society for the Cognitive Sciences of Music ESCOM 2009, Jyväskylä (2009)

Liikkanen, L.A.: Music in everymind: commonality of involuntary musical imagery. Paper Presented at the 10th International Conference on Music Perception and Cognition, Sapporo (2008)

Liikkanen, L.A.: New directions for understanding involuntary musical imagery. Paper Presented at the ICMPC-ESCOM 12, Thessaloniki (2012)

Liikkanen, L.A.: Inducing involuntary musical imagery: an experimental study. In: EDITOR? Music. Sci. **16**(2), 217–234 (2012)

Liikkanen, L.A.: Involuntary Musical Imagery – Everyday But Ephemeral (University of Helsinki. Department of Digital Humanities, Studies in Cognitive Science 10). Digital edition (2018). https://creativecommons.org/licenses/by-nc-nd/4.0/

Lin, K.-S., Yang, Y.-H., Lee, C.-T., Chen, H.-H.: Automatic highlights extraction for drama video using music emotion and human face features. Neurocomputing **119**, 11–17 (Nov 2013)

Lin, Y.-C., Yang, Y.-H., Chen, H.-H.: Exploiting online tags for music emotion classification. ACM Trans. Multimedia Comput. Commun. Appl. 7S(1). Special Issue on Social Media October 2011

Lin, Y.-P., Yang, Y.-Hs., Jung. Tz.-P.: Fusion of Electroencephalogram dynamics and musical contents for estimating emotional responses in music listening. Front. Neurosci. **8**(94), 1–14, May 2014

Lindner, G.: Grundlagen und Anwendung der Phonetik (Sammlung Akademie-Verlag 36. Sprache), Berlin 1981. Akademie-Verlag (1981)

Linke, A., Nussbaumer, M., Portmann, P.: Studienbuch Linguistik (Germanistische Linguistik, Bd. 121. Kollegbuch), ergänzt um das Kapitel Phonetik und Phonologie von Urs Willi, Tübingen 3. Aufl. 1996. Max Niemeyer (1996)

Liu, C., Yang, Y., Wu, P., Chen, H.: Detecting and classifying emotion in popular music. In: Proceedings of the 9th Joint Conference on Information Sciences (JCIS)/CVPRIP (2006)

Liu, C., Yang, Y., Wu, P., Chen, H.: Detecting and classifying emotion in popular music. In: Proceedings of the 9th Joint Conference on Information Sciences (JCIS)/CVPRIP (2006)

Liu, D., Lu, L., Zhang, N.: Automatic mood detection from acoustic music data. In: Proceedings of the Fifth International Symposium on Music Information Retrieval (2003)

Lock, G., Murray, D. (eds.): The Hearing Eye. Jazz & Blues Influences in African American Visual Art. Oxford University Press, Oxford and New York (2009)

Locke, D.L.: The metric matrix: simultaneous multidimensionality in African music. Anal. Approaches World Music **1**(1), 48–72 (2011)

Logothetis, A.: Kurze Musikalische Spurenkunde. Eine Darstellung des Klanges. In: Melos 1970, No. 2, pp. 39–43. Schott Verlag, Mainz (1970)

Logothetis, A.: Über die Darstellung des Klanges im Schriftbild. In: Impulse für Spielmusikgruppen, pp. 3–9. Universal Edition, Wien (1973)

Logothetis, A.: Zeichen als Aggregatzustand der Musik. Jugend & Volk, Wien (1974)

Lombardi, D.: Il rumore del tempo. Un' antologia di segno/gesto/suono. Prato, Palazzo Novellucci, luglio 1983. Centro Di, catalogo 163, Firenze (1983)

Löther, R.: Die Beherrschung der Mannigfaltigkeit. Philosophische Grundlagen der Taxonomie, Digitalisat Max Stirner Archiv Leipzig - 17.5.2015 (2017)

Lowell, N.: Eugène Ysaÿe and Musical Dialogue. dornsife.usc.edu (not dated)

Lu, L., Liu, D., Zhang, H.J.: Automatic mood detection and tracking of music audio signals. IEEE Trans. Audio Speech Lang. Proc. **14**(1) (2006)

Lubben, J.: Metric vibration: acoustic perspectives on metric dissonance. In: Utz, C., Gnadenstätter, C. (ed.): Musiktheorie als interdisziplinäres Fach. Music Theory and Interdisciplinarity. 8. Kongress der Gesellschaft für Musiktheorie Graz 2008 (musik.theorien der gegenwart 4). Pfau, Saarbrücken (2010)

Manning, P.: Electronic Electronic and Computer Music, 4th edn. Oxford University Press, Oxford and New York (2013)

Markuse, B., Schneider, A.: Ähnlichkeit, Nähe, Distanz: zur Anwendung multidimensionaler Skalierung in musik-wissenschaftlichen Untersuchungen. In: Systematische Musikwissenschaft / Systematic Musicology/Musicologie syst[['e]]matique 4, pp. 53–89 (1996)

Mathias, B., Zamm, A., Gianferrara, P.G., Ross, B., Palmer, C.: Rhythm complexity modulates behavioral and neural dynamics during auditory-motor synchronization. J. Cogn. Neurosci. **32**(10), 1864–1880 (2020). Doi:https://doi.org/10.1162/jocn_a_01601

Mauch 2011: Estimation of Harpsichord Inharmonicity and Temperament from Musical Recordings, 28 MÄRZ 2011. https://schall-und-mauch.de/artificialmusicality/2011/03/estimation-of-harpsichord-inharmonicity-and-temperament-from-musical-recordings/. Accessed 30 Jan 2014

Mauch, M., Fujihara, M., Yoshii, K., Goto, M.: Timbre and melody features for the recognition of vocal activity and instrumental solos in polyphonic music. International Society for Music Information Retrieval Conference: Proceedings of the 12th International Society for Music Information Retrieval Conference, ISMIR 2011, Miami, Florida, USA, October 24–28 (2011)

Mauch, M., Müllensiefen, D., Dixon, S., Wiggins, G.: What's a Word in Harmony? Paper Given at 10th International Conference for Music Perception and Cognition, Hokaido University, Sapporo, 25–30th of August 2008. Full paper (2008)

Mayer, G.: Über das „Rauschen“ des Irrealen. In: Mayer 2005, pp. 193–214 (1996)

Mayer, G.: Zur Theorie des Ästhetischen. Texte aus vier Jahrzehnten (Zwischen/Töne, hg. v. H.-W. Heister, Neue Folge Bd. 5), Berlin 2005. Weidler Buchverlag (2005)

Mazzola, G. (with the collaboration of Muzzulini, D., with a contribution of Hofmann, G.R.): Geometrie der Töne: Elemente der mathematischen Musiktheorie. Birkhäuser, Basel, Boston and Berlin (1990)

Mazzola, G.B., Cherlin, P.B.: Flow, Gesture, and Spaces in Free Jazz. (Computational Music Science). Springer, Berlin, Heidelberg (2009)

Mazzola, G., Mannone, M., Pang, Y.: Cool Math for Hot Music. A First Introduction to Mathematics for Music Theorists (Computational Music Science, series edited by G. Mazzola and M. Andreatta). Springer, Berlin, Heidelberg (2016)

Mazzola, G., Park, J., Thalmann, F.: Strategies and Tools in Composition and Improvisation (Computational Music Science). Springer, Berlin, Heidelberg (2011)

Mazzola, G.: Musical Performance. A Comprehensive Approach: Theory, Analytical Tools, and Case Studies (Computational Music Science). Springer, Berlin, Heidelberg (2011)

Mazzola, G.: Review of Hesse's Glass Bead Game. In: Castalia. The End of Games and Means (2016). https://www.glass-bead.org/article/melting-glass-beads/?lang=enview. Accessed 20 Feb 2017

Mazzola, G.: The topos geometry of musical logic. In: Assayag, G., Feichtinger, H., Rodrigues, J.F. (eds.): Mathematics and Music. A Diderot Mathematical Forum, pp. 199–214. Springer, Berlin (2002)

McInerney, M.J.: Performance and the Page: An artist's investigation of the dialogue between the musical event and the written score, A thesis submitted to the University of Plymouth in partial fulfilment for the degree of Doctor of Philosophy, vol. 1: Written Inquiry. Dartington College of Arts. Creative and Performing Arts (2007)

McKay, C., Fujinaga, I.: Automatic genre classification using large high-level musical feature sets. In: Proceedings of the International Conference on Music Information Retrieval, pp. 525–530 (2004)

McLean, A., Griffiths, D., Collins, N., Wiggins, G.: Visualisation of live code (https://yaxu.org/visualisation-of-live-code/). In: Electronic Visualisation and the Arts, London 2010. Ed. by. Alan Seal, A., Jonathan P. Bowen, J.P., Kia Ng, K. Springer Series on Cultural Computing (2013). https://yaxu.org/writing/visualisation-of-live-code.pdf

Messing, J.: Allgemeine Theorie des menschlichen Bewusstseins. Weidler Buchverlag, Berlin (1999)

Metzger, H.-K., Riehn, R. (eds.): Bruckners Neunte im Fegefeuer der Rezeption (Musik-Konzepte, vols. 120–122). text+kritik, München (2011)

Meyer-Eppler, W.: Grundlagen und Anwendungen der Informationstheorie. Springer-Verlag, Berlin, Göttingen. Heidelberg (1959)

Meyer-Eppler, W.: Statistische und psychologische Klangprobleme. In: die reihe 1: elektronische Musik. Informationen über serielle Musik, pp. 22–28. Universal-Edition, Wien (1955)

Meyer, L.B.: Emotion and Meaning in Music. University of Chicago Press, Chicago (1956)

Meyer, P.M. (ed.): Sound History. Acoustic Turn. Wilhelm Fink-Verlag, München (2008)

MGG[1] (Die Musik in Geschichte und Gegenwart) (The Music in History and Present Day), ed. F. Blume, 14 vols. +2 supplement-vols. + 1 vol. register. Bärenreiter-Verlag, Kassel etc. (1949–1979)

MGG[2] (Die Musik in Geschichte und Gegenwart) 2nd, new revised edition, ed. L. Finscher. Subject part, 9 vols. Bärenreiter/Metzler, Kassel and Stuttgart (1994–1998)

Miller, D.: Are Scores Maps? A Cartographic Response to Goodman. Computer Science (2017). https://www.semanticscholar.org/paper/Are-scores-maps%3A-a-cartographic-response-to-goodman-Miller/b2a0e29fd3856282706ca3639a85d00649ed9682

Milmeister, G.: The Rubato Composer Music Software. Component-Based Implementation of a Functorial Concept Architecture (Computational Music Science). Springer, Berlin, Heidelberg (2009)

Mindern, Missbräuche verhindern. In: Sitzungsberichte der Leibniz-Sozietät der Wissenschaften zu Berlin, vol. 112, pp. 19–41. trafo Verlag, Berlin (2011)

Moles, A.A.: Informationstheorie und ästhetische Wahrnehmung. DuMont, Köln (1971)

Möller, H.: Angst, Entmutigung und Perfektionismus bei Musikern. Mit Fallbeispielen aus der Praxis. Self-publishing, Berlin (2018)

Monson, K.: La Relazione con Hanna Fuchs-Robettin e la «Lyrische Suite». In: Karen Monson, Berg, pp. 237–263. Rusconi, Milano (1982)

Montiel, M., Peck, R.W. (Eds.): MATHEMATICAL MUSIC THEORY. Algebraic, Geometric, Combinatorial,Topological and Applied Approaches to Understanding Musical Phenomena. World Scientific, New Jersey (2018)

Moorey, G.: Everyday Reveries: Recorded Music, Memory & Emotion. A thesis submitted in partial fulfilment of the requirements of the University of the West of England, Bristol for the degree of Doctor of Philosophy, Faculty of Arts, Creative Industries and Education). University of the West of England, Bristol (2012)

Morat, D., Ziemer, H. (with the collaboration of R. Rutz) (eds.): Handbuch Sound. Geschichte – Begriffe – Ansätze. J.B. Metzler Verlag, Stuttgart (2018)

Mörchen, F., Ultsch, A., Thies, M., Löhken, I., Nöcker, M., Stamm, C., Efthymiou, N., Kümmerer, M.: MusicMiner: visualizing timbre distances of music as topographical maps. Tech. rep., Department of Mathematics and Computer Science, University of Marburg, Germany (2005)

Moritz, K.Ph.: Über den Begriff des in sich selbst Vollendeten [1785], In: Werke in 2 Bänden, Berlin und Weimar: Aufbau 1973, pp. 203–210 (1973)

Morris, D.: Biologie der Kunst. Ein Beitrag zur Untersuchung bildnerischer Verhaltensweisen bei Menschenaffen und zur Grundlagenforschung der Kunst. Karl Rauch-Verlag, Düsseldorf (1962)

Mukarovský, J.: Kapitel aus der Ästhetik. Edition Suhrkamp, Frankfurt a. M. (1970)

Müllensiefen, D.: Dealing with ambiguity in empirical models of music cognition. Talk at the Music and Brain Club, Wellcome Centre, University College, London (2008)

Müllensiefen, D., Lewis, D., Wiggins, G.A.: Feature description and corpus-based musicology: tools for modelling music cognition. Paper at the Conference of International Musicological Society (IMS), University of Zürich (2007)

Müllensiefen, D., Lewis, D., Wiggins, G. A.: Where's the chorus? A computational approach for the automatic segmentation of pop songs. Paper at the Conference of International Association for the Study of Popular Music (IASPM), Mexico City (2007)

Müllensiefen, D., Pearce, M.T., Wiggins, G.A., Frieler, K.: Segmenting pop melodies: a model comparison approach. Presented at the Conference for the Society for Music Perception and Cognition. Concordia University, Montreal, Canada. August 2007 (2007)

Müllensiefen, D., Wiggins, G.: Representing melodic contour by polynomials. Paper given at 10th International Conference for Music Perception and Cognition, Hokaido University, Sapporo, 25–30th of August 2008 (2008)

Müllensiefen, D., Wiggins, G.: Polynomial functions as a representation of melodic phrase contour. In: Schneider, A., von Ruschkowski, A. (eds.) Systematic Musicology: Empirical and Theoretcial Studies, pp. 63–88. Peter Lang, Frankfurt a. M. (2011)

Müllensiefen, D., Wiggins, G., Lewis, D.: High-level feature descriptors and corpus-based musicology: techniques for modelling music cognition. In: Schneider, A. (ed.) Hamburger Jahrbuch für Musikwissenschaft, 24, pp. 133–155). Peter Lang, Frankfurt (2008)

Müllensiefen, D.: Using R for data analysis. Skills Session given as part of Goldsmiths' staff training, August 2009, Goldsmiths, University of London (2009)

Müllensiefen, D.: Can you predict how courts decide in plagiarism cases? Talk given at the public event Striking Your Own Chord: Journeys into Musical Plagiarism, at the The Science Museum's Dana Centre (2010)

Müllensiefen, D.: Similarity algorithms for music plagiarism. Invited talk given at the 4th International Plagiarism Conference, University of Newcastle, 22nd of June 2010 (2010)

Müllensiefen, D.: A Psychological Perspective on Similarity and Distance Measures. Invited talk given at the Symposium of The German Classification Society (GfKl), 1st of September 2011, University of Frankfurt (2011)

Müllensiefen, D.: Review of Noyce, Küssner and Sollich: quantifying shapes. Empirical Musicol. Rev. **8**(2), 155–157 (2013)

Müllensiefen, D., Wiggins, G.: Sloboda & Parker's recall paradigm for melodic memory: a new computational perspective. In: Deliege, I., Davidson, J. (eds.) Music and the Mind: Essays in Honour of John Sloboda, pp. 161–188. University Press, Oxford (2011)

Müllensiefen, D., Frieler, K.: Evaluating different approaches to measuring the similarity of melodies. In: Batagelj, V., Bock, H.-H., Ferligoj, A., Ziberna, A. (eds.) Data Science and Classification, pp. 299–306. Springer, Berlin (2006)

Müllensiefen, D., Halpern, A.R.: What's behind hits and false alarms in musical memory? Poster presented at the 4th Conference on the Neurosciences and Music, Edinburgh, 9–12. of June 2011 (2011)

Müllensiefen, D., Skiada, N.: Comment on Hennig and Liao's discussion paper on mixed-type clustering for social stratification. J. Royal Stat. Soc. Series C **62**(3), 354–355 (2013)

Müllensiefen, D., Cason, R.: Singing from the same sheet: a new approach to measuring tune similarity and its legal implications. Invited speaker seminar, 5th of December, School of Law, University of Greenwich (2012)

Müllensiefen, D., Frieler, K.: Modelling expert's notions of melodic similarity. Musicae Scientiae Discussion Forum **4A**, 183–210 (2007)

Müllensiefen, D., Pendzich, M.: Court decisions on music plagiarism and the predictive value of similarity algorithms. Musicae Scientiae Discussion Forum **4B**, 257–295 (2009)

Müllensiefen, D., Wiggins, G.A.: Sloboda's & Parker's recall paradigm for melodic memory: a new computational persepctive, p. 2009. Around the Contribution of John Sloboda, IRCAM, Talk at the Symposium Music and Cognition (2009)

Müllensiefen, D., Bonometti, M., Stewart, L., Wiggins, G.: Testing different models of melodic contour. In: 7th Triennial Conference of the European Society of the Cognitive Sciences of Music (ESCOM 2009), University of Jyäskylä, Finland (2009)

Müllensiefen, D., Davies, C., Dossman, L., Hansen, J.L., Pickering, A.: Implicit and explicit effects of music on brand perception in TV ads. Presentation given at the Audio Branding Congress 2012, 11th of December, University of Oxford (2012)

Müllensiefen, D., Frieler, K.: Cognitive adequacy in the measurement of melodic similarity. Algorithmic vs. human judgments. Comput. Musicol. **13**, 147–176 (2004)

Müllensiefen, D., Frieler, K.: Optimizing measures of melodic similarity for the exploration of a large folk song database. In: 5th International Conference on Music Information Retrieval, Audiovisual Institute, Universitat Pompeu Fabra, Barcelona, Spain, pp. 274–280 (2004b)

Müllensiefen, D., Fry, J., Jones, R., Jilka, S., Stewart, L., Williamson, V.J.: Individual differences predict patterns in spontaneous involuntary musical imagery. Music Perception **31**(4), 323–338 (2014)

Müllensiefen, D., Hennig, C.: Modeling memory for melodies. In: Spiliopoulou, M., Kruse, R., Borgelt, C., Nürnberger, A., Gaul, W. (eds.) From Data and Information Analysis to Knowledge Engineering, pp. 732–739. Springer, Berlin (2006)

Müllensiefen, D., Lewis, D., Rhodes, C., Wiggins, G. A.: Evaluating a Chord Labeling Algorithm. In Proceedings of the Eighth International Conference on Music Information Retrieval, pp. 317–318, Vienna (2007)

Müllensiefen, D., Stewart, L., Cooper, A.: Predicting item difficulty in a melodic discrimination task using item response theory and computational structure analysis. Talk given at the Manchester meeting of the Experimental Psychology Society (EPS), University of Manchester, 7th of July 2010 (2010)

Müllensiefen, D., Pfleiderer, M., Frieler, K.: The perception of accents in pop music melodies. J. New Music Res. **38**(1), 19–44 (2009). https://dx.doi.org/10.1080/09298210903085857. Accessed 11 Sept 2012

Müllensiefen, D.: Variabilität and Konstanz von Melodien in der Erinnerung: ein Beitrag zur musikpsychologischen Gedächtnisforschung. Doctoral Dissertation, Institute of Musicology, University of Hamburg, Hamburg, Germany (2004)

Müller, G.: 2. Mathematik der Fuzzy-Mengen (n.d). https://ejb-ressourcen.de/docs/FuzzyLogic/node5.html

Müller, G.K., Christa: Geheimnisse der Pflanzenwelt. Waltrop und Leipzig (2003)
Müller, H.-C.: Zur Theorie und Praxis indeterminierter Musik. Aufführungspraxis zwischen Experiment und Improvisation. G. Bosse, Kassel (1994)
Mundhenke, F., Arenas, F.R., Wilke, Th. (eds.). Mashups. Neue Praktiken und Ästhetiken in populären Medienkulturen. Springer VS, Wiesbaden (2014)
Murcott, D.: Mechanical troubles: performing nancarrow's player piano studies today. Pianola J. **23**, 3–10 (2013)
Murphy, K.R., Jako, R.A., Anhalt, R.L.: Nature and Consequences of Halo Error. A critical analysis. J. Appl. Psychol. **78**(2), 218–225 (1993)
Musil, J., El-Nusairi, B., Müllensiefen, D.: Perceptual dimensions of short audio clips and corresponding timbre features. In: From Sounds to Music and Emotions. Lecture Notes in Computer Science, 7900, pp. 214–227. Springer, Berlin (2013)
Nagel, F., Kopiez, R., Grewe, O., Altenmüller, E.: EMuJoy. Software for the continuous measurement of emotions in music. Behav. Res. Methods **39**(2), 283–290 (2007)
Nägele, R.: Musikwissenschaftsdämmerung, Anmerkungen zu einem unzeitgemäßen Fach. In: Merkur, issue (842), July 2019 (2019)
Naskar, S., Victor, R., Nath, K., Choudhury, H. A.: "Radio inside my head": a curious case of early onset "stuck song syndrome" or obsessive-compulsive disorder with predominant musical obsession. Indian J. Psychol. Med. **39**(3), 373–375 (2017). Doi:https://doi.org/10.4103/0253-7176.207333
Neue Gesellschaft für Bildende Kunst Berlin (ed.): Absolut modern sein: Zwischen Fahrrad und Fließband – culture technique in Frankreich 1889–1937. Elefantenpress, Berlin (1986)
Neuhoff, H.: Mensch – Musik – Maschine. Das digitale Credo und seine Probleme. Eine Antwort auf Reinhard Kopiez. In: Musica 1/1996, pp. 27–32 (1996)
Nicolas, F.: Questions of logic: writing, dialectics and musical strategies. In: Assayag, G., Feichtinger, H.G., Rodrigues, J.F. (eds.) Mathematics and Music. A Diderot Mathematical Forum, pp. 89–112. Springer, Berlin, Heidelberg, New York, Barcelona, Hong Kong, London, Milan, Paris, Tokyo (2002)
Nolte, J.: Kollaps der Moderne. Traktat über die letzten Bilder. Droemersche Verlagsanstalt Th. Knaur Nachf., München (1989)
Nonnenmann, R.: Invention durch Notation – Von der Verfertigung musikalischer Gedanken beim Schreiben. In: NZfM, 169. Jg. 2008, issue 5, pp. 20–25 (2008)
Nono-Schoenberg, N. (ed.): Arnold Schönberg. 1874–1951. Lebensgeschichte in Begegnungen. Verlag Ritter, Klagenfurt (1992)
Nowak, A.: Musikalische Logik, HmT - 38. Delivery, winter 2004/05
Nyre, L.: Sound Media. From live journalism to music recording. Routledge, Oxon and New York (2008)
Oberhaus, L., Oetken, M. (unter Mitarbeit von Susanne Stamm) (Ed.): Farbe, Klang, Reim, Rhythmus. Interdisziplinäre Zugänge zur Musik im Bilderbuch, Bielefeld 2017. transcript Verlag, Einleitung 7–22 (2017)
Oberhoff, B. (Ed.): Das Unbewußte in der Musik. Psychosozial-Verlag, Gießen (2002)
Oesch, H. (with collaboration of Haas, M. and Haller, P.): Außereuropäische Musik, vol. 2 (Neues Handbuch der Musikwissenschaft, vol. 9). Laaber-Verlag, Laaber (1987)
Orjuela-Rojas, J.M., Lizeth, I., Rodríguez, L.: The stuck song syndrome: a case of musical obsessions. Am. J. Case Rep. **19**, 1329–1333 (2018). doi:https://doi.org/10.12659/AJCR.912402
Oswald, J.: Plunderphonics, or Audio Piracy as a Compositional Prerogative - as presented by to the Wired Society Electro-Acoustic Conference in Toronto (1985)
Pace, I.: Graphic Notation. Paper presented at the Lecture, 13–06–2007, Hochschule für Musik, Freiburg i. Br., 35 p. (2007). https://openaccess.city.ac.uk/6476/
Pampalk, E., Dixon, S., Widmer, G.: On the evaluation of perceptual similarity measures for music. In: Proceedings of the International Conference on Digital Audio Effects, pp. 6–12 (2003)

Pampalk, E., Flexer, A., Widmer, G.: Hierarchical organization and description of music collections at the artist level. In: Proceedings of the 9th European Conference on Research and Advanced Technology for Digital Libraries, pp. 37–48 (2005)

Pampalk, E.: Audio-based music similarity and retrieval: Combining a spectral similarity model with information extracted from fluctuation patterns. In: 3rd Annual Music Information Retrieval eXchange (MIREX'06), n.p. (2006a). https://pampalk.at/publications/

Pampalk, E.: Computational models of music similarity and their application in music information retrieval. PhD thesis, Computer Science Department, Technical University Vienna, Austria (2006b)

Pang, H., Yoon, D.: Automatic detection of vibrato in monophonic music. Pattern Recogn. **38**(7), 1135–1138 (2005)

Pareyon, G.: On musical self-similarity. Intersemiosis as Synecdoche and Analogy. Acta Semiotica Fennica XXXIX. Approaches to Musical Semiotics 13. International Semiotics Institute at Imatra. Semiotic Society of Finland, Helsinki (2011)

Parncutt, R.A.: A perceptual model of pulse salience and metrical accent in musical rhythms. Music Perception **11**(4), 409–464 (1994). https://doi.org/10.2307/40285633

Paul, G., Schock, R.: Sound des Jahrhunderts. Geräusche, Töne, Stimmen 1889 bis heute, pp. 313–319. Bundeszentrale für politische Bildung bpb, Bonn (2013)

Pawley, A., Müllensiefen, D.: An empirical field study on sing-along behaviour in the North of England. Presentation given at the 12th International Conference on Music Perception and Cognition (ICMPC), 27th of July, Thessaloniki (2012)

Pawley, A., Müllensiefen, D.: An empirical field study on sing-along behaviour in the North of England. Presentation given at the 7th International Conference on Music Since 1900, 30th of July 2011, University of Lancaster (2011)

Pawley, A., Müllensiefen, D.: The science of singing along: a quantitative field study on sing-along behaviour in the north of England. Music Perception **30**(2), 129–146 (2012)

Pearce, M.T., Müllensiefen, D., Wiggins, G.A.: A comparison of statistical and rule-based models of melodic segmentation. In Proceedings of the Ninth International Conference on Music Information Retrieval, pp. 89–94. Drexel University, Philadelphia, USA (2008)

Pearce, M.T., Müllensiefen, D., Wiggins, G.A.: The role of expectation and probabilistic reasoning in auditory boundary perception: a model comparison. Perception **39**(10), 1367–1391 (2010)

Pearce, M.T., Müllensiefen, D., Lewis, D., Rhodes, C.S.: David Temperley, Music and Probability. Cambridge, Massachusetts: MIT Press, 2007, ISBN-13: 978–0–262–20166–7 (hardcover) $40.00. In: Empirical Musicol. Rev. **2**(4), 155–163 (2007)

Pearce, M.T., Müllensiefen, D., Wiggins, G.A.: Melodic grouping in music information retrieval: new methods and applications. In: Zbigniew, R., Wieczorkowska, A. (eds.) Advances in Music Information Retrieval (Studies in Computational Intelligence, vol. 274), pp. 365–390. Springer, Berlin (2010)

Pearce, M.T., Wiggins, G.A., Müllensiefen D.: Statistical Models of Music Cognition. Invited talk at the symposium Around Fred Lerdahl & Ray Jackendoff's Generative Theory of Tonal Music. IRCAM, Paris. January 2008 (2008)

Pearce, M., Conklin, D., Wiggins, G.: Methods for combining statistical models of music. In: Wiil, U.K. (ed.): CMMR 2004. LNCS, vol. 33, no. 10, pp. 295–312. Springer, Heidelberg (2005)

Pearce, M., Müllensiefen, D., Wiggins, G.: Perceptual Segmentation of Melodies: Ambiguity, Rules, and Statistical Learning. Paper given at 10th International Conference for Music Perception and Cognition, Hokaido University, Sapporo, 25–30th of August 2008 (2008)

Pearce, M.T., Wiggins, G.A.: Expectation in melody: the influence of context and learning. Music Perception: Interdisc. J. **23**(5), 377–405 (June 2006). University of California Press Oakland, CA (2006)

Pendzich, M., Müllensiefen, D.: Court decisions on music plagiarism and the predictive value of similarity algorithms. In: 7th Triennial Conference of the European Society of the Cognitive Sciences of Music (ESCOM 2009), University of Jyäskylä, Finland (2009)

Perle, G.: The Operas of Alban Berg: Vol. 2 Lulu. University of California Press, Berkeley (1989)

Perle, G.: Twelve-Tone Tonality. University of California Press, Berkeley (1977)
Petzer, A.: Farbe Klang Körper. Zur Bewegungsästhetik in Wassily Kandinskys Bühnenkomposition „Der gelbe Klang". Diplomarbeit im Studiengang Pädagogische Ausbildung der Fakultät Musik der Universität der Künste Berlin (2017)
Petzer, T., Steiner, St. (eds.): Synergie. Kultur- und Wissensgeschichte einer Denkfigur (Trajekte-Buchreihe). Wilhelm Fink Verlag, München (2016)
Phleps, T.: "In meinen Tönen spreche ich" - Biographische Chiffren in Kompositionen von Brahms. In: Heister, H.-W. (ed.): Johannes Brahms oder Die Relativierung der "absoluten" Musik (Zwischen/Töne, vol. 5), pp. 147–198. von Bockel, Hamburg (1997)
Phleps, T.: »Complex, but simple«. Conlon Nancarrows tempo-dissonierende Boogie Woogies und Canons für Player Piano. In: Floros, C., Geiger, F., Schäfer, Th. (eds.) Komposition als Kommunikation. Zur Musik des 20. Jahrhunderts (Hamburger Jahrbuch für Musikwissenschaft, vol. 17), pp. 175–205. Lang, Frankfurt a. M (2000)
Phleps, T.: Affectionen einer lebhaft begehrenden Sinnlichkeit. Verschlüsselte Botschaften in Schubert-Liedern. In: Heister, H.-W. (ed.): Kunstwerk und Biographie. Gedenkschrift Harry Goldschmidt (Zwischen/Töne. Neue Folge, vol. 1), pp. 335–360. Weidler Buchverlag, Berlin (2002)
Phleps, T.: Das Schaffen des Künstlers ist triebhaft. Über das Bewusstsein von Zahlen in Schönbergs kleinem Klavierstück op. 19,2. In: Neue Zeitschrift für Musik vol. 162, No. 1/2001, pp. 17–23. Schott International, Mainz (2001)
Phleps, Th., Reich, W. (eds.): Musik-Kontexte Festschrift für Hanns-Werner Heister. Monsenstein und Vannerdat, Münster (2011)
Pias, C.: „Hollerith 'gefiederter' Kristalle." Kunst, Wissenschaft und Computer in Zeiten der Kybernetik. In: Hagner, M., Hörl, E. (eds.): Die Transformation des Humanen. Beiträge zur Kulturgeschichte der Kybernetik, pp. 72–106. suhrkamp taschenbuch, Frankfurt a. M. (2008)
Piatigorsky, G., Mein Cello und ich und unsere Begegnungen. dtv, München (1975)
Polth, M.: Zum Verhältnis von Harmonik und Instrumentation ›vor Wagner‹. ZGMTH **11**(1), 37–61 (2014)
Popper, K.R.: Of clouds and clocks. In: Popper, K.R. (eds.) Objective Knowledge: An Evolutionary Approach, pp. 2–66. Clarendon, Oxford (1972)
Por, P.: Die Entstehung des Fragments (Mallarmé: L'après-Midi d'un Faune) und die Entstehung des Fragments aus dem Geist des Fragments (Hofmannsthal: Der Tod des Tizian). In: TRANS. Internet-Zeitschrift für Kulturwissenschaften, no. 14/2002
Prümm, K.: Die Montage als alles durchdringendes Prinzip der Stadtsinfonie. Anmerkungen zu Walter Ruttmanns Berlin. Die Sinfonie der Grossstadt und die Folgen. In: montage AV 20/1/2011, pp. 61–71 (2011)
Puckette, M.: The Theory and Technique of Electronic Music. DRAFT, September 24, 2006, sine loco. msp.ucsd.edu › techniques (2006). https://www.researchgate.net/publication/242020079_The_Theory_and_Technique_of_Electronic_Music
Quinn, I.: .Fuzzy extensions to the theory of contour. Music Theory Spectrum **19**, 232–263 (1997). https://doi.org/10.1525/mts.1997.19.2.02a00050
Rabinovich, G., Slonimski, J.: Towards a galant pedagogy: partimenti and schemata as tools in the pedagogy of eighteenth-century style improvisation. MTO J. Soc. Music Theory **21**(3) (2015). https://mtosmt.org/issues/mto.15.21.3/mto.15.21.3.rabinovitch.php
Ras, Z.W., Dardzinska, A.: Solving failing queries through cooperation and collaboration. Special Issue on Web Resources Access. World Wide Web J. **9**(2), 173–186 (2006)
Ratcliffe, R.: A proposed typology of sampled material within electronic dance music. Dancecult: J. Electron. Dance Music Cult. **6**(1), 97–122 (2014)
Red'ko, V.G., et al.: Project "Animat brain": Designing the animat control system on the basis of the functional systems theory. In: Butz, M.V., Sigaud, O., Pezzulo, G., Baldassarre, G. (eds.): Anticipatory Behavior in Adaptive Learning Systems. ABiALS 2006. Lecture Notes in Computer Science, vol. 4520. Springer, Berlin, Heidelberg (2007). https://doi.org/10.1007/978-3-540-742 62-3_6

Redlich, H.F.: Alban Berg. Versuch einer Würdigung. Universal-Edition, Wien, Zürich, London (1957)
Reich, St.: Einige optimistische Voraussagen über die Zukunft der Musik (1970) In: Danuser et al., p. 291 (1993)
Reicher-Marek, M.E.: Werk und Autorschaft. Eine Ontologie der Kunst. mentis. s.l. (2018/2019)
Reicher-Marek, M.E.: Works and realizations. In: Guarino, N. (ed.): Formal Ontology in Information Systems. Proceedings of the First International Conference (FOIS 98, June 6–8), pp. 121–132. IOS
Reik, T.: The Haunting Melody: Psychoanalytic Experiences in Life and Music. Farrar, Straus and Young, New York (1953)
Reish, G.N.: Una nota sola: Giacinto Scelsi and the genesis of music on a single note. J. Musicol. Res. **25**, 149–189 (2006). https://doi.org/10.1080/01411890600613827
Reith, D.: "Elektronische Musik und Algorithmische Komposition". In: Neue Musiktechnologie, ed. B. Enders, Schott International, Mainz (1993)
Rense, R.: Music, Deciphered: A Demonic Prelude by an Ailing Chopin, NYT May 12 (2002). https://www.nytimes.com/2002/05/12/arts/music-deciphered-a-demonic-prelude-by-an-ailing-chopin.html
Reuter, C., Enders, B., Jacobi, R.: Lexikon Musikautomaten. CD-Rom. Schott International, Mainz (2000)
Rhodes, C., Lewis, D., Müllensiefen, D.: Bayesian model selection for harmonic labelling. In: Klouche, T., Noll, Th. (eds.) Mathematics and Computation in Music: First International Conference, MCM 2007, Berlin, Germany, May 18–20, 2007. Revised Selected Papers (Communications in Computer and Information Science), pp. 107–116. Springer, Berlin (2009)
Riemann, H.: Geschichte der Musik seit Beethoven (1800–1900). Peters, Leipzig (1901)
Riemann, H.: Ideen zu einer "Lehre von den Tonvorstellungen". In: Jahrbuch der Musikbibliothek, vol. 21/22, pp. 1–26. Peters, Leipzig (1914/15), C.F. Peters, Leipzig (published 1916)
Risi, S., Mörchen, F., Ultsch, A., Lewark, P.: Visual mining in music collections with emergent SOM. In: Proceedings workshop on self-organizing maps (WSOM), pp. 3–6 (2007). https://pure.itu.dk/portal/en/publications/visual-mining-in-music-collections-with-emergent-som(4207cb96-55e6-45c9-a6ef-9a5eb1ca642f)/export.html
Risset, J.-C.: Computing musical sound. In: Assayag et al. (eds.) Mathematics and Music. A Diderot Mathematical Forum, pp. 215–231. Springer-Verlag, Berlin (2002)
Rodriguez Zivic, P.H., Shifres, F., Cecchi, G.A.: Perceptual basis of evolving Western musical styles. In: Sejnowski, T.J. (ed.) PNAS, vol. 110, no. 24, pp. 10034–10038 (2013)
Rohrhuber, J.: Artificial, Natural, Historical in Transdisciplinary Digital Art. Sound, Vision and the New Screen. Springer. Berlin and Heidelberg (2008). https://akustik.hfbk.net/publications/rohrhuber_artificial_natural_historical.pdf
Rosenbaum, I.: Scored Movement: Incorporation of Prescribed Movement of Musicians in Musical Scores, Doctor of musical arts in composition, University of California Santa Cruz (2014). https://escholarship.org/uc/item/0735k2n4
Rossignol, S., Depalle, P., Soumagne, J., Rodet, X., Collette, J.L.: Vibrato: detection, estimation, extraction, modification. In: Proceedings of the COST-G6 Workshop on Digital Audio Effects (DAFx-99 (1999a)
Röver, C., Klefenz, F., Weihs, C.: Identification of musical instruments by means of the Hough-transformation. In: Weihs, C., Gaul, W. (eds.): Classification - the Ubiquitous Challenge, pp. 608–615. Springer, Berlin (2005)
Russo, F., Thompson, W.: The subjective size of melodic intervals over a two-octave range. Psychon. Bull. Rev. **12**, 1068–1075 (2005)
Sacks, O.: Der einarmige Pianist. Über Musik und Gehirn. Rowohlt, Reinbek (2009)
Sacks, O.: Musicophilia: Tales of Music and the Brain. Knopf, New York (2007)
Saffran, J.R., Johnson, E.K., Aslin, R.N., Newport, E.L.: Statistical learning of tone sequences by human infants and adults. In: Cognition **70**(1), 27–52 (1999)

Sandkühler, H.J.: Analogie. In: Sandkühler, H.J., et al. (eds.) Europäische Enzyklopädie zu Philosophie und Wissenschaften, vol. 1, pp. 101–108. Felix Meiner Verlag, Hamburg (1990)
Sandvold, V., Herrera, P.: Towards a semantic descriptor of subjective intensity in music. In: Proceedings of the International Computer Music Conference (2005). https://mtg.upf.edu/files/publications/Sandvold-ICMC-2005.pdf
Sanguinetti, G.: The Art of Partimento. History, Theory, KOMMA and Practice. Oxford University Press, Oxford, New York etc. (2012)
Schacter, D.L., Chiao, J.Y., Mitchell, J.P.: The seven sins of memory. Implications for self. In: Annals of the New York Academy of Sciences. 2003;1001, pp. 226–239 (2003). https://scholar.harvard.edu/files/schacterlab/files/schacterchiaomitchell2003.pdf
Schacter, D.L.: Wir sind Erinnerung. Gedächtnis und Persönlichkeit. Rowohlt, Reinbek (2001)
Schaeffer, P.: Musique concrète. Von den Pariser Anfängen um 1948 bis zur elektronischen Musik von heute. E. Klett, Stuttgart (1974)
Schäfer, A., Kursell, J.: Kräftespiel. Zur Dissymmetrie von Schall und Wahrnehmung. In: Zeitschrift für Medienwissenschaft. issue 2: Materialität/Immaterialität, vol. 2, No. 1, pp. 24–40 (2010). https://doi.org/10.25969/mediarep/1365
Schedel, M., Pohle, T.P., Knees, P., Widmer, G.: Assigning and visualizing music genres by web-based co-occurance analysis. In: Proceedings of the 7th International Conference on Music Information Retrieval, pp. 260–265 (2006)
Schedel, M., Bressan, F.: Notation for an electric stage: twenty years of writing about notation and a thought experiment. with additional commentary about preservation. In: Musica/Tecnologia, vol. 13, pp. 85–101. Firenze University Press and Fondazione Ezio Franceschini, Florence (2019)
Scheirer, E.D.: Tempo and beat analysis of acoustic musical signals. J. Acoust. Soc. Am. **103**(1), 588–601 (1998)
Schellenberg, E.G.: Simplifying the implication-realization model of melodic expectancy. Music Percept. **14**, 295–318 (1997)
Scheytt, J.: Estampes, No. 1 Pagodes (n.d.). https://www.jochenscheytt.de/debussy/debussywerke/pagodes.html
Schillinger, J.: Schillinger System of Musical Composition. C. Fischer, Inc., New York (1946)
Schillinger, J.: The Mathematical Basis of the Arts. Philosophical Library, New York (1948)
Schippan, Th.: Einführung in die Onomasiologie. VEB Bibliographisches Institut, Leipzig (1972)
Schlothauer, B.: Analyse von Arnold Schönbergs Orchesterstück „Farben", Opus 16 / 3. Hausarbeit an der Technischen Universität Berlin im Wintersemester 2005/06
Schmidt, M.: Klangfarbenmelodie. In: Oesterreichisches Musiklexikon online (2001). https://www.musiklexikon.ac.at/ml/musik_K/Klangfarbenmelodie.xml. Accessed 06 May 2001
Schmitt-Thomas, R.: Die Entwicklung der deutschen Konzertkritik im Spiegel der Leipziger Allgemeinen Musikalischen Zeitung (1798–1848). Phil. Diss., Johann Wolfgang Goethe-Universität zu Frankfurt am Main (1969)
Schmitz, R., Ure, B.: Wie Mozart in die Mozartkugel kam. Kurioses und Überraschendes aus der Welt der klassischen Musik. Pantheon-Verlag, Munich (2019)
Schnebel, D.: Anschläge – Ausschläge. Texte zur neuen Musik. Carl Hanser, München (1993)
Schnebel, D.: Denkbare Musik. Schriften 1952–1972 (edited by Zeller, H.R.). DuMont Schauberg, Köln (1972)
Schnebel, D.: Mauricio Kagel Musik Theater Film. DuMont-Verlag, Cologne (1970)
Schnebel, D.: MO-NO. Musik zum Lesen. DuMont-Verlag, Köln (1969)
Schneck, D.J., Berger, D.S.: The Music Effect. Music Physiology and Clinical Applications. Jessica Kingsley Publishers, London and Philadelphia (2006)
Schönberg, A.: Harmonielehre. Universal-Edition, Wien (1911)
Schönberg, A.: Stil und Gedanke. In: Vojtěch, I. (ed.): Aufsätze zur Musik, pp. 409–436. Fischer-Verlag, Frankfurt a. M. (1976)
Schönberg, A.: War-clouds diary. JJ. Arnold Schoenberg Inst. **9/1**(1), 55–77 (1986)

Schuller, B., Eyben, F., Rigoll, G.: Tango or Waltz? Putting ballroom dance style into tempo detection. In: EURASIP Journal on Audio, Speech, and Music Processing, volume 2008, Article ID 846135 2008, 12 p. (2008). https://link.springer.com/article/10.1155/2008/846135

Schuller, G.: Early Jazz: Its Roots and Musical Development. Oxford University Press, New York (1968)

Schultz, C.M., Seedorf, M.: loop – ein Ensemble virtueller Musikinstrumente zur Vermittlung elektronischer Musik und ihrer technischen Grundlagen (Masterthesis Technische Universität Berlin, Fakultät I: Geisteswissenschaften. Institut für Sprache und Kommunikation, Fachgebiet Audiokommunikation) (2016)

Schulze, H.: Das aleatorische Spiel: Erkundung und Anwendung der nichtintentionalen Werkgenese im 20. Jahrhundert. Hochschule der Künste Berlin (Phil. Diss. 1999). Fink, Munich and Berlin (2000)

Schulze, S., Conrad, D. (eds.): Tiere: Respekt - Harmonie - Unterwerfung (Ausstellungskatalog Museum für Kunst und Gewerbe Hamburg). Hirmer-Verlag, München (2017)

Schwartz, L.R.: Repair and Self-Formation through Verbal Notation: Analyzing Self in Works by Jennifer Walshe and Pauline Oliveros and An Anatomical Study on Escape [Original Composition], Submitted to the Graduate Faculty of the Kenneth P. Dietrich School of Arts and Sciences in partial fulfillment of the requirements for the degree of Doctor of Philosophy University of Pittsburgh (2019)

Schwarz, W.M.: Grafische Notation menschlicher rhythmischer Bewegungen in einem interaktiven multimedialen System. In: Reichel, H. (ed.): Informatik – Wirtschaft – Gesellschaft. Informatik aktuell. Springer, Berlin and Heidelberg (1993). https://doi.org/10.1007/978-3-642-78486-6_77

Searchinger, C.: Artur Schnabel. A Biography, with a Tribute by Clifford Curzon. Dodd, Mead & Co, New York (1957)

Seidel, H.: Kurze Autobiographie, In: Biese, Alfred: Fritz Reuter, Heinrich Seidel und der Humor in der neueren deutschen Dichtung, Kiel und Leipzig: Lipsius und Tischer 1891, pp. 49–55. (Deutsche Schriften für Litteratur und Kunst. 1. Reihe, Heft 5) (1891)

Seidel, W.: Rhythmus, metrum, Takt. In: MGG^2, vol. 8, 1998, col. 257–317 (1998)

Seifert, U., Kim, J.U., Moore, A. (eds.): Paradoxes of Interactivity. Perspectives for Media Theory, Human-Computer Interaction, and Artistic Investigations. transcript Verlag, Bielefeld (2008)

Seising R.: Information – ein Begriff im Fluss. 60 Jahre 'Mathematische Theorie der Kommunikation'", Vortrag zum Symposion 1948–2008: 60 Jahre Kybernetik - eine nach wie vor junge Wissenschaft?! Berlin, 21.-22.11.2008 (not published)

Seising, R., Trillas, E., Moraga, C., Termini, S. (eds.): On Fuzziness. A Homage to Lotfi A. Zadeh – Volume 1 and 2, Studies in Fuzziness and Soft Computing, vol. 298. Springer, Berlin and Heidelberg (2013)

Seising, R., Zadeh, L.A.: On fuzzy sets and the precisiation of meaning - an interview with Prof. Dr. Lotfi A. Zadeh. APHSC 1:2013 (2013). https://www.aphsc.org

Seising, R.: Die Entstehung der Fuzzy Set Theorie und ihrer ersten Anwendungen – Ihre Entwicklung bis in die 70er Jahre des 20. Jahrhunderts (Boethius: Texte und Abhandlungen zur Geschichte der Mathematik und der Naturwissenschaften Band 54). Franz Steiner, Stuttgart (2005)

Seising, R.: The Fuzzification of Systems: The Genesis of Fuzzy Set Theory and Its Initial Applications – Developments Up to the 1970s. Springer-Verlag, Berlin etc. (2007)

Seising, R., Sanz, V. (eds.): Soft Computing in Humanities and Social Sciences, Volume I (Springer series Studies in Fuzziness and Soft Computing). Springer, Berlin etc. (2012)

Seneta, E.: Lewis Caroll as a probabilist and a mathematician. Math. Scientist **1984**(9), 79–94 (1984)

Seneta, E.: Lewis Carroll's "Pillow Problems": on the 1993 centenary. Stat. Sci. **8**(2), 180–186 (1993)

Serébrinnikow, B.A.: Allgemeine Sprachwissenschaft, vol. 1: Existenzformen, Funktion und Geschichte der Sprache, 2., revised ed. (russian 1970). Akademie-Verlag, Berlin (1975)

Serres, M., Farouki, N. (ed.): Thesaurus der exakten Wissenschaften, Frankfurt a. M. (2001). (französisch 1997)

Shafer, S.: Performance practice of real-time notation. In: Proceedings of the 2016 International Conference on Technologies for Music Notation & Representation (2016)

Siebert, C.: Qualia. Das Phänomenale als Problem philosophischer und empirischer Bewußtseinstheorien. Phil. Diss., Humboldt-Universität zu Berlin (1998)

Siegele, U.: Vortrag, MGG[1], vol. 14 , col. 16–31 (1968)

Sinclair, J.B.: A Descriptive Catalogue of the Music of Charles Ives. Yale University Press (1999)

Skupin, R.: Giacinto Scelsi – homo viator and his musical itinerary. The Academy of Music, January 2012 (2012). https://www.researchgate.net/publication/316172352

Slonimsky, N.: Thesaurus of Scales and Melodic Patterns, Archived 2017–01–09 at the Wayback Machine (1975)

Smaragdis, P., Brown, J.: Non-negative matrix Factorization for polyphonic music transcription. In: IEEE Workshop on Applications of Signal Processing to Audio and Acoustics, pp. 177–180 (2003)

Snyder, B.: Music and Memory: An Introduction. MIT Press, New York (2000)

Soleymani, M., Yang, Y.-Hs., Irie, G., Hanjalic, A.: Challenges and perspectives for affective analysis in multimedia. IEEE Trans. Affective Comput. **6**(3), 206–208 (2015)

Solis, A., Panoutsos, G.: Granular computing neural-fuzzy modelling: A neutrosophic approach. https://doi.org/10.1016/j.asoc.2012.09.002 (2013)

Solis, A., Panoutsos, G.: An ensemble data-driven fuzzy network for laser welding quality prediction. IEEE International Conference on Fuzzy Systems (FUZZ-IEEE). https://doi.org/10.1109/FUZZ-IEEE.2017.8015496 (2017)

Solms, M., Turnbull, O.: Das Gehirn und die innere Welt. Neurowissenschaft und Psychoanalyse (English 2002). Walter-Verlag, Düsseldorf (2002)

Sondereguer, C.: Sistemas compositivos Amerindos. Morfoproporcionalidad/El concepto arquitectonico-escultorico en Amerindia. Praxis de un pensmiento morfoespacial, urbanistico y simbolico. Corregidor, Buenos Aires (2010)

Sonnenfeld, A., Falkenberg Hansen, K.: S-notation: a complete musical notation system for scratching and sample music derived from 'Theory of Motions' (TENOR, in Proceedings of the International Conference on Technologies for Music Notation and Representation) (2016)

Sparrer, W.-W.: Helmut Lachenmann: Pression für Violoncello solo (1969/1970), In: Heister 2005 (ed.): Geschichte der Musik im 20. Jahrhundert: 1945–1975 (Handbuch der Musik im 20. Jahrhundert Band 3), pp. 340–344. Laaber-Verlag, Laaber (2005)

Sparrer, W.-W.: Isang Yun. In: Heister, H.-W., Sparrer, W.-W. (eds.) Komponisten der Gegenwart. Loose-Leaf lexikon, vol. 1 [40 pp.]. edition text + kritik, München (1992)

Spitzer 2005, Musik im Kopf. Hören, Musizieren, Verstehen und Erleben im neuronalen Netzwerk, Stuttgart und New York: Schattauer 5th ed. 2005 (1st 2002)

Spitzer, M.: Geist im Netz. Spectrum Akademischer Verlag, Heidelberg (1996)

Spohr, L.: Lebenserinnerungen [written between 1847 and 1858, first edition Kassel 1860/1861], edited by Folker Göthel. Schneider, Tutzing (1968)

Steinbeck, W.: Struktur und Ähnlichkeit. Methoden automatisierter Melodieanalyse (Kieler Schriften zur Musikwissenschaft, vol. XXV). Bärenreiter, Kassel (1982)

Stelzner, W.: 3.2.2 Mehrwertige Logik, from: Logik. In: Enzyklopädie Philosophie. Three volumes, plus one CD-ROM. In cooperation with Borchers, D., Regenbogen, A., Schürmann, V. and Stekeler-Weithofer, P., ed. by Sandkühler, H.J., pp. 1442ff. Felix Meiner Verlag, Hamburg (2010)

Stengård, E., van den Berg, R.: Imperfect Bayesian inference in visual perception. PLOS Computat. Biol. **15**(4) (2019). doi: https://doi.org/10.1371/journal.pcbi.1006465

Stern, D.: Filmmusik als Musik. Musikbehandlung im frühen Tonfilm als Modell für Grundsatzfragen der Filmmusik. Manuscript, (May 2020). Introduction to the book Musik und Film: Aneignung der Wirklichkeit. Weidler Buchverlag, Berlin (will be published 2021) (2020)

Sterne, J. (ed.): The Sound Studies Reader. Routledge, Abingdon (2012)

Sterneck, W.: Das künstliche Geräusch – Futuristische und konkrete musik, revised passage from the book Der Kampf um die Träume – Musik und Gesellschaft (1998). https://www.sterneck.net/musik/futurismus/index.php

Stewart, L., Müllensiefen, : MSc in Music, Mind and Brain at Goldsmiths. University of London. In Psychomusicology **20** (1–2), 177–179 (2010)

Stoll, R.W.: Belle bonne … Amour curtois als Herrscherlob? (2020). https://www.sinustext.com/single-post/belle-bonne

Stopka, K.: Rauschen. Morat Ziemer Rainer Rutz **2018**, 205–209 (2018)

Stowell, R.: Violin Technique and Performance Practice in the Late Eighteenth and Early Nineteenth Centuries. Cambridge University Press, Cambridge (1990)

Straebel, V.: Musikalische Repräsentation geometrischer Objekte in Alvin Luciers Kammermusik. In: Tadday, U. (ed.) Alvin Lucier (Musik-Konzepte, vol. 180/181), pp. 49–71. Edition Text + Kritik, München (2018)

Straebel, V.: Aspects of conceptual composing. The Open Space Magazine, issue 10, fall 2008, pp. 69–78 (2008)

Streich, S., Herrera, P.: Detrended fluctuation analysis of music signals: danceability estimation and further semantic characterization. In: Proceedings of the 118th AES Convention (2005). https://www.researchgate.net/publication/238677665_Detrended_Fluctuation_Analysis_of_Music_Signals_Danceability_Estimation_and_further_Semantic_Characterization/link/0046352d3b8a4599cf000000/download

Stroh, W. M.: Midi-Experimente und Algorithmisches Komponieren – eine Anleitung zum kreativen Programmieren und Komponieren am Computer, mit Diskette(n). midi-pädagogische schriftenreihe, vol. 3. musiklabor Berlin, Berlin (1990)

Stürzbecher, U.: Werkstattgespräche mit Komponisten. Gesprächsteilnehmer u.a.: Boris Blacher, Pierre Boulez, Günter Bialas, K. Höller, Hans-Werner Henze, Karlheinz Stockhausen, D. Erdmann. dtv, München (1973)

Su, L., Yang, Y.-Hs.: Combining spectral and temporal representations for multipitch estimation of polyphonic music. In: IEEE/ACM Transactions on Audio, Speech, and Language Processing, vol. 23, no. 10, pp. 1600–1612, October 2015

Supper, M.: Elektroakustische Musik – Live-Elektronik (Bibliography with E. Ungeheuer). In: MGG^2, vol. **2**, col. 1717–1765 (1995)

Supper, M.: Elektroakustische Musik und Computer Musik, Darmstadt (1997)

Takeda, H., Nishimoto, T., Sagayama, S.: Rhythm and tempo analysis toward automatic music transcription. In: IEEE International Conference on Acoustics, Speech and Signal Processing, ICASSP 2007, vol. **4** (2007). DOI: https://doi.org/10.1109/ICASSP.2007.367320

Tamagnan, F., Yang, Y.-Hs.: Drum fills detection and generation. In: Proceedings of the International Symposium on Computer Music Multidisciplinary Research (CMMR), October 2019

Teich Geertinger, A.: Werkbegriffe in Gesamtausgabe und Werkverzeichnis, oder: Wann ist ein Werk keins? Definitorische Probleme dargestellt am Beispiel Carl Nielsens. In: Richts, K. and Stadler, P. (eds,) „Ei, dem alten Herrn zoll' ich Achtung gern'" (Festschrift für Joachim Veit zum 60. Geburtstag, for the Virtuellen Forschungsverbund Edirom), pp. 265–282. Allitera Verlag, München (2016)

Temperley, D., Sleator, D.: Modeling meter and harmony: A preference-rule approach. Comp. Music J. **15**(1), 10–27 (1999)

Temperley, D.: The Cognition of Basic Musical Structures. MIT Press, Cambridge, MA (2001)

Temperley, D.: Music and Probability. MIT Press, Cambridge, MA (2006)

Temperley, D.: A probabilistic model of melody perception. Cogn. Sci.. A Multidisciplinary J. **32**(2), 418–444 (2018). https://doi.org/10.1080/03640210701864089. Accessed 2 Apr 2018

Teng, Y.-Ch., Kuo, Y.-Sh., Yang, Y.-Hs.: A large in-situ dataset for context-aware music recommendation on smartphones. IEEE International Conference on Multimedia and Expo, July 2013

Tenney, J.: From Scratch Writings in Music Theory, Edited by Larry Polansky, Lauren Pratt, Robert Wannamaker, and Michael Winter, University of Illinois Press. Chicago, and Springfield, Urbana (2015)

Thompson, W.F., Balkwill, L.L., Vernescu, R.: Expectancies generated by recent exposure to melodic sequences. Memory Cogn. **28**(4), 547–555 (2000)

Thym, J.: Song as Memory, Memory as Song. In: Archiv für Musikwissenschaft, vol. **69**, no. 3, pp. 263-273. Franz Steiner Verlag, Stuttgart (2012)

Tischhauser, K.J.: A Survey of the Use of Extended Techniques and Their Notations in Twentieth Century String Quartets Written since 1933 by American Composers with a Selected Annotated Bibliography and Discography (A Treatise submitted to the School of Music in partial fulfillment of the requirements for the degree of Doctor of Music) (2002)

Toiviainen, P.: Musikalische Wahrnehmung und Kognition im Computermodell. In: Bruhn, H., Kopiez, R. Lehmann, A. (eds.): Musikpsychologie. Das neue Handbuch, pp. 476–489. Rowohlt, Reinbek (2006)

Tomás, E.: Musical instruments as scores: a hybrid approach. In: Proceedings of the International Conference on Technologies for Music Notation and Representation TENOR (2016). https://www.tenor-conference.org/proceedings.html

Tomatis, A.A.: Introduction and editing S. Manassi: Der Klang des Lebens: vorgeburtliche Kommunikation - die Anfänge der seelischen Entwicklung. Rowohlt, Reinbek (1990)

Tzanetakis, G., Cook, P.: Musical genre classification of audio signals. IEEE Trans Speech Audio Process **10**(5), 293–302 (2002)

Tzanetakis, G., Ermolinskyi, A., Cook, P.: Beyond the query-by-example paradigm: new query interfaces for music. In: Proceedings of the International Computer Music Conference, pp. 177–183 (2002a)

Tzanetakis, G., Ermolinskyi, A., Cook, P.: Pitch histograms in audio and symbolic music information retrieval. In: Proceedings of the 3rd International Conference on Music Information Retrieval, pp. 31–38 (2002b)

Tzanetakis, G., Essl, G., Cook, P.: Human perception and computer extraction of beat strength. In: Proceedings of the international conference on digital audio effects (DAFx-02), pp. 257–261 (2002). https://www.researchgate.net/publication/228769561_Human_perception_and_computer_extraction_of_musical_beat_strength

Ultsch, A.: Self organizing neural networks perform different from statistical k-means clustering. In: BMBF Statusseminar Künstliche Intelligenz, Neuroinformatik und Intelligente Systeme, München, pp. 55–74. Bielefeld: transcript (1996)

Ultsch, A.: Self-organizing neural networks for visualization and classification. In: Opitz, O., Lausen, B., Klar, R. (eds.) Information and Classification - Concepts, Methods, and Applications, pp. 307–313. Springer, Berlin (1993)

Ungeheuer, E. (ed.) Elektroakustische Musik (Handbuch der Musik im 20. Jahrhundert, Bd. 5). Laaber-Verlag, Laaber (2002)

Ungeheuer, E.: Elektroakustische Musik 1945–75: Prisma musikalischer Originalität. In: Heister (ed.) Geschichte der Musik im 20. Jahrhundert, Bd. III: 1945–1975, Laaber 2005 (Handbuch der Musik des 20. Jahrhunderts, Bd. 3), pp. 96–106 (2005)

Vakhania, N.: On a probability problem of Lewis Carroll. Bull. Georgian Natl. Acad. Sci. **3**(3), 8–11 (2009)

Veitl, A.: Musique, causalité et écriture: Mathews, Risset, Cadoz et les recherches en synthèsenumérique des sons. In: Musique, Instruments, Machines, Paris, OMF-Paris IV Sorbonne, pp. 235–251 (2006)

Vogler, H.: Polyphonie: Komplexität der Komposition versus Kapazität der rezeptiven Verarbeitung, Staatsexamensarbeit Hochschule für Musik und Theater Hamburg (2010)

Völz, H.: Computer und Kunst, Leipzig, Jena und BerlIn: Urania 2nd ed. (1990). (1st 1988)

von Bertalanffy, L.: … aber vom Menschen wissen wir nichts. Robots, Men and Minds, Düsseldorf und Wien 1970 (englisch 1967)

von Blumröder, Ch.: Formel-Komposition—Minimal Music—Neue Einfachheit: Musikalische Konzeptionen der siebziger Jahre. In: Henck, H. (ed.) Neuland Jahrbuch 2 (1981/82), pp. 183–205. Neuland Verlag, Bergisch Gladbach (1982)

von Dadelsen, G.: Parodie und Kontrafaktur seit 1600. MGG[1] **10**, pp. 826–834 (1962)

von Dadelsen, G.: Verzierungen. In: MGG[1], vol. **13**, col., pp. 1526–1566 (1966) von Falkenhausen, L.: Suspended Music: Chime-Bells in the Culture of Bronze Age China. University of California Press, Berkeley (1993)

von Maur, K. (ed.): Vom Klang der Bilder. Die Musik in der Kunst des **20**. Jahrhunderts. Prestel, München (1985)

von Weber, M.M.: Carl Maria von Weber. Ein Lebensbild, vol. **2**. Ernst Keil, Leipzig. https://www.zeno.org/nid/20007797583 (1866)

Waeltner, E.L.: "O du lieber Augustin". Der Scherzo-Satz im II. Streichquartett von Arnold Schönberg. In: Bericht über den 1. Kongreß der Internationalen Schönberg-Gesellschaft, edited by Stephan, R. (Publikationen der Internationalen Schönberg-Gesellschaft, vol. 1), pp. 246–262. Lafite, Wien (1978)

Wagner, R.: My Life. Dodd, Mead and Company, New York 1911. PART III. 1850–1861 (In German written 1865/1880) (1911)

Walker, M.: Hitchcock's Motifs. Amsterdam University Press, Amsterdam (2005)

Walton, C.: Lies and Epiphanies: composers and their inspiration from Wagner to Berg (Eastman studies in music, v. 111), University of Rochester Press (2014)

Wang, G., Cook, P.: "On-the-fly programming: using code as an expressive musical instrument. In: Proceedings of the 2004 International Conference on New Interfaces for Musical Expression (NIME). NIME, New York (2006). https://soundlab.cs.princeton.edu/publications/on-the-fly_nime2004.pdf

Wang, J.-CH., Yang, J.-Hs., Jhuo, I.-H., Lin, Y.-Y., Wang. Hs.-M.: The acousticvisual emotion Gaussians model for automatic generation of music video. In: ACM International Conference on Multimedia (MM), Grand Challenge solution paper (First Prize), November 2012

Wang, M., Zhang, N., Zhu, H.: User-adaptive music emotion recognition. In: 7th Inter-national Conference on Signal Processing, ICSP (2004)

Wang, Y.-Ch., Yang, Y.-Hs., Chang, K., Wang, Hs.-M., Jeng, Sh.-K.: Exploring the relationship between categorical and dimensional emotion semantics of music. In: International Workshop on Music Information Retrieval with User-Centered and Multimodal Strategies (MIRUM), in Conjunction with ACM Multimedia (MM), November 2012

Weber, P.: Zyklusbildung in Beethovens letzten Sinfonien. In: Archiv für Musikwissenschaft, vol. **71**, iss. 4, pp. 282–290. Franz Steiner-Verlag, Stuttgart (2014)

Wehnert, M.: Thema und Motiv. In: MGG^1, vol. 13, col 282–311 (1966)

Weihs, C., Szepannek, G., Ligges, U., Luebke, K., Raabe, N.: Local models in register classification by timbre. In: Batagelj, V., Bock, H.H., Ferligoj, A., Ziberna, A. (eds.) Data science and classification, pp. 315–322. Springer, Berlin (2006)

Weihs, C., Berghoff, S., Hasse-Becker, P., Ligges, U.: Assessment of purity of intonation in singing presentations by discriminant analysis. In: Kunert, J., Trenkler, G. (eds.) Mathematical Statistics and Biometrical Applications, pp. 395–410. Josef Eul, Bergisch-Gladbach (2001)

Weihs, C., Ligges, U., Mörchen, F., Müllensiefen, D.: Classification in music research. Adv. Data Anal. Classif. **1**(3), 255–291 (2008)

Weinberg, S.: The First Three Minutes. A Modern View of the Origin of the Universe. Basic Books, New York (1977)

Weinstein-Reiman, M.D.: Young man's fancy: enlightenment taxonomy and the feminization of the free musical fantasia, thesis School of Music and Dance and Graduate School of the University of Oregon for the degree of Master of Arts (2014). https://hdl.handle.net/1794/18409. Accessed 29 Sept 2014

Weiss, A.S.: Experimental Sound & Radio. MIT Press, Cambridge (1996)

Wells Hall, R.: The sound of numbers. A tour of mathematical music theory, Department of Mathematics and Computer Science. Saint Joseph's University Philadelphia (2008). https://www.sju.edu/~rhall. Accessed 17 June 2020

Welsh, C.: Sirene. In: Morat/Ziemer 2018, pp. 211–216 (2018)

Wescher, H.: Die Geschichte der Collage. Vom Kubismus bis zur Gegenwart (1st ed. 1968). DuMont, Köln (1987)

Westby, S.K., Glynn, R.: Fabricating performance: reciprocal constructs of dance notation. In: Nexus Network Journal 20.01.2018, pp. 75–94 (2018)

Weyde, T., Dalighaus, K.: Recognition of musical rhythm patterns based on a neuro-fuzzy-system. In: Dagli, C.H. (ed.) Proceedings of the ANNIE 2001 Conference, St. Louis, USA (2001)

Weyde, T.: Lern- und wissensbasierte Analyse von Rhythmen - zur Entwicklung eines Neuro-Fuzzy-Systems für Erkennung und Vergleich rhythmischer Muster und Strukturen. Dissertation, Universität Osnabrück (2002)

Whatley, K.: "Rough Romance: Sequenza II for Harp as Study and Statement". In: Halfyard, J.K. (ed.) I: Berio's Sequenzas: Essays on Performance, Composition and Analysis, pp. 39–52. Ashgate Publishing Limited, Aldershot, Burlington, VT (2007)

Wicke, P.: Ästhetische Dimensionen technisch produzierter Klanggestalten. In: Phleps, T., Reich, W. (eds.) Musik-Kontexte. Festschrift für Hanns-Werner Heister, vol. 2, pp. 1062–1085. Monsenstein und Vannerdat, Münster (2011)

Wicke, P., Ziegenrücker, K.-E.: Handbuch der populären Musik. Geschichte – Stile – Praxis – Industrie, Erweiterte Neuausg. Mainz 2007. Schott Music GmbH (2007)

Wickler, W.: Mimikry. Nachahmung und Täuschung in der Natur. Fischer-Verlag, Frankfurt a. M. (1973)

Widmer, G., Dixon, S., Goebl, W., Pampalk, E., Tobudic, A.: In search of the Horowitz factor. AI Mag. **2**(3), 111–130 (2003)

Wieczorkowska, A., Ras, Z.W., Zhang, X., Lewis, R.: Multi-way hierarchic classification of musical instrument sounds. In: IEEE CS International Conference on Multimedia and Ubiquitous Engineering (MUE 2007), Seoul, Korea, April 26–28 (2007)

Wieczorkowska, A., Raś, Z.W.: Do we need automatic indexing of musical instruments? In: Bolc, L., Michalewicz, Z., Nishida, T. (eds.) IMTCI 2004. LNCS (LNAI), vol. 3490, pp. 239–245. Springer, Heidelberg (2005)

Wieczorkowska, A., Synak, P., Ras, Z.: Multi-label classification of emotions in music. In: Intelligent Information Processing and Web Mining, Advances in Soft Computing, Proceedings of IIS 2006 Symposium, Ustron, Poland, vol. 35, pp. 307–315 (2006)

Wiener, N.: Kybernetik. Regelung und Nachrichtenübertragung in Lebewesen und Maschine. Rowohlt, Reinbek 1968 (English 1948)

Wieser, W.: Organismen, Strukturen, Maschinen. Zu einer Lehre vom Organismus. Fischer-Taschenbuch, Frankfurt a. M. (1959)

Wiggins, G.A., Pearce M.T., Müllensiefen, D.: Computational modelling of music cognition and musical creativity. In: Dean, R. (ed.) The Oxford Handbook of Computer Music, pp. 383–420. University Press, Oxford (2009)

Wiggins, G., et al.: A framework for the evaluation of music representation systems. Comput. Music J. 17.03.1993, 31–42 (1993)

Wild, Ch.: Charles Ives' three places in New England: an interpretation and a conductor's guide, A Doctoral Thesis, Presented to the Faculty of the Bienen School of Music of Northwestern University in Candidacy for the Degree of Doctor of Musical Arts (2018)

Williamson, V., Jilka, S., Fry, J., Finkel, S., Müllensiefen, D., Stewart, L.: How do 'earworms' start? Classifying the everyday circumstances of involuntary musical imagery. Psychol. Music **40**(3), 259–284 (2012)

Williamson, V., Müllensiefen, D.: Earworms from three angles. In: Presentation Given at the 12th International Conference on Music Perception and Cognition (ICMPC), 26th of July, Thessaloniki (2012). https://www.doc.gold.ac.uk/~mas03dm/papers/VickyDaniel_earworms_ICMPC2012.pdf

Wilson, E.O.: Die Einheit des Wissens. Siedler Verlag, Berlin (1998)

Wilson, P.N.: Reduktion. Zur Aktualität einer musikalischen Strategie (edition neue zeitschrift für musik, edited by Stoll, R.W.). Schott International, Mainz (2003)

Wit, E.: 'All models are wrong...': an introduction to model uncertainty. Statistica Neerlandica (2012). doi: https://doi.org/10.1111/j.1467-9574.2012.00530.x1, quoted afterhttps://www.philos.rug.nl/~romeyn/paper/2012_wit_et_al_-_all_models_are_wrong.pdf. Accessed 1 Mar 2019

Wojcik, J., Kostek, B.: Intelligent technologies for inconsistent processing. In: Nguyen, N.T. (ed.) Intelligent Methods for Musical Rhythm Finding Systems (Int. Series on Advanced Intelligence, vol. 10), pp. 187–202 (2004)

Wojcik, J., Kostek, B.: Representations of music in ranking rhythmic hypotheses. advances in music information retrieval. In: Studies in Computational Intelligence, vol. 274, pp. 39–64. Springer, Berlin (2010)

Wojcik, J.: Methods of forming and ranking rhythmic hypotheses in musical pieces, Ph.D. Thesis, Faculty of Electronics, Telecommunications and Informatics, Gdansk University of Technology, Gdansk (2006)

Wolf, A., Müllensiefen, D.: The perception of similarity in court cases of melodic plagiarism and a review of measures of melodic similarity. In: Wewers, J., Seifert, U. (eds.) Under Construction: Trans- and Interdisciplinary Routes in Music Research: Proceedings of SysMus 11, Cologne 2011, pp. 215–230. Osnabrück, EpOs (2012)

Wolf, G. (ed.): BI-Lexikon Neurobiologie, Leipzig (1988)

Wolf, R.: Die Musikmaschinen von Kaufmann, Mälzel und Robertson. Eine Quellenedition. Preprint 5. Deutsche Museum, München (2012)

Wright, R., Uchanski, R.M.: Music perception and appraisal: cochlear implant users and simulated CI listening. J. Am. Acad. Audiol. **2**(5), 350–379 (2012)

Wrzeciono, P.: Statistical evaluation of violinist's performance based on recordings from the AMATI multimedia database. In: 55th Open Seminar on Acoustics, Wrocław-Piechowice, Poland (2008)

Xenakis, I.: Formalized music. Thought and mathematics in composition, Revised Edition, Additional material compiled and edited by Sharon Kanach. Harmonologia Series No. 6, Pendragon Press, Stuyvesant NY (1992)

Yang, Y.-H. Chen, H. H.: Machine recognition of music emotion: a review. ACM Trans. Intell.

Yang, Y.-H. Chen, H.-H.: Music emotion ranking. In: IEEE International Conference on Acoustics, Speech and Signal Processing (ICASSP), April 2009

Yang, Y.-H., Chen, H.-H.: iMR: Interactive music recommendation via active interactive genetic algorithm. In: Workshop on Computer Music and Audio Technology (WOCMAT) (2009)

Yang, Y.-H., Chen, H.-H.: Machine recognition of music emotion: a review. ACM Trans. Intell. Syst. Technol. **3**(3), 40 (2012)

Yang, Y.-H., Chen, H.-H.: Prediction of the distribution of perceived music emotions using discrete samples. IEEE Trans. Audio Speech Lang. Process. **19**(7), 2184–2196 (September 2011)

Yang, Y.-H., Liu J.-Y.: Quantitative study of music listening behavior in a social and affective context. IEEE Trans. Multimedia 15(6), 1304–1315, October 2013, Special Issue on Social Media as Sensors (2013)

Yang, Y.-Hs., Hu, X.: Cross-cultural music mood classification: a comparison of English and Chinese songs. In: International Society for Music Information Retrieval Conference (ISMIR), October 2012

Yang, Y., Su, Y., Lin, Y., Chen, H.: Music emotion recognition: the role of individuality. In: Proceedings of the International Workshop on Human-Centered Multimedia, HCM 2007 (2007)

Yankovskii, B.A.: Method for the objective appraisal of violin tone quality. Soviet Phys.-Acoust. 11(269) (1965)

Yeh, Y.-Ch, Hsiao, W.-Y.,Fukayama, S., Kitahara, T., Genchel, B., Min Liu, H.-M., Dong, H.-W., Chen, Y., Leong, T., Yang, Y.-Hs.: Automatic melody harmonization with triad chords: a comparative study. In: ArXiv e-prints. abs/2001.02360, January 2020

Yeh, Y.-Ch., Yu Liu, J.-Y., Hsiao, W.-Hs., Huang, Y.-S.,Yang, Y.-Hs.: Learning to generate Jazz and Pop piano music from audio via MIR techniques. In: ISMIR demo paper, November 2019.

Yoon, H. J.: Stochastische und fraktale Modelle in der Algorithmischen Komposition. epOs Electronic Publishing Osnabrück Music, Osnabrück (2002)

You, M., Liu, J., Guo-Zheng Li, G.-Zh., Chen, J.: Embedded feature selection for multi-label classification of music emotions, pp. 668–678 (2012). https://www.tandfonline.com/doi/abs/10.1080/18756891.2012.718113

Yun-Ning Hung, Y.-N., Yang, Y.-Hs.: Frame-level instrument recognition by timbre and pitch. In: Proceedings of the International Society for Music Information Retrieval Conference (ISMIR), September 2018

Yun, Sh.-H.: Die Hauptklangkomposition Isang Yuns unter dem Aspekt der bildenden Kunst. In: Umakhak 11, ed. by Korean Musicological Society, pp. 129–161. Seoul (2004)

Yun, Sh.-H.: Zwischen zwei Musikwelten. Studien zum musikalischen Denken Isang Yuns, Dissertation zur Erlangung des akademischen Grades Dr. phil., Seoul Oktober 2000. Ms. (2000)

Zadeh, L.A.: Foreword to: Kruse, R., Gebhardt J., Klawonn, F. (eds.) Foundations of Fuzzy Systems, p. VI. J. Wiley & Sons, Chichester (1994)

Zadeh, L.A.: Precisiated natural language (PNL). In: AI-Magazine (AI MAG), vol. 25/3, pp. 74–91. Association for the Advancement of Artificial Intelligence (2004)

Zadeh, L.A.: Precisication of meaning via translation into PRUF. In: Vaina, L., Hintikka, J. (eds.) Cognitive constraints on communication, pp. 373–402. Kluwer Academic Publishers B.V., Alphen an den Rijn (1984)

Zadeh, L.A., Polak, E. (ed.): System Theory. New York (1969)

Zadeh, L.A.: Linguistic cybernetics. In: Proceedings of the International Symposium on Systems Sciences and Cybernetics, Oxford (1972)

Zadeh, L.A.: From computing with numbers to computing with words – from manipulation of measurements to manipulation of perceptions. IEEE Trans. Circ. Syst. **45**(1), 105–119 (1999)

Zeller, M.: Planal Analysis and the Emancipation of Timbre: Klangfarbenmelodie and Timbral Function in Mahler, Schoenberg, and Webern, Dissertation submitted in partial fulfillment of the requirements for the degree of Doctor of Philosophy in the Department of Music in the Graduate School of Duke University (2020). https://dukespace.lib.duke.edu/dspace/handle/10161/21042

Zender, H.: Happy New Ears. Herder, Freiburg i. Br. et al. (1991)

Zender, H,: Messiaens Aktualität – das Haiku-Denken. In: Positionen **13**(1992), pp. 46–48 (1992)

Zender, H.: Wir steigen niemals in denselben Fluß. Wie Musikhören sich wandelt. Herder, Freiburg i. Br. u.a (1996)

Zender, H.: Die Sinne denken. Texte zur Musik 1975–2003. (Ed.) J. P. Hiekel. Breitkopf & Härtel, Wiesbaden et al. (2004)

Zhang, J., Harbottle, G., Wang, C., Kong, Z.: Oldest Playable Musical Instrument Found At Jiahu Early Neolithic Site In China (2007)

Zillien, N.: Ludwik Fleck und die »Verehrung der Zahl«. Beitrag zu einer Soziologie der Quantifizierung (vol. 6 Robuste Vagheit: Quantifizierung in der Wissenschaftsgesellschaft). In: Endreß, M., Lichtblau, K., Moebius, S. (eds.) Zyklos 3. Jahrbuch für Theorie und Geschichte der Soziologie, pp. 15–51. Springer, Wiesbaden (2017)

Zoltai, D.: Ethos und Affekt. Geschichte der philosophischen Musikästhetik von den Anfängen bis zu Hegel. Akademie-Verlag, Berlin and Budapest (1970)